W9-DJB-559

www.wadsworth.com

wadsworth.com is the World Wide Web site for Wadsworth Publishing Company and is your direct source to dozens of online resources.

At *wadsworth.com* you can find out about supplements, demonstration software, and student resources. You can also send e-mail to many of our authors and preview new publications and exciting new technologies.

wadsworth.com
Changing the way the world learns®

POPULATION

An Introduction to Concepts and Issues
Seventh Edition

John R. Weeks
San Diego State University

Wadsworth Publishing Company

I(T)P® An International Thomson Publishing Company

Belmont, CA • Albany, NY • Boston • Cincinnati • Johannesburg • London • Madrid • Melbourne
Mexico City • New York • Pacific Grove, CA • Scottsdale, AZ • Singapore • Tokyo • Toronto

Sociology Editor: Halee Dinsey
Assistant Editor: Barbara Yien
Editorial Assistant: Jennifer Jones
Marketing Manager: Christine Henry
Project Editor: Jerilyn Emori
Print Buyer: Karen Hunt
Permissions Editor: Robert Kauser
Manuscript Editor: Deanna Weeks
Production: Marcia Craig, Graphic World Publishing Services
Copy Editor: Lois Stagg
Illustrations: Graphic World Illustration Services
Cover Design: Margarite Reynolds
Compositor: Graphic World, Inc.
Printer: RR Donnelley & Sons/Crawfordsville

The cover illustration shows countries in proportion to population. Adapted by the author from Population Reference Bureau data.

COPYRIGHT © 1999 by Wadsworth Publishing Company
A Division of International Thomson Publishing Inc.
I(T)P® The ITP logo is a registered trademark under license.

Printed in the United States of America   5   6   7   8   9   10

For more information, contact Wadsworth Publishing Company, 10 Davis Drive, Belmont, CA 94002, or electronically at http://www.wadsworth.com

International Thomson Publishing Europe
Berkshire House
168-173 High Holborn
London, WC1V 7AA, United Kingdom

International Thomson Editores
Seneca, 53
Colonia Polanco
11560 México D.F. México

Nelson ITP, Australia
102 Dodds Street
South Melbourne
Victoria 3205 Australia

International Thomson Publishing Asia
60 Albert Street
#15-01 Albert Complex
Singapore 189969

Nelson Canada
1120 Birchmount Road
Scarborough, Ontario
Canada M1K 5G4

International Thomson Publishing Japan
Hirakawa-cho Kyowa Building, 3F
2-2-1 Hirakawa-cho, Chiyoda-ku
Tokyo 102 Japan

International Thomson Publishing Southern Africa
Building 18, Constantia Square
138 Sixteenth Road, P.O. Box 2459
Halfway House, 1685 South Africa

All rights reserved. No part of this work covered by the copyright hereon may be reproduced or used in any form or by any means—graphic, electronic, or mechanical, including photocopying, recording, taping, or information storage and retrieval systems—without the written permission of the publisher.

Library of Congress Cataloging-in-Publication Data
Weeks, John Robert
    Population : an introduction to concepts and issues /
John R. Weeks. — 7th ed.
        p.   cm.
    Includes bibliographical references and index.
    ISBN 0-534-55305-2
    1. Population.   I. Title.
HB871.W43   1998
    304.6—dc21   98-39201

This book is printed on acid-free recycled paper.

To Deanna

BRIEF TABLE OF CONTENTS

DETAILED TABLE OF CONTENTS

PREFACE

When I think about population growth in the world, I conjure up an image of a bus hurtling down the highway toward what appears to be a cliff. The bus is semi-automatic and has no driver who is really in charge of its progress. Some of the passengers on the bus are ignorant of what seems to lie ahead and are more worried about whether or not the air conditioning is turned up high enough or wondering how many snacks they have left for the journey. Other more alert passengers are looking down the road, but some of them think that what seems like a cliff is really just an optical illusion and is nothing to worry about; some think it may just be a dip, not really a cliff. Those who think it is a cliff are trying to figure out how to apply the brakes, knowing that a big bus takes a long time to slow down even after the brakes are put on. Are we headed toward a disastrous scenario? We don't really know for sure, but we simply can't afford the luxury of hoping for the best. The population bus is causing damage and creating vortexes of change as it charges down the highway, whether or not we are on the cliff route; and the better we understand its speed and direction, the better we will be at steering it and managing it successfully.

Over the years I have found that most people are either blissfully unaware of the enormous impact that population growth and change have on their lives, or else they have heard so many horror stories about impending doom that they are nearly overwhelmed whenever they think of population growth. My purpose in this book is to shake you out of your lethargy (if you are one of those types), without necessarily scaring you in the process. I will introduce you to the basic concepts of population studies and help you develop your own demographic perspective, enabling you to understand some of the most important issues confronting the world. My intention is to sharpen your perception of population growth and change, to increase your awareness of what is happening and why, and to help prepare you to cope with (and help shape) a future that will be shared with billions more people than there are today.

I wrote this book with a wide audience in mind because I find that students in my classes come from a wide range of academic disciplines and bring with them an incredible variety of viewpoints and backgrounds. No matter who you are, demographic events are influencing your life, and the more you know about them, the better off you will be.

What Is New in this Seventh Edition

This edition represents a major overhaul from the sixth edition, with several of the chapters reordered to conform to the evolution in the way that I and many other users of previous editions teach demography. In that process much of the book has been completely rewritten and, of course, throughout the book I have scoured the literature and data to incorporate the latest trends and thinking in the field.

Chapters Have Been Reordered and Rewritten

Chapter 1 now introduces the field of population by providing an overview of global population trends. This previously was done in the second chapter, but many users, including me, had been starting their classes by giving students an introduction to what is going in the world and thus why they should care. The book now does that too.

Chapter 2 is now the chapter dealing with demographic sources—where did all of that information in the first chapter come from? This chapter includes a discussion of the Census 2000 in the United States, and also reviews census activities in Canada and Mexico.

Chapter 3 has had several sections revised, but remains essentially the same.

Chapter 4 is now the one dealing with mortality—the cause of world population growth. This chapter used to follow the chapters on fertility, but many users told me that it made more sense to discuss mortality first, since its decline, in essence, created the whole field of demography. I agree.

Chapters 5 and 6 are now the chapters dealing with fertility concepts (Chapter 5) and trends and levels (Chapter 6). Both chapters have been substantially updated.

Chapters 7 and 8 deal with migration and the age/sex structure, as before. They have both been updated.

Chapter 9 is now the chapter on population aging and the life course. Its new placement reflects the logical continuation of the discussion of the age and sex structure (Chapter 8) by discussing the impact of aging on societies.

Chapter 10 is now family demography and life chances and is a combination of what previously had been Chapters 9 and 10, covering population characteristics and household and family structure. In previous editions there had been quite a bit of overlap in these two chapters and they are now more efficiently and parsimoniously combined into a single discussion. This has had the added bonus of reducing the size of the book to 15 chapters, which I know from experience is easier to deal with in the usual academic term.

Chapter 11 is now the chapter on the urban transition, and has been extensively rewritten.

Chapter 12 is now the chapter on population growth and development. It had undergone a major revision in the sixth edition, and has been updated appropriately for this edition.

Chapter 13 is now the chapter on population growth, food, and the environment and has been extensively rewritten to reflect the explosion of literature in this field of intense interest to the entire global community.

Chapters 14 and 15 are now the chapters on population policy, and demographics, respectively. They have both been updated for this seventh edition.

Websites have been incorporated as resources at the end of every chapter, following the suggested additional readings. I find myself searching the web as a matter of course in almost everything I do, and I have incorporated some of my favorite sites into the book. These suggested internet resources are tied to classroom and homework exercises that I have prepared and are available to readers at the Wadsworth sociology home page: **http://www.sociology.wadsworth.com**.

The focus of the book is even more global than previous editions and I have incorporated a broader North American perspective by expanding the coverage of demographic change in Canada and Mexico, in addition to the United States.

Finally, let me note that the Glossary and the Appendix have both been revised and updated, and the book includes three indexes (subject, author, and geographic).

Organization of the Book

The book is organized into five parts, each building on the previous one.

Part One "A Demographic Perspective" (Chapters 1–3) begins with an introduction to world population trends, so that you have a good idea of what population studies are all about. The second chapter reviews the sources of the data that we use to understand demographic trends, and the third chapter in this section introduces you to the major perspectives or ways of thinking about population growth and change.

Part Two "Population Processes" (Chapters 4–7) discusses the three basic demographic processes: mortality (Chapter 4), fertility (Chapters 5 and 6), and migration (Chapter 7). Knowledge of the three population processes provides you with the foundation you need to understand why changes occur and what might be done about them.

Part Three In "Population Structure and Characteristics" (Chapters 8–11), I discuss the interaction of the population processes and societal change according to demographic characteristics such as age and sex (Chapter 8), aging and the life course more specifically (Chapter 9), family and household structure, as well as sociodemographic characteristics including marital status, race, ethnicity, education, labor force participation, occupation, and religion (Chapter 10), and the urban transition (Chapter 11).

Part Four In "Population, Development, and the Environment" (Chapters 12 and 13), I explore with you the demographic underpinnings of several massively important issues confronting the global community—can economic growth and development be sustained (Chapter 12), especially given the need to feed billions more people at a time when we are perhaps irreparably degrading the environment (Chapter 13)?

Part Five I conclude with two chapters that make up the section on "Using the Demographic Perspective." In Chapter 14 I discuss various ways to use your new-found demographic perspective to try to alter the course of demographic events in order to achieve the kinds of social and economic goals that you might have in mind for yourself and the world, and this chapter includes an extensive discussion of the United Nations International Conference on Population and Development (ICPD) held in Cairo in 1994. Chapter 15, "Demographics," is a chapter in which I review the useful and potentially profitable applications of demographic information to business decision-making, social policy implementation, and political planning.

Special Features of the Book

To help increase your understanding of the basic concepts and issues of population studies, the book contains the following special features.

Short Essays Each chapter contains a short essay on a particular population concept, designed to help you better understand current demographic issues.

Main Points A list of main points appears at the end of each chapter, following the summary, to aid in your review of chapter highlights.

Suggested Readings At the end of each chapter I have listed five of the most important and more readable references for additional review of the topics covered in that chapter.

Websites of Interest The internet has become a useful supplement to published reading material, and there is a great deal of information available on the world wide web that is of relevance to population studies. At the end of each chapter I list and annotate five websites that I have found to be particularly interesting and helpful to students.

Glossary A glossary in the back of the book defines key population terms. These terms are in boldface type when introduced in the text to signal their appearance in the glossary.

Complete Bibliography This is a fully-referenced book and all of the publications and data sources that I have utilized are included in the bibliography at the end of the text.

Thorough Indexes To help you find what you need in the book, we have built three different indexes for you—a subject index, a geographic index, and a name index.

Personal Acknowledgments

Like most authors, I have an intellectual lineage that I feel is worth tracing. In particular I would like to acknowledge the late Kingsley Davis, whose standards as a teacher and scholar will always keep me reaching, Eduardo Arriaga, the late Judith

Blake, Thomas Burch, Carlo Cipolla, Murray Gendell, Nathan Keyfitz, and Samuel Preston. In small and large ways, they have helped me unravel the mysteries of how the world operates demographically. Thanks is due also to Steve Rutter, formerly of Wadsworth Publishing Company, whose idea this book was in the beginning. Eve Howard and Denise Simon have provided key insights for the updating of this seventh edition, and I am very grateful for the terrific production work of Marcia Craig at Graphic World Publishing Services. Special thanks go to John, Gregory, Jennifer, and Suzanne for teaching me the costs and benefits of children. They have instructed me, respectively, in the advantages of being first-born, in the coziness of the middle child, in the joys that immigration can bring to a family, and in the wonderful gifts (including Andrew and Sophie) that a daughter-in-law can bring to a family.

However, the one person who is directly responsible for the fact that the first, second, third, fourth, fifth, the updated fifth, sixth, and now the seventh edition were written, and who deserves credit for the book's strengths, is my wife Deanna. Her creativity, good judgment, and hard work in editing the manuscript benefited virtually every page, and I have dedicated the book to her.

Other Acknowledgments

I would also like to thank the users of the earlier editions, including professors and their students and my own students, for their comments and suggestions. In particular, for the seventh edition, I thank Etienne van de Walle, University of Pennsylvania; Daniel Krymkowski, University of Vermont; David Sly, Florida State University; and Thomas Espenshade, Princeton University.

For their helpful comments and suggestions, I would also like to thank the following: John L. Anderson, University of Louisville; Roger Avery, Brown University; Phillips Cutright, Indiana University; Lillian Rogers Daughaday, Murray State University; Saad Gadalla, San Diego State University; Shoshana Grossbard-Schechtman, San Diego State University; Larry J. Halford, Washburn University of Topeka; Roberto Ham-Chande, El Colegio de la Frontera Norte (Mexico); Nan Johnson, Michigan State University; Jerry N. Judy, Shippensburg University of Pennsylvania; J. William Leasure, San Diego State University; Tim Futing Liao, University of Illinois at Urbana–Champaign; Jerome McKibben, Indiana University; Gail McMillen, Arizona State University; William D. Mangold, University of Arkansas–Fayetteville; Glenna Spitze, State University of New York, Albany; Gillian Stevens, University of Illinois at Urbana–Champaign; Teresa Sullivan, University of Texas, Austin; Russell Thornton, University of California, Berkeley; Kwaku Twumasi-Ankrah, Fayetteville State University; John Wardwell, Washington State University; DeeAnn Went, University of Oklahoma; Charles Westoff, Princeton University; Lynn White, University of Nebraska; and Howard Wineberg, Portland State University.

Part One
A Demographic Perspective

Population growth is the single most important set of events ever to occur in human history. It has changed—and continues to alter—the way of life in even the most remote corners of the earth. The number of people added to the world each day is unprecedented in history and unparalleled in its consequences. We now live in a world crowded not only with people but with contradiction. There are more highly educated people than ever before yet also more illiterates, more rich people but also more poor, more well-fed children and also more hunger-ravaged babies whose images haunt us. We have better control over the environment than ever before, but we are damaging our living space in ways we are afraid to imagine.

Control of the environment is the key to understanding why the population is growing. Our partial mastery of the environment has allowed us to conquer many of the diseases that once routinely killed us. It is not too much to say that the rapid, dramatic drop in mortality all over the world is one of the greatest stories ever told. However, no good deed goes unpunished, even an altruistic one like conquering (or at least delaying) death. Birth rates almost never go down in tandem with the decline in the death rate, and the result is rapid population growth. This relentless increase in the population is fuel for both environmental damage and social upheaval.

Not all demographic change is bad, of course, but population growth does make implacable demands on natural and societal resources that will certainly be detrimental to the long-term health of the planet and its inhabitants, human and otherwise. A baby born this year may not create an immediate stir, but in a few years that child will be eating more, needing clothes, an education, and then a job and place of her own to live.

Population growth is indeed an irresistible force. Virtually every social, political, and economic problem that faces the world has demographic change as one of its root causes. Furthermore, I can guarantee you that it is a force that will increasingly affect your life personally in ways both large and small.

What are the causes of population growth, and what are its consequences? Population change is not just something that happens to other people. It is taking place all around us, and we each make our own contribution to it. Whether our concern with demography is personal or global, unraveling the "whys" of population growth and change will provide you with a better perspective on population issues and what can (or cannot) be done about them.

I begin in Chapter 1 with an overview of the world's population situation, describing current trends in the size and rate of growth of countries in various parts of the world and explaining along the way how we got ourselves into this mess. I do

1

not just make up these numbers, of course, and in Chapter 2, I discuss the various demographic resources that form the bedrock of our analysis. In Chapter 3, I turn your attention to the major theoretical perspectives that have been developed to organize population information into a meaningful framework so that we can better understand how the world works.

CHAPTER 1
Introduction to the World's Population

Introduction

"The increase of population is the most revolutionary phenomenon of our times." With that, Spanish philosopher Ortega y Gasset neatly summarized the situation more than 50 years ago, and it is still difficult to quarrel with his assessment. Consider that at this moment you are sharing the planet, vying for space and resources, with 6 billion other people, and by the year 2010 yet another billion souls will have joined you at the world's table. This in and of itself is pretty impressive, but it becomes truly alarming when you realize that this is a huge leap up from the "only" 2.5 billion in residence as recently as 1950. This phenomenon is obviously without precedent, so we are sailing in uncharted territory.

Primary among our concerns, of course, is that grappling with this upward spiral of growth will consume an increasing amount of the world's resources. This massive increase in population density is THE major contributor to concerns about energy sources, housing shortages, rampant hunger, environmental degradation, and the global shift of manufacturing jobs to less developed nations.

In order to deal intelligently with a future that will be shared with billions more people than today, we have to understand why the populations of most countries are growing (and why some are not) and what happens to societies as their patterns of birth, death, or migration change. In recent years we have heard pessimists (the "doomsters") argue that population growth is a bomb that is exploding with soon-to-be-felt catastrophic consequences. On the other hand, optimists (the "boomsters") say that population growth is not really as threatening as it seems and that problems could be solved with new technologies and better distribution of resources. Either way, people are concerned about a future that they know will be altered by the consequences of population change—be it growth (as in most places) or decline (as in a few places). Understanding these and a wide range of related issues is the business of **demography**, the science of population.

Demography: The Science of Population

Demography is concerned with virtually everything that influences or can be influenced by population size, distribution, processes, structure, or characteristics. We are all contributors to the world demographic scene. Taken together, the sum of millions, indeed billions, of individual decisions made by each of us become the substance of demographic reality. At the most basic level, everyone experiences at least two of the demographic processes: we are all born, and we all die. In between, most of us will have children of our own, and almost everybody migrates from one place to another at least once. In addition, the lines at the supermarket, the price of gas at the pump, the chances that you will marry and maybe divorce, have children, the kind of job you will land, the housing you find, the choices that you will have for a mid-life career change, and the kind of social support you can expect in old age are all dependent on demographic forces.

The demographic foundation of our lives is deep and broad. As you will see in this book, demography affects nearly every facet of your life in some way or another. Population change is one of the prime forces behind social and technological change all over the world. As population size and composition changes in an area—whether it be growth or decline—people have to adjust, and from those adjustments radiate innumerable alterations to the way society operates. Very few acres of this earth await human discovery as more and more people with more and better machines traverse the planet, coming into contact (and often conflict) with one another. A century and a half ago the American philosopher and naturalist Henry David Thoreau wrote, "I would rather sit on a pumpkin and have it all to myself than be crowded on a velvet cushion." But population growth has nearly swallowed up his famous Walden Pond in the Boston metropolitan area, and his ghost must now share even his pumpkin with thousands of visitors each year, never mind a velvet cushion.

You are probably well aware of the substantial migration (legal and undocumented) from Mexico to the United States, and you can no doubt readily grasp the idea that people from a developing nation would prefer to earn the higher wages that prevail in the highly industrialized nations. Less obvious are the demographic trends in Mexico and the United States that are channeling that migration stream. In Mexico, high rates of population growth have made it impossible for the economy to generate enough jobs for each year's crop of new workers. The unemployment in Mexico naturally increases the attractiveness of migrating to where the jobs are. That happens to be the United States, because low rates of population growth have left many jobs open, especially at the lower end of the economic ladder, into which foreign laborers often fit. One of the selling points of the North American Free Trade Agreement (NAFTA), which went into effect in 1994, was that it encouraged economic development in Mexico and promised eventually to dampen the flow of migrants into the United States.

The population of the world is increasing by nearly a quarter of a million people per day, as I will discuss in more detail later, and that rapid growth is reflected in the political instability of several regions of the world. Population growth is, of course, not the only source of trouble in the world, and its impact often is largely incendiary, igniting other dilemmas that face human society (Davis, 1984). Increased scarcity of resources and economic inequality among nations are frequent causes of conflict, but these problems are worsened by population growth (Choucri, 1984).

Throughout the Middle East, for example, we can see with special clarity the crucial role that demography has played. It was, in part, pressure from population growth that encouraged oil-producing nations to think about raising prices (or, in the case of Iraq, to invade a neighbor) to ensure a larger share of resources for their own citizens. The migration of poor rural peasants to the cities of Iran, especially Tehran, contributed to the political revolution in that country in 1978 by creating a pool of young, unemployed men who were ready recruits to the cause of overthrowing the existing government (Kazemi, 1980). Further, one reason for the increased visibility of the Palestine Liberation Organization over time has been the very high birth rate among Palestinian refugees living in the West Bank and Gaza. The fact that there are more young Palestinians each year than there are new jobs being created means that there is a pool of discontent waiting to be tapped by any group interested in overturning the present social order (Fargues, 1995).

Likewise, population growth in Central America has been a major contributor to political instability (Diaz-Briquets, 1986), providing a spark for violence at every downturn in the economy and any hint of societal injustice. In the 1970s a period of economic instability in Central America coincided with the coming of age of a large number of people needing work but finding that a rigid social structure and a failing economy were not providing it. The result was more than a decade of guerilla warfare and instability in the region. Sub-Saharan Africa is another part of the world where population growth has been increasing faster than resources can be generated to support it. Rwanda and Burundi, two land-locked countries in central sub-Saharan Africa, both reached the exploding point in the mid-1990s. These two countries, which in decades past were likened to the countryside of Switzerland, have now been almost completely deforested, as a population that doubles every 30 years harvests firewood and seeks land on which to grow food. The vast majority of people are in grinding poverty with little hope of improvement and it is a recipe for disaster that has produced interethnic violence and mass murders. In 1993 a genocidal civil war erupted between the rival Tutsi and Hutu ethnic groups, leaving at least 100,000 Burundis dead. The following year a similar eruption in Rwanda killed an estimated 1 million Rwandans, and it continues.

Now let us trace the history of population growth in the world, to give you a clue as to how we got ourselves into the current situation. Then I will take you on

WHY STUDY DEMOGRAPHY?

One of the most compelling reasons to study demography is that population growth can compound and magnify, if not create, a wide variety of social, economic, and political problems. Among the more significant are the following:

Food Security None of the basic resources required to expand food output—land, water, energy, fertilizer—can be considered abundant today. This especially affects third world countries with rapidly rising food demands and small energy reserves. Even now in sub-Saharan Africa, food production is not keeping pace with population growth, and this raises the fear that the world may have surpassed its ability to sustain current levels of food production. Furthermore, the annual fish catch is leveling off and seems unlikely to keep up with the pace of population growth in the future.

Energy Every person added to the world's population requires energy to prepare food, to provide clothing and shelter, and to fuel economic life.

Each increment in demand is another claim on energy resources (and creates problems of what to do with the byproducts of energy use), forcing further global adjustments in the use of energy.

Environmental Degradation As the human population has increased, its potential for disrupting the earth's biosphere has grown. We are polluting the atmosphere (producing problems such as acid rain, global warming, and holes in the ozone layer); the hydrosphere (contaminating the fresh water supply, destroying coral reefs, and depleting the fish population of the ocean); and the lithosphere (degrading the land with toxic waste and permitting top soil loss, desertification, and deforestation).

Urbanization The rural population worldwide is growing so fast that people are forced to leave the rural areas and search for jobs in cities, adding to the size of slums and generating problems of infrastructure, health, and public safety.

Sources: Adapted from L. Brown, C. Flavin, and H. French, 1997, State of the World 1997 (New York: W.W. Norton); Population Action International, 1996, Why Population Matters (Washington, D.C.: Population Action International); and Population Institute, 1997, Toward the 21st Century: Resource Conservation, Population and Sustainable Development (Washington, D.C.: The Population Institute).

a brief guided tour through each of the world's major regions, highlighting current patterns of population size and rates of growth, with a special emphasis on the world's 10 most populous nations.

World Population Growth

A Brief History

Human beings have been around for at least 200,000 years (Wilson and Cann, 1992). For almost all of that time, their presence on the earth was scarcely noticeable. Humans were hunter-gatherers living a primitive existence marked by only very slow population growth. Given the amount of space required by a hunting-gathering society, it seems unlikely that the earth could support more than several million people living like that (Biraben, 1979a; Coale, 1974), so it is no surprise that the population of the world on the eve of the Agricultural Revolution 10,000 years ago is estimated at about 4 million (see Table 1.1). Many people argue that the Agricultural Revolution occurred slowly but pervasively across the face of the

---

International Migration Population growth is a direct contributor to international migration, as people in rapidly growing poorer areas seek opportunities elsewhere, and an indirect contributor when population growth fuels instability and violence within a region, which forces people to flee for their lives.

Housing As a result of the swelling demand for houses, the costs of land, lumber, cement, and fuel have risen beyond the financial means of the majority of the world's 6 billion people.

Infrastructure A safe and healthy environment requires the infrastructure necessary to provide clean water and to treat sewage and waste, and a growing economy requires transportation and communications infrastructure. Rapidly growing countries find that their "demographic overhead" is too great to keep up with the demand not only for new infrastructure, but just to maintain what they already have.

Income Population growth may offset all economic growth in some countries, aggravating the very unequal global distribution of income and wealth.

Unemployment With current technologies, economists estimate that countries experiencing a 3 percent rate of population growth require a 9 percent rate of economic growth just to maintain employment at the current level. In many countries of the world, population is still growing rapidly while economic growth is not, thus intensifying problems of unemployment.

Generation Gap Differences in birth rates produce differences in age structure, and the wealthier, slower growing countries are struggling with how to care for an aging population, while the emerging nations are struggling with how to deal with hordes of young people.

Individual Freedom As more and more people require space and resources on this planet, more and more rules and regulations are required to supervise individual use of the earth's resources for the common good.

Table 1.1 Population Growth Was Very Slow in the Earlier Years of Human Existence but Has Accelerated in the Past 200 Years

Year	Population (in millions)	Average Annual Rate of Growth (%)	Doubling Time (years to double population)	Average Annual Increase in Population
−8000	4			
−5000	5	0.01	9277	372
−3000	14	0.05	1340	7,209
−2000	27	0.07	1051	17,739
−1000	50	0.06	1120	30,819
−500	100	0.14	498	138,726
1	211	0.15	463	314,708
500	198	−0.01	−5414	−25,231
1000	290	0.08	906	220,868
1100	311	0.07	1000	214,297
1200	380	0.20	342	768,328
1300	396	0.04	1673	163,356
1400	362	−0.09	−769	−324,822
1500	473	0.27	258	1,266,737
1600	562	0.17	400	969,760
1700	645	0.14	504	883,396
1750	764	0.34	203	2,603,446
1800	945	0.42	163	4,018,906
1850	1,234	0.53	129	6,603,265
1900	1,654	0.59	118	9,717,345
1910	1,750	0.57	122	9,943,830
1920	1,860	0.61	113	11,373,321
1930	2,070	1.07	65	22,262,088
1940	2,300	1.05	65	24,361,028
1950	2,556	1.47	47	37,768,237
1960	3,039	1.33	52	40,629,803
1970	3,707	2.07	33	77,395,382
1980	4,454	1.69	41	76,035,404
1990	5,278	1.56	44	82,903,255
2000	6,081	1.25	55	76,753,814
2010	6,840	1.10	63	75,755,042
2020	7,570	0.90	77	68,403,637
2030	8,225	0.76	91	62,407,650
2040	8,820	0.62	111	54,564,430
2050	9,309	0.44	157	41,049,843

Sources: The population data for each year are drawn from the U.S. Bureau of the Census, International Programs Center, "Historical Estimates of the World Population," (http://www.census.gov/ipc/www/worldhis.html). The numbers from -8000 through 1940 reflect the average of the estimates made by Biraben, 1979a; Durand, 1967; and McEvedy and Jones, 1978, whereas the estimates from 1950 through 1990 and the projections from 2000 through 2050 are those of the Bureau of the Census.

earth precisely because the hunting-gathering populations were growing just enough to push the limit of the **carrying capacity** of that way of life (Boserup, 1965; Cohen, 1977; Harris and Ross, 1987; Sanderson, 1995). *Carrying capacity* refers to the number of people that can be supported in an area given the available physical resources and the way that people use those resources (Miller, 1998). Hunting and gathering uses resources *extensively* rather than *intensively*, so it was natural that over tens of thousands of years humans moved into the remote corners of the earth in search of sustenance. Eventually, people in most of those corners began to use the environment more intensively, leading to the more sedentary, agricultural way of life that has characterized most of human society for the past 10,000 years.

The population began to grow more noticeably after the Agricultural Revolution, as Table 1.1 shows. Between 8000 B.C. and 5000 B.C. about 372 people on average were being added to the world's total population each year, but by 500 B.C., as major civilizations were being established in China and Greece, the world was adding nearly 139,000 people each year to the total. By the time of Christ (the Roman Period, 1 A.D.) there may well have been more than 200 million people on the planet, increasing by more than 300,000 each year. Plagues and invasions produced some backsliding in the rate of growth after that, but on the eve of the Industrial Revolution in the middle of the eighteenth century (about 1750), the population of the world was approaching 1 billion people and was increasing by 2.6 million every year.

It is quite likely that the Industrial Revolution occurred in part because of this growth. It is theorized that the Europe of 300 or 400 years ago was reaching the carrying capacity of its agricultural society, so Europeans first spread out looking for more room and then began to invent more intensive uses of their resources to meet the needs of a growing population. The major resource was energy, which, along with the discovery of fossil fuels (first coal, then oil, and more recently natural gas), helped to light the fire under industrialization (Harrison, 1993).

Since the beginning of the Industrial Revolution approximately 200 years ago, the size and rate of world population growth have increased even more dramatically, as you can see graphically in Figure 1.1. So, for tens of thousands of years the population of the world grew slowly, and then within little more than 200 years, the number of people mushroomed to 6 billion. There can be little question why the term *population explosion* was coined to describe recent demographic events. The world's population did not reach 1 billion until after the American Revolution (the United Nations fixes the year at 1804 [United Nations Population Division, 1994]), but since then we have been adding each additional billion people at an accelerating pace. The 2 billion mark was hit in 1927, just before the Great Depression and 123 years after the first billion. In 1960, only 33 years later, came 3 billion; and 4 billion came along only 14 years after that, in 1974. We then hit 5 billion in another 13 years, in 1987, a milestone that the United Nations celebrated by declaring Matej Gaspar, a boy born to a Yugoslav nurse in Zagreb (now the capital of Croatia), to be the symbolic 5-billionth person. Current projections (see Table 1.1) suggest that each of the next few billions will come in quick succession, with barely a decade separating one additional billion from the next. This can certainly be labeled the "population deluge."

Figure 1.1   The World's Population is Expoding in Size

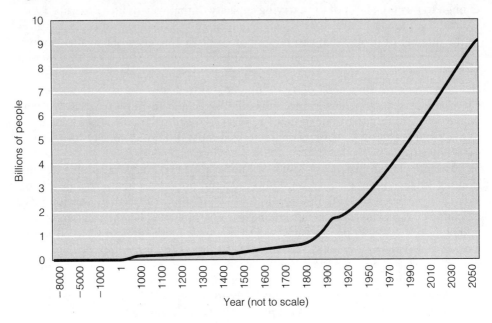

Source: Data in Table 1.1

How Fast Is the World's Population Growing Now?

Though it is good news that the rate of population growth in the world has declined since hitting its peak (2.19 percent per year) in 1962–63, dropping to the current rate of approximately 1.25 percent (see Table 1.1 and Figure 1.2), when you build on a base of 6 billion, even that lower rate of growth translates into about 77 million more humans next year than there are right now. Put another way, during the next 12 months, approximately 130 million babies will be born in the world while 53 million of all ages will die, resulting in that net addition of 77 million people. Back in 1950, as the world's population was taking off, the rate of growth was higher than it is now (1.47 percent, as shown in Table 1.1), yet because the total population size was only 2.5 billion, we were adding "only" 38 million new humans each year, as opposed to the current 77 million. In Table 1.1 you can see that the rate of growth in 1960 was lower than in 1950. This was due to a terrible famine in China in 1959–60, which was produced by Mao Zedong's Great Leap Forward program of 1958, in which the Chinese government "leapt forward" into industrialization by selling "surplus" grain to finance industrial growth. Unfortunately, the grain was not surplus, and the confiscation of food amounted to a self-imposed disaster that led to the deaths of 30 million Chinese in the following 2 years (1959 and 1960) (Becker, 1997). Though the Chinese famine may have been the largest single disaster in human history, world population growth quickly rebounded, and the growth rate hit its record high shortly after that.

Figure 1.2   Nearly 80 Million People Are Still Being Added to the World's Total Population Each Year, Despite the Drop in the Rate of Growth

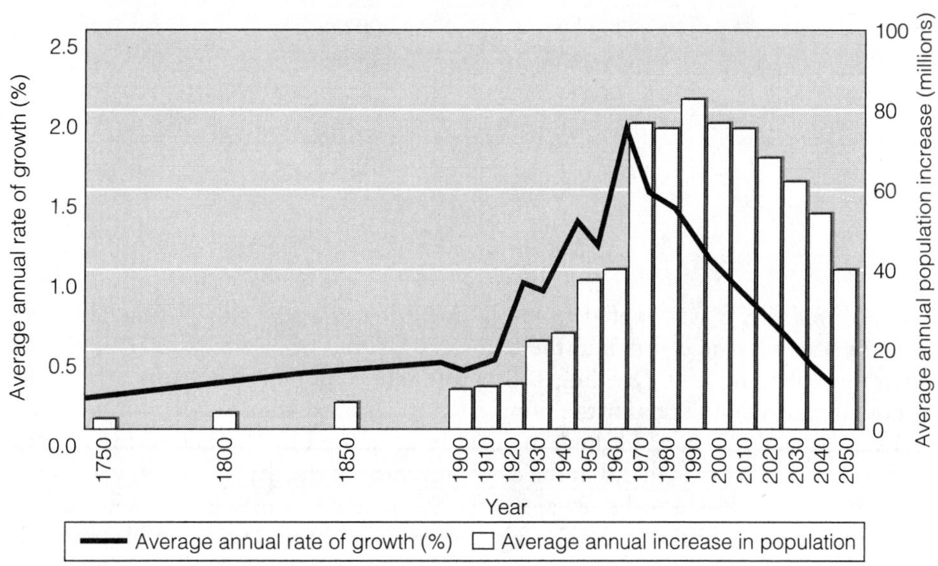

Source: Data in Table 1.1.

Even computed on a daily basis the volume of population growth in the world to-day is astounding. You can see in Table 1.2 that more than 200,000 people are added to the total each day, the net result of 356,000 births less the 145,000 people who die every day (of whom about 23,000 are infants). The vast majority of this increase is taking place in the less developed nations, as I will discuss later in this chapter.

How Fast Can Populations Grow? The Power of Doubling

Human populations, like all living things, have the capacity for exponential in-crease. A common way of measuring the growth potential of any combination of birth and death rates is to calculate the doubling time, the time required for a pop-ulation to double in number if the current rate of growth continues. You can calcu-late this easily for yourself by remembering the "rule of 69." The doubling time is approximately equal to 69 divided by the growth rate (in percent per year). So, if we estimate the world's rate of growth in the year 2000 to be 1.25 percent per year, we can calculate that the doubling time is 55 years. Where does the 69 come from in the doubling formula? Exponential growth is expressed mathematically by nat-ural logarithms. Thus, to find out how long it would take a population to double in size, we first must find the natural logarithm (ln) of 2. This turns out to be 0.69, which we multiply by 100 to get rid of the decimal point. Then dividing the rate of growth into 69 tells us how many years would be required for a population to

Table 1.2 In 1 Day the World's Population Increases by More than 200,000 People

Time Period	Births	Deaths	Natural Increase
Year	130,000,000	53,000,000	77,000,000
Day	356,000	145,000	211,000
Hour	14,833	6,041	8,792
Minute	247	101	146
Second	4.1	1.7	2.4

double. Similarly, if we wanted to know how long it would take a population to triple in size, we would first find the natural logarithm of 3, which is 1.10, or 110 when multiplied by 100. Dividing 110 by the rate of population growth then tells us how long it would take for the population to triple in size.

The incredible power of doubling can be illustrated by the tale of the Persian chessboard. The story is told that the clever inventor of the game of chess was called in by the King of Persia to be rewarded for this marvelous new game. When asked what he would like his reward to be, his answer was that he was a humble man and deserved only a humble reward. Gesturing to the board of 64 squares that he had devised for the game, he asked that he be given a single grain of wheat on the first square, twice that on the second square, twice *that* on the third square, and so on, until each square had its complement of wheat. The king protested that this was far too modest a prize, but the inventor persisted and the king finally relented. When the Overseer of the Royal Granary began counting out the wheat, it started out small enough: 1, 2, 4, 16, 32 . . . , but by the time the sixty-fourth square was reached, the number was staggering—nearly 18.5 quintillion grains of wheat (about 75 billion metric tons!) (Sagan, 1989). This, of course, exceeded the "carrying capacity" of the royal granary in the same way that successive doublings of the human population threaten to exceed the carrying capacity of the planet. Looking back at Table 1.1 you can see that it took several thousand years for the population to double from 4 million to 8 million, a few thousand years to double from 8 to 16 million, about a thousand to double from 16 to 32, less than a thousand to double to 54. About 400 years elapsed between the European renaissance and the Industrial Revolution, and the world's population doubled in size during that time. But from 1750 it took only a little more than a 100 years to double again, and the next doubling occurred in less than 100 years. The most recent doubling (from 3 to 6 billion) took only about 40 years.

Will we double again in the future? Most projections suggest not (Lutz, 1994; United Nations Population Division, 1998), but we will probably come perilously close to it by the middle of the twenty-first century, and we don't really know at this point how we will feed, clothe, educate, and find jobs for nearly twice as many people as we currently have on the planet. The social, economic, and political implications of this growth in human numbers are to say the least immense.

Once you realize how rapidly a population actually can grow, it is reasonable to wonder why early growth of the human population was so slow.

Why Was Early Growth So Slow?

The reason the population grew so slowly during the first 99 percent of human history was that death rates were very high, and at the same time very few populations have ever tried to maximize the number of children born (Abernethy, 1979). During the hunting-gathering phase of human history, it is likely that life expectancy at birth averaged about 20 years (Livi-Bacci, 1992; Petersen, 1975). At this level of mortality, more than half of all children born will die before age 5, and the average woman who survives through the reproductive years will have to bear nearly seven children in order to assure that two will survive to adulthood. This is still well below the biological limit of fertility, however, as I discuss in Chapter 5. Research among the very few remaining hunting-gathering populations in sub-Saharan Africa suggests that a premodern woman might have deliberately limited the number of children born by spacing them a few years apart to make it easier to nurse and to carry her youngest child and to permit her to do her work (Dumond, 1975). She may have accomplished this by abstinence, abortion, or possibly even **infanticide** (Howell, 1979; Lee, 1972). Similarly, sick and infirm members of society were at risk of abandonment once they were no longer able to fend for themselves. For most groups of people, the balance between a large number of births and an almost equally large number of deaths led over time to only slight increases in population size.

It was once believed that the Agricultural Revolution increased growth rates, because when people settled down in stable farming communities, death rates were lowered. Sedentary life was thought to have improved living conditions because of the more reliable supply of food. The theory went that birth rates remained high but death rates declined slightly, and thus the population grew. However, archaeological evidence combined with studies of extant hunter-gatherer groups have offered another explanation for growth during this period (Spooner, 1972). Possibly the sedentary life and the high-density living associated with farming actually raised death rates by creating sanitation problems and heightening the exposure to communicable diseases. Growth rates probably went up because fertility rates rose as new diets improved the ability of women to conceive and bear children (see Chapter 5). Also, it became easier to wean children from the breast earlier because of the greater availability of soft foods, which are easily eaten by babies. This would have shortened the birth intervals, and the birth rate could have risen on that account alone.

It should be kept in mind, of course, that only a small difference between birth and death rates is required to account for the slow growth achieved after the Agricultural Revolution. Between 8000 B.C. and 1750 A.D., the world was adding an average of only 67,000 people each year to the population. By the end of the twentieth century, that many people were being added every 7.5 hours.

Why Are Recent Increases So Rapid?

The rapid acceleration in growth after 1750 was due almost entirely to the declines in the death rate that accompanied the Industrial Revolution. First in Europe and North America and more recently in less developed countries, death rates have decreased sooner and much more rapidly than have fertility rates. The result has been that many fewer people die than are born each year. In the industrialized countries, declines in mortality at first were due to the effects of economic development and a rising standard of living—people were eating better, wearing warmer clothes, living in better houses, bathing more often, drinking cleaner water, and so on (McKeown, 1976). These improvements in the human condition helped to lower the exposure to disease and also to build up resistance to illness. Later, after 1900, much of the decline in mortality was due to improvements in medical technology, especially vaccination against infectious diseases.

Declines in the death rates first occurred in only those countries experiencing economic development. In each of these areas, primarily Europe and North America, fertility also began to decline within at least one or two generations after the death rate began its drop. However, since World War II, medical and public health technology has been available to virtually all countries of the world regardless of their level of economic development. In the underdeveloped countries, although the risk of death has been lowered dramatically, as yet the birth rates have gone down less significantly, and the result is rapid population growth. As you can see in Table 1.3, 98 percent of the growth of the world's population is taking place in less developed nations.

As a population increases, the same rate of growth will produce a larger absolute increase in size from year to year. In fact, the age structure of the world is now sufficiently young (as I detail in Chapter 8), and the population already so large, that for the next few decades the same number of people will be added each decade despite the fact that the rate of growth is expected to decline, as illustrated in Figure 1.2.

How Many Humans Have Ever Lived?

In a burst of poetic license, not to mention perhaps unprecedented ethnocentricity, William Matthews wrote in 1980 that "there are now more of us alive than ever have been dead" (Matthews, 1980). It is certainly true that the population is increasing each year in numbers that strain credulity—could it be that today's 6 billion people exceed in number all the people who have ever been born and died over the entire history of human beings?

Table 1.3 Less Developed Regions Are the Sites of Future Population Growth (To the Year 2025)

	Area		
	More Developed Nations	Less Developed Nations	World
Population in 2000 (in millions)	1,181	4,900	6,081
Projection to year 2025 (in millions)	1,213	6,692	7,905
Increase between 2000 and 2025 (in millions)	32	1,792	1,824
Percent of increase attributable to each area	2%	98%	

No, actually not. As a matter of fact, our contribution to history's total represents only a small fraction of all humans who have ever lived. This question has intrigued scholars and has been examined with some regularity over the decades. In 1960, for example, William Deevey estimated that "a cumulative total of about 110 billion individuals seem to have passed their days, and left their bones, if not their marks, on this crowded market" (1960:197).

Others have made similar types of calculations (Desmond, Wellemeyer, and Lorimer, 1962; Haub, 1995; Westing, 1981), but the most analytic of the estimates has been made by Nathan Keyfitz (1966, 1985). I have employed Keyfitz's formulas to estimate the number of people who have ever lived and as you can see in Table 1.4, the results of my calculations suggest that a total of 60 billion humans have been born over the past 200,000 years, of whom the 6 billion alive today constitute 10 percent.

As Keyfitz (1985) has pointed out, however, the dramatic drop in infant and childhood mortality over the last two centuries means that babies are now far more likely than ever to grow up to be adults. Thus the adults alive today represent a considerable fraction of all humans who have ever lived to adulthood. Furthermore, if we look at specific categories of people, such as engineers or college professors, then it is probable that the vast majority of such individuals that have ever lived on earth are still alive today.

The increase in size of the world's population is not the only important demographic change to occur in the past few hundred years. In addition, there has been a massive redistribution of population.

Redistribution of the World's Population Through Migration

As populations have grown disproportionately in different areas of the world, the pressures or desires to migrate have also grown. Migration streams generally flow from rapidly growing areas into less rapidly growing ones, for example, from Mexico to the United States. In earlier decades, as population grew dense in a particular

Table 1.4   How Many People Have Ever Lived?

Historical Period	Number of People Born During the Period (in Billions)	Cumulative Total Ever Born (in Billions)
200,000 B.C. to 8,000 B.C.	2.1	2.1
8,001 B.C. to 0 A.D.	16.7	18.8
1 A.D. to 1799	29.4	48.2
1800 to 1899	3.6	51.8
1900 to 1949	2.6	54.4
1950 to 1999	6.0	60.4

Source: Calculated by the author based on formulas from Keyfitz (1966) for estimates from 200,000 B.C. to 1949 and based on population size and birth rate estimates from the United Nations and the Population Reference Bureau for the period 1950–1999.

region, people were able to move to some other less populated area, much as high-pressure storm fronts move into low-pressure weather systems. The most notable illustration of this kind of migration is the expansion of European populations into other parts of the world. This phenomenon is especially notable because as Europeans moved around the world, they altered patterns of life, including their own, wherever they went. Indeed, European expansion has been such an important part of world history that it deserves additional comment.

European Expansion   Beginning in the fourteenth century, migration out of Europe started gaining momentum, and this expansion of the European population virtually revolutionized the entire human population. With their gun-laden sailboats, Europeans began to stake out the less developed areas of the world in the fifteenth and sixteenth centuries, and this was only the beginning. Migration of Europeans to other parts of the world on a massive scale did not occur until the nineteenth century, when the European nations began to industrialize and swell in numbers. As Kingsley Davis has put it:

> Although the continent was already crowded, the death rate began to drop and the population began to expand rapidly. Simultaneous urbanization, new occupations, financial panics, and unrestrained competition gave rise to status instability on a scale never known before. Many a bruised or disappointed European was ready to seek his fortune abroad, particularly since the new lands, tamed by the pioneers, no longer seemed wild and remote but rather like paradises where one could own land and start a new life. The invention of the steamship (the first one crossed the Atlantic in 1827) made the decision less irrevocable [1974:98].

Before the great expansion of European people and culture, Europeans represented about 18 percent of the world's population, with almost 90 percent of these people of European origin living in Europe itself. By the 1930s, at the peak of European dominance in the world, people of European origin in Europe, North America, and Oceania accounted for 35 percent of the world's population (Durand, 1967). At the end of the twentieth century, the percentage has declined to 16, and it is projected to drop to 11 percent by the middle of the twenty-first century (U.S. Bureau of the Census, 1998). Even that may be a bit of an exaggeration, since the rate of growth in North American and European countries is increasingly influenced by immigrants and births to immigrants from developing nations.

Since the 1930s the outward expansion of Europeans has ceased. Until then, European populations had been growing more rapidly than the populations in Africa, Asia, and Latin America, but since World War II that trend has been reversed. The less developed areas now have by far the most rapidly growing populations. Indeed, demographer Judith Blake once commented that "population growth used to be a reward for doing well; now it's a scourge for doing badly" (1979). This change in the pattern of population has resulted in a shift in the direction of migration. On balance, there is now more migration from less developed to developed areas than the reverse. Furthermore, since migrants from less developed areas generally have higher levels of fertility than natives of the developed regions, their migration makes a disproportionate contribution over time to the overall population increase in the developed area to which they have migrated. As a result, the proportion of the population whose origin is one of the modern world's less developed nations tends to be on the rise in nearly every developed country. Within the United States, for example, it is projected that the population of Hispanics and those of Asian origin in California will represent a majority of the total by early in the twenty-first century (Campbell, 1996).

When Europeans migrated, they were generally filling up territory that had very few people. Those seemingly empty lands or frontiers have largely disappeared today, and as a consequence migration into a country now results in more noticeable increases in population density. Closely associated with this increasing population density in the modern world is another major form of population redistribution associated with migration, the urban revolution.

The Urban Revolution   Until very recently in world history, almost everyone lived in basically rural areas. Large cities were few and far between. For example, Rome's population of 650,000 in 100 A.D. was probably the largest in the ancient world (Chandler and Fox, 1974). It is estimated that as recently as 1800, less than 1 percent of the world's population lived in cities of 100,000 or more. By the first decade of the twenty-first century, more than one third of all humans will be living in cities of that size.

The redistribution of people from rural to urban areas is most marked in the industrialized nations. For example, in 1800 about 10 percent of the English population lived in urban areas, primarily London. By the 1990s, 90 percent of the British were in cities. Similar patterns of urbanization have been experienced in other European countries, the United States, Canada, and Japan as they have industrialized.

In the less developed areas of the world, urbanization was initially associated with a commercial response to industrialization in Europe, America, and Japan. In other words, in many areas where industrialization was not occurring, Europeans had established colonies or trade relationships. The principal economic activities in these areas were not industrial but commercial in nature, associated with buying and selling. The wealth acquired by people engaging in these activities naturally attracted attention, and urban centers sprang up all over the world as Europeans sought populations to whom they could sell their goods. Thus, urban populations began to grow in some countries even without industrialization, as places where goods were exchanged and where services were developed to meet the needs of merchants. For example, Europeans sailing along the African coast traded cloth and other European products for African slaves. The slaves were transported to the Caribbean, where they were traded for sugar and coffee, which were then shipped back to Europe and traded for cloth and other European goods. So the trade evolved, with each trade generating a profit for merchants as they went, and these profits attracting people to the urban areas where the money was spent.

Geographic Distribution of the World's Population

The five largest countries in the world account for nearly half the world's population (48 percent as of the year 2000) but only 21 percent of the world's land surface. These countries include China, India, the United States, Indonesia, and

Table 1.5 The 20 Most Populous Countries in the World: 1950, 2000, and 2025

Rank	1950 Country	1950 Population (in Millions)	1950 Area (in Square Miles)	2000 Country	2000 Population (in Millions)	2000 Area (in Square Miles)	2025 Country	2025 Population (in Millions)	2025 Area (in Square Miles)
1	China	563	3,601,310	China	1256	3,601,310	India	1408	1,147,950
2	India	370	1,147,950	India	1016	1,147,950	China	1408	3,601,310
3	Soviet Union	180	8,649,500	United States	275	3,536,340	United States	335	3,536,340
4	United States	152	3,536,340	Indonesia	219	705,190	Indonesia	288	705,190
5	Japan	84	145,370	Brazil	174	3,265,060	Pakistan	212	297,640
6	Indonesia	83	705,190	Russia	146	6,520,660	Brazil	210	3,265,060
7	Germany	68	134,850	Pakistan	141	297,640	Nigeria	203	351,650
8	Brazil	53	3,265,060	Bangladesh	132	50,260	Bangladesh	181	50,260
9	United Kingdom	50	93,280	Japan	126	145,370	Mexico	142	736,950
10	Italy	47	113,540	Nigeria	117	351,650	Russia	139	6,520,660
11	Bangladesh	46	50,260	Mexico	102	736,950	Philippines	121	115,120
12	France	42	212,390	Germany	82	134,850	Japan	120	145,370
13	Pakistan	39	297,640	Philippines	81	115,120	Iran	112	631,660
14	Nigeria	32	351,650	Vietnam	78	125,670	Zaire (Congo)	106	875,310
15	Mexico	28	736,950	Iran	72	631,660	Vietnam	104	125,670
16	Spain	28	192,830	Egypt	68	384,340	Ethiopia	99	386,100
17	Vietnam	26	125,670	Turkey	67	297,150	Egypt	97	384,340
18	Poland	25	117,540	Thailand	61	197,250	Turkey	90	297,150
19	Ethiopia	22	386,100	Ethiopia	61	386,100	Germany	75	134,850
20	Egypt	21	384,340	France	59	212,390	Thailand	70	197,250
Top 20		1,959	24,247,760		4,333	22,842,910		5,520	23,505,830
World		2,556	57,900,000		6,081	57,900,000		7,905	57,900,000
% top 5		53%	37%		48%	34%		46%	35%
% top 10		65%	42%		59%			57%	
% top 20		77%			71%	39%		70%	41%

Source: U.S. Bureau of the Census, International Population Data Base.

Brazil, as you can see in Table 1.5. Rounding out the top 10 are Russia, Pakistan, Bangladesh, Japan, and Nigeria. Within these most populous 10 nations reside 59 percent of all humans. It turns out that you have to visit only 15 countries in order to shake hands with two thirds of all the people in the world. In doing so, you would travel across 37 percent of the earth's land surface. The rest of the population is spread out among 180 or so other countries, accounting for the remaining 63 percent of the earth's terrain.

If you set a goal to be as efficient as possible in maximizing the number of people you could visit while minimizing the distance you would travel, your best bet would be to schedule a trip to China and the Indian subcontinent. As of the year 2000, 42 percent of all the world's people will live in those two contiguous regions of Asia, and you can see how these areas stand out in the map of the world drawn with country size proportionate to population (see Figure 1.3). Throughout this area, the average population density is a whopping 530 people per square mile, compared with the overall world average of 105 per square mile and the American average of 78 per square mile.

Population growth in Asia is not a new story. In Table 1.6 on page 22 you can see that in 1500, as Europeans were venturing beyond their shores, China and India (or more technically the Indian subcontinent, including the modern nations of India, Pakistan, and Bangladesh) were already the most populous places on earth, and all of Asia accounted for 53 percent of the world's 461 million people. Five centuries later, Asians account for 60 percent of all the people on earth. Sub-Saharan Africa, on the other hand, had nearly 100 million people in 1500, almost as many as in Europe. However, contact with Europeans tended to be deadly for Africans (as it was for Indians in America) because of disease, violence, and slavery. Only in this century has sub-Saharan Africa rebounded in population size, and it is expected to comprise 11 percent of the total world population in 2000, compared with 17 percent in 1500. Note that the population rebound in Africa is taking place in spite of the high prevalence of HIV infection (which has primarily affected countries of sub-Saharan Africa but also has affected Brazil, Haiti, and Thailand) (Way and Stanecki, 1994). The HIV/AIDS pandemic is having the most devastating effect in Africa, but even after taking AIDS into account, virtually all African countries have such high levels of fertility that their populations will continue to grow. Only Thailand seems to face the potential for population decline as a result of AIDS. I discuss AIDS in greater detail in Chapter 4.

Because differential rates of population growth help explain the shifting distribution of people around the globe, it is worth our while to examine them briefly.

Current Differences in Growth Rates

The world's population is estimated to be multiplying at a rate approximating 1.25 percent per year as of 2000. At this rate (if it continued unchecked) the world's population would double in size in about 55 years. Five of the ten most populous countries in the world are growing even more rapidly than the world average, especially Nigeria and Pakistan but also Bangladesh, India, and Indonesia. Nigeria is growing

Figure 1.3 Cartogram of the Countries of the World by Population Size

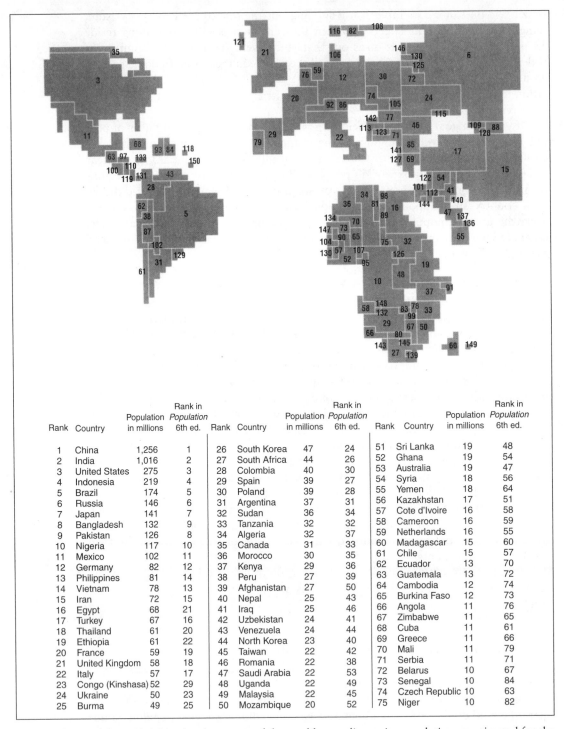

Rank	Country	Population in millions	Rank in *Population* 6th ed.	Rank	Country	Population in millions	Rank in *Population* 6th ed.	Rank	Country	Population in millions	Rank in *Population* 6th ed.
1	China	1,256	1	26	South Korea	47	24	51	Sri Lanka	19	48
2	India	1,016	2	27	South Africa	44	26	52	Ghana	19	54
3	United States	275	3	28	Colombia	40	30	53	Australia	19	47
4	Indonesia	219	4	29	Spain	39	27	54	Syria	18	56
5	Brazil	174	5	30	Poland	39	28	55	Yemen	18	64
6	Russia	146	6	31	Argentina	37	31	56	Kazakhstan	17	51
7	Japan	141	7	32	Sudan	36	34	57	Cote d'Ivoire	16	58
8	Bangladesh	132	9	33	Tanzania	32	32	58	Cameroon	16	59
9	Pakistan	126	8	34	Algeria	32	37	59	Netherlands	16	55
10	Nigeria	117	10	35	Canada	31	33	60	Madagascar	15	60
11	Mexico	102	11	36	Morocco	30	35	61	Chile	15	57
12	Germany	82	12	37	Kenya	29	36	62	Ecuador	13	70
13	Philippines	81	14	38	Peru	27	39	63	Guatemala	13	72
14	Vietnam	78	13	39	Afghanistan	27	50	64	Cambodia	12	74
15	Iran	72	15	40	Nepal	25	43	65	Burkina Faso	12	73
16	Egypt	68	21	41	Iraq	25	46	66	Angola	11	76
17	Turkey	67	16	42	Uzbekistan	24	41	67	Zimbabwe	11	65
18	Thailand	61	20	43	Venezuela	24	44	68	Cuba	11	61
19	Ethiopia	61	22	44	North Korea	23	40	69	Greece	11	66
20	France	59	19	45	Taiwan	22	42	70	Mali	11	79
21	United Kingdom	58	18	46	Romania	22	38	71	Serbia	11	71
22	Italy	57	17	47	Saudi Arabia	22	53	72	Belarus	10	67
23	Congo (Kinshasa)	52	29	48	Uganda	22	49	73	Senegal	10	84
24	Ukraine	50	23	49	Malaysia	22	45	74	Czech Republic	10	63
25	Burma	49	25	50	Mozambique	20	52	75	Niger	10	82

Note: The map shows the size of each country of the world according to its population as estimated for the year 2000 (based on data from the U.S. Bureau of the Census International Data Base).

Figure 1.3 (continued)

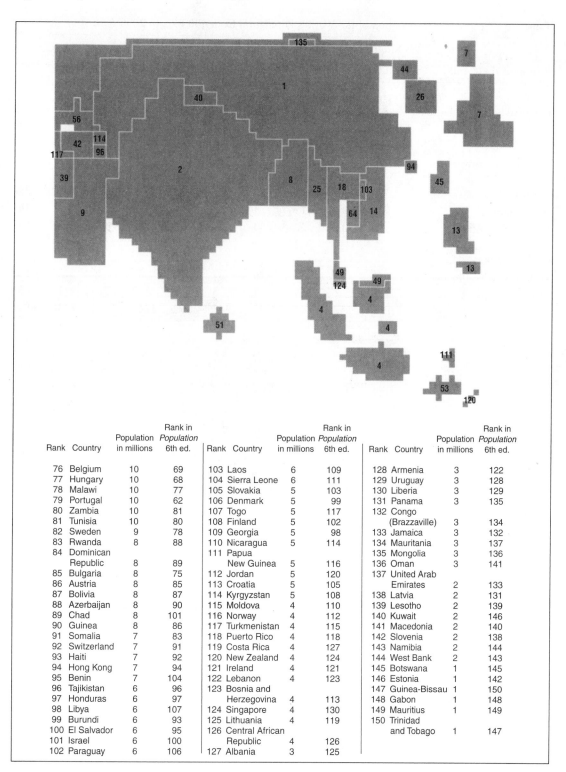

Rank	Country	Population in millions	Rank in *Population* 6th ed.	Rank	Country	Population in millions	Rank in *Population* 6th ed.	Rank	Country	Population in millions	Rank in *Population* 6th ed.
76	Belgium	10	69	103	Laos	6	109	128	Armenia	3	122
77	Hungary	10	68	104	Sierra Leone	6	111	129	Uruguay	3	128
78	Malawi	10	77	105	Slovakia	5	103	130	Liberia	3	129
79	Portugal	10	62	106	Denmark	5	99	131	Panama	3	135
80	Zambia	10	81	107	Togo	5	117	132	Congo		
81	Tunisia	10	80	108	Finland	5	102		(Brazzaville)	3	134
82	Sweden	9	78	109	Georgia	5	98	133	Jamaica	3	132
83	Rwanda	8	88	110	Nicaragua	5	114	134	Mauritania	3	137
84	Dominican			111	Papua			135	Mongolia	3	136
	Republic	8	89		New Guinea	5	116	136	Oman	3	141
85	Bulgaria	8	75	112	Jordan	5	120	137	United Arab		
86	Austria	8	85	113	Croatia	5	105		Emirates	2	133
87	Bolivia	8	87	114	Kyrgyzstan	5	108	138	Latvia	2	131
88	Azerbaijan	8	90	115	Moldova	4	110	139	Lesotho	2	139
89	Chad	8	101	116	Norway	4	112	140	Kuwait	2	146
90	Guinea	8	86	117	Turkmenistan	4	115	141	Macedonia	2	140
91	Somalia	7	83	118	Puerto Rico	4	118	142	Slovenia	2	138
92	Switzerland	7	91	119	Costa Rica	4	127	143	Namibia	2	144
93	Haiti	7	92	120	New Zealand	4	124	144	West Bank	2	143
94	Hong Kong	7	94	121	Ireland	4	121	145	Botswana	1	145
95	Benin	7	104	122	Lebanon	4	123	146	Estonia	1	142
96	Tajikistan	6	96	123	Bosnia and			147	Guinea-Bissau	1	150
97	Honduras	6	97		Herzegovina	4	113	148	Gabon	1	148
98	Libya	6	107	124	Singapore	4	130	149	Mauritius	1	149
99	Burundi	6	93	125	Lithuania	4	119	150	Trinidad		
100	El Salvador	6	95	126	Central African				and Tobago	1	147
101	Israel	6	100		Republic	4	126				
102	Paraguay	6	106	127	Albania	3	125				

Table 1.6 Geographic Distribution of the World's Population from 400 B.C. to 2000 A.D.

Region of the World	Number of People (in Millions) in the Region as of Year:									
	−400	0	500	1000	1500	1750	1850	1900	1950	2000
China	19	70	32	56	84	220	435	415	558	1,256
India/Pakistan/Bangladesh	30	46	33	40	95	165	216	290	431	1,289
Japan	1	2	5	4	10	26	30	45	84	126
Rest of Asia	45	52	49	52	56	89	109	153	320	956
Europe (w/o former U.S.S.R.)	19	31	30	30	67	111	209	295	395	508
Former USSR	13	12	11	13	17	35	79	127	180	290
North Africa	10	14	11	9	9	10	12	43	52	147
Sub-Saharan Africa	7	12	20	30	78	94	90	95	167	650
North America (U.S./Canada)	1	2	2	2	3	3	25	90	166	306
Central and South America	7	10	13	16	39	15	34	75	164	523
Oceania	1	1	1	1	3	3	2	6	13	30
TOTAL	153	252	207	253	461	771	1241	1634	2530	6081
Percentage:										
China	12.4	27.8	15.5	22.1	18.2	28.5	35.1	25.4	22.1	20.7
India/Pakistan/Bangladesh	19.6	18.3	15.9	15.8	20.6	21.4	17.4	17.7	17.0	21.2
Japan	0.7	0.8	2.4	1.6	2.2	3.4	2.4	2.8	3.3	2.1
Rest of Asia	29.4	20.6	23.7	20.6	12.1	11.5	8.8	9.4	12.6	15.7
Europe (w/o former U.S.S.R.)	12.4	12.3	14.5	11.9	14.5	14.4	16.8	18.1	15.6	8.4
Former U.S.S.R.	8.5	4.8	5.3	5.1	3.7	4.5	6.4	7.8	7.1	4.8
North Africa	6.5	5.6	5.3	3.6	2.0	1.3	1.0	2.6	2.1	2.4
Sub-Saharan Africa	4.6	4.8	9.7	11.9	16.9	12.2	7.3	5.8	6.6	10.7
North America (U.S./Canada)	0.7	0.8	1.0	0.8	0.7	0.4	2.0	5.5	6.6	5.0
Central and South America	4.6	4.0	6.3	6.3	8.5	1.9	2.7	4.6	6.5	8.6
Oceania	0.7	0.4	0.5	0.4	0.7	0.4	0.2	0.4	0.5	0.5
TOTAL	100.0	100.0	100.0	100.0	100.0	100.0	100.0	100.0	100.0	100.0

Sources: Adapted from J-N. Biraben, 1979, "Essai Sur L'Evolution Du Nombre Des Hommes." Population, 34:13; and updated by the author to 2000 using data from the U.S. Bureau of the Census International Data Base.

at a rate that doubles the population in only 24 years, while in India the population will double in 43 years at the current rate.

Among the other five of the world's ten largest countries, Russia is growing most slowly (indeed, Russia and much of Eastern Europe and the Baltic states appear to be depopulating), followed by Japan, China, the United States, and then Brazil. In Figure 1.4 you can see that the most rapidly growing regions in

Figure 1.4   Rates of Population Growth are Highest in the Middle Latitudes

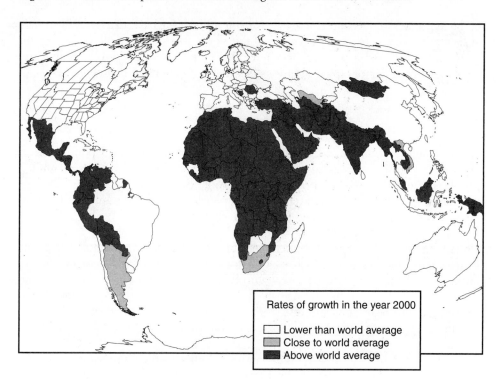

Note: In this map the world has been divided into those countries that have rates of population growth that are (a) clearly below the world average (estimated to be 1.25 percent in 2000); (b) near the world average; and (c) clearly above the world average. Virtually all of the countries growing at a rate above the world average are in the middle latitudes—Mexico and Central America, the northwestern part of South America, almost all of Africa, and western, south and southeast Asia. The slower growing southern hemisphere countries are in southern and eastern South America and Oceania.

the world tend to be those that are least developed economically, which tend to be located in the mid-latitudes, and the slowest growing are the most advanced industrially, which tend to be more northerly. It has not always been that way, however.

Before the Great Depression of the 1930s, the populations of Europe and especially North America were the most rapidly growing in the world. During the decade of the 1930s, growth rates declined in those two areas to match approximately that of most of the rest of the world, which was about 0.75 percent per year—a doubling time of 93 years. Since the end of World War II, the situation has changed again, and now Europe and North America rank among the more slowly growing populations, with rapid growth in the less developed countries of Asia, Latin America, and Africa now responsible for most of the world's population increase. Let's examine these trends in more detail.

Global Variation in Population Size and Rates of Growth

In the remainder of the chapter, I review for you some of the basic demographic trends in each of the major geographic regions of the world, focusing particular attention on the 10 most populous nations, with a few other countries included to help illustrate the variability of demographic situations in which countries find themselves.

North America

North America—Canada and the United States—have a combined population of about 300 million, representing 5 percent of the world's total. Canada's 31 million people account for about 10 percent of North America's total, and the United States, with about 270 million, has the remaining 90 percent. The demographic histories of the two countries are intertwined but are not identical.

United States   It does not take a demographer to notice that the population of the United States has undergone a total transformation since John Cabot (an Italian hired by the British to search the new world) landed in Newfoundland in 1497 and claimed North America for the British. As was true throughout the western hemisphere, the European guns and diseases rather quickly decimated the native American Indian population, making it easier to establish a new culture. It is estimated that the population of North America in 1650 consisted of about 50,000 European colonists and 2 to 3 million native American Indians—the Europeans were outnumbered by as much as 60 to 1. By 1850, disease and warfare had reduced the Indian population to perhaps as few as 250,000, while the European population had increased to 25 million. Indeed, it was widely assumed that the Indian population was on the verge of disappearance (Snipp, 1989).

Although early America was a model of rapid population growth (at least for the European-origin population), it was also a land of substantial demographic contrasts. Among the colonies existing in the seventeenth century, for example, those in New England seem to have been characterized by very high birth rates (women had an average of seven to nine children) yet relatively low mortality rates (infant mortality rates in Plymouth colony may have been lower than in some of today's less developed nations, apparently a result of the fairly good health of Americans even during that era) (Demos, 1965; Wells, 1982). Demos notes that "the popular impression today that colonial families were extremely large finds the strongest possible confirmation in the case of Plymouth. A sample of some ninety families, about whom there is fairly reliable information, suggests that there was an average of seven to eight children per family who actually grew to adulthood" (1965:270). In the southern colonies during the same time period, however, life was apparently much harsher, probably because the environment was more amenable to the spread of disease. In the Chesapeake Bay colony of Charles Parish, higher mortality meant that few parents had more than two or three living children at the time of their death (Smith, 1978).

Despite the regional diversity, the American population grew rather steadily during the seventeenth and eighteenth centuries, and though much of the increase in the number of Europeans in America was attributable to in-migration, the greater percentage actually was due to natural increase. The nation's first census, taken in 1790, shortly after the American Revolution, counted 3.9 million Americans, and though the population was increasing by nearly 120,000 a year, only about 3 percent of the increase was a result of immigration. With a crude birth rate of about 55 births per thousand population (comparable to the highest national birth rates in the world today) and a crude death rate of about 28 deaths per thousand, there were twice as many people being born each year as were dying. At this rate the population was doubling in size every 25 years.

Although Americans may picture foreigners pouring in seeking freedom or fortune, it was not until the second third of the nineteenth century that migration became a substantial factor in American population growth. In fact, during the first half of the nineteenth century in-migrants accounted for less than 5 percent of the population increase in each decade, whereas in every decade from the 1850s through the 1920s in-migrants accounted for at least 20 percent of the growth of population (see Chapter 7 for more details).

Throughout the late nineteenth and early twentieth centuries the birth rate in the United States was falling. There is evidence that fertility among American Quakers began to be limited at about the time of the American Revolution (Wells, 1971), and the rest of the nation was only a few decades behind their pace. By the 1930s, fertility actually dropped below the level required to replace the next generation (as I discuss more thoroughly in Chapter 6). Furthermore, restrictions on immigration had all but halted the influx of foreigners, and Americans were facing the prospect of potential depopulation.

The early post-World War II era upset forecasts of population decline; they were replaced by the realities of a population explosion. The period from the mid-1940s to the mid-1960s is generally known as the "Baby Boom" era, a time when the United States experienced a rapid rate of increase in population, accomplished almost entirely by increases in fertility. The Baby Boom, in turn, was followed in the late 1960s and early 1970s by a "Baby Bust." An echo of the Baby Boom was experienced as the "Baby Boomlet" of the 1980s, but fertility has since remained at or below the replacement level (Morgan, 1996). Nonetheless, population growth, rather than population decline, has continued to be the order of the day, because in the 1960s and again in 1990 a rewriting of the nation's immigration laws opened the nation's doors wider, and the result has been renewed high levels of migration into the United States. Indeed, the 900,000 immigrants added each year in the 1990s accounted for more than one third of the total annual population increase. The long-term implication of this continued immigration is that variations in fertility levels in the United States are increasingly determined by fertility differences among the various racial and ethnic groups (see Chapter 6 for additional discussion of this).

By current world standards the United States today is one of the slower growing countries, but its rate of growth is still one of the highest among the more developed nations. With a total rate of growth of about 0.8 percent per year, the U.S. population is growing at a rate that is similar to the sub-Saharan African nation of Zimbabwe and the Asian nation of Thailand. Fertility in the United States is low (2.1 children per woman) but not as low as in virtually all European countries. On the other hand, mortality is also very low (female life expectancy in the United

States is now 79 years) but not as low as the countries of Western Europe nor as low as Japan (which has the highest life expectancy in the world). The only major demographic category in which the United States leads the world is in the number of immigrants received each year. Canada is also a nation of immigrants.

Canada   The French were the first Europeans to settle the area that has become Canada, and by 1760 the population of New France included about 70,000 persons of European (largely French) origin, along with a diminishing population of indigenous peoples. In 1763 the French government ceded control of the region to the British. The French-origin population was thus "cut off from intercourse with the French, and they received no further demographic reinforcements" (Overbeek, 1980:10). Twenty years later, in 1783, the establishment of American independence led about 30,000 "loyalists" to leave the United States and migrate to Canada, the forerunner of a much more massive migration from England to Canada during the nineteenth century (Kalbach and McVey, 1979). The conflict between the British-origin and French-origin populations helped to stimulate the British North America Act of 1867, which united all of the provinces of Canada into the Dominion of Canada, and every census since then has asked about ethnicity as a way of keeping track of the numerical balance between these historically rival groups (Kralt, 1990). In 1931, when Canada was granted independence from Great Britain, the Province of Québec, in which most of French-speaking Canadians reside, accounted for 28 percent of the population of Canada. By 1996 that figure had dropped slightly to 24 percent.

In the seventeenth and eighteenth centuries the high fertility of women in New France was legendary, and French speakers in Canada maintained higher-than-average levels of fertility until the 1960s. In the rest of Canada, fertility began to drop in the nineteenth century and reached very low levels in the 1930s, before rebounding after World War II in a Baby Boom that was similar in its impact on Canadian society to that experienced in the United States. This boom was similarly followed by a Baby Bust and then a small echo of the baby boom (Foot, 1996). Canada now has a fertility level (1.6 children per woman) that is below replacement, and women in Québec are bearing children at a rate even below the national level (1.4 children per women), making Francophone (French-speaking) Canada one of the lowest fertility regions anywhere in the western hemisphere.

Just as fertility is lower in Canada than the United States, so is mortality, making life expectancy in Canada about 3 years longer than in the United States. Indeed, in both of these respects the demographic profile of Canada is more like Europe than it is the United States. However, when it comes to immigration, Canada reflects the Northern American history of being a receiving ground for people from other nations. Despite its lower fertility, Canada's overall rate of population growth exceeds that of the United States because it accepts more immigrants per person than the United States does. For example, in the 1990s Canada was admitting about 235,000 immigrants per year (about 8 per thousand population) and the foreign-born population represented 17 percent of the Canadian total in 1996 (Statistics Canada, 1997). In the same period there were 900,000 immigrants annually entering the United States (about 3 per thousand population), and in 1995 the foreign born accounted for 9 percent of the United States population (U.S. Bureau of the Census, 1997). Immigrants to Canada in recent years have come especially from countries of Asia and the Middle East. By contrast, immigrants to the United States tend to come from Asia and Latin America, especially Mexico.

Mexico and Central America

Mexico and Central America had developed more advanced agricultural societies than had North America at the time of European contact. The Aztec civilization in central Mexico and the remnants of the Mayan civilization farther south (centered near Guatemala) encompassed many millions more people than lived on the northern side of what is now the United States-Mexico border. This fact, combined with the Spanish goal of extracting resources (a polite term for plundering) from the New World rather than colonizing it, produced a very different demographic legacy from what we find in Canada and the United States.

Mexico and the countries of Central America have been growing rapidly since the end of World War II as a result of rapidly dropping death rates and birth rates that have only more recently begun to drop but which remain above the world average. Mexico's estimated population size of 102 million as of the year 2000 represents three fourths of the population of the region, with the remaining 25 percent distributed among (in order of size) Guatemala, El Salvador, Honduras, Nicaragua, Costa Rica, Panama, and Belize. The combined regional population of 136 million as estimated for the year 2000 is a little more than 2 percent of the world's total.

Mexico was the site of a series of agricultural civilizations as far back as 2,500 years before the invasion by the Spanish in 1519. Within a relatively short time the population of several million was cut by as much as 80 percent as a result of disease and violence. This population collapse was precipitated by contact with European diseases, but it reflects the fact that mortality was already quite high before the arrival of the Europeans, and it did not take much to upset the demographic balance (Alchon, 1997). By the beginning of the twentieth century, life expectancy was still very low in Mexico, less than 30 years (Morelos, 1994), and fertility was very high in response to the high death rate. However, since the 1930s the death rate has dropped dramatically, and nearing the beginning of the twenty-first century, life expectancy in Mexico is more than 72 years, several years above the world average.

For several decades this decline in mortality was not accompanied by a change in the birth rate, and the result was a massive explosion in the size of the Mexican population. In 1920, before the death rate began to drop, there were 14 million people in Mexico (Mier y Terán, 1991). By 1950 that had nearly doubled to 26 million, and by 1970 it had nearly doubled again to 49 million. In the 1970s the birth rate finally began to decline in Mexico. Mexican women had been bearing an average of 6 children each for many decades (if not centuries), but within just two decades—by the mid-1990s—it had dropped to about 3.5 children per woman. However, the massive build-up of young people has strained every ounce of the Mexican economy, encouraging outmigration, especially to the United States.

The other countries of Central America have experienced similar patterns of rapidly declining mortality, leading to population growth and its attendant pressures for migration to other countries where the opportunities might be better. Not every region has experienced the same fertility decline as Mexico, however. In particular, those areas in which a high proportion of the population is indigenous (rather than being of mixed Indian and European origin) have birth rates that remain much higher than the world average.

South America

The 330 million inhabitants of South America represent about 5 percent of the world's total population, with Brazil alone accounting for about half of that. The modern history of Brazil began when Portuguese explorers found a hunter-gatherer indigenous population in that region and tried to enslave them to work on plantations. These attempts were largely unsuccessful, and the Portuguese wound up populating the colony largely with African slaves, with the 4 million slaves taken to Brazil representing more than one third of all slaves transported from Africa to the western hemisphere between the sixteenth and nineteenth centuries (Thomas, 1997). The Napoleonic Wars in Europe in the early part of the nineteenth century allowed Brazil, like most Latin American countries, to gain their independence from Europe, and the economic development that followed ultimately led to substantial migration into Brazil from Europe during the latter part of the nineteenth century. The result is a society that is now about half European-origin and half African-origin or mixed race.

Brazil boasts a land area nearly equal in size to the United States, upon which an estimated 173 million people currently reside. Since 1965 Brazil has experienced a reduction in fertility that has been described as "nothing short of spectacular" (Martine, 1996:47). In 1960 the average Brazilian woman was giving birth to 6 children, which dropped to 2.3 by the end of the 1990s. This decline was largely unexpected, since the country was not experiencing dramatic economic improvement nor was there a big family planning campaign. Indeed, for many years the influence of the Catholic Church was strong enough to cause the government to forbid the dissemination of contraceptive information or devices (Martine, 1996). Rather, for a variety of reasons that we will delve into in Chapter 6, women have been using abortion and sterilization to limit family size.

The infant mortality rate of just under 40 deaths per 1,000 live births in Brazil is just below the world average, as is the female life expectancy of 69 years. Nearly 90 percent of Brazilian baby girls can expect still to be around at age 50, but the odds are slimmer for boys, for whom life expectancy at birth (59 years) is 10 years shorter than for females.

The other countries of South America that are predominantly European-origin, such as Argentina, Chile, and Uruguay, also tend to have fertility levels as low as Brazil's but slightly higher life expectancy. On the other hand, the countries with a large indigenous population—principally the descendants of the Incan civilization—such as Peru, Bolivia, and Ecuador, tend to have higher fertility, higher mortality, and higher rates of population growth.

Europe

The combined population of western, southern, northern, central, and eastern Europe is about 730 million, or about 12 percent of the world's total. Russia is the most populous, accounting for 20 percent of Europe's population, followed by Germany, the United Kingdom, France, and Italy.

At the end of the twentieth century, several European nations, especially the former Eastern Bloc nations, are experiencing negative population growth. Foremost among these is Russia, in which the breakup of the Soviet Union was foreshadowed by a rise in death rates (Feshbach and Friendly, 1992; Shkolnikov, Meslé, and Vallin, 1996), which has been followed by a Baby Bust that has lowered the average number of children being born per woman to 1.3. These trends in Russia mirror demographic events in Eastern Germany just before and after the fall of the Berlin Wall, and in both cases the situation has been described as "a society convulsed by its stresses" (Eberstadt, 1994:149). At the same time, it appears that the low fertility in these countries is not just a temporary phenomenon caused by political and economic upheaval. Rather, it appears that the motivation to have large families has disappeared and has been replaced by a propensity to try to improve the family's standard of living by limiting the number of children (Avdeev and Monnier, 1995).

The rest of Europe has experienced very low birth rates but without the drop in life expectancy. Thus, while Eastern and Central Europe are actually depopulating, the rest of Europe is either not growing or is growing only very slowly. Where slight population growth is occurring, such as in France and the United Kingdom, it is largely attributable to the immigration of people from less developed nations (Hall and White, 1995).

It should not be a surprise that fertility and mortality are both low in Europe, since that is the part of the world where mortality first began its worldwide decline approximately 200 years ago and where fertility began *its* worldwide decline about 100 years ago. What is surprising, however, is how low the birth rate has fallen. It is especially low in the Mediterranean countries of Italy and Spain, where fertility has dropped well below replacement level—in predominantly Catholic societies where fertility for most of history has been higher than in the rest of Europe. Sweden and the other Scandinavian countries have emerged as having fertility rates that are now among the highest in Europe, although scarcely at the replacement level, and Chesnais (1996) has offered the intriguing thesis that an improvement in the status of women may be required to push fertility levels in Europe back up to the replacement level. We return to this theme in Chapters 6 and 9.

Northern Africa and Western Asia

Egypt and western Asia includes the area most often referred to as the Middle East, while the remainder of Northern Africa to the west of Egypt is usually referred to as the Maghrib. The entire region is characterized especially by the presence of Islam (with the obvious notable exception of Israel) and by the fact that it includes a globally disproportionate share of countries with the highest rates of population growth. Total population in the region is 350 million, representing 6 percent of the world's total. Egypt is the most populous of the countries in northern Africa and western Asia, followed by Turkey, and together they account for about one third of the regional total population.

There are 68 million Egyptians crowded into the narrow Nile Valley. With its rate of growth of 1.8 percent per year, Egypt's population could double in less than 40 years, and this rapid growth constantly hampers even the most ambitious strategies for economic growth and development. This explosion in numbers is due, of course, to the dramatic drop in mortality since the end of World War II. In 1937 the life expectancy at birth in Egypt was less than 40 years (Omran, 1973; Omran and Roudi, 1993), whereas by the 1990s it had risen to 65 for females. Even with such a drop, however, death rates are above the world average, and there is considerable room for improvement.

As mortality declined, fertility remained almost intransigently high until very recently, save for brief dips during World War II and again during the wars with Israel in the late 1960s and early 1970s. For as long as statistics have been kept, Egyptian women had been bearing an average of six children, until the late 1970s. Massive family planning efforts were initiated in the 1970s under President Sadat and then reinvigorated in the 1980s under President Mubarak. These programs, especially in combination with increasing levels of education among women (Fargues, 1997), have had an effect, and the estimated fertility level is now an average of 3.4 children per woman, much lower than it used to be but still very high. Because of the high fertility, a very high proportion (40 percent) of the population is under age 15.

It is the size and rate of increase in the youthful population that has been especially explosive throughout Northern Africa and western Asia. The rapid drop in mortality after World War II, followed by a long delay in the start of fertility decline, have produced a very large population of young people in need of jobs. They have spread throughout the region looking for work, and many have gone on to Europe and North and South America. But the economies have not been able to keep up with the demand for jobs, and this has produced a generation of young people who, paradoxically, are better educated than their parents but who face an uncertain future in an increasingly crowded world. The declining birth rates hold out hope that the situation will ease in a few decades (Fargues, 1994), but in the meantime the demographic situation has fueled discontent and has threatened the political stability of the region (see, for example, Huntington, 1996).

Sub-Saharan Africa

Sub-Saharan Africa is the place from which all human life originated, according to most evidence (Wilson and Cann, 1992), and the 650 million people living there now comprise 11 percent of the world's total. Nigeria accounts for about 1 in 6 of those 650 million, followed by Ethiopia and Zaire. If you do a few quick calculations in Table 1.6, you will discover that between 1000 A.D. and 1500 A.D., before extensive European contact with sub-Saharan Africa, that part of the world was increasing in population at a higher rate than anywhere else. However, Europeans brought their diseases to Africa and, much more significantly, they commercialized slavery to an extent never previously known in human history. More than

11 million Africans were enslaved and sent off to the western hemisphere between the sixteenth and nineteenth centuries (Thomas, 1997), contributing to a decline in population size in sub-Saharan Africa between 1600 and 1850.

In recent decades, mortality has declined so much throughout the world that it is increasingly difficult to find countries with both high mortality and high fertility rates. Before World War II it was quite easy, because death rates in Africa were frequently as high as 40 deaths per 1,000 population and were associated with a life expectancy in the range of 30 to 40 years, which is lower than the United States has experienced at any time since the American Revolution. Currently there is perhaps only one country with such a high death rate (Sierra Leone, on the west coast of sub-Saharan Africa), but Africans still tend to have the highest risk of death in the world, aggravated, of course, by the high incidence of AIDS (Bongaarts, 1996; Goliber, 1997).

If we rather arbitrarily define a high-mortality country as one in which the life expectancy for females is 50 years or less, then we find that there are currently 21 such countries in the world. All have accompanying high birth rates, and 19 of the 21 are located in sub-Saharan Africa. In the past several years the tremendous suffering of people in some of these high-mortality countries, such as Somalia, Angola, Rwanda, and Uganda, has received worldwide attention.

In the first edition of this book I suggested that the demographics of high-fertility, high-mortality countries could be illustrated by Cameroon, a middle-African country just north of the equator and facing the Atlantic Ocean. I further indicated that "countries like Cameroon all represent potentially explosive populations in terms of future growth if mortality conditions improve" (Weeks, 1978:49). By now this set of events is in full swing. The death rate in Cameroon began to drop in the mid-1970s, and the rate of population growth has jumped from 1.8 percent per year (a doubling in 39 years) to 2.8 percent per year (a doubling in a mere 25 years).

Somalia is a country whose images were brought to television viewers all over the world in the early 1990s when the United States, under the auspices of the United Nations, sent a military contingent to provide emergency relief and to attempt to restore civil order. Somalia is a predominantly agricultural nation, and about half of the more than 10 million inhabitants of the country are nomadic pastoralists. Islam has been the major religion in the region for more than 1,000 years. Modern difficulties began with European colonization. The area was divided up between England, Italy, and France from the late 1800s until 1960, when the pieces were put together again to create a newly independent state. Women in Somalia are bearing an average of 7 children each, and although the infant mortality rate of 122 deaths per 1,000 live births is one of the highest in the world, it is still lower than it used to be. Life expectancy for a Somali female baby is only 49 years (about the same as the United States in 1900), but a mother can still expect that three fourths of her children (5 of the 7) will survive to adulthood, helping to explain why 47 percent of the population is under the age of 15. The problem is that the economic infrastructure does not exist to accommodate all of these people who are growing into adulthood.

The drop in the death rate in the past few decades, unaccompanied by a drop in the birth rate, has put tremendous pressure on the relatively fragile environment. The population of Somalia has doubled in just 20 years, dramatically increasing the demand for food, and this has led to the overgrazing of land and desertification, as well as contributing to an overall collapse of organized society.

So despite being the region of the world where mortality is highest, the fact that mortality is dropping and fertility is not (at least not by very much) explains how it is that between 1950 and the year 2000 sub-Saharan Africa regained its place as the most rapidly growing region in the world. This situation is not likely to change in the foreseeable future, since "the onset of the fertility transition in sub-Saharan Africa has been slow and limited, and even in cases where evidence of fertility decline has been found, the results have been subject to debate" (Shapiro, 1996:89).

East Asia

There are 1.5 billion people living in East Asia, with the region dominated demographically by China (accounting for 85 percent of the population in this part of the world) and Japan (with 8 percent of the region's population). East Asia includes 25 percent of the world's total population, but its share will diminish a bit in the future as China brakes its population growth.

China   The People's Republic of China has a population of more than 1.2 billion people, maintaining for the time being its long-standing place as the most populous country in the world. With more than one fifth of all human beings, China dominates the map of the world drawn to scale according to population size (see Figure 1.3). If we add in the Chinese in Taiwan (which the government of mainland China still claims as its own), Singapore, and the overseas Chinese elsewhere in the world (Poston, Mao, and Yu, 1994), closer to one out of every four human beings is of Chinese origin. Nonetheless, China's share of the world's total population actually peaked in the middle of the nineteenth century, as you can see in Table 1.6. In 1850 more than one in three humans was living in China, and that fraction has steadily declined over time.

For years the Chinese tried to ignore their demographic bulk, perhaps fearful that outside reports of population size and growth rates would not take sufficient account of the fact that the Communist revolution inherited a very large problem. In 1982 China took stock of the magnitude of its problem with its first national census since 1964 (which had been taken shortly after the terrible famine that I mentioned earlier). A total of just slightly more than 1 billion people were counted, and the results seemed to reinforce the government's belief that it was on the right track in vigorously pursuing a fairly coercive one child policy to cut the birth rate (which I will discuss in detail in Chapter 15). Fertility decline actually began in China's cities in the 1960s and spread rapidly throughout the rest of the country in the 1970s, when the government introduced the family planning program known as *wan xi shao*, meaning "later" (marriage), "longer" (birth interval), "fewer" (children) (Banister, 1987; Goldstein and Feng, 1996). In 1979 this was transformed into the one child policy, but fertility was already on its way down by that time. Between 1963 and 1983 China experienced what Blayo (1992) has called a "breathtaking drop in TFRs [total fertility rates, see Chapter 5] from 7.5 to 2.5 children" (p. 213), and what Riley and Gardner (1996) suggest is "the most rapid sustained fertility decline ever seen in a population of any size" (p. 1).

China's fertility has continued to decline, although at a slower pace. China's birth rate has now dropped to 1.9 children per woman, which is below the replacement rate, although as I will discuss later in the book, that does not yet mean that the population has stopped growing. Furthermore, it is well above the government goal of 1 child per woman. Life expectancy at birth for females is 71 years, nearly 3 years above the world average, but there is still room for improvement. The infant mortality of 45 deaths per 1,000 live births is several times higher than the rate in the United States and Canada (6 per 1,000).

Despite its low birth rate, the number of births each year in China is more than twice the number of deaths just because China is paying for its previous high birth rate. There are so many young women of reproductive age (women born 20 to 40 years ago when birth rates were much higher) that their babies vastly outnumber the people who are dying each year. As a result, the rate of natural increase in China is quite a bit higher than in the United States. However, the United States has essentially the same total rate of population growth as does China, because the United States accepts more immigrants each year than any other country in the world. Of course, since China has more than four times as many inhabitants as does the United States, it is adding 14 million people each year to its territory, whereas the United States is accumulating "only" an additional 2.5 million (and that includes immigrants). Thus population growth remains a serious concern in China, reflected in the following comment by a Chinese demographer:

> Despite the outstanding achievements made in population and development, China still confronts a series of basic problems including a large population base, insufficient cultivated land, under-development, inadequate resources on a per capita basis and an uneven social and economic development among regions. At present, the annual new-borns number 20 million resulting in a net increase of population of about 14 million. Too many people has impeded seriously the speed of social and economic development of the country and the rise of the standard of living of the people. Many difficulties encountered in the course of social and economic development are directly attributable to population problems [Peng, 1996:7].

Japan   Let me also briefly mention Japan, which has the lowest level of mortality in the world, with a female life expectancy at birth of 83 years. Japan's health (accompanied by its wealth) translates demographically into very high probabilities of survival to old age—indeed, more than half of all Japanese born in the year 2000 will likely still be alive at age 80. This very low mortality is accompanied by very low fertility. Japanese women are bearing an average of 1.5 children each, of whom virtually all will live to adulthood. This is a level well below that required to replace each generation (2.1), and Japan is currently growing in number only because of the fact that there are still enough women of reproductive age that, even at 1.5 each, they produce more babies than there are people dying each year. Japan's low mortality and low fertility have produced a population in which only 16 percent are under age 15 and 14 percent are 65 or older, with a forecasted rise to 26 percent aged 65 or older by the year 2025. The resulting "graying" of Japan has caused a new set of problems for that island nation to worry about (Ogawa and Retherford, 1997), but thus far there have been only a few suggestions that the nation should encourage a higher birth rate.

Japan has an extremely restrictive immigration policy, so migrants contribute little to demographic change in Japan. By contrast, in several European countries the sustained low birth rates would have produced an even steeper increase in the percent of the population that is elderly were it not for the impact of in-migrants from less developed regions of the world.

South and Southeast Asia

South and southeast Asia is home to about 2 billion people, one third of the world's total. The Indian subcontinent dominates this area demographically, with India, Pakistan, and Bangladesh encompassing two thirds of the region's population. But Indonesia, the world's fourth most populous nation (and the one with the largest Muslim population in the world) is also part of southeast Asia, as are 3 other countries on the top 20 list—the Philippines, Vietnam, and Thailand.

India, Pakistan, and Bangladesh   Second in population size in the world is India, with the current population estimated to be just beyond 1 billion. Mortality is somewhat higher in India than in China, and the birth rate is quite a bit higher than in China. Indian females have a life expectancy at birth of 64 years—below the world average, but a substantial improvement over the life expectancy of less than 27 years that prevailed back in the 1920s (Adlahka and Banister, 1995). The infant mortality rate of 63 per 1,000 is above the world average, but it is also far lower than it was just a few decades ago. Women are bearing children at a rate of 3.2 each, and most children in India now survive to adulthood. With an annual growth rate of 1.6 percent, the Indian population is adding 17 million people to the world's total each year. As a result, the population of India is expected to exceed that of China by the year 2025, as you can see in Table 1.5. In fact, the population of the Indian subcontinent is already more populous than mainland China (see Table 1.6), and that does not take into account the 15 million people of Indian origin who are estimated to be living elsewhere in the world (Visaria and Visaria, 1995).

It appears that fertility prior to 1961 was high and remained virtually constant (Jain and Adlahka, 1982), but since then it has declined, although not nearly as rapidly as in China. Furthermore, these declines have been nearly matched by declining mortality, thus maintaining the rate of population growth in India at a level consistently above the world's average. In fact, given its birth rate, if its death rate were to drop to a level equal to that of China, India would now be adding 20 million people a year, instead of its present 17 million.

India's population is culturally diverse, and this is reflected in rather dramatic geographic differences in fertility and rates of population growth within the country. For example, in the southern states of Kerala and Tamil Nadu, fertility has dropped below the replacement level. By comparison, in the 4 most populous states in the north (Bihar, Madhya Pradesh, Rajasthan, and Uttar Pradesh), where

40 percent of the Indian population lives, the average woman is bearing 5 children (Adlahka and Banister, 1995).

At the end of World War II, as India was granted its independence from British rule, the country was divided into predominantly Hindu India and predominantly Muslim Pakistan, with the latter having territory divided between West Pakistan and East Pakistan. In 1971 a civil war erupted between the two disconnected Pakistans, and, with the help of India, East Pakistan won the war and became Bangladesh. Although Pakistan and Bangladesh are both Muslim, Bangladesh has a demographic profile that now looks more like India than Pakistan. The average woman in Bangladesh now gives birth to 3.3 children (compared with 3.4 in India), whereas fertility in Pakistan has remained much higher (4.9 children per woman). The overall rate of population growth in Bangladesh is 1.7 percent per year (compared with 1.6 in India), but it is 2.2 percent per year in Pakistan.

Indonesia   Indonesia is a string of islands in Southeast Asia and is the world's most populous Muslim nation, with an estimated 219 million people. A former Dutch colony, it has experienced a substantial decline in fertility in recent years, and Indonesian women now bear an average of 2.6 children each. Life expectancy for women at birth is 65 years (just below the world average), while the infant mortality rate of 59 deaths per 1,000 live births is just at the world average.

Oceania

Oceania is home to a wide range of indigenous populations, including Melanesian and Polynesian, but European influence has been very strong, and the region is generally thought of as being "overseas European." Its population of 30 million is about the same number as Canada, less than 1 percent of world's total. Australia accounts for two thirds of the region's population, followed by Papua New Guinea and New Zealand. In a pattern repeated elsewhere in the world, the lowest birth rates and lowest death rates (and thus the lowest rates of population growth) are found in countries whose populations are largely European-origin (Australia and New Zealand, in this case), whereas the countries with a higher fraction of the population that is of indigenous origin have higher birth rates, higher mortality, and substantially higher rates of population growth (exemplified in Oceania by Papua New Guinea).

Summary and Conclusion

High death rates kept the number of people in the world from growing rapidly until approximately the time of the Industrial Revolution. Then improved living conditions and medical advancements accelerated the pace of growth dramatically. As populations have grown, the pressure or desire to migrate has also increased. The vast European expansion into the less developed areas of the world, which began in the fifteenth and sixteenth centuries, is a notable illustration of massive migration and population redistribution. Today migration patterns have shifted, and more people move from less developed areas to industrialized nations. Closely associated with migration and population density is the urban revolution—that is, the movement from rural to urban areas.

The current world situation finds China and India as the most populous countries, followed by the United States, Indonesia, and Brazil. The United States is growing more slowly than the other populous nations because its fertility is lower. Everywhere that the population is growing we find that death rates have declined more rapidly than have the birth rates, but there is considerable global and regional variability in both the birth and the death rates and thus in the rate of population growth. Dealing with the pressure of an expanding young population is the task of the developing countries, whereas the more developed countries, with slowly growing (and thus aging) populations, are coping with the fact that the demand for labor in their economies is most readily met by immigrants from the more rapidly growing countries. Thus, demographic dynamics represent the leading edge of social change in the modern world.

These are only some among many ways in which population influences people's lives, and as I explore with you the causes and consequences of population growth and the uses to which such knowledge can be applied, you will need to know about the sources of demographic data. What is the empirical base of our understanding of the relationship between population and society? We turn to that topic in the next chapter.

Main Points

1. Pessimists argue that population growth is like an exploding bomb, whereas optimists hope for new technologies to take care of ever-increasing numbers.

2. Regardless of whether you are a pessimist or an optimist, you will be sharing a future with millions more people than there are today. Analyzing these changes is the business of demography, the science of population.

3. Demography is concerned with virtually everything that influences or can be influenced by population size, distribution, processes, structure, or characteristics.

4. The cornerstones of population studies are the processes of mortality (a deadly subject), fertility (a well-conceived topic), and migration (a moving experience).

5. During the first 90 percent of human existence, the population of the world had grown only to the size of New York City today.

6. Between 1750 and 1950 the world's population mushroomed from 800 million to 2.5 billion, and since 1950 it has expanded to 6 billion.

7. The doubling time is a convenient way to summarize the rate of population growth. It is found by dividing the average annual rate of population growth into 69.

8. Early growth of population was slow not because birth rates were low but because death rates were very high.

9. On the other hand, recent rapid increases are due to dramatic declines in mortality without a commensurate decline in fertility.

10. World population growth has been accompanied by migration from rapidly growing areas into less rapidly growing regions. Initially, that meant an outward expansion of the European population, but more recently it has meant migration from less developed to more developed nations.

11. Migration has also involved the shift of people from rural to urban areas, and urban regions on average are currently growing more rapidly than ever before in history.

12. Although migration is crucial to the demographic history of the United States and Canada, both countries have grown largely as a result of natural increase—the excess of births over deaths—after the migrants arrived.

13. At the time of the American Revolution, fertility levels in North America were among the highest in the world. Now they are among the lowest, even though there are still more babies born each year than there are people dying.

14. The world's 10 most populous countries are the People's Republic of China, India, the United States, Indonesia, Brazil, Russia, Pakistan, Japan, Bangladesh, and Nigeria. Together they account for more than half of the world's population.

15. At its present rate of growth the world's population could double in size in about 50 years. Five of the ten largest countries are growing even more rapidly than that because of their relatively low death rates combined with relatively high birth rates.

Suggested Readings

1. Joel Cohen, 1995, How Many People Can the Earth Support? (New York: W.W. Norton).

   Cohen shows how to approach an answer to this question, since there is not a single correct solution. In doing so, he has literally jam-packed his analysis with useful and interesting information about past, present, and future population growth trends and what they might mean for human society.

2. Massimo Livi-Bacci, 1992, A Concise History of World Population (translated by Carl Ipsen) (Cambridge, MA: Blackwell).

   This is a brief but very thoughtful and well-documented account of population growth from ancient times to the early 1990s.

3. United Nations, Demographic Yearbook (New York: United Nations), published annually.

   No matter how boring it first may seem, you will find yourself becoming absorbed by (and, in a way, into) this annual compilation of demographic data for nearly all of the countries on the globe.

4. Lester R. Brown and associates, State of the World: A Worldwatch Institute Report on Progress Toward a Sustainable Society (New York: W.W. Norton), published annually.

   This annual set of research papers by scholars at the Worldwatch Institute in Washington, D.C. is a nice companion piece to the Demographic Yearbook, because its goal is to analyze what these trends in population growth (and related issues) mean for the future of the globe.

5. Thomas McKeown, 1976, The Modern Rise of Population (London: Edward Arnold).

   This is probably the single most important (and still widely discussed and reevaluated) study of why mortality began to decline in Europe—the event that set off the chain reaction around the world known as the population explosion.

✪ Websites of Interest

Remember that websites are not as permanent as books and journals, so I cannot guarantee that each of the following websites still exists at the moment that you are reading this.

1. http://www.slate.com/redirect/announce.asp?gotoT=/Code/DDD/DDD.
   asp?file=population

   Slate magazine is a Microsoft Network–owned on-line magazine that features regular debates on various subjects of interest. This dialogue, published in late 1997, features Ben Wattenberg (a "boomster" at the American Enterprise Institute) and Kenneth Hill (a professor at The Johns Hopkins University who is probably more of a "realist" than necessarily a "doomster").

2. http://www.census.gov/ipc/www

   The web site of the International Programs Center at the U.S. Bureau of the Census has several very useful features, including world population information, a world population clock (where you can check the latest estimate of the total world population), and an international database of population information that you can search on-line or download to your own computer.

3. http://www.popexpo.net/eMain.html

   Would you like to know not only how many people there are at this moment in the world but how many there were when you were born? This website offers you that bit of interesting information (no, not trivia—this is serious) and a lot more.

4. http://www.un.org/Pubs/CyberSchoolBus/infonation/e_infonation.htm

   The United Nations has developed this website that allows you to choose several countries at a time and look at demographic, economic, and social indicators for those nations.

5. http://www.popnet.org

   The Population Reference Bureau in Washington, D.C. has created this very useful website that provides you not only with their information but with linkages to a wide range of other population-related websites located all over the world.

CHAPTER 2
Demographic Resources

In order to analyze the demography of a particular society, we need to know how many people live there, how they are distributed geographically, how many are being born, how many are dying, how many are moving in, and how many are moving out. That, of course, is only the beginning. If we want to unravel the mysteries of why things are as they are and not just describe what they are, we have to know about the social, psychological, economic, and even physical characteristics of the people being studied. Furthermore, we need to know these things not just for the present, but for the past as well. Let me begin the discussion, however, with the sources of basic information about the numbers of living people, births, deaths, and migrants.

Sources of Demographic Data

The primary source of data on population size and distribution, as well as on demographic structure and characteristics, is the **census of population,** and after an overview of the history of population censuses, I will take a closer look at census taking in North America—the United States, Canada, and Mexico. The major source of information on the population processes of births and deaths is the registration of **vital statistics,** although in a few countries this task is accomplished by **population registers.** These sources are often supplemented with data from *sample surveys* as well as **historical sources. Administrative data** provide much of the information about population changes at the local level and about geographic mobility and migration.

Overview of Population Censuses

For centuries, governments have wanted to know how many people were under their rule. Rarely has their curiosity been piqued by scientific concern, but rather governments wanted to know who the taxpayers were, or they wanted to identify potential laborers and soldiers. The most direct way to find out how many people there are is to count them, and when you do that you are conducting a population census. The United Nations defines a census of population more specifically as "the total process of collecting, compiling and publishing demographic, economic and social data pertaining, at a specified time or times, to all persons in a country or delimited territory" (United Nations 1958:3). In practice this does not mean that every person actually is seen and interviewed by a census taker. In most countries it means that one adult in a household answers questions about all the people living in that household. These answers may be written responses to a questionnaire sent by mail or verbal responses to questions asked in person by the census taker.

The term *census* comes from Latin, and for Romans it meant a register of adult male citizens and their property for purposes of taxation, the distribution of military obligations, and the determination of political status (Starr 1987). Thus, in 119 A.D. a person named Horos from the village of Bacchias left behind a letter on papyrus in which he states: "I register myself and those of my household for the house-by-house census of the past second year of Hadrian Caesar our Lord. I am

Horos, the aforesaid, a cultivator of state land, forty-eight years old, with a scar on my left eyebrow, and I register my wife Tapekusis, daughter of Horos, forty-five years old. . ." (Winter 1936:187).

As far as we know, the earliest governments to undertake censuses of their populations were those in the ancient civilizations of Egypt, Babylonia, China, India, and Rome (Shryock, et al. 1973). For about 800 years, citizens of Rome were counted every 5 years for tax and military purposes, and this enumeration was extended to the entire Roman Empire in 5 B.C. The Bible records this event as follows: "In those days a decree went out from Caesar Augustus that all the world should be enrolled. This was the first enrollment, when Quirinius was governor of Syria. And all went to be enrolled, each to his own city" (Luke 2:1-3). You can, of course, imagine the deficiencies of a census that required people to show up at their birthplaces rather than paying census takers to go out and do the counting.

In the seventh century A.D. the Prophet Mohammed led his followers to Medina (in Saudi Arabia), and upon establishing a city-state there, one of his first activities was to conduct a written census of the entire Muslim population in the city (the returns showed a total of 1,500). William of Normandy employed a similar strategy in 1086 after having conquered England in 1066. William ordered an enumeration of all the landed wealth in the newly conquered territory in order to determine how much revenue the landowners owed. Data were recorded in the Domesday Book, the term *domesday* being a corruption of the word *doomsday,* which is the day of final judgment. The census document was so named because once entered, the information was considered final and irrevocable. The Domesday Book was not really what we think of today as a census, because it was an enumeration of "hearths," or household heads and their wealth, rather than of people. In order to calculate the total population of England in 1086 from the Domesday Book you would have to multiply the number of "hearths" by some estimate of household size. More than 300,000 households were included, so if they averaged five persons per household, the population of England at the time was approximately 1.5 million (Hinde 1995).

The European Renaissance began in northern Italy in the 14th century, and the Venetians and then the Florentines were interested in counting the wealth of their region, as William had been after conquering England. They developed a *catasto* that combined a count of the hearth and individuals. Thus, unlike the Domesday Book, the Florentine Catasto of 1427 recorded not only the wealth of households but also data about each member of the household. In fact, so much information was collected that most of it went unexamined until the modern advent of computers (Herlihy and Klapisch-Zuber 1985).

A few other sporadic censuses were taken in Europe between the Domesday Book and the arrival in the 18th century of modern nation-states, which gave rise to the quest for population information (Hollingsworth 1969). Indeed the term *statistic* is derived from the German word meaning "facts about a state." Sweden was one of the first of the European nations to regularly keep track of its population, with the establishment in 1749 of a combined population register and census administered in each diocese by the local clergy (Statistika Centralbyran [Sweden] 1983). Denmark and several Italian states (before the uniting of Italy) also conducted censuses during the 18th century (Carr-Saunders 1936), as did the United States (where the first census was conducted in 1790). England launched its first census in 1801.

By the latter part of the 19th century the statistical approach to understanding business and government affairs had started to take root in the Western world (Cassedy 1969). The population census began to be viewed as a potential tool for finding out more than just how many people there were and where they lived. Governments began to ask questions about age, marital status, whether people were employed and what their occupation was, literacy, and so forth.

Census data (in combination with other statistics) have become the "lenses through which we form images of our society." Frederick Jackson Turner announced this famous view on the significance of the closing of the frontier on the basis of data from the 1890 census. Our national self-image today is confirmed or challenged by numbers that tell of drastic changes in the family, the increase in ethnic diversity, and many other trends. Winston Churchill observed that "first we shape our buildings and then they shape us. The same may be said of our statistics" (Alonso and Starr 1982:30). The potential power behind the numbers that censuses produce can be gauged by public reaction to a census. In Germany the enumeration of 1983 was postponed to 1987 because of public concern that the census was prying unduly into private lives. In the past few decades protests have occurred in England, Switzerland, and the Netherlands as well. In the latter case the census was actually canceled after a survey indicating that the majority of the urban population would not cooperate (Robey 1983), and no census has been taken since. However, Netherlands is a country that maintains a population register, which I discuss later, and so they do have good demographic information even in the absence of a census.

Since the end of World War II, the United Nations has encouraged countries to enumerate their populations in a census, often providing financial as well as technical aid. As a result, between 1953 and 1964, 78 percent of the world's population (including that of mainland China) was enumerated by census. Between 1965 and 1974 the United Nations calculates that 86 percent of the countries of the world conducted at least one census (Suharto and Volkov 1993). However, since mainland China was not one of those nations taking a census, only 55 percent of the world's population was enumerated in that decade. Between 1975 and 1984, 89 percent of countries conducted a census, encompassing 96 percent of the world's population. In the most recent census decade (1985 through 1994), 77 percent of countries took a census, and in the process 90 percent of the world's population was enumerated (United Nations 1996). Throughout the world, there has been a fairly clear relationship between a country's level of economic advancement and the taking of censuses. Reasons for this relationship are numerous, but the most obvious is cost (the Census 2000 in the United States carries an estimated price tag of $3.9 billion).

In the less developed nations, there is less pressure on the government to conduct a census, because the benefits of the undertaking may not appear to outweigh the costs. Nevertheless, in the early 1990s the world's two largest less developed nations (in fact, the two largest nations regardless of stage of development), China and India, each conducted a census. In October 1990 the Chinese government undertook the most ambitious census in world history when it counted its 1.1 billion inhabitants. The task itself involved nearly 7 million enumerators, more people at the time than the entire population of Switzerland. The Chinese were aided in their census by the United Nations Population Fund, and, like the previous census in 1982, computers aided the number crunching. Before that, data from Chinese censuses had been tabulated using abacuses and manual methods.

The year 1991 began the second 100 years of census taking in India, the first census having been taken in 1881 under the supervision of the British. The census counted 844 million Indians as of March 1991, although many people believed that this number was too low. On the other hand, the 1981 census of India counted 12 million more people than the government had expected to find. These results called into question some of the optimism that Indian officials had expressed during the 1970s over the potential success of the nation's family planning program (Visaria and Visaria 1981).

In contrast to India's regular census-taking, another of England's former colonies, Nigeria (the world's tenth most populous nation), has been less successful in these efforts. Nigeria's population is divided among three broad ethnic groups: the Hausa-Fulani in the north, who are predominately Muslim; the Yorubas in the southwest (of various religious faiths); and the largely Christian Ibos in the southeast. The 1952 census of Nigeria indicated that the Hausa-Fulani had the largest share of the population, and so they dominated the first post-colonial government set up after independence in 1960. The newly independent nation ordered a census to be taken in 1962, but the results showed that northerners accounted for only 30 percent of the population. A "recount" in 1963 led somewhat suspiciously to the north accounting for 67 percent of the population. This exacerbated the underlying ethnic tensions, culminating in the Ibos declaring independence. The resulting 3-year Biafran war (1967–70) saw at least 3 million people lose their lives before the Ibo rejoined the rest of Nigeria. A census in 1973 was never accepted by the government, and it was not until 1991 that the nation felt stable enough to try its hand again at an enumeration, after agreeing that there would be no questions about ethnic group, language, or religion, and that population numbers would not be used as a basis for distributing government expenditures. The official census count was 88.5 million people, well below the 110 million that many population experts had been guessing in the absence of any real data.

Lebanon has not been enumerated since 1932, when the country was under French colonial rule (Domschke and Goyer 1986). The nearly equal number of Christians and Muslims in the country and the political strife between those groups have combined to make a census a very sensitive political issue. Before the nation was literally torn apart by civil war in the 1980s, the Christians had held a slight majority with respect to political representation. But Muslims almost certainly now hold a demographic majority (Faour 1991), and a census could either bolster or deny a group's claim to power and thus create the potential for further divisiveness.

I should note that censuses historically have been unpopular in that part of the world. The Old Testament of the Bible tells us that in ancient times King David ordered a census of Israel in which his enumerators counted "one million, one hundred thousand men who drew the sword. . . . But God was displeased with this thing [the census], and he smote Israel. . . . So the Lord sent a pestilence upon Israel; and there fell seventy thousand men of Israel" (1 Chronicles 21). Fortunately, in modern times the advantages of census taking seem more clearly to outweigh the disadvantages. This has been especially true in the United States, where records indicate that no census has been followed directly by a pestilence.

The Census of the United States

Population censuses were part of colonial life prior to the creation of the United States. A census had been conducted in Virginia in the early 1600s, and most of the northern colonies had conducted a census prior to the Revolution (U. S. Bureau of the Census 1978). A census of population has been taken every 10 years since 1790 in the United States as a part of the constitutional mandate that seats in the House of Representatives be apportioned on the bases of population size and distribution. But even in 1790 the government used the census to find out more than just how many people there were. The census asked for the names of head of family, free white males aged 16 years and older, free white females, slaves, and other persons (Shryock, et al. 1973:22). The census questions were reflections of the social importance of those categories.

For the first 100 years of census taking in the United States, the population was enumerated by U.S. marshals. In 1880 special census agents were hired for the first time, and by 1902 the Census Bureau had become a permanent part of the government bureaucracy (Francese 1979). Beyond a core of demographic and housing information, the questions asked on the census have fluctuated according to the concerns of the time. Interest in international migration, for example, peaked in 1920 just before the passage of a restrictive immigration law (see Chapter 14), and the census in that year added a battery of questions about the foreign-born population. In 1970, as the number of disabled people in the country increased—especially young men disabled in the Vietnam War—questions about disabilities were asked for the first time since 1910. Questions are added and deleted by the Census Bureau by consultation with other government officials and census statistics users.

One of the more controversial items for the Census 2000 questionnaire was the question about race and ethnicity. The growing racial and ethnic diversity of the United States has led to a larger number of interracial/interethnic marriages and relationships producing children of mixed origin (also called multiracial). Previous censuses had asked people to choose a single category of race to describe themselves, but there was a considerable public sentiment that people should be able to identify themselves as being of mixed or multiple origins if, in fact, they perceived themselves in that way. Late in 1997 the government accepted the recommendation from a federally appointed committee that people of mixed racial heritage will be able to choose more than one race category when filling out the Census 2000 questionnaire. Thus, a person whose mother is white and whose father is African-American could check both "White" and "Black or African American" whereas in the past the choice would had to have been made between one or the other. There is still a separate question on "Hispanic/Latino/Spanish Origin" identity.

If questions on the U.S. Census do not have a clear legal mandate (even if they are generally useful), then they are susceptible to being dropped, and that is what happened to several items that were asked in the 1990 Census but are not included in the Census 2000. In particular, Census 2000 does not have a question on fertility of American women (vital statistics and surveys now take over that role).

The census is designed as a complete enumeration of the population, but only a few of the questions are actually asked of everyone. For reasons of economy, most items in the census questionnaire are administered to a sample of households. From 1790 through 1930, all questions were asked of all applicable persons, but as the

American population grew and Congress kept adding new questions to the Census, the savings involved in sampling grew, and in 1940 the Census Bureau began its practice of asking only a small number of items of all households, and using a sample of households to gather data that are more detailed. Fortunately, there has been no significant loss in accuracy (after all, a sample of 1 out of every 6 households in 1990 still yielded detailed data for more than 40 million people).

The items of information obtained from everyone are often called the long-form items and include basic demographic and housing characteristics. The draft of the questionnaire for Census 2000 is reproduced as Figure 2.1. The first page asks that everyone in the household be listed by name, which allows the Census Bureau to check for duplicate listings of people (such as college students away from home). Then, information is requested for each person in the household regarding his or her relationship to the head of household (who is to be listed in column 1), sex, racial and ethnic identification, age and year of birth, marital status, and Hispanic/Latino origin. These and the questions relating to characteristics of the housing unit comprise the short-form. Approximately 5 in 6 of all households will receive only the short form to fill out, whereas 1 in 6 (about 17 percent of households) will be asked to complete the longer, more detailed questionnaire shown in Figure 2.1. Table 2.1 on pages 52 and 53 lists the items included on the U.S. Census 2000 questionnaire, in comparison with a list of the data obtained by the 1996 Census of Canada and the 1990 Census of Mexico. The table indicates which items are asked of every household and which are asked of a sample of households.

In theory, a census obtains accurate information from everyone. In reality, however, a number of possible errors can creep into the enumeration process. We can divide these into the two broad categories of **nonsampling error** and **sampling error**.

Nonsampling Error   Inaccuracies can occur in a census from a variety of sources, but the two most common sources of error are *coverage error* and *content error.* Although in theory a census counts everyone, in reality there are always people who are missed, as well as a few who are counted more than once (this is coverage error). In the 1970 U.S. census, for example, it is estimated that about 5 million people were missed (Siegel 1974), representing a little more than 2 percent of the total. The 1980 U.S. census was even better, missing only an estimated 1.4 percent of the total population (Passell, Cowan, and Wolter 1983). The undercount in 1990 was estimated to be 2.1 percent, higher than in 1980 (Choldin 1994). Still, this was a major accomplishment given the fact that only 74 percent of households mailed back their census returns, a number much lower than had been expected. However, the U.S. Bureau of the Census sent enumerators into the field to collect completed forms, to interview people who had not completed forms, and, in some cases, to find out about people whom they were unable to contact. About 3 percent of households were counted on this "last resort" basis. After at least six unsuccessful attempts to track down people in person or on the phone, enumerators contacted neighbors, mail carriers, building superintendents, or anyone else who might have knowledge of the otherwise missing person. When you combine this with the fact that one member of a household may have filled in the information for all household members, it is easy to see why so many people think they have not been counted in the census—someone else answered for them.

Text continues on page 53.

46

Figure 2.1 Draft of U.S. Census 2000 Questionnaire

United States
Census 2000
Dress Rehearsal

U.S. Department of Commerce
Bureau of the Census

This is the official form for all the people at this address. It is quick and easy, and your answers are protected by law. Complete the census and help your community get what it needs — today and in the future!

This booklet shows the content of the two main questionnaires being used in the U.S. Census 2000 Dress Rehearsal. See the explanatory notes on page 2.

Start Here

1. **If possible, this form should be filled out by one of the people who live here who owns, is buying, or rents this house or apartment.** A black or blue pen is preferred.

2. **If this house or apartment is a vacation or seasonal home or a temporary residence for your household, please call the Census Bureau at 1-888-421-1998 before you fill out this form.** The telephone call is free.

➜ **Next, please turn the page and print the names of all the people living or staying here on April 18, 1998.**

If you need help completing this form, call 1-888-421-1998 between 8:00 a.m. and 9:00 p.m., 7 days a week. The telephone call is free.

TDD — Telephone display device for the hearing impaired. Call 1-800-582-8330 between 8:00 a.m. and 9:00 p.m., 7 days a week. The telephone call is free.

¿NECESITA AYUDA? Si usted habla español y necesita ayuda para completar su cuestionario o si requiere un cuestionario en español, llame sin cargo alguno al 1-888-427-1998 entre las 8:00 a.m. y las 9:00 p.m., 7 días a la semana (la llamada telefónica es gratis).

The Census Bureau estimates that, for the average household, this form will take about 38 minutes to complete, including the time for reviewing the instructions and answers. Comments about the estimate should be directed to the Associate Director for Administration/Controller, Attn: Paperwork Reduction Project 0607-0848, Room 3104, Federal Building 3, Bureau of the Census, Washington, DC 20233.

Respondents are not required to respond to any information collection unless it displays a valid approval number from the Office of Management and Budget.

OMB No. 0607-0848. Approval Expires 12/31/98

Form IDX-61 (I-32-98)

INFORMATIONAL COPY

List of Persons

1 **Please print the names of all the people living or staying here on April 18, 1998, as shown in this example:**

Example — Last Name
J O H N S O N

First Name | MI
R O B I N | J

BE SURE TO INCLUDE anyone who is:
- a foster child, roomer, or housemate
- staying here on April 18, 1998 and has no other permanent place to stay
- staying here most of the time while working even if he or she has another place to live

DO NOT INCLUDE anyone who:
- is living away while attending college
- was in a correctional facility, nursing home, or mental hospital on April 18, 1998
- is in the Armed Forces and living somewhere else
- lives or stays at another place most of the time

Start with the person, or one of the people living here in whose name this house or apartment is owned, being bought, or rented. If there is no such person, start with any adult living or staying here.

Person 1 — Last Name

First Name | MI

Person 2 — Last Name

First Name | MI

Person 3 — Last Name

First Name | MI

Person 4 — Last Name

First Name | MI

Form IDX-61 (I-12-98)

Person 5 — Last Name

First Name | MI

Person 6 — Last Name

First Name | MI

Person 7 — Last Name

First Name | MI

Person 8 — Last Name

First Name | MI

Person 9 — Last Name

First Name | MI

Person 10 — Last Name

First Name | MI

Person 11 — Last Name

First Name | MI

Person 12 — Last Name

First Name | MI

➜ **Next, answer questions about Person 1.**

EXPLANATORY NOTES

This "Informational Copy" shows the basic content of the two questionnaires that will be mailed to households for the United States Census 2000 Dress Rehearsal. Each household will receive either a short form (100-percent questions) or a long form (100-percent and sample questions). The content of the 1990 census, consulting with federal and non-federal data users, and conducting tests.

Short form – This questionnaire contains 6 population questions and 1 housing question. On average, about 5 in every 6 households will receive this form. For the average household, this form will take an estimated 10 minutes to complete.

Long form – This questionnaire has all the short form questions, plus 26 population questions, and 20 housing questions. A statistical sample of approximately 1 in every 6 households will receive this form. For the average household, this form will take an estimated 38 minutes to complete.

Population Questions
- Questions 1–6 are asked of all persons on both forms.
- Questions 7–32 are asked of persons in a sample of house-holds on the long form.

Housing Questions
- Question 33 is asked of all households on both forms.
- Questions 34–53 are asked of a sample of households on the long form.

2

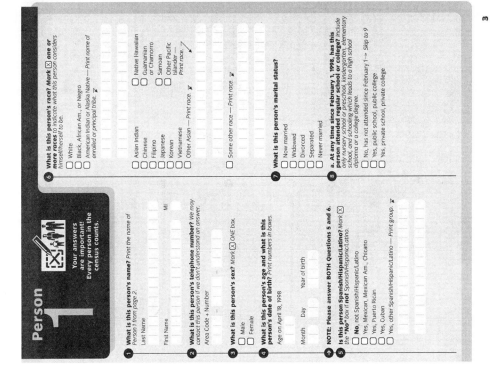

Person 1

Your answers are important!
Every person in the census counts.

1. What is this person's name? *Print the name of Person 1 from page 2.*

Last Name

First Name ⬚ MI

2. What is this person's telephone number? *We may contact this person if we don't understand an answer.*

Area Code + Number

3. What is this person's sex? *Mark ⊠ ONE box.*
☐ Male
☐ Female

4. What is this person's age and what is this person's date of birth? *Print numbers in boxes.*

Age on April 18, 2008

Month ⬚ Day ⬚ Year of birth

→ **NOTE: Please answer BOTH Questions 5 and 6.**

5. Is this person Spanish/Hispanic/Latino? *Mark ⊠ the "No" box if not Spanish/Hispanic/Latino.*

☐ **No,** not Spanish/Hispanic/Latino
☐ Yes, Mexican, Mexican Am., Chicano
☐ Yes, Puerto Rican
☐ Yes, Cuban
☐ Yes, other Spanish/Hispanic/Latino — *Print group.* ➚

6. What is this person's race? *Mark ⊠ one or more races to indicate what this person considers himself/herself to be.*

☐ White
☐ Black, African Am., or Negro
☐ American Indian or Alaska Native — *Print name of enrolled or principal tribe.* ➚

☐ Asian Indian   ☐ Native Hawaiian
☐ Chinese   ☐ Guamanian or Chamorro
☐ Filipino   ☐ Samoan
☐ Japanese   ☐ Other Pacific Islander — *Print race.* ➚
☐ Korean
☐ Vietnamese
☐ Other Asian — *Print race.* ➚

☐ Some other race — *Print race.* ➚

7. What is this person's marital status?
☐ Now married
☐ Widowed
☐ Divorced
☐ Separated
☐ Never married

8. a. At any time since February 1, 1998, has this person attended regular school or college? *Include only nursery school or preschool, kindergarten, elementary school, and schooling which leads to a high school diploma or a college degree.*

☐ No, has not attended since February 1 → *Skip to 9*
☐ Yes, public school, public college
☐ Yes, private school, private college

3

<hr>

Person 1 (continued)

8. b. What grade or level was this person attending? *Mark ⊠ ONE box.*
☐ Nursery school, preschool
☐ Kindergarten
☐ Grade 1 to grade 4
☐ Grade 5 to grade 8
☐ Grade 9 to grade 12
☐ College undergraduate years (freshman to senior)
☐ Graduate or professional school (for example: medical, dental, or law school)

9. What is the highest degree or level of school this person has COMPLETED? *Mark ⊠ ONE box. If currently enrolled, mark the previous grade or highest degree received.*
☐ No schooling completed
☐ Nursery school to 4th grade
☐ 5th grade or 6th grade
☐ 7th grade or 8th grade
☐ 9th grade
☐ 10th grade
☐ 11th grade
☐ 12th grade, **NO DIPLOMA**
☐ **HIGH SCHOOL GRADUATE** — high school DIPLOMA or the equivalent (for example: GED)
☐ Some college credit, but less than 1 year
☐ 1 or more years of college, no degree
☐ Associate degree (for example: AA, AS)
☐ Bachelor's degree (for example: BA, AB, BS)
☐ Master's degree (for example: MA, MS, MEng, MEd, MSW, MBA)
☐ Professional degree (for example: MD, DDS, DVM, LLB, JD)
☐ Doctorate degree (for example: PhD, EdD)

10. What is this person's ancestry or ethnic origin?

(For example: Italian, Jamaican, African Am., Cambodian, Cape Verdean, Norwegian, Dominican, French Canadian, Haitian, Korean, Lebanese, Polish, Nigerian, Mexican, Taiwanese, Ukrainian, and so on.)

11. a. Does this person speak a language other than English at home?
☐ Yes
☐ No → *Skip to 12*

b. What is this language?

(For example: Korean, Italian, Spanish, Vietnamese)

c. How well does this person speak English?
☐ Very well
☐ Well
☐ Not well
☐ Not at all

12. Where was this person born?
☐ In the United States — *Print name of state.*

☐ Outside the United States — *Print name of foreign country, or Puerto Rico, Guam, etc.*

13. Is this person a CITIZEN of the United States?
☐ Yes, born in the United States → *Skip to 15a*
☐ Yes, born in Puerto Rico, Guam, the U.S. Virgin Islands, or Northern Marianas
☐ Yes, born abroad of American parent or parents
☐ Yes, U.S. citizen by naturalization
☐ No, not a citizen of the United States

14. When did this person come to live in the United States? *Print numbers in boxes.*

Year

15. a. Did this person live in this house or apartment 5 years ago (on April 18, 1993)?
☐ Person is under 5 years old → *Skip to 33*
☐ Yes, this house → *Skip to 16*
☐ No, outside the United States — *Print name of foreign country, or Puerto Rico, Guam, etc., below, then skip to 16.*

☐ No, different house in the United States

47

Figure 2.1   (continued)

48

Person 1 (continued)

15 b. Where did this person live 5 years ago?

Name of city, town, or post office
[ ]

Did this person live inside the limits of the city or town?
☐ Yes
☐ No, outside the city/town limits

Name of county
[ ]

Name of state
[ ]

ZIP Code
[ ]

16 Does this person have any of the following long-lasting conditions:

	Yes	No
a. Blindness, deafness, or a severe vision or hearing impairment?	☐	☐
b. A condition that substantially limits one or more basic physical activities such as walking, climbing stairs, reaching, lifting, or carrying?	☐	☐

17 Because of a physical, mental, or emotional condition lasting 6 months or more, does this person have any difficulty in doing any of the following activities:

	Yes	No
a. Learning, remembering, or concentrating?	☐	☐
b. Dressing, bathing, or getting around inside the home?	☐	☐
c. (Answer if this person is 16 YEARS OLD OR OVER) Going outside the home alone to shop or visit a doctor's office?	☐	☐
d. (Answer if this person is 16 YEARS OLD OR OVER) Working at a job or business?	☐	☐

18 Was this person under 15 years of age on April 18, 1998?
☐ Yes → Skip to 33
☐ No

19 a. Does this person have any of his/her own grandchildren under the age of 18 living in this house or apartment?
☐ Yes
☐ No → Skip to 20a

b. Is this grandparent currently responsible for most of the basic needs of any grandchild(ren) under the age of 18 who live(s) in this house or apartment?
☐ Yes
☐ No → Skip to 20a

c. How long has this grandparent been responsible for the(se) grandchild(ren)? *If the grandparent is financially responsible for more than one grandchild, answer the question for the grandchild for whom the grandparent has been responsible for the longest period of time.*
☐ Less than 1 month
☐ 1 to 6 months
☐ 7 to 12 months
☐ More than 12 months
☐ Don't know

20 a. Has this person ever served on active duty in the U.S. Armed Forces, military Reserves, or National Guard? *Active duty does not include training for the Reserves or National Guard, but DOES include activation, for example, for the Persian Gulf War.*
☐ Yes, now on active duty
☐ Yes, on active duty in past, but not now
☐ No, training for Reserves or National Guard only → Skip to 21
☐ No, never served in the military → Skip to 21

b. When did this person serve on active duty in the U.S. Armed Forces? *Mark ☒ a box for EACH period in which this person served.*
☐ April 1995 or later
☐ August 1990 to March 1995 (including Persian Gulf War)
☐ September 1980 to July 1990
☐ May 1975 to August 1980
☐ Vietnam era (August 1964–April 1975)
☐ February 1955 to July 1964
☐ Korean conflict (June 1950–January 1955)
☐ World War II (September 1940–July 1947)
☐ Some other time

c. In total, how many years of active-duty military service has this person had?
☐ Less than 2 years
☐ 2 years or more

5

Person 1 (continued)

21 LAST WEEK, did this person do ANY work for either pay or profit? *Mark ☒ the "Yes" box even if the person worked only 1 hour, or helped without pay in a family business or farm for 15 hours or more, or was on active duty in the Armed Forces.*
☐ Yes
☐ No → Skip to 25a

22 At what location did this person work LAST WEEK? *If this person worked at more than one location, print where he or she worked most last week.*

a. Address (Number and street name)
[ ]

(If the exact address is not known, give a description of the location such as the building name or the nearest street or intersection.)

b. Name of city, town, or post office
[ ]

c. Is the work location inside the limits of that city or town?
☐ Yes
☐ No, outside the city/town limits

d. Name of county
[ ]

e. Name of U.S. state or foreign country
[ ]

f. ZIP Code
[ ]

23 a. How did this person usually get to work LAST WEEK? *If this person usually used more than one method of transportation during the trip, mark ☒ the box of the one used for most of the distance.*
☐ Car, truck, or van
☐ Bus or trolley bus
☐ Streetcar or trolley car
☐ Subway or elevated
☐ Railroad
☐ Ferryboat
☐ Taxicab
☐ Motorcycle
☐ Bicycle
☐ Walked
☐ Worked at home → Skip to 27
☐ Other method

→ If "Car, truck, or van" is marked in 23a, go to 23b. Otherwise, skip to 24a.

b. How many people, including this person, usually rode to work in the car, truck, or van LAST WEEK?
☐ Drove alone
☐ 2 people
☐ 3 people
☐ 4 people
☐ 5 or 6 people
☐ 7 or more people

24 a. What time did this person usually leave home to go to work LAST WEEK?
☐ a.m. ☐ p.m.

b. How many minutes did it usually take this person to get from home to work LAST WEEK?
Minutes [ ]

→ Answer questions 25–26 for persons who did not work for pay or profit last week. Others skip to 27.

25 a. LAST WEEK, was this person on layoff from a job?
☐ Yes → Skip to 25c
☐ No

b. LAST WEEK, was this person TEMPORARILY absent from a job or business?
☐ Yes, on vacation, temporary illness, labor dispute, etc. → Skip to 26
☐ No → Skip to 25d

c. Has this person been informed that he or she will be recalled to work within the next 6 months OR been given a date to return to work?
☐ Yes → Skip to 25e
☐ No

d. Has this person been looking for work during the last 4 weeks?
☐ Yes
☐ No → Skip to 26

e. LAST WEEK, could this person have started a job if offered one, or returned to work if recalled?
☐ Yes, could have gone to work
☐ No, because of own temporary illness
☐ No, because of all other reasons (in school, etc.)

26 When did this person last work, even for a few days?
☐ 1993 to 1998
☐ 1992 or earlier, or never worked → Skip to 31

Form D-61-D2/98

6

Person 1 (continued)

27 Industry or Employer — *Describe clearly this person's chief job activity or business last week. If this person had more than one job, describe the one at which this person worked the most hours. If this person had no job or business last week, give the information for his/her last job or business since 1993.*

a. For whom did this person work? *If now on active duty in the Armed Forces, mark ☒ this box → ☐ and print the branch of the Armed Forces.*

Name of company, business, or other employer

☐☐☐☐☐☐☐☐☐☐☐☐☐☐☐☐☐☐☐☐
☐☐☐☐☐☐☐☐☐☐☐☐☐☐☐☐☐☐☐☐

b. What kind of business or industry was this? *Describe the activity at location where employed. (For example: hospital, newspaper publishing, mail order house, auto repair shop, bank)*

☐☐☐☐☐☐☐☐☐☐☐☐☐☐☐☐☐☐☐☐
☐☐☐☐☐☐☐☐☐☐☐☐☐☐☐☐☐☐☐☐

c. Is this mainly — *Mark ☒ ONE box.*
☐ Manufacturing?
☐ Wholesale trade?
☐ Retail trade?
☐ Other *(agriculture, construction, service, government, etc.)?*

28 Occupation

a. What kind of work was this person doing? *For example: registered nurse, personnel manager, supervisor of order department, auto mechanic, accountant*

☐☐☐☐☐☐☐☐☐☐☐☐☐☐☐☐☐☐☐☐
☐☐☐☐☐☐☐☐☐☐☐☐☐☐☐☐☐☐☐☐

b. What were this person's most important activities or duties? *(For example: patient care, directing hiring policies, supervising order clerks, repairing automobiles, reconciling financial records)*

☐☐☐☐☐☐☐☐☐☐☐☐☐☐☐☐☐☐☐☐
☐☐☐☐☐☐☐☐☐☐☐☐☐☐☐☐☐☐☐☐

29 Was this person — *Mark ☒ ONE box.*
☐ Employee of a PRIVATE FOR PROFIT company or business or of an individual, for wages, salary, or commissions
☐ Employee of a PRIVATE NOT-FOR-PROFIT, tax-exempt, or charitable organization
☐ Local GOVERNMENT employee *(city, county, etc.)*
☐ State GOVERNMENT employee
☐ Federal GOVERNMENT employee
☐ SELF-EMPLOYED in own NOT INCORPORATED business, professional practice, or farm
☐ SELF-EMPLOYED in own INCORPORATED business, professional practice, or farm
☐ Working WITHOUT PAY in family business or farm

30 a. LAST YEAR, 1997, did this person work at a job or business at any time?
☐ Yes
☐ No → *Skip to 31*

b. How many weeks did this person work in 1997? *Count paid vacation, paid sick leave, and military service.*

Weeks ☐☐

c. During the weeks WORKED in 1997, how many hours did this person usually work each WEEK?

Usual hours worked each WEEK ☐☐

31 INCOME IN 1997 — *Mark ☒ the "Yes" box for each income source received during 1997 and enter the total amount received during 1997 to a maximum of $999,999. Otherwise, mark ☒ the "No" box.*

If net income was a loss, enter the amount and mark ☒ the "Loss" box next to the dollar amount.

For income received jointly, report, if possible, the appropriate share for each person; otherwise, report the whole amount for only one person and mark ☒ the "No" box for the other person. If exact amount is not known, please give best estimate.

a. Wages, salary, commissions, bonuses, or tips from all jobs — *Report amount before deductions for taxes, bonds, dues, or other items.*
☐ Yes Annual amount — *Dollars*
$ ☐☐☐☐☐☐ .00
☐ No

b. Self-employment income from own nonfarm businesses or farm businesses, including proprietorships and partnerships — *Report NET income after business expenses.*
☐ Yes Annual amount — *Dollars*
$ ☐☐☐☐☐☐ .00 ☐ Loss
☐ No

Person 1 (continued)

31 *(continued)*

c. Interest, dividends, net rental income, royalty income, or income from estates and trusts — *Report even small amounts credited to an account.*
☐ Yes Annual amount — *Dollars*
$ ☐☐☐☐☐☐ .00 ☐ Loss
☐ No

d. Social Security or Railroad Retirement
☐ Yes Annual amount — *Dollars*
$ ☐☐☐☐☐☐ .00
☐ No

e. Supplemental Security Income (SSI)
☐ Yes Annual amount — *Dollars*
$ ☐☐☐☐☐☐ .00
☐ No

f. Any public assistance or welfare payments from the state or local welfare office
☐ Yes Annual amount — *Dollars*
$ ☐☐☐☐☐☐ .00
☐ No

g. Retirement, survivor, or disability pensions — *Do NOT include Social Security.*
☐ Yes Annual amount — *Dollars*
$ ☐☐☐☐☐☐ .00
☐ No

h. Any other sources of income received regularly such as Veterans' (VA) payments, unemployment compensation, child support, or alimony — *Do NOT include lump-sum payments such as money from an inheritance or sale of a home.*
☐ Yes Annual amount — *Dollars*
$ ☐☐☐☐☐☐ .00
☐ No

32 What was this person's total income in 1997? *Add entries in questions 31a – 31h; subtract any losses. If net income was a loss, enter the amount and mark ☒ the "Loss" box next to the dollar amount.*
Annual amount — *Dollars*
☐ None OR $ ☐☐☐☐☐☐ .00 ☐ Loss

→ Now, please answer questions 33 – 53 about your household.

33 Is this house, apartment, or mobile home —
☐ Owned by you or someone in this household with a mortgage or loan?
☐ Owned by you or someone in this household free and clear (without a mortgage or loan)?
☐ Rented for cash rent?
☐ Occupied without payment of cash rent?

34 Which best describes this building? *Include all apartments, flats, etc., even if vacant.*
☐ A mobile home
☐ A one-family house detached from any other house
☐ A one-family house attached to one or more houses
☐ A building with 2 apartments
☐ A building with 3 or 4 apartments
☐ A building with 5 to 9 apartments
☐ A building with 10 to 19 apartments
☐ A building with 20 to 49 apartments
☐ A building with 50 or more apartments
☐ Boat, RV, van, etc.

35 About when was this building first built?
☐ 1996 to 1998
☐ 1990 to 1995
☐ 1980 to 1989
☐ 1970 to 1979
☐ 1960 to 1969
☐ 1950 to 1959
☐ 1940 to 1949
☐ 1939 or earlier

36 When did this person move into this house, apartment, or mobile home?
☐ 1997 or 1998
☐ 1996
☐ 1990 to 1995
☐ 1980 to 1989
☐ 1970 to 1979
☐ 1969 or earlier

37 How many rooms do you have in this house, apartment, or mobile home? *Do NOT count bathrooms, porches, balconies, foyers, halls, or half-rooms.*
☐ 1 room   ☐ 6 rooms
☐ 2 rooms   ☐ 7 rooms
☐ 3 rooms   ☐ 8 rooms
☐ 4 rooms   ☐ 9 or more rooms
☐ 5 rooms

Form DX-01 (9-30-98)

Figure 2.1   (continued)

Person 1 (continued)

38. How many bedrooms do you have; that is, how many bedrooms would you list if this house, apartment, or mobile home were on the market for sale or rent?
- No bedroom
- 1 bedroom
- 2 bedrooms
- 3 bedrooms
- 4 bedrooms
- 5 or more bedrooms

39. Do you have COMPLETE plumbing facilities in this house, apartment, or mobile home; that is, 1) hot and cold piped water, 2) a flush toilet, and 3) a bathtub or shower?
- Yes, have all three facilities
- No

40. Do you have COMPLETE kitchen facilities in this house, apartment, or mobile home; that is, 1) a sink with piped water, 2) a range or stove, and 3) a refrigerator?
- Yes, have all three facilities
- No

41. Is there telephone service available in this house, apartment, or mobile home from which you can both make and receive calls?
- Yes
- No

42. Which FUEL is used MOST for heating this house, apartment, or mobile home?
- Gas: from underground pipes serving the neighborhood
- Gas: bottled, tank, or LP
- Electricity
- Fuel oil, kerosene, etc.
- Coal or coke
- Wood
- Solar energy
- Other fuel
- No fuel used

43. How many automobiles, vans, and trucks of one-ton capacity or less are kept at home for use by members of your household?
- None
- 1
- 2
- 3
- 4
- 5
- 6 or more

44. Answer ONLY if this is a ONE-FAMILY HOUSE OR MOBILE HOME — All others skip to 45.
a. Is there a business (such as a store or barber shop) or a medical office on this property?
- Yes
- No
b. How many acres is this house or mobile home on?
- Less than 1 acre → Skip to 45
- 1 to 9.9 acres
- 10 or more acres
c. In 1997, what were the actual sales of all agricultural products from this property?
- None
- $1 to $999
- $1,000 to $2,499
- $2,500 to $4,999
- $5,000 to $9,999
- $10,000 or more

45. What are the annual costs of utilities and fuels for this house, apartment, or mobile home? If you have lived here less than 1 year, estimate the annual cost.
a. Electricity
Annual cost — Dollars
$ ____ .00
OR
- Included in rent or in condominium fee
- No charge or electricity not used
b. Gas
Annual cost — Dollars
$ ____ .00
OR
- Included in rent or in condominium fee
- No charge or gas not used
c. Water and sewer
Annual cost — Dollars
$ ____ .00
OR
- Included in rent or in condominium fee
d. Oil, coal, kerosene, wood, etc.
Annual cost — Dollars
$ ____ .00
OR
- Included in rent or in condominium fee
- No charge or these fuels not used

Person 1 (continued)

46. Answer ONLY if you PAY RENT for this house, apartment, or mobile home — All others skip to 47.
a. What is the monthly rent?
Monthly amount — Dollars
$ ____ .00
b. Does the monthly rent include any meals?
- Yes
- No

47. Answer questions 47a – 53 if you or someone in this household owns or is buying this house, apartment, or mobile home; otherwise, skip to questions for Person 2.
a. Do you have a mortgage, deed of trust, contract to purchase, or similar debt on THIS property?
- Yes, mortgage, deed of trust, or similar debt
- Yes, contract to purchase
- No → Skip to 48a
b. How much is your regular monthly mortgage payment on THIS property? Include payment only on first mortgage or contract to purchase.
Monthly amount — Dollars
$ ____ .00
OR
- No regular payment required → Skip to 48a
c. Does your regular monthly mortgage payment include payments for real estate taxes on THIS property?
- Yes, taxes included in mortgage payment
- No, taxes paid separately or taxes not required
d. Does your regular monthly mortgage payment include payments for fire, hazard, or flood insurance on THIS property?
- Yes, insurance included in mortgage payment
- No, insurance paid separately or no insurance

48. a. Do you have a second mortgage or a home equity loan on THIS property? Mark ⊠ all boxes that apply.
- Yes, a second mortgage
- Yes, a home equity loan
- No → Skip to 49
b. How much is your regular monthly payment on all second or junior mortgages and all home equity loans on THIS property?
Monthly amount — Dollars
$ ____ .00
OR
- No regular payment required

49. What were the real estate taxes on THIS property last year?
Yearly amount — Dollars
$ ____ .00
OR
- None

50. What was the annual payment for fire, hazard, and flood insurance on THIS property?
Annual amount — Dollars
$ ____ .00
OR
- None

51. What is the value of this property; that is, how much do you think this house and lot, apartment, or mobile home and lot would sell for if it were for sale?
- Less than $10,000
- $10,000 to $14,999
- $15,000 to $19,999
- $20,000 to $24,999
- $25,000 to $29,999
- $30,000 to $34,999
- $35,000 to $39,999
- $40,000 to $49,999
- $50,000 to $59,999
- $60,000 to $69,999
- $70,000 to $79,999
- $80,000 to $89,999
- $90,000 to $99,999
- $100,000 to $124,999
- $125,000 to $149,999
- $150,000 to $174,999
- $175,000 to $199,999
- $200,000 to $249,999
- $250,000 to $299,999
- $300,000 to $399,999
- $400,000 to $499,999
- $500,000 to $749,999
- $750,000 to $999,999
- $1,000,000 or more

52. Answer ONLY if this is a CONDOMINIUM —
What is the monthly condominium fee?
Monthly amount — Dollars
$ ____ .00

53. Answer ONLY if this is a MOBILE HOME —
a. Do you have an installment loan or contract on THIS mobile home?
- Yes
- No
b. What was the total cost for installment loan payments, personal property taxes, site rent, registration fees, and license fees on THIS mobile home and its site last year? Exclude real estate taxes.
Yearly amount — Dollars
$ ____ .00
OR
- No regular payment required

→ Are there more people living here? If yes, continue with Person 2.

Form D-61 (1-0-98)

Person 3

Information about children helps your community plan for child care, education, and recreation.

Short form: for Persons 3–5, repeat questions 1–6 of Person 2.

Long form: for Persons 3–5, repeat questions 1–32 of Person 2.

NOTE – The content for Question 2 on both forms varies between Person 1 and Persons 2–5.

THANK YOU

for completing your U.S. census form. If there are more than five people at this address, the Census Bureau will contact you.

For additional information about the Census 2000 Dress Rehearsal, visit our website at **www.census.gov/2000** or write the Director, Bureau of the Census, Washington, DC 20233.

Form DX-6113 (2-98)

Person 2

Census information helps your community get financial assistance for roads, hospitals, schools, and more.

Short form: for Person 2, repeat questions 3–6 of Person 1.

Long form: for Person 2, repeat questions 3–32 of Person 1.

1 What is this person's name? *Print the name of Person 2 from page 2.*

Last Name

First Name   MI

2 How is this person related to Person 1?
Mark ⊠ ONE box.

☐ Husband/wife
☐ Natural-born son/daughter
☐ Adopted son/daughter
☐ Stepson/stepdaughter
☐ Brother/sister
☐ Father/mother
☐ Grandchild
☐ Parent-in-law
☐ Son-in-law/daughter-in-law
☐ Other relative — *Print exact relationship.*

If NOT RELATED to Person 1:
☐ Roomer, boarder
☐ Housemate, roommate
☐ Unmarried partner
☐ Foster child
☐ Other nonrelative

Table 2.1   Comparison of Items Included in the U.S. Census 2000
Questionnaire, the 1996 Census of Canada, and the 1990 Census of Mexico

Census Item	U.S. Census 2000	Canada 1996	Mexico 1990
Population Characteristics:			
Age	XX	XX	XX
Gender	XX	XX	XX
Relationship to householder (family structure)	XX	XX	XX
Race	XX	X	
Hispanic origin	XX		
Marital status	X	XX	XX
Fertility			XX
Income	X	X	XX
Sources of income	X	X	
Household activities		X	
Labor force status	X	X	XX
Industry, occupation, and class of worker	X	X	XX
Work status last year	X		
Veteran status	X		
Grandparents as caregivers	X		
Place of work and journey to work	X	X	
Vehicles available	X		
Ancestry	X	X	
Place of birth	X	X	XX
Citizenship	X	X	
Year of entry if not born in this country	X	X	
Language spoken at home	X	XX	XX
Religion			XX
Educational attainment	X	X	XX
School enrollment	X	X	
Residence 1 year ago (migration)		X	
Residence 5 year ago (migration)	X	X	XX
Disability	X	X	
Housing characteristics:			
Tenure (rent or own)	XX		XX
Type of housing	XX		XX
Material used for construction of walls			XX
Material used for construction of roof			XX
Material used for construction of floors			XX
Repairs needed on structure		X	

continued

Table 2.1   (continued)

Census Item	U.S. Census 2000	Canada 1996	Mexico 1990
Housing characteristics: (continued)			
Year structure built	X	X	
Units in structure	X		
Rooms in unit	X	X	XX
Bedrooms	X	X	
Kitchen facilities	X		XX
Plumbing facilities	X		XX
Telephone in unit	X		
House heating fuel	X		XX
Year moved into unit	X		
Farm residence	X		
Value	X		
Selected monthly owner costs	X	X	
Rent	X	X	

Note: XX = Included and asked of every household. X = Included but asked of only a sample of households. Questions asked on each census may be different; similar categories of questions asked does not necessarily mean strict comparability of data.

   Despite the Census Bureau's best efforts, there were nearly 5 million people who were not included in the 1990 U.S. census count. Data from this and previous censuses indicate that these individuals are disproportionately minority group members, especially in inner-city or rural areas that could suffer financially from an undercount, since many government resources are distributed on the basis of the census information. An undercount could also have a political effect on the way seats in the U.S. House of Representatives are reapportioned (see Chapter 15). The Census Bureau has estimated that the 1990 under-enumeration was 5.2 percent for Hispanics, 5 percent for American Indians, and 4.8 percent for African-Americans. This issue created a series of political crises for the Census Bureau as it prepared for Census 2000 in the United States, as I discuss in the essay that accompanies this chapter.

   About two thirds of people who were missed in the 1990 U.S. Census were not enumerated because whoever filled out the census form did not, in fact, include them. The remaining third were missed because the Census Bureau did not know that their address existed and so they never received a questionnaire to fill out (U.S. Bureau of the Census 1997). Of course that raises the question of who should be included in the census. There are several ways to answer that question, and each produces a potentially different total number of people. At one extreme is the concept of the **de facto population,** which counts people who are in a given territory on the census day. At the other extreme is the **de jure population,** which represents people who legally "belong" to a given area in some way or another, regardless of whether they were there on the day of the census. For countries with few foreign workers and where working in another area is rare, the distinction makes little difference. But some countries, such as Switzerland or Germany, with large numbers of alien workers, have a larger de facto than de jure population. On the other hand, a country such as Mexico, from which migrants regularly leave temporarily to go to the United States, the de jure population is actually greater than the de facto.

Most countries (including the United States, Canada, and Mexico) have now adopted a concept that lies somewhere between the extremes of de facto and de jure, and they include people in the census on the basis of **usual residence,** which is roughly defined as the place where a person usually sleeps. College students who live away from home, for example, are included at their college address rather than being counted in their parents' household. People with no usual residence (the homeless, including migratory workers, vagrants, and "street people") are counted where they are found. On the other hand, visitors and tourists from other countries who "belong" somewhere else are not included, even though they may be in the country when the census is being conducted. At the same time, the concept of usual residence means that undocumented immigrants (who technically do not "belong" where they are found) will be included in the census along with everyone else.

Although coverage error is a concern in any census, there can also be problems with the accuracy of the data obtained in the census (content error). Content error includes nonresponses to particular questions on the census or inaccurate responses if people do not understand the question. Errors can also occur if information is inaccurately recorded on the form or if there is some glitch in the processing (coding, data entry, or editing) of the census return. In comparison with other censuses, the United Nations rates the American census as highly accurate, especially with respect to recording age, one of the most important demographic characteristics (United Nations 1996). By and large, content error is not a problem in the U.S. census, although the data are certainly not 100 percent accurate. In general, data from the United Nations suggest that the more highly developed a country is, the more accurate the content of its census data will be. This fact is probably accounted for especially by the level of education of a population.

Sampling Error   If any of the data in a census are collected on a sample basis (as is done in both the United States and Canada), then sampling error is introduced into the results. With any sample, scientifically selected or not, differences are likely to exist between the characteristics of the sampled population and the larger group from which the sample was chosen. However, in a scientific sample, such as is employed in most census operations, sampling error is readily measured based on the mathematics of probability. To a certain extent, sampling error can be controlled— samples can be designed to ensure comparable levels of error across groups or across geographic areas (U.S. Bureau of the Census 1997). However, nonsampling error and biases that it introduces in terms of some people being more likely to included than others are present throughout the census process and can reduce the quality of results more than sampling error.

Measuring Coverage Error   Right now you are probably asking yourself how the Census Bureau could ever begin to estimate how many people are missed in a census. This is not an easy task, and statisticians in the U.S. and other countries have experimented with a number of methods over the years. There are two principal methods that are typically employed: (1) **demographic analysis (DA)**, and (2) **dual system estimation (DSE).**

The demographic analysis approach uses the demographic balancing equation to estimate what the population at the latest census should have been, and then that number is compared with the actual count. Thus, if we know the number of people from the previous census, we can add the number of births since then, subtract the number of deaths since then, add the number of in-migrants since then, and subtract the number of out-migrants since then to estimate what the total population count should have been. A comparison of this number with the actual census count provides a clue to the accuracy of the census. Thus, by comparing the number of males aged 20 to 24 in 2000 with the number of males aged 10 to 14 in 1990, we can determine whether there are fewer males aged 20 to 24 in 2000 than we should have expected, taking death and migration into account. Using these methods, in conjunction with birth and death records and administrative data from the U.S. Immigration and Naturalization Service, the Census Bureau is able to piece together a composite rendering of what the population "should" look like. Differences from that picture and the one painted by the census can be used as estimates of under-enumeration. By making similar calculations for all ages, we can arrive at an estimate of the possible undercount in certain age and sex categories.

Of course, if we do not have an accurate count of births, deaths, and migrants, then our demographically-derived estimate may itself be wrong, so this method requires a lot of careful attention to the quality of the noncensus data. And, you say, why should we even take a census if we think that we can estimate the number of people more accurately without it? The answer is that the demographic analysis approach usually only produces an estimate of the total number of people in any age and sex group, without providing a way of knowing the details of the population—which is what we obtain from the census questionnaire.

The dual system estimation method involves taking a carefully constructed sample survey right after the census is finished and then matching up people in the sample survey with their responses in the census. Experienced enumerators conduct the interviews to determine whether households and the individuals within the households had been counted in both the census and the survey (the ideal situation); in the census but not in the survey (possible but not likely); or in the survey but not in the census (the usual measure of under-enumeration). Obviously, some people may be missed by both the census and the survey, but the logic underlying the method is analogous to the capture-recapture method used by biologists tracking wildlife (Choldin, 1994). The strategy is to capture a sample of animals, mark them, and release them. Later, another sample is captured, and some of the marked animals will wind up being recaptured. The ratio of recaptured animals to all animals caught in the second sample is assumed to represent the ratio of the first group captured to the whole population, and on this basis the wildlife population can be estimated. Although some humans are certainly "wild," a few adjustments are required to apply the method to human populations. After the 1990 Census in the United States, a post-enumeration survey (PES) of 150,000 households was conducted to measure the extent of coverage error. A much larger survey (approximately 750,000

TO SAMPLE OR NOT TO SAMPLE: THAT IS THE QUESTION

In anticipation of the undercount of minority group members in the 1990 U.S. census, New York City, Los Angeles, and several other cities, states, and civil rights groups filed suit in federal court in New York in 1988 to try to force the Census Bureau to make a postcensal adjustment. The procedure would have statistically estimated the under-enumeration and added people to the totals in those areas where the Census Bureau thought underenumeration had occurred. A court settlement in 1989 required that the Secretary of Commerce review the pros and cons of such an adjustment and make a decision by July 15, 1991. As it turned out, the Census Bureau calculated the adjustments, but the Department of Commerce (the bureaucracy within which the Census Bureau is housed) decided not to make those data available. Had the adjustments been made, the states of Wisconsin and Pennsylvania would each have lost a seat in Congress and California and Arizona would each have gained a seat. The decision not to adjust sparked a new flurry of lawsuits, of course, and in 1996 the U.S. Supreme Court finally ruled unanimously that the Census Bureau was under no legal obligation to adjust the final figures, despite its acknowledged undercount of urban minority group members (Barrett 1996).

You might well think that a unanimous Supreme Court decision would settle an issue, but you would be wrong in this case, because in the meantime the Bureau of the Census had decided to take the advice offered by the National Academy of Sciences (Edmonston and Schultze 1995; National Research Council 1993; White and Rust 1997) and announced that for Census 2000 it intended to use statistical sampling techniques to estimate and include those people who were otherwise missed by the census enumeration (Rosewicz, 1996), in a process known as sampling for nonresponse follow-up (SNRFU for you lovers of acronyms). The basic idea is that rather than having a postcensal survey to estimate how many people were missed by the census, the survey process would become part of the census, and thus those who were estimated to have been

missed would be included in the final count instead of being lamented and fought over later. The traditional methods of enumeration would be used to obtain the first 90 percent of responses, but the last 10 percent (the really hard to find people) would be estimated from a large (750,000 households) sample survey. This idea sparked a new flurry of Congressional mandates and lawsuits.

On one side of the debate were those members of Congress who wanted no sampling whatsoever and who introduced bills into Congress to forbid the Census Bureau from spending any money even on planning for sampling (Duff 1997). Some of these members of Congress were, in fact, proposing that the decennial census be reduced to a few questions that could fit on a postcard. On the other side of the debate was the Census Bureau, arguing that its Congressional mandate to make Census 2000 more accurate and yet less costly than the 1990 Census could only be met by the judicious use of sampling (Riche 1997).

Is the uproar over sampling genuinely an issue about how best to conduct a census? Not really. Virtually everyone in the community of professional demographers and statisticians agrees that properly implemented statistical sampling techniques will improve the accuracy of the final population estimates, because it is simply impossible in a large country actually to physically locate and enumerate every single person. The real issue was politics: those people missed by the census in the United States are perceived to be more sympathetic to the Democratic political party. It has been estimated that including all those people who have typically been missed by the census would cause Democrats to gain two dozen seats in the House of Representatives, at the expense, of course, of the Republican party (Spar 1998). A search for the truth takes a back seat in such negotiations, causing the Director of the Census Bureau, Martha Riche, to resign in 1998, and reminding us that the United States is as susceptible to playing politics with a census as is any country.

households) is planned for the Census 2000 evaluation (U.S. Bureau of the Census 1997). Note that the dual system estimation approach also provides a way of testing for content error by comparing the responses of people when they answered the census questionnaire with their answers to the post-enumeration survey questionnaire.

Continuous Measurement   Almost all of the detailed data about population characteristics obtained from the decennial censuses in the United States come from the "long form," which is administered to only about 1 in 6 households. The success of survey sampling in obtaining reliable demographic data (which I discuss later in the chapter) has led the U.S. Bureau of Census to propose that in the twenty first century it may phase in a process of "continuous measurement" and thereby delete the long form from subsequent decennial censuses. The vehicle for this would be the monthly American Community Survey, which the Census Bureau has already implemented on an experimental basis (Pollard 1997). Were this to be implemented, the decennial census would then consist only of the short form in order to meet the minimal Constitutional requirement for population data to apportion the House of Representatives. The advantages to this would be that it would provide a better on-going picture of social change in the United States. However, the major disadvantage is that fewer households would be included than are currently enumerated by the long form, and the level of geographic detail would be much less. Interestingly enough, many big businesses (who might usually be on the side of cutting government expenses) have voiced rather vociferous opposition to eliminating the long form because they rely on those data for their marketing efforts (Peyser and Alexander 1997; Sherwood 1995) (see Chapter 15).

The Census of Canada

The first census in Canada was taken in 1666 when the French colony of New France was counted on orders of King Louis XIV. This turned out to be a door-to-door enumeration of all 3,215 settlers in Canada at that time. A series of wars between England and France ended in France ceding Canada to England in 1763, and the British undertook censuses on an irregular but fairly consistent basis (Statistics Canada 1995). The several regions of Canada were united under the British North America Act of 1867, and that Act specified that censuses were to be taken regularly to establish the number of representatives that each province would send to the House of Commons. The first of these was taken in 1871, although similar censuses had been taken in 1851 and 1861. In 1905 the census bureau became a permanent government agency, now known as Statistics Canada.

Canada began using sampling in 1941, the year after the United States experimented with it. In 1956 they conducted the first quinquennial census (every 5 years, as opposed to every 10 years—the decennial census), and in 1971 Canada made it a law that the census was to be conducted every 5 years. The U.S. Congress passed similar legislation in the 1970s but never funded those efforts, so the United States has stayed with the decennial census.

Two census forms are employed in Canada, as in the United States—a short form with just a few key items (see Table 2.1) and a more detailed long form. The long form in 1996 went to a sample of 20 percent of the Canadian households. Among the few questions asked on the short form is, not surprisingly, language—the split between English and French speakers nearly tore the country apart in the 1990s.

Public opinion influences census activities in Canada, as it does in the United States, and so the 1996 Census of Canada included a question on unpaid household activities because of the protest of a Saskatoon housewife. She refused to fill out the 1991 Census form (and risked going to jail as a result) because the census form's definition of work did not include household work or child care. This helped to galvanize public opinion to include a set of questions on this type of activity in the 1996 census (De Santis 1996). On the other hand, the Canadian government decided that the number of children born to a woman might be too private a question to be asked any longer (it had been asked on every decennial census since 1941), and it was not included in the 1996 census (but it may be included in the 2001 Census).

Statistics Canada estimates coverage error by conducting post-enumeration surveys and then matching those randomly sampled individuals to their responses (or lack thereof) in the census. Based on this method, the underenumeration in the 1991 census was estimated to be 2.9 percent (Statistics Canada 1996).

The Census of Mexico

Like Canada and the United States, Mexico has a long history of census taking. Spain conducted several censuses in Mexico during the colonial years, including a general census of New Spain (Nueva España as they knew it) in 1790. Mexico gained independence from Spain in 1821, but it was not until 1895 that the first of the modern series of national censuses was undertaken. A second enumeration came in 1900, but since then they have been taken every 10 years (with the exception of the one in 1921, which was 1 year out of sequence). From 1895 through the 1970s, the census activities were carried out by the General Directorate of Statistics (Dirección General de Estadística), and there were no permanent census employees. However, the bureaucracy was reorganized for the 1980 census, and in 1983 the Instituto Nacional de Estadística, Geografía e Informática (INEGI) became the permanent government agency in charge of the census and other government data collection.

There were fewer questions asked in Mexico in the 1990 census than in the other two North American nations, as you can see in Table 2.1, and no sampling was involved—all households received the same questionnaire. Considerably less detail is obtained in Mexico about income than in Canada or the United States, and socioeconomic categories are more often derived from the outward manifestations of income, such as the quality of housing, for which there are several detailed questions. Since most Mexicans are "mestizos" (Spanish for mixed race, in this case mainly European and Amerindian), no questions are asked about race or ethnicity. The only allusion to diversity within Mexico is found in the question about language, in which people are asked if they speak an Indian language. If so, they are also asked if they speak Spanish.

In 1995, INEGI conducted an innovative quinquennial count (called a "Conteo," signifying an enumeration that was less extensive than the decennial censuses). A short household questionnaire was administered to every household in the nation, asking about very basic information, including number of people in the household, their age and sex, literacy, and whether they speak an Indian language, along with a few questions about the type of housing and whether it was connected to a sewer and a water line. A much more detailed questionnaire was administered to a sample of households in the country (Instituto Nacional de Estadística 1996).

In Mexico the evaluation of coverage error in the census has generally been made using the method of demographic analysis (discussed earlier). On this basis Corona (1991) has estimated that underenumeration in the 1990 census of Mexico was somewhere between 2.3 and 7.3 percent.

Registration of Vital Events

When you were born, a birth certificate was filled out for you, probably by a clerk in the hospital where you were born. When you die, someone (again, typically a hospital clerk) will fill out a death certificate on your behalf. Standard birth and death certificates used in the United States are shown in Figure 2.2. Births and deaths, as well as marriages, divorces, and abortions, are known as vital events, and when they are recorded by the government and compiled for use they become vital statistics. These statistics are the major source of data on births and deaths, and they are most useful when combined with census data.

Registration of vital events in Europe began as a chore of the church. Priests often recorded baptisms, marriages, and deaths, and historical demographers have used the surviving records to try to reconstruct the demographic history of parts of Europe (see Wrigley 1969 and Wrigley and Schofield 1981). Among the more demographically important tasks that befell the clergy was that of recording burials that occurred in England during the many years of the plague. In the early 16th century the city of London ordered that the number of people dying be recorded in each parish, along with the number of Christenings. Beginning in 1592 these records (or "bills") were printed and circulated on a weekly basis during particularly rough years, and so they were called the London Bills of Mortality (Laxton 1987; Lorimer 1959). Between 1603 and 1849 these records were published weekly (on Thursdays, with an annual summary on the Thursday before Christmas) in what amounts to one of the most important series of vital statistics before the establishment in the 19th century of official government bureaucracies to collect and analyze such data. To begin with, the information on deaths indicated the cause (since one goal was to keep track of the deadly plague), but starting in the 18th century the age of those dying was also noted.

In 1662 a Londoner named John Graunt, who is sometimes called the father of demography, analyzed the series of Bills of Mortality in the first known statistical analysis of demographic data (Sutherland 1963). Although he was a haberdasher by trade, Graunt used his spare moments to conduct studies that were truly remarkable for his time. He discovered that for every 100 people born in London, only 16 were still alive at age 36 and only 3 at age 66 (Dublin, Lotka, and Spielgeman 1949;

Figure 2.2 Standard Birth and Death Certificates Used in the United States

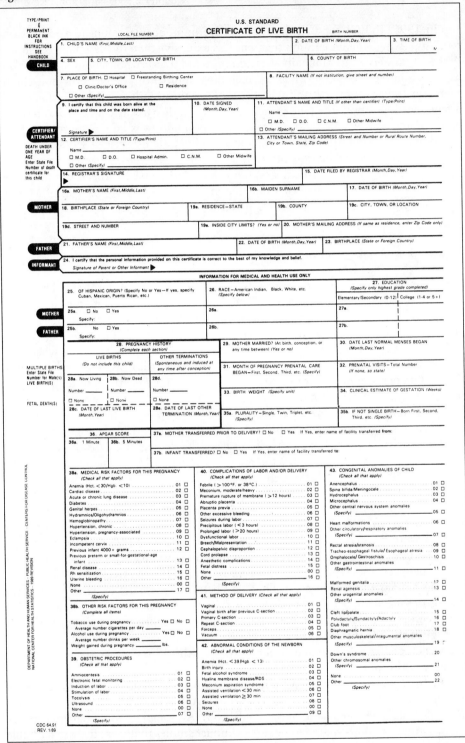

Note: Although each state in the United States may design its own birth and death certificates, these are the standard forms suggested by the U.S. National Center for Health Statistics.

Figure 2.2 (continued)

LOCAL FILE NUMBER

U.S. STANDARD
CERTIFICATE OF DEATH

STATE FILE NUMBER

1. DECEDENT'S NAME (First, Middle, Last)	2. SEX	3. DATE OF DEATH (Month, Day, Year)

DECEDENT

4. SOCIAL SECURITY NUMBER	5a. AGE—Last Birthday (Years)	5b. UNDER 1 YEAR Months / Days	5c. UNDER 1 DAY Hours / Minutes	6. DATE OF BIRTH (Month, Day, Year)	7. BIRTHPLACE (City and State or Foreign Country)

8. WAS DECEDENT EVER IN U.S. ARMED FORCES? (Yes or no)	9a. PLACE OF DEATH (Check only one; see instructions on other side)

HOSPITAL: ☐ Inpatient ☐ ER/Outpatient ☐ DOA OTHER: ☐ Nursing Home ☐ Residence ☐ Other (Specify)

9b. FACILITY NAME (If not institution, give street and number)	9c. CITY, TOWN, OR LOCATION OF DEATH	9d. COUNTY OF DEATH

NAME OF DECEDENT: For use by physician or institution
SEE INSTRUCTIONS ON OTHER SIDE

10. MARITAL STATUS—Married, Never Married, Widowed, Divorced (Specify)	11. SURVIVING SPOUSE (If wife, give maiden name)	12a. DECEDENT'S USUAL OCCUPATION (Give kind of work done during most of working life. Do not use retired.)	12b. KIND OF BUSINESS/INDUSTRY

13a. RESIDENCE—STATE	13b. COUNTY	13c. CITY, TOWN, OR LOCATION	13d. STREET AND NUMBER

13e. INSIDE CITY LIMITS? (Yes or no)	13f. ZIP CODE	14. WAS DECEDENT OF HISPANIC ORIGIN? (Specify No or Yes—If yes, specify Cuban, Mexican, Puerto Rican, etc.) ☐ No ☐ Yes Specify:	15. RACE—American Indian, Black, White, etc. (Specify)	16. DECEDENT'S EDUCATION (Specify only highest grade completed) Elementary/Secondary (0-12) / College (1-4 or 5+)

PARENTS

17. FATHER'S NAME (First, Middle, Last)	18. MOTHER'S NAME (First, Middle, Maiden Surname)

INFORMANT

19a. INFORMANT'S NAME (Type/Print)	19b. MAILING ADDRESS (Street and Number or Rural Route Number, City or Town, State, Zip Code)

DISPOSITION

20a. METHOD OF DISPOSITION ☐ Burial ☐ Cremation ☐ Removal from State ☐ Donation ☐ Other (Specify)	20b. PLACE OF DISPOSITION (Name of cemetery, crematory, or other place)	20c. LOCATION—City or Town, State

21a. SIGNATURE OF FUNERAL SERVICE LICENSEE OR PERSON ACTING AS SUCH ▶	21b. LICENSE NUMBER (of Licensee)	22. NAME AND ADDRESS OF FACILITY

SEE DEFINITION ON OTHER SIDE

PRONOUNCING PHYSICIAN ONLY

Complete items 23a-c only when certifying physician is not available at time of death to certify cause of death.

23a. To the best of my knowledge, death occurred at the time, date, and place stated. Signature and Title ▶	23b. LICENSE NUMBER	23c. DATE SIGNED (Month, Day, Year)

ITEMS 24-26 MUST BE COMPLETED BY PERSON WHO PRONOUNCES DEATH

24. TIME OF DEATH ___ M	25. DATE PRONOUNCED DEAD (Month, Day, Year)	26. WAS CASE REFERRED TO MEDICAL EXAMINER/CORONER? (Yes or no)

SEE INSTRUCTIONS ON OTHER SIDE

27. PART I. Enter the diseases, injuries, or complications that caused the death. Do not enter the mode of dying, such as cardiac or respiratory arrest, shock, or heart failure. List only one cause on each line.

Approximate Interval Between Onset and Death

IMMEDIATE CAUSE (Final disease or condition resulting in death) ➤ a. _____
DUE TO (OR AS A CONSEQUENCE OF):

Sequentially list conditions, if any, leading to immediate cause. Enter **UNDERLYING CAUSE** (Disease or injury that initiated events resulting in death) **LAST**

b. _____
DUE TO (OR AS A CONSEQUENCE OF):

c. _____
DUE TO (OR AS A CONSEQUENCE OF):

d. _____

CAUSE OF DEATH

PART II. Other significant conditions contributing to death but not resulting in the underlying cause given in Part I.	28a. WAS AN AUTOPSY PERFORMED? (Yes or no)	28b. WERE AUTOPSY FINDINGS AVAILABLE PRIOR TO COMPLETION OF CAUSE OF DEATH? (Yes or no)

29. MANNER OF DEATH ☐ Natural ☐ Accident ☐ Suicide ☐ Homicide ☐ Pending Investigation ☐ Could not be Determined	30a. DATE OF INJURY (Month, Day, Year)	30b. TIME OF INJURY ___ M	30c. INJURY AT WORK? (Yes or no)	30d. DESCRIBE HOW INJURY OCCURRED
	30e. PLACE OF INJURY—At home, farm, street, factory, office building, etc. (Specify)		30f. LOCATION (Street and Number or Rural Route Number, City or Town, State		

SEE DEFINITION ON OTHER SIDE

CERTIFIER

31a. CERTIFIER (Check only one)

☐ CERTIFYING PHYSICIAN (Physician certifying cause of death when another physician has pronounced death and completed Item 23)
To the best of my knowledge, death occurred due to the cause(s) and manner as stated.

☐ PRONOUNCING AND CERTIFYING PHYSICIAN (Physician both pronouncing death and certifying to cause of death)
To the best of my knowledge, death occurred at the time, date, and place, and due to the cause(s) and manner as stated.

☐ MEDICAL EXAMINER/CORONER
On the basis of examination and/or investigation, in my opinion, death occurred at the time, date, and place, and due to the cause(s) and manner as stated.

31b. SIGNATURE AND TITLE OF CERTIFIER ▶	31c. LICENSE NUMBER	31d. DATE SIGNED (Month, Day, Year)

32. NAME AND ADDRESS OF PERSON WHO COMPLETED CAUSE OF DEATH (ITEM 27) (Type/Print)

REGISTRAR

33. REGISTRAR'S SIGNATURE ▶	34. DATE FILED (Month, Day, Year)

Graunt 1662 [1939]). With these data he uncovered the high incidence of infant mortality in London and found, somewhat to the amazement of people at the time, that there were regular patterns of death in different parts of London. Graunt "opened the way both for the later discovery of uniformities in many social or volitional phenomena like marriage, suicide and crime, and for a study of these uniformities, their nature and their limits; thus he, more than any other man, was the founder of statistics" (Willcox 1936:xiii).

One of Graunt's close friends (and possibly the person who coaxed him into this work) was William Petty, a member of the Royal Society in London. Petty circulated Graunt's work to the Society (which would not have otherwise paid much attention to a "tradesman"), and this brought it to the attention of the scientific world of 17th century Europe. Several years later, in 1693, Edmund Halley (of Halley's comet fame) became the first scientist to elaborate on the probabilities of death. Although Halley, like Graunt, was a Londoner, he came across a list of births and deaths kept for the city of Breslau in Silesia (now Poland). From these data, Halley used the life-table technique (discussed in Chapter 4) to determine that the expectation of life in Breslau between 1687 and 1691 was 33.5 years (Dublin, et al. 1949). Yet despite the interest created by the work of Graunt, Petty, Halley, and others, people remained skeptical about the quality of the data and unsure of what could be done with them (Starr 1987). So it was not until the middle of the 19th century that civil registration of births and deaths became compulsory in all of England, and an office of vital statistics was officially established by the government in that country, mirroring events in much of Europe and North America (Linder 1959).

Today we find the most complete vital registration system in countries that are most highly developed and the least complete (often nonexistent) in the least developed. Such systems seem to be tied to literacy (there must be someone in each area to record events), to adequate communication, and to the cost of the bureaucracy required for such record keeping, all of which is associated with economic development. Among countries where systems of vital registration do exist, there is wide variation in the completeness with which events are recorded. Even in the United States the registration of births is not yet 100 percent complete.

Although most nations have separate systems of birth and death registration, there are dozens of countries that maintain **population registers**. These are lists of all people in the country. Alongside each name are recorded all vital events for that individual, typically birth, death, marriage, divorce, and change of residence. Such registers are kept primarily for administrative (that is, social control) purposes such as legal identification of people, an election roll, and a call for military service, but they are also extremely valuable for demographic purposes, since they provide a demographic life history for each individual. However, population registers vary widely in quality. The most complete and demographically usable ones are probably those found in the Netherlands (where they now serve as a substitute for the census), Denmark, Norway, Sweden, Finland, and Japan. Registers are expensive to maintain, but many countries that could afford them, such as the United States, tend to avoid them because of the perceived threat to personal freedom that can be inherent in a system that compiles and centralizes data.

Combining the Census and Vital Statistics

Although the recording of vital events provides information about the number of births and deaths (along with other events) according to such characteristics as age and sex, we also need to know how many people there are at risk of these events. Thus, vital statistics data can be teamed up with census data, which do include that information. For example, you may know from the vital statistics that there were 3,899,589 births in the United States in 1995, but that figure tells you nothing about whether birth rates were high or low. In order to draw any conclusion you must relate those births to the 262,890,000 Americans who were alive in 1995, and only then do you discover a low birth rate of 14.8 births per 1,000 population, down from 16.7 in 1990.

Since censuses are not conducted every year, you may wonder how an estimate of the population can be produced for an intercensal years such as 1995. Again, the answer is that census data are combined with vital statistics data. For example, the population in any year after a census should be equal to the census population plus all the people who have been born since the census, minus the people who have died in the interim, plus the people who have migrated in, minus the people who have migrated out. Naturally, deficiencies in any of these data sources will lead to inaccuracies in the estimate of the number of people alive at any time.

Administrative Data

Knowing that censuses and the collection of vital statistics were not originally designed to provide data for demographic analysis has alerted demographers everywhere to keep their collective eyes open for any data source that might yield important information. For example, an important source of information about immigration to the United States is the compilation of **administrative records** filled out for each person entering the country from abroad. These forms are collected and tabulated by the Immigration and Naturalization Service (INS), which is part of the U.S. Department of Justice. Of course, we need other means to estimate the number of people who enter without documents and avoid detection by the INS, and I discuss that more in Chapter 7. Data are not routinely gathered on people who permanently leave the United States, but the administrative records of the U.S. Social Security Administration provide some clues about the number and destination of such individuals. Migration within the United States is estimated on an annual basis by the Census Bureau using data provided to them by the Internal Revenue Service (IRS). Although no personal information is ever divulged, the IRS can match social security numbers of taxpayers each year and see if their address has changed, thus providing a clue about geographic mobility from that type of administrative record.

At the local level a variety of administrative data are tapped into regularly to determine demographic patterns. School enrollment data provide clues to patterns of population growth and migration. Utility hookups (connections and disconnections) can also be used to discern local population trends, as can the number of people signing up for various government-sponsored programs such as health (Medicaid and Medicare) and income assistance (various forms of welfare).

Sample Surveys

There are two major difficulties with using data collected in the census or by the vital statistics registration system or derived from administrative records: (1) they are usually collected for purposes other than demographic analysis (this is true even for the censuses in most countries) and thus do not necessarily reflect the theoretical concerns of demography; and (2) they are collected by a lot of different people using many different methods and may be prone to numerous kinds of error. For these two reasons, in addition to the cost of big data-collection schemes, sample surveys have been used increasingly often to gather demographic data. Sample surveys may provide the social, psychological, economic, and even physical data that I referred to earlier as being necessary to an understanding of why things are as they are.

By using a carefully selected sample of even a few thousand people, demographers have been able to ask questions about births, deaths, migration, and other subjects that reveal aspects of the "why" of demographic events rather than just the "what." In some poor or remote areas of the world, sample surveys can provide reasonably accurate estimates of the levels of fertility, mortality, and migration in the absence of census or vital registration data.

In the United States, one of the most important sample surveys is the Current Population Survey (CPS) conducted every month by the U.S. Bureau of the Census. Since 1943, thousands of households (currently over 50,000) have been queried each month about a variety of things, although a major thrust of the survey is to gather information on the labor force. Each year, detailed questions are also asked about fertility and migration and such characteristics as education, income, marital status, and living arrangements. These data are a major source of demographic information about the American population. In 1983 the Census Bureau launched the Survey on Income and Program Participation (SIPP), which is a companion to the Current Population Survey. Using a rotating panel of about 20,000 households, the SIPP gathers detailed data on sources of income and wealth, disability, and the extent to which household members participate in government assistance programs. The Census Bureau also conducts the American Housing Survey on a regular basis, providing important data on mobility and migration patterns in the United States. The National Center for Health Statistics in the United States generates data about reproductive health in the National Survey of Family Growth, which it conducts every 5 years or so.

Canada has a monthly Labor Force Survey (LFS), initiated in 1947 to track employment trends after the end of World War II. It too is a rotating panel of more than 50,000 households, and although its major purpose is to produce data on the labor force (hence the name), it also obtains data on most of the core sociodemographic characteristics of people in each sampled household, so it provides a continuous measure of population trends in Canada, just as the CPS does in the United States. Since 1985 Statistics Canada has also conducted an annual

General Social Survey, which is designed to elicit detailed data about various aspects of life in Canada, such as health and social support, families, time use, and related topics.

Mexico conducts a monthly survey comparable in many ways to the CPS and the LFS. The National Survey of Urban Employment (Encuesta Nacional de Empleo Urbano [ENEU]) is a large stratified sample of households undertaken by INEGI in the 43 largest urban areas (places with 100,000 people or more) in the country. As with the Current Population Survey, the goal is to provide a way of continuously measuring and monitoring the sociodemographic and economic characteristics of the population. Most of the population questions asked in the census (see Table 2.1) are asked in the ENEU, along with a detailed set of questions about labor force activity of everyone in the household who is 12 years of age or older.

One of the largest social scientific projects ever undertaken was the World Fertility Survey, conducted under the auspices of the International Statistical Institute in the Netherlands. Between 1972 and 1982, a total of nearly 350,000 women of childbearing age were interviewed, encompassing 42 developing nations and 20 developed countries (Lightbourne, Singh, and Green 1982). These data have contributed substantially to our knowledge of how and why people in different parts of the world control their fertility, as will be apparent in Chapter 6. Concurrent with the World Fertility Survey was a series of Contraceptive Prevalence Surveys, conducted in Latin America, Asia, and Africa with funding from the U.S. Agency for International Development. In 1984 the work of the World Fertility Survey and the Contraceptive Prevalence Surveys was combined into one large project called the Demographic and Health Surveys Program, which continues to gather data on reproductive behavior in developing nations.

Historical Sources

I have already alluded to the lengthy history of demographic data gathering. The flip side is that our modern understanding of population processes is shaped not only by our perception of current trends but by historical events as well. Historical demography requires that we almost literally dig up information about the patterns of mortality, fertility, and migration in past generations—to reconstruct "the world we have lost," as Peter Laslett (1971) has called it. You may prefer to whistle past the graveyard, but researchers at England's Cambridge University Group for the History of Population and Social Structure have spent the past few decades pioneering ways in which to recreate history by reading dates on tombstones and organizing information contained in parish church registers and other local documents (Reher and Schofield 1993; Wrigley and Schofield 1981).

Historical sources of demographic information include censuses and vital statistics, but the general lack of good historical vital statistics is what typically necessitates special detective work to locate birth records in church registers and death records in the graveyards. Even in the absence of a census, a complete set of good local records for a small village may allow a researcher to reconstruct the demographic profile of families by matching entries of births, marriages, and deaths in the community over a period of several years. Yet another source of such information is family genealogies, the compilation of which has become somewhat more common in recent years.

The results of these labors can be of considerable importance in testing our notions about how the world used to work. For example, through historical demographic research we now know that the conjugal family (parents and their children) is not a product of industrialization and urbanization, as was once thought (Wrigley 1974). In fact, such small family units were quite common throughout Europe for several centuries before the Industrial Revolution and may actually have contributed to the process of industrialization by allowing the family more flexibility to meet the needs of the changing economy. In the United States, however, extended families may have been more common prior to the 20th century than has generally been thought (Ruggles 1994). Conclusions such as this come from the Integrated Public Use Microdata Series of the Historical Census Projects at the University of Minnesota, an innovative database built by selecting samples of household data from a long series of censuses (1850 to 1990) in the United States. This project has including information on "over 50 million individuals spread over 140 years of extraordinary social and economic change" (Ruggles and Menard 1995:2).

By quantifying (and thereby clarifying) our knowledge of past patterns of demographic events, we are also better able to interpret historical events in a meaningful fashion. Wells (1985) has reminded us that the history of the struggle of American colonists to survive, marry, and bear children may tell us more about the determination to forge a union of states than would a detailed recounting of the actions of British officials. In subsequent chapters we will also have numerous occasions to draw upon the results of the Princeton European Fertility Project, which gathered and analyzed data on marriage and reproduction throughout 19th- and early 20th- century Europe.

Who Uses Population Data?

As the world has become more aware of its demographic underpinnings, population studies have been used in a wide range of planning strategies. In both government and business, people are aware that demography has important implications for social, economic, and political planning. Indeed, the Baby Boom generations in the United States and Canada have repeatedly reminded society about how important population change can be. In Mexico, declining fertility may be the key factor in helping economic growth to lift greater fractions of people into the middle class. Political leaders may use demographic data to provide insights into the forces producing social and economic change in their area, whether that area is a local region, a state, or a nation. Pakistan's Minister of Population Welfare put it all on the table in 1997 when she said "If we achieve success in lowering our population growth substantially, Pakistan has a future. But if, God forbid, we should not—no future" (H. E. Syeda Abida Hussain, quoted in Population Institute 1997:3)

Local governments use demographic data for more mundane but still extremely important purposes, such as planning for changes in municipal services and infrastructure, establishing voting district boundaries, estimating future tax revenue, and designing public programs. Politicians use population information to decide what their constituents are like and who will vote for them and to evaluate the issues likely to be important in the areas they represent. They also may use demographic data to provide insights into the forces producing social and economic change in their area as a way of formulating appropriate policy or legislation.

Demographic data are also crucial to business and social planners. Just as legislators need to consider the future ratio of retirees to wage earners when determining the tax rate for old-age insurance, insurance companies need to know the probabilities of death before insuring a life. Building contractors need to know the age and economic status of prospective buyers before embarking on a residential construction project. Educators need to project the future number of students in an area before hiring (or firing) faculty and staff. A highway engineer needs to know how a proposed freeway interchange might alter the population of a nearby village or town. Sales efforts are increasingly being based on demographic trends and characteristics. Such data are used in picking sites for setting up new businesses, for targeting particular marketing strategies, and for forecasting long-term sales trends (see Chapter 15).

Most demographers use population data to improve the understanding of human society; that is my major focus in this book. Using data for this purpose, of course, also aids in the other uses of population data I have mentioned. People in business who can interpret population changes in human terms, rather than in mere statistical terms, are likely to use the data more effectively for profitable purposes. Similarly, a social understanding of the data used by government planners will lead to better decision making for the public welfare. For you as an individual, understanding the social, economic, and political causes and consequences of population growth will improve your ability to cope with a future that will, without any doubt, be significantly influenced by demographic events.

Demographic Uses of Geographic Information Systems

Demographers have been using maps as a tool for analysis for a long time, and some of the earliest analyses of disease and death relied heavily on maps showing, for example, where it was that people were dying from particular causes (Cliff and Haggett 1996). In the middle of the 19th century, London physician John Snow used maps to trace a local cholera epidemic. He was able to show that cholera occurred much more frequently in customers of a water company that drew its water from the lower Thames River (downstream), where it had become contaminated with London sewage, whereas another company obtained water from the upper Thames (Snow 1936). Today a far more sophisticated version of this same idea is available to demographers (see, for example, Ricketts, et al. 1997). The advent of powerful desktop computers has created a "GIS revolution" (Longley and Batty 1996) that allows demographers (and others, of course) to bring maps together with data in a **geographic information system (GIS).**

A GIS is a computer-based system that allows us to combine maps with data that refer to particular places on those maps and then to analyze those data and display the results as thematic maps or some other graphic format. The computer allows us to transform a map into a set of areas (such as a country or a state or a census tract), lines (such as streets or highways or rivers), and points (such as a house or a school or a health clinic). Our demographic data must then be **geo-referenced** so that the computer will associate them with the correct area, line, or point. If the computer knows that a particular set of latitude and longitude coordinates represents the map of Mexico, then our data for Mexico must be "referenced" to that particular location. Or, if we have a survey of households, then each variable for the household would be geo-referenced to the specific address (the point) of that household. The "revolutionary" aspect of GIS is that the geo-referencing of data to places on the map means that we can combine different types of data (such as census and survey data) for the same place, and we can do it for more than one time (such as data for 1990 and 2000). This greatly increases our ability to visualize and thus understand the kinds of demographic changes that are taking place over time and space.

The computer—the revolutionary element that has created GIS (the maps and the demographic data have been around for a long time)—has vastly expanded our capacity to process and analyze data. It is no coincidence that census data are so readily amenable to being "crunched" by the computer; the history of the computer and the U.S. Bureau of the Census goes back a long way together. Prior to the 1890 census, the U.S. government held a contest to see who could come up with the best machine for counting the data from that census. The winner was Herman Hollerith, who had worked on the 1880 census right after graduating from Columbia University. His method of feeding a punched card through a tabulating machine proved to be very successful, and in 1886 he organized the Tabulating Machine Company, which in 1911 was merged with two other companies and became the International Business Machine (IBM) Corporation (Kaplan and Van Valey 1980). Then, after World War II, the Census Bureau sponsored the development of the first computer designed for mass data processing—the UNIVAC-I—which was used to help with the 1950 census and led the world into the computer age.

Another really useful innovation was the creation for the 1980 U.S. census of the DIME (Dual Independent Map Encoding) files. This was the first step toward computer mapping, in which each piece of data could be coded in a way that could be matched electronically to a place on a map. In the 1980s there were several private firms that latched onto this technology, improved it, and made it available to other companies for their own business uses.

By the early 1990s the pieces of the puzzle had come together. The data from the 1990 U.S. census were made available for the first time on CD-ROM and at prices affordable to a wide range of users. Furthermore, the Census Bureau reconfigured its geographic coding of data, creating what it calls the TIGER (Topologically Integrated Geographic Encoding and Referencing) system, a "digital (computer-readable) geographic database that automates the mapping and related geographic activities required to support the Census Bureau's census and survey programs" (U.S. Bureau of the Census 1993:A11). At the same time, and certainly in response to increased demand, personal computers came along that were powerful enough and had enough memory to be able to store and manipulate huge census files, including both the geographic database and the actual population and housing data. Not far behind was the software to run those computers, and there is now a growing number of firms making software for desktop computers that allow interactive spatial analysis of census and other kinds of data and then the production of high-quality color maps of the results of the analysis. Two of these firms— Environmental Systems Research Institute (ESRI) and Geographic Data Technology (GDT)—were awarded contracts in 1997 to help the U.S. Bureau of the Census update its computerized Master Address File (the information used to continuously update the TIGER files) in order to improve accuracy in the Census 2000.

Knowledge and understanding are based on information, and our information base grows by being able to tap more deeply into rich data sources such as censuses and surveys. GIS is an effective tool for doing this, and you will see numerous examples of GIS at work in the remaining chapters.

Where Can You Go for Information?

The power of the internet and the world wide web have greatly expanded the scope of demographic information available to people all over the world. I could be camped out in a remote valley in the Andes Mountains in Peru, but if I have my laptop computer with a modem and a cellular phone, I can log onto the internet and discover the latest demographic information for almost any country in the world, using some of the websites listed at the end of each chapter in this book. Nonetheless, there is still a lot more population information, especially the details, that is published on paper than is published on the internet, although that may well change in your lifetime. The U.S. Bureau of the Census has been moving toward the day when virtually all of its data will be made available through the internet, rather through traditional printing on paper. Statistics Canada also has a great deal of its data on the internet, whereas INEGI has been making much of its data available for sale on CD-ROM. The addresses for the websites of these agencies are given at the end of this chapter, by the way.

The most convenient and widely available source of published international data is the Demographic Yearbook, published by the United Nations every year since 1948. The Population Reference Bureau regularly publishes information on the world's population in its Population Bulletin, and it also publishes an annual World Population Data Sheet that is a very handy reference for up-to-date estimates of basic demographic facts. The single best source for fairly detailed information about the population of the United States is the Statistical Abstract of the United States, which is published annually by the U.S. Department of Commerce. Tables from the Abstract are now available through the website of the U.S. Bureau of the Census. Many countries of the world have similar kinds of statistical yearbooks that summarize their demographic, social, and economic data, and these are often available at libraries.

There is a wide range of periodicals that tend to concentrate on issues specific to population studies. At the more popular level is *American Demographics,* which is explicitly aimed at interpreting population statistics for applied users. Journals such as *Population and Development Review, Population Research* and *Policy Review, Family Planning Perspectives,* and *International Family Planning Perspectives* are aimed especially at field practitioners and policy planners. The more technical journals include *Demography, Population Studies, Studies in Family Planning, Population and Environment, Genus, European Journal of Population, International Journal of Population Geography, Social Biology,* and *International Migration*

Review. Keep in mind that many important articles dealing with demographic issues are published in nondemographic journals, such as the *American Sociological Review, American Journal of Sociology, Social Forces,* and *The Professional Geographer.* Although all of the above journals publish articles that may refer to any geographic area in the world, demographic studies of Canada may also be found in *Canadian Studies in Population,* and research about Mexico is often published in *Estudios Demográficos y Urbanos.* To find the articles that you need you are always well advised to consult the *Population Index* (Princeton University, Office of Population and Research). The *Population Index* has been the single most important bibliographical tool for demographers since it was first published in the 1930s. It is now available online at http://popindex.princeton.edu/.

Summary and Conclusion

The working bases of any science are facts and theory. In this chapter I have discussed the major sources of demographic information, the wells from which population data are drawn. The census is the most widely known source of data on populations, and humans have been counting themselves in this way for a long time. However, the modern series of more scientific censuses date only from late 18th and early 19th centuries. The high cost of censuses, combined with the increasing knowledge that we have about the value of surveys, has meant that even in so-called complete enumerations there is often some kind of sampling. That is certainly true in North America, as the United States, Canada, and Mexico all have employed sampling techniques in recent censuses. Even vital statistics can be estimated using sample surveys, although the usual pattern is for births and deaths (and often marriages, divorces, and abortions) to be registered with the civil authorities. Some countries take this a step farther and maintain a complete register of life events for everybody.

Knowledge can also be gleaned from administrative data that are gathered for nondemographic purposes. These are particularly important sources of information for helping us to measure migration. It is not just the present that we attempt to measure; historical sources of information can add much to our understanding of current trends in population growth and change. Our understanding is increasingly enhanced by incorporating our demographic data into a geographic information system, permitting us to ask questions that were not really answerable before the advent of the computer.

In this chapter and the preceding one, I have introduced you first to where the world currently stands demographically and then to the sources of data that permit us to know these things. In the next chapter I will introduce you to the major theories that attempt to explain how population growth is related to the social system.

Main Points

1. In order to study population processes and change, you need to know how many people are alive, how many are being born, how many are dying, how many are moving in and out, and why these things are happening.

2. A basic source of demographic information is the population census, in which information is obtained about all people in a given area at a specific time.

3. Not all countries regularly conduct censuses. For example, between 1965 and 1974 only about 55 percent of the world's population was enumerated in a census, although 90 percent was enumerated between 1985 and 1994.

4. In the United States, censuses have been taken every 10 years since 1790.

5. In Canada, censuses have been taken every 10 years since 1851 and every 5 years since 1951.

6. In Mexico, censuses have been taken every 10 years since 1900.

7. Errors in the census typically come about as a result of nonsampling errors (the most important source of error) or sampling error.

8. The principal types of nonsampling error are coverage error and content error.

9. Coverage error is typically measured by demographic analysis and/or a dual system of estimation based on a postcensal survey.

10. Information about births and deaths usually comes from vital registration records—data recorded and compiled by government agencies. The most complete vital registration systems are found in the most highly developed nations, while they are often nonexistent in less developed areas.

11. Most of the estimates of the magnitude of population growth and change are derived by combining census data with vital registration data.

12. Sample surveys are sources of information for places in which census or vital registration data do not exist or where reliable information can be obtained less expensively by sampling than by conducting a census.

13. There is a wide range of users of population data, including people in business, government, and academics.

14. Geographic information systems combine maps and demographic data into a computer system for analysis.

Suggested Readings

1. John A. Ross (ed.), 1985, International Encyclopedia of Population (New York: The Free Press).

   This is a resource book that will help you to define basic demographic concepts and issues from A to Z.

2. William Petersen and Renee Petersen, 1985, Dictionary of Demography (Westport, CT: Greenwood Press).

   This five-volume set includes biographies of population researchers from throughout the world; terms, concepts, and institutions; and a multilingual glossary. Think of this as the dictionary that complements the encyclopedia mentioned above.

3. Harvey Choldin, 1994, Looking for the Last Percent: The Controversy Over Census Undercounts (New Brunswick, NJ: Rutgers University Press).

The author takes you inside the Census Bureau to compare the way in which the 1980 and 1990 censuses of the United States were conducted.

4. Daedalus 97(2), 1968 (special issue on historical population studies).

Roger Revelle organized this fascinating set of papers authored by some of the foremost researchers in the field of historical demography, providing glimpses of the "world we have lost."

5. Thomas J. Goliber, 1997, "Population and Reproductive Health in Sub-Saharan Africa," Population Bulletin 52(4).

This overview of population issues in sub-Saharan Africa provides you with an excellent example of how diverse sources of demographic information can be put together to tell an important story.

Websites of Interest

Remember that websites are not as permanent as books and journals, so I cannot guarantee that each of the following websites still exists at the moment that you are reading this.

1. **http://www.census.gov**

   The home page of the U.S. Bureau of the Census. From here you can locate an amazing variety of information, including updates on the Census 2000. You can also download data for the United States. This is one of the most accessed websites in the world.

2. **http://www.statcan.ca**

   The home page of Statistics Canada, which conducts the censuses in Canada. From here you can obtain data from the 1996 Census and can check information about the 2001 census of Canada.

3. **http://inegi.gob.mx**

   The home page of INEGI (Instituto Nacional de Estadística, Geografía, y Informática), which is the government agency in Mexico that conducts the decennial censuses and related demographic surveys.

4. **http://www.ciesin.org**

   The Consortium for International Earth Science Information Network (CIESIN) has been building an impressive collection of downloadable demographic data and maps (so that you can create your own GIS). For example, they have 1990 census data and maps for Mexico that can be accessed at **http://sedac.ciesin.org/home-page/mexico.html**, and data and maps for the 1982 census of China that can be obtained from **http://plue.sedac.ciesin.org/china.**

5. **http://www.cdc.gov/nchswww/nchshome.htm**

   In many countries (including Canada and Mexico) there is a central statistical agency that conducts censuses and also collects vital statistics. Not so in the United States, where these functions are divided up. The vital statistics data are collected from each state, tabulated, analyzed, and disseminated by the National Center for Health Statistics (NCHS), which is part of the U.S. Centers for Disease Control and Prevention (CDC).

CHAPTER 3
Demographic Perspectives

In order to get a handle on population problems and issues, you have to put the facts of population together with the "whys" and "wherefores." In other words, you need a demographic perspective—a way of relating basic information to theories about how the world operates demographically. A **demographic perspective** will guide you through the sometimes tangled relationships between population factors (such as size, distribution, age structure, and growth) and the rest of what is going on in society. As you develop your own demographic perspective, you will acquire a new awareness about your community or your job, for example, or national and world political or social problems. You will be able to ask yourself about the influences that demographic changes have had (or might have had), and you will consider the demographic consequences of events.

In this chapter I discuss several theories of how population processes are entwined with general social processes. There are actually two levels of population theory. At the core of demographic analysis is the technical side of the field—the mathematical and biomedical theories that predict the kinds of changes taking place in the more biological components of demography: fertility, mortality, and the distribution of a population by age and sex. Demography, for example, has played a central role in the development of the fields of probability, statistics, and sampling (see Kraeger 1993). This hard core is crucial to our understanding of human populations (and we will dive into it in the next chapter), but there is a "softer" (but no less important) outer wrapping of theory that relates demographic processes to the real events of the social world (Schofield and Coleman 1986). The linkage of the core with its outer wrapping is what produces your demographic perspective.

There are two broad questions that have to be answered before you will be able to develop your own perspective: (1) what are the causes of population growth (or, at least, population change); and (2) what are the consequences of population growth or change? In this chapter I discuss several perspectives that provide broad answers to these questions and that also introduce you to the major lines of demographic theory. The purpose of this review is to give you a start in developing your own demographic perspective by taking advantage of what others have learned and passed on to us.

I begin the chapter with a brief review of premodern thinking on the subject of population. Most of these ideas are what we call doctrine, as opposed to theory. Early thinkers were certain they had the answers and certain that their proclamations represented the truth about population growth and its implications for society. By contrast, the essence of modern scientific thought is to assume that you do not have the answer and to acknowledge that you are willing to consider evidence regardless of the conclusion to which it points. In the process of sorting out the evidence, we develop tentative explanations (theories) that help to guide our thinking and our search for understanding. In demography, as in all of the sciences, theories replace doctrine when new systematically collected information (censuses and other sources discussed in Chapter 2) becomes available, allowing people to question old ideas and formulate new ones.

Premodern Population Doctrines

Until about 2,500 years ago, human societies probably shared a common concern about population: they valued reproduction as a means of replacing people lost through the universally high mortality. Ancient Judaism, for example, provided the prescription to "be fruitful and multiply" (Genesis 1:28). Indeed, reproductive power was often deified, as in ancient Greece, where it was the job of a variety of goddesses to help mortals successfully bring children into the world and raise those children to adulthood. By the 5th century B.C., the writings of the school of Confucius in China showed an awareness of the relationship between population and resources, and it was suggested that the government should move people from over-populated to under-populated areas. Still, the idea of promoting population growth was clear in the doctrine of Confucius (Keyfitz 1973).

Plato, writing in The Laws in the 3rd century B.C., emphasized the importance of population stability rather than growth. Specifically, Plato proposed keeping the ideal community of free citizens (much of the work was done by indentured laborers or slaves who had few civil rights) at a constant 5,040, which is 7 factorial. He felt that too many people led to anonymity, which would undermine democracy, and that too few people would prevent an adequate division of labor and would not allow a community to be properly defended. Population size would be controlled by late marriage, infanticide, and migration (in or out as the situation demanded) (Plato 1960). Plato was an early exponent of the doctrine that quality in humans is more important than quantity.

Plato's most famous student, Aristotle, was especially concerned that the population of a city-state not grow beyond the means of the families to support themselves. He advocated that the number of children be limited by law, and that if a woman became pregnant after already having all of the children that the law allowed, an abortion would be appropriate (Stangeland 1904). Echoes of this idea are found today in the one child policy in China.

In the Roman Empire, the reigns of Julius and Augustus Caesar were marked by clearly **pronatalist** doctrines. Cicero noted that population growth was seen by emperors as a necessary means of replacing war casualties and of ensuring enough people to help colonize the empire. Several scholars have speculated, however, that by the 2nd century A.D., as the old pagan Roman Empire was waning in power, the birth rate in Rome may have been declining (Stangeland 1904; Veyne 1987). In a thoroughly modern sentiment, Pliny ("the younger") complained that "in our time most people hold that an only son is already a heavy burden and that it is advantageous not to be overburdened with posterity" (quoted in Veyne 1987:13).

The Middle Ages in Europe, which followed the decline of Rome and its transformation from a pagan to a Christian society, was characterized by a combination of both pronatalist and **antinatalist** Christian doctrines. Christianity condemned polygamy, divorce, abortion, and infanticide, practices that had kept earlier Roman growth rates lower than they otherwise might have been. But the writings of Paul in the New Testament led the influential Christian leader, mystic, and writer Augustine (354–430) to argue that virginity is the highest form of human existence. His doctrine of otherworldliness held that if all men would abstain from intercourse, then so much sooner would the City of God be filled and the end of the world hastened (Keyfitz 1973). This was an economically stagnant, fatalistic period of European history, and for centuries Europeans were content with the idea that population was a matter best regulated by God.

Intellectual stagnation in Europe was countered by the flowering of Islamic scholarship in the Middle East and North Africa. By the 14th century, one of the great Arab historians and philosophers, Ibn Khaldun, was in Tunis writing about the benefits of a growing population. In particular, he argued that population growth creates the need for specialization of occupations, which in turn leads to higher incomes, concentrated especially in cities: "Thus, the inhabitants of a more populous city are more prosperous than their counterparts in a less populous one. . . . The fundamental cause of this is the difference in the nature of the occupations carried on in the different places. For each town is a market for different kinds of labour, and each market absorbs a total expenditure proportionate to its size" (quoted in Issawi 1987:93). Ibn Khaldun was not a utopian. His philosophy was that societies evolved and were transformed as part of natural and normal processes. One of these processes was that "procreation is stimulated by high hopes and resulting heightening of animal energies" (quoted in Issawi 1987:268).

While Europe muddled through the Middle Ages, Islam (which had emerged in the 7th century A.D.) was expanding throughout the Mediterranean. Muslims took control of southern Italy and the Iberian peninsula and, under the Ottoman Empire, controlled the Balkans and the rest of southeastern Europe. Europe's reaction had been the Crusades, a series of wars launched by Christians to wrestle control away from Muslims. These expeditions were largely unsuccessful from a military perspective, but they did put Europeans into contact with the Muslim world, which ultimately led to the Renaissance—the rebirth of Europe:

> The Islamic contribution to Europe is enormous, both of its own creations and of its borrowings—reworked and adapted—from the ancient civilizations of the eastern Mediterranean and from the remoter cultures of Asia. Greek science and philosophy, preserved and improved by the Muslims but forgotten in Europe; Indian numbers and Chinese paper; oranges and lemons, cotton and sugar, and a whole series of other plants along with the methods of cultivating them—all these are but a few of the many things that medieval Europe learned or acquired from the vastly more advanced and more sophisticated civilization of the Mediterranean Islamic world [Lewis 1995:274].

The cultural reawakening of Europe took place in the context of a growing population, as I noted in the first chapter. Not surprisingly then, new murmurings were heard about the place of population growth in the human scheme of things. The Renaissance began with the Venetians, who had established trade with Muslims and others as the eastern Mediterranean ceased to be a Crusade war zone in the 13th century. In that century, an influential Dominican monk, Thomas Aquinas, argued that marriage and family building were not inferior to celibacy. By the 14th century the plague had largely disappeared from Europe; by the 16th century the Muslims (and Jews) had been expelled from southern Spain, and the Europeans had begun their discovery and exploitation of Africa, the Americas, and south Asia. Cities began to grow noticeably, and Giovanni Botero, a 16th century Italian statesman, wrote that "the powers of generation are the same now as one thousand years ago, and, if they had no impediment, the propagation of man would grow without limit and the growth of cities would never stop" (quoted in Hutchinson 1967:111). The 17th and 18th centuries witnessed an historically unprecedented trade (the so-called

Columbian Exchange) of food, manufactured goods, people, and disease between the Americas and most of the rest of the world (see Crosby 1972), undertaken largely by European merchants with their "guns and sails" (Cipolla 1965).

This rise in trade, prompted at least in part by population growth, generated the doctrine of **mercantilism** among the new nation-states of Europe. Mercantilism maintained that a nation's wealth was determined by the amount of precious metals it had in its possession, which were acquired by exporting more goods than were imported, with the difference (the profit) being stored in precious metals. The catch here was that a nation had to have things to produce to sell to others, and the idea was that the more workers you had, the more you could produce. Thus population growth was seen as essential to an increase in national revenue, and mercantilist writers sought to encourage it by a number of means, including penalties for nonmarriage, encouragements to get married, lessening penalties for illegitimate births, limiting out-migration, and promoting immigration of productive laborers. It is important to keep in mind that these doctrines were concerned with the wealth and welfare of a specific country, not all of human society. "The underlying doctrine was, either tacitly or explicitly, that the nation which became the strongest in material goods and in men would survive; the nations which lost in the economic struggle would have their populations reduced by want, or they would be forced to resort to war, in which their chances of success would be small. The war, however, would kill many and make the condition of the survivors more tolerable" (Stangeland 1904:183).

Mercantilist doctrines were supported by the emerging demographic analyses of people like John Graunt, William Petty, and Gregory King (all English) in the 17th century (see Chapter 2) and Johann Peter Süssmilch, an 18th century chaplain in the army of Frederick the Great of Prussia (now Germany). Süssmilch built on the work of Graunt and others and added his own analyses to the observation of the regular patterns of marriage, birth and death in Prussia and believed that he saw in these the divine hand of God ruling human society (Hecht 1987; Keyfitz 1973). His view, widely disseminated throughout Europe, was that a larger population was always better than a smaller one, and, in direct contradistinction to Plato, he valued quantity over quality. He believed that indefinite improvements in agriculture and industry would postpone overpopulation so far into the future that it wouldn't matter.

The issue of population growth was more than idle speculation, because we now know with a fair amount of certainty that the population of England, for example, doubled during the 18th century (Petersen 1979). More generally, during the period from about 1650 to 1850, Europe as a whole experienced rather dramatic population growth as a result of the agricultural "evolution" that preceded (and almost certainly stimulated) the industrial revolution (Cohen 1995). The increasing interest in population stimulated the publication of two important essays on population size, one by David Hume (1752 [1963]) and the other by Robert Wallace (1761 [1969]), which were then to influence Malthus, whom I discuss later. These essays stimulated considerable debate and controversy, because there were large issues at stake: "Was a large and rapidly growing population a sure sign of a society's good health? On balance, were the growth of industry and cities, the movement of larger numbers from one social class to another—in short, all of what we now term 'modernization'—a boon to the people or the contrary? And in society's efforts to resolve such dilemmas, could it depend on the sum of individuals' self-interest or was considerable state control called for?" (Petersen 1979:139). These are questions that we are still dealing with more than 200 years later.

So the population had, in fact, increased during the Mercantilist era, although probably not as a result of any of the policies put forth by its adherents. However, it was less obvious that the population was better off. Rather, the Mercantilist period had become associated with a rising level of poverty (Keyfitz 1972). Mercantilism relied upon a state-sponsored system of promoting foreign trade, while inhibiting imports and thus competition. This generated wealth for a small elite but not for most people. One of the more famous reactions against mercantilism was that mounted by François Quesnay, a physician to King Louis XV of France (and an economist when not "on duty"). While mercantilists argued that wealth depends upon the number of people, Quesnay turned that around and argued that the number of people depends upon the means of subsistence (a general term for level of living). The essence of this view, called **physiocratic** thought, was that land, not people, is the real source of wealth of a nation. In other words, population went from being an independent variable, causing change in society, to a dependent variable, being altered by societal change. As you will see throughout the book, both perspectives have their merits.

Physiocrats also believed that free trade (rather than the import restrictions demanded by mercantilists) was essential to economic prosperity. This concept of "laissez-faire" was picked up by Adam Smith, one of the first modern economic theorists. Central to Smith's view of the world was the idea that, if left to their own devices, people acting in their own self-interest would produce what was best for the community as a whole. Smith differed slightly from the physiocrats, however, on the idea of what led to wealth in a society. Smith believed that wealth sprang from the labor applied to the land (we might now say the "value added" to the land by labor), rather than it being just in the land itself. From this idea sprang the belief that there is a natural harmony between economic growth and population growth, with the latter depending always on the former. Thus, he felt that population size is determined by the demand for labor, which is determined by the productivity of the land. These ideas are important to us because the work of Smith served as an inspiration for the Malthusian theory of population, as Malthus himself acknowledges (see the preface to the sixth edition of Malthus 1872) and as I discuss later.

The 18th century was the Age of Enlightenment in Europe, a time when the goodness of the common person was championed. This perspective that the rights of individuals superseded the demands of a monarchy inspired the American and French Revolutions and was generally very optimistic and utopian, characterized by a great deal of enthusiasm for life and a belief in the perfectibility of humans. In France, these ideas were well expressed by Marie Jean Antoine Nicolas de Caritat, marquis de Condorcet, a member of the French aristocracy who forsook a military career to pursue a life devoted to mathematics and philosophy. His ideas helped to shape the French Revolution, but despite his sympathy with that cause, he died in prison at the hands of revolutionaries. In hiding before his arrest, Condorcet had written a *Sketch for an Historical Picture of the Progress of the Human Mind* (Condorcet 1795 [1955]). He was a visionary who "saw the outlines of liberal democracy more than a century in advance of his time: universal education; universal suffrage; equality before the law; freedom of thought and expression; the right to freedom and self-determination of colonial peoples; the redistribution of wealth; a system of national insurance and pensions; equal rights for women" (Hampshire 1955:x).

Condorcet's optimism was based on his belief that technological progress has no limits. "With all this progress in industry and welfare which establishes a happier proportion between men's talents and their needs, each successive generation will have larger possessions, either as a result of this progress or through the preservation of the products of industry, and so, *as a consequence of the physical constitution of the human race, the number of people will increase*" (Condorcet 1795 [1955]:188; emphasis added). He then asked whether it might not happen that eventually the happiness of the population would reach a limit. If that happens, Condorcet concluded, "we can assume that by then men will know that . . . their aim should be to promote the general welfare of the human race or of the society in which they live or of the family to which they belong, rather than foolishly to encumber the world with useless and wretched beings" (p. 189). Condorcet thus saw prosperity and population growth increasing hand in hand, and he felt that if the limits to growth were ever reached, the final solution would be birth control.

On the other side of the English Channel, similar ideas were being expressed by William Godwin (father of Mary Wollstonecraft Shelley, author of Frankenstein, and father-in-law of the poet Percy Bysshe Shelley). Godwin's *Enquiry Concerning Political Justice and Its Influences on Morals and Happiness* appeared in its first edition in 1793, revealing his ideas that scientific progress would enable the food supply to grow far beyond the levels of his day, and that such prosperity would not lead to overpopulation because people would deliberately limit their sexual expression and procreation. Furthermore, he believed that most of the problems of the poor were due not to overpopulation but to the inequities of the social institutions, especially greed and accumulation of property (Godwin 1793 [1946]).

Thomas Robert Malthus had recently graduated from Cambridge and was a country curate and a nonresident fellow of Cambridge as he read and contemplated the works of Godwin, Condorcet, and others who shared the utopian view of the perfectibility of human society. Although he wanted to be able to embrace such an openly optimistic philosophy of life, he felt that intellectually he had to reject it. In doing so, he unleashed a controversy about population growth and its consequences that rages to this very day.

The Malthusian Perspective

The **Malthusian** perspective derives from the writings of Thomas Robert Malthus, an English clergyman and college professor. His first *Essay on the Principle of Population as it affects the future improvement of society; With remarks on the speculations of Mr. Godwin, M. Condorcet, and other writers* was published anonymously in 1798. Malthus's original intention was not to carve out a career in demography, but only to show that the unbounded optimism of the physiocrats and philosophers was misplaced. He introduced his essay by commenting that "I have read some of the speculations on the perfectibility of man and society, with great pleasure. I have been warmed and delighted with the enchanting picture which they hold forth. I ardently wish for such happy improvements. But I see great, and, to my understanding, unconquerable difficulties in the way to them" (Malthus 1798 [1965]:7).

These "difficulties," of course, are the problems posed by his now-famous **principle of population.** He derived his theory as follows:

> I think I may fairly make two postulata. First, that food is necessary to the existence of man. Secondly, that the passion between the sexes is necessary, and will remain nearly in its present state . . . . Assuming then, my postulata as granted, I say, that the power of population is indefinitely greater than the power in the earth to produce subsistence for man. Population, when unchecked, increases in a geometrical ratio. Subsistence increases only in an arithmetical ratio . . . . By the law of our nature which makes food necessary to the life of man, the effects of these two unequal powers must be kept equal. This implies a strong and constantly operating check on population from the difficulty of subsistence. This difficulty must fall somewhere; and must necessarily be severely felt by a large portion of mankind . . . . Consequently, if the premises are just, the argument is conclusive against the perfectibility of the mass of mankind [Malthus 1798 (1965):11].

Malthus demolished the Utopian optimism by suggesting that the laws of nature, operating through the principle of population, essentially prescribed poverty for a certain segment of humanity. Malthus was a shy person by nature (James 1979; Petersen 1979), and he seemed ill prepared for the notoriety created by his essay. Nonetheless, after owning up to its authorship, he proceeded to document his population principles and to respond to critics by publishing a substantially revised version in 1803, slightly but importantly retitled to read *An Essay on the Principle of Population; or a view of its past and present effects on human happiness; with an inquiry into our prospects respecting the future removal or mitigation of the evils which it occasions.* In all, seven editions of Malthus's essay on population were published, and as a whole they have undoubtedly been the single most influential work relating population growth to its social consequences. Although Malthus drew heavily on earlier writers, he was the first to draw out in a systematic way a picture that links the consequences of growth to its causes.

Causes of Population Growth

Malthus believed that human beings, like plants and nonrational animals, are "impelled" to increase the population of the species by what he called a powerful "instinct," the urge to reproduce. Further, if there were no checks on population growth, human beings would multiply to an "incalculable" number, filling "millions of worlds in a few thousand years" (Malthus 1872[1971]:6). We humans, though, have not accomplished anything nearly that impressive. Why not? Because of the checks to growth that Malthus pointed out—factors that have kept population growth from reaching its biological potential for covering the earth with human bodies.

According to Malthus, the ultimate check to growth is lack of food (the "means of subsistence"). In turn, the means of subsistence are limited by the amount of land

available, the "arts" or technology that could be applied to the land, and "social organization" or land ownership patterns. A cornerstone of his argument is that populations tend to grow more rapidly than does the food supply (a topic I return to in
Chapter 13), since population has the potential for growing geometrically—two
parents could have four children, sixteen grandchildren, and so on—while he believed that food production could be increased only arithmetically, by adding one
acre at a time. In the natural order, then, population growth will outstrip the food
supply, and the lack of food will ultimately put a stop to the increase of people (see
Figure 3.1).

Of course, Malthus was aware that starvation rarely operates directly to kill
people, since something else usually intervenes to kill them before they actually die
of starvation. This "something else" represents what Malthus calls **positive checks,**
primarily those measures "whether of a moral or physical nature, which tend prematurely to weaken and destroy the human frame" (Malthus 1872[1971]:12). Today we would call these the causes of mortality. There are also **preventive checks**—
limits to birth. In theory, the preventive checks would include all possible means of
birth control, including abstinence, contraception, and abortion. However, to
Malthus the only acceptable means of preventing a birth was to exercise **moral restraint;** that is, to postpone marriage, remaining chaste in the meantime, until a man
feels "secure that, should he have a large family, his utmost exertions can save them
from rags and squalid poverty, and their consequent degradation in the community"

Figure 3.1   Over Time, Geometric Growth Overtakes Arithmetic Growth

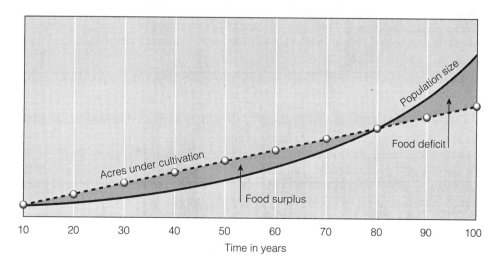

Note: If we start with 100 acres supporting a population of 100 people and then add
100 acres of cultivated land per decade (arithmetic growth) while the population is increasing by 3 percent per year (geometric growth), the result is a few decades of food surplus before population growth overtakes the increase in the acres under cultivation, producing a
food deficit, or "misery," as Malthus called it.

(1872[1971]:13). Any other means of birth control, including contraception (either before or after marriage), abortion, infanticide, or any "improper means," was viewed as a vice that would "lower, in a marked manner, the dignity of human nature." Moral restraint was a very important point with Malthus, because he believed that if people were allowed to prevent births by "improper means" (that is, prostitution, contraception, abortion, or sterilization), then they would expend their energies in ways that are, so to speak, not economically productive.

I should point out that as a scientific theory, the Malthusian perspective leaves much to be desired, since he constantly confuses moralistic and scientific thinking (see Davis 1955). Despite its shortcomings, however, which were evident even in his time, Malthus's reasoning led him to draw some important conclusions about the consequences of population growth.

Consequences of Population Growth

Malthus believed that a natural consequence of population growth was poverty. This is the logical end result of his arguments that (1) people have a natural urge to reproduce, and (2) the increase in the supply of food cannot keep up with population growth. In his analysis, Malthus turned the argument of Adam Smith upside down. Instead of population growth being dependent on the demand for labor, as Smith argued, Malthus believed that the urge to reproduce always forces population pressure to precede the demand for labor. Thus, "overpopulation" (as measured by the level of unemployment) would force wages down to the point where people could not afford to marry and raise a family. At such low wages, with a surplus of labor and the need for each person to work harder just to earn a subsistence wage, cultivators could employ more labor, put more acres into production, and thus increase the means of subsistence. Malthus believed that this cycle of increased food resources leading to population growth leading to too many people for available resources leading then back to poverty was part of a natural law of population. Each increase in the food supply only meant that eventually more people could live in poverty.

As you can see, Malthus did not have an altogether high opinion of his fellow creatures. He figured that most of them were too "inert, sluggish, and averse from labor" (1798[1965]:363) to try to harness the urge to reproduce and avoid the increase in numbers that would lead back to poverty whenever more resources were available. In this way he essentially blamed poverty on the poor themselves. There remained only one improbable way to avoid this dreary situation.

Avoiding the Consequences

Borrowing from John Locke, Malthus argued that "the endeavor to avoid pain rather than the pursuit of pleasure is the great stimulus to action in life" (1798[1965]:359). Pleasure will not stimulate activity until its absence is defined as being painful. Malthus suggested that the well-educated, rational person would perceive in advance the pain of having hungry children or being in debt and would postpone marriage and sexual intercourse until he was sure that he could avoid that pain. If that motivation existed and the preventive check was operating, then the miserable consequences of population growth could be avoided. You will recall that Condorcet had suggested the possibility of birth control as a preventive check but that Malthus objected to this solution: "To remove the difficulty in this way, will, surely in the opinion of most men, be to destroy that virtue, and purity of manners, which the advocates of equality, and of the perfectibility of man, profess to be the end and object of their views" (1798:154). So the only way to break the cycle is to change human nature. Malthus felt that if everyone shared middle-class values, the problem would solve itself. He saw that as impossible, though, since not everyone has the talent to be a virtuous, industrious, middle-class success story, but if most people at least tried, poverty would be reduced considerably.

To Malthus, material success is a consequence of human ability to plan rationally—to be educated about future consequences of current behavior—and he was a man who practiced what he preached. He planned his family rationally, waiting to marry and have children until he was 39, shortly after getting a secure job in 1805 as a college professor. Also, although his detractors later attributed 11 children to him, he and his wife, 11 years his junior, had only 3 children (Nickerson 1975).

To summarize, the major consequence of population growth, according to Malthus, is poverty. Within that poverty, though, is the stimulus for action that can lift people out of misery. So, if people remain poor, it is their own fault for not trying to do something about it. For that reason, Malthus was opposed to the English Poor Laws (welfare benefits for the poor), because he felt they would actually serve to perpetuate misery. They permitted poor people to be supported by others and thus not feel that great pain, the avoidance of which might lead to birth prevention. Malthus argued that if every man had to provide for his own children, he would be more prudent about getting married and raising a family. In his own time, this particular conclusion of Malthus brought him the greatest notoriety, because the number of people on welfare had been increasing and English parliamentarians were trying to decide what to do about the problem. Although the Poor Laws were not abolished, they were reformed largely because Malthus had given legitimacy to public criticism of the entire concept of welfare payments (Himmelfarb 1984). The Malthusian perspective that blamed the poor for their own poverty endures, contrasted with the equally enduring view of Godwin and Condorcet that poverty is the creation of unjust human institutions. Two hundred additional years of debate have only sharpened the edges of the controversy.

Critique of Malthus

The single most obvious measure of Malthus's importance is the number of books and articles that have attacked him, and the attacks began virtually the moment his first essay appeared in 1798. The three most strongly criticized aspects of his theory have been (1) the assertion that food production could not keep up with population growth, (2) the conclusion that poverty was an inevitable result of population growth , and (3) the belief that moral restraint was the only acceptable preventive check. Malthus was not a firm believer in progress; rather, he accepted the notion that each society had a fixed set of institutions that established a stationary level of living. He was aware, of course, of the Industrial Revolution, but he was skeptical of its long-run value. He was convinced that the increase in manufacturing wages that accompanied industrialization would promote population growth without increasing the agricultural production necessary to feed those additional mouths. Although it is clear that he was a voracious reader (Petersen 1979) and that he was a founder of the Statistical Society of London (Starr 1987), it is also clear that Malthus paid scant attention to the economic statistics that were available to him. "There is no sign that even at the end of his life he knew anything in detail about industrialization. His thesis was based on the life of an island agricultural nation, and so it remained long after the exports of manufacturers had begun to pay for the imports of large quantities of raw materials" (Eversley 1959:256). Thus, Malthus either failed to see or refused to acknowledge that technological progress was possible, and that its end result was a higher standard of living, not a lower one.

The crucial part of Malthus's ratio of population growth to food increase was that food (including both plants and animals) would not grow exponentially. Yet when Charles Darwin acknowledged that his *Origin of the Species* was inspired by Malthus's essay, he implicitly rejected this central tenet of Malthus's argument. "Darwin described his own theory as 'the doctrine of Malthus applied with manifold force to the whole animal and vegetable kingdoms; for in this case there can be no artificial increase of food, and no prudential restraint from marriage.' Thus plants and animals, even more than men, would increase geometrically if unchecked" (Himmelfarb 1984:128).

Malthus's argument that poverty is an inevitable result of population growth is also open to scrutiny. For one thing, his writing reveals a certain circularity in logic. In Malthus's view, a laborer could achieve a higher standard of living only by being prudent and refraining from marriage until he could afford it, but Malthus also believed that you could not expect prudence from a laborer until he had attained a higher standard of living. Thus, our hypothetical laborer seems squarely enmeshed in a catch-22. Even if we were to ignore this logical inconsistency, there are problems with Malthus's belief that the Poor Laws contributed to the misery of the poor by discouraging them from exercising prudence. Historical evidence has revealed that between 1801 and 1835 those English parishes that administered Poor Law allowances did not have higher birth, marriage, or total population growth rates than those in which Poor Law assistance was not available (Huzel 1969; 1980; 1984). Clearly, problems with the logic of Malthus's argument seemed to be compounded by his apparent inability to see the social world accurately. "The results of the 1831 Census were out before he died, yet he never came to interpret them. Statistics apart, the main charge against him must be that he was a bad observer of his fellow human beings" (Eversley 1959:256).

Neo-Malthusians

Those who criticize Malthus's insistence on the value of moral restraint, while accepting many of his other conclusions, are typically known as **neo-Malthusians.**

Specifically, neo-Malthusians favor contraception rather than simple reliance on moral restraint. During his lifetime, Malthus was constantly defending moral restraint against critics (many of whom were his friends) who encouraged him to deal more favorably with other means of birth control. In the fifth edition of his Essay he did discuss the concept of **prudential restraint,** which meant the delay of marriage until a family could be afforded without necessarily refraining from premarital sexual intercourse in the meantime. He never fully embraced the idea, however, nor did he ever bow to pressure to accept anything but moral restraint as a viable preventive check.

Ironically, the open controversy actually helped to spread knowledge of birth control among people in 19th century England and America. This was aided materially by the trial and conviction (later overturned on a technicality) in 1877–78 of two neo-Malthusians, Charles Bradlaugh and Annie Besant, for publishing a birth control handbook (*Fruits of Philosophy: The Private Companion of Young Married People,* written by a physician in Massachusetts and originally published in 1832). The publicity surrounding the trial enabled the English public to become more widely knowledgeable about those techniques (Chandrasekhar 1979; Himes 1976). Eventually, the widespread adoption of birth control broke the connection between intercourse and fertility that had seemed so natural to Malthus: "In time fertility behavior came to be better epitomized by inverting the Malthusian assumption than by adhering to it . . . . The supplanting of the social control of fertility *within* marriage completely undermined the validity of the assumption that population growth rates would rise in step with prosperity" (Wrigley 1988:44; emphasis in original).

Criticisms of Malthus do not, however, diminish the importance of his work:

> There are good reasons for using Malthus as a point of departure in the discussion of population theory. These are the reasons that made his work influential in his day and make it influential now. But they have little to do with whether his views are right or wrong. . . . Malthus' theories are not now and never were empirically valid, but they nevertheless were theoretically significant [Davis 1955:541].

As I noted earlier, part of Malthus's significance also lies in the storm of controversy his theories stimulated. Particularly vigorous in their attacks on Malthus were Karl Marx and Friedrich Engels.

The Marxist Perspective

Karl Marx and Friedrich Engels were both teenagers in Germany when Malthus died in England in 1834, and by the time they had independently moved to England and met, Malthus's ideas already were politically influential in their native land. Several German states and Austria had responded to what they believed was overly rapid growth in the number of poor people by legislating against marriages in which the applicant could not guarantee that his family would not wind up on welfare (Glass 1953). As it turned out, that scheme backfired on the German states, because people continued to have children, but out of wedlock. Thus, the welfare rolls grew as the illegitimate children had to be cared for by the state (Knodel 1970). The laws were eventually repealed, but they had an impact on Marx and Engels, who saw the Malthusian point of view as an outrage against humanity. Their demographic perspective thus arose in reaction to Malthus.

WHO ARE THE NEO-MALTHUSIANS?

"Picture a tropical island with luscious breadfruits hanging from every branch, toasting in the sun. It is a small island, but there are only 400 of us on it so there are more breadfruits than we know what to do with. We're rich. Now picture 4,000 people on the same island, reaching for the same breadfruits: Number one, there are fewer to go around; number two, you've got to build ladders to reach most of them; number three, the island is becoming littered with breadfruit crumbs. Things get worse and worse as the population gradually expands to 40,000. Welcome to a poor, littered tropical paradise" (Tobias 1979:49). This scenario would probably have drawn a nod of understanding from Malthus himself, and it is typical of the modern neo-Malthusian view of the world.

One of the most influential of recent neo-Malthusians is the biologist Garrett Hardin. In 1968 he published an article that raised the level of consciousness about population growth in the minds of professional scientists. Hardin's theme was simple and had been made in a previous article by Kingsley Davis (1963): personal goals are not necessarily consistent with societal goals when it comes to population growth. Hardin's metaphor is "the tragedy of the commons." He asks us to imagine an open field, available as a common ground for herdsmen to graze their cattle. "As a rational being, each herdsman seeks to maximize his gain. Explicitly or implicitly, more or less consciously, he asks, 'What is the utility to me of adding one more animal to my herd?'" (Hardin 1968:1244). The benefit, of course, is the net proceeds from the eventual sale of each additional animal, whereas the cost lies in the chance that an additional animal may result in overgrazing of the common ground. Since the ground is shared by many people, the cost is spread out over all, so for the individual herdsman, the benefit of an-

other animal exceeds its cost. "But," notes Hardin, "this is the conclusion reached by each and every rational herdsman sharing a commons. Therein is the tragedy. Each man is locked into a system that compels him to increase his herd without limit—in a world that is limited" (1968:1244). The moral, as Hardin puts it, is that "ruin is the destination toward which all men rush, each pursuing his own best interest in a society that believes in the freedom of the commons. Freedom in a commons brings ruin to all" (1968:1244).

Hardin reminds us that most societies are committed to a social welfare ideal. Families are not completely on their own. We share numerous things in common: education, public health, and police protection, and in the United States we are guaranteed a minimum amount of food and income at the public expense. This leads to a moral dilemma that is at the heart of Hardin's message: "To couple the concept of freedom to breed with the belief that everyone born has an equal right to the commons is to lock the world into a tragic course of action" (Hardin 1968:1246). He is referring, of course, to the ultimate Malthusian clash of population and resources, and Hardin is no more optimistic than Malthus about the likelihood of people voluntarily limiting their fertility before it is too late.

Meanwhile, the public was becoming keenly aware of the population crisis through the writings of the person who is arguably the most famous of all neo-Malthusians, Paul Ehrlich. Like Hardin, Ehrlich is a biologist (at Stanford University), not a professional demographer. His *Population Bomb* (Ehrlich 1968) was an immediate sensation when it came out in 1968 and continues to set the tone for public debate about population issues. In the second edition of his book (Ehrlich 1971), he phrased

Causes of Population Growth

Neither Marx nor Engels ever directly addressed the issue of why and how populations grew. They seem to have had little quarrel with Malthus on this point, although they were in favor of equal rights for men and women and saw no harm in

---

the situation in three parts: "too many people," "too little food," and, adding a wrinkle not foreseen directly by Malthus, "environmental degradation" (Ehrlich called Earth "a dying planet").

In 1990 Ehrlich, in collaboration with his wife Anne, followed with an update titled *The Population Explosion* (Ehrlich and Ehrlich 1990), reflecting their view that the bomb they worried about in 1968 had detonated in the meantime. The level of concern about the destruction of the environment has grown tremendously since 1968. Ehrlich's book had inspired the first Earth Day in the spring of 1970, and the twentieth anniversary of Earth Day in 1990 generated worldwide publicity. Yet in his 1990 book Ehrlich rightly questions why, in the face of the serious environmental degradation that has concerned him for so long, people have regularly failed to grasp its primary cause as being rapid population growth: "Arresting population growth should be second in importance only to avoiding nuclear war on humanity's agenda. Overpopulation and rapid population growth are intimately connected with most aspects of the current human predicament, including rapid depletion of nonrenewable resources, deterioration of the environment (including rapid climate change), and increasing international tensions" (Ehrlich and Ehrlich 1990:18).

Ehrlich thus argues that Malthus was right—dead right. But the death struggle is more complicated than that foreseen by Malthus. To Ehrlich, the poor are dying of hunger, while rich and poor alike are dying from the by-products of affluence—pollution and ecological disaster. Indeed, this is part of the "commons" problem. A few benefit, all suffer. What does the future hold? Ehrlich suggests that there are only two solutions to the population problem: the birth rate solution (lowering the birth rate) and the death-rate solution (a rise in the death rate). He views the death-rate solution as being the most likely to happen, because, like Malthus, he has little faith in the ability of humankind to pull its act together. The only way to avoid that scenario is to bring the birth rate under control, perhaps even by force.

The population of the world is growing rapidly and somebody has to do something. A major part of Ehrlich's contribution has been to encourage people to take some action themselves, to spread the word and practice what they preach. Like Hardin, Ehrlich feels that population growth is outstripping resources and ruining the environment. If we sit back and wait for people to react to this situation, disaster will occur. Therefore we need to act swiftly to push people to limit their fertility by whatever means possible.

Neo-Malthusians thus differ from Malthus because they reject moral restraint as the only acceptable means of birth control and because they face a situation in which they see population growth as leading not simply to poverty but also to widespread calamity. For neo-Malthusians, the "evil arising from the redundancy of population" (Malthus 1872[1971]: preface to the fifth edition) has broadened in scope, and the remedies proposed are thus more dramatic.

Gloomy they may seem, but the messages of Ehrlich and Hardin are important and impressive and have brought population issues to the attention of the entire globe. One of the ironies of neo-Malthusianism is that if the world's population does avoid future calamity, people will likely claim that the neo-Malthusians were wrong. In fact, however, much of the stimulus to action to bring down birth rates and to find new ways to feed people and protect the environment has come as a reaction to the concerns raised by neo-Malthusians.

preventing birth. Nonetheless, they were skeptical of the eternal or natural laws of nature as stated by Malthus (that population tends to outstrip resources), preferring instead to view human activity as the product of a particular social and economic environment. The basic **Marxist** perspective is that each society at each point in history has its own law of population that determines the consequences of population growth. For **capitalism,** the consequences are overpopulation and poverty, whereas for **socialism,** population growth is readily absorbed by the economy with no side effects. This line of reasoning led to a vehement rejection of Malthus, because if Malthus was right about his "pretended 'natural law of population'" (Marx 1890 [1906]:680), then Marx's theory would be wrong.

Consequences of Population Growth

Marx and Engels especially quarreled with the Malthusian idea that resources could not grow as rapidly as population, since they saw no reason to suspect that science and technology could not increase the availability of food and other goods at least as quickly as population grows. Engels argued in 1865 that whatever population pressure existed in society was really pressure against the means of employment rather than against the means of subsistence (Meek 1971). Thus, they flatly rejected the notion that poverty can be blamed on the poor. Instead, they said, poverty is the result of a poorly organized society, especially a capitalist society. Implicit in the writings of Marx and Engels is the idea that the normal consequence of population growth should be a significant increase in production. After all, each worker obviously was producing more than he or she required, or else how would all the dependents (including the wealthy manufacturers) survive? In a well-ordered society, if there were more people, there ought to be more wealth, not more poverty (Engels 1844[1953]).

Not only did Marx and Engels feel generally that poverty was not the end result of population growth, they argued specifically that even in England at the time there was enough wealth to eliminate poverty. Engels had himself managed a textile plant in Manchester, and he believed that in England more people had meant more wealth for the capitalists rather than for the workers because the capitalists were skimming off some of the workers' wages as profits for themselves. Marx argued that they did that by stripping the workers of their tools and then, in essence, charging the workers for being able to come to the factory to work. For example, if you do not have the tools to make a car but want a job making cars, you could get hired at the factory and work 8 hours a day. But, according to Marx, you might get paid for only 6 hours, the capitalist (owner of the factory) keeping part of your wages as payment for the tools you were using. The more the capitalist keeps, of course, the lower your wages and the poorer you will be.

Furthermore, Marx argued that capitalism worked by using the labor of the working classes to earn profits to buy machines that would replace the laborers, which, in turn, led to unemployment and poverty. Thus, the poor were not poor because they overran the food supply, but only because capitalists had first taken away part of their wages and then taken away their very jobs and replaced them with machines. Thus, the consequences of population growth that Malthus discussed were really the consequences of capitalist society, not of population growth per se. Over-population in a capitalist society was thought to be a result of the capitalists' desire for an industrial reserve army that would keep wages low through competition for jobs and, at the same time, would force workers to be more productive in order to keep their jobs. To Marx, however, the logical extension of this was that the growing population would bear the seeds of destruction for capitalism, because unemployment would lead to disaffection and revolution. If society could be reorganized in a more equitable (that is, socialist) way, then the population problems would disappear.

It is noteworthy that Marx, like Malthus, practiced what he preached. Marx was adamantly opposed to the notion of moral restraint, and his life repudiated that concept. He married at the relatively young age (compared with Malthus) of 25, proceeded to father 8 children, including one illegitimate son, and was on intimate terms with poverty for much of his life.

In its original formulation, the Marxist (as well as the Malthusian) perspective was somewhat provincial, in the sense that its primary concern was England in the 19th century. Marx was an intense scholar who focused especially on the historical analysis of economics as applied to England, which he considered to be the classic example of capitalism. However, as his writings have found favor in other places and times, revisions have been forced upon the Marxist view of population.

Critique of Marx

Not all who have adopted a Marxist world view fully share the original Marx-Engels demographic perspective. Marxist countries have had trouble because of the lack of political direction offered by the Marxist notion that different stages of social development produce different relationships between population growth and economic development. Indeed, much of the so-called Marxist thought on population is in fact attributable to Lenin, one of the most prolific interpreters of Marx. For Marx, the Malthusian principle operated under capitalism only, whereas under pure socialism there would be no population problem. Unfortunately, he offered no guidelines for the transition period. At best, Marx implied that the socialist law of population should be the antithesis of the capitalist law. If the birth rate were low under capitalism, then the assumption was that it should be high under socialism; if abortion seemed bad for a capitalist society, it must be good for a socialistic society. Thus, it was difficult for Soviet demographers to reconcile the fact that demographic trends in the former Soviet Union were remarkably similar to trends in other developed nations. Furthermore, Soviet socialism was unable to alleviate one of the worst evils that Marx attributed to capitalism, higher death rates among people in the working class than those in the higher classes (Brackett 1967). The flip side of that is that birth rates dropped to such low levels throughout pre-1990 Marxist eastern Europe that it was no longer possible to claim (as Marx had done) that low birth rates were bourgeois.

In China the empirical reality of having to deal with the world's largest national population has led to a radical departure from Marxist ideology. As early as 1953 the Chinese government organized efforts to control population by relaxing regulations concerning contraception and abortion. Ironically, after the terrible demographic disaster that followed the "Great Leap Forward" in 1958 (see Chapter 1), a Chinese official quoted Chairman Mao as having said, "A large population in China is a good thing. With a population increase of several fold we still have an adequate solution. The solution lies in production" (Ta-k'un 1960:704). Yet by 1979 production no longer seemed to be a panacea, and the interpretation of Marx took an about-face as another Chinese official wrote that under Marxism the law of production "demands not only a planned production of natural goods, but also the planned reproduction of human beings" (Muhua 1979:724).

Thus, despite Marx's denial of a population problem, the Marxist government in China is dealing with one by rejecting its Marxist-Leninist roots and embracing instead one of the most aggressive and coercive government programs ever launched to reduce fertility through restraints on marriage (the Malthusian solution), contraception (the neo-Malthusian solution), and abortion (a remnant of the Leninist approach) (Teitelbaum and Winter 1988). In a formulation such as this, Marxism is being revised in the light of new scientific evidence about how people behave in the same way that Malthusian thought has been revised. Bear in mind that although the Marxist and Malthusian perspectives are often seen as antithetical, they both originated in 19th century Europe in the midst of that particular milieu of economic, social, and demographic change.

The population growth controversy, initiated by Malthus and fueled by Marx, emerged into a series of 19th century and early 20th century reformulations that have led directly to prevailing theories in demography. In the next section I briefly discuss three individuals who contributed to those reformulations: John Stuart Mill, Arsène Dumont, and Emile Durkheim.

Other Early Modern Population Theories

Mill

An extremely influential writer of the 19th century was the English philosopher and economist John Stuart Mill. Mill was not so quarrelsome about Malthus as Marx and Engels had been; his scientific insights were greater than those of Malthus at the same time that his politics were less radical than those of Marx and Engels. Although Mill accepted the Malthusian calculations about the potential for population growth to outstrip food production as being axiomatic (a self-truth), he was more optimistic about human nature than was Malthus. Mill believed that although a person's character is formed by circumstances, one's own desires can do much to shape circumstances and modify future habits (Mill 1873 [1924]).

Mill's basic thesis was that the standard of living is a major determinant of fertility levels. "In proportion as mankind rises above the condition of the beast, population is restrained by the fear of want, rather than by want itself. Even where there is no question of starvation, many are similarly acted upon by the apprehension of losing what have come to be regarded as the decencies of their situation in life" (1848:Book I, Chap. 10 [1929]). The belief that people could be and should be free to pursue their own goals in life led him to reject the idea that poverty is inevitable (as Malthus implied) or that it is the creation of capitalist society (as Marx argued). One of Mill's most famous comments is that "the niggardliness of nature, not the injustice of society, is the cause of the penalty attached to overpopulation" (1848:Book 1, Chap. 13 [1929]). This is a point of view conditioned by Mill's reading of Malthus, but Mill denies the Malthusian inevitability of a population growing beyond its available resources. Mill believed that people do not "propagate like swine, but are capable, though in very unequal degrees, of being withheld by prudence, or by the social affections, from giving existence to beings born only to misery and premature death" (1848:Book I, Chap. 7 [1929]). In the event that population ever did overrun the food supply, however, Mill felt that it would likely be a temporary situation with at least two possible solutions: import food or export people.

The ideal state from Mill's point of view is that in which all members of a society are economically comfortable. At that point he felt (as Plato had centuries earlier) that the population should stabilize and people should try to progress culturally, morally, and socially instead of attempting continually to get ahead economically. It does sound good, but how do we get to that point? It was Mill's belief that before reaching the point at which both population and production are stable, there is essentially a race between the two. What is required to settle the issue is a dramatic improvement in the living conditions of the poor. If social and economic development is to occur, there must be a sudden increase in income, which could give rise to a new standard of living for a whole generation, thus allowing productivity to outdistance population growth. According to Mill, this was the situation in France after the revolution:

> During the generation which the Revolution raised from the extremes of hopeless wretchedness to sudden abundance, a great increase of population took place. But a generation has grown up, which, having been born in improved circumstances, has not learnt to be miserable; and upon them the spirit of thrift operates most conspicuously, in keeping the increase of population within the increase of national wealth [1848:Book II, Chap. 7(1929)].

Mill was further convinced that an important ingredient in the transformation to a nongrowing population is that women do not want as many children as men do, and if they are allowed to voice their opinions, the birth rate will decline. Mill, like Marx, was a champion of equal rights for both sexes, and one of Mill's more notable essays, *On Liberty,* was co-authored with his wife. He reasoned further that a system of national education for poor children would provide them with the "common sense" (as Mill put it) to refrain from having too many children.

Overall, Mill's perspective on population growth was significant enough that we find his arguments surviving today in the writings of many of the 20th century demographers whose names appear in the pages that follow. However, before getting to those people and their ideas, it is important to acknowledge at least two other 19th century individuals whose thinking has an amazingly modern sound: Arsène Dumont and Emile Durkheim.

Dumont

Arsène Dumont was a late 19th century French demographer who felt that he had discovered a new principle of population, which he called "social capillarity" (Dumont 1890). *Social capillarity* refers to the desire of a person to rise on the social scale, to increase one's individuality as well as one's personal wealth. The concept is drawn from an analogy to a liquid rising into the narrow neck of a laboratory flask. The flask is like the hierarchical structure of most societies, which is broad at the bottom and diminishes as you near the top. To ascend the social hierarchy often requires that sacrifices be made, and Dumont argued that having few or no children was the price that many people paid to get ahead. Dumont recognized that such ambitions were not possible in every society. In a highly stratified aristocracy, few people outside of the aristocracy could aspire to a career beyond subsistence. However, in a democracy (such as late 19th century France), opportunities to succeed existed at all social levels. Spengler (1979) has succinctly summarized Dumont's thesis: "The bulk of the population, therefore, not only strove to ascend politically, economically, socially, and intellectually, but experienced an imperative urge to climb and a palsying fear of descent. Consequently, since children impeded individual and familial ascension, their number was limited" (p. 158).

Notice that Dumont added an important ingredient to Mill's recipe for fertility control. Mill argued that it was fear of social slippage that motivated people to limit fertility below the level that Malthus had expected. Dumont went beyond that to suggest that social aspiration was the root cause of a slowdown in population growth. Dumont was not happy with this situation, by the way. He was upset by the low level of French fertility and used the concept of social capillarity to propose policies to undermine it. He believed that socialism would undercut the desire for upward social mobility and would thus stimulate the birth rate.

Durkheim

While Dumont was concerned primarily with the causes of population growth, another late 19th century French sociologist, Emile Durkheim, was basing an entire social theory on the consequences of population growth. In discussing the increasing complexity of modern societies, characterized particularly by increasing divisions of labor, Durkheim proposed that "the division of labor varies in direct ratio with the volume and density of societies, and, if it progresses in a continuous manner in the course of social development, it is because societies become regularly denser and more voluminous" (Durkheim 1933:262). Durkheim proceeded to explain that population growth leads to greater societal specialization, because the struggle for existence is more acute when there are more people.

If you compare a primitive society with an industrialized society, the primitive society is not very specialized. By contrast, in industrialized societies there is a lot of differentiation; that is, there is an increasingly long list of occupations and social classes. Why is this? The answer is in the volume and density of the population. Growth creates competition for society's resources, and in order to improve one's advantage in the struggle, each person specializes. This thesis of Durkheim that population growth leads to specialization was derived (he himself acknowl-edges) from Darwin's theory of evolution. In turn, Darwin acknowledged his own debt to Malthus. You will notice that Durkheim also clearly echoes the words of Ibn Khaldun, although it is uncertain whether Durkheim knew of Ibn Khaldun's work.

The critical theorizing of the 19th and early 20th centuries set the stage for more systematic collection of data to test aspects of those theories and to examine more carefully those that might be valid and those that should be discarded. As popula-tion studies became more quantitative in the 20th century, a phenomenon called the demographic transition took shape and took the attention of demographers.

The Theory of the Demographic Transition

Although it has dominated recent demographic thinking, the **demographic transition** theory actually began as only a description of the demographic changes that had taken place over time in the advanced nations. In particular, it described the transition from high birth and death rates to low birth and death rates. The idea emerged as early as 1929, when Warren Thompson gathered data from "certain countries" for the period 1908–27 and showed that the countries fell into three main groups, according to their patterns of population growth:

> Group A (northern and western Europe and the United States): From the latter part of the nineteenth century to 1927 they had moved from having very high rates of natural increase to having very low rates of increase "and will shortly become stationary and start to decline in numbers" (Thompson 1929:968).
>
> Group B (Italy, Spain, and the "Slavic" peoples of central Europe): Thompson saw evidence of a decline in both birth rates and death rates but suggested that "it appears probable that the death rate will decline as rapidly or even more rapidly than the birth rate for some time yet. The condition in these Group B countries is much the same as ex-isted in the Group A countries thirty to fifty years ago" (1929:968).
>
> Group C (the rest of the world): In the rest of the world Thompson saw little evi-dence of control over either births or deaths.

As a consequence of this relative lack of voluntary control over births and deaths (a concept which we will question later), Thompson felt that the Group C countries (which included about 70 to 75 percent of the population of the world at the time) would continue to have their growth "determined largely by the opportu-nities they have to increase their means of subsistence. Malthus described their processes of growth quite accurately when he said "'that population does invariably increase, where there are means of subsistence. . . .'" (Thompson, 1929:971).

Thompson's work, however, came at a time when there was relatively little con-cern about overpopulation. Indeed, by 1936 birth rates in the United States and Eu-rope were so low that Enid Charles published a widely read book called *The Twi-light of Parenthood*, which was introduced with the comment that "in place of the Malthusian menace of overpopulation there is now real danger of underpopulation" (Charles 1936:v). Furthermore, Thompson's labels for his categories had little charisma (it is difficult to build a theory around categories called A, B, and C).

Sixteen years after Thompson's work, in 1945, Frank Notestein picked up the threads of his thesis and provided labels for the three types of growth patterns that Thompson had simply called A, B, and C. Notestein called the Group A pattern **incipient decline,** the Group B pattern **transitional growth,** and the Group C pattern **high growth potential** (Notestein 1945). That same year, Kingsley Davis (1945) edited a volume of The Annals of the American Academy of Political and Social Sciences titled *World Population in Transition,* and in the lead article (titled *The Demographic Transition*) he noted that "viewed in the long-run, earth's population has been like a long, thin powder fuse that burns slowly and haltingly until it finally reaches the charge and explodes" (Davis 1945:1). The term *population explosion,* alluded to by Davis, refers to the phase that Notestein called transitional growth. Thus was born the term *demographic transition.* It is that period of rapid growth when a country is moving from high birth and death rates to low birth and death rates, from high growth potential to incipient decline (see Figure 3.2).

At this point in the 1940s, however, the demographic transition was merely a picture of demographic change, not a theory. But each new country studied fit into the picture, and it seemed as though some new universal law of population growth—an evolutionary scheme—was being developed. The apparent historical uniqueness of the demographic transition (all known cases have occurred within the last 200 years) has spawned a host of alternative names, such as the "vital revolution" and the "demographic revolution."

Figure 3.2   The Demographic Transition

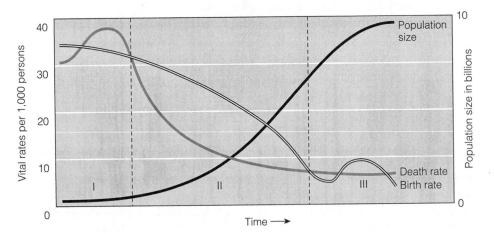

Note: The demographic transition is divided roughly into three stages. In the first stage there is high growth potential because both birth and death rates are high. The second stage is the transition from high to low birth and death rates. During this stage the growth potential is realized as the death rate drops before the birth rate drops, resulting in rapid population growth. Finally, the last stage is a time when death rates are as low as they are likely to go, while fertility may continue to decline to the point that the population might eventually decline in numbers. In the developed countries, the full transition took place essentially as schematized. However, most less developed nations have not yet followed the full pattern of change.

Between the mid-1940s and the late 1960s rapid population growth became a worldwide concern, and demographers devoted a great deal of time to the demographic transition perspective. By 1964 George Stolnitz was able to report that "demographic transitions rank among the most sweeping and best-documented trends of modern times . . . based upon hundreds of investigations, covering a host of specific places, periods and events" (1964:20). As the pattern of change took shape, explanations developed for why and how countries pass through the transition. These explanations tended to be cobbled together in a somewhat piecemeal fashion from the 19th and early 20th century writers I discussed earlier in this chapter, but overall they were derived from the concept of **modernization.**

The modernization theory is based on the idea that in premodern times human society was generally governed by "tradition," and that the massive economic changes wrought by industrialization forced societies to alter traditional institutions: "In traditional societies fertility and mortality are high. In modern societies fertility and mortality are low. In between, there is demographic transition" (Demeny 1968:502). In the process, behavior has changed and the world has been permanently transformed. It is a macro-level theory that sees human actors as being buffeted by changing social institutions. Thus, individuals did not deliberately lower their risk of death to precipitate the modern decline in mortality. Rather, society-wide increases in income and improved public health infrastructure brought about this change. Similarly, people did not just decide to move from the farm to town to take a job in a factory. Economic changes took place that created those higher wage urban jobs while eliminating many agricultural jobs. These same economic forces improved transportation and communication and made it possible for individuals to migrate in heretofore unheard of numbers.

Modernization theory provided the vehicle that allowed the demographic transition to move from a mere description of events to a demographic perspective. In its initial formulations this perspective was perhaps best expressed by the sentiments "take care of the people and population will take care of itself" or "development is the best contraceptive" (Teitelbaum 1975). It drew on the available data for most countries that had gone through the transition. Death rates declined as the standard of living improved, and birth rates almost always declined a few decades later, eventually dropping to low levels, although rarely as low as the death rate. It was argued that the decline in the birth rate typically lagged behind the decline in the death rate because it takes time for a population to adjust to the fact that mortality really is lower, and because the social and economic institutions that favored high fertility require time to adjust to new norms of lower fertility that are more consistent with the lower levels of mortality. Since most people value the prolongation of life, it is not hard to lower mortality, but the reduction of fertility is contrary to the established norms of societies that have required high birth rates to keep pace with high death rates. Such norms are not easily changed, even in the face of poverty.

Birth rates eventually declined, it was argued, as the importance of family life was diminished by industrial and urban life, thus weakening the pressure for large families. Large families are presumed to have been desired because they provided parents with a built-in labor pool, and because children provided old-age security for parents. The same economic development that lowered mortality is theorized to transform a society into an urban industrial state in which compulsory education lowers the value of children by removing them from the labor force, and people come to realize that lower infant mortality means that fewer children need to be born to achieve a certain number of surviving children. Finally, as a consequence of the many alterations in social institutions, "the pressure from high fertility weakens and the idea of conscious control of fertility gradually gains strength" (Teitelbaum 1975:421).

Critique of the Demographic Transition Theory

It has been argued that the concept underlying the demographic transition is that population stability (sometimes called "homeostasis"—see, for example, Lee 1987) is the normal state of affairs in human societies and that change (the "transition") is what requires explanation (Kraeger 1986). But that is only partially true. In its original formulation, the demographic transition theory explained high fertility as a reaction to high mortality. As mortality declines, the need for high fertility lessens, and so birth rates go down. There is a spurt of growth in that transition period, but presumably the consequences will not be serious if the decline in mortality was produced by a rise in the standard of living, which, in its turn, produces a motivation for smaller families. But what will be the consequences if mortality declines and fertility does not? That situation presumably is precluded by the theory of demographic transition, but the demographic transition theory has not been capable of predicting levels of mortality or fertility or the timing of the fertility decline. This is because the initial explanation for the demographic behavior during the transition tended to be ethnocentric. It relied almost exclusively on the sentiment that "what is good for the goose is good for the gander." In other words, if this is what happened to the developed countries, why should it not also happen to other countries that are not so advanced? One reason might be that the preconditions for the demographic transition are considerably different now from what they were when the industrialized countries began their transition.

For example, prior to undergoing the demographic transition, few of the currently industrialized countries had birth rates as high as those of most currently less developed countries, nor indeed were their levels of mortality so high. Yet when mortality did decline, it did so as a consequence of internal economic development, not as a result of a foreign country bringing in sophisticated techniques of disease prevention, as is the case today. A second reason might be that the factors leading to the demographic transition were actually different from what for years had been accepted as true. These problems with the usual explanations of the demographic transition have led to new research and a reformulation of the perspective.

Reformulation of the Demographic Transition Theory

One of the most important social scientific endeavors to cast doubt on the classic explanation was the European Fertility Project, directed by Ansley Coale at Princeton University. In the early 1960s J. William Leasure, then a graduate student in Economics at Princeton, was writing a doctoral dissertation on the fertility decline in Spain, using data for each of that nation's 49 provinces. Surprisingly, his thesis revealed that the history of fertility change in Spain was not explained by a simple version of the demographic transition theory. Fertility in Spain declined in contiguous areas that were culturally similar, even though the levels of urbanization and economic development might be different (Leasure 1962). At about the same time, other students began to uncover similarly puzzling historical patterns in European data (Coale 1986). A systematic review of the demographic histories of Europe was thus begun in order to establish exactly how and why the transition occurred. The focus was on the decline in fertility, because it is the most problematic aspect of the classic explanation. These new findings have been used to revise the theory of the demographic transition.

One of the important clues that a revision was needed was the discovery that the decline of fertility in Europe occurred in the context of widely differing social, economic, and demographic conditions (van de Walle and Knodel 1967). Economic development emerges, then, as a sufficient cause of fertility decline, though not a necessary one (Coale 1973). For example, many provinces of Europe experienced a rapid drop in the birth rate even though they were not very urban, infant mortality rates were high, and a low percentage of the population was in industrial occupations. The data suggest that one of the more common similarities in those areas that have undergone fertility declines is the rapid spread of **secularization.** Secularization is an attitude of autonomy from otherworldly powers and a sense of responsibility for one's own well-being (Leasure 1982; Lesthaeghe 1977). It is difficult to know exactly why such attitudes arise when and where they do, but we do know that industrialization and economic development are virtually always accompanied by secularization. Secularization, however, can occur independently of industrialization. It might be thought of as a modernization of thought, distinct from a modernization of social institutions. Some theorists have suggested that secularization is part of the process of Westernization (see, for example, Caldwell 1982). In all events, when it pops up, secularization often spreads quickly, being diffused through social networks as people imitate the behavior of others to whom they look for clues to proper and appropriate conduct.

Education has been identified as one potential stimulant to such altered attitudes, especially mass education that tends to emphasize modernization and secular concepts (Caldwell 1980). Education facilitates the rapid spread of new ideas and information, which would perhaps help explain another of the important findings from the Princeton European Fertility Project, that the onset of long-term fertility decline tended to be concentrated in a relatively short period of time (van de Walle and Knodel 1980). The data from Europe suggest that once marital fertility had dropped by as little as 10 percent in a region, the decline spread rapidly. This occurred whether or not infant mortality had already declined (Watkins 1986). Some areas of Europe that were similar with respect to socioeconomic development did not experience a fertility decline at the same time, whereas other provinces that were less similar socioeconomically experienced nearly identical drops in fertility. The data suggest that this riddle is solved by examining cultural factors rather than socioeconomic ones. Areas that share a similar culture (same language, common ethnic background, similar lifestyle) were more likely to share a decline in fertility than areas that were culturally less similar (Watkins 1991). The principal reason for this is that the idea of family planning seemed to spread quickly only until it ran into a barrier to its communication. Language is one such barrier (Leasure 1962; Lesthaeghe 1977), and social and economic inequality in a region is another (Lengyel-Cook and Repetto 1982). Social distance between people turns out to be a very effective inhibitor of communication of new ideas and attitudes.

What kinds of ideas and attitudes might encourage people to rethink how many children they ought to have? To answer this kind of question we must shift our focus from the macro (societal) level to the micro (individual) level and ask how people actually do respond to the social and economic changes taking place around them (see Kertzer and Hogan 1989, for an interesting Italian case study). A prevailing individual-level perspective is that of **rational choice theory** (sometimes referred to as RAT) (Coleman and Fararo 1992). The essence of rational choice theory is that human behavior is the result of individuals making calculated cost-benefit analyses about how to act and what to do. For example, Caldwell (1976; 1982) has suggested that "there is no ceiling in primitive and traditional societies to the number of children who would be economically beneficial" (1976:331). Children are a source of income and support for parents throughout life, and they produce far more than they cost in such societies. The **wealth flow**, as Caldwell calls it, is from children to parents. But the process of modernization eventually results in the tearing apart of large, extended family units into smaller, nuclear units that are economically and emotionally self-sufficient. As that happens, children begin to cost parents more (including the cost of educating them as demanded by a modernizing society), and the amount of support that parents get from children begins to decline (starting with the income lost because the children are in school rather than working). As the wealth flow reverses and parents begin to spend their income on children, rather than deriving income from them, the economic value of children declines, and people no longer derive any economic advantage from children. Economic rationality would dictate having zero children, but in reality, of course, people continue having children for a variety of social reasons (which I detail in Chapter 6).

I would be remiss if I did not also add that motivation to limit children needs to be linked to the means to do so. In the demographic transitions of the now developed societies, motivation to control children had to be particularly strong in order for fertility to drop, because the methods available to prevent conception or birth were not very sophisticated. In our modern world, safe, effective, and fairly easy to use methods of fertility control are widely available (albeit unevenly so, as I discuss in Chapter 14). Thus, fertility can be readily influenced with lower levels of motivation for family limitation than in the past, or, more optimistically, birth rates can now decline much more rapidly as the motivation for fertility control grows. The existence of modern contraceptive methods has, of course, encouraged the development of programs designed to deliver those methods to couples wanting them, and those programs have the potential to independently influence couples to limit fertility.

Overall, then, the principle ingredient in the reformulation of the demographic transition perspective is to add "ideational" factors to "demand" factors as the likely causes of fertility decline (Cleland and Wilson 1987). The original version of the theory suggested that modernization reduces the demand for children and so fertility falls—if people are rational economic creatures, then this is what should happen. But the real world is more complex, and the diffusion of ideas can shape fertility behavior along with, or even in the absence of, the usual signs of modernization.

A strength of the reformulation of the demographic transition is that nearly all other perspectives can find a home here. Malthusians note with satisfaction that fertility first declined in Europe primarily as a result of a delay in marriage, much as Malthus would have preferred. Neo-Malthusians can take heart from the fact that rapid and sustained declines occurred simultaneously with the spread of knowledge about family planning practices. Marxists also find a place for themselves in the reformulated demographic transition perspective, because its basic tenet is that a change in the social structure is necessary to bring about a decline in fertility. This is only a short step away from agreeing with Marx that there is no universal law of population, but rather that each stage of development and social organization has its own law.

However, I should caution you that many theorists would argue that the reason we have so much trouble understanding human behavior is that not all behavior (demographic or otherwise) is governed by social institutions or by rational choice or by the diffusion of new ideas and attitudes. Some portion of behavior may be driven by nonrational influences such as "human nature" (Wilson 1993), or even more specifically, by hormones (Udry 1994).

Another attempt to go beyond the usual explanation offered by the demographic transition theory is the theory of demographic change and response.

The Theory of Demographic Change and Response

The theory of **demographic change and response** was put forward by Kingsley Davis in 1963 as an adjunct, not really an alternative, to the demographic transition theory. Davis's concern is with the interaction of the causes and consequences of population growth, on the assumption that in order to do anything about the consequences, you have to know the causes. The basic problem Davis attempts to deal with is the central issue of the demographic transition theory: How (and under what conditions) can a mortality decline lead to a fertility decline?

To answer that question, Davis asked what happens to individuals when mortality declines. The answer is that more children survive through adulthood, putting greater pressure on family resources, and people have to reorganize their lives in an attempt to relieve that pressure; that is, people respond to the demographic change. But note that their response will be in terms of personal goals, not national goals. It rarely matters what a government wants. If individual members of a society do not stand to gain by behaving in a particular way, they probably will not behave that way. Indeed, that was a major argument made by the neo-Malthusians against moral restraint. Why advocate postponement of marriage and sexual gratification rather than contraception when you know that few people who postpone marriage are actually going to postpone sexual intercourse too? In fact, Ludwig Brentano (1910) quite forthrightly suggested that Malthus was insane to think that abstinence was the cure for the poor.

In all events, Davis argued that the response that individuals make to the population pressure created by more members joining their ranks is determined by the means available to them. A first response, nondemographic in nature, is to try to increase resources by working harder—longer hours perhaps, a second job, and so on. If that is not sufficient or there are no such opportunities, then migration of some family members (typically unmarried sons or daughters) is the easiest demographic response. This is, of course, the option that people have been using forever (undoubtedly explaining in large part why humans being have spread out over the planet). Already in the early 18th century, Richard Cantillon, an Irish-French economist, was pointing out what happened in Europe when families grew too large (and this was even before mortality began markedly to decline):

> If all the labourers in a village breed up several sons to the same work, there will be too many labourers to cultivate the lands belonging to the village, and the surplus adults must go to seek a livelihood elsewhere, which they generally do in cities. . . . If a tailor makes all the clothes there and breeds up three sons to the same, yet there is work enough for but one successor to him, the two others must go to seek their livelihood elsewhere; if they do not find enough employment in the neighboring town they must go further afield or change their occupation to get a living . . . [Cantillon 1755 (1964):23].

But what will be the response of that second generation, the children who now have survived when previously they would not have, and who have thus put the pressure on resources? Davis argues that if (and this is a big if) there is in fact a chance for social or economic improvement, then people will try to take advantage of those opportunities by avoiding the large families that caused problems for their parents. Davis suggests that the most powerful motive for family limitation is not fear of poverty or avoidance of pain as Malthus argued; rather it is the prospect of rising prosperity that will most often motivate people to find the means to limit the number of children they have (see Chapter 5 for a fuller discussion of these means). Davis here echoes the themes of Mill and Dumont, but adds that at the very least, the desire to maintain one's relative status in society may lead to an active desire to prevent too many children from draining away one's resources. Of course, that assumes the individuals in question have already attained some status worth maintaining.

One of Davis's most important contributions to our demographic perspective is, as Cicourel put it, that he "seems to rely on an implicit model of the actor who makes everyday interpretations of perceived environmental changes" (Cicourel 1974:8). For example, people will respond to a decline in mortality only if they notice it, and then their response will be determined by the social situation in which they find themselves. Davis's analysis is important in reminding us of the crucial link between the everyday lives of individuals and the kinds of population changes that take place in society.

Another demographer who has extended his work to this kind of analysis is Richard Easterlin, whose ideas have been called the relative income hypothesis.

The Relative Income Hypothesis

The **relative income hypothesis** is based on the idea that the birth rate does not necessarily respond to absolute levels of economic well-being but rather to levels that are relative to those to which one is accustomed (Easterlin 1968; 1978). Easterlin assumes that the standard of living you experience in late childhood is the base from which you evaluate your chances as an adult. If you can easily improve your income as an adult relative to your late childhood level, then you will be more likely to marry early and have several children. On the other hand, if you perceive that it is going to be rough sledding as an adult even to match the level of living you were accustomed to as a child, that fear will probably lead you to postpone marriage or at least postpone childbearing.

So far the theory of relative income is strikingly similar to Mill's writing a hundred years earlier. But Easterlin goes on to ask what factors might cause you to be in a relatively advantageous or disadvantageous position as you begin adulthood. He argues that the answer lies in the relationship between the business cycles and the demographic responses to those cycles. In a society unencumbered by government intervention, a long-term (say, 15-year) upswing in business will encourage immigration, and it may also make it easier for people to marry and have children. The 'may' in this case depends on another demographic variable that has not yet entered the picture, the age structure—the number and proportion of people at each age in a society. If young people are relatively scarce in society and business is good, they will be in relatively high demand. In nearly classic Malthusian fashion, they will be able to command high wages and thus be more likely to feel comfortable about getting married and starting a family. Of course, how comfortable they are will depend on how much those wages can buy relative to what they are accustomed to. If young people are in relatively abundant supply, then even if business is good, the competition for jobs will be stiff and it will be difficult for people to maintain their accustomed level of living, much less marry and start a family.

You should be asking yourself why there might be a relative scarcity or abundance of young people in the age structure. Although I discuss this in more detail in Chapters 6 and 8, suffice it here to note that primarily it is due to fluctuations in the birth rate, which are influenced by the pattern of people marrying and having children. Thus, Easterlin's thesis presents a model of society in which demographic change and economic change are closely interrelated. Economic changes produce demographic changes, which in turn produce economic changes, and so on. The model, however, has a certain middle-class bias. What about the people at the bottom end of the economic ladder, for whom relative deprivation does not necessarily apply because they already have so little? Are they caught in a constant Malthusian cycle of overpopulation and poverty? Mill suggested in 1848 that they were, unless one entire generation could be catapulted into the middle class. It seems that Easterlin, like many American theorists, has ignored the importance of social class as an independent factor influencing demographic behavior (Lesthaeghe and Surkyn 1988).

Empirical tests of the Easterlin hypothesis have produced mixed results. Several studies provide evidence that, in general, changes in cohort size and the birth rate tend to move in directions predicted by Easterlin (Butz and Ward 1979; Crenshaw, Ameen, and Christenson 1997; Lee 1976; Pampel 1993). On the other hand, tests of specific aspects of the Easterlin hypothesis have generally failed to support it (Behrman and Taubman 1989; Kahn and Mason 1987; and Wright 1989) or have supported it only partially (Pampel 1996). The Easterlin hypothesis is intuitively appealing, but its apparent lack of solid empirical support may be due to the fact that, as Lesthaeghe and Surkyn (1988) suggest, it is a "building stone" for a theory rather than the edifice of the theory itself. The idea of a demographic feedback cycle, which is at the core of Easterlin's thinking, is compelling, and it is certain that relative cohort size is one factor that may influence some kinds of social change. Whether these cycles can in fact explain fluctuations in fertility, however, is still up in the air (Wachter 1991).

Theories about the Consequences of Population Growth

As you can see from the doctrines and theories discussed in this chapter, most demographic thinking before the 20th century focused on the consequences of population growth and, to be sure, that is the principle element of interest in the Malthusian perspective. But you cannot really be sure of consequences if you do not understand the causes (and complexities) of population growth. The modern field of population studies came about largely to encourage and inspire deeper insight into the causes of changes in fertility, mortality and migration, age and sex structure, and population characteristics and distribution. Demographers spent most of the 20th century doing that, but always with an eye toward what new things would be learned about what demographic change meant for human society. Unlike in Malthus's day, population growth is no longer viewed as being caused by one set of factors, nor as having a simple prescribed set of consequences.

Perhaps the closest that we can come at present to "big" theories are those that try to place demographic events and behavior into the context of other global change, especially political change, economic development, and westernization. One of the more ambitious and influential of these writers is Jack Goldstone (1986; 1991), whose theories incorporate population growth as a precursor of change in the "early modern world" (defined by him as 1500 to 1800). He argues that population growth in the presence of rigid social structures produced dramatic political change in England and France, in the Ottoman Empire, and in China. Population growth led to increased government expenditures, which led to inflation, which led to fiscal crisis. In these societies with no real opportunities for social mobility, population growth (which initially increases the number of younger persons) led to disaffection and popular unrest and created a new cohort of young people receptive to new ideas. The result in the four cases studies he analyzes were rebellion and revolution.

Stephen Sanderson (1995) promotes the idea that population growth has been an important stimulus to change throughout human history, but especially since the Agricultural Revolution. Thus, his scope is much broader than that of Goldstone. Sanderson largely synthesizes the work of others (such as Boserup 1965; Cohen 1977; and Johnson and Earle 1987). "Had Paleolithic hunter-gatherers been able to keep their populations from growing, the whole world would likely still be surviving entirely by hunting and gathering" (Sanderson 1995:49). Instead, population growth generated the Agricultural Revolution (an idea that I discussed in Chapter 1) and then the Industrial Revolution. The sedentary life associated with the Agricultural Revolution increased social complexity (a very Durkheimian idea), which leads to the rise of civilization (cities) and the state (city-states and then nation-states).

These and other writers have encouraged social scientists to realize that population change is largely imperceptible to us as it is occurring, so it requires an analytic observer to tell us what is going on. Thus, if we can look back and see that in the past there were momentous historical consequences of population growth, can we look forward into the future and project similar kinds of influences? Many people (including me) would say yes, and the rest of this book will show you why.

There Are Many Other Theories

Demography is often accused of being a science that is weak on theory but strong on evidence. Part of the reason for this perception is that as more and more evidence has been gathered about demographic behavior, the world has appeared increasingly complex, and no one has stepped forward to offer an overarching, all-encompassing theory that explains every aspect of demographic behavior. The studies by Goldstone and Sanderson, just discussed, move in that direction, as do general theories of the relationship between population growth and economic development. In Chapter 12, I discuss several such theories, including attempts to incorporate a demographic perspective into the broader framework of a world-systems theory. Coming up sooner than that, however, are theories that focus more narrowly on specific aspects of the causes and consequences of population change. In the next chapter (4) I discuss the epidemiological transition as a theory of mortality decline; in Chapter 6 I discuss another theory put forward by Richard Easterlin, the supply-demand framework for fertility determinants; and in Chapter 7 I review the century-old Laws of Migration developed by Ravenstein, along with other more recent theoretical approaches to the causes and consequences of human migration. In Chapter 8, I discuss the critical role that the age structure plays in determining the nature of society. Population growth is not some unitary phenomenon—rather it occurs differently at different ages and in different segments of society. In each of the other chapters I evaluate a variety of perspectives on the consequences of population growth. Thus, the remainder of this book is devoted to examining both theories and facts in order to increase your understanding of the interrelationships between population and society.

Summary and Conclusion

I have traced for you the progression of demographic thinking from ancient doctrines to contemporary systematic perspectives. Malthus was not necessarily the first, but he was certainly the most influential of the early modern writers. Malthus believed that a biological urge to reproduce was the cause of population growth, and that its natural consequence was poverty. Marx, on the other hand, did not openly argue with the Malthusian causes of growth, but he vehemently disagreed with the idea that poverty is the natural consequence of population growth. Marx denied that population growth was a problem per se—it only appeared that way in capitalist society. It may have seemed peculiar to you to discuss a person who denied the importance of a demographic perspective in a chapter dedicated to that very importance. However, the Marxist point of view is sufficiently prevalent today among political leaders and intellectuals in enough countries that this attitude becomes in itself a demographic perspective of some significance.

The perspective of Mill, who seems very contemporary in many of his ideas, was somewhere between that of Malthus and Marx. He believed that increased productivity could lead to a motivation for having smaller families, especially if the influence of women was allowed to be felt and if people were educated about the possible consequences of having a large family. Dumont took these kinds of individual motivations a step further and suggested in greater detail the reasons why prosperity and ambition, operating through the principle of social capillarity, generally lead to a decline in the birth rate. Durkheim's perspective emphasized the consequences more than the causes of population growth. He was convinced that the complexity of modern societies almost entirely is due to the social responses to population growth—more people lead to higher levels of innovation and specialization.

More recently developed demographic perspectives have implicitly assumed that the consequences of population growth are serious and problematic, and they move directly to explanations of the causes of population growth. The theory of the demographic transition suggests that growth is an intermediate stage between the more stable conditions of high birth and death rates and low birth and death rates. If you accept this perspective, you view the world in an evolutionary way—a decline in mortality will almost necessarily be followed by a decline in fertility. The theory of demographic change and response considers the kind of individual decision making that has to take place before fertility will decline from previous high levels. The theory of relative income builds on the idea that reproductive behavior is not based solely on what is happening in the rest of society but also on one's relative status in society. It is a perspective that specifically ties together the interaction between the causes and consequences of demographic change.

With a bit of theory in hand (and hopefully in your head as well), we are now ready to probe more deeply into the analysis of population processes; to come to an appreciation of how important the decline in the death rate is, yet why it is still so much higher in some places than in others; why birth rates are high in some places and low in others; and why some people move and others do not.

Main Points

1. A demographic perspective is a way of relating basic population information to theories about how the world operates demographically.

2. Population doctrines and theories prior to Malthus tended to be pronatalist and often utopian. Condorcet and Godwin were two such theorists, and they stimulated Malthus to write his *Essay on Population*.

3. The Malthusian perspective is based on the writings of Thomas Robert Malthus, whose first *Essay on Population* appeared in 1798 and has been one of the most influential works ever written on population growth and its societal consequences.

4. According to Malthus, population growth is generated by the urge to reproduce, although growth is checked ultimately by the means of subsistence.

5. The natural consequences of population growth according to Malthus are misery and poverty because of the tendency for populations to grow faster than the

food supply. Nonetheless, he believed that misery could be avoided if people practiced moral restraint—a simple formula of chastity before marriage and a delay in marriage until you can afford all the children that God might provide.

6. Karl Marx and Friedrich Engels strenuously objected to the Malthusian population perspective because it blamed poverty on the poor rather than on the evils of social organization.

7. Marx and Engels believed that overpopulation was a product of capitalism and that in a socialist society either there would be enough resources per person or else people would be motivated to keep families small.

8. Not suprisingly, very few people have bought the Malthusian idea of moral restraint, although there are many who agree that population growth tends to outstrip food production. Such people are usually called neo-Malthusians and believe in the use of birth control.

9. Revisions of Marxist ideology frequently include a more active government role in trying to influence birth limitation.

10. John Stuart Mill argued that the standard of living is a major determinant of fertility levels, but he also felt that people could influence their own demographic destinies.

11. Arsène Dumont argued that personal ambition generated a process of social capillarity, which induced people to limit the number of children in order to get ahead socially and economically.

12. Emile Durkheim built an entire theory of social structure on his conception of the consequences of population growth.

13. The theory of the demographic transition is a perspective that emphasizes the importance of economic and social development, which leads first to a decline in mortality and then, after some time lag, to a commensurate decline in fertility. It is based on the experience of the developed nations.

14. Reformulations of the demographic transition theory suggest that secularization and the cultural diffusion of ideas may be important keys to declining fertility.

15. The theory of demographic change and response emphasizes that people must perceive a personal need to change behavior before a decline in fertility will take place, and that the kind of response they make will depend on what means are available to them.

16. The relative income hypothesis views changes in the birth rate as being a response to levels of economic well-being that are relative to those to which one is accustomed, particularly when combined with birth cohorts of unequal size.

Suggested Readings

1. United Nations, 1973, The Determinants and Consequences of Population Trends (New York: United Nations).

   This volume was written by well-known and highly expert demographers and is the best single source for a comprehensive review of major demographic doctrines and theories.

2. Thomas Robert Malthus, 1798, An Essay on the Principles of Population, 1st Ed. (available from a variety of publishers).

   Despite the fact that numerous good summaries of Malthus have been written, you owe it to yourself to sample the real thing.

3. David Coleman and Roger Schofield, (eds.) 1986, The State of Population Theory: Forward from Malthus (Oxford: Basil Blackwell).

   A set of serious contemplations of how far demographic theory has or has not come since Malthus's first essay was published in 1798.

4. Ansley Coale and Susan Cotts Watkins, (eds.) 1986, The Decline of Fertility in Europe: The Revised Proceedings of a Conference on the Princeton European Fertility Project (Princeton: Princeton University Press).

   The most comprehensive statement in the literature of ways in which the Princeton European Fertility Project caused a rethinking of the theory of the demographic transition.

5. Paul R. Ehrlich and Anne H. Ehrlich, 1990, The Population Explosion (New York: Simon and Schuster).

   Published more than 20 years after his great neo-Malthusian classic, *The Population Bomb*, this book reassesses the worldwide consequences of continued population growth.

🌐 Websites of Interest

Remember that websites are not as permanent as books and journals, so I cannot guarantee that each of the following websites still exists at the moment that your are reading this.

1. **http://www-groups.dcs.st-and.ac.uk/~history/Mathematicians/Condorcet.html**

   The marquis de Condorcet, who helped to inspire Malthus's essay, is the subject of this website, located at the University of St. Andrews in Scotland. It includes biographical information and a list of his publications

2. **http://www.ecn.bris.ac.uk/het/malthus/popu.txt**

   The beauty of this website, located in the Department of Economics at the University of Bristol in England, is that it contains the full text of Malthus's first (1798) *Essay on Population.*

3. **http://www.emagazine.com/november-december_1996/1196conv.html**

   This is a conversation with the neo-Malthusians Paul and Anne Ehrlich, published as an article from E/The Environmental Magazine and archived on-line.

4. **http://opr.princeton.edu/archive/eufert/eufert.html**

   At this site you will find a description of the Princeton European Fertility Project, along with links to some of the data and resources used in the study and a bibliography of publications of findings from the project.

5. **http://www.popcouncil.org/pdr/pdrabs.html**

   Population and Development Review is a journal published by the Population Research Council in New York City, and its articles tend to focus on issues that directly or indirectly test demographic theories and perspectives. At their website you can peruse abstracts of articles published in the journal to stay up to date on recent research.

Part Two
Population Processes

Mortality, fertility, and migration are the dynamic elements of demographic analysis. They are the population processes that lead to change in the demographic structure and usually in the social, economic, and political structure of society as well. In the following four chapters that make up Part Two, I will analyze each of these population processes in turn. My purpose is to improve your understanding of how and why population change occurs. If you believe that a population is growing too rapidly and want to implement a remedy, you first have to understand why the population is growing to recognize what kinds of changes or policies are likely to work.

Population growth occurs as a result of the combination of mortality, fertility, and migration. I begin with mortality (Chapter 4). Bringing death under control has been one of the most significant achievements in human history, yet it has also been the cause of many problems. Specifically, declining death rates are the cause of increased rates of population growth.

Although I will discuss each process in some detail, I devote the greatest amount of time to fertility (Chapters 5 and 6), since the lowering of fertility is, and should be, a very high-priority item on the world agenda of contemporary issues. Migration (Chapter 7) is not a factor at the global level, because interplanetary migration has never been documented. At national and international levels, however, migration is of considerable importance, since it relieves population pressure in some places while contributing to growth in other areas—sometimes with beneficial results, sometimes with negative consequences.

An important point for you to keep in mind as you read these four chapters is that there is incredible variety within the human experience, and the explanations for behavior in one society do not necessarily work in another. We must constantly be aware of the different social environments in which people live as we try to explain why some people are more likely to die younger than others, why those who survive have as many children as they do, and why some people migrate when others do not.

CHAPTER 4
Mortality

Declining mortality, not rising fertility, is the root cause of current world population growth. It is not that people breed like rabbits; rather, they no longer die like flies. Virtually within our lifetime, mortality has been brought under control to the point that most of us now are able to take a long life pretty much for granted. In fact, long life is enjoying widespread popularity all over the world, and we are surviving in record numbers. Human triumph over disease and early death surely represents one of the most significant improvements ever made in the condition of human life, and we can rightly be proud of ourselves. Nevertheless, one by-product of our triumph is current world growth and the problems that have grown along with population size. Furthermore, the problems will continue to grow, for although mortality is increasingly under control, there are still wide variations in life expectancy in different parts of the world as well as among different groups of people within countries. These differences represent a potential reservoir of population growth, since further declines in death rates will trigger even higher rates of growth unless we are able to curtail fertility.

It was once widely believed that differences in death rates were genetic or biological in nature and thus difficult to change, but we now know that most variations are due to social, not biological, causes. I will pursue this point by beginning the chapter with a discussion of the differences between the biological and social components of mortality. Then I will explore with you why people die—what are the specific causes of death? Next, after a brief explanation of how to measure mortality, I will turn to an examination of who dies. The answer to this question has two parts, reflecting (1) the changes over time in the societal risk of death, which we know as the **epidemiological transition;** and (2) the differences in death rates in the same society at any given time (mortality differentials). After examining these issues, the concepts are used to help you understand the mortality situation in developed, low mortality areas such as Europe, East Asia, and North America, as well as the situation in the developing countries in the rest of the world.

Components of Mortality

There are two biological aspects of mortality. The first is **lifespan,** and refers to the oldest age to which human beings can survive. The second is **longevity** and, refers to the ability to remain alive from one year to the next—the ability to resist death. Lifespan is almost entirely a biological phenomenon, whereas longevity has both biological and social components.

Lifespan

Lifespan, remember, refers to how long a person can possibly live. Since it is, of course, impossible accurately to predict how long a person *could* live, we must be content to assume that the oldest age to which a human actually *has* lived (a figure that may change from day to day) is the oldest age to which it is possible to live. Claims of long human lifespan are widespread, but confirmation of those claims is more difficult to find. The oldest authenticated age to which a human has ever lived is 122 years, an age achieved by a French woman, Jeanne Louise Calment, who died in August, 1997. Her authenticated birth date was February 21, 1875, and on her 120th birthday in 1995, she was asked what kind of future she expected. "A very short one," she replied (Wallis 1995:85).

You probably have heard reports of people who claim to have lived longer than that, but none of those reports has ever been verified, principally because birth records were not so scrupulously kept prior to the 20th century, or else they have been destroyed (by fire or flood) or just plain lost. One such example is the case of an ex-slave named Charlie Smith who claimed to have been born in Liberia in 1842 and brought to the United States in 1854 and who died at a reported age of 137. The usual way to check on the reported age of an older person for whom no birth certificate exists is to look back through the census returns for information about those individuals when they were younger (these data are available on microfilm at the National Archives in Washington, D.C.). Unfortunately, Charlie Smith is such a common name, and he had moved so often, that he was impossible to trace (Meyers 1978). However, a marriage record in Florida that appears to refer to him has led investigators to believe that his age was exaggerated by at least 33 years, and that he may not even have been quite 100 years old when he died (McWhirter 1983).

Age exaggeration is often suspected among very old people. In a piece of fascinating detective work, Meyers (1978) was able to dig through old census data and show that a Pennsylvania man who died at a reported age of 112 in 1866 had, in fact, "aged" 27 years in the 10-year period between the 1850 and 1860 censuses and had "aged" 8 years in the 6-year period from the 1860 census to his death. Clearly, there was some exaggeration going on. Other evidence of age exaggeration among American centenarians has been discovered by Rosenwaike (1979), who found significant discrepancies between ages recorded on death certificates and ages recorded in the census. In 1992 in the United States, for example, there were three people who died with age 122 or over recorded on their death certificates (National Center for Health Statistics 1996), but their ages have not been verified by birth certificates.

So we know that humans can live to age 122 and maybe even beyond, yet very few people come close to achieving that age. Most, in fact, can expect to live only about half that long (life expectancy at birth for the world as a whole is estimated to be about 66 years). It is this concept, the age to which people *actually* survive, their demonstrated ability to stay alive as opposed to their potential, that we refer to as longevity.

Longevity

Biological Factors Longevity is usually measured by life expectancy, the statistically average length of life, and is greatly influenced by the society in which we live (as I discuss later in connection with the epidemiological transition) and by the genetic characteristics with which we are born. The strength of vital organs, predisposition to particular diseases, metabolism rate, and so on, are biological factors over which we presently have little control. Many of the most severe biological weaknesses, though, tend to display themselves rather soon after birth, and, as a result, mortality is considerably higher in the first year or so of life than it is in the remainder of childhood and early adulthood.

After the initial year of life there is a period of time, usually lasting at least until middle age, when risks of death are relatively low. Beyond middle age, mortality increases at an accelerating rate. This pattern of death by age is illustrated in Figure 4.1, where you can see that the pattern is similar whether the actual death rates are high or low. The genetic or biological aspects of longevity have led many theorists over time to believe that the age patterns of longevity shown in Figure 4.1 could be explained by a simple mathematical formula similar perhaps to the law of gravity and other laws of nature. The most famous of these was put forward in 1825 by Benjamin Gompertz and describes a simple geometric relationship between age and death rates from the point of sexual maturity to the extreme old ages (Olshanskey and Carnes 1997). These mathematical models are interesting, but they have so far not been able to capture the actual variability in the human experience with death. More promising are newly developed genetic models of human frailty that attempt to combine demographic analysis with quantitative genetics and genetic epidemiology (Weiss 1990; Yashin and Iachine 1997).

Figure 4.1   The Very Young and the Old Have the Highest Death Rates

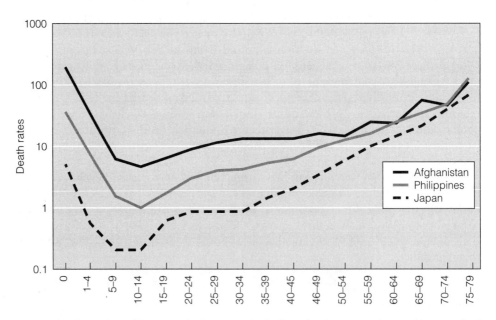

Note: In the 1990s Afghanistan had among the highest death rates in the world, Japan had the lowest, and the Philippines was in between. Yet all three countries exhibited the virtually universal age pattern of mortality—high at both ends and lowest in the middle. Data are for males, but the pattern is the same for females.

Source: United Nations, 1997, Demographic Yearbook, 1995 (New York: United Nations), Table 20.

Despite our individual biological strengths and weaknesses, the actual levels of mortality for each sex at each age in each society appear to be due primarily to social factors that influence when and why death occurs.

Social Factors   The social world influences the risk of death in a variety of ways that can be reasonably reduced to two broad categories: (1) the social and economic infrastructure (how much control we exercise over nature) and (2) lifestyle (how much control we exercise over ourselves). The infrastructure of society refers to the way in which wealth is generated and distributed and reflects the extent to which water and milk are purified, diseases are vaccinated against, rodents controlled, waste eliminated, and food, shelter, clothing, and acute care medical assistance are made available to members of society. Most of the rest of this chapter will examine this theme in detail. Within any particular social setting, however, death rates may also be influenced by lifestyle. An increasing body of evidence has implicated such things as excesses in fatty food, salt, and alcohol, tobacco, drugs, and too little exercise as lifestyle factors that may shorten longevity.

Although one key to a long life may be your "choice" of long-lived parents, prescriptions for a long life are most often a brew of lifestyle choices. A typical list of ways to maximize longevity includes regular exercise, daily breakfast, normal weight, no smoking, only moderate drinking, 7 to 8 hours of sleep, regular meal-taking, and maintaining an optimistic outlook on life. These suggestions, by the way, are not unique to the western world, nor are they particularly modern. A group of medical workers studying older people in southern China concluded that the important factors for long life are fresh air, moderate drinking and eating, regular exercise, and an optimistic attitude (Associated Press 1980). Similarly, note the words of a Dr. Weber, who was 83 in 1904 when he published an article in the British Medical Journal outlining his prescriptions for a long life:

> Be moderate in food and drink and in all physical pleasures; take exercise daily, regardless of the weather; go to bed early, rise early, sleep for no more than 6-7 hours; bathe daily; work and occupy yourself mentally on a regular basis—stimulate the enjoyment of life so that the mind may be tranquil and full of hope; control the passions; be resolute about preserving health; and avoid alcohol, narcotics, and soothing drugs [quoted in Metchnikoff 1908].

Other examples of the way in which social and psychological processes can apparently influence death have been given by David Phillips. In the first of a series of studies, Phillips (1974) found that mortality from suicide tends to increase right after a famous person commits a well-publicized suicide (see also Stack 1987; Wasserman 1984). Thus, some people seem to "follow the leader" when it comes to dying. Further, people do so in more invidious ways than just simple suicide. In follow-up studies, Phillips has found that the number of fatal automobile crashes (especially single-car, single-person crashes) goes up after publicized suicides (Phillips 1977) and (incredibly enough) that private airplane accident fatalities also increase just after newspaper stories about a murder-suicide. It appears "that murder-suicide stories trigger subsequent murder-suicides, some of which are disguised as airplane accidents" (Phillips 1978:748). In a 1983 study, Phillips demonstrated that mass-media violence can also trigger homicides. He discovered that in the United States between 1973 and 1978, homicides regularly increased right after championship prize fights. Furthermore, the more heavily publicized the fight, the greater the rise in homicides (Phillips 1983).

It is easier, of course, to die than to resist death, adding interest to another angle of Phillips's research. He has found that there is a tendency for people who are near death to postpone dying until after a special event, especially a birthday. The story is often told that Thomas Jefferson lingered on his death-bed late on the evening of July 3, 1826, until his physician assured him that it was past midnight and was now the 4th of July, whereupon Jefferson died. Phillips and Smith (1990) found in two large samples of nearly 3 million people that women especially are indeed more likely to die in the week right after their birthday than in any other week of the year, suggesting the prolongation of life until after their birthday. Men, on the other hand, show a peak mortality just before their birthday, suggesting a "deadline" for death. The flip-side of that is the intriguing finding that Chinese-Americans who are born in a year that Chinese astrology considers ill-fated and who have a disease that Chinese medicine considers to be ill-fated have significantly lower life expectancy than normal (Phillips 1993). It may be, of course, that Chinese astrology and medicine are correct about the fates, but Phillips thinks it is more likely that people succumb to an earlier death because of psychosomatic processes—their *belief* that they are fated to die.

These studies illustrate some of the more extreme ways in which social factors can influence death. Normally, however, social factors have less direct influence. In general, different patterns of social organization produce different levels of environmental protection against disease and death. For example, in 1990 the life expectancy for white males in the United States was 72.7 years. In that same year, in the Central Harlem section of New York City, an African-American male had a life expectancy of 53.8 years (about the same as Bangladesh, as seen in Figure 4.2) (Findley and Ford 1993). At ages 35 to 44, African-American males in Harlem were more than 10 times as likely to die as the average male in the United States. This especially was due to the effect of AIDS, discussed later in this chapter, compounded by homicide, suicide, and the spread of infectious diseases in a poverty- and crime-ridden area.

There is considerable variation in longevity around the world (see Figure 4.2), and social factors affect those differences to the extent that they influence specific causes of death. In the next section I review the important causes of death, so we can then examine the social factors that influence why and when you might die.

Causes of Death

In general, there are three major reasons why people die: (1) they are killed by diseases that can be transmitted from one person to another (infectious and parasitic diseases), (2) they degenerate, or (3) they are killed by products of the social and economic environment.

Infectious and Parasitic Diseases

For much of human history, infectious diseases have been the major causes of death, killing people before they had a chance to die of something else. These diseases are parasites that survive by feasting on the host. They are spread in different ways (by different vectors), and have varying degrees of severity. Malaria, for example, typically is spread by mosquitoes first biting an infected person. Then the blood from

Figure 4.2 Variations in Longevity Around the World

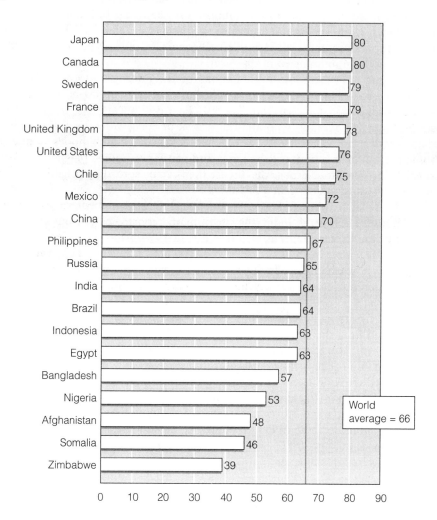

Note: Less developed nations in Africa and South Asia tend to be below the world average (66 years) in terms of life expectancy. Latin American and Western Asian countries tend to be close the the world average, while the more developed countries in North America, Europe, and East Asia are well above average.
Source: U.S. Bureau of the Census, International Data Base, accessed in 1998.

the malarial person spends a week or more in the mosquito's stomach, where the malarial spores develop and enter the mosquito's salivary gland. The disease is passed along with the mosquito's next bite to a human. The probability of dying from untreated malaria is more than 10 percent (Benenson 1990).

Infectious diseases are usually either a virus (a noncellular organism) or a bacterium (a single-celled organism). Measles is an example of an acute viral disease that is severe in infancy and adulthood but less so in childhood. It is usually spread by droplets passed through the air when an infected person coughs or sneezes. If left untreated in an infant or adult, the chance of death is 5 to 10 percent. Vaccinations now protect most people in the developed world from measles, and the United Nations has been working to increase immunization elsewhere. In India in 1985 only 1 percent of children had been immunized against measles, but by 1995 that figure had jumped to 78 percent (UNICEF 1997). By contrast, only 40 percent of children in Nigeria had been immunized in that year.

The plague is perhaps the most famous of the infectious diseases. Plague is caused by the bacterium *Yersinia pestis,* which lives mainly in flea-bearing rodents, especially wild ground squirrels, but also rats. That is important because rats live in closer proximity to humans, who are thus at risk of being bitten by a flea living on a plague-ridden rat. The disease can either infect the lymph nodes and cause bubonic plague, or it can attack the lungs and cause pneumonic plague. The latter is especially problematic because the disease then can be spread by coughing, an easier method of transmission than a flea bite.

The same plague that was known as the "Black Death" in medieval Europe and Asia still pops up not just in India (the site of many reported cases), but in the rural southwest of the United States as well (see Ewald 1994; McNeill 1976). Untreated bubonic plague has a case fatality rate of about 50 percent, whereas untreated pneumonic plague is nearly always fatal—if you get it, you die if you do not seek medical attention. Fortunately, tetracycline and other antibiotics are effective treatments.

A major victory against infectious disease was scored in 1979 when smallpox was eliminated from the world. Two years had passed without a reported case (and the World Health Organization had offered a $1,000 reward to ferret out any cases), and WHO officials thus declared the disease eradicated (except for sample vials contained in laboratories in the United States and Russia).

Infectious and parasitic diseases are far less important as a cause of death in the United States, Canada and other developed societies now than it was in the past, as I discuss later in more detail in describing the epidemiological transition. As seen in Figure 4.3, as of the late 1990s only two of the top 10 killers in the United States and Canada were infectious and parasitic diseases: pneumonia and influenza (separate but related respiratory diseases) and HIV/AIDS (human immunodeficiency virus infection). The latter disease has literally exploded on the scene since the early 1980s to become a worldwide pandemic.

HIV/AIDS   At about the same time that the world was apparently rid of smallpox, another deadly viral disease came along to remind us of how tenuous is our control over mortality. I refer to acquired immunodeficiency syndrome (AIDS).

Figure 4.3   The Leading Causes of Death in the United States, Canada, and Mexico

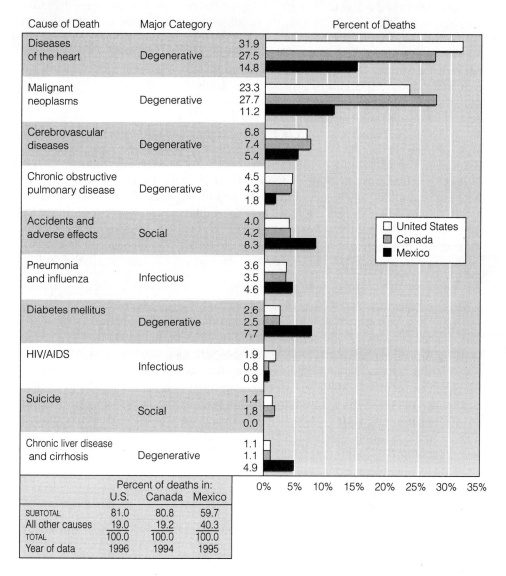

		Percent of deaths in:		
		U.S.	Canada	Mexico
SUBTOTAL		81.0	80.8	59.7
All other causes		19.0	19.2	40.3
TOTAL		100.0	100.0	100.0
Year of data		1996	1994	1995

Sources: United States data are from National Center for Health Statistics, 1998, Monthly Vital Statistics Report, 46(1)(S2), "Births and Deaths: United States, 1996," Canada data are from Statistics Canada, 1998 (http//www.statcan.ca/english/Pgdb/People/Health/death.htm); Mexico data are from INEGI, 1998 (**http://www.inegi.gob.mx/homeing/estadistica/ sociodem/sal-3.htm**).

Unfortunately, this newer disease is worse than the one eliminated. Classic smallpox (variola major) had a case fatality rate as high as 40 percent (Benenson 1990), which is certainly terrible, but AIDS appears to kill virtually everyone who develops its symptoms. AIDS is the most severe of several manifestations of infection with the human immunodeficiency virus (HIV), and technically AIDS is not a single disease but rather a complex of diseases and symptoms whose underlying causes are still not fully understood. AIDS kills by attacking the immune system and rendering the victim susceptible to various kinds of infections and malignancies that then lead to death in a degenerative, disabling fashion.

AIDS initially was recognized as a specific cause of death in 1981 (Quinn 1987), after entering the United States sometime in the 1970s. On December 10, 1981, a headline in the *Wall Street Journal* announced that a "Mysterious Ailment Plagues Drug Users, Homosexual Males." Only 6 years later the United Nations declared AIDS to be a global emergency, and by the end of 1997 the World Health Organization reported that there were more than 30 million people in the world infected with HIV (World Health Organization 1997). The U.S. Bureau of the Census estimates that 2.8 million in the world will die of AIDS in the year 2000 and that could rise to a peak of 4.5 million in 2010, after which the number should decline. (McDevitt 1996)

There is some disagreement about the exact origin of the disease, although it appears that various strains of HIV have been around for decades in both humans and monkeys (Ewald 1994). The evidence suggests that the virus circulating in the world today (HIV-1) probably jumped from a chimpanzee to a human in the Congo (formerly Zaire) in the central region of sub-Saharan Africa (Zhu, et al. 1998). The transfer may have come from humans killing infected animals for food or possibly through the use of animal blood for some magical potion (Gould 1993). Regardless of the mechanism, it must have happened as early as the 1950s, since an Englishman who had been to Africa died in 1959 of what we now know was AIDS (Gould 1993), and a reexamination of a 1959 blood sample (the so-called ZR59 sample) from a man in the Congo tested positive for HIV (Zhu, et al. 1998).

> Around a year after ZR59 was collected, the Congo erupted into one of the bloodiest and most disruptive civil wars in African history. War, and the refugees and starvation which result from it, provide ideal circumstances for any disease to spread, and the activities of armies composed largely of young men are particularly likely to give sexually transmitted infection a boost. Without the Congolese war, HIV-1 might, like its cousin HIV-2, still be confined to a small area of Africa [The Economist 1998:82].

An HIV infection typically produces no clinical symptoms for 2 to 14 years, although it appears that there is a "massive covert infection" taking place internally (Pantaleo, et al. 1993). The transition from HIV to AIDS occurs when an HIV-infected person has at least one of the following conditions: opportunistic infections, malignancies, dementia, weight loss of 15 percent of body weight, or a T-helper cell count below 200.

The progression from HIV infection to AIDS appears to be fairly slow in the first 5 years after initial infection (Schinaia, et al. 1991), but within 10 years after initial infection, it is estimated that 50 percent of HIV-infected persons will have developed AIDS symptoms (Heymann, Chin, and Mann 1990). It is this slow progression that has made AIDS so deadly, because infected people have many years in which to spread the disease. Many new viruses, like Ebola, kill their hosts so quickly that the epidemic quite literally dies out after a short period of time. On the other hand, HIV has a fairly low probability of being transmitted in any single sexual encounter, but the host lives long enough to have multiple encounters, thereby increasing the odds of passing the infection along (Goudsmit 1997).

There are several known methods of transmission of HIV. Heterosexual vaginal intercourse with an infected person seems to have a transmission rate of about 1 percent, although there is considerable variability around this average (Heymann, Chin, and Mann 1990). An open sore or lesion (common in a person already infected with a sexually transmitted disease such as gonorrhea) seems especially to heighten the risk of infection. Anal intercourse, including oroanal intercourse, appears to carry higher risks of infection, which is one reason why homosexual males have a higher infection rate (Grulich, et al. 1997). Homosexual activity also carries a greater risk because of the greater number of sexual partners typically involved. Contaminated blood is another mode of transmission. There is about a 90 percent chance of being infected with HIV if you are given HIV-infected blood. Sharing contaminated needles is a method of transmission among intravenous (IV) drug users, and needle sharing has also been known to infect patients in third-world medical settings. It is estimated that each contaminated needle exposure carries a transmission risk of about 1 percent. Finally, and very importantly, an infected pregnant woman has a 25 to 50 percent chance of passing the disease along to her child. Research conducted in Thailand does suggest, however, that giving an HIV-infected woman doses of AZT during the late stages of pregnancy can dramatically reduce the transmission rate to her baby (Gayle 1998).

In theory, anyone could become infected with HIV, but reality is much different, because transmission of the disease is highly dependent upon the pattern of sexual networking in a society. Among all known cases of AIDS in the United States through 1996, 86 percent have been men. More specifically, in 58 percent of the cases the method of transmission was men having sex with men, most of whom were not IV drug users. IV drug use among men and women accounted for 25 percent of AIDS cases, and only 4 percent were attributable to people who were only heterosexual and were not IV drug users (National Center for Health Statistics 1997).

In North America and western Europe the openness with which AIDS has been discussed has helped to slow down the spread of the disease by encouraging the use of condoms or even abstinence and by increasing the chance that someone infected with HIV will seek treatment that may forestall or perhaps even prevent HIV from progressing to AIDS and premature death. But in much of the rest of world, especially Africa and Asia, the situation is dramatically different because the vast majority of people (perhaps 90 percent) who are infected with HIV do not know of their infection, and so they continue to spread the disease to others (World Health Organization 1997). Table 4.1 provides estimates from the World Health Organization of the worldwide distribution of HIV/AIDS cases as of the beginning of 1998.

The incredibly high rate of infection in sub-Saharan Africa jumps out from Table 4.1, which shows that 2 out of 3 people estimated to be infected with HIV as

Table 4.1 Worldwide Distribution of HIV/AIDS

Region	Number of People Estimated to Be Living With HIV/AIDS (in Millions)	Adult Prevalence Rate	Percent Women Among Those Infected	Cumulative Number of Orphans (in Millions)	Main Mode of Transmission (In Order of Importance)*
Sub-Saharan Africa	20.800	7.40%	50	7.800	Hetero
North Africa, Middle East	0.210	0.13%	20	0.140	IDU, Hetero
South and Southeast Asia	6.000	0.60%	25	0.200	Hetero
East Asia, Pacific	0.440	0.05%	11	0.002	IDU, Hetero, MSM
Latin America	1.300	0.50%	19	0.090	MSM, IDU, Hetero
Caribbean	0.310	1.90%	33	0.050	Hetero, MSM
Eastern Europe and Central Asia	0.150	0.07%	25	0.000	IDU, MSM
Western Europe	0.530	0.30%	20	0.009	IDU, MSM
North America	0.860	0.60%	20	0.070	MSM, IDU, Hetero
Australia and New Zealand	0.012	0.10%	5	0.000	MSM, IDU
TOTAL	30.612	1.00%	41	8.361	

*Hetero = heterosexual intercourse; IDU = intravenous drug use; MSM = men having sex with men
Source: World Health Organization, 1997, Report on the Global HIV/AIDS Epidemic, December 1997.

of the late 1990s were living in that region of the world, and more than 7 percent of all adults were probably infected, including more than one third of pregnant women in Zimbabwe. In sub-Saharan Africa women are as likely as men to be infected, since the disease is spread largely through heterosexual intercourse. The difference between sub-Saharan Africa and much of the rest of the world appears to be related to the fundamentally different pattern of sexual networking prevalent throughout the region. Premarital and extramarital sexual intercourse have long been fairly common in sub-Saharan Africa, without the attendant moral or religious guilt with which it is associated in Western societies (National Center for Health Statistics 1997). Factors that seem to encourage these sexual patterns include long periods of postpartum abstinence on the part of women, which may encourage males to seek sexual encounters elsewhere; polygyny, which institutionalizes male sexual adventurousness (Ahonsi 1991; Orubuloye, Caldwell, and Caldwell 1991), and the low status of women, which forces them to be sexually submissive (Ankrah 1991). In such an environment, any sexually related disease has increased opportunities for transmission through heterosexual channels. Other sexually transmitted diseases already are much more prevalent in sub-Saharan Africa than elsewhere in the world, and the open sores associated with those diseases increase the risk of HIV transmission, as discussed previously.

The above-average prevalence of HIV/AIDS in the Caribbean is accounted for especially by the high rate in Haiti, where the sexual networking patterns are similar to those in sub-Saharan Africa. The rates for south and southeast Asia are influenced by Thailand, where the frequent use of commercial sex workers without frequent use of condoms increases the spread of the disease within the heterosexual population (VanLandingham, et al. 1993).

AIDS is also found disproportionately among young people, which is not surprising since the young are more likely to be sexually promiscuous and to be IV drug users. However, the lag of several years between infection and death means that it is not until ages 35 to 44 in the United States that AIDS becomes the single most important cause of death, exceeding even deaths from accidents and homicides combined.

Without a cure the outlook is bleak, especially for most areas of sub-Saharan Africa, where many people surveyed prefer to risk HIV infection rather than use a condom. In many parts of Africa, the use of condoms is stigmatized because it is traditionally associated with commercial sex workers (Anderson 1994). The U.S. Bureau of the Census has made projections of population growth for those countries hardest hit by AIDS to estimate the impact of AIDS deaths on overall rates of population growth (U.S. Bureau of the Census 1996; Way and Stanecki 1994). Despite the severity of the pandemic, their projections suggest that AIDS will not noticeably slow the rate of population growth at the global level. As can be seen in Figure 4.4, sub-Saharan Africa will be the region hardest hit, but the fact that people typically die after the prime reproductive ages means that AIDS does not affect fertility; rather it produces many more orphans than there would otherwise be (see Table 4.1). As a result, AIDS in Africa has been called the "grandmother's disease" because it falls upon the older women to care for these orphans, as well for the sick and dying (Møller 1997). Every country but Thailand is projected to continue growing (albeit more slowly because of the higher death rate), even in the face of the AIDS pandemic.

Figure 4.4 Most Countries with High Percentages of HIV-Infected Adults Are Still Projected to Increase in Size, but More Slowly

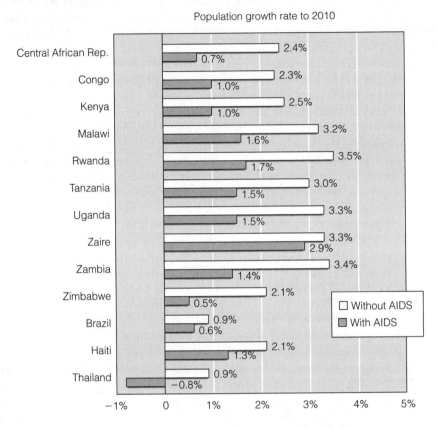

Population growth rate to 2010

Sources: Adapted from U.S. Bureau of the Census, 1994, "AIDS takes its toll on sub-Saharan Africa," *Census and You*, 29(6):1.

Pneumonia and Influenza   The impact of AIDS has overshadowed the quiet deadliness of influenza, which has killed millions of people throughout history. Influenza is much like a cold but more severe. It is an acute viral infection of the respiratory tract that produces a fever, headache, sore throat, cough, and sometimes nausea, vomiting, and diarrhea. In August 1918, as World War I was ending, a particularly virulent form of the flu apparently mutated almost spontaneously in West Africa (Sierra Leone), although it was later called the "Spanish Influenza." For the next year it spread quickly around the world, killing about 20 million people in its path, including more than 500,000 in the United States and Canada—disproportionately healthy young adults (Crosby 1989).

Pneumonia is a family of bacterial respiratory infections and is lumped together with influenza in government summaries of causes of death. It is often called "opportunistic" (like a vulture), because it is a disease that preys especially on physically weak individuals (infants, older people, and others with suppressed immune systems). For a reasonably healthy person, the pneumonia case fatality rate is about 5 to 10 percent, even with treatment, but that can rise to 20 to 40 percent for people who have pneumonia on top of another underlying disease (Benenson 1990). The data in Figure 4.3 show that twice as many people die of pneumonia and influenza as die of AIDS in the United States, and the figure is four times as many in Canada, and five times as many in Mexico.

Degeneration from Chronic Diseases

Degeneration refers to the biological deterioration of a body. In American medicine, "[C]hronic diseases have been referred to as chronic illnesses, noncommunicable diseases, and degenerative diseases. They are generally characterized by uncertain etiology, multiple risk factors, a long latency period, a prolonged course of illness, noncontagious origin, functional impairment or disability, and incurability" (Taylor, et al. 1993). The major chronic diseases associated with degeneration, in order of importance as a cause of death in the United States, Canada (with one exception), and Mexico (with two exceptions) are heart disease, cancer, cerebrovascular disease (stroke), chronic lung disease, diabetes mellitus, and chronic liver disease and cirrhosis (see Figure 4.3). The exceptions are that in Canada cancer is just slightly ahead of heart disease as the leading cause of death, and in Mexico diabetes and chronic liver disease are more common causes of death than in either Canada or the US.

Deaths from coronary heart disease occur as a result of a reduced blood supply to the heart muscle, most often caused by a narrowing of the coronary arteries, which can be a consequence of atherosclerosis, "a slowly progressing condition in which the inner layer of the artery walls become thick and irregular because of plaque—deposits of fat, cholesterol, and other substances. As the plaque builds up, the arteries narrow, the blood flow is decreased, and the likelihood of a blood clot increases" (Smith and Pratt 1993).

The second leading cause of death is cancer, a group of diseases that kill by generating uncontrolled growth and spread of abnormal cells. These cells, if untreated, may then metastasize (invade neighboring tissue and organs) and cause dysfunction and death by replacing the normal tissue in vital organs (Brownson, et al. 1993). In the United States, lung cancer is responsible for more cancer deaths than any other type (28 percent), followed by colon and rectal cancer (11 percent), breast cancer (9 percent), and prostate cancer (7 percent).

Cerebrovascular disease (stroke) is the third leading cause of death. It is actually part of the family of cardiovascular diseases, along with coronary heart disease, but whereas heart disease produces death by the failure to get enough blood to the heart muscle, stroke is the result of the rupture or clogging of an artery in the brain. This causes a loss of blood supply to nerve cells in the affected part of the brain, and these cells die within minutes (Smith and Pratt 1993).

Fourth on the list of most important causes of death in both the United States and Canada (but much further down the list in Mexico) is chronic lung disease. This also is a family of problems, including bronchitis, emphysema, and asthma. The underlying functional problem is difficulty breathing, symptomatic of inadequate oxygen delivery.

Diabetes mellitus is a disease that inhibits the body's production of insulin, a hormone needed to convert glucose into energy (Bishop, et al. 1993). Like most of the other degenerative diseases, diabetes is part of a group of related diseases, all of which can lead to further health complications such as heart disease, blindness, and renal failure. It is the third most common cause of death among degenerative diseases in Mexico and the fourth leading cause of death in that country.

In the United States chronic liver disease and cirrhosis account for 1 percent of all deaths, compared with 2 percent in Canada and 5 percent in Mexico. The liver is a very forgiving organ with tremendous reserves, so it takes considerable abuse to die from liver failure. The single biggest cause of liver disease and cirrhosis is excessive alcohol. Viral hepatitis can also cause deaths from liver disease (Stroup, Dufour, Hurwitz, and Desenclos 1993).

Products of the Social and Economic Environment

Despite the widespread desire of humans to live as long as possible, we have devised myriad ways to put ourselves at risk of accidental death as a result of the way in which we organize our lives and deal with products of our technology. Furthermore, we are the only known species of animal that routinely kills other members of the same species (homicide) for reasons beyond pure survival, and we seem to be unique in being able to kill ourselves intentionally (suicide). These uniquely human causes of death are included among the top killers in North America. Of course, other aspects of our lifestyle also contribute to our risk of infectious and chronic disease, as discussed in the accompanying essay.

Accidents   Every time a person is killed in an automobile accident, slips and is killed in the bathtub, puts a gun to the temple and pulls the trigger, or is murdered by a mugger, that death is attributable to the social and economic environment. Accidental deaths generally occur in the absence of degeneration or transmissible disease, although an otherwise sick person may be more susceptible to having an accident. The only types of accidents not directly attributable to the social and economic environment are those due to natural phenomena such as floods, tornadoes, avalanches, earthquakes, and others. However, in most of these instances deaths can be attributed to human risk taking. For instance, people die in an earthquake not from the quake itself, but from the building that falls on top of them because it was unable to withstand the shaking. Nevertheless, buildings continue to be built and occupied very near, and in some cases right on top of, known fault lines. Likewise, floods generally kill those who unwisely build homes in flood plains or who venture away from secure areas. Tornadoes, too, tend to kill those who do not take precautions and, for whatever reason, have not found a basement, ravine, or ditch to hide in. Also intriguing is the finding that tornadoes are more likely to touch down in urban counties than in rural counties, raising the possibility that human urban settlements may themselves increase the odds of being victimized by a tornado (Aguirre, et al. 1993).

About half of all accidental deaths in the United States are attributable to motor vehicles, compared with one third in Canada and less than one fifth in Mexico.

Yet, despite the differences in the reasons for accidental deaths, in all three countries they are in the top five causes of death (in fact, third in Mexico). In 1995 there were more than 50,000 lives lost in traffic accidents in Canada, the United States, and Mexico; tens of thousands more were injured and will face permanent disability. These victims are disproportionately young males, and alcohol is involved in a high, although declining, fraction of cases. Alcohol also plays a role in other accidents, including bicycle fatalities. A study in Baltimore, for example, found that nearly one half of the males aged 25 to 34 who had died in bike crashes between 1987 and 1991 had alcohol in their blood (Englehart 1994).

Suicide   By world standards, the suicide rate is not unusually high in the United States and Canada—nearly every Scandinavian country, for example, has a rate at least twice as high. Interestingly enough, countries that are culturally similar to the United States and Canada, such as the United Kingdom and New Zealand, have suicide rates similar to North American rates (United Nations 1997). The suicide rate in the United States rises through the teen years (a phenomenon that receives considerable publicity), peaks in the young adult ages, plateaus in the middle years, and then rises for males while it drops for females in the older ages. More specifically, the rate goes up among white males; it does not follow so strong a pattern among African-American males. Although the reasons for this racial difference are not yet well defined, one explanation offered is that women and minority elders experience more adversity during their lives than do white males and thus may be better equipped to cope with the losses and frustrations that sometimes accompany old age (Miller 1979). These losses and frustrations probably explain why in almost every nation for which data are available the suicide rate is higher for older men than for any other group (World Health Organization 1995).

It is worth noting that official suicide rates do not take into consideration the kinds of death, such as auto accidents (discussed earlier in the chapter), that may be disguised suicides. Nor do they show that at all ages women, at least in the United States, attempt more suicides than men, nor that young men in the United States attempt to take their lives more often than do older men but use less lethal means and thus are proportionately less successful (Kushner 1989).

Homicide   Men are not only more successful at killing themselves, they are also more likely to be killed by someone else. Homicide rates are highest for young adult males in virtually every country for which data are available (Garner 1990; World Health Organization 1990). In the United States in 1995 males were three times more likely than females to be homicide victims, but the homicide rate peaks at ages 15 to 24. For white males at this age, the homicide rate in 1995 was 17 per 100,000 compared with a staggering 132 deaths per 100,000 African-American males aged 15 to 24 (National Center for Health Statistics 1997).

Homicide death rates in the United States are higher than for any other industrialized nation (Anderson, Kochanek, and Murphy 1997), possibly reflecting the cultural acceptance of violence as a response to conflict (Straus 1983) combined with the ready availability of guns (which are used in two thirds of homicides in the United States). The remarkable contrast between African-American and white homicide rates within the United States has existed for decades and appears to be

THE "REAL" CAUSES OF DEATH

The causes of death discussed in this chapter reflect those items listed on a person's death certificate. The World Health Organization has worked diligently over the years to try to standardize those causes under a set of guidelines called the International Classification of Diseases, so that the **pathological** conditions leading to death will be identified consistently from one person to the next and from one country to the next. This enhances comparability, but it ignores the actual things going on that contribute to that death. Thus, when public concern first arose over the role of alcohol in traffic fatalities, there were no data available to suggest whether a person who died in an accident was a victim of his or her own alcohol use or the alcohol use of someone else. Similarly, a person who dies of lung cancer or heart disease may really be dying of smoking, no matter what the actual pathological condition that led immediately to death.

There is a vast amount of literature in the health sciences tracing the etiology (origins) of the diseases listed on death certificates, and in 1993 two physicians (McGinnis and Foege) culled those studies in order to estimate the "real" or "actual" causes of death in the United States in 1990, in comparison with the 10 leading causes of death as shown in Figure 4.3 in this chapter. The actual causes of death, as traced by McGinnis and Foege (1993), are detailed in the accompanying table. Of the 2,148,000 people who died in the United States in 1990, this table suggests that 400,000 (19 percent) died as a result of tobacco use. Tobacco has been traced to cancer deaths (especially cancers of the lung, esophagus, oral cavity, pancreas, kidney, and bladder), cardiovascular deaths (coronary heart disease, stroke, and high blood pressure), chronic lung disease, low birth weight and other problems of infancy as a result of mothers who smoke, and to accidental deaths from burning cigarettes.

The second most important "real" cause of death is diet and activity patterns of the United States population, which account for 300,000 deaths annually (14 percent of the total in 1990). The major dietary abuses identified in the literature include high cholesterol, high sodium, and a high consumption of animal fat, while the principle activity pattern of concern is the lack of activity—a sedentary lifestyle. Poor diet and inactivity contribute to heart disease and stroke, cancer (especially colon, breast, and prostate), and diabetes mellitus.

Alcohol misuse is the third (albeit a distant third) real cause of death in the United States, although McGinnis and Foege note that the consequences of alcohol misuse extend well beyond death and include the ruination of lives due to alcohol dependency. Alcohol contributes to death from cirrhosis, vehicle accidents, injuries in the home, drowning, fire fatalities, job injuries, and some cancers. Health care practitioners have been more cautious in their assessment of the positive effects of limited alcohol consumption. A great deal of publicity has been given to the beneficial effects of red wine in keeping heart disease rates lower in France than in the United States, but closer inspection of death rates suggests that red wine has not protected the French against other causes of death, and the French have death rates that are almost identical to those in the United States. Klatsky and Friedman (1995) have noted "a J-shaped alcohol-mortality curve, with the lowest risk among drinkers who take less than three drinks daily" (p. 16).

Less equivocal is the danger from microbial agents—infectious diseases (beyond HIV or infections associated with tobacco, alcohol, or drug use). These represent the fourth leading cause of death and are, in theory at least, largely preventable through appropriate vaccination and sanitation.

Toxic agents are responsible for an estimated 60,000 deaths annually in the United States. This topic is discussed in Chapter 13, but for now note that these agents include occupational hazards, environmental pollutants, contaminants of food and water supplies, and components of commercial products. Toxins are known to contribute to cancer and to diseases of the heart, lungs, liver, kidneys, bladder, and the neurological system.

The role of firearms is obvious. They contribute to an estimated 36,000 deaths annually, including 16,000 homicides, 19,000 suicides, and 1,400 unintentional deaths. The United States is unique in the world in the number of deaths from firearms, and the most pressing problem is that of guns in the hands of teenagers and young adults, who disproportionately use the weapons to kill themselves and/or someone else.

The 30,000 deaths included in the table as being the result of sexual behavior include 21,000 from sexually acquired HIV infection, 5,000 from infant deaths resulting from unintended pregnancies, 4,000 from cervical cancer, and 1,600 from sexually acquired hepatitis B infection. The deaths directly attributable to motor vehicles are those not already accounted for by the use of alcohol or drugs. Beyond the need for safer driving in general, deaths can be avoided by use of lap and seat belts, air bags, child passenger restraints, and helmets for motorcyclists.

Illicit use of drugs is estimated to cause 20,000 deaths annually by contributing to infant deaths (through the mother's use of drugs), as well as to drug overdose, suicide, homicide, motor vehicle deaths, HIV infection, pneumonia, hepatitis, and heart disease.

Researchers have begun exploring the real causes of death in other countries as well. For example, Nizard and Muñoz-Pérez (1994) have found that 16 percent of deaths in France in 1986 were attributable to tobacco use (a little lower than in the United States), and that 10 percent of deaths in France are traceable to alcohol consumption (higher than in the United States).

In tracing the social origins of death, the implication is that these deaths are preventable, and there can be little doubt that each of us takes life-threatening risks (whether big or small) that are not only unnecessary but also are potentially very expensive, since health care is one of the biggest societal expenditures in the wealthier countries. McGinnis and Foege point out that prevention of disease and death accounts for only 5 percent of health care costs, despite the fact that their analysis suggests that fully 50 percent of deaths can be attributed to preventable causes.

These data feed into another debate currently underway among health specialists about how likely it is that life expectancy can be raised substantially in the United States and other low mortality societies (see, for example, Manton, Stallard, and Corder 1997; Rogers 1995). Can we anticipate that life expectancy at birth will ever reach 100 years? On the surface, it would seem easy to suggest that if 50 percent of deaths are preventable, then preventing those deaths will lower the death rate substantially. Life is more complicated than that, however, because the person who died yesterday from a heart attack might have died soon from prostate cancer had he survived the heart attack. People are at multiple risk of death at any given moment, and elimination of one cause of death does not necessarily lower the risks from the other causes. Nonetheless, Rogers has concluded that healthy behaviors do indeed add years to life expectancy, even if the gains may not be as big as it seems they should be.

The "Real" Causes of Death: United States, 1990

Cause of death	Estimated number of deaths	Percentage of all deaths
Tobacco	400,000	19
Diet/activity patterns	300,000	14
Alcohol	100,000	5
Microbial agents	90,000	4
Toxic agents	60,000	3
Firearms	35,000	2
Sexual behavior	30,000	1
Motor vehicles	25,000	1
Illicit use of drugs	20,000	<1
Total	**1,060,000**	**50**

Source: J. M. McGinnis and W. H. Foege, 1993, "Actual Causes of Death in the United States," Journal of the American Medical Association, 270(18):2208.

most readily explained by the proposition that "economic stress resulting from the inadequate or unequal distribution of resources is a major contribution to high rates of interpersonal violence" (Gartner 1990:95). Put another way, a "subculture of exasperation" (Harvey 1986) promotes a "masculine way of violence" (Staples 1986).

If we put together all of the above comments about homicide, we can more readily understand the differences in death rates from homicide in Canada, the United States, and Mexico as shown in Figure 4.5. Canada clearly has the lowest rates, due especially to the laws that make it very difficult to own a gun in Canada, while fairly high levels of taxation and government spending have kept the income distribution somewhat more even than in the United States or Mexico. On the other hand, Mexican males are far more likely to be homicide victims than are men in Canada and the United States (although the rates for Mexican males are not as high as for African-American males in the United States). Mexico not only has a supply of weapons with which people can kill each other, the other kinds of conditions also prevail in which the society is wracked by a noticeably unequal distribution of income, and violence has been an historically more acceptable way for people to settle their differences. The rates for the United States fall between those of Canada and Mexico.

Figure 4.5 Death Rates from Homicide Are Highest for Young Adult Men, Especially in Mexico and the United States

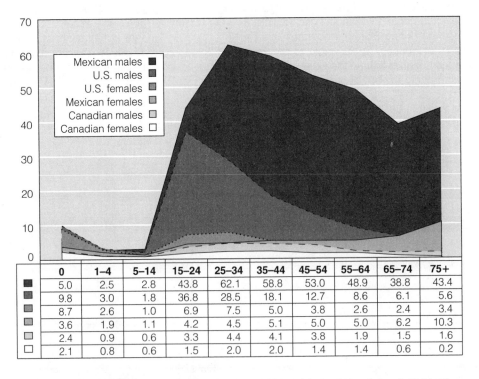

	0	1–4	5–14	15–24	25–34	35–44	45–54	55–64	65–74	75+
■	5.0	2.5	2.8	43.8	62.1	58.8	53.0	48.9	38.8	43.4
■	9.8	3.0	1.8	36.8	28.5	18.1	12.7	8.6	6.1	5.6
■	8.7	2.6	1.0	6.9	7.5	5.0	3.8	2.6	2.4	3.4
■	3.6	1.9	1.1	4.2	4.5	5.1	5.0	5.0	6.2	10.3
□	2.4	0.9	0.6	3.3	4.4	4.1	3.8	1.9	1.5	1.6
□	2.1	0.8	0.6	1.5	2.0	2.0	1.4	1.4	0.6	0.2

Source: World Health Organization, 1995, World Health Statistics Annual 1994 (Geneva: World Health Organization), Table B-1.

Thus far I have mentioned life expectancy and death rates, but I have not actually defined those rates for you. So, before proceeding to an analysis of differences and trends in mortality levels around the world, it will be useful to consider the methods by which mortality is usually measured.

Measuring Mortality

In measuring mortality, we are attempting to estimate the **force of mortality,** the extent to which people are unable to live to their biological maximum. The ability to measure accurately varies according to the amount of information available, and as a consequence, the measures of mortality differ considerably in their level of sophistication. The least sophisticated and most often quoted measure of mortality is the crude death rate.

Crude Death Rate

The **crude death rate** (CDR) is the total number of deaths in a year divided by the average total population. In general form:

$$CDR = \frac{\text{Total deaths in a year}}{\text{Average total population in that year}} \times 1{,}000$$

It is called crude because it does not take into account the differences by age and sex in the likelihood of death. Nonetheless, it is frequently used because it requires only two pieces of information, total deaths and total population, which often can be estimated with reasonable accuracy even in the absence of costly censuses or vital registration systems.

Differences in the CDR between two countries could be due entirely to differences in the distribution of the population by age, even though the force of mortality was actually the same. Thus, if one population has a high proportion of old people, its crude death rate will be higher than that of a population with a high proportion of young adults, even though at each age the probabilities of death are identical. For example, in 1997 Mexico had a crude death rate of 5 per 1,000, exactly half the 10 per 1,000 in Poland in that year. Nonetheless, the two countries actually had an identical life expectancy at birth of 72 years. The difference in crude death rates was accounted for by the fact that only 4 percent of Mexico's population was aged 65 and older, whereas the elderly accounted for 11 percent of Poland's population. Mexico's crude death rate was also lower than the level in the United States (9 per 1,000 in 1997). Yet in Mexico a baby at birth could expect to live 4 years less than a baby in the United States. The younger age structure in Mexico puts a smaller fraction of the population at risk of dying each year, even though the actual probability of death at each age is higher in Mexico than in the United States.

In order to account for the differences in dying by age (and sex), we must calculate age/sex-specific death rates.

Age/Sex-Specific Death Rates

To measure mortality at each age and for each sex we must have a vital registration system in which deaths by age and sex are reported, along with census data that provide estimates of the number of people in each age and sex category. The **age/sex-specific death rate** (ASDR) is measured as the number of deaths in a year of people of a particular age (usually aged x to $x + 5$) divided by the average number of people of that age in the population. It is usually multiplied by 100,000 to get rid of the decimal point. In general:

$$\text{ASDR} = \frac{\text{Number of deaths in a year of people aged } x \text{ to } x + 5}{\text{Average number of people in that year aged } x \text{ to } x + 5} \times 100{,}000$$

In the United States in 1995, the ASDR for males aged 65 to 69 was 2,650 per 100,000, while for females it was 1,566, almost half that for males. In 1900 the ASDR for males aged 65 to 69 was 5,000 per 100,000, and for females 5,500. Thus, we can see that in just less than a century the death rates for males aged 65 to 69 dropped by 47 percent, while for females the decline was 72 percent. To be sure, in 1900 the death rate for females was actually a bit higher than for males, whereas by 1990 it was well below that for males.

Life Tables

Often it is awkward or inconvenient to compare mortality on an age-by-age basis. We would like to have a single index that sums up the mortality experience of a population, while also taking into account the age and sex structure. A frequently used index is the expectation of life at birth. This measure is derived from a **life table,** which is a fairly complicated statistical device that I have saved for the Appendix, if you want to examine it. The important thing for you to know at this point is how to interpret a life expectancy at birth. It is the average age at death for a hypothetical group of people born in a particular year and being subjected to the risks of death experienced by people of all ages in that year. The expectation of life at birth for U.S. females in 1995 of 78.9 years (Anderson, Kochanek, and Murphy 1997) does not mean that the average age at death in that year for females was 78.9. What it does mean is that if all the females born in the United States in 1995 had the same risks of dying throughout their lives as those indicated by the age-specific death rates in 1995, then their average age at death would be 78.9. Of course, some of them would have died in infancy while others might live to be 120, but the age-specific death rates in 1995 implied an average of 78.9. Note that life expectancy is based on a hypothetical population, so the *actual* longevity of a population would be measured by the average age at death. Yet, since we do not usually want to wait decades to find out how long people are actually going to live, the hypothetical situation set up by life expectancy provides a useful, quick comparison between populations. One of the more glaring limitations of basing the life table on rates for a given year is that the death rates of older people in that year will almost certainly be higher than will be experienced by today's babies when they reach that age. For this reason, life tables typically underestimate the actual life expectancy of an infant by as much as 10 or 15 years (Murray 1996).

Having discussed the more prominent causes of death and the usual methods for measuring mortality levels, we are ready to analyze the trends and levels in mortality. We must first explain how and why mortality has declined over time across different regions of the globe. This phenomenon is best described by the epidemiological transition, discussed in the next section. Then, after looking at the mortality transition, we explore why it is that everywhere we go in the world there are some fairly predictable differences within each society in the risks of death—differences between urban and rural places, differences by social strata, and differences by age and sex.

The Epidemiological Transition

Most of us take our long life expectancy for granted. Yet scarcely a century ago, and for virtually all of human history before that, death rates were very high and early death was commonplace. Within the past 200 years, and especially during the 20th century, country after country has experienced an epidemiological transition (Omran 1971; 1977)—a long-term shift in health and disease patterns that has brought death rates down from very high levels in which people die young, primarily from communicable disease, to low levels with deaths concentrated among the elderly, who die from degenerative diseases. This is mortality's contribution to the overall demographic transition. We can perhaps best appreciate the epidemiological transition by following it through time, from premodern societies to the present.

Premodern Mortality

In much of the world and for most of human history, life expectancy probably fluctuated between 20 and 30 years (United Nations 1973; Weiss 1973). At this level of mortality, only about two thirds of babies would survive to their first birthday, and only about one half would still be alive at age 5, as seen in Table 4.2. This means that one half of all deaths occurred before age 5. At the other end of the age continuum, around 10 percent of people made it to age 65 in a premodern society (the consequences of that are discussed in Chapter 9). So in the premodern world about one half the deaths were to children under age 5 and only about 1 in 10 were to a person aged 65 or older.

In hunter-gatherer societies, it is likely that the principal causes of death were due to poor nutrition—people literally starving to death—combined perhaps with selective infanticide and geronticide (the killing of older people) (McKeown 1988), although there is too little evidence to do more than speculate about this (Livi-Bacci 1991). As humans gained more control over the environment by domesticating plants and animals (the Agricultural Revolution), both birth and death rates probably went up, as I mentioned in Chapter 1. It was perhaps in the sedentary, more densely settled villages occasioned by agriculture that infectious diseases became a

Table 4.2   The Meaning of Improvements in Life Expectancy

Period	Life Expectancy (Females)	% Surviving to Age:				% of Deaths:		No. of Births Required for ZPG
		1	5	25	65	<5	65+	
Premodern	20	63	47	34	8	53	8	6.1
	30	74	61	50	17	39	17	4.2
Lowest in world today (comparable to Europe/US in 19th century)	40	82	73	63	29	27	29	3.3
World average circa 2000	68	97	96	94	74	4	74	2.1
Mexico	76	98	98	98	83	2	83	2.1
United States	80	99	99	98	86	1	86	2.1
Canada	83	99	99	98	88	1	88	2.1
Japan (highest in the world)	83	99	99	98	89	1	89	2.1

Source: Based on stable population models in United Nations, 1967, Methods of Estimating Basic Demographic Measures from Incomplete Data, Population Studies No. 42 (New York: United Nations), updated with the use of software provided by Eduardo E. Arriaga, 1994, Population Analysis with Microcomputers (Washington, DC: US Bureau of the Census).

more prevalent cause of death. People were almost certainly better fed, but closer contact with one another, with animals, and with human and animal waste would encourage the spread of disease, a situation that prevailed for thousands of years.

The Roman Era to the Industrial Revolution

Life expectancy in the Roman era is estimated to have been 22 years (Petersen 1975), but by the Middle Ages nutrition had probably improved enough to raise life expectancy to more than 30 years. The plague, or Black Death, hit Europe in the 14th century, having spread west from Asia (probably Mongolia [McNeill 1976]). It is estimated that one third of the population of Europe may have perished from the disease between 1346 and 1350. The plague then made a home for itself in Europe and, as Cipolla says, "[F]or more than three centuries epidemics of plague kept flaring up in one area after another. The recurrent outbreaks of the disease deeply affected European life at all levels—the demographic as well as the economic, the social as well as the political, the artistic as well as the religious" (Cipolla 1981:3).

I mentioned in Chapter 1 that Europe's increasing dominance in oceanic shipping and weapons gave it an unrivaled ability not only to trade goods with the rest of the world, but to trade diseases as well. Most famous of these disease transfers is the so-called Columbian Exchange, involving the diseases that Columbus and other European explorers took to the Americas (and a few that they took back to Europe).

The relative immunity to the diseases they brought with them, at least in comparison with the devastation those diseases wrought on the indigenous populations, is one explanation for the relative ease with which Spain was able to dominate much of Latin America after arriving there around 1500. The populations in Middle America at the time of European conquest were already living under conditions of "severe nutritional stress and extremely high mortality" (McCaa 1994:7), but contact with the Spaniards turned a bad situation into what McCaa (1994) has called a "demographic hell" with high rates of orphanhood and life expectancy probably dipping below 20. Spain itself was hit by at least three major plague outbreaks between 1596 and 1685, and McNeill (1976) suggests that this may have been a significant factor in Spain's decline as an economic and political power.

Industrial Revolution to the Present

The plague had been more prevalent in the Mediterranean area (where it is too warm for the fleas to die during the winter) than farther north or east, and the last major sighting of the plague in Europe was in the south of France, in Marseilles, in 1720. It is no coincidence that this was the eve of the Industrial Revolution. The plague retreated (rather than disappeared) probably as a result of "changes in housing, shipping, sanitary practices, and similar factors affecting the way rats, fleas, and humans encountered one another" (McNeill 1976:174).

By the early 19th century, after the plague had receded and as increasing income improved nutrition, housing, and sanitation, life expectancy in Europe and the United States was approximately 40 years. As Table 4.2 shows, this is a transitional stage at which there are just about as many deaths to children under age 5 as there are deaths at ages 65 and over. Infectious diseases (including influenza, acute respiratory infections, pneumonia, enteric fever, malaria, cholera, and smallpox) were still the dominant reasons for death, but their ability to kill was diminishing.

Although death rates began to decline in the middle of the 19th century, at first improvements were fairly slow to develop for various reasons. Famines were frequent in Europe as late as the middle of the 19th century—the Irish potato famine of the late 1840s and Swedish harvest failures of the early 1860s are prominent examples. These crop failures were widespread, and it was common for local regions to suffer greatly from the effects of a poor harvest, because poor transportation made relief very difficult. Epidemics and pandemics of infectious diseases, including the 1918 influenza pandemic, helped to keep death rates high even into this century.

Until recently, then, increases in longevity primarily were due to environmental improvements—better nutrition and increasing standards of living—not to better medical care:

> Soap production seems to have increased considerably in England, and the availability of cheap cotton goods brought more frequent change of clothing within the economic feasibility of ordinary people. Better communication within and between European countries promoted dissemination of knowledge, including knowledge of disease and the ways to avoid it, and may help to explain the decline of mortality in areas which had neither an industrial nor an agricultural revolution at the time [Boserup 1981:124–125].

McKeown and Record (1962), who did the pioneering research in this area, argue that the factors most responsible for 19th century mortality declines were improved diet and hygienic changes, with medical improvements largely restricted to smallpox vaccinations. Preston and Haines (1991), on the other hand, have concluded that public health knowledge was more important in controlling infectious disease than was nutrition.

> In 1900, the United States was the richest country in the world (Cole and Deane 1965:Table IV). Its population was also highly literate and exceptionally well-fed. On the scale of per capita income, literacy, and food consumption, it would rank in the top quarter of countries were it somehow transplanted to the present. Yet 18 percent of its children were dying before age 5, a figure that would rank in the bottom quarter of the contemporary countries. Why couldn't the United States translate its economic and social advantages into better levels of child survival? Our explanation is that infectious disease processes . . . were still poorly understood . . .[Preston and Haines 1991:208]

Of course, the epidemiological transition was already under way in North America and Europe, but genuine public health knowledge emerged only in the late 19th and early 20th centuries and then began to be widely diffused. Along with this came a new assumption that child health was a public responsibility, not just a private family matter.

The experience of the North American population has been generally comparable to that of Europeans, although there is some evidence that during the 18th and 19th centuries, mortality was slightly lower in the United States than in Europe. Nonetheless, by 1900 the expectations of life at birth were nearly identical in England and the United States, and they have remained that way since.

Life expectancy has increased enormously in the United States since the mid-19th century, as seen in Figure 4.6. In 1850, the numbers were 40.5 years for females and 38.3 years for males. Referring to Table 4.2, this meant that about 72 babies out of 100 would survive to age 5 and about 30 percent of people born would still be alive at age 65. Figure 4.6 also shows that life expectancy began to increase more rapidly as we moved into the 20th century. The data for Canada from 1920 on show that life expectancy for females was generally a bit higher than in the United States, whereas Mexico began the century with a life expectancy that was not quite 30 years, and the improvement has been rapid to levels that are now very close to those of Canada and the United States.

Mortality in Europe was changing in a pattern similar to that in North America. For example, an analysis of data from Italy concluded that in the 100 years between 1887 and 1986, life expectancy had gone from 38 years to 79 years (for females), and that 65 percent of that gain was due to the fact that mortality for infectious diseases had been nearly wiped out (Caselli and Egidi 1991).

Currently in the United States there is an 86 percent chance that a female baby will survive to age 65. In fact, at current levels of mortality, with a life expectancy of nearly 80 years for women (for men it is lower than that), more than half of all women born will still be alive at age 80, and 4 out of 10 will still be alive at age 85. Trivial perhaps, but interesting, is the fact that mortality is currently low enough that more than 10 percent of all babies born in the United States could have a living great-great-grandmother (Keyfitz 1985).

Figure 4.6 Life Expectancy Has Improved Substantially in the United States, Canada, and Mexico in the Last Century

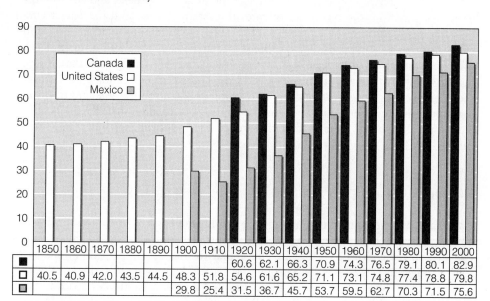

	1850	1860	1870	1880	1890	1900	1910	1920	1930	1940	1950	1960	1970	1980	1990	2000
■ (Canada)								60.6	62.1	66.3	70.9	74.3	76.5	79.1	80.1	82.9
□ (United States)	40.5	40.9	42.0	43.5	44.5	48.3	51.8	54.6	61.6	65.2	71.1	73.1	74.8	77.4	78.8	79.8
▨ (Mexico)						29.8	25.4	31.5	36.7	45.7	53.7	59.5	62.7	70.3	71.5	75.6

Sources: Data for the United States 1850 through 1970 are from U.S. Bureau of the Census, 1975, Historical Statistics of the United States, Colonial Times to the 1970 Bicentennial Edition, Part I (Washington, D.C.: Government Printing Office); Tables B107–115 and B126–135 (data for 1850 through 1880 refer only to Massachusetts); Data for Mexico from 1900 to 1950 are from Marta Mier y Terán, 1991, "El Gran Cambio Demográfico," Demos 5:4–5; Data for Canada from 1920 to 1970 are from Statistics Canada, Catalogue no. 82-221-XDE; all other data are from U.S. Bureau of the Census, International Data Base. All data for Canada and the U.S. are for females, data from 1950 to 2000 are for females for Mexico, but 1900 through 1940 refers to both sexes combined.

Japan has the highest life expectancy in the world (taking over a few years ago from Sweden, which had been in first place for a long time). This probably is due to diet and culturally based health practices, in combination with the societal wealth, which engenders access to clean water, good sewage systems, a high standard of living, and excellent medical care. These characteristics are not peculiar to Japan. East Asians, in general, have lower than average death rates, and Asians in the United States have the highest life expectancy of any group, as discussed later.

A genuine difficulty in drawing conclusions about the exact reasons for mortality decline in Europe (as elsewhere in the world) is that we know less than you might think about the factors that promote or retard the decline of mortality. As Schofield and Reher (1991) have noted in a review of the European mortality decline, "there is no simple or unilateral road to low mortality, but rather a combination of many different elements ranging from improved nutrition to improved education" (p. 17).

On the other hand, Eastern Europe is one area of the world where mortality has experienced a backslide. Data for the former Soviet Union, Hungary, and other eastern European nations all exhibit similar downward trends (Carlson and Watson 1990). To the extent that death rates reflect the underlying social and economic situation of a nation, the rise in death rates in eastern Europe and the Soviet Union perhaps foreshadowed the massive political and economic upheaval in these countries in the early 1990s. In Russia, life expectancy had been falling behind other Western nations since at least the 1970s, and it has been suggested that Russia's health system was unable to move beyond the control of communicable disease to the control of the degenerative diseases that kill people in the later stages of the epidemiological transition. With the collapse of the Soviet Union in 1989 the health of Russians undoubtedly was further threatened by "political instability and human turmoil" (Chen, Wittgenstein, and McKeon 1996:526).

Developing Countries

As mortality has declined throughout the world, the control of communicable diseases has been the major reason, although improved control of degenerative disease has also played a part (Gage 1994). This is true for the less developed nations of the world today, just as it was for Europe and North America before them. However, there is a big difference between the developed and developing countries in what precipitated the drop in death rates. Whereas socioeconomic development was a precursor to improving health in the developed societies, the less developed nations have been the lucky recipients of the transfer of public health knowledge and medical technology from the developed world. Much of this has taken place since World War II.

World War II conjures up images of German bombing raids on London, desert battles in Egypt, D-Day, and the nuclear explosion in Hiroshima. It was a devastating war costing more lives than any previous combat in history. Yet it was also the staging ground for the most amazing resurgence in human numbers ever witnessed. To keep their own soldiers alive, each side in that war spent millions of dollars figuring out how to control disease, clean up water supplies, and deal with human waste. All of the knowledge and technology was transferred to the rest of the world at the war's end, leading immediately to significant declines in the death rate. Not every area of the world has benefited equally, however, so there is still considerable unevenness in the pattern of mortality decline throughout the underdeveloped areas of the world. Among less developed nations, mortality tends to be lowest in Latin America, followed by Asia, with most African nations trailing behind in terms of their success at battling disease.

Prior to the Spanish invasion in the 16th century, the area now called Latin America was dotted with primitive civilizations in which medicine was practiced as a magic, religious, and healing art. In an interesting reconstruction of history, Ortiz de Montellano (1975) made chemical tests of herbs used by the Aztecs in Mexico and claimed by them to have particular healing powers. He found that a majority of those remedies he was able to replicate were, in fact, effective. Most of the remedies were for problems very similar to those for which Americans spend millions of dollars a year on over-the-counter drugs: coughs, sores, nausea, and diarrhea. Unfortunately, these remedies were not sufficient to combat most diseases and mortality remained very high throughout Latin America until the 1920s, when it started to decline at an accelerating rate.

Since the 1920s, death rates have been declining so rapidly that by now, Mexico, as an example of the region as a whole, has reduced mortality to the level that the United States had achieved in the early 1970s.

It took only half a century in Latin America for mortality to fall to a point that had taken at least five centuries in European countries. Countries no longer have to develop economically to improve their health levels if public health facilities can be emulated and medical care imported from European countries. As Arriaga has noted: "Because public health programs in backward countries depend largely on other countries, we can expect that the later in historical time a massive public health program is applied in an underdeveloped country previously lacking public health programs, the higher the rate of mortality decline will be" (Arriaga 1970).

Included among those techniques that can be used to lower mortality even in the absence of economic development are eradication of disease-carrying insects and rodents, chlorination of drinking water, good sewage systems, vaccinations, dietary supplements, use of new drugs, better personal hygiene, and oral rehydration therapy to control diarrhea in infants.

The decade of the 1950s was the period in which mortality declined most rapidly in the less developed nations. More recently, however, that improvement has slowed:

> This should not be surprising, because the pace of mortality decline during the 1950s was rather unique. It would be hard to maintain such a rapid rate of mortality change over a long period of time. The 1950s were the years of the application of low cost and massive health programs which produced a large decline in mortality, as the simpler causes of death were reduced. For continuation of rapid mortality decline, more expensive programs, including implementation of additional public health programs, would have been required [Arriaga 1982:9].

This is another way of noting that communicable diseases generally are easier to bring under control than are the other causes of death, and this is one of the reasons why mortality differentials still persist in country after country.

What are the "routes" to low mortality? Caldwell (1986) has pointed out that a high level of national income is nearly always associated with higher life expectancies, but the bigger question is whether it is possible for poorer countries to lower their mortality levels. The answer is at least a qualified "yes," and there are several ways to do it. Geography makes some difference. Caldwell notes that islands and other countries with very limited territories seem to be able to lower their mortality more readily than territorially larger nations. Being "in the path" of European expansion has also been fortuitously beneficial to some countries (such as Costa Rica and Sri Lanka) because it increases the opportunities for the transfer of death control technology.

The implementation of a basic primary health care system is probably the single most important element in lowering mortality, but it is not enough to provide a health care system. It must be used appropriately, and for this reason much of the improvement in health is associated with education. Since children account for a large share of the population at risk of death in a poorer country, the education of the caregivers—the women in the population—will have an important bearing on the societal death rate.

The education of women is a cultural variable more than an economic one and is closely related to the status of women in society. Caldwell points out that the poorer countries of the world with higher than expected life expectancies all share in common an egalitarian attitude toward women, equal access of women to education, and a political openness (his "star" examples are Sri Lanka, Costa Rica, and the Indian state of Kerala). On the other hand, wealthier countries with lower than average life expectancies share in common a lower status for women, lower levels of education for women than for men, and less democracy in the political system. Most of the examples of such countries are Muslim, such as Saudi Arabia, Oman, and Iran.

Some developing countries have high mortality because they simply have too few resources to overcome the risks of death. This is especially true of nations in parts of sub-Saharan Africa, where high death rates continue to prevail, although they have declined over time. Throughout most of Africa, life expectancies hover around 50 years, a level equivalent to the United States in about 1910. Africa remains the furthest behind of the world's regions in terms of the mortality transition. It has been estimated that between 1985 and 1990, 59 percent of all deaths on the African continent were under the age of 15, and this was a time before the real impact of AIDS (Heligman, Chen, and Babakol 1993). The gains in life expectancy experienced in Africa are being eroded by AIDS and that will continue for the foreseeable future in that part of the world (Bongaarts 1996).

The comparison of life expectancies in developing countries with those in the developed nations of the world can be seen in Figure 4.7. In this graph each box represents the middle 50 percent of all countries in the region, and the line through the box shows the median life expectancy in the region. The lines ("whiskers") extending to the right and left of the box show, respectively, the highest and lowest life expectancies found in the region. Sub-Saharan Africa is clearly set off from other regions of the world in terms of life expectancy, with half of all countries in the region falling between 46 and 56 years and the median life expectancy in the year 2000 expected to be only 50. The lowest figure is for Zimbabwe (reflecting the impact of AIDS) and the highest figure is for Reunion, an island off of Africa in the Indian Ocean that is controlled by France.

The region with the next highest average level of life expectancy is South Asia, where half of the countries in the region have a life expectancy somewhere between 58 and 64, with the median being 63, which is three years below the world average (as shown Figure 4.3). In that region the lowest life expectancy is found in Afghanistan (which had been the highest mortality country in the world for some time until the AIDS pandemic spread through Africa), whereas the highest is found in Sri Lanka (as discussed above). North Africa's median life expectancy is right at the world average, with the extremes in the region being represented by Tunisia at the high end and Western Sahara at the low end. Southeast Asia is the only other region in which a large fraction of countries in the region are below the world average. The median of 68 years of life is just above the average, but the middle half of countries fall between 55 and 71 years. The lowest life expectancy is found in

Figure 4.7 There is Considerable Variability in Life Expectancy Among Developing Countries

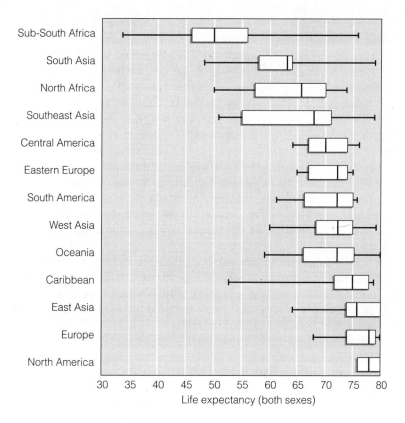

Note: In this graph each box shows the middle 50 percent of all countries in a region arranged by life expectancy for men and women combined. The line through the box represents the median life expectancy among countries in the region and the lines to the right and left of the box stretch to the highest and lowest life expectancies for countries in the region. *Source:* Data refer to the year 2000 and are drawn from the International Data Base of the U.S. Bureau of the Census, accessed in 1998.

Cambodia, while the highest is found in Singapore. You can probably deduce for yourself that the very low life expectancy in the Caribbean region belongs to Haiti, another victim of the AIDS pandemic.

Urban and Rural Differentials in Mortality

Until a few decades ago, cities were deadly places to live. Mortality levels were invariably higher there than in surrounding areas, since the crowding of people into small spaces, along with poor sanitation and contacts with travelers who might be carrying disease, helped to maintain fairly high levels of communicable diseases in cities. For example, life expectancy in 1841 was 40 years for native English males and 42 years for females, but in London it was 5 years less than that (see, for example, Landers 1993). In Liverpool, the port city for the burgeoning coal regions of Manchester, life expectancy was only 25 years for males and 27 years for females. In probability terms, a female child born in the city of Liverpool in 1841 had less than a 25 percent chance of living to her 55th birthday, while a rural female had nearly a 50 percent chance of surviving to age 55. Sanitation in Liverpool at that time was atrociously bad. Pumphrey notes that "pits and deep open channels, from which solid material (human wastes) had to be cleared periodically, often ran the whole length of streets. From June to October cesspools were never emptied, for it was found that any disturbance was inevitably followed by an outbreak of disease" (Pumphrey 1940:141).

In general, we can conclude that the early differences in urban and rural mortality were due less to favorable conditions in the countryside than to decidedly unfavorable conditions in the cities. Over time, however, medical advances and environmental improvements have benefited the urban population more than the rural, leading to the current situation of better mortality conditions in urban areas.

As the world continues to urbanize (see Chapter 11), a greater fraction of the population in each country will be in closer contact with systems of prevention and cure. At the same time, the sprawling slums of many third-world cities blur the distinction between urban and rural and may produce their own unhealthy environments.

Social Status Differentials in Mortality

Differences in mortality by social status are among the most pervasive inequalities in modern society, and they are especially apt to show up in cities. The connection between income and health has been obvious for centuries, and whatever characteristic lowers your status in society may therefore put you at greater risk of death. Marx attributed the higher death rate in the working classes to the evils of capitalism and argued that mortality differentials would disappear in a socialist society. That may have been overly optimistic, but data do clearly suggest that by nearly every index of status, the higher your position in society, the longer you are likely to live.

I should note here that good data are not always easy to come by, since death certificates rarely contain information on occupation and virtually never on income or education, and when they do have occupation data, many indicate "retired" or "housewife," giving no further clues to social status. Thus, more circumlocutious means must be devised to obtain data on the likelihood of death among members of different social strata. An important method used is record linkage, such as in the study by Kitagawa and Hauser (1973) for the United States, in which death certificates for individuals in a census year were linked with census information obtained for that individual prior to death. Age and cause of death were ascertained from the death certificate, while the census data provided information on occupation, education, income, marital status, and race.

Occupation

Among white American men aged 25 to 64 when they died in 1960, mortality rates for laborers were 19 percent above the average, while those for professional men were 20 percent below the average (Kitagawa and Hauser 1973). Similar kinds of results have been reported more recently for all Americans (not just males) followed in the National Longitude Mortality Study, in which the risk of death was clearly lowest for the highest occupational positions and highest for those in the lowest occupational strata (Gregario, Walsh, and Paturzo 1997).

In England researchers followed a group of 12,000 civil servants in London who were first interviewed in 1967–69 when they were aged 40 to 64. They were tracked for the next 10 years, and it was clear that even after adjusting for age and sex, the higher the pay grade, the lower the death rate. Furthermore, within each pay grade, those who owned a car (a more significant index of status in England than in the United States) had lower death rates than those without a car (Smith, Shipley, and Rose 1990). The importance of this study is that it relates to a country that has a highly egalitarian National Health Service. Even so, equal access to health services does not necessarily lead to equal health outcomes.

Use of vehicles on the job represents a significant risk factor for occupational fatalities in the United States, as seen in Table 4.3. Among all occupations in 1993, people involved in fishing had the highest rate of fatal work injuries, caused especially by boat mishaps. Airplane pilots and navigators were at risk of air crashes, while taxi drivers and chauffeurs were at risk of being killed in their vehicles by robbers. Overall, 7 of the 11 deadliest occupations involved vehicles in some way or another.

Table 4.3   The Deadliest Occupations in the United States, 1993

Occupation	Fatalities per 100,000 Employed	Major Deadly Event
Fishers	155	Boating mishap
Timber cutting and logging	133	Struck by tree
Airplane pilots and navigators	103	Air crash
Structural metal workers	76	Fall
Taxicab drivers and chauffeurs	50	Homicide
Electrical power installers and repairers	38	Electrocution
Farm operators, managers and supervisors	36	Vehicular
Construction laborers	33	Fall, contact with objects
Truck drivers	26	Highway crash
Driver–sales workers	23	Highway crash
Farm workers	21	Vehicular

Note: The national average fatality rate per occupation is 5 per 100,000.

Source: U.S. Department of Labor, Bureau of Labor Statistics, 1995, "Outdoor occupations exhibit high rates of fatal injuries," Issues in Labor Statistics, Summary 95-6 (March).

By the way, socialism did not generally reduce occupational differences in mortality, no matter what Marx may have hoped for. Carlson and Tsvetarsky (1992) found that manual laborers in Bulgaria under socialism not only had higher death rates than nonmanual workers, but the difference had been growing over time. This mirrored trends in other Eastern European nations. Indeed, it has been said that "communism could seriously damage your health; but post-communism is proving even worse" (The Economist 1995:41).

The whole family is affected, of course, by the social status of the household head. Fertility surveys have consistently generated data showing an inverse relationship between infant and childhood mortality and the father's occupation. For example, the Kenya Fertility Study in 1978 found that death rates for children aged 1 to 4 whose fathers were employed in professional or clerical occupations were nearly half the level of children whose fathers were farmers or unskilled workers (United Nations 1986). A study in Sweden suggests that social status differentials in childhood continue to affect mortality later in adulthood as well (Östberg and Vägero 1991).

It has to be admitted that "no one knows precisely why people with high status are more healthy" (Shweder 1997:E). Nonetheless, the most important aspects of occupation and social class that relate to mortality are undoubtedly income and education—income to buy protection against and cures for diseases, and education to know the means whereby disease and occupational risks can be minimized.

Income and Education

There is a striking relationship between income and mortality in the United States. Kitagawa and Hauser's data for 1960 showed clearly that as income went up, mortality went down, and more recent studies have confirmed that conclusion. Menchik (1993) used data from the National Longitudinal Survey of Older Men to suggest that, at least for older men in the United States, income is more important than anything else in explaining social status differences in mortality. McDonough and her associates (McDonough, et al. 1997) used data from the Panel Study of Income Dynamics to show that among people under 65, regardless of race or sex, income was a strong predictor of mortality levels.

As with income, there is a marked decline in the risk of death as education increases. A white male in 1960 with an eighth-grade education had a 6 percent chance of dying between the ages of 25 and 45, whereas for a college graduate the probability was only half as high (Kitagawa and Hauser 1973). For women, education makes an even bigger difference, especially at the extremes. A white female with a college education could expect, at age 25, to live 10 years longer than a woman who had 4 or fewer years of schooling.

For virtually every major cause of death, white males with at least 1 year of college had lower risks of death than those with less education. The differences appear to be least for the degenerative chronic diseases and greatest for accidental deaths. This is consistent with the way you might theorize that education would affect mortality, since it should enhance an individual's ability to avoid dangerous, high-risk situations.

Kitagawa and Hauser's work has been partially replicated by Duleep (1989) using 1973 Current Population Survey data matched to 1973–78 Social Security records. This study showed a persistence of the pattern of mortality differentials by education. Overall, death rates were 45 percent higher for white males aged 25 to 64 who had less than a high school education than those who had at least some college education. Pappas and his associates used the 1986 National Mortality Followback Survey and the 1986 National Health Interview Survey to further replicate Kitagawa and Hauser's analysis. They found that "despite an overall decline in death rates in the United States since 1960, poor and poorly educated people still die at higher rates than those with higher incomes and better education, and the disparity increased between 1960 and 1986" (Pappas, Queen, Hadden, and Fisher 1993:103). Data from Canada (Choiniére 1993), as well as from the Netherlands (Doornbus and Kromhout 1991) confirm that these patterns are not unique to the United States.

Race and Ethnicity

In most societies in which more than one racial or ethnic group exists, one group tends to dominate the others. This generally leads to social and economic disadvantages for the subordinate groups, and such disadvantages frequently result in lower life expectancies for the racial or ethnic minority group members. Some of the disadvantages are the obvious ones in which prejudice and discrimination lead to lower levels of education, occupation, and income and thus to higher death rates. But an increasing body of evidence suggests that there is a psychosocial component to health and mortality, causing marginalized peoples in societies to have lower life expectancies than you might otherwise expect (see, for example, Kunitz 1994; Ross and Wu 1995). In the United States, African-Americans and native Americans have been particularly marginalized and experience higher than average death rates. In Canada and Mexico, the indigenous populations are similarly disadvantaged with respect to the rest of society and suffer from higher than average death rates.

The data for the United States in 1995 from the National Center for Health Statistics show that at every age up to 70, African-American mortality rates are nearly double the rates for the white population (Anderson, Kochanek, and Murphy 1997). Between ages 20 and 59, death rates among blacks are more than double those for whites. In 1900 African-Americans in the United States had a life expectancy that was 15.6 years less than for whites, and that differential had been reduced to 6.9 years by 1995. Yet that is a larger gap than exists, for example, between the United States as a whole and the population of Mexico and, perhaps even more significantly, it is a larger gap than had existed 5 years earlier.

African-Americans have higher risks of death from almost every major cause of death than do whites. There are three causes of death, in particular, that help to account for the overall difference. Most important is the higher rate of heart disease for African-Americans of both sexes (Keith and Smith 1988; Poednak 1989), which may be explained partly by the higher rates of unemployment among African-Americans (Guest, Almgren, and Hussey 1998; Potter 1991). Cancer and homicide rates are also important factors in the African-American mortality differential (Keith and Smith 1988). Especially disturbing is the fact that life expectancy for African-Americans actually declined between 1984 and 1988 and again between 1990 and 1991, and by 1995 it was only back to where it had been in 1984. This was particularly noticeable for males and was apparently caused by an increase in deaths from HIV/AIDS and by a rise in the already high homicide rates. Neither Canada nor Mexico publishes mortality data by race and ethnicity, so we have no comparable data for those two countries.

The gap in mortality in the United States between whites and African-Americans could be closed completely with increased standards of living and improved lifestyle, according to a study by Rogers (1992). His analysis of data from the 1986 National Health Interview Survey suggests that sociodemographic factors such as age, sex, marital status, family size, and income account for most of the racial differences in life expectancy. This finding generally is consistent with the pattern for other racial/ethnic groups. For example, over time the income and social status gap has narrowed between "Anglos" (non-Hispanic whites) and the Mexican-origin population in the United States. As this happened, differences in life expectancy between the groups essentially disappeared (Bradshaw and Frisbie 1992; Rosenwaike 1991).

Asians in the United States are among the more economically well-off groups (as I will discuss later in Chapter 10), and data from California indicate that life expectancy is higher for Asians than for any other group. In San Diego County in 1990, Asian females had a life expectancy at birth of 84.0 years, followed by Hispanics (largely of Mexican origin) at 81.6 years, then non-Hispanic whites (79.1), and African-Americans (75.1) (San Diego Association of Governments 1991). For the United States as a whole for 1990, Edmonston and Passel (1994) estimate that Asians and non-Hispanic Whites have virtually the same life expectancy (80.0 for females), with Hispanics just slightly behind that level (79.9 for females), while black females lagged behind at 75.6 years.

American Indians tend to be at the lower end of the income continuum in the United States, and mortality rates consistently have been higher than for whites, although the gap has been narrowing. In fact, as recently as World War II the Inuit in Canada had a life expectancy of only 29 years (Young 1994), and given the health consequences for many tribes of having been herded onto reservations in the second half of the 19th and early in the 20th centuries, the American Indian population actually has experienced a fairly rapid epidemiological transition (Trafzer 1997).

The United States is one of the few countries for which reliable data are available on ethnic differences in mortality, but earlier studies revealed that in India in 1931, members of a religious sect called Parsis had (and probably still have) an expectation of life at birth that was 20 years longer than that of the total population of India. The difference apparently is due to the relatively high economic position of the Parsis (Axelrod 1990). Similarly, in the 1940s before Palestine was partitioned to make Israel, the death rate among Muslims was two or three times that of the Jewish population, probably due to the higher economic status of the Jews (United Nations 1953). The mortality gap between Jews and Muslims within the modern state of Israel has narrowed considerably, but Jews continue to have a slightly higher life expectancy (Israel Center Bureau of Statistics 1998).

Marital Status

It has long been observed that married people tend to live longer than unmarried people. This is true not only in the United States but in other countries as well (Hu and Goldman 1990). A long-standing explanation for this phenomenon is that marriage is selective of healthy people; that is, people who are physically handicapped or in ill health may have both a lower chance of marrying and a higher risk of death. At least some of the difference in mortality by marital status certainly is due to this (Kisker and Goldman 1987).

Another explanation is that marriage is good for your health; protective, not just selective. In 1973, Gove examined this issue, looking at cause-of-death data for the United States in 1959–61. His analysis indicates that differences in mortality of married and unmarried people "are particularly marked among those types of mortality where one's psychological state would appear to affect one's life chances" (1973:65). As examples, Gove notes that among men aged 25 to 64, suicide rates for single men are double those of married men. For women, the differences are in the same direction, but they are not as large. Unmarried males and females also have higher death rates from what Gove calls "the use of socially approved narcotics" such as alcohol and cigarettes. Cirrhosis of the liver is associated with heavy drinking, and single males have three times the death rate of married males from that disease; for divorced males, the rates are nine times higher. Again, the differences are less extreme for females. Lung cancer is associated with smoking, and single males have death rates 45 percent higher than married men; divorced men have three times the rate.

Finally, Gove notes that for mortality associated with diseases requiring "prolonged and methodical care" unmarried people are also at a disadvantage. In these cases, the most extreme rates continue to be among divorced men, who have death rates nine times higher than married men. In general, Gove's analysis suggests that married people, especially men, have lower levels of mortality than unmarried people because their levels of social and psychological adjustment are higher. Researchers more recently found that economic factors also may play a protective role. Married women are healthier than unmarried women partly because they have higher incomes (Hahn 1993), and unmarried men are more likely to die than married men, particularly if they are below the poverty line (Smith and Waitzman 1994).

Does this mean that if you are currently single you should jump into a marriage just because you think it might prolong your life? Not necessarily, but Lillard and Panis (1996) have found evidence from the Panel Study of Income Dynamics in the United States that relatively unhealthy men tend to marry early and to remain married longer, presumably using marriage as a protective mechanism. The flip side of that is that international comparisons of data suggest that it is the ending of a marriage that especially elevates the risk of death. Divorced males have a noticeably higher death rate than people in any other marital category (Fenwick and Barresi 1981; Hu and Goldman 1990).

Gender Differentials in Mortality

Women generally live longer than men do, and the gap had been widening until recently. In 1900 women could expect to live an average of 2 years longer than men in the United States, and by 1975, the difference had peaked at 7.8 years. Since then the difference has dropped a bit to 6.9 years, but the survival advantage of women is nearly universal among the nations of the world. Canada has followed a nearly identical pattern, with a 1.8 year advantage in 1920 expanding to a 7.2 year advantage in 1980 before dropping back to a 6.4 year advantage in 1990. Mexico is still in the phase of a widening gender gap. In 1960, for example, women in Mexico had a life expectancy that was about the same as in Canada and the United States in 1920, and their advantage over men was 3.2 years. By 1990 female life expectancy in Mexico was at the level reached in Canada and the United States in 1970, and the advantage for women had grown to 6.0 years.

This phenomenon has attracted curiosity for a long time, and it has been suggested facetiously that the early death of men is nature's way of repaying those women who have spent a lifetime with demanding, difficult husbands. However, the situation has been more thoroughly investigated by a variety of researchers.

There is some evidence that a widening gender gap is a normal accompaniment to the epidemiological transition because, for reasons not yet well understood, women are more likely than men to experience a decline in degenerative disease, which takes over as the main reason for death in the later stages of the transition (Gage 1994; Lopez 1983). This argues for a biological interpretation of the difference, and Retherford (1975) has pointed out that a large number of studies show that throughout the animal kingdom females survive longer than males, which could indicate a basic biological superiority in the ability of females to survive relative to males. However, in human populations, although the survival advantage of women is widespread, it is not quite universal. This implies that there are social factors that are also operating. One such factor is the status of women. It is in those countries where women are most dominated by men (especially south Asia) that women have been least likely to outlive men.

In Bangladesh, the preference of parents for sons means that the health of girls may be neglected relative to that of boys (Muhuri and Preston 1991), repeating a pattern found in India (Langford and Storey 1993). Still, this does not mean that girls are universally discriminated against (Basu 1991). In the rice-growing regions of India, where the work of women is more highly valued, life expectancy for women is higher than in the wheat-growing areas, where men do more of the work (Kishor 1993).

In North America, as in European and other countries, the trend toward greater independence of women has been accompanied by a large gender gap in life expectancy, perhaps because women have been gaining in their knowledge of preventive health measures (such as proper nutrition and exercise) and because greater independence enhances the ability to seek proper medical care. The Conference on Gender and Longevity sponsored by the National Institute on Aging in 1987 concluded that the higher rate of heart disease for men combined with men's higher-risk behaviors accounted for most of the difference in the gender gap (Gerontology Center Newsletter 1988). Noteworthy in the high-risk category are smoking, alcohol use, accidents, suicides, and homicides.

You recall from the essay accompanying this chapter that tobacco use is the top "real" killer of people in the United States. Since 1900 males have smoked cigarettes much more than have females, and this has helped to elevate male risks of death from cancer, degenerative lung diseases (such as chronic bronchitis and emphysema), and cardiovascular diseases (Preston 1970). However, cigarette smoking by women increased after World War II and in 1995 women aged 45 to 64 were almost as likely as men of that age to be smokers (the smoking version of gender equality), but fortunately the percentage of people who smoke has been steadily declining since the 1960s (National Center for Health Statistics 1997). Of course, deaths associated with smoking tend to occur many years after smoking begins, and so the result of women's post-World War II smoking habits has been a predictable recent rise in death rates from lung cancer.

Using data for 1986, Rogers and Powell-Griner (1991) have calculated that males and females who smoke heavily have similar (and lower than average) life expectancies, but nonsmoking males still have lower life expectancies than nonsmoking females (although still higher than smokers). Smoking thus accounts for some, but not all, of the gender gap. It is also possible that a real biological superiority exists for women in the form of an immune function, perhaps imparted by the hormone estrogen (Waldron 1983; Waldron 1986).

Age Differentials in Mortality

Infant Mortality

There are few things in the world more frightening and awesome than the responsibility for a newborn child, fragile and completely dependent on others for survival. In many societies, that fragility and dependency are translated into high **infant mortality rates** (the number of deaths during the first year of life per 1,000 live births). In some of the less developed nations, especially in equatorial Africa, infant death rates are as high as 123 deaths per 1,000 live births. That figure is for Sierra Leone, plagued by drought and famine and site of some of the highest mortality ever recorded for a human population (McDaniel 1992), but it is indicative of trends in that region of the world.

Infant death rates are closely correlated with life expectancy, and Figure 4.8 shows the infant mortality rates for the same countries for which life expectancy was listed in Figure 4.2. Japan is tied with Finland, Sweden, and Singapore for the honor of having the world's lowest rate of infant mortality, at 4 deaths per 1,000 live births. Canada's rate of 5 per 1,000 puts it in the top 20, whereas the United States has a rate of 7 per 1,000, placing it in the top 30. Mexico's level of 23 per 1,000 is well down the list among all countries in the world but is still about half of the world average of 45 per 1,000.

Why do babies have higher death rates in some countries than in others? The answer is perhaps best summed up by mentioning the two characteristics common to people in places where infant death rates are low—high levels of education and income. These are key ingredients at both the societal level and the individual level. In general, those countries with the highest levels of income and education are those with enough money to provide the population with clean water, adequate sanitation, food and shelter, and, very importantly, access to health care services, especially oral rehydration therapy. Even in an impoverished country such as

Figure 4.8   Variations in Infant Mortality Around the World

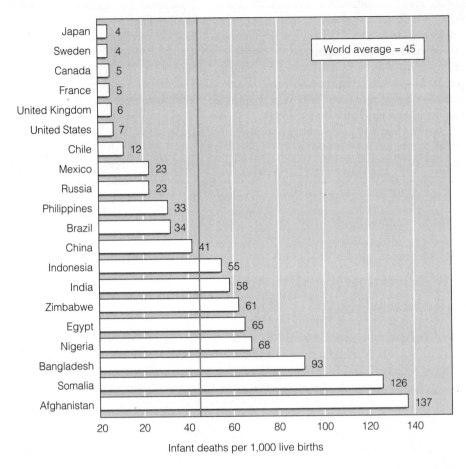

Source: Data are from U.S. Bureau of the Census, International Data Base, accessed in 1998 and refer to the year 2000.

Bangladesh, the odds of an infant dying decrease the closer the family lives to a physician (Paul 1991). At the individual level, education here can refer simply to knowledge of a few basic rules that would avoid unnecessary infant death. For example, one study in a rural Indian village revealed that tetanus was the major cause of infant death. Further investigation ascribed this to the fact that umbilical cords were often cut with instruments such as unsanitary farm implements, and the cord was dressed with ash from the cow dung fire typical of that part of the country (Bouvier and van der Tak 1976).

Income is important in order to provide babies with a nutritious, sanitary diet that prevents diarrhea, an important cause of death among infants. Nursing mothers can best provide this service if their diet is adequate in amount and quality. Income is also frequently associated with the ability of a nation to provide, or an individual to buy, adequate medical protection from disease. In places where infant death rates are high, communicable diseases are a major cause of death, and most of those deaths could be prevented with medical assistance. For example, between 1861 and 1960, the infant death rate in England and Wales dropped from 160 to 20, with more than two thirds of that decline due to the control of communicable diseases. Since disease in infants often manifests itself in diarrhea and dehydration, rehydration therapy can be an inexpensively effective way to save the life of a sick baby. This was an important reason why the infant death rate in the United Nations–supervised Palestinian refugee camps in the West Bank (the Israeli-occupied part of Jordan) plummeted from 120 in 1962 to 25 in 1987 (United Nations 1988).

Resistance to disease is, of course, closely related to the overall health of the child, which in turn is associated with the health of the mother. Mothers who are healthy while pregnant and who maintain that good health after giving birth are more likely to have healthy babies. Since levels of health are generally higher in the more advanced nations, infant mortality is generally lowest there. Although the relationship between income and health is not perfect, a glance at Figure 4.8 shows that the countries with the lowest infant death rates are all highly developed nations, whereas those with the highest rates are among the world's poorest countries.

In advanced nations like the United States and Canada, prematurity accounts for a vast majority of deaths among infants, and in many cases prematurity results from lack of proper care of the mother during pregnancy. Pregnant women who do not maintain an adequate diet, who smoke or take drugs, or in general do not care for themselves have an elevated chance of giving birth prematurely, thus putting their baby at a distinct disadvantage in terms of survival after birth. Rumbaut and Weeks (1996) have found that immigrants to the United States have lower levels of infant mortality than do native-born whites, even when they come from countries with high infant mortality and even when they themselves have fairly low levels of education and income. Again, the explanation seems to be found at least partly in the fact that immigrant women are much less likely than U.S.-born women to smoke, drink, or take drugs during pregnancy.

Prior to its collapse (and perhaps symptomatic of the underlying problems) the Soviet Union had difficulty keeping its infant mortality rate low, and in fact the rate had actually risen, a trend that was unique in the history of developed countries (Feshbach and Friendly 1992). Apparently to hide this health slippage from the rest of the world, the USSR stopped publishing detailed mortality data in 1974, but other data collected by Murray Feshbach at Georgetown University made it clear that infants were at increasingly high risks of death, with a rate perhaps as high as 33 per 1,000. A 1983 editorial in the *Wall Street Journal* detailing Feshbach's findings drew an irate denial from the Soviet Ministry of Health, but they provided little data to rebut Feshbach's argument. In 1986, the Soviets resumed the publication of data on infant mortality after a 10-year hiatus, and it is estimated that Russia's infant mortality rate in the year 2000 will be 23 per 1,000—the same as Mexico (see Figure 4.8).

Although infant mortality measures infant deaths from birth through the first year of life, the most dangerous time for infants is just before and just after birth. There are special measures of infant death that take these risks into account. For example, **late fetal mortality** refers to fetal deaths that occur after at least 28 weeks of gestation. **Neonatal** mortality refers to deaths of infants within 28 days after birth.

Postneonatal mortality then covers deaths from 28 days to 1 year after birth. In addition, there is an index called **perinatal** mortality, which includes late fetal deaths plus deaths within the first 7 days after birth.

In general, neonatal mortality is less preventable than is postneonatal mortality, especially in countries where most babies are born in a hospital. Thus, as infant mortality declines, postneonatal mortality will likely decline more than will the neonatal component. When infant mortality is very low, as in the United States or Canada, neonatal deaths will comprise a large fraction of all babies who die. Thus, in the United States in 1995, the neonatal mortality rate was 491 deaths per 100,000 live births during the first month of life, while the postneonatal mortality rate, reflecting the deaths occurring from then until the first birthday, was 267 per 100,000. Together, they combined for an infant mortality rate of 758 per 100,000 (or a little more than 7 per 1,000).

Throughout the world, the infant mortality rate is a fairly sensitive indicator of societal development because as the standard of living goes up, so does the average level of health in a population, and the health of babies typically improves earlier and faster than at other ages. One interesting (and still controversial) explanation for the relative sensitivity of infant mortality to development is that in a poor society some infant deaths may be the result of a response by parents to too few resources (Scrimshaw 1978; Simmons, et al. 1982). This is the rational choice model—the supply-demand framework—applied to the analysis of infant mortality. If parents lack the foresight or the ability to control their fertility, the option always exists of controlling mortality by reducing the amount of food given to an infant or by selective inattention to the health and medical needs of an unwanted child.

Data from the Princeton European Fertility Project have led to the conjecture that infant and child abuse in the 19th century may have been a result of the undesirability of large families (Cherfas 1980), and an example of this is provided by an examination of family life in Paris. In that city in the 19th century, there was an increase in the percentage of women working outside the home (especially among middle-class artisans and shopkeepers). Before bottle-feeding, a woman with a baby who wished to continue working had to hire a wet nurse. Most wet nurses were peasant women who lived in the countryside, and a mother would have to give up the child for several months if she wished to keep working.

Infant mortality was as high as 250 deaths per 1,000 infants among those placed with wet nurses (Sussman 1977), yet in the early 1800s nearly one fourth of all babies born in Paris were placed with wet nurses. Infants were a bother, and the risk was worth taking. Lynch and Greenhouse (1994) would argue that though this might be true for some families, the historical data for Sweden suggest that death at all ages tended to cluster in families. Rather than higher-order births being targeted for neglect, within families all children tended to have similar risk of infant mortality, but in some families the risk was higher than in others. This could be thought of as a micro-level "cultural" approach to understanding infant mortality.

Birth can be a traumatic and dangerous time not only for the infant, but for the mother as well. Until the last few decades, pregnancy and childbearing were the major reasons for death among young female adults in North America, and this is still true in many developing nations. Let us turn to an examination of mortality during that age period.

Mortality in Young Adulthood

Currently in North America and throughout the developed world, young adults have very low risks of death. In the United States and Canada, for example, there are fewer than two chances in a hundred that a 20-year-old will die before reaching age 35—about one tenth the risk experienced in 1900. For females at these ages, the decline in deaths associated with childbearing has been especially important and in 1995 in the United States, death rates associated with complications of pregnancy and childbirth were less than one tenth the level of 1950 and just a tiny fraction of the rate that prevailed in 1900. Age still plays a role, however. The maternal death rate for women aged 30 to 34 is more than twice the rate for women under age 20. Furthermore, African-American women have five times the risk of dying in childbirth as do white women, almost certainly reflecting the differential access to health services in the United States.

Women are only at risk of a maternal death if they become pregnant, so maternal death rates need to take that risk into account. Thus, the maternal mortality ratio measures the number of maternal deaths per 100,000 live births. In the United States in 1991 this rate for all groups combined was 7.1 (Anderson, Kochanek, and Murphy 1997), whereas the world average is about 430 (Salter, Johnston, and Hengen 1997). More than a half million women die each year from childbirth—the equivalent of three jumbo jets crashing each day and killing all passengers. These deaths leave a trail of tragedy throughout the world, but disproportionately in developing nations, where pregnancies are more frequent among women and health care systems are less adequate.

Despite the steep drop in maternal mortality as a cause of death in young adulthood in the now developed countries, it was the decline in communicable disease that actually led the drop in mortality over time at the young adulthood ages. Rarely do young adults, male or female, now die from such causes. Instead, those few who do die are much more likely to be victims of accidents, automobile or otherwise. Among women, accidents account for one third of all deaths between ages 20 and 35, and for males they account for two thirds of all deaths in this age range. This is a pattern that actually exists from early childhood on through the middle years of life, but by old age the pattern has changed considerably.

Mortality at Older Ages

It has been said that in the past parents buried their children; now, children bury their parents. This describes the mortality transition in a nutshell. The postponement of death until the older ages means that the number of deaths among friends and relatives in your own age group is small in the early years and then accumulates more rapidly in the later decades of life. By the time people move into their fifties, their chances of dying start to increase at an accelerating rate. At ages above 60, by far the largest number of people of either sex die of cardiovascular diseases. For example, between two thirds and three fourths of all people of that age in the United States die from heart attack and stroke. It is probable that many of the deaths from cardiovascular causes represent biological degeneration, although social factors, especially stress and smoking, are frequently implicated, as discussed earlier in the chapter. Running a distant second as a cause of death among the elderly is cancer, although in the past few decades there has been considerable progress in treating and thus delaying death from cancer, and this has added to life expectancy in the later years of life.

Summary and Conclusion

The control of mortality has vastly improved the human condition but has produced worldwide population growth in its wake. There are, however, wide variations between and within nations with respect to both the probabilities of dying and the causes of death. The differences between nations exist because countries are at different stages of the epidemiological transition, the shift from high mortality (largely from infectious diseases and associated with most deaths occurring at young ages) to low mortality (with most deaths occurring at the older ages and largely caused by degenerative diseases). On the other hand, differences within a society tend to be due to social status inequalities. Differences in mortality by social status are among the most pervasive inequalities in modern society. For example, as prestige goes up, death rates go down. Likewise, as education and income levels go up, death rates go down. The social and economic disadvantages felt by minority groups often lead to lower life expectancies. Marital status is also an important variable, with married people tending to live longer than unmarried people. Females have a survival advantage over males at every age in most of the world, and a widening gender gap seems to be a feature of the epidemiological transition.

In general, infancy—birth to the first birthday—is the riskiest age until you are well into the retirement years. The vast majority of infant deaths in North America are due to prematurity and low birth weight, whereas infectious diseases kill many babies in developing countries. The risk of death in young adulthood is relatively low in all societies, and then it accelerates at older ages.

Although mortality rates are low in the developed nations and are declining in most developing nations, diseases that can kill us still exist if we relax our vigilance. The emergence of HIV/AIDS is one reminder of that, as was the outbreak in 1995 in Zaire of the extremely deadly Ebola virus (the same virus featured in the book and movie *The Hot Zone*). Less deadly but more pernicious is tuberculosis. Most forms of the disease are avoidable and until recently it had been nearly wiped out in the United States, but the number of cases reported increased by 20 percent between 1984 and 1992. Malaria has also been making a resurgence in many tropical regions of the world, reminding us that death control cannot be achieved and then taken for granted, for as Zinsser has so aptly put it:

> However secure and well-regulated civilized life may become, bacteria, protozoa, viruses, infected fleas, lice, ticks, mosquitoes, and bedbugs will always lurk in the shadows ready to pounce when neglect, poverty, famine, or war lets down the defenses. And even in normal times they prey on the weak, the very young, and the very old, living along with us, in mysterious obscurity waiting their opportunities [1935:13].

If the thought of those lurking diseases scares you to death, then I suppose that too is part of the mortality transition. Also lurking around the corner is Chapter 5, in which we examine fertility concepts and measurements.

Main Points

1. Lifespan refers to the oldest age to which members of a species can survive.

2. Longevity is the ability to resist death from year to year.

3. Although biological factors affect each individual's chance of survival, social factors are important overall determinants of longevity.

4. Most deaths can be broadly classified as a result of infectious disease, degeneration, or a product of the social and economic environment.

5. Infectious diseases account for the worldwide drop in mortality and historically were the most important causes of death.

6. AIDS is now among the 10 most important causes of death in the United States and has spread rapidly in the world, especially in sub-Saharan Africa.

7. Degeneration is associated with chronic diseases, which account for the majority of all deaths in developed nations such as the United States and Canada. Heart disease is the most important of the chronic diseases as a cause of death.

8. Accidents and homicides and, increasingly, AIDS, are the most significant causes of death among young adults.

9. The most important "real" cause of death in the United States is the use of tobacco.

10. Mortality is measured with the crude death rate, the age-specific death rate, and life expectancy.

11. The changes over time in death rates and life expectancy are captured by the perspective of the epidemiological transition.

12. Significant widespread improvements in the probability of survival date back only to the 19th century and have been especially impressive since the end of World War II. The drop in mortality, of course, precipitated the massive growth in the size of the human population.

13. The role played by public health preventive measures in bringing down death rates is exemplified by the saying at the turn of this century that the amount of soap used could be taken as an index of the degree of civilization of a people.

14. Living in a city used to verge on being a form of latent suicide, but now cities tend to have lower death rates than rural areas.

15. Rich people live longer than poor people on average (and they have more money, too).

16. Married people tend to live longer than unmarried people, and females generally live longer than males.

17. The single most vulnerable age in a person's life (at least until well beyond retirement age) is the first year. Reductions in infant mortality often account for a major part of the early mortality decline in a previously high-mortality society.

Suggested Readings

1. Abdel Omran, 1977, "Epidemiologic transition in the United States," Population Bulletin 32(2).

   Omran was the first to develop the concept of the epidemiological transition, and it is well summarized in this report.

2. James N. Gribble and Samuel H. Preston, editors, 1993, The Epidemiological Transition: Policy and Planning Implications for Developing Countries (Washington, D.C.: National Academy Press).

   A report from a National Research Council Workshop exploring the ways in which developing countries can best complete the epidemiological transition.

3. Murray Feshbach and Alfred Friendly, Jr., 1992, Ecocide in the USSR: Health and Nature Under Siege (New York: Basic Books).

   It is almost unprecedented for a modern nation to backslide with respect to the mortality transition, but it happened in the former Soviet Union, and this book describes the process.

4. Christopher J. L. Murray and Alan D. Lopez, editors, 1996, The Global Burden of Disease: A Comprehensive Assessment of Mortality and Disability From Diseases, Injuries, and Risk Factors in 1990 and Projected to 2020 (Cambridge: Harvard University Press).

   This volume is part of a massive project undertaken by the Harvard School of Public Health in collaboration with the World Health Organization and the World Bank to assess the societal impact of disease and death on human societies. Although it is not an "easy read," there is a wealth of information packed into the book.

5. S. Jay Olshansky, Bruce Carnes, Richard G. Rogers, and Len Smith, 1997, "Infectious Diseases—New and Ancient Threats to World Health," Population Bulletin, 52(2).

   Zinsser's worry in 1935 that infectious diseases are lurking around waiting to strike back has, in fact, become reality in the last part of the 20th century, and the issues are reviewed in the monograph.

Websites of Interest

Remember that websites are not as permanent as books and journals, so I cannot guarantee that each of the following websites still exists at the moment that you are reading this.

1. http://www.northwesternmutual.com/noframes/games/longevity

   How long can you expect to live? Play the Longevity Game and calculate your life expectancy. This innovative "game" from Northwestern Mutual Life Insurance Company lets you calculate your own expected age at death, given the data that you enter about yourself, your health, and your lifestyle characteristics.

2. http://www.who.ch/whosis

   The World Health Organization is the United Nations agency that monitors—you guessed it—the world's health, and their website includes their Statistical Information System (SIS), which allows you to tap into their latest data, including tables and graphs from their annual publication, the World Health Report.

3. http://www.cihi.com/cihidesc.htm

   The Center for International Health (CIHI) is an information base funded by the U.S. Agency for International Development and maintained by Information Management Consultants in conjunction with the Futures Group and the International Science and Technology Institute. The data are organized into country health profiles, drawing data not only from the World Health Organization but from a wide range of other sources. Data are limited largely to developing nations.

4. http://www.cdc.gov/nchswww/fastats/deaths.htm

   The National Center for Health Statistics is the source of vital statistics for the United States, and this web page takes you directly to data on mortality in the United States.

5. http://www.statcan.ca/english/Pgdb/People/health.htm

   Statistics Canada is the source of vital statistics for Canada, and this web page takes you directly to data on mortality in Canada.

CHAPTER 5
Fertility Concepts and Measurements

Control over human reproduction is one of the most important and revolutionary changes that has ever occurred, comparable in many ways to the remarkable progress in postponing death, which I discussed in the previous chapter. It frees women and men alike from the bondage of unwanted parenthood and helps to time and space those children that are desired. To control fertility does not necessarily mean to limit it, yet almost everywhere you go in the world, the two concepts are nearly synonymous. This suggests, of course, that as mortality declines and the survival of children and their parents is assured, people generally want relatively small families, and the wider the range of means available to accomplish that goal, the greater the chance of success.

Fertility control means much more than taking "the pill" or using a condom. As with most things, life is more complex. How do we define fertility? What are all of the ways in which it may be brought under our control? If we know how to control it, why does it vary from one area to another? In this chapter I deal with these questions by discussing the concept of fertility and then analyzing the means by which fertility levels can be altered. I conclude with a brief discussion of how fertility is measured. In the next chapter, I discuss trends in fertility and offer explanations for differences in fertility levels.

What Is Fertility?

Fertility refers to the number of children born to women. Note that though our concern lies primarily with the total impact of childbearing on a society, we have to recognize that the birth rate is the accumulation of millions of individual decisions to have or not have children. Thus, when we refer to a "high-fertility society," we are referring to a population in which most women have several children, whereas a "low-fertility society" is one in which most women have few children. Naturally, some women in high-fertility societies have few children, and vice versa.

Fertility, like mortality, is composed of two parts, one biological and one social. The biological component refers to the capacity to reproduce, and while obviously a necessary condition for parenthood, it is not sufficient alone. Whether children will actually be born and if so, how many, given the capacity to reproduce, is largely a result of the social environment in which people live.

The Biological Component

The physical ability to reproduce is usually called **fecundity** by demographers. A fecund person can produce children; an infecund (sterile) person cannot. The term *fertility* is typically reserved to describe reproductive performance, that is, the actual birth of children, rather than the mere capacity to do so. However, since people are rarely tested in the laboratory to determine their level of fecundity, most estimates are based on levels of fertility. Couples who have tried unsuccessfully for at least 12 months to conceive a child are usually called "infertile" by physicians (demographers would say "infecund"). The 1995 National Survey of Family Growth (NSFG) showed that 7 percent of American couples (where the wife was aged 15 to 44) are infertile by that criterion (Abma, et al. 1997)—a decline from 11 percent in 1965 (Mosher and Pratt 1990). A more general concept is the idea of **impaired fecundity**, measured by a woman's response to survey questions about her fecundity status. A woman is classified as having impaired fecundity if she believes that it is impossible for her to have a baby, if a physician has told her not to become pregnant because the pregnancy would pose a health risk for her or her baby, or if she has been continuously married for at least 36 months, has not used contraception, and yet has not gotten pregnant. In 1995, 10 percent of American women fell into that category—an increase from 8 percent in 1982 (Chandra and Stephen 1998).

For most people, fecundity is not an all-or-none proposition and varies according to age. Among women it tends to increase from **menarche** (the onset of menstruation, which usually occurs in the early teens), peaks in the twenties, and then declines to **menopause** (the end of menstruation) (Kline, Stein, and Susser 1989). Male fecundity increases from puberty to young adulthood, and then gradually declines, but men are generally fecund to a much older age than are women.

At the individual level, very young girls occasionally become mothers (indeed, in 1995 there were more than 12,000 babies born to mothers under 15 years of age in the United States), whereas at the other end of the age continuum, hormone treatment of postmenopausal women suggests that a woman of virtually any age might be able to bear a child. Until the mid-1990s, the oldest verified mother had been an American named Ruth Kistler, who gave birth to a child in Los Angeles, California in 1956 at the age of 57 years, 129 days (McFarlan, et al. 1991). It is now possible, however, to induce pregnancy by implanting in a woman an embryo created from a donated egg impregnated with sperm, as was done for a 63-year-old Filipino-American woman, also in Los Angeles, in 1996. Nonetheless, older mothers are rare enough that the U.S. National Center for Health Statistics is likely to treat as erroneous any age listed on a birth certificate as being over 49. The world's verified most prolific mother was a Russian woman who in the eighteenth century gave birth to 69 children. She actually had "only" 27 pregnancies, but experienced several multiple births (Mathews, et al. 1995).

Let us put the extremes of individual variation aside. Assuming that most couples are normally fecund, how many babies could be born to women in a population that uses no method of fertility control? If we assume that an average woman can bear a child during a 35-year span between ages 15 and 49, that each pregnancy lasts a little less than 9 months (accounting for some pregnancy losses such as miscarriages), and that, in the absence of fertility limitation, there would be an average of about 18 months between the end of one pregnancy and the beginning of the next, then the average woman could bear a child every 2.2 years for a potential total of 16 children per woman (see Bongaarts 1978). This can be thought of as the maximum level of reproduction for an entire group of people.

No known society has averaged as many as 16 births per woman, and there are biological reasons why such high fertility is unlikely. For one thing, pregnancy is dangerous (in the previous chapter I noted the high rates of maternal mortality in many parts of the world), and many women in the real world would die before (if not while) delivering their sixteenth child. The other problem with the above calculation was the assumption that all couples are "normally" fecund. The principal control that a woman has over her fecundity is to provide herself with a good diet and physical care. Without such good care, of course, the result may well be lower fertility. In some sub-Saharan African countries, such as Cameroon and Nigeria, more than one third of women of reproductive age are infertile (Larsen 1995). Studies of fertility rates of U.S. blacks also suggest that the drop in fertility between the late 1880s and the early 1930s was due in part to the deteriorating health conditions of black women, and that part of the post-World War II Baby Boom among blacks was due to improved health conditions, especially the eradication of venereal disease and tuberculosis (Farley 1970; McFalls and McFall 1984; Tolnay 1989).

Naturally, disease is not the only factor that can lower the level of fecundity in a population. Nutrition seems also to play a role, and Frisch (1978) has suggested that a certain amount of fat must be stored as energy before menstruation and ovulation can occur on a regular basis. Thus, if a woman's level of nutrition is too low to permit fat accumulation she may experience **amenorrhea** and **anovulatory** cycles. For younger women, the onset of puberty may be delayed until an undernourished girl reaches a certain critical weight (Komlos 1989), as exemplified by a study in Bangladesh that found that girls who were better fed began menstruating at an earlier age than their less fortunate counterparts (Haq 1984).

The maximum level of fertility described above is thus not the level of fertility that we would expect to find in premodern societies. A slightly different concept, **natural fertility,** has historically been defined as the level of reproduction that exists in the absence of deliberate fertility control (Coale and Trussell 1974; Henry 1961). This is clearly lower than the maximum possible level of fertility, and it may be that the secret of human success lies in the very fact that as a species we have not actually been content to let nature take its course; that rather than there being some "natural" level of fertility, humans have always tried to exercise some control over reproduction:

> The genius of the species has not been to rely on a birth rate so high that it can overcome almost any death rate, no matter how high. The genius of the species is rather to have few offspring and to invest heavily in their care and training, so that the advantages of a cultural adaptation can be realized. Throughout 99 percent of hominid history, then, fertility was kept as low as it could be, given the current mortality [Davis 1986:52].

The clear implication of this idea is that the social component of human reproductive behavior is at least as important as the biological capacity to reproduce.

The Social Component

Opportunities and motivations for childbearing vary considerably from one social environment to another, and the result is great variability in the average number of children born to women. As I mentioned in the first chapter, hunter-gatherer societies were probably motivated to space children several years apart, thus keeping fertility lower than it might otherwise have been. Agricultural societies provide an environment in which more children may be advantageous, whereas modern industrial societies reduce the demand for children well below anything previously imagined in human existence.

The most famous high-fertility group is the Hutterites, an Anabaptist (opposed to childhood baptism) religious group, who live in agrarian communes in the northern plains states of the United States and the western provinces of Canada. In the late nineteenth century about 400 Hutterites migrated to the United States from Russia, having originally fled there from eastern Europe (Kephart 1982), and in the span of about 100 years they have doubled their population more than six times to a current total of more than 30,000. In the 1930s the average Hutterite woman who survived through her entire reproductive years could have expected to have given birth to more than 12 children (Eaton and Mayer 1954).

The secret to Hutterite fertility, until recently, has been a fairly early age at marriage, a good diet, good medical care, and a passion to follow the biblical prescription to "be fruitful and multiply." Also, of course, they engage regularly in sexual intercourse without using contraception or abortion, believing as they do that any form of birth control is a sin. In recent years, however, population growth has slowed among the Hutterites. Each Hutterite farming colony typically grows to a size of about 130 people. Then, in a manner reminiscent of Plato's Republic, the division of labor becomes unwieldy and the colony branches off to form a new group. Branching requires that additional land be purchased to establish the new colony, and the gobbling up of vacant land by Hutterites has caused considerable alarm, especially in Canada, where most Hutterites now live. Laws have been passed in Canada restricting the Hutterites' ability to buy land and, at the same time, new technological changes in farming methods (which the Hutterites tend to keep up with) have changed the pattern of work in the colonies. These social dynamics have apparently had the effect of raising the average age at marriage for women by as much as 4 or 5 years. Furthermore, access to modern health care has led women at the other end of their reproductive years to agree to sterilization for "health reasons" (Peter 1987). In an extensive updating of earlier analyses, Peter (1987) has found that overall fertility levels among Hutterites in 1980 were only half what they had been in 1930.

Lest you think that the decline of fertility among Hutterites ends the story of extremely high fertility in the modern world, I offer you the case of the Shipibo Indians of the eastern Amazon region of Peru. This is a population that has been studied for more than three decades by Hern (1977; 1992), whose analysis of data collected in 1964–69 and again in 1982–84 revealed a level of reproduction even higher than that of the Hutterites. Among the Shipibo, Hern found that the average age at marriage was 14 years and the average age at delivery of the first child was only 15.6 years. This is, by the way, a pattern similar to that found in the United States among Hmong refugees from the highlands of Laos. Upon arrival in the United States in the early 1980s, the Hmong continued their pattern of early and frequent motherhood, achieving levels of fertility that, at least temporarily, also exceeded the Hutterites (Weeks, et al. 1989).

In Figure 5.1, I have contrasted for you the maximum or "full" fertility rates by age that the Hutterites could have attained (but did not), leading to an average of 16 children per woman, along with the birth rates that produced their achieved levels of 11 children per woman earlier in this century. For comparison, the figure also offers you the age pattern of fertility for women in the United States and Mexico in 1995.

It is obvious that fertility levels in modern societies such as the United States and Mexico are vastly lower than among the Hutterites, who themselves had levels of reproduction well below the potential biological maximum for a group. Why the variation? There are two things that must be considered in answering that question. The first is *how* can variations in fertility be accomplished—what are the means available to limit births or encourage them? The second is *why* would people be motivated to use or not use the various means that exist to control fertility? The means of fertility regulation (the *how*) in fact mediate the relationship between motivation (the *why*) and actual fertility behavior. Once we explore how it can be done in this chapter, we will proceed in the next chapter to try to understand why a couple decides to limit fertility. For now, though, let's assume that you have already made your decision to keep your own family small. What means are available to help you do that? In the remainder of this chapter I will examine what these are, but keep in mind that not all possible avenues of fertility control are open to all people. Abortion is a good example. Although it is a legal method of birth control in the United States, it is not "available" to people who object to it for religious or personal reasons.

Figure 5.1   Maximum and Natural Hutterite Fertility Compared with Contemporary United States and Mexico Fertility Levels

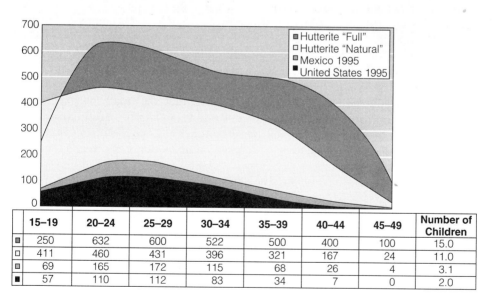

	15–19	20–24	25–29	30–34	35–39	40–44	45–49	Number of Children
▪	250	632	600	522	500	400	100	15.0
▫	411	460	431	396	321	167	24	11.0
▪	69	165	172	115	68	26	4	3.1
▪	57	110	112	83	34	7	0	2.0

Note: Birth rates are per 1,000 women.

Sources: Data for Hutterites were drawn from Warren Robinson, 1986, "Another Look at the Hutterites and Natural Fertility," Social Biology, 33:65-76; Table 6; data for Mexico are from the International Data Base of the US Bureau of the Census; data for the US are from Stephanie Ventura, Joyce Martin, Sally Curtin, and T. J. Matthews, 1997, Report of Final Natality Statistics, 1995, Monthly Vital Statistics Report, 45(11): Supplement, Table A.

How Can Fertility Be Controlled?

The means for regulating fertility have been popularly labeled (popular, at least, in population studies) the **intermediate variables** (Davis and Blake 1956). These represent 11 variables through which any social factors influencing the level of fertility must operate. Davis and Blake point out that there are actually three phases to fertility: **intercourse, conception,** and **gestation.** Intercourse is required if conception is to occur; if conception occurs, successful gestation is required if a baby is to be born alive. Table 5.1 lists the eleven intermediate variables according to whether they influence the likelihood of intercourse, conception, or gestation.

Although each of the eleven intermediate variables plays a role in determining the overall level of fertility in a society, the relative importance of each varies considerably. Bongaarts (1978) has been instrumental in refining our understanding of fertility control first by calling these variables the **proximate determinants of fertility** instead of intermediate variables, and secondly by suggesting that differences in fertility from one population to the next are largely accounted for by only four of those variables: proportion married, use of contraceptives, incidence of abortion, and involuntary infecundity (especially postpartum fecundity as affected by breast-feeding practices) (Bongaarts 1982). These variables are noted with a checkmark in Table 5.1. Although it gets ahead of our story a bit, it is worth noting that among countries with generally high birth rates, Bongaarts found that differences largely were due to lactation; countries where breast-feeding is common have somewhat lower levels of fertility than those in which breast-feeding is less common. Furthermore, the transition from high to low fertility is usually found to be accomplished with a combination of declines in breast-feeding (a seemingly contradictory pattern that I discuss later in the chapter), later marriage, increased use of contraception (as I discuss in the essay in this chapter), and increased incidence of abortion.

Bongaarts does not mean to imply, however, that the other intermediate or proximate determinants are irrelevant to our understanding of fertility among humans, only that they are relatively less important. Let us discuss each one, in the order listed in Table 5.1, considering first those variables that can affect a woman's exposure to sexual intercourse. As we review these variables, you will discover, by the way, that there is heavier emphasis on the behavior of women than of men. That is simply because if a woman never has intercourse, she will never have a child, whereas a man will never have a child no matter what he does. Of course, avoiding conception by sterilization or contraceptives can be done by either sex, but if conception occurs, it is only the woman who bears the burden of either pregnancy or abortion.

Intercourse Variables

Age at Entry into a Sexual Union   Permanent virginity is rare, but the longer past puberty a woman waits to begin engaging in sexual unions, the fewer children she will probably have because of the shorter time she will be at risk of bearing children (Variable 1 in Table 5.1). Cross-cultural data on age at menarche have revealed that women with an earlier onset of puberty tend to begin sexual activity and have their first child sooner than do women whose menarche occurs later (Udry and Cliquet 1982).

In pre-1960 American and Canadian societies, the age at marriage and the age at entry into a sexual union were essentially the same. But times have changed, and now we recognize a distinction between the two events. Since the 1970s, for example, the percentage of teenage girls aged 15 to 17 in the United States who have engaged in sexual intercourse has risen substantially over time, although it leveled off at 38 percent between 1988 and 1995 (Abma, et al. 1997). However, the average

Table 5.1   The Proximate Determinants of Fertility—the Intermediate Variables through Which Social Factors Influence Fertility

Most Important of the Proximate Determinants	Proximate Determinants or Intermediate Variables
	I. Factors affecting exposure to intercourse ("intercourse variables").
	A. Those governing the formation and dissolution of unions in the reproductive period.
	1. Age of entry into sexual unions (legitimate and illegitimate).
✔	2. Permanent celibacy: proportion of women never entering sexual unions.
	3. Amount of reproductive period spent after or between unions.
	a. When unions are broken by divorce, separation, or desertion.
	b. When unions are broken by death of husband.
	B. Those governing the exposure to intercourse within unions.
	4. Voluntary abstinence.
	5. Involuntary abstinence (from impotence, illness, unavoidable but temporary separations).
	6. Coital frequency (excluding periods of abstinence).
	II. Factors affecting exposure to conception ("conception variables").
✔	7. Fecundity or infecundity, as affected by involuntary causes, but including breast-feeding.
✔	8. Use or non-use of contraception.
	a. By mechanical and chemical means.
	b. By other means.
	9. Fecundity or infecundity, as affected by voluntary causes (sterilization, medical treatment, and so on).
	III. Factors affecting gestation and successful parturition ("gestation variables").
	10. Fetal mortality from involuntary causes (miscarriage).
✔	11. Fetal mortality from voluntary causes (induced abortion).

Sources: Kingsley Davis and Judith Blake, 1956, "Social Structure and Fertility: An Analytic Framework," Economic Development and Cultural Change 4 (April): no. 3. Used by permission; modified using information from John Bongaarts, 1982, "The Fertility-Inhibiting Effects of the Intermediate Fertility Variables," Studies in Family Planning 13:179–189.

age at first marriage has also been rising in the United States, suggesting that women are sexually active for an increasingly longer time prior to marriage.

Modern contraception has altered the relationship between intercourse and having a baby, but it remains true that an effective way to postpone childbearing is to postpone engaging in sexual activity, particularly on the regular basis implied in marriage. Historically, those societies with a later age at marriage have been the ones in which the fertility was lower, and it is still true that more traditional populations in sub-Saharan Africa and south Asia are characterized by early marriage for women and consequent higher than average levels of fertility (Singh and Samara 1996).

Permanent Celibacy   Permanent **celibacy** (Variable 2) refers to those women who never marry. It will not surprise you to learn that the greater the proportion of woman in a society who never marry, the lower the level of fertility tends to be. The idea of celibacy has not caught on, though, and the proportion of women who never marry is consistently low, never reaching even 10 percent outside of Europe.

The highest level of permanent celibacy occurs in northwestern Europe, which has historically also had among the lowest levels of fertility in the world. For decades, Ireland had the highest rate of nonmarriage in the world, reaching a level of 18 percent of women aged 40 to 44 never having been married in 1971 (Ireland Central Statistics Office 1978). However, by the late 1980s, the rate had dropped to half that level and Finland's 11 percent was higher than Ireland's 9 percent (United Nations 1992).

Time between Unions   Under normal circumstances, once a woman is involved in a sexual union such as marriage, sexual intercourse will be more or less regularly engaged in. However, not all sexual unions are permanent, since some are broken up by divorce, separation, or desertion (Variable 3a), while some are broken up by the death of a spouse (Variable 3b). During that fecund period of life when a woman has had a sexual union broken and has not entered into a new one, months or years of pregnancy risk are lost, never to be recouped. Data for the United States and nearly every other country of the world reveal that regardless of the age at which a woman first married, if her marriage has broken up, her fertility will be lower than if she had remained continuously married (see, for example, Lee and Pol 1988). While I sincerely doubt that people deliberately break up a marriage just to keep from having children, the result is the same.

Voluntary Abstinence   Voluntary abstinence (Variable 4) will clearly eliminate the risk of pregnancy within a marriage, but this is not a popular option. It is uncommon in industrialized countries except shortly after the birth of a child, when abstinence does not have much effect on fertility anyway, since childbirth is normally followed by a period of 1 or 2 months of temporary sterility (called **postpartum amenorrhea,** and referring to the fact that a woman does not ovulate during this time). In less developed societies, postpartum (that is, after childbirth) taboos on intercourse occasionally extend to several months or even to a few years in societies in which intercourse is forbidden while a mother is nursing a child. The rationale is based generally on superstitions that intercourse will somehow be harmful to either mother or child (Davis and Blake 1956) or on the generally correct notion that a pregnancy will be harmful to the nursing child because it will reduce the quantity of mother's milk. Voluntary abstinence after childbirth may help to keep birth rates below the potential maximum in many premodern societies by lengthening the interval between pregnancies.

In areas of the world where more sophisticated birth control techniques are not regularly available, abstinence may be an effective means of fertility limitation for highly motivated people. For example, in 1970 Brandes (1975) studied a small Spanish peasant village in which there was considerable pressure to limit family size to two children. He relates the story of a 35-year-old man with 5 young children who "is constantly greeted with the refrain, 'iAtate al catre!'—Tie yourself to the bedpost!" Brandes goes on to explain:

> Through a combination of coitus interruptus and abstention, most villagers manage to keep births down. Though they openly complain about the frustrations which these methods entail, they are willing to make the sacrifice in order to assure their ability to provide their children the best possible opportunities. Those husbands who cannot exercise control are seen as somewhat animalistic [1975:43].

A similar story is related by Reining and her associates (1977) about behavior in a small Mexican village. In examining the life history of a 37-year-old woman with 5 children, it was discovered that the woman "doesn't want to have any more children, for the benefit of the children she already has, and in order not to deplete the economic resources of the family." But since her husband does not accept the use of contraceptives she has refused to sleep with him.

Thus, if the motivation exists to limit family size, peasants will do so, turning to the time-honored methods of abstinence and coitus interruptus (discussed later in the chapter). Even in the 1980s, an American woman wrote to "Dear Abby" explaining that "the most effective method yet found for birth control is a large dog sleeping in the middle of the bed on top of the covers. Try it. It works. That's what we've been doing for 17 years" (Universal Press Syndicate 1982).

Involuntary Abstinence Involuntary abstinence (Variable 5) could be a result of either impotence or involuntary separation. It is not easy to tell how important the role played by impotence is in involuntary birth control. On the other hand, we do know that temporary separations are fairly common in many occupations, such as sales and transportation, and for various other reasons—separate vacations, hospitalization, military duty, and so on. In underdeveloped countries, where migratory labor is more common than in the United States, we might also expect to find involuntary abstinence to be lowering fertility somewhat. Contrary to the typical pattern in the United States, in many parts of the world migratory labor involves the laborer only, not his or her family. The *Bracero* program in the United States, which brought thousands of Mexican men into the United States in the 1950s and early 1960s to work on American farms, is a typical example. Many of the men were married and left their wives and children behind. Some countries, such as Kuwait, which are heavily dependent upon guest workers, actively discourage workers from bringing a spouse along, precisely because they do not want to absorb the expense of the children that will be produced (United Nations 1988).

Coital Frequency   Within marriage (or any sexual union) the regularity of sexual intercourse (Variable 6) will influence the likelihood of a pregnancy, particularly if no contraception is used. In general, the more often a couple has intercourse, the more likely it is that the woman will get pregnant, primarily because it increases the chance that intercourse will occur during the fertile period of the menstrual cycle, as shown in Figure 5.2. It is probably true that if a woman engages in sexual intercourse with the same man more than once a day, every day, the risk of pregnancy is lower than expected because the male's sperm does not have the opportunity to fully mature between ejaculations. This rarely appears to be a problem, however.

In a study done in Lebanon, Yaukey (1961) found that during the first year of marriage, among women who were not using contraceptives, the more frequently intercourse was engaged in, the less time it took for a woman to get pregnant. Studies in the United States have produced similar results. MacLeod and Gold (1953). found that 51 percent of married women engaging in intercourse 3 times per week

Figure 5.2   Conception Occurs Near the Middle of the Menstrual Cycle

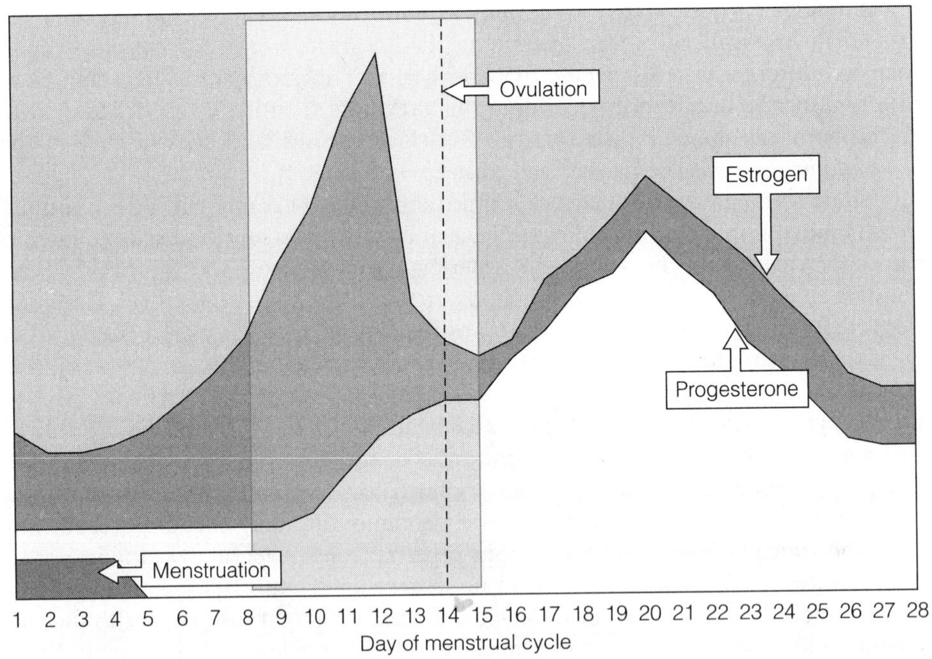

Note: In this model of a typical menstrual cycle, days 8 through 15 represent the time of the cycle when the chance of conception is greatest. Sperm can survive for several days in the uterus, so intercourse during the few days prior to ovulation (day 14 of the typical 28-day cycle) has the highest chance of producing a fertilized egg and thus conception. Since an unfertilized ovum dies very quickly, the risk of conception drops off quickly after ovulation.
Sources: Robert Hatcher, et al., 1994, Contraceptive Technology, Sixteenth Revised Edition (New York: Irvington Publishers), and Wilcox et al., 1995, "Timing of Sexual Intercourse in Relation to Ovulation," The New England Journal of Medicine 333:1517-1521.

were pregnant within 6 months, compared with only 32 percent who engaged in intercourse only once per week. Also in the 1950s, Kinsey found that among stable married couples, coital frequency decreased steadily as age increased. He found that women aged 41 to 45 had sexual intercourse only half as often as women aged 21 to 25. Of course, that was in an era before the modern contraceptive revolution, when couples may have deliberately slowed down the pace of sexual activity precisely to lower the risk of conception. Married women now exhibit little change in sexual activity with age (Bachrach and Horn 1988; Jasso 1985), because high levels of contraception and voluntary sterilization have dramatically loosened the connection between intercourse and pregnancy.

Conception Variables

Breast-feeding   Even if a woman is regularly sexually active, she will not necessarily conceive a child because she may be involuntarily infecund, as I mentioned previously (Variable 7), and one of the major ways in which human societies differ in this regard is with respect to breast-feeding practices.

Breast-feeding prolongs the period of postpartum amenorrhea and suppresses ovulation, thus producing in most women the effect of temporary subfecundity. In fact, nature provides the average new mother with a brief respite from the risk of conception after the birth of a baby whether she breast-feeds or not; however, the period of infecundity is typically only about 2 months among women who do not nurse their babies, compared with 10 to 18 months among lactating mothers (Konner and Worthman 1980). Research suggests that stimulation of the nipple during nursing sets up a neuroendocrine reflex that reduces the secretion of the luteinizing hormone (LH) and thus suppresses ovulation (Jones 1988). A study in Indonesia found that women whose babies nurse intensively (several nursing bouts per day of fairly long duration) delay the return of menses by an average of 21 months, nearly twice the delay of those women who breast-feed with low intensity (Jones 1988). On the other hand, the cessation of lactation signals a prompt return of menstruation and the concomitant risk of conception in most women (Prema and Ravindranath 1982).

Although breast-feeding is a natural contraceptive and otherwise keeps fertility lower than it would otherwise be, the percentage of women who breast-feed their babies has tended to decline with modernization, and so we find the seemingly contradictory fact that breast-feeding declines as fertility declines. In the earlier decades of modernization, bottle-feeding was seen as preferable to breast-feeding because it gave the mother more flexibility to work and to have someone else care for the child, and there was a certain cachet attached to being able to bottle-feed rather than engaging in the more "peasant" activity of breast-feeding. Over the past few decades there has been concern that women in less developed nations were abandoning breast-feeding in favor of bottle-feeding. In the absence of some method of contraception, a decline in breast-feeding would, of course, increase fertility by spacing children closer to one another. Closer spacing poses a threat to the health of both mother and child, and bottle-feeding also is likely to raise the infant death rate, since in the less developed nations bottle-feeding often is accompanied by watered-down formulas that are less nutritious than a mother's milk. Bacteria growing in unsterilized bottles and reused plastic liners can also lead to disease, especially diarrhea, which is often fatal to infants.

In response to these concerns, the World Health Organization approved a voluntary code in 1981 to regulate the advertising and marketing of infant formula. However, baby food companies lobbied strongly against the code, the United States opposed it, and a year later only 6 percent of the countries who voted for it had actually adopted it (Adelman 1982). The pressure continued to mount, however, and in 1991 the Swiss firm Nestlé finally decided to limit its supply of free baby formula to third world hospitals (Freedman 1991), and other baby formula suppliers quickly followed suit. Of some interest is the publication in that same year of a study suggesting that the marketing practices of the infant food industry may have a smaller impact on a mother's decision to breast- or bottle-feed than is commonly assumed (Stewart, et al. 1991).

The women most likely to lead the movement back to breast-feeding in any given country are, somewhat ironically, the better educated (Akin, et al. 1981; Hirschman and Butler 1981), the very same women whose fertility is apt to be kept low by deliberate use of other means of contraception rather than by the influence of lactation. In the United States in 1995, 81 percent of women with a college degree breast-fed their most recent baby, compared with less than half of the mothers with only a high school education or less (Abma, et al. 1997).

Contraception   The first thing we usually think of when we consider ways to control fertility is **contraception** (Variable 8). There are so many different types of contraceptives, not to mention new research currently being conducted on contraceptives, that it is useful to categorize methods before discussing them. In Table 5.2, I have divided contraceptive methods into three main categories, according to whether they are primarily female, male, or couple methods. Under each of those categories I have listed methods according to whether their action is primarily barrier, chemical, natural, or surgical. Since individuals or couples often combine methods, this classification scheme is not mutually exclusive, but it is illustrative. Surgical contraception is actually captured by the category of voluntary infecundity (Variable 9), but it has become so common that I have included it in the table of contraceptive methods.

Female Barrier Methods   Barrier methods include the **diaphragm; cervical cap;** a disposable vaginal **sponge;** the intrauterine device (IUD); the female condom; and vaginal **spermicidal** foams, jellies, suppositories, and film.

The diaphragm is a rubber disk that is inserted deep into the vagina and over the mouth of the uterus sometime before sexual intercourse. Used alone, it is a fairly ineffective physical barrier. As with all barrier methods, it is normally used in conjunction with a spermicide which is spread in and around the diaphragm before insertion into the vagina. The spermicide acts to kill the sperm, and the diaphragm operates to keep the spermicide in place, preventing the passage of sperm into the uterus and on into the fallopian tubes where fertilization occurs. Since sperm can live in the vagina for several hours, it is necessary for a woman to leave the diaphragm in place for awhile after intercourse. Used in this way, the diaphragm can be very effective, and until the 1960s when oral contraceptives were placed on the market, the diaphragm (invented in 1883) was a fairly common method of birth control for married women. Because it requires a great deal of forethought, it is obviously not useful for those occasions, especially outside of marriage, when intercourse occurs spontaneously.

Variations on the diaphragm theme have appeared in recent years. One of these is a vaginal sponge called "Today" that contains a spermicide and fits over the cervix. The sponge works by releasing a spermicide, providing a barrier between the sperm and the cervix, and by trapping sperm in the sponge. Studies suggest that it is about equally effective as the diaphragm. Unfortunately, the sole manufacturer of the sponge, Whitehall-Robins Healthcare (a division of American Home Products), announced in 1995 that they were going to stop producing the contraceptive because they could not afford to upgrade their manufacturing plant to meet new Food and Drug Administration safety standards and, as of this writing, they have not changed their mind. Cervical caps, which are similar to the diaphragm, have been available in Europe for several decades, and a model called the Prentif Cavity Rim cervical cap is now widely available in the United States.

A female condom called Reality went on sale in the United States in the fall of 1994 after a series of tests in England (where the device is manufactured). It is a

Table 5.2   Contraceptive Methods

Primary Characteristic	Primary Use:		
	Female	Male	Couple
Barrier	Diaphragm	Condom	
	Cervical cap		
	Vaginal sponge		
	Female condom		
	Vaginal spermicides		
	IUD		
Chemical			
Precoital	Pill		
	Mini-pill		
	Implants		
	Injectables		
Postcoital	Morning-after pill		
	RU486		
Natural	Breast-feeding	Withdrawal	Fertility awareness
			Oral/anal sex and other forms of incomplete intercourse
Surgical	Tubal ligation	Vasectomy	

lubricated polyurethane sheath that is inserted like a diaphragm into the vagina just before intercourse. The marketability of the female condom is based not only on its protection against pregnancy, but also on its ability to permit a woman to limit her exposure to AIDS and other sexually transmitted diseases. Since its introduction it has been distributed widely in developing countries where women are attracted to the fact that they no longer have to worry about whether a male sex partner will have a condom or, even if he does, will use it properly.

The IUD was first designed in 1909, but it was not widely manufactured and distributed until the 1960s. At that time it was the contraceptive technique that many family planners believed would bring an end to the world's population explosion. Its success has been, however, much less spectacular than that, and although it is still used by millions of women around the world, it has not enjoyed the wholesale acceptance that many had hoped. No one knows for certain how it works to prevent a birth, but it is generally believed that since it is a foreign object, it sets up a chemical reaction inside the uterus that may destroy the fertilized egg and/or prevent implantation of a fertilized egg from taking place (Sivin 1989). Although it comes in several different shapes and can be made of various materials, the IUD most commonly used is a nylon plastic coil.

About 2 or 3 percent of the women who have an IUD inserted expel it, but for the remainder the IUD theoretically will stay in place for long periods of time and is very effective in preventing pregnancy. In the less developed areas of the world, however, high proportions of women who have an IUD inserted choose to have it removed even if they do not want to get pregnant. The principal reasons are infections and excessive bleeding caused by the IUD, although fear and superstition may also be causes.

One brand of IUD, the Dalkon Shield, was responsible for the deaths of at least 13 women prior to its removal from the market in 1975. Most IUDs have a string attached to them to aid in removal and serve as a sign that the device is in place; the string of the Dalkon Shield turned out to be a "ladder for bacteria" (Roberts 1978) and caused infection in some users. Lawsuits stemming from those cases caused virtually all IUD manufacturers to withdraw the device from the U.S. market in 1986. Since then only two IUDs have been available in the United States—TCu-380A (known as Paragard) and the progesterone T device (known as Progestasert) (Hatcher, et al. 1994). Other IUDs continue to be available elsewhere in the world. Indeed, Poston (1986) estimates that 70 percent of all IUD use occurs in China.

Female Chemical Methods   The oral contraceptive, or "the pill" as it is popularly known, has revolutionized birth prevention for millions of women all over the world. The pill is a compound of synthetic hormones that suppress ovulation by keeping the estrogen level high in a female. This prevents the pituitary gland from sending a signal to the ovaries to release an egg. In addition, the progestin content of the pill makes the cervical mucus hostile to implantation of the egg if it is indeed released. The combination pill is virtually 100 percent effective when taken as directed. It has estrogen and progestin in each pill and typically is taken for 21 days, followed by 7 days when it is not taken and the woman has a menstrual flow. The sequential, or triphasic, pill varies the dosages of estrogen and progestin over the menstrual cycle (see Figure 5.2), trying to mimic the "natural" change in a woman's hormones while still protecting against pregnancy (Hatcher, et al. 1994).

In the 1960s and 1970s the pill was the method of choice for women of all ages, and it continues to be the most popular nonsurgical method of birth control in the United States (see Table 5.3). All things considered, the pill would be just about the perfect solution except for the problem of side effects. When first taking the pill, women often report nausea, breast tenderness, and spotting, although these problems usually do not persist. More important by far are the risks of death that might be associated with the pill. Although the evidence is far from conclusive, it appears that oral contraceptives can aggravate problems such as high blood pressure, blood clotting, diabetes, and migraine headaches. Women with histories of these problems are normally advised not to use the pill. Much of the early scare about increased risk of cancer, however, has subsided as researchers have been able to lower the dosages of chemicals in the pills, and as they have learned that oral contraceptives do not, in fact, increase the risk of breast cancer, even after prolonged use (Wharton and Blackburn 1988). More encouraging are the studies suggesting that using the pill may actually reduce a woman's risk of ovarian cancer, endometrial cancer, molar

Table 5.3 Sterilization Is the Most Commonly Used Method of Fertility Control among Women in the United States (1995)

	15–44	15–19	20–24	25–29	30–34	35–39	40–44
Not currently using contraception	35.8	70.2	36.6	30.7	27.3	27.1	28.5
Among those using contraception:							
Surgical methods:							
Female sterilized	27.7	0.3	3.9	17.0	29.4	40.9	49.8
Male sterilized	10.9	0.0	1.1	4.5	10.5	18.7	20.3
Nonsurgical methods							
Pill	26.9	43.6	52.2	39.0	28.5	11.1	5.9
Implant	1.4	2.7	3.8	2.0	0.7	0.3	0.1
Injectable	3.0	9.7	6.2	4.2	1.8	1.1	0.3
IUD	0.8	0.0	0.3	0.7	0.8	0.1	1.3
Diaphragm	1.9	0.0	0.6	0.9	2.3	0.3	2.7
Condom	20.4	36.6	26.3	24.2	18.4	16.9	12.3
Female condom	0.0	0.0	0.2	0.0	0.0	0.0	0.0
Periodic abstinence	2.3	1.3	0.9	1.7	3.2	2.9	2.5
NFP	0.3	0.0	0.2	0.3	0.4	0.5	0.3
Withdrawal	3.1	0.4	3.3	3.8	2.9	3.2	2.0
Other methods	1.6	0.1	1.4	1.7	1.8	1.2	2.5

Note: Although the pill is the most popular form of fertility control for younger women, from ages 30 on, sterilization (both male and female) is the most widely used method.

Source: J. Abma, A. Chandra, W. Mosher, L. Peterson, L. Piccinino, 1997, "Fertility, Family Planning, and Women's Health: New Data From the 1995 National Survey of Family Growth," National Center for Health Statistics, Vital Health Statistics 23(19): Table 41.

pregnancies, fibroids, and ovarian cysts (Hatcher, et al. 1994). One study has even concluded that a woman who takes the pill for 5 years before the age of 30 can, in effect, add 4 days to her life expectancy (Fortney, Harper, and Potts 1986).

The pill has been embroiled in a storm of controversy about its potentially harmful side effects ever since its introduction. It is the most widely studied drug ever invented, and its principal inventor, Carl Djerassi of Stanford University, is understandably defensive. In 1981 he wrote:

> The point that these opponents of the Pill really miss is that the Pill as well as all other methods of fertility control should be available for any woman who is willing and able to use them given her particular circumstances. . . . The reality is that for many women throughout the world, the Pill is the best contraceptive method currently available [1981:47].

Researchers and manufacturers have been particularly active in the development of progestin-only contraceptives, including the "mini-pill," Norplant implants, and Depo-Provera injectables. The mini-pill, which contains only progestin, may not prevent ovulation, but rather seems to work primarily by producing a hostile environment for sperm and egg. It is taken every day without stopping. Mini-pills are especially useful for breast-feeding women, because they offer additional protection against pregnancy without affecting lactation.

The pill represents one way of introducing synthetic hormones into the body, and implants represent another method. Norplant is a progestin-only implant developed by the Population Council in New York. It is a one eighth inch long silicone rubber capsule containing a small amount of levonorgestrel, a synthetic progestin. The capsule is implanted just beneath the skin of a woman's upper arm and releases a tiny amount of progestin every day for up to 5 years before needing to be replaced. After years of field tests and an endorsement by the World Health Organization, Norplant was approved by the Food and Drug Administration in December 1990 for use in the United States.

Another variation on the theme is for a woman to receive a periodic injection of chemicals. Without doubt the most widely debated of these injectable contraceptives is Depo-Provera, a highly effective contraceptive that requires a woman to be injected with a massive dose of synthetic progestin every 3 months. Although it had been tested in the United States since 1968 and has been approved for contraceptive use in at least 80 countries, The U.S. Food and Drug Administration delayed its approval for use in the United States until 1992.

Another category of chemical contraceptive encompasses those methods that essentially back up all of the previously discussed techniques. **Postcoital contraception,** or "emergency contraception" as it is more popularly known, is designed to avert pregnancy within a few days after intercourse (Ellertson 1996). The so-called "morning-after" pill as currently used is actually four of the combination birth control pills marketed in the United States as Ovral (two doses of two pills taken 12 hours apart). Clinic tests in Europe and in several developing countries suggest that this regimen has a very low failure rate (Hatcher, et al. 1994).

The more controversial of the emergency contraceptive methods is RU486 (technically known as mifepristone), which interferes with the hormone progesterone, causing the uterine lining to be sloughed off and any embryo thus to be expelled. It appears to be most effective when taken within 10 days of intercourse. The drug was invented in France and has been approved for use in several countries, including China. It has been tested throughout the United States and has received conditional approval from the U.S. Food and Drug Administration to be made available to health care providers.

Female Natural Methods   I discussed breast-feeding earlier in the context of involuntary infecundity. In the 1980s, however, breast-feeding was elevated to the status of a natural contraceptive, especially through the issuance of the so-called "Bellagio Consensus." In 1988 a group of experts met in Bellagio, Italy, to review the evidence regarding the use of breast-feeding as a deliberate method of preventing conception. "The consensus of the group was that the maximum birth spacing effect of breastfeeding is achieved when a mother 'fully' or nearly fully breastfeeds and remains amenorrheic. When these two conditions are fulfilled, breast feeding provides more than 98 percent protection from pregnancy in the first six months" (Kennedy, Rivera, and McNeilly 1989). Other analyses have confirmed the important fertility-reducing impact of lactation for those mothers who are willing to breast-feed on demand, whenever or wherever that might be (Guz and Hobcraft 1991).

One natural method that is not listed in Table 5.2 is vaginal douching. There have been references to douching throughout recorded history, stretching back to ancient Egypt (Baird, et al. 1996). Over time it has been recommended as a means of treating specific gynecological conditions and also as a contraceptive, on the theory that washing sperm out of the vagina right after intercourse (the "dash for the douche") might prevent conception. Unfortunately for the one doing the douching, the sperm take only about 15 seconds to travel through the vagina into the cervical canal, so the effectiveness of douching is very limited. There is also some evidence that douching may increase the risk of pelvic inflammatory disease (Hatcher, et al. 1994).

Male Methods   Men have very few options available to them in contraceptives. In fact, they are basically limited to the **condom.** This is a rubber or latex sheath inserted over the erect penis just prior to intercourse. During ejaculation the sperm are trapped inside the condom, which is then removed immediately after intercourse while the penis is still erect, to avoid spillage. The condom is very effective, and when used properly in conjunction with a spermicidal foam, it is virtually 100 percent effective. The condom is also useful in preventing the spread of sexually transmitted diseases, including venereal disease and, more popularly, AIDS. Although use of the condom dropped off considerably during the 1960s and 1970s, by 1995 (the most recent year for which data are available) it had regained its place just after the pill in popularity among younger Americans (Piccinino and Mosher 1998) (see Table 5.3).

Withdrawal is also an essentially (although not exclusively) male method of birth control. It has a long history (for example, it is referenced in the Bible), but its effectiveness is relatively limited. It is, in fact, a form of incomplete intercourse (thus its formal name "coitus interruptus") because it requires the male to withdraw his erect penis from his partner's vagina just before ejaculation. The method leaves little room for error, especially since there may be an emission of semen just before ejaculation, but it is one of the more popular methods historically for trying to control fertility.

Couple Methods   Until recently the principal method of contraception to require couple cooperation was rhythm. Users of this method, however, are jokingly referred to as parents. Recent developments have altered this view.

In an era of highly sophisticated chemical technology, the search for alternative natural methods of doing things has been accelerating, and methods of fertility control have not been exempted from this search. Indeed, new couple-oriented methods of periodic abstinence have emerged as extremely effective means for limiting births, at least among highly motivated couples. No longer does Natural Family Planning (NFP) mean the rhythm method of periodic abstinence. It now refers more broadly to a set of fertility awareness methods, including (1) the ovulation method, (2) the basal body temperature method, and (3) the combination of ovulation and temperature methods and other new tests into what is known as the symptothermal method of fertility control.

The ovulation method (OM) is based on the discovery that changing estrogen levels produce changes in the consistency of a woman's cervical mucus as ovulation approaches. With appropriate training most women can easily learn to identify these mucus changes and thus determine when ovulation is about to occur (Keefe 1962). In a typical woman the vaginal area feels dry immediately after menstruation, followed in a few days by the secretion of a small amount of cervical mucus. As ovulation approaches, the mucus changes to a slippery, raw eggwhite–like consistency. This is an indication that ovulation is about to take place (McCarthy 1977). Research suggests that when used properly, the ovulation method is virtually as effective as the pill (Trussell and Grummer-Strawn 1990), but proper use requires considerable self-control.

The basal body temperature method gives evidence that ovulation has already occurred. Basal body temperature (BBT) is a person's resting temperature taken the same time daily under similar conditions (such as shortly after awakening in the morning). Prior to ovulation, the BBT is relatively low; however, it typically begins to shift upward as ovulation occurs and remains high until the next menses. Most experts feel that 3 days of a sustained upward shift in the BBT indicates that ovulation has occurred. Accurate recording of temperature is important in using this method, because the temperature shift that we're talking about is generally no more than 0.5 degrees Fahrenheit.

The symptothermal (S-T) method may be more effective in practice than either OM or BBT, because S-T combines evidence about the timing of ovulation from both sources. OM tells you when ovulation is about to occur, while BBT tells you that it has occurred. Since the secret of success with NFP is abstaining from intercourse during the fertile (ovulatory) period, it is obvious that precise identification of the fertile period is crucial. For couples using fertility awareness methods, Unipath Corporation (a subsidiary of Unilever) has a device they call Persona, which uses an at-home urine test to monitor changes in hormone levels to spot the hormonal surge that usually signals the onset of ovulation (see Figure 5.2). So far the product is sold only in Europe. Quidel Corporation in the United States is working on a similar product, except that theirs detects when a woman has entered the in-fertile phase of the menstrual cycle. This would give couples a clear indication that it is safe to resume sexual intercourse without risk of conception. Presumably the use of both tests would provide a way of bracketing the fertile period when inter-course should be avoided.

Despite the apparent effectiveness of the newer methods of NFP, most women who practice periodic abstinence still use the old calendar rhythm method (Abma, et al. 1997; Sheon and Stanton 1989), which continues to be plagued by a high rate of accidental pregnancy.

Other methods of birth control that explicitly require cooperation between both partners are mutual masturbation and oral-genital sex. In Davis and Blake's 1956 article, these forms of incomplete intercourse were listed as "perversions," but in more recent years, with a significant change in openness about sex and with the in-troduction of best-selling "how-to" manuals, they have become more openly ac-ceptable techniques for engaging in sexual activity. These methods are sometimes practiced in combination with coitus interruptus.

New Methods Being Tested   The increasing worldwide demand for contracep-tion encourages the development of new methods, and there are female, male, and couple methods currently being tested in various parts of the world. For example, vaginal rings are small devices that release either levonorgestrel (the active ingredi-ent in Norplant) or progesterone when inserted in the vagina. Although they are a physical object in the vagina, they can be removed prior to intercourse, since the progestin will already have been absorbed into the body. Vaginal rings are still in the experimental stage, but they appear to offer a high rate of effectiveness with few side effects (Hatcher, et al. 1994).

An additional progestin-only contraceptive being tested is Capronar, developed by the Research Triangle Institute in North Carolina. It also uses the same chemi-cals as Norplant but, unlike Norplant, it does not have to be removed. The hor-mones are encased in a biodegradable capsule, which is inserted under the skin (usu-ally the arm, as with Norplant) and emits pregnancy-protecting hormones for a period of up to 18 months.

Chinese scientists have developed a birth control pill for men derived from cot-tonseed oil and called gossypol. It appears to work by temporarily interfering with the production of sperm. Serious side effects have been observed in some men, and the method is still being tested in several parts of the world, including the United States. Research in the United States on male contraceptives has focused on the use of luteinizing hormone–releasing hormone (LH-RH), a substance that can suppress sperm production and reduce the level of the male hormone testosterone. However, studies of hormonal contraceptives have shown fairly consistently that these drugs may adversely affect a wide range of sexuality measures, including frequency of orgasm and intensity of desire. Nonetheless, research is continuing and in 1990 the Population Council (which had developed Norplant) was authorized by the U.S. Food and Drug Administration to begin tests of an LH-RH male birth control vaccine.

Primitive or "Traditional" Methods   It is probable that at least some people in most societies throughout human history have pondered ways to control fertility

(Himes 1976). Abstinence and withdrawal are the most ancient of such primitive means, but there is some historical evidence that various plants were used in the premodern world to produce "oral contraceptives" and early-stage abortifacients (Riddle 1992). We do not know much about the actual effectiveness of such methods, but they were almost certainly far less effective in preventing conception or birth and far riskier for a woman's health than are modern methods. The Shipibo Indians in the Amazon basin of Peru, whose extraordinarily high fertility was noted earlier, nonetheless (or maybe because of this) have knowledge of an herb that is believed to have contraceptive powers. It is made from the root of the cyperus plant (a bog plant, not the same as the cypress), and they believe that it induces temporary sterility (Maxwell 1977). Its real level of effectiveness, however, appears to be nil (Hern 1976).

How Effective Are Different Contraceptive Techniques?   Contraceptive effectiveness refers to how well a contraceptive works to prevent a pregnancy. There are two ways to measure this. The first is to look at **theoretical effectiveness,** meaning the percentage of the time that a method should prevent pregnancy after sexual intercourse if everything else is perfect. Thus, if we say that oral contraceptives are 100 percent effective, we mean that if a woman is in normal health and if she takes the pills exactly as directed, she will never get pregnant. Theoretical effectiveness is very difficult to assess and, in addition, it ignores the human failure that can occur with any method.

The more usual measure of a contraceptive's effectiveness is called **use-effectiveness,** which measures the actual pregnancy performance associated with a particular method. There are several specific ways by which use-effectiveness can be measured. The most common is called the Pearl method or the pregnancy-failure rate, the number of contraceptive failures per 100 woman-years of contraceptive exposure. For example, if 200 women used the pill for a year, they have had a combined total of 200 woman-years of exposure. If four of the women became pregnant while using the pill, then the pregnancy rate is two. In other words, there were 2 failures (accidental pregnancies) for every 100 woman-years of exposure. The lower the failure rate, the more effective the contraceptive.

Use-effectiveness combines two kinds of failures: (1) method failure, such as when a woman gets pregnant even though she has an IUD in place, and (2) user failure, such as when a woman on the pill goes on a weekend trip and leaves her pills at home accidentally, or when a man is not sufficiently motivated to always use a condom.

The most precise way to measure use-effectiveness is to study couples who are using a specific method of contraception for some period of time (such as a year) and see what proportion of the couples wind up with an accidental pregnancy. Such data are presented in Table 5.4. If a couple were to use no method at all and rely simply on chance to avoid a pregnancy, there is an 85 percent chance of a pregnancy during the year. Implants and injectables improve on those odds considerably, with failure rates of .09 percent and 0.03 percent, respectively. The data in Table 5.4 show very clearly that, regardless of the particular method chosen, even the least effective contraceptive dramatically reduces the odds of a pregnancy compared with not using any method at all. Data from a variety of surveys also have shown clearly that the more highly motivated couples (that is, those who do not want any more pregnancies, as opposed to those merely spacing their children) have higher use-effectiveness rates than the less motivated, regardless of the method chosen.

Voluntary Infecundity   The last of the conception variables that can be manipulated to achieve low fertility is voluntary infecundity or **sterilization** (Variable 9). These methods are sometimes known as **surgical contraception.** For females, voluntary sterilization procedure spans a spectrum from the removal of the ovaries (the most drastic) to a tubal ligation (the least drastic), but only the latter is usually employed as means of contraception. The other more extreme types of surgical sterilization are usually required for health reason. Removal of the ovaries (called oophorectomy) not only removes all egg follicles and prevents further ovulation but also changes the hormonal balance of a woman. This operation is fairly serious and is generally done for medical reasons unrelated to a desire not to have children. The next most serious operation is a hysterectomy, the removal of the uterus. This too requires major surgery. Each year more than 400,000 American women have a hysterectomy, making it the fourth most frequently performed operation on women. Nevertheless, since most hysterectomies are performed on women over 35, its use as a birth control method is limited.

The type of female sterilization most often recommended and utilized for contraceptive purposes are minilaparotomy and laparoscopy (generically labeled as

Table 5.4   Theoretical and Use-Effectiveness for Selected Contraceptive Techniques

Method	Theoretical Effectiveness	Use-Effectiveness (Percentage of Couples Experiencing an Accidental Pregnancy During the First Year of Use)
		Typical Findings
None	Chance	85.0
Implants (Norplant)	Virtually no failures	0.09
Injectables (Depo-Provera)	Virtually no failures	0.3
IUD	Virtually no failures	2.0
Pill	Virtually no failures	3.0
Male condom	Very few failures	12.0
Diaghragm/cap	Some failures	18.0
Withdrawal	Some failures	19.0
Fertility awareness	Some failures	20.0
Female condom	Some failures	21.0
Spermicides	Some failures	21.0
Vaginal sponge	Some failures	24.0

Source: Use-effectiveness data from R. Hatcher, et al., 1994, Contraceptive Technology, Sixteenth Revised Edition (New York: Irvington): Table 27-1.

tubal ligation). A minilaparotomy involves an incision just above the pubic hairline, through which a surgeon grasps the fallopian tubes and "ties" them off with rings or by some other method. Laparoscopy utilizes a smaller incision, through which a laparoscope (a tiny viewing device) is inserted to assist the surgeon to "tie" the tubes. In both cases, the result is that an egg is released normally, but it is prevented from entering the uterus, and sperm are prevented from reaching the egg. The menstrual cycle continues as usual, except that the egg produced is absorbed into the body. Biologically there is no effect on a woman's sexual response, although psychologically the virtual absence of a reason to fear pregnancy probably enhances a woman's sexuality.

As is true for females, there are drastic as well as simple means of sterilization for males. The drastic means is castration, which is removal or destruction of the testes. This generally eliminates sexual responsiveness in the male, causing him to be impotent (incapable of having an erection). However, a vasectomy does not alter a male's sexual response. A vasectomy involves cutting and tying off the vasa deferens, which are the tubes leading from each testicle to the penis. The male continues to generate sperm, but they are unable to leave the testicle and are absorbed into the body. Vasectomy is quite popular in the United States, as you can see in Table 5.3. In fact, at ages 35 to 44 it is the second most common method of fertility control among American couples, trailing only female sterilization. As early as 1982 sterilization of either sex had stolen the lead away from the pill as the method of choice among married women of reproductive age. By 1995 nearly two thirds of women aged 35 to 44 who used some method of contraception relied on sterilization (of themselves or their partner).

Gestation Variables: Abortion

Assuming that conception has occurred, a live birth may still be prevented. This could happen as a result of involuntary fetal mortality (Variable 10), which is either a spontaneous **abortion** (miscarriage) or a stillbirth. More important for our discussion, though, is voluntary fetal mortality or induced abortion (Variable 11). Abortions became legal in Canada in 1969 and in the United States in 1973, and they are legal in all three of the world's most populous nations (China, India, and the United States) as well as in Japan and virtually all of Europe, except Ireland (although Irish women may travel to England for an abortion). Worldwide, the demand for abortion remains high, even in places where it is not legal, and it has been estimated that one of three pregnancies in the world may end in abortion (Henshaw 1990), with China and Russia leading the list. During the 1970s, abortion in many countries "changed from a largely disreputable practice into an accepted medical one, from a subject of gossip into an openly debated public issue" (Tietze and Lewit 1977:21). Figure 5.3 provides illustrative data.

Abortion is perhaps the single most often used form of birth control in the world and is resorted to when methods of contraception are otherwise not available or have failed. Abortions have played a major role in fertility declines around the world, and they are an important reason for the continued low birth rate in many countries, including the United States and Canada. They have played a role in Mexico's fertility decline, as well, despite the fact that elective abortion is not legally available to women in Mexico.

The number of legally induced abortions reported in the United States increased steadily from 1973 until 1980, since which time the number has leveled off. The abortion ratio (the number of abortions per 100 live births) has also remained quite constant at around 35 abortions per 100 live births since 1980 (Koonin, et al. 1996). Abortion ratios are highest for unmarried women, teenagers, nonwhites, and women with no prior pregnancies, a profile that has not changed much over time. The abortion rate (the number of abortions per 1,000 women aged 15 to 44) has remained fairly constant in Canada since abortion laws were liberalized, and Canadian women are less than half as likely as women in the United States to use abortion (Millar, Wadhers, and Henshaw 1997). Since the fertility rate in Canada is slightly lower than in the United States, the implication is that Canadian women are more efficient users of other methods of contraception than are women in the United States. In Mexico the public appears to favor the broader legalization of

Figure 5.3   The Number of Abortions per 100 Pregnancies Varies Throughout the World

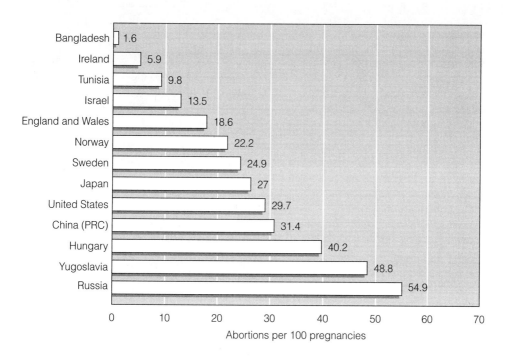

Note: The abortion ratio (abortions per 100 pregnancies) is low in predominantly Muslim Bangladesh and predominantly Catholic Ireland, while it is high in Russia and other eastern European nations.

Source: Adapted from S. Henshaw, 1990, "Induced Abortion: a World Review, 1990," *Family Planning Perspectives* 22(2):76–89, Table 2.

THE IMPACT OF CONTRACEPTION ON FERTILITY

Of all proximate determinants of fertility, contraception is by far the most important. Without modern contraception, fertility can be maintained at levels below the biological maximum, perhaps even as low as three children per woman, but it is very difficult to achieve low levels of fertility without a substantial fraction of reproductive-age couples using some form of modern fertility control. The exact nature of the relationship between fertility and contraception has been measured with increasing accuracy and frequency over the past few decades, beginning with the Knowledge, Attitude, and Practice (KAP) surveys of the 1960s and 1970s and followed by the World Fertility Surveys and the Contraceptive Prevalence Surveys of the 1970s and 1980s and then by the Demographic and Health Surveys of the 1980s and 1990s. In North America, these kinds of data have come to us through the National Surveys of Family Growth in the United States, through the General Social Survey in Canada, and through the National Fertility Studies in Mexico.

Contraceptive utilization is usually measured by calculating the rate of **contraceptive prevalence,** which is the percentage of "at risk" women of reproductive age (15 to 44 or 15 to 49) who are using a method of contraception. Being at risk of a pregnancy means that a woman is in a sexual union, is fecund, but is not currently pregnant. There are approximately 900 million women in the world right now who are at risk of pregnancy, and about 56 percent of them are using some contraceptive method (Population Reference Bureau 1997). Contraceptive methods tend to be lumped into two broad categories: modern and traditional. The modern methods include the pill, IUD, implants, injectables, sponge and cervical cap, diaphragm, condom (male and female), and voluntary sterilization. Thus, the definition of contraception in most surveys encompasses two of the intermediate variables listed in Table 5.1: contraception (Variable 8) and voluntary infecundity (Variable 9). The traditional methods include total abstinence, periodic abstinence, withdrawal, and other methods. Periodic abstinence is not really a traditional method, as I have discussed in this chapter. In fact, the timing of ovulation in the menstrual cycle (which led to the development of the rhythm method of periodic abstinence) was unknown until the 1930s when Kyusaku Ogino and Herman Knaus independently discovered the fact that peak fecundity in women occurs at the approximate midpoint between menses and that, despite the variability in the amount of time between the onset of menses and ovulation, the interval between ovulation and the next menses is fairly constant at about 14 days (Population Information Program 1985). Nonetheless, the high failure rate of the rhythm method (in contrast to newer methods of periodic abstinence) removes it from the modern category.

As of 1998 data on contraceptive prevalence were available for 115 countries. The data come from surveys conducted principally in the early 1990s, although the dates are obviously not the same for all countries. In the accompanying graph I have charted the percentage of women using contraception (including both modern and traditional types) against the total fertility rate (TFR, an index similar to the total number of children born to women, as I discuss in this chapter). Each triangle on the graph represents one of those 115 countries in terms of its total fertility rate and contraceptive prevalence. Thus, Yemen, in the upper left of the graph, has a very high fertility rate (7.7 children per woman) and a very low 10 percent of women using any method of contraception. In the middle of the graph is India with a TFR of 3.4 and a middling 41 percent of women using contraception. Mexico is a little farther down the line with a TFR of 3.1 and a fairly high 65 percent of women using contraception. Toward the lower right-hand side of the graph are the United States and Canada, both with a low TFR (2.0 in the United States and 1.6 in Canada) and a high percentage of women using contraception (71 and 73, respectively). You do not need a degree in statistics to see that there is a close relationship between these two variables, and the R^2 of .83 confirms that fact for those of you familiar with correlation coefficients.

The regression equation calculated from these data shows that replacement-level fertility (equivalent to a total fertility rate of 2.1 children per woman) is associated with 70 percent of women of reproductive age using contraception. In a similar analysis, Bongaarts (1986) has calculated that 75 percent of *married* women in a society need to be contraceptive users in order to produce a level of fertility that just replaces each generation.

You can see in the graph that not every country is exactly on the line (which would represent a perfect relationship between fertility and contraceptive prevalence). For example, Togo (not labeled on the graph), with a total fertility rate of 6.9 children, has higher fertility than might be expected given that 34 percent of women appear to be using contraception. The explanation is that most women reported their method as periodic abstinence, a method with a high failure rate. On the other hand, most eastern European nations have lower fertility than might be expected just on the basis of contraceptive prevalence. In these countries, legal abortions tend to account for most of the difference (Mauldin and Segal 1988). There are several countries in Latin America where the evidence suggests that illegal abortions also play an important auxiliary role in limiting fertility (Frejka and Atkin 1996).

Two important points need to be kept in mind as you examine the accompanying graph: (1) Increases in contraceptive utilization over time have been overwhelmingly due to the use of female methods, and (2) contraception is the means by which changing motivations for family size are implemented. In order to understand these changes, we must examine a wide range of social factors, as we will begin to do in the next chapter.

Higher Levels of Contraceptive Use Lead to Lower Levels of Fertility

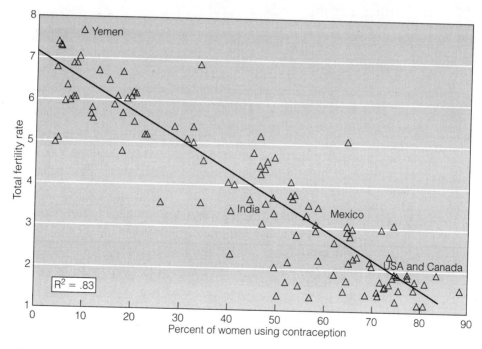

Source: Adapted from data in Population Reference Bureau, 1997, World Population Data Sheet 1997 (Washington, D.C.: Population Reference Bureau).

abortion (Reider and Pick 1992), but most abortions remain clandestine. Despite that fact, the abortion rate in Mexico is just slightly lower than that in Canada (Singh and Sedgh 1997).

There are several different types of abortion techniques, varying primarily according to how far along the pregnancy is before an abortion is attempted. In the United States the majority of reported induced legal abortions are performed during the first 12 weeks of pregnancy. Most of these are done using vacuum curettage, in which gentle suction removes the contents of the uterus. A less common method is the traditional dilation and curettage (D & C), in which a sharp metal instrument is used instead of vacuum curettage. The most common method of abortion between 13 and 16 weeks of pregnancy in the United States is dilation and evacuation (D & E), involving a gentle scraping of the lining of the uterus with a long instrument (Hatcher, et al. 1994). An abortion later in the pregnancy might require a hypertonic saline abortion. In this technique the amniotic fluid is replaced by a concentrated salt solution, which induces uterine contractions. After a few hours the fetus and placenta are expelled. This requires a 2- to 3-day hospital stay. After 20 weeks, or if the saline method fails, the typical abortion procedure is a hysterotomy, which is similar to a cesarean section, in which the fetus is removed through a small abdominal incision.

Measuring Fertility

Having discussed the numerous means by which fertility can be controlled, we are almost ready to discuss and try to explain recent trends in fertility around the world. However, in that discussion I will, of necessity, use several different measures of fertility, and each of them deserves some comment ahead of time so that you can understand their various weaknesses and strengths. In some cases, the entire trend in fertility over time in a country will appear a bit different just by using different measures of fertility. Since many of these measures regularly find their way into newspapers and popular magazines, familiarizing yourself with them will make you a better consumer of such information, in addition to the contribution it will make to your understanding of fertility trends and levels.

The basic goal in measuring fertility is to estimate the number of children that women are having. The data used for measuring this usually come from a combination of census and vital statistics sources and from sample surveys, as I discussed in Chapter 2. The measurement of fertility is often complicated by differences between one country (or state) and another in the amount and quality of data available; the less information available, the less sophisticated and accurate will be the estimates of fertility. Nowhere is this more apparent than in the contrast between period and cohort data.

Period Versus Cohort Data

Period data refer to a particular calendar year and represent a cross section of the population at one specific time. **Cohort** measures of fertility, on the other hand, "are designed to follow the fertility of groups of women as they proceed through the childbearing years of life. In other words, period rates focus on the experience of specific calendar years, but cohort rates focus on the experience of groups of women over a number of years" (Kiser, Grabill, and Campbell 1968). In general, we learn more about human behavior by following people through time (the cohort approach) than we do by taking a snapshot (the period approach). But in the interest of assessing the demographic situation based on data that are available to us at the moment, we often base our calculations on a **synthetic cohort,** meaning that we treat our period data as though they referred to a cohort. Thus, if we have data for women aged 20 to 24 and 25 to 29 in the year 2000, they really represent the period data for two different cohorts. If we assume that the women who are now 20 to 24 will have just the same experience 5 years from now as the women who are currently 25 to 29, then we have constructed a synthetic cohort from our period data.

In the following discussion, we will look first at the usual period measures of fertility, including the simplest and most frequently used fertility measure—the crude birth rate—as well as the general fertility rate, the child-woman ratio, Coale's Index of Fertility, age-specific fertility rates, the total fertility rate, gross reproduction rate, and net reproduction rate. Note that the latter three measures, in particular, are based on synthetic cohorts. Then we will examine real cohort methods, including completed fertility and fertility expectations.

Period Measures of Fertility

Crude Birth Rate   The **crude birth rate** (CBR) is the number of live births in a year divided by the mid-year population. It is usually multiplied by 1,000 to eliminate the decimal point:

$$CBR = \frac{\text{Number of live births in year } x}{\text{Mid-year population in year } x} \times 1,000$$

The CBR is "crude" because (1) it does not take into account which people in the population were actually at risk of having the births; and (2) it ignores the age structure of the population, which can greatly affect how many live births can be expected in a given year. Thus, the CBR (which is sometimes called simply "the birth rate") can mask significant differences in actual reproductive behavior between two populations and, on the other hand, can indicate differences that do not really exist. For example, if a population of 1,000 people contained 300 women who were of childbearing age and one tenth of them (30) had a baby in a particular year, the crude birth rate would be (30 births/1,000 total people) = 30 births per 1,000 population. However, in another population, one tenth of all women may also have had a child that year. Yet, if out of 1,000 people there were only 150 women of childbearing age, then only 15 babies would be born, and the crude birth rate would be 15 per 1,000. The range of crude birth rates in the world is from a low of 9 per 1,000 (in Italy and Russia) to a high of 52 per 1,000 in the Gaza Strip. As of 1998 the CBR in Canada was 12, compared with 14 in the United States, and 25 in Mexico.

Despite its shortcomings, the crude birth rate is often used because it requires only two pieces of information: the number of births in a year and the total population size. If, in addition, we have a distribution of the population by age and sex, usually obtained from a census (but also obtainable from a large survey), then we can be more sophisticated in our measurement of fertility and calculate the general fertility rate.

General Fertility Rate   The **general fertility rate** (GFR) uses information about the age and sex structure of a population to be more specific about who actually has been at risk of having the births that are recorded in a given year. The GFR (which is sometimes called simply "the fertility rate") is the total number of births in a year divided by the number of women in the childbearing ages:

$$GFR = \frac{\text{Total number of births in year } x}{\text{Midyear population of women aged 15–44 in year } x} \times 1{,}000$$

Let me give you an example of the differences between CBR and GFR in measuring fertility. In 1935, during the Depression, fertility was low in the United States. The CBR was 19 births per 1,000 population. In 1967, in the middle of the fertility decline that followed the Baby Boom, the CBR was 18, suggesting that fertility was actually lower than during the Depression. However, the lower CBR in 1967 than in 1935 was a result of differences in the number of people at each age in those 2 years rather than lower levels of childbearing. In 1967 the total population was swelled by young Baby Boom children who were not of childbearing age but whose inclusion in the total population (the denominator of the CBR) lowered its value. This problem is corrected by calculating the GFR. In 1935 the GFR was 78 births per 1,000 women aged 15 to 44. In 1967 it was higher than that, 88 per 1,000. Thus, by relating the births more closely to the people having the babies, we get a more accurate sense of the actual level of reproduction.

Although the CBR and GFR may tell different tales, in general they are reasonably consistent and Smith (1992) has noted that the GFR tends to be equal to about 4 1/2 times the CBR. Thus, in 1995 the GFR in the United States was 65.6, which was roughly 4 1/2 times the CBR of 14.8 that year.

Child–Woman Ratio   Thus far, the two measures we have discussed rely on vital statistics data for their information about births; however, even if such data are not available, all is not lost. The **child–woman ratio** (CWR) provides an index of fertility that is conceptually similar to the general fertility rate but relies solely on census data. The CWR is measured by the ratio of young children (aged 0 to 4) enumerated in the census to the number of women of childbearing ages (15 to 49):

$$CWR = \frac{\text{Total number of children aged 0–4}}{\text{Women aged 15–49}} \times 1{,}000$$

Notice that we use an older upper limit on the age of women than that used with the GFR, because some of the children aged 0 to 4 will have been born up to 5 years prior to the census date. In 1990 there were 18,354,000 children aged 0 to 4 in the

United States and 65,698,000 women aged 15 to 49; thus, the CWR was 279 children aged 0 to 4 per 1,000 women of childbearing age. Interestingly enough, this was virtually the same CWR as in 1980. By contrast, the child-woman ratio for Mexico in 1990 was 489, nearly double the level in the United States.

The child-woman ratio can be affected by the underenumeration of infants, by infant and childhood mortality (some of the children born will have died before being counted), and by the age distribution of women within the childbearing years. The latter problem is usually considered to be the most serious, and it is taken into account with a technique called age standardization (see the Appendix). As an example we can note that if the distribution of women by age in 1990 in Mexico had been identical to that in the United States, Mexico's CWR would have been less than 489. The younger age structure of Mexico placed a higher fraction of women in the prime reproductive years (ages 20 to 34) and "inflated" the CWR.

Just as the GFR is roughly 4 1/2 times the CBR, so it is that the CWR is approximately 4 1/2 times the GFR (Smith 1992). The CWR for the United States in 1990, as I noted above, was 279, which was just a bit less than 4 1/2 times the GFR that year of 70.9.

Coale's Index of Fertility   As part of the Princeton European Fertility Project (first discussed in Chapter 3), Ansley Coale produced an index of fertility that has been useful in making historical comparisons of fertility levels. The overall index of fertility (I_f) is the product of the proportion of the female population that is married (I_m) and the index of marital fertility (I_g). Thus:

$$I_f = I_m \times I_g$$

Marital fertility is calculated in an interesting way, however. It is the ratio of marital fertility (live births per 1,000 married women) in a particular population to the marital fertility rates of the Hutterites in the 1930s. Since they were presumed to have had the highest overall level of "natural" fertility, any other group might come close to, but not likely exceed, that level (see Figure 5.1). Thus, the Hutterites represent a good benchmark for the upper limit of fertility. An I_g of 1.0 would mean that a population's marital fertility was equal to that of the Hutterites, whereas an I_g of 0.5 would represent a level of childbearing only half that.

Calculating marital fertility as a proportion, rather than as a rate, allows the researcher to readily estimate how much of a change in fertility over time is due to the proportion of women who are married and how much is due to a shift in reproduction within marriage. Much of the work leading to the reformulation of the demographic transition (see Chapter 3) has been based on analyses of I_f, I_m, and I_g.

Age-Specific Fertility Rate   One of the most precise ways of measuring fertility is the **age-specific fertility rate**. These rates, like the other measures still to be discussed, require a rather complete set of data: births according to the age of the mother and a distribution of the total population by age and sex. An age-specific fertility rate (ASFR) is the number of births occurring annually per 1,000 women of a specific age (usually given in 5-year age groups):

$$\text{ASFR} = \frac{\text{Births in a year to women aged } x \text{ to } (x + 5)}{\text{Total mid-year population of women aged } x \text{ to } (x + 5)} \times 1,000$$

For example, in the United States in 1995 there were 110 births per 1,000 women aged 20 to 24. In 1955 in the United States, childbearing activity for women aged 20 to 24 was more than twice that, reflected in an ASFR of 242. In 1995 the ASFR for women aged 25 to 29 was 112, compared with 191 in 1955. Thus, we can conclude that between 1955 and 1989, fertility dropped more for women aged 20 to 24 than for women aged 25 to 29.

It is apparent that ASFRs require analysis of fertility on an age-by-age basis. For a quick comparison of an entire population, this can be a pain in the neck. Therefore, demographers have devised a method for combining ASFRs into a single fertility index covering all ages—the total fertility rate.

Total Fertility Rate   The **total fertility rate** employs the synthetic cohort approach and approximates being able to ask women how many children they have had when they are all through with childbearing by using the age-specific fertility rates at a particular date to project what could happen in the future if all women went through their lives bearing children at the same rate that women of different ages were bearing them at that date. For example, I just mentioned that in 1995 American women aged 25 to 29 were bearing children at a rate of 112 births per 1,000 women per year. Thus, over a 5-year span (from ages 25 through 29), for every 1,000 women we could expect 560 ($= 5 \times 112$) births if everything else remained the same. By applying that logic to all ages, we can calculate the TFR as the sum (signified by the symbol Σ) of the ASFRs through all ages:

$$\text{TFR} = \Sigma \, (\text{ASFR} \times 5)$$

The ASFR for each age group is multiplied by 5 only if the ages are grouped into 5-year intervals. If we had data by single year of age, we would not have to make that adjustment. The calculation of the TFR for period data is illustrated in Table 5.5.

The TFR can be readily compared from one population to another because it takes into account the differences in age structure, and, although the calculation of the TFR may seem slightly complicated, its interpretation is simple and straightforward. The total fertility rate is an estimate of the average number of children born to each woman, assuming that current birth rates remain constant. In 1995, the TFR in the United States was 2,019 children per 1,000 women, or 2.0 children per woman. This was a bit more than half of the 1955 figure of 3.6 children per woman. The TFR calculated from period data was almost identical to the completed fertility rate for women aged 40 to 44 in 1995, according to the Current Population Survey (Bachu 1997). These women, who were wrapping up their childbearing by 1995, had given birth to an average of 1.96 children during their lifetimes.

For a "quick and dirty" calculation of the TFR (measured per 1,000 women) you can multiply the GFR by 30 or, if you have only the CBR, multiply it by 4 1/2 and then again by 30. So, in the United States in 1995, the TFR of 2.019 (or 2019 per 1,000 women) is very close to 30 times the GFR of 65.6.

Gross Reproduction Rate   A further refinement of the total fertility rate is to look at female births only (since it is only the female babies who eventually bear children). Thus, if we multiply the TFR by the proportion of all births that are girls, we have the **gross reproduction rate** (GRR). Specifically,

$$GRR = TFR \times \frac{\text{Female births}}{\text{All births}}$$

In the United States in 1995, 48.8 percent of all births were girls. Since the TFR was 2.019, we multiply that figure by 0.488 (the percentage converted to a proportion) to obtain a GRR of 0.985.

The gross reproduction rate is generally interpreted as the number of female children that a female just born may expect to have during her lifetime, assuming that birth rates stay the same and ignoring her chances of survival through her reproductive years. A value of 1 indicates that women will just replace themselves, while a number less than 1 (as for the United States in 1995) indicates that women will not quite replace themselves, and a value greater than 1 indicates that the next generation of women will be more numerous than the present one. The GRR is called "gross" because it assumes that a girl will survive through all her reproductive years. Actually, some women will die before reaching the oldest age at which they might bear children. To take account of mortality risks, we can calculate the net reproduction rate.

Net Reproduction Rate   The **net reproduction rate** (NRR), like the TFR and GRR, is based on the use of period data to construct a synthetic cohort. With a real

Table 5.5   Calculation of Total Fertility Rate for the United States, 1995

Age	Women in Age Group (Midyear Population in Thousands)	Live Births to Women in Age Group	Age-Specific Fertility Rates (Live Births Per 1,000 Women)	ASFR $\times$ 5
10–14	9,417	12,242	1.3	6.5
15–19	8,801	499,873	56.8	284.0
20–24	8,794	965,547	109.8	549.0
25–29	9,479	1,063,539	112.2	561.0
30–34	10,966	904,666	82.5	412.5
35–39	11,188	383,745	34.3	171.5
40–44	10,189	67,250	6.6	33.0
45–49	9,090	2,727	0.3	1.5
Totals				2,019.0*

*TFR = Σ (ASFR $\times$ 5) = 2,019

Source of data: Stephanie J. Ventura, Joyce A. Martin, Sally C. Curtin, and T.J. Mathews 1997, Report of Final Natality Statistics, 1995, Monthly Vital Statistics Report, 45(11):supplement, Tables 2 and 4.

cohort, the net reproduction rate is typically referred to as the *generational replacement rate*. Either way, the NRR represents the number of female children that a female child just born can expect to bear, taking into account her risk of dying before the end of her reproductive years. Thus, the NRR is always less than the GRR, since some people always die before the end of their reproductive periods. How much before, of course, depends on death rates. In a low-mortality society such as the United States, the NRR is only slightly less than the GRR (the ratio of the NRR of .965 to the GRR of 0.985 is .98). By contrast, in a high-mortality society such as Ethiopia, the difference can be substantial (the ratio of the NRR of 2.6 to the GRR of 3.7 is .70). Since calculation of the NRR is somewhat complicated, I have spared you having to deal with it at this point, putting it instead in the Appendix. The most important point is that you understand how to interpret it, for it is a fairly often cited measure of fertility.

As an index of generational replacement, an NRR of 1 indicates that each generation of females has the potential to just replace itself. This indicates a population that will eventually stop growing if fertility and mortality do not change. A value less than 1 indicates a potential decline in numbers, and a value greater than 1 indicates the potential for growth unless fertility and mortality change. It must be emphasized that the NRR is not equivalent to the rate of population growth in most societies. For example, in the United States the NRR in 1995 was less than 1 (0.965, as I just mentioned), yet the population was still increasing by 2.2 million people each year. The NRR represents the future potential for growth inherent in a population's fertility and mortality regimes. However, peculiarities in the age structure (such as large numbers of women in the childbearing ages), as well as migration, affect the actual rate of growth at any point in time.

Cohort Measures of Fertility

Completed Fertility   The basic measure of cohort fertility is births to date, eventually resulting in a final measure of completed fertility, the cohort total fertility rate. For example, women born in 1915 began their childbearing during the Depression. By the time those women had reached age 25 in 1939, they had given birth to 890 babies per 1,000 women (Heuser 1976). By age 44 in 1958 those women had finished their childbearing in the Baby Boom years with a total fertility rate of 2,429 births per 1,000 women. We can compare those women with another cohort of women who were raised during the Depression and began their childbearing right after World War II. The cohort born in 1930 had borne a total of 1,415 children per 1,000 women by the time they were age 25 in 1954. This level is 60 percent greater than the 1915 cohort. By age 44 in 1973 the 1930 cohort had borne 3,153 children per 1,000 women, 30 percent higher than the 1915 cohort. Indeed, examining cohort data for the United States, it turns out that the women born in 1933 were the most fertile of any group of American women since the cohort born in 1881. Although such information is illuminating, we cannot always wait until women complete their childbearing to estimate their level of fertility; which is why we typically use the synthetic cohort approach to calculate the TFR.

Fertility Intentions   It is useful to have estimates of what the women who are presently of childbearing age might do in the future, and period data cannot do that any better than the cohort births-to-date information, so demographers estimate future cohort fertility by asking women about the number of births they expect in the future. For example, in June 1971, the Current Population Survey asked women how many births they expected to have had by 1976. Women aged 18 to 24 (the birth cohorts of 1946 to 1953) responded with an average of 1.99 births. Overall, they expected to have 2.38 births on average during their lifetime. In 1976, when they were reinterviewed, it appeared that their expectations had exceeded reality; they were averaging only 1.76 births per woman, and it seemed unlikely that they would reach their lifetime expectations. But, by 1981, when interviewed a third time (now aged 28 to 34) their behavior had nearly caught up with their expectations of a decade earlier, and they were averaging 2.21 children each. What had happened? Women who were single in 1971 failed to anticipate their average delay in getting married and, within marriage, a postponement of children. Thus, their short-run expectations missed the mark (O'Connell and Rogers 1983). Yet ultimately they made up for lost time, and by 1986 they had reached their expected number of lifetime births.

The usefulness of data on fertility intentions has been the subject of considerable debate for a number of years (see, for example, Masnick 1981; Westoff and Ryder 1977; Demeny 1988). Westoff (1990) used data from 134 different fertility surveys conducted in 84 different countries to conclude that "the proportion of women reporting that they want no more children has high predictive validity and is therefore a useful tool for short-term fertility forecasting" (p. 84).

What do such measurements suggest about future fertility in the United States? In 1992 women aged 18 to 24 expected to have 2.053 children in their lifetime, nearly identical to the TFR for that year. So solid is the preference for just 2 children that 80 percent of the women who already had 2 children in 1992 did not expect to have another baby (Bachu 1993).

Summary and Conclusion

Fertility has both a biological and a social component. The capacity to reproduce is biological (although it can be influenced to a certain extent by the environment), but we have to look to the social environment to find out why women are having a particular number of children. Fertility control can be accomplished by a variety of means. In general, they include ways by which intercourse can be prevented, conception can be prevented if intercourse takes place, and successful gestation can be prevented if conception occurs. An understanding of these proximate determinants or intermediate variables is an important preliminary to a good analysis of fertility trends and levels, as is the knowledge of how fertility is measured. In this chapter I examined the measures of fertility most commonly used in demographic analysis. In the next chapter I will build on the concepts and measures of fertility as I examine and attempt to explain reproductive behavior in different parts of the world.

Main Points

1. Fertility refers to the number of children born to women (or fathered by men).

2. Fecundity, on the other hand, refers to the biological capacity to produce children.

3. The group of people who are always pointed out when high fertility is discussed is the Hutterites, who live in the northern plains of the United States and in Canada. In the 1930s each married woman averaged 11 children.

4. Fertility can be controlled by manipulating exposure to intercourse, trying to prevent conception, or interrupting pregnancy.

5. Researchers have discovered that women who do not engage in sexual intercourse do not have babies. This may be the cheap and effective means of birth control that the world has been looking for.

6. The longer a woman delays marriage, the fewer children she will likely have.

7. Women whose marriages are interrupted by the death of a spouse, divorce, or separation tend to have fewer children than those who remain continuously in a marriage.

8. Implants, injections, and the pill are among the most effective means of contraception, but the IUD is also effective, as is the condom, especially when used in conjunction with a spermicidal foam, jelly, or film.

9. The douche is almost useless as a contraceptive, as is the rhythm method. However, the symptothermal method appears to be an effective natural method for highly motivated couples.

10. Sterilization is an increasingly popular method of avoiding conception and, furthermore, governments around the world are increasingly allowing abortion as a legal means to end unwanted pregnancies.

11. The principal measures of fertility used by demographers are the crude birth rate, general fertility rate, child-woman ratio, Coale's Index of Fertility, age-specific fertility rate, total fertility rate, gross reproduction rate, net reproduction rate, and birth expectations. Don't let people tell you what the birth rate is without finding out what rate they are referring to and whether the rate refers to period or cohort data.

Suggested Readings

1. Kingsley Davis and Judith Blake, 1955, "Social structure and fertility: an analytic framework," Economic Development and Cultural Change 4:211–35.

   The chapter that you have just finished reading has leaned heavily on this classic statement of the role of "intermediate variables" as determinants of fertility levels in human societies.

2. John Bongaarts, 1978, "A Framework for Analyzing the Proximate Determinants of Fertility," Population and Development Review 4(1):105–32.

This is a very important updating of the analysis of Davis and Blake's "intermediate variables," which Bongaarts has redefined as the "proximate determinants of fertility."

3. Robert Hatcher, et al., Contraceptive Technology (New York: Irvington Publishers), published regularly.

This basic reference work provides updated information on the prevalence and effectiveness of various fertility control methods.

4. Stanley Henshaw, 1990, "Induced Abortion: a World Review, 1990," Family Planning Perspectives 22(2):76–89.

As the Deputy Director of Research at the Alan Guttmacher Institute, Henshaw periodically updates the world data on abortion and publishes it in this journal.

5. Linda Piccinino and William D. Mosher, 1998, "Trends in Contraceptive Use in the United States," Family Planning Perspectives 30(1):4–11.

By combining data from the National Surveys of Family Growth from 1982 through 1995 the authors are able to draw a very detailed picture of trends in contraceptive use in the United States.

⊕ Websites of Interest

Remember that websites are not as permanent as books and journals, so I cannot guarantee that each of the following websites still exists at the moment that your are reading this.

1. http://www.jhuccp.org/poprpts.stm

The Center for Communication Programs at the Johns Hopkins University has for several years published an excellent series of detailed reports called Population Reports that provide background on a wide range of fertility and family planning topics. You can download many of the reports from this website.

2. http://www.agi-usa.org/home.html

The Alan Guttmacher Institute in New York City is a not-for-profit organization devoted to research and dissemination of research results on issues of family planning and reproductive health. Among their many activities are the publication of the journal's Family Planning Perspectives and International Family Planning Perspectives. Information about the organization and their publications is available at this website.

3. http://www.ippf.org

The International Planned Parenthood Federation is based in London, England. The organization coordinates a wide range of research, policy, education, and dissemination programs.

4. http://www.macroint.com/dhs/

Most of the information that we have about fertility and reproductive health in developing countries comes from the Demographic and Health Surveys, conducted by Macro International. At their website you can get a summary of results and check on the status of current surveys.

5. http://www.yahoo.com/Health/Reproductive_Health/Birth_Control

Yahoo is an internet company that specializes in searches and they have packaged one set of searches already for you, this one on birth control methods.

CHAPTER 6
Fertility Trends, Levels, and Explanations

I am inclined to think that the most important of Western values is the habit of a low birth-rate. If this can be spread throughout the world, the rest of what is good in Western life can also be spread. There can be not only prosperity, but peace. But if the West continues to monopolize the benefits of low birth-rate(s), war, pestilence, and famine must continue, and our brief emergence from those ancient evils must be swallowed in a new flood of ignorance, destitution and war.

When Bertrand Russell, whose words these are (1951:49), died in 1970 at age 89, he had witnessed almost the entire demographic transition in the Western nations; birth rates in many less developed areas, however, seemed stubbornly high and few people would have predicted much of a drop. In Chapter 3, I mentioned that most of the 20th century theorizing in population studies has focused on the demographic transition and its reformulation. The inability of demographic transition theory to adequately predict what has been happening from one country to the next has led demographers to disaggregate the demographic transition into its three logical component parts—the mortality transition (or the epidemiological transition, as it is usually called—the shift from high to low mortality, as discussed in Chapter 4), the **fertility transition** (the shift from "natural" fertility, as discussed in Chapter 5, to fertility limitation), and the **migration transition** (the shift of people from rural to urban places, and the shift to higher levels of international migration, which I will discuss in the next chapter). The fertility transition is my focus in this chapter.

From the original formulation of the demographic transition comes the "supply–demand framework" and from the reformulation of the demographic transition comes the "diffusion-innovation" perspective. Both perspectives seem to explain at least part of the variation in fertility over time and place, and so I will review some of the attempts to synthesize these two contrasting perspectives into a single unified theory of fertility. I begin this chapter with a discussion of these perspectives on fertility and then use them to see how they help us to understand high fertility and the transition to low fertility.

Explanations for the Fertility Transition

The Supply–Demand Framework

The demographic transition envisions a world in which the normal state of affairs is a balance between births and deaths. Mortality is assumed to decline for reasons that are often beyond the control of the average person (**exogenous factors**), but a person's reproductive behavior is dominated by a rational calculation of the costs and benefits to himself or herself (**endogenous factors**) of maintaining high fertility in the face of declining mortality. The theory implies that people will eventually perceive that lower mortality has produced a situation in which more children are going to survive than can be afforded and, at that point, fertility will decline.

The economist Richard Easterlin (mentioned earlier in Chapter 3) is most notable for his work in this regard, and the resulting perspective is somewhat clumsily called "the theory of supply, demand, and the costs of regulation" or, in shorthand, "**the supply–demand framework.**" It is also known as "the new household economics" because the household, rather than the individual or the couple, is often taken as the unit of analysis. As I discuss later in this chapter, high fertility may help households to avoid risk in the context of low economic development and weak institutional stability, especially when children generate a positive net flow of income to the parents.

The essence of the supply–demand framework is that the level of fertility in a society is determined by the choices made by individual couples within their cultural (and household) context (see Bongaarts 1993; Bulatao and Lee 1983; Easterlin and Crimmins 1985; Easterlin 1978; McDonald 1993; and Robinson 1997). Couples strive to maintain a balance between the potential supply of children (which is essentially a biological phenomenon determined especially by fecundity, as noted in Chapter 5) and the demand for children (which refers to a couple's ideal number of surviving children). If mortality is high, the number of surviving children may be small, and the supply may essentially equal the demand. In such a situation, there is no need for fertility regulation. However, if the supply begins to exceed the demand, either because infant and child survival has increased, or because the **opportunity costs** of children are rising, then couples may adjust the situation by using some method of fertility regulation. The decision to regulate fertility will be based on the couple's perception of the costs of doing so, which include the financial costs of the method and the social costs (such as stigmas attached to the use of family planning).

What are the opportunity costs of children? Let us assume that people are rational and that they make choices based on what they perceive to be their best self-interest. One set of choices that most people make in life revolves around children. The idea that children might be thought of as "commodities" was introduced in 1960 by University of Chicago economist Gary Becker, whose work on the economic analysis of households and fertility earned him a Nobel prize in 1992. Becker's theory treated children as though they were consumer goods that require both time and money for parents to acquire. Then he drew on classic microeconomic theory to argue that for each individual a utility function could be found that would express the relationship between a couple's desire for children and all other goods or activities that compete with children for time and money (Becker 1960). It is important to note that time as well as money is being considered, for if money were the only criterion, then one would expect that (in a society where there are social pressures to have children) the more money a person had, the more children he or she would want to have. Yet we know that in virtually every industrialized nation, those who are less well-off financially tend to have more children than do those who are more well-off.

With the introduction of time into the calculations, along with an implicit recognition that social class determines a person's tastes and lifestyle, Becker's economic theory turns into a trade-off between quantity and quality of children. For the less well-off, the expectations that exist for children are presumed to be low and thus the cost is at its minimum. In the higher economic strata the expectations for the children are presumed to be greater, both in terms of money and especially in terms of time spent on each child. The theory asserts that parents in the higher strata also are exposed to a greater number of opportunities to buy goods and engage in time-consuming activities. Thus, to produce the kind of child desired, the number must be limited.

Later in the chapter I will return to the question of what specific kinds of opportunities compete with children, implicitly elbowing out some children in the competition for parents' time and money. For now, let me note that the opportunity costs theory is an intuitively appealing explanation for differences in fertility within a particular society, but its scope needs to be broadened to provide a more complete explanation of why fertility might be high in some societies and lower in others.

The Innovation/Diffusion and "Cultural" Perspective

The supply–demand framework draws its concepts largely from the field of neo-classical economics, which assumes that people make rational choices about what they want and how to go about getting it. Not all social scientists agree that human behavior works like that. In particular, sociologists, anthropologists and cultural geographers have often been drawn to the idea that many changes in society are the result of the diffusion of innovations (Brown 1981; Rogers 1995). We know, for example, that much of human behavior is driven by fads and fashions. Last year's style of clothing will go unworn by some people this year, even though the clothes may be in very good repair, just because that is not what "people" are wearing. These "people" are important agents of change in society—those who, for reasons that may have nothing to do with money or economic factors, are able to set trends. You may call it charisma, or karma, or just plain influence, but some people set trends and others do not. We see it happen many times in our lives. Often these change agents are members of the upper strata of society. They may not be the inventors of the innovation, but when they adopt it, others follow suit. Notice, too, that the innovation may be technological, such as the cellular telephone, or it may be attitudinal and behavioral, such as deciding that two children is the ideal family size and then utilizing the most popular means to achieve that number of children.

In Chapter 3, I mentioned that the fertility history of Europe suggests a pattern of geographic diffusion of the innovation of fertility limitation within marriage. The practice seemed to spread quickly across regions that shared a common language and ethnic origin, despite varying levels of mortality and economic development. This finding led to speculation that fertility decline could be induced in a society, even in the absence of major structural changes such as economic development, if the innovation could be properly packaged and adopted by the appropriately influential change agents. However, this is where the concept of "culture" comes into play, because some societies are more prone to accept innovations than are others (see Pollack and Watkins 1993).

To accept an innovation and to change your behavior accordingly, you must be "empowered" (to use an increasingly overworked term) to believe that it is within your control to alter your behavior. Not all members of all societies necessarily feel this way. In many premodern and "traditional" societies, people accept the idea that their behavior is governed by God, or multiple Gods, or more generally by "Fate," or more concretely by their older family members (dead or alive). In such a society, an innovation is likely to be seen as an evil intrusion and is not apt to be tolerated.

You can perhaps appreciate, then, that the diffusion of an innovation requires that people believe that they have some control over their life, which is the essence of the rational choice model that underlies the economic approach to the fertility transition. In other words, the supply–demand model, and the innovation-diffusion model may not be so different after all. In fact, both approaches can be helpful in explaining why fertility declines if we examine the three preconditions for a fertility decline as laid out by Ansley Coale.

Three Preconditions for a Fertility Decline

In 1973, in response to the findings emerging from the Princeton European Fertility Project, Ansley Coale tried to deduce how an individual would have to perceive the world on a daily basis if fertility were to be consciously limited. In this revised approach to the demographic transition, he states that there are three preconditions for a substantial fertility decline: (1) the acceptance of calculated choice as a valid element in marital fertility, (2) the perception of advantages from reduced fertility, and (3) knowledge and mastery of effective techniques of control (Coale 1973). Although the societal changes that produced mortality declines may also induce fertility change, they will do so, Coale argues, only if the three preconditions exist. The preconditions, though, may exist even in the absence of mortality decline. The important causal factors determining whether these preconditions will exist include "unmeasured traditions and habits of mind" (Burch 1975), especially secularization, as I discussed earlier and will mention again later.

Coale's first precondition thus goes to the very philosophical foundation of individual and group life: who is in control? If a supernatural power is believed to control reproduction, then it is unlikely that people will run the risk of offending that deity by impudently trying to limit fertility—even when they know that means of fertility control exist. On the other hand, the more secular people are (even if still religious), the more likely it is that they will believe that they and other humans have the ability to control important aspects of life, including reproduction.

The second precondition recognizes that more is required than just the belief that you can control your reproduction. You must have some reason, in particular, to want to limit fertility. Otherwise, the natural attraction between males and females will lead to unprotected intercourse and, eventually, to numerous children. What kinds of changes in society might motivate people to want fewer children? Davis (1963; 1967) has suggested that people will be motivated to delay marriage and limit births within marriage if economic and social opportunities make it advantageous for them to do so. He argues further that having children per se rarely satisfies an end in itself but is generally a means to some other end, such as expanding the ego, proving gender roles, maintaining a descent group, continuing the inheritance of property, securing economic welfare in old age, assuring future family labor, or satisfying the believed demand of a deity. If the goals that are important change, or if the means available to achieve those goals change (for example, money instead of children), then the desire to have children may change.

As an entire social structure changes or as a person's position within the social structure changes, the ends or goals that individuals have in mind change and their motivations to have children change. We know, for example, that as education increases, wealth and prestige tend to increase (see Chapter 10) and the number of children born in a family declines. Apparently, education leads to a greater ability to acquire wealth and prestige and this competes with having children, since children are for many years resource consumers rather than resource producers. This principle seems to work for individuals as they attempt to be upwardly mobile socially within a society, or as they try to prevent social slippage—a loss of social and economic status relative to others.

In 1942 Joseph Schumpeter discussed the emerging attitudes of middle-class families toward childbearing and rearing. Heralding yuppie ideas, but 40 years ahead of his time, he posed the question "that is so clearly in many potential parents' minds: 'why should we stunt our ambitions and impoverish our lives in order to be insulted and looked down upon in our old age?'" (Schumpeter 1942:157). This is a very different view of the world from that in which family members work primarily for the mutual survival of all other family members or the view from the bottom of the social heap in which the only thing that matters is day-to-day survival.

Though motivations for limiting fertility implicitly incorporate the supply–demand framework into the cost-benefit calculations that couples are making about their reproductive behavior, one must also recognize that motivations for low fertility do not appear magically just because one aspires to wealth, or has received a college education, or is an only child and also prefers a small family. Motivations for low fertility arise out of our communication with other people and other ideas. Fertility behavior, like all behavior, is in large part determined by the information we receive, process, and then act on. The people with whom, and the ideas with which, we interact in our everyday lives shape our existence as social creatures. Leon Tabah has gone so far as to suggest that:

> Motivations for childbearing cannot in themselves explain behavior without reference to the social environment. Thus, American sociologists and demographers have for many years observed that changes in the social climate of the United States at the time of the Great Depression and equally during the postwar baby boom had more influence on fertility than the so-called personal variables (education, income, religion of the individual). Through this reasoning, one comes to appeal to the 'collective conscience,' that overriding force that operates on our lives at the same time that we believe we are controlling them ourselves [1980:364].

These social influences on what otherwise seem like private decisions remind us of the importance of the diffusion of innovations. In any social situation in which influential couples are able to improve their own or their children's economic and social success by concentrating resources on a relatively smaller number of children, other parents may feel called upon to follow suit if they and their offspring are to be socially competitive (Caldwell 1982; Handwerker 1986; Turke 1989).

The importance of the "influential couples" is sometimes ignored by North Americans who prefer the ideal of a classless society. European demographers offer the reminder that two enduring theories of social stratification have strong implications for fertility behavior: (1) cultural innovation typically takes place in higher social strata as a result of privilege, education, and concentration of resources; lower social strata adopt new preferences through imitation; and (2) rigid social stratification or closure of class or caste inhibits such downward cultural mobility (Lesthaeghe and Surkyn 1988). Thus, the innovative behavior of influential people will be diffused downward through the social structure, as long as there are effective means of communication among and between social strata. From our perspective, the innovation of importance is the preference for smaller families, implemented through the innovations of delayed marriage and/or fertility limitation.

Coale's third precondition involves the knowledge and mastery of effective means of fertility control. Specific methods of fertility control may be thought of as technological innovations, the spread of which is an example of diffusion. As I mentioned in Chapter 5, women in the US and Canada now typically use the pill to space children and then use sterilization to end reproduction after the desired number of children are born. This is a different set of techniques than prevailed 30 years ago, and 30 years from now the mix will doubtless be different still. At the same time that knowledge of methods is being diffused, part of the decision about what method of fertility regulation to use is based on the individual's cost-benefit calculation about the "costs of fertility regulation" as laid out in the supply–demand framework. The economic and psychosocial costs of various methods may well change over time and cause people to change their fertility behavior accordingly.

Explanations for High Fertility

The theories of the fertility transition try to explain why a society might go from a high level of reproduction to a lower level, begging the question of why fertility is high in the first place. All of the above perspectives accept the idea that high fertility is a societal response to high mortality. Since very high mortality was the normal situation for virtually every human society until the past two hundred years or so (as discussed in Chapter 4), it is easy to argue that only those societies with sufficiently high fertility managed to survive over the years. There may have been lower fertility societies in the past, but we know nothing about them because they did not replenish themselves over time.

Need to Replenish Society

A crucial aspect of high mortality is that a baby's chances of surviving to adulthood are not very good. Yet if a society is going to replace itself, an average of at least two children for every woman must survive long enough to be able to produce more children. So, under adverse conditions, any person who limited fertility might be threatening the very existence of society. In this light it is not surprising that societies have generally been unwilling to leave it strictly up to the individual or to chance to have the required number of children. Societies everywhere have developed social institutions to encourage childbearing and reward parenthood in various ways. For example, among the Kgatla people in South Africa, mortality was very high during the 1930s when they were being studied by Schapera. He discovered that "to them it is inconceivable that a married couple should for economic or personal reasons deliberately seek to restrict the number of its offspring" (Schapera 1941:213). Several social factors encouraged the Kgatla to desire children:

> A woman with many children is honored. Married couples acquire new dignity after the birth of their first child. Since the Kgatla have a patrilineal descent system (inheritance passes through the sons), the birth of a son makes the father the founder of a line that will perpetuate his name and memory . . . [the mother's] kin are pleased because the birth saves them from shame [Nag 1962:29].

In a 1973 study of another African society, the Yoruba of western Nigeria, families of fewer than four children were (and still are) looked upon with horror. Ware reports that "even if it could be guaranteed that two children would survive to adulthood, Yoruban parents would find such a family very lonely, for many of the features of the large family which have come to be negatively valued in the West, such as noise and bustle, are positively valued by them" (Ware 1975:284).

The 1989 Kenya Demographic and Health Survey discovered that in the coastal province of that eastern African nation, the average ideal family size of 5.6 children was higher than the actual level of reproduction. Women wanted larger families than they were having, but they were slowed down by involuntary infecundity (Cross, Obungu, and Kizito 1991).

Social encouragements to fertility have been discussed in a more general form by Davis:

> We often find for example that the permissive enjoyment of sexual intercourse, the ownership of land, the admission to certain offices, the claim to respect, and the attainment of blessedness are made contingent upon marriage. Marriage accomplished, the more specific encouragements to fertility apply. In familistic societies where kinship forms the chief basis of social organization, reproduction is a necessary means to nearly every major goal in life. The salvation of the soul, the security of old age, the production of goods, the protection of the hearth, and the assurance of affection may depend upon the presence, help, and comfort of progeny. . . . [T]his articulation of the parental status with the rest of one's statuses is the supreme encouragement to fertility [Davis 1949:561].

The supply–demand theory of fertility suggests that couples try to balance the supply of children with their demand. Thus, when mortality declines, people should adjust their fertility downward. Yet, you may notice from these examples that social pressures are not actually defined in terms of the need to replace society, and an individual would likely not recognize them for what they are. By and large the social institutions and norms that encourage high fertility are so taken for granted by the members of society that anyone who consciously said, "I am having a baby in order to continue the existence of my society" would be viewed as a bit weird. Further, if people really acted solely on the basis that they had to replace society, then higher fertility societies would now actually have much lower levels of fertility, since in all such countries the birth rates exceed the death rates by a substantial margin.

The societal disconnection between infant and child mortality and reproductive behavior probably explains why there is not much evidence of a direct connection between infant mortality in a family and the fertility level of that couple (see especially Montgomery and Cohen 1998; Preston 1975). Thus, the more culturally oriented perspectives on fertility begin with the assumption that fertility is maintained at a high level in the premodern setting by institutional arrangements that bear no logical relationship to mortality. Premodern groups accepted high mortality, especially among children, as given and devised various ways to ensure that fertility would be high enough to ensure group survival. Pronatalist pressures encourage family members to bring power and prestige to themselves and to their group by having children, and this may have no particular relationship to the level of mortality.

Children as Security and Labor

In a premodern society, human beings are the principal economic resource. Even as youngsters they can help in many tasks, and as they mature they provide the bulk of the labor force that supports those, such as the aged, who are no longer able to support themselves. Boserup (1981) reports, for example, that:

> In most of Africa, a large share of the agricultural work was and is done by women and the children, even very young ones, perform numerous tasks in rural areas. A man with many children can have his land cleared for long-fallow cultivation by young sons, and all, or nearly all, other agricultural work done by women and smaller children. He need not pay for hired labor or fear for lack of support in old age. A large family is an economic advantage—a provider of social security, and of prestige in the local community. Therefore, the large family is the universally agreed on ideal in most African communities [1981:180].

More broadly speaking, children can be viewed as a form of insurance that rural parents, in particular, have against a variety of risks, such as a drought or a poor harvest (Cain 1981). Though at first blush it may seem as though children would be a burden under such adverse conditions, many parents view a large family as providing a safety net—at least one or two of the adult children may be able to bail them out of a bad situation. One important way that this may happen is that one or more children may migrate elsewhere and send money home. (I will return to that theme in Chapter 7.) Although children may clearly provide a source of income for parents until they themselves become adults (and parents), it is less certain that children will actually provide for parents in their old age. Despite an almost world-wide norm that children should care for parents in old age, a relatively small fraction of parents in a premodern setting actually survive to a dependent old age. As Vlassoff (1990) and Dharmalingam (1994) have noted, there is very little empirical evidence to suggest a positive relationship between fertility and the perceived need for old-age security. For now, it is most noteworthy that in a premodern setting, the quantity of children may matter more than the quality, and the nature of parenting is more to bear children than to rear them (Gillis, Tilly, and Levine 1992). Still, the noneconomic, nonrational part of society intrudes by often suggesting that male children are more desirable than female children.

Desire for Sons

Although it is clear in many countries today that the status of women is steadily improving, it is nonetheless true that in many societies around the world, desired social goals can be achieved only by the birth and survival of a son—indeed, most known societies throughout human history have been dominated by men. Since in most societies males have been valued more highly than females, it is easy to understand why many families would continue to have children until they have at least one son. Furthermore, if babies are likely to die, a family may have at least two sons in order to increase the likelihood that one of them will survive to adulthood (an "heir and a spare"). For example, the total fertility rate of 5.6 children per woman in Pakistan in the mid-1990s was just the level of fertility required to ensure an average of two surviving sons per woman. Analogously, in Angola, where mortality is even higher, the total fertility rate of 6.6 in the mid-1990s was the level at which the average woman had two sons surviving to adulthood.

India is an example of a country where the desire for a surviving son is relatively strong, since the Hindu religion requires that parents be buried by their son (Mandelbaum 1974). Malthus was very aware of this stimulus to fertility in India and, in his Essay on Population, quoted an Indian legislator who wrote that under Hindu law a male heir is "an object of the first importance. `By a son a man obtains victory over all people; by a son's son he enjoys immortality; and afterwards by the son of that grandson he reaches the solar abode'" (Malthus 1872:116). The deep cultural significance of sons continues to this day in India (Vlassoff 1990). Such beliefs, of course, also serve to ensure that society will be replaced in the face of high mortality.

In North America and Europe, the preference for sons has abated over time, and male preference has become more subtle. Parents often express a desire for a son first, and then a daughter. In Asia the preference manifests itself more openly, especially among the Chinese. In Malaysia, for example, ethnic Malay (who are predominantly Muslim) and ethnic Indians (who are predominantly Hindu) show relatively little son preference, whereas ethnic Chinese are far more likely to stop having any more children if they have only sons, and are far more likely to keep having children if they have only daughters (Pong 1994).

Yet, as Arnold (1988) reminds us, the desire for sons cannot alone account for high birth rates. In Korea and China, two Asian societies with very strong preferences, the drop in fertility has been rapid (Park and Cho 1995) and Vietnam, another nation with a marked son preference, is also in the midst of a rapid fertility decline (Goodkind 1995b). The major impact of son preference in the midst of a fertility decline is to increase the chance that a female fetus may be aborted, leading to the phenomenon of the "missing females" in China (Coale and Bannister 1994).

In the 1980s there was a great deal of speculation and concern that the missing females were victims of female infanticide. Since female infanticide had been fairly common during the precommunist era, the probability seemed great that the one-child policy in China would lead a couple to kill or abandon a newborn female infant, reserving their one-child quota for the birth of a boy (Mosher 1983). Recent analyses of data in both China and Korea (where a similar pattern of fertility decline has occurred without a coercive one-child policy) suggest that sex-selective abortion, combined with the nonregistration of some female births, accounts for almost all of the "missing" females, and that the role of infanticide probably was exaggerated (Coale and Banister 1994; Park and Cho 1995). Infanticide in most cases is probably a result of the abandonment of children, and there is evidence of a large number of orphans in China who are foundlings, suggesting that at least some children who are abandoned are found and do survive (Johnson 1996).

Family Control and Fertility Control

In Chapter 5 I discussed the fact that "natural" fertility is rarely as high as the maximum level that would be possible. In most societies families are trying to have the number of surviving children that will be most beneficial to them. But people for most of human history have lived close to the subsistence level and have lived in the shadow of high death rates. Thus, it is not surprising that in such circumstances couples are unlikely to have a preference for a specific number of children (van de Walle 1992). The vagaries of both child mortality and the food supply are apt to cause people to "play things by ear" rather than plan in advance the number of children that are desired or wanted. With high mortality it is less important how many children are born than how many survive—which is how I defined the net reproduction rate for you in Chapter 5.

When we realize that it is net reproduction (surviving children, not just children ever born) that is of importance, we can see that human beings have been very clever at dealing with family size by controlling the family, rather than by controlling fertility. For example, higher than desired fertility in terms of live births can be responded to after a child is born by what Skinner (1997) has called "**child control**," or by what Mason (1997) has labeled "postnatal control." There are at least three ways of dealing with a child that is not wanted or cannot be cared by its parents after it is born, some of which were discussed above with reference to China: (1) infanticide (known to have been practiced in much of Asia); (2) **fosterage** (sending, or even selling, an "excess" child to another family that needs or can afford it—a relatively common practice in sub-Saharan Africa and parts of Africa, and not uncommon in pre-transition Europe); and (3) **orphanage** (a pre-transition practice in much of Europe, as documented by Kertzer [1993]). The levels of infant and child mortality may condition the need for these practices and, at the same time, they contribute to higher levels of infant and child mortality. Child control is not one of the proximate determinants of fertility in the usual demographic framework because it follows a live birth, rather than precedes it. Nonetheless, it is a potentially powerful mechanism for obtaining the desired level of net reproduction.

Family control does not end at that point, of course. As more children survive through childhood, the child control options become more difficult because the number of children that can no longer be afforded may stretch the limits of what families can get away with in terms of infanticide, fosterage, and orphanage. This hearkens back to the theory of demographic change and response, which I discussed in Chapter 3. As declining infant and child mortality impinges on familial resources, the family's reaction may be to work harder, especially if "child control" is not available to them. Then, as children become teenagers (sometimes even before that), they may be sent elsewhere in search of work, either to reduce the current burden on the family or more specifically to earn wages to send back to the parents. These decisions are typically made by the family, not by the individual young person, and they are made in direct, albeit belated, response to the drop in the death rate that permitted these "excess" children to survive to adulthood.

Family control, rather than fertility control, permits couples to "keep their options open," so to speak, because the encouragements to high fertility often persist even after mortality declines. Childbearing is rarely an end in itself, but rather a means to achieve other goals; so if the attainment of the other goals is perceived as being more important than limiting fertility, a woman may continue to risk pregnancy even though she may be **ambivalent** about having a child. Some of the factors that may enhance feelings of ambivalence at various points in a woman's life include the exclusive identification of a woman's role with reproduction (that is, males do not help with child rearing), lack of participation of women in work outside of the immediate family, low levels of education, lack of communication between husband and wife, lack of potential for social mobility, and an extended family system in which couples need not be economically independent to afford children. Most of these factors are related to the domination of women by men and may help to account for the persistence of higher than average fertility in Muslim societies.

Despite all of the publicity in the modern world about women needing fertility drugs and other help to get pregnant, remember that for most women in most of human history it was easier to have several children than to limit the number to one or two, regardless of the level of motivation. For the average person, a high level of desire and access to the means of fertility control are required to keep families small.

Higher Fertility Countries

Having analyzed why high fertility might persist even after mortality has declined, let me now briefly review three countries where we find that very situation today: Jordan, India and Mexico (see Figure 6.1).

Jordan

Jordan is a small Middle Eastern nation sandwiched between Israel to the west and Saudi Arabia, Iraq, and Syria on its other sides, and it has the distinction of being home to the world's largest number of Palestinians. Like so many countries in the region, Jordan's modern boundaries were set somewhat arbitrarily at the end of World War I when the remnants of the Ottoman Empire were put under the "protectorate," or "mandate," of various European powers—England, in the case of Jordan. In 1946 Jordan was granted independence, and its area at that time included territory on both sides of the Jordan River. However, Jordan was part of the Arab coalition that attacked Israel in 1967, and Israel has militarily occupied the West Bank since that date. Many Arabs who had fled to the West Bank from Palestine in 1948, when it was partitioned to create the modern state of Israel, fled further east into Jordan as a result of the 1967 war.

Figure 6.1   The Contrast between High and Low Fertility

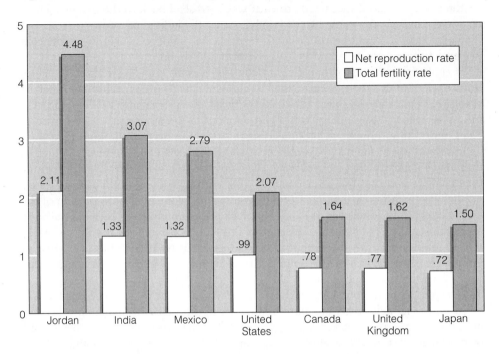

Note: The difference between the higher and lower fertility countries discussed in this chapter is vividly portrayed in this graph. Higher fertility countries like Jordan, India, and Mexico have total fertility rates that are clearly above those of lower fertility countries such as the United States, Canada, the United Kingdom, and Japan. These differences are still marked when we look at the net reproduction rate, which takes into account the difference in child survival between countries.

Source: Adapted from the International Data Base of the U.S. Bureau of the Census. Data are estimates for the year 2000.

Fertility in Jordan is high, but declining. The estimate of the total fertility rate for the year 2000, shown in Figure 6.1, is 4.48, and given the declining mortality in Jordan that translates into a net reproduction rate of 2.11, indicating that the next generation could be more than twice the size of the current generation. Still, that rate for 2000 is less than the 5.6 children per woman found in the 1990 Demographic and Health Survey and is considerably lower than the rate of 7.7 measured in a 1976 fertility survey (Zou'bi, Poedjastoeti, and Ayad 1992). Part of the decline is due to an increase in the age at marriage among women, but part is due as well to an increase in the percentage of married women who are using methods of fertility control. In 1976, 17 percent of women were estimated to have been using a modern method of contraception and that had increased to 27 percent by 1990. Nonetheless, a segment of the Jordanian population may not yet be convinced that reproductive decisions are within their personal control. In 1985 a survey of husbands in Jordan revealed that over half of that nation's married men believed that family-size decisions were up to God, and more than 40 percent explicitly did not believe in practicing contraception (Warren, et al. 1990).

As is true in most countries, rural husbands were more likely to think that God would determine family size than were urban husbands, and the rural men also expected larger families. Similarly, the more educated a man, the more likely he was to believe that he, rather than God, would choose, and the fewer children he expected to have.

There are more than 4 million people in Jordan, and the population is currently growing at a rate that will double in size in scarcely more than 20 years. Still, the government has no official population policy, although contraceptives are available in government-run clinics. Since most Jordanians are Muslim, it is tempting to think that Islam is responsible for high fertility. As I have argued elsewhere (Weeks 1988), it is unlikely that religion per se is the motivating force. The low level of education contributes to the maintenance of traditional beliefs (which tend to support high fertility regardless of the specific religious persuasion), and the flames of high fertility are fanned by the Palestine Liberation Organization (PLO), which views the high birth rate as a "secret weapon" in its fight to regain the land conquered by Israel (Waldman 1991). Jordan is one of the few Arab nations that has granted citizenship to Palestinians, and a large fraction of the Jordanian population is of Palestinian origin.

The counterpoint to Jordan's high birth rate is the area's limited supply of water (Richards and Waterbury 1990), and there is a serious question as to how long the region's ecological resources can sustain the current rate of population growth. To avoid further disaster it seems necessary not simply for the birth rate to decline, but for it to decline at even a more rapid rate than it is currently doing.

India

India is especially interesting, since virtually every explanation I have discussed for high fertility applies to at least some area of that country. You will recall from Chapter 1 that India is a large, culturally diverse society, where the pattern of fertility has been one of consistently high levels. Davis (1951) estimated that in 1881–91 the crude birth rate was about 49 per 1,000 population. After 1921 there is some evidence of a decline, such that by 1941 the birth rate was about 45 per 1,000. Data were not reliable enough to ensure that the decline was real, and if it was, there is no apparent explanation for it, since historically there had been little use of contraceptives in India. In fact, Davis reports that in a study done in 1941 in Kolhapur City with a population of nearly 100,000 people, in a sample of 1,661 married women, only 3 practiced contraception. After an exhaustive analysis of the population of India and Pakistan, Davis reached "the melancholy conclusion that an early and substantial decline of fertility in India seems unlikely unless rapid changes not now known or envisaged are made in Indian life" (1951:82).

In 1958, Coale and Hoover concluded that low fertility had become established among only a negligible fraction of the Indian population—a very small group of urban, highly educated people (Coale and Hoover 1958). Between 1941 and 1961, estimates of the birth rate in India indicate that it remained at a level of about 45 per 1,000 population. Data for the period 1961–71 indicated a small decline in the crude birth rate to 42 per 1,000. An analysis by Adlakha and Kirk (1974) suggested that this decline represented a real drop in marital fertility among Indian women, which may have been a result of the concerted government efforts to promote family planning techniques, especially sterilization. However, the decline in fertility was outweighed by declines in the death rate, and in fact the population of India was growing faster than ever before. By 1995 the crude birth rate had dropped to 29 per 1,000, but the steady drop in the death rate has maintained the pace of Indian population growth.

As has been alluded to earlier in this chapter, India is a nation in which sons are often preferred to daughters, and this fact is reflected in the marriage patterns. A bride usually lives with her husband's family and brings a dowry with her, which means that a son's marriage will provide his family with a young woman, who often is expected to do much of the domestic work, along with money or goods from the girl's family. On the other hand, a family with several daughters faces a very expensive prospect as the girls reach puberty.

In a review of studies on India, Mandelbaum reports that "typically a woman knows of no acceptable alternative for herself than that of wife-mother. . . . For all but a relative few, a woman's destiny lies mainly in her procreation; the mark of her success as a person is in her living, thriving children" (1974:16). The status of women in India is as low as anywhere in the world, and thus a failure to conceive is routinely blamed solely on the woman (Jeffery, Jeffery, and Lyon 1989). Women know that husbands may die before they do and children, especially sons, may offer a source of economic security. "Parents look to sons for support in old age and, in keeping with tradition, maintain that it is only with sons that parents can share a home" (Mehta 1975:134). Further, fathers often argue that they need the sons to help work the land, even if their farm is small (Jejeebhoy and Kulkarni 1989). Above all, in a nation where there is still only limited material wealth to which most people can aspire, children help fulfill aspiration for some status in society and for expressing their parents' creativity.

From 1954 through 1960 an intensive birth control program was conducted in the Indian state of Punjab. Known as the Khanna study, it was sponsored by the Harvard School of Public Health, and it cost well over $1 million. A follow-up study in 1969 showed that the program had been a failure. Why did it fail? In analyzing the data from the program, Mamdani concluded that "no program would have succeeded, because birth control contradicted the vital interests of the majority of the villagers. To practice contraception would have meant to willfully court economic disaster" (1972:21). The point is that if a person does not see a clear advantage to limiting family size, it is not likely to happen.

Something has been happening, however, at least in parts of India. Between 1960–64 and 1980–84, the total fertility rate declined from 5.81 to 4.75. About one fourth of this was due to a delay in marriage (during this period the legal age at marriage for girls in India was increased from 15 to 18). The remaining three fourths was accounted for by a decline in the average number of children being born to married women, led especially by women in their thirties (Retherford and Rele 1989). The trend carried into the 1990s and the National Family Health Survey in India in 1993 reported a total fertility rate of 3.4 (East-West Center 1995). In the north of

India, the most populous state in the country (Uttar Pradesh, with 139 million people according to the 1991 census) still had a total fertility rate of 4.8, but one state, Kerala (in the southwest of India) has already reached replacement level fertility. What has been happening is that both precondition 2 (motivation to limit fertility) and precondition 3 (availability of means) have been kicking into gear. Between the early 1980s and the early 1990s the economy of India was improving. Roads and electricity reached most rural villages during this period (Economist 1994). Electricity brought television, and the roads brought access to consumer goods, as well as to family planning clinics. By the mid-1990s it is estimated that 45 percent of Indian couples were using some form of contraception, primarily condoms, to space children and then sterilization to stop reproduction (Rajan, Mishra, and Ramanathan 1993; Ross, Mauldin, and Miller 1993).

It is of some importance to note that although fertility may now be yielding in parts of India, its level even a few decades ago was still well below the biological maximum. Why wasn't it any higher? The answer to this question is related to the social environment in which reproduction occurs. Childbearing is only one of the activities necessary for the ongoing conduct of society. There are many institutions, such as religion, government, and education, that are also important, and activities involving other aspects of social life often compete or conflict with sexual intercourse or childbearing. It is because of this fact that most premodern societies have been able to maintain a relative balance between resources and population (Lesthaeghe 1980). Thus, virtually every known society has social barriers to maximizing fertility and India is no exception. In different societies, these barriers may involve late age at marriage, restrictions against remarriage, special times when intercourse is taboo, or a host of other customs that keep fertility lower than it would otherwise be. Individuals may not be consciously motivated to limit family size, but social institutions prevent them from reaching their maximum potential.

These competing pressures operate through the proximate determinants of fertility. For example, in India it is believed (correctly so) that the health of an infant is endangered if the next child comes too quickly. Thus, in many regions there is a taboo on the mother having sexual intercourse for several months (sometimes longer) after the birth of a child. As a result, Indian children are spaced an average of 3 to 4 years apart (Mandelbaum 1974). This of course lowers the overall number of children that a woman can have in her lifetime. Hutterite women space their children very close together—less than 2 years on the average, according to Sheps, (1965)—and that fact alone could help to explain why Indians have fewer children than Hutterites do. It is also reported that there is a "pregnant grandmother complex" in India; that is, a woman is openly criticized and disgraced if she becomes pregnant after she is already a grandmother (Caldwell, Reddy, and Caldwell 1988).

The Indian government has become increasingly concerned about how the nation's welfare is being affected by population growth and has adopted several policies to try to promote a two-child norm. In Chapter 14, I discuss their policies in more detail, but it will suffice here to note that government attempts to limit births were met by opposition leading to violence in 1976 and the subsequent ouster of the antinatalist government of Indira Gandhi. After she regained power, Gandhi carefully sidestepped the family limitation issue, setting an example followed by several of her successors.

Mexico

Mexico is the eleventh most populous nation in the world and it arrived there through a classic pattern of declining mortality that for several decades was not matched by a drop in fertility. Life expectancy began to increase in Mexico in the 1930s, but there was virtually no sign of a change in reproductive behavior until the 1960s (Mier y Terán 1991). Therein lies the reason for the population of Mexico to have tripled in size (from 17 to 49 million) between 1930 and 1970. Women continued to bear an average of more than 6 children each until the women born just after World War II reached adulthood in the 1960s. They were the ones that led Mexico toward lower fertility (Juárez and Quilodrán 1996; Zavala de Cosío 1989), although as you can see in Figure 6.1 the estimated total fertility rate for the year 2000 is still nearly 3 children per woman.

How did these younger members of Mexican society bring about this change? Partly through a delay in marriage, but largely through the use of modern contraception (Moreno and Singh 1996), facilitated by the government's sudden reversal in 1974 of its policy on birth control. In 1974, President Luis Echeverria, who had campaigned in 1970 on a pronatalist platform, changed his mind and pushed through a change in Mexico's General Law of Population that not only made it legal to buy and sell contraceptives, but also set up government agencies to make contraceptives available. In doing so, Echeverria was responding to the needs of younger adults who were already looking for ways to limit their family size.

Throughout the world these dynamics of fertility change are being played out differently, but there are some obvious regional similarities, as you can see in Figure 6.2. The examples that I gave you above of higher fertility countries come from the Middle East, South Asia, and Latin America where, as you can from Figure 6.2, almost all countries have fertility levels that are well above the replacement level. Still, in most of these countries there are signs that fertility is declining. Will all countries join Europe, East Asia, and North America in bringing fertility down to the level where it once again matches the mortality level? And, if so, how will that come about? Just what is it that motivates people to have small families or no families at all? These are among the most important questions facing the world.

Explanations for Low Fertility

All of the perspectives on the fertility transition discussed earlier in the chapter assume that fertility will not decline until people see it as being in their interest to limit fertility. The supply–demand framework assumes that people are making rational economic choices between the quantity and quality of children, whereas the innovation-diffusion perspective argues that social pressure is the motivation, regardless of the underlying economic circumstances. Coale's three preconditions do not specify what the motivating factors might be, leaving open the possibility of some combination of economic and social motivations. In the real world, a combination of economic and social forces seems to predominate, illustrated by the way in which wealth (an economic factor) and prestige (a social variable) influence fertility in nearly every society.

Wealth, Prestige, and Fertility

Historically the most persistent socioeconomic factors related to fertility are wealth and prestige (both of which are further related to power). A study of primitive societies has shown that population control was related to a competition for power and prestige (Douglas 1966), which often led to an emphasis on high fertility (there may be power in numbers) rather than low fertility. However, as Benedict (1972) has noted, in urban industrial societies of the 19th and 20th centuries, prestige and wealth tend to be linked with low fertility rather than high fertility. This reversal may seem puzzling at first glance. After all, it would seem on an a priori basis that as people acquired wealth and prestige they would acquire more children, since presumably they could afford more rather than fewer. The new home economics approach suggests that the key is in the availability of resources. In most countries, wealth and prestige are scarce economic and social commodities, and it may require sacrifices of one kind or another if an individual is to beat the competition. One sacrifice is the large family. In 1938, an Englishman put it rather succinctly: ". . . in our existing economic system, apart from luck, there are two ways of rising in the economic system; one is by ability, and the other by infertility. It is clear that of two equally able men—the

Figure 6.2   Total Fertility Rates are Highest In Africa and Western Asia

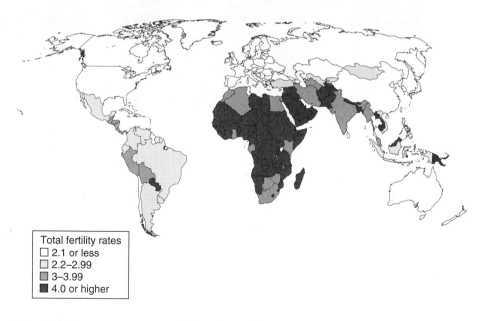

Total fertility rates
☐ 2.1 or less
☐ 2.2–2.99
▨ 3–3.99
■ 4.0 or higher

Note: Fertility rates are at or below the replacement (2.1 or less) in North America, Europe, East Asia, and Oceania, but remain above that level in the rest of the world, with the highest rates (4.0 or higher) in Africa and Western Asia.

Source: Mapped by the author from data in the U.S. Bureau of the Census International Data Base. Data are estimates for the year 2000.

one with a single child, and the other with eight children—the one with a single child will be more likely to rise in the social scale" (Daly 1971:33).

Thus, acquiring wealth may require that a family be kept small, whereas already having wealth may permit, and even encourage, the growth of families. Often people who have kept families small in order to acquire wealth and prestige are past their reproductive years when they reach their goal (if they reach it), or they may have grown comfortable with a small family and decide not to have more children even though finally they might be able to afford them. This difference in the timing of fertility and wealth is one important point to keep in mind in reviewing explanations for low fertility.

Income and Fertility

In June of each year, the Current Population Survey of the U.S. Census Bureau asks American women about their reproductive behavior. I have reproduced some of the findings from the 1995 survey in Figure 6.3. From these data we may infer that, for wives of any age, whether they are working or not, the more money the family has, the fewer children are likely to have been born to date. For example, women aged 25 to 34 who were working and had a family income of $75,000 or more had given birth to an average of 1.008 children, compared with 1.609 for those whose family income was less than $20,000. The idea that opportunity costs can lower fertility is seen by comparing the fertility of women who work with those who do not, at the same income level. Among women aged 25 to 34 in a marriage where the family income is $50,000 to $74,999, those whose work contributes to that income have an average of 1.148 children each. This is nearly one child less than the women at that income level who are not in the labor force, who average 1.804 children each. Overall, the data in Figure 6.3 show that the highest levels of fertility in the United States are found among poor women who do not work, whereas the lowest levels of fertility are among those who do work and are well paid.

Education and Fertility

Although wealth and prestige are characteristics of premodern as well as contemporary societies, education is a new dimension that has been added to the mix by the Renaissance and the Industrial Revolution that it engendered. As I mentioned previously in Chapter 3, education (and its closely related cousin, secularization) is the single best clue to a person's attitude toward reproduction. An increase in education (probably to a level beyond primary) is most apt to be associated with Coale's first precondition for a decline in fertility (the acceptance of calculated choice) and is most strongly associated with the kind of rational decision-making implied in the supply–demand framework. Furthermore, the better educated members of society are most likely to be the agents of change that will encourage the diffusion of an innovation such as fertility limitation.

I have never seen a set of data from anywhere in the world that did not show that more educated women had lower fertility than the less educated women in that society. It is nearly axiomatic, and it is the identification of this kind of **fertility**

differential that helps to build our understanding of reproductive dynamics in human societies, because it causes us to ask what it is about education that makes reproduction so sensitive to it. In general terms, the answer is that education offers to people (men and women) a view of the world that expands their horizon beyond the boundaries of traditional society, and causes them to reassess the value of children, and to reevaluate the role of women in society. Education also increases the opportunity for social mobility which, in turn, sharpens the likelihood that people will be in the path of innovative behavior, such as fertility limitation, which they may try themselves. I will return to these ideas later in Chapter 10.

Figure 6.3   Higher Income Women Have Fewer Children (United States, 1995)

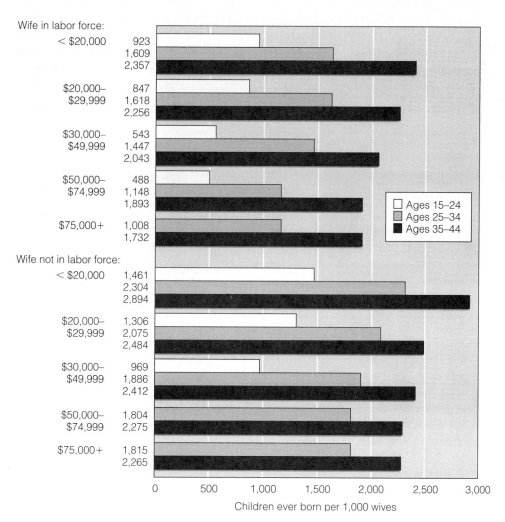

Source: Adapted by the author from Amara Bachu, U.S. Bureau of the Census, 1997, "Fertility of American Women: June 1995 (Update)," Current Population Reports, Series P-20, No. 499, Table 3.

Canada and Mexico, like virtually all other societies, offer us examples of the relationship between fertility and education, as shown in Figure 6.4. Data are shown for women who are old enough to have completed their childbearing (aged 40 to 49), so that any babies whose arrival might have been delayed by a woman's education at a younger age have now had a chance to be born. The differences in the number of children ever born by education are more striking in the case of Mexico because it is still in the midst of a fertility decline, led especially by those women with the higher levels of education. Yet, even in Canada, where fertility has now dropped to below replacement level, it is still true that among these women aged 40 to 49 interviewed in the National Fertility Study, the better educated had fewer children than the less well educated. An important difference between Canada and Mexico that you can see from Figure 6.4, of course, is that at each level of education, women in Mexico have had more children than the women in Canada. So, education always leads to fertility differentials within a population, but similar levels of education do not always lead to similar levels of fertility in different populations. Part of the reason for this difference may be economic: education will buy you more in some societies than in others, and the more it will buy you, the higher might be the opportunity costs of children.

Figure 6.4   Fertility Goes Down as Education Goes Up: Canada and Mexico as Examples

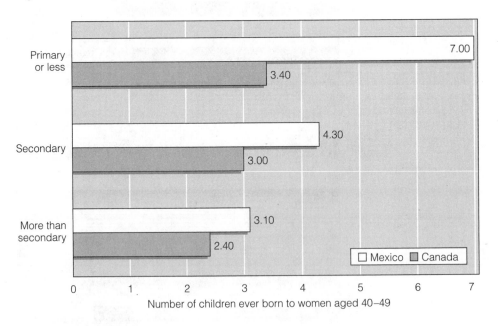

Sources: Adapted from data for Canada from T.R. Balakrishnan, Evelyne LaPierre-Adamcyk, and Karol J. Krótke, 1993, Family and Childbearing in Canada: A Demographic Analysis (Toronto: University of Toronto Press), Table 3.5; and for Mexico from Dirección General de Planificación Familiar, México: Encuesta Nacional Sobre Fecundidad y Salud, 1987 (Columbia, MD: IRD/Macro International), 1989, Table 6.4.

Other Factors

There are other factors that help to explain what motivates people to limit their fertility. These include the social characteristics such as race/ethnicity and religion, as well as more blatantly economic characteristics such as occupational status. In various parts of the book, but especially in Chapter 10, I discuss some of these factors in more detail. Remember also that the means available is an important issue in determining the birth rate, once the motivation exists to limit fertility. Availability often is a reflection of public policy, and I discuss this more in Chapter 14. For now, let us examine some of the factors involved in declining fertility by examining four countries that have experienced that phenomenon—England, Japan, the United States, and Canada. Note, by the way, that the fertility decline in China is discussed in considerable detail in Chapter 14.

Lower Fertility Countries

Let me now put the fertility transition perspectives to work in describing and accounting for changes in fertility in four impressive cases of long-run fertility declines accompanying improvements in the standard of living—England, Japan, the United States, and Canada. As a preface to that discussion, however, I should note that in England and other parts of Europe the beginnings of a potential fertility decline may well have existed before the Industrial Revolution touched off the dramatic rise in the standard of living. In English parishes there is evidence that withdrawal (coitus interruptus) was used to reduce marital fertility during the late 17th and early 18th centuries, and it was apparently also a major reason for a steady decline in marital fertility in France during the late 18th and early 19th centuries. Abortion was quite probably also fairly common (Wrigley 1974). Furthermore, the fact that preindustrial birth rates were much higher in the European colonies of America than in Europe points to the fact that fertility limitation in Europe was widely accepted and practiced, especially through the mechanism of deliberately delayed marriage, but through other more direct means as well.

England

The enormous economic and social upheaval of industrialization took place earlier in England than anywhere else, and by the first part of the 19th century, England was well into the Machine Age. For the average worker, however, it was not until the latter half of the 19th century that sustained increases in real wages actually occurred. During the first part of the century the Napoleonic Wars were tripling the national debt in England, increasing prices by as much as 90 percent without an increase in production. Thus, during almost the entire professional life of Malthus, his country was experiencing substantial inflation and job insecurity. These relatively adverse conditions undoubtedly contributed to a general decline in the birth rate during the first half of the 19th century (see Figure 6.5), brought about largely by delayed marriage (Wrigley 1987).

Figure 6.5   England's Demographic Transition

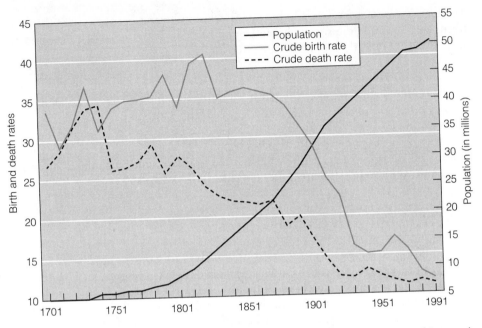

Note: Between 1701 and 1991 England experienced all stages of the demographic transition. The death rate began to drop late in the eighteenth century during the preliminary period of the Industrial Revolution. Birth rates did not experience a sustained drop until almost 100 years later. Since the late 19th century, fertility has dropped fairly steadily to the point where both birth and death rates are currently low.
Sources: Adapted from data in B. R. Mitchell, 1988, British Historical Statistics (Cambridge: Cambridge University Press), Population and Vital Statistics Tables 1, 2, 10, 11, and 13; and Central Statistical Office (U.K.), 1991, Annual Abstract of Statistics 1991 (London: HMSO), Tables 2.1, 2.16, and 2.22.

After about 1850, economic conditions improved considerably, and the first response was a rise in the birth rate (as the marriage rate increased), followed by a long-run decline, as can be seen in Figure 6.5. This was a period in which all of Ansley Coale's preconditions for a fertility decline existed: (1) people had apparently accepted calculated choice as a valid element, (2) there were perceived advantages from lowered fertility, and (3) they were aware of effective means of birth control. As I mentioned earlier, the British were accustomed to thinking in terms of family limitation, and delayed marriage, abstinence, and coitus interruptus within marriage were known to be effective means to reduce fertility. In the second half of the 19th century, then, motivation to limit family size came in the form of larger numbers of surviving children combined with aspiration for higher standards of living.

As you can see in Figure 6.5, fertility actually went up in England before it went down. With fewer families broken by death and with increasing real income, which made earlier marriage more feasible, birth rates rose before people could adjust their reproductive behavior to meet the new demographic and economic situations. This is reflected in the net reproduction rate, which measures each woman's average number of female children who will survive to adulthood (a figure of 1 means exact generational replacement). In 1841, the net reproduction rate was 1.35, and it had increased to 1.51 by 1881. By 1911, however, it was down to 1.12 (Wrigley 1974:195).

It is important to remember that the restriction of fertility was in many ways a return to preindustrial patterns, in which an average of about two children survived to adulthood in each generation. Thus, as we discussed earlier in relation to the theories of the fertility transition, mortality declines produced changes in the lives of individuals to which they had to respond. The English reacted in ways that were consistent with the theory of demographic change and response. They responded to population growth by migrating, by delaying marriage, and then, only when those options were played out, did marital fertility clearly decline (Friedlander 1983).

The best-known explanation of the fertility decline in England in the latter half of the 19th century is that offered by J. A. Banks (1954) in *Prosperity and Parenthood* and its sequel *Feminism and Family Planning in Victorian England* (Banks and Banks 1964). Banks's thesis is by now a familiar one: that the rising standard of living in England, especially among the middle classes, gave rise to a declining fertility by (1) raising expectations of upward social mobility, (2) creating fears of social slippage (you had to "keep up with the Joneses"), and (3) redefining the roles of women from housewife to a fragile luxury of a middle-class man.

While it is true that birth rates dropped more quickly in the upper social strata of English society than in the lower social strata, by 1880 all segments of English society were experiencing fertility declines. From about 1880 to 1910 England shared with much of continental Europe in what Knodel and van de Walle (1979) have called "the momentous revolution of family limitation." Fertility has continued on a slow downward trend since then, interrupted by a post-war Baby Boom, which peaked in the mid-1960s. Since about 1964, however, birth rates have resumed their downward trend—England's Baby Bust (Hobcraft 1996)—and England has settled into a consistent pattern of below-replacement-level fertility, with a TFR for the year 2000 of only 1.6 children per woman (see Figure 6.1).

Japan

During the 19th century, under the influence of the Tokugawa Shogunate, Japan developed an isolated, self-sufficient economy based on commerce rather than landholdings. The Japanese borrowed ideas and technology from China and Korea to produce a commercial economy similar to, but independent from, that which had developed in Europe. The real "take-off" period for Japan's economic development came, however, between 1878 and 1900 after the Meiji Restoration and perhaps not coincidentally, this happened after Japan had reestablished relations with Europe. In 1920 both birth and death rates started to decline (Muramatsu 1971). The decline in fertility was accomplished primarily through abortion and the use of condoms, which the Japanese were producing themselves. Nonetheless, mortality was decreasing more rapidly than fertility, and the population was experiencing fairly rapid growth as well as high rates of urbanization: the migration of people from the countryside to the city. Growth was fostered in the early 20th century by the pronatalist, imperialistic policy of the Japanese government.

Japan, like England, has had a long history of demographic consciousness. There is substantial evidence that mortality and fertility were both low in Japan by world standards as far back as the 17th century. Fertility was apparently kept low by a combination of delayed marriage and abortion (Hanley and Yamamura 1977). But our demographic interest in Japan actually lies less in what happened before World War II than in the dramatic drop in fertility after the war. Between 1947 and 1957 the crude birth rate went down 50 percent, from 34.3 to 17.2 per 1,000 population. As measured by the total fertility rate, the drop was equally sensational. In 1947 the TFR was estimated to have been 4.5 children per woman, and by 1957 it was down to 2.0. Thus, in only a decade, reproductive performance was cut by more than half, principally by means of induced abortion. Previously, abortions had been illegal, but legal abortions were made possible by the Eugenics Protection Act of 1948. This law was not really aimed at reducing population growth but rather at protecting women's health by eliminating the need for illegal abortions, which had been increasing in number (see Table 6.1). During this time, the condom continued as a popular contraceptive, and the age at marriage for females also increased slightly (Kobayashi 1969). Since 1955 the use of contraceptives has been increasing steadily (although not until 1997 was the pill approved

Table 6.1   Abortion Played an Important Role in Japan's Fertility Decline

Year	Annual Totals (in Thousands)			Sum per 1,000 Population
	Births	Abortions	Sum	
1949	2,697	102	2,799	34.4
1950	2,338	320	2,658	32.1
1951	2,138	459	2,597	30.8
1952	2,005	798	2,803	32.8
1953	1,868	1,067	2,935	33.9
1954	1,770	1,143	2,913	33.1
1955	1,727	1,170	2,897	32.6
1956	1,665	1,159	2,825	31.4
1957	1,563	1,122	2,685	29.6
1958	1,653	1,128	2,781	30.4
1959	1,636	1,099	2,725	29.5

Note: Between the late 1940s and the late 1950s the birth rate fell dramatically in Japan, largely as a result of the rise in induced abortion. You can see that there was little variation from year to year in the number of conceptions (births plus abortions), with the number of abortions going up as the number of births went down. Note that due to rounding the births and abortions may not sum exactly.

Source: K. Davis, 1963, "The theory of change and response in modern demographic history," Population Index 294:347, Table 1. Used by permission.

for use by Japanese women) and is currently the major source of maintaining low fertility, along with voluntary sterilization.

An interesting sidelight is the incredible social impact of an ancient superstition on a modern, rational population. I refer to the Year of the Fiery Horse. In 1966 the birth rate made a sudden 1-year dip—this was the Year of the Fiery Horse. According to a widely held Japanese superstition, girls born in the Year of the Fiery Horse (which occurs every 60 years) will have troublesome characters, such as a propensity to murder their husbands. Thus, girls born in that year are hard to marry off, and many couples avoided having children in 1966. Again, this was accomplished mainly by contraception rather than abortion. However, in 1906—another Fiery Horse year—fertility had also declined dramatically without the availability of modern contraception. The Chinese follow a different cultural calendar, by the way, and the year of the Dragon in the Chinese calendar is thought to be a good omen for the birth of a child. Thus, among the Chinese in Taiwan, Hong Kong, Singapore, and Malaysia (although not in mainland China) two recent years of the Dragon (1976 and 1988) produced a spike in birth rates, interrupting fertility declines in those populations (Goodkind 1991; 1995a). While at first glance this might seem like nonrational "cultural" behavior, Goodkind (1991) has explained it in terms of the supply–demand framework, suggesting that as couples "plan for ever-smaller families, parents become more concerned with child quality, and, for some parents, the first investment in that quality concerns choosing the right zodiacal sign for the child" (p. 673).

I have already mentioned that Japan has the highest life expectancy in the world, and that has been matched by one of the lowest levels of fertility in the world. As you can see in Figure 6.1, the total fertility rate in the year 2000 is estimated to be 1.50, with a net reproduction rate of only 0.72. At the current rate, Tokyo could eventually be quite a bit less crowded than it is now. Retherford and his associates (Retherford, Ogawa, and Sakamoto 1996) have attributed this low fertility to changing attitudes of women toward marriage and family-building. Although surveys in Japan suggest that the average Japanese woman thinks that more than two children is the ideal family size, women are nonetheless staying single longer, and are delaying their first birth once they get married (out-of-wedlock births are very rare in Japan).

The status of women in Japan is climbing, but it is still demonstrably inferior to that of men, and fertility is so low partly because women are reluctant to yield to the more traditional pressures inherent in being a wife and mother in Japan. These findings for Japan are consistent with analyses of two European countries—Italy and Spain—where fertility is also well below replacement for similar reasons. Chesnais (1996) has offered the seemingly paradoxical suggestion that in highly developed countries, a *rise* in the status of women may be necessary to bring fertility back up to replacement level. Increased education and occupational opportunities for women in these countries have raised their expectations for independence, but the maintenance of traditional family roles, as in Japan, discourage women from marriage and childbearing while, at the same time, traditional attitudes toward women punish them socially for childbearing outside of marriage. The result has been very low fertility, which may only rise when women are more assured that family-building activities will not be seen as precluding a career and economic and social independence.

The United States

Around 1800, when Malthus was writing his "Essay on Population," he found the growth rate in America to be remarkably high and commented on the large frontier families about which he had read. Indeed, it is estimated that the average number of children born per woman in colonial America was about eight. It is probably no exaggeration to say that early in the history of the United States, American fertility was higher than any European population had ever experienced. Early data are not very reliable, but in 1963 Ansley Coale and Melvin Zelnick made new estimates of crude birth rates in the United States going back as far as 1800; these indicate that the crude birth rate of nearly 55 per 1,000 population was higher than the rate in any less developed country today (Coale and Zelnick 1963). Even in 1855 the crude birth rate in America was 43 per 1,000, higher than the current levels in any of the higher fertility countries discussed earlier in this chapter. However, the birth rate had clearly begun a rapid decline, and by 1870 the American birth rate had reached the lower levels of European countries. This decline continued virtually unabated until the Great Depression of the 1930s, during which time it bottomed out at a low level only recently reapproached. Why the precipitous drop?

As I discussed in Chapter 1, almost all voluntary migrants to the North American continent were Europeans. The people who made up the population of the early United States came from a social environment in which fertility limitation was known and practiced. Despite the frontier movement westward, America in the century after the Revolution was urbanizing and commercializing rapidly. Furthermore, the United States was experiencing the process of secularization, and people's lives were increasingly loosened from the control of Church and State. Malthus had commented that, with respect to all aspects of life, including reproduction, "despotism, ignorance, and oppression produced irresponsibility; civil and political liberty and an informed public gave grounds for expecting prudence and restraint" (quoted by Wrigley 1988:39). Bolton and Leasure (1979) have shown that throughout Europe, the early decline in fertility occurred near the time of revolution, democratic reform, or the growth of a nationalist movement. Analogously, Leasure (1989) has found that the decline in fertility in the United States in the 19th century was closely associated with a rise in what he calls the "spirit of autonomy," measured early in the century by the proportion of the population in an area belonging to the more tolerant Protestant denominations (Congregational, Presbyterian, Quaker, Unitarian, and Universalist) and measured later in the century by educational level.

Lower fertility was accomplished by a rise in the average age at marriage and by various means of birth control within marriage, including coitus interruptus, abortion (even though it was illegal), and breast-feeding (Sanderson 1979). Nineteenth-century America also witnessed the secret spread of knowledge about douching and periodic abstinence, neither of which is necessarily very effective on the face of it, but the fact that women were searching for ways to prevent pregnancy is clear evidence of the motivation that women had to limit fertility (Brodie 1994).

After World War I the use of condoms became widespread in the United States (and in Europe as well) and, along with withdrawal and abstinence, contributed to the very low levels of fertility during the Depression (Himes 1976). Estimates of the percentage of women who were using contraceptive methods at that time range from 42 to 95 percent (Himes, 1976:343). Note that the condom method, like coitus interruptus and abstinence, requires the initiative or cooperation of the male, whereas modern chemical or surgical methods do not. As a consequence, they are also methods that provide women with a way of negotiating sexuality (by offering a condom to her partner, or forcing the "interruptus" during coitus, or insisting upon abstinence). Thus, the adoption of these methods of fertility limitation can be seen as part of the long-term change in gender relations in the United States (Brodie 1994).

During the Depression, fertility fell to levels below generational replacement. The United States was not unique in this respect, but that bottoming out did cap the most sustained drop in fertility that the world has yet seen. It was undoubtedly a response to the economic insecurity of the period, especially since that insecurity had come about as a quick reversal of increasing prosperity. Fear of social slippage was thus a very likely motive for keeping families small. The American demographic response was for many couples to defer marriage and to postpone having children, hoping to marry later on and have a larger family. Gallup polls starting in 1936 indicate that the average ideal family size was three children, and that most people felt that somewhere between two and four was what they would like. Thus, people were apparently having fewer children than they would liked to have had under ideal circumstances.

In 1933, the birth rate hit rock bottom because women of all ages, regardless of how many children they already had, lowered their level of reproduction. But from 1934 on, the birth rates for first and second children rose steadily (reflecting people getting married and having small families), while birth rates for third and later children continued to decline (reflecting the postponement of larger families) until about 1940 (Grabill, Kiser, and Whelpton 1958). Just as the United States was entering World War II in late 1941 and 1942, there was a momentary rise in the birth rate as husbands went off to war, followed by a lull during the war. However, the end of World War II signaled one of the most dramatic demographic phenomena in North American history—the **Baby Boom.**

The Baby Boom   Most of you can probably appreciate that immediately after the end of a war, families and lovers are reunited and the birth rate will go up temporarily as people make up for lost time. This occurred in the United States as well as in England and Canada and several of the other countries actively involved in World War II. Surprisingly, these Baby Booms lasted not for 1 or 2 years but for several years after the war. Birth rates in the United States continued to rise through the 1950s, as the total fertility rate went from 2.19 in 1940 to 3.58 in 1957, an increase of nearly 1.5 children per woman. Note that in the United States the term "baby boomers" is usually applied to people born between 1946 and 1964, but the "boom" peaked in 1957—12 years after the war ended.

An important contribution to the Baby Boom was the fact that after the war, women started marrying earlier and having their children sooner after marriage. For example, in 1940 the average first child was born when the mother was 23.2 years old, whereas by 1960, the average age had dropped to 21.8. This had the effect of bunching up the births of babies, which in earlier times would have been more spread out. Further, not only were young women having children at younger ages, but older women were having babies at older than usual ages, due at least in part to their having postponed births during the Depression and war. After the war, many women stopped postponing and added to the crop of babies each year. These somewhat mechanical aspects of a "catching up" process explain only the early part of the Baby Boom. What accounts for its prolongation?

We do not have a definitive answer to this question (and remember that it occurred in other countries besides just the US), but a widely discussed explanation is offered by Easterlin (1968; 1978), which I mentioned in Chapter 3 as the relative income hypothesis—a spin-off of the supply–demand perspective. Easterlin begins his analysis by noting that the long-term decline in the birth rate in the United States was uneven, sometimes declining more rapidly than at other times. In particular, it declined less rapidly during times of greater economic growth. If a young man could easily find a well-paying job, he could get married and have children; if job hunting were more difficult, marriage (and children within marriage) would be postponed. Thus, it was natural that the postwar Baby Boom occurred, because the economy was growing rapidly during that time. What was unusual was that economic growth was more rapid than in previous decades, and the resultant demand for labor was less easily met by large numbers of immigrants, because in the 1920s the United States had passed very restrictive immigration laws (see Chapter 14). Furthermore, the number of young people looking for work was rather small because of the low birth rates in the 1920s and 1930s. Finally, the demand for labor was not easily met by females, since there was a distinct bias against married women working in the United States, particularly a woman who had any children. In some states legislation existed actually restricting married women from working in certain occupations. To be sure, women did work, especially single women, but their opportunities were limited (see Chapter 10 for further discussion). Thus, economic expansion, restricted immigration, a small labor force, and discrimination against women in the labor force meant that young men looking for jobs could find relatively well-paying positions, marry early, and have children. Indeed, income was rising so rapidly after the war and on into the 1950s that it was relatively easy for couples to achieve the life-style to which they were accustomed, or even to which they might moderately aspire, and still have enough money left over to have several children.

As Campbell (1969) has pointed out, Easterlin's thesis is consistent with Davis's theory of demographic change and response (see Chapter 3). Davis argued that fear of relative deprivation, rather than the threat of famine or absolute deprivation, is a subjective stimulus to limiting fertility. The other side of the coin is that when you feel more secure, the desire for children may resurface if pronatalist pressures still exist, as they did and still do.

In 1958 the crude birth rate and the general fertility rate in the United States registered clear declines—a downward change that carried into the late 1970s—as you can see in Figure 6.6. At first the decline was due merely to the fact that the trend of earlier marriage and closer spacing of children had bottomed out. The number of women in the childbearing ages had also declined as the relatively small number of babies born during the Depression reached childbearing age. At this point in the early 1960s there was no discernible trend toward smaller families. The ideal family size among Americans had remained quite stable between 1952 and 1966, ranging only between 3.3 and 3.6 children. But in 1967, Blake discovered in a national sample taken the year before that "young women (those under age 30) gave 'two' children as their ideal more frequently than they had in any surveys since the early nineteen-fifties" (Blake 1967:20). This was the first solid evidence that the desired family size might be on the way down—that the Baby Bust period had arrived.

The Baby Bust and the Baby Boomlet   There were social and economic factors that suggested fertility might continue to decline for awhile. The rate of economic growth had slackened off, and there was no longer a labor shortage. As Norman Ryder very presciently noted:

> In the United States today the cohorts entering adulthood are much larger than their predecessors. In consequence they were raised in crowded housing, crammed together in schools, and are now threatening to be a glut on the labor market. Perhaps they will have to delay marriage, because of too few jobs or houses, and have fewer children. It is not entirely coincidental that the American cohorts whose fertility levels appear to be the highest in this century were those with the smallest number [Ryder 1960:845].

Figure 6.6   The Baby Boom, the Baby Bust, and the Baby Boomlet in the United States

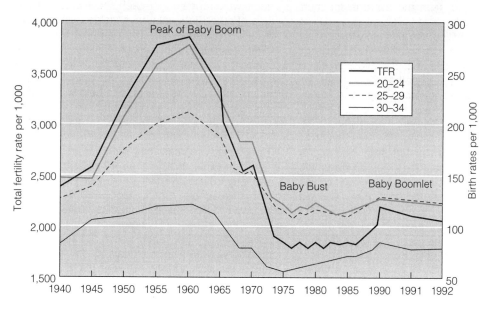

Note: The total fertility rate shows the overall pattern of reproductive behavior in the United States, while the age-specific birth rates show the contribution of individual age groups to that overall pattern.

Sources: National Center for Health Statistics, 1990, Vital Statistics of the United States, 1988, Vol. I, Natality (Hyattsville, MD: NOHS), Table 1-6; S, Ventura, et al., 1994, Advance Report of Final Natality Statistics, 1992, Monthly Vital Statistics Report 43(5): Table 4.

IS STUDENT ACHIEVEMENT RELATED TO FAMILY SIZE?

The issue of quality versus quantity of children finds expression in the relationship between student achievement and family size. With a larger family the measured intelligence of an average child in the family will likely be lower; the explanation for which may be what Blake (1989) has called the "dilution factor." The more children there are in a family, the more diluted is the amount of attention that each child will receive from parents. The lower level of adult interaction seems to affect verbal performance, in particular, which in its turn is related to student achievement and ultimately to educational attainment.

Belmont and Marolla (1973) were able to utilize data gathered for nearly 400,000 men in the Netherlands born in 1944–47 who were examined at age 19 to determine their fitness for military service. Intelligence was measured for the study using the Raven Progressive Matrices, which in this case was a 40-item written test. They found that the average IQ for men from families with two children was higher than the average for men from families of other sizes. The average IQ for men from large families was invariably lower than the average for men from small families. Belmont and Marolla also found that the average IQ of firstborn children was higher than the average for second children, and so forth. Curiously enough, however, they discovered that average intelligence was highest for firstborn men in a two-child family, followed by the firstborn in a three-child family. "Only" children ranked sixth overall. As you might have guessed, the lowest average IQs were for the ninth child in families of nine (families of nine were the largest for which they reported data).

Let me stress that these averages do not necessarily reflect what you, as an individual, might achieve in terms of measured intelligence. You may well be an only child and be "smarter" than the firstborn of a two-child family. The average reduces a lot of human variability into a single index, so don't take it personally one way or another.

More recent analyses have called some of those conclusions into question because they did not control for the fact that only children are more likely than others to come from a broken home, which fact is likely to lower test scores independently of family size. Studies that have taken broken or intact families into account suggest that only children actually do better than children of other birth orders on achievement tests (Blake 1981).

Overall, it seems certain that larger families are associated with poorer student achievement (Retherford and Sewell 1991; Zajonc, et al. 1991). Since it is often true that the larger families are found in the lower social classes and vice versa, Belmont and Marolla looked at their data for each social class to see if that could explain the relationship they had found. It couldn't. The same facts were repeated for each social class. Analyses of other data sets have yielded similar conclusions (Blake 1989). The probable reason for this goes back to the resource issue. Within any given family, each additional child dilutes not only the interpersonal resources of the parents, but the financial resources available to help with education, as well, thus leaving each child a little less well off than he or she might otherwise be. However, the relationship between interpersonal and economic resources may be different. Analyses of longitudinal data indicate that parents spend more time with the first child, but spend more money (especially for college) on the last child (Downey 1995; Steelman and Powell 1989).

One of the implications of the relationships among birth order, family size, and intelligence is that they might help to explain the fluctuations in the average Scholastic Aptitude Test (SAT) score of high school seniors in the United States. The drop in SAT scores between 1962 and 1979 caused considerable alarm and was blamed on television, a desire to get away from disciplined learning, and the rising proportion of minority group students taking the test. But, as Zajonc (1976) has pointed out, there is no evidence to support any of those contentions. What was happening, though, was that the average birth order of children graduating from high school between 1964 and 1979 was going up. These were the children of the large families of the Baby Boom, and in each year that went by during that time, the average birth order of graduating seniors was higher. Since younger siblings tend to do less well on intelligence tests, it seemed reasonable to infer that they would do less well on the SAT.

The parallel changes between SAT scores and average birth order for the time period 1965–79 were striking, but Blake (1989) discovered that the relationship weakened if one went back in time to 1951 when the SAT was first administered. She suggested that family size, not birth order, was the more important correlate of the long-term trend in the SAT scores. Her analysis shows a close relationship between the average SAT score in a given year and the average number of children in the family of the student taking the test. As family size goes up, SAT scores go down and vice versa. The relationship correctly predicted that 1980 was the year that SAT scores bottomed out and that they would begin to rise in 1982. In 1982 the report was that "nationwide scores by high school seniors on the Scholastic Aptitude Test inched upward this year, reversing a 19-year decline. . . . A jubilant George H. Hanford, president of the College Board, said, `This year's rise, however slight . . . is a welcome sign for educators, parents, and students that serious efforts by the nation's schools and their students to improve the quality of education are tak-

ing effect'" (Colvin 1982). 1983 results were exactly the same as those of 1982, and then "average national scores on the Scholastic Aptitude Test rose a total of four points in 1984, the largest combined gain since 1963. . . . In Washington, Education Secretary Terrel Bell said that `the gain in SAT scores reflects the movement toward excellence in our schools that is sweeping the nation'" (Mackey-Smith 1984:60). But, you might well ask, did the quality of education improve or did the quality of the students improve through the family-size effect?

The answer to that question was muddled a bit in the mid-1980s by the changing sociodemographic characteristics of students taking the test, as well as by changes made to the test itself. Nationally, SAT scores dropped between 1988 and 1991, but hidden within that pattern were rising scores for both whites and nonwhites separately, correlated still with a continuing decline in the average size of test takers' families. Since nonwhites tend to have lower SAT scores than whites, their increasing fraction of all test takers probably contributed to the decline in the average SAT score between 1988 and 1991 (Stout 1992). Between 1992 and 1997, on the other hand, the SAT steadily inched up each year (College Board 1998). Of course, the reasons for this increase were variously ascribed to better teaching, better student preparation, and improved parental support. All of these things were probably made possible by the fact that the groups of high school seniors taking the SAT between 1992 and 1997 were products of the end of the Baby Bust (born between 1976 and 1980: see Figure 6.6) and had few siblings with which to compete for resources. Another glance at Figure 6.6 suggests that the family size of children taking the test in the early part of the 21st century will rise, giving rise to lower SAT scores and another round of national angst.

In fact, between the early 1960s and the mid-1970s the average age at marriage among American women went up slightly and the number of children ever born to women aged 20 to 24 went steadily down, declining by 27 percent between 1960 and 1971. The period 1967 to 1971 was especially dramatic. Between those two dates the total number of lifetime births expected by currently married American women aged 18 to 24 dropped from 2,852 to 2,375 per 1,000. By 1977 the birth expectations had continued to drop at a slower pace to 2,137 births per 1,000 women, and then the number rose in 1978 to 2,166 (U.S. Bureau of the Census 1979). Expectations for older women also dropped steeply between 1967 and 1977, but the decline came later than for the younger women. It was, of course, the younger couples who were most influenced by the cohort squeeze as outlined by Easterlin.

Almost all the fertility decline was due to a drop in marital fertility (Gibson 1976), mainly as a result of more efficient use of contraception, and to a rise in the use of abortion. As fertility dropped, family size ideals dropped as well. Gallup surveys reported by Blake (1974) indicated, for example, that the proportion of white women under age 30 saying that two children was an ideal number rose dramatically from a low of 16 percent in 1957 to 57 percent in 1971. The decline in fertility following the Baby Boom peak seemed thus to signal a major shift in the norms surrounding parenthood in American society. "Motherhood is becoming a legitimate question of *preferences*. Women are now entitled to seek rewards from the pursuit of activities other than childrearing" (Ryder 1990:477, emphasis added). I have already commented on Caldwell's theory of the reversal of income flow between parents and children. The Baby Boomers have been trapped in the final turn of that transition—sandwiched between parents who feel that children owe them something and children who feel that parents owe them something.

The rewards of parenthood in American society have certainly become less tangible, but they do exist. In fact, pronatalism may be encouraged by the fact that even in a low-fertility society, a disproportionate share of people have themselves grown up in a large family (Blake 1989). Since this is not necessarily an intuitively obvious fact, let me give you an example. Suppose that we had five mothers giving birth to an average of two children, but with some variability around that average. Two of the women have one child, two have two children, and the fifth mother has four children. Despite the average of two children per mother and despite the fact that only one mother (20 percent of the mothers) has exceeded the average, 40 percent of the children in this example came from a family of four. Using census and vital statistics data, Blake (1989) has shown that the Baby Boom mothers born in 1931–35 had an average of 3.45 children, but the average child of the Baby Boom came from a family with 4.49 children. In fact, less than 14 percent of Baby Boomers grew up in families with only two children. We can conclude then, that growing up in a large family is not yet a thing of the past.

The Baby Bust troughed in the mid-1970s and was followed by a baby boomlet, as you can see in Figure 6.6. In a much-publicized fashion, the boomlet was sparked by women in their early thirties whose birth rate had risen almost uninterruptedly since 1975. Birth rates also rose for women in their late thirties and early forties, but I have not shown those data in Figure 6.6 in order to keep the graph more readable and, besides, 90 percent of all births in the United States are to women under the age of 35.

The Baby Boomers have been having little boomers, but that phenomenon appears to have peaked in 1990 and, as you can discern from Figure 6.6 (and Figure 6.1), birth rates have dropped slightly since then. Furthermore, these general trends hide a great deal of complexity in American fertility patterns, and two trends, in particular, are worth noting at this point: (1) a rise in illegitimacy, which has consequences for household structure and well-being that are discussed later in Chapter 10; and (2) an increase in childlessness, which also has consequences for household structure and well-being.

Rise in Illegitimacy   As fertility has generally declined since the peak of the Baby Boom, and as the Boomers themselves have moved through the childbearing years, a steadily increasing fraction of babies has been born to unmarried women. In 1970 in the United States, only 11 percent of babies were born out of wedlock. By 1980 this had increased to 18 percent; it hit 28 percent in 1990; and in 1995 it was up to 32 percent (National Center for Health Statistics 1994 #79; (Ventura, et al. 1997). There are substantial differences by race/ethnicity in these percentages. At one extreme, 70 percent of all African-American babies born in 1995 were out-of-wedlock, whereas at the other extreme, only 8 percent of Chinese-American babies fell into that category.

These data may say something about the relative status of women within ethnic groups and they have certainly fueled enormous public debate about the social cost of out-of-wedlock births (measured monetarily by welfare benefits and socioculturally by the deprivations suffered by fatherless children) and they have been interpreted as signs of imminent cultural decay. But, as McLanahan and Casper (1995) point out, the situation is not quite what it seems:

> The increase in the illegitimacy ratio [the ratio of illegitimate births to all births] between 1960 and 1975 was due to two factors: the decline in the fertility of married women and the delaying of marriage. But not to an increasing birthrate among unmarried women. Beginning in the mid-1970s, marital fertility rates stopped declining, nonmarital fertility rates began to rise, and the age at first marriage continued to rise. After 1975, the rise in the illegitimacy ratio was due to increases in nonmarital fertility as well as to increases in the number of women at risk of having a nonmarital birth [p. 11].

Although in the past few years the probability has increased that an unmarried woman will bear a child, part of the rise in the illegitimacy ratio was actually because married women were having fewer or no children, thus tilting the ratio of births in the direction of those born out-of-wedlock.

Rise in Childlessness   It is to be expected that about 7 percent of couples may remain childless because of impaired fecundity, as was discussed in Chapter 5. In the 1970s, data from the Current Population Survey suggested that a slightly higher percentage of women (10 percent) was reaching age 40 to 44 without having had a child. Since then, however, the percentage has steadily increased to 18 percent by 1995 (Bachu 1997). Some fraction of these women probably "drifted" toward childlessness by continually postponing the first child, whereas other women may have intended at an earlier age to remain childless but changed their mind somewhere along the way. We know that most reproductive decisions, in both high and low fertility societies, are made sequentially from one birth to the next (Morgan 1996).

One of the more important aspects of childlessness is that it illustrates the control that now exists over fertility in the United States. Brodie (1994) suggests that a substantial fraction of women even in 19th century America intended not to have children, but were thwarted, so to speak, by their inability to prevent pregnancy. The vastly improved fertility control technology that is now available makes it much easier for a woman to implement such a desire. It can also be said that, given the set of high monetary costs and opportunity costs of children in advanced societies, it would be very surprising, indeed, if there were not a noticeable fraction of women who chose not to have children.

No matter how important the diffusion of innovative ideas and technologies may have been in initiating the fertility transition, most of the evidence seems clearly to suggest that some variant of the rational choice model now dictates the reproductive decisions made by most Americans (see, for example, Yamaguchi 1995; Rindfuss, Morgan, and Offutt 1996). That raises the question again, of course, of why people have children at all in a society like the United States? Why doesn't everyone opt for childlessness? The answer, as I alluded to earlier in the chapter, is that children do provide benefits to their parents, even though the return to the parents is rarely economic. Rather, children provide a source of "social capital" for parents, as Schoen and his associates (Schoen, et al. 1997) have suggested, building on the rational choice theory of James Coleman (which I discussed in Chapter 3). Children establish for their parents a network with other relatives, as well as with other people in the community, and they do that not just while they are young, but potentially for the rest of the parents' lives. Humans are social creatures, and through children we assure ourselves of a social network.

Canada

It is tempting to look at Canada and assume that its demographic pattern is a clone of what is happening in the United States. That would be a mistake. Many of the demographic trends that have been highly publicized in the United States either emerged first (albeit to less fanfare) in Canada, or followed a distinctive path. The different cultural history and ethnic composition of Canada has produced its own response to the demographic transition, although there are, to be sure, many similarities with other low fertility, advanced societies.

Canada has always been a more multicultural society than has the United States and so its history of fertility has been marked by significant regional and ethnic differences. Fertility was declining in parts of Canada in the late 19th and early 20th centuries, but fertility was generally higher in Canada than in the United States (Overbeek 1980), and in the 1920s the fertility level of the French-speaking population in Quebec was among the highest anywhere in North America (a total fertility rate over 4), whereas fertility in British Columbia was already at a fairly low level (with a total fertility rate under 3). As a result, in the 1930s, when fertility in the United States dropped below replacement level, Canadian fertility was close to an average of 3 children per woman, and only British Columbia dropped below replacement level (Foot 1982).

After World War II, Canada's Baby Boom started a year later than the United States, in 1947 (compared to 1946 in the United States), because Canadian soldiers were largely deployed in Europe during the war, and soldiers were returned home more slowly from there than were American soldiers from areas such as Asia (Foot 1996). Because it was starting from a higher level than in the US, the Baby Boom in Canada was also "louder" than in the United States—reaching a higher fertility level when it peaked (in 1959) than did the United States. In a variation on the "the higher you are, the farther you fall" theme, the birth rate in Canada has plummeted to a level below that in the United States, and the decline was led, to the surprise of most observers, by the province of Quebec, which within just three decades went from a total fertility rate of nearly 4 children per woman in 1956 down to level of only 1.4 children per woman in 1986 (Statistics Canada 1998).

The decline in fertility was probably a response to the opportunity costs of children (the economic explanations) more than anything else (Merrigan and St.-Pierre 1998). With respect to the proximate determinants, the decline has been made possible by an increase in the age at marriage, and by use of the pill among younger women, and surgical sterilization as a contraceptive measure at the older reproductive ages (Balakrishnan, Lapierre-Adamczyk, and Krotki 1993). Canada legalized abortion 4 years earlier than did the United States, but abortion rates have generally been lower in Canada than in the United States (Millar, Wadhers, and Henshaw 1997). Canada did experience a minor Baby Boomlet in the early 1990s, fueled largely by women in their thirties deciding to have children that they had previously postponed. The birth rate has since resumed its level at well below replacement.

Summary and Conclusion

The fertility transition is viewed by many as having an essentially economic interpretation, emphasizing the relationship between the supply of children (which is driven by biological factors) and the demand for children (based on a couple's calculations about the costs and benefits of children), given the costs (monetary and psychosocial) of fertility regulation. This is known as the supply–demand framework. It is countered by those who argue that fertility limitation is an innovation that is diffused through societies along social strata that may be independent of economic factors. Both viewpoints seem to find a home in the company of Ansley Coale's three preconditions for a fertility decline. These include the acceptance of calculated choice about reproductive behavior, a motivation to limit fertility, and the availability of means by which fertility can be limited.

All of these perspectives on the fertility transition can be used to help understand the fertility situation in the modern world, ranging from those countries where fertility remains high, to those where it has dropped to or below the replacement level. In higher fertility societies such as Jordan, India and Mexico, there is evidence that some fraction of the population still does not believe that fertility is, or should be, under the control of humans. Furthermore, the motivations to limit fertility are still underdeveloped in all of these countries, although there is evidence in virtually all higher fertility societies that the diffusion of family planning methods contributes to fertility differentials (such as the lower fertility of the more educated segments of the population), which may point the way to lower fertility.

People expect (and receive) pressures from others to have children, even in lower fertility countries such as the United States. This is why it is often difficult for fertility to decline when mortality declines, especially if no other economic or social changes have preceded the mortality decline. If marriage and childbearing are prestigious when mortality is high, why should it be any different when mortality is low? Social norms may change slowly in response to the new environment, because the link between mortality levels and fertility levels is not given conscious recognition.

Theories of low fertility emphasize the role of wealth and economic development in lowering levels of fertility, although it is clear that these are not sufficient reasons for fertility to decline. You must also assess the overall social environment in which change is occurring. When there are desired and scarce resources, wealth, prestige, status, education, and other related factors often help to lower fertility because they change the way people perceive and think about the social world and their place in it. Human beings are amazingly adaptable when they want to be. When people believe that having no children or only a few children is in their best interest, they behave in that way. Sophisticated contraceptive techniques make it easier, but they are not necessary, as the history of fertility declines in England, Japan, the United States and Canada illustrates.

One of the most important ways in which societies change in the modern world is through migration. Migrants bring not only themselves, but their ideas with them when they move, and as communication and transportation get increasingly easier, they are more apt to diffuse ideas and innovations backwards to their place of origin. In the next chapter we turn our attention, then, to this third aspect of the demographic transition: migration.

Main Points

1. The fertility transition represents the shift from "natural fertility" to more deliberate fertility limitation.

2. The supply–demand perspective on the fertility transition suggests that couples strive to maintain a balance between the potential supply of children and the demand (desired number of surviving children), given the cost of fertility regulation.

3. An important concept in the "new home economics" approach to the supply–demand framework is the tradeoff between the quantity and quality of children, expressed typically as the opportunity costs of children.

4. The innovation-diffusion model of fertility draws on sociological and anthropological evidence that much of human behavior is driven by the diffusion of new innovations—both technological and attitudinal—that may have little to do with a rational calculus of costs and benefits.

5. Ansley Coale's three preconditions for a fertility decline include the following sequence: (1) acceptance of calculated choice in reproductive decision-making; (2) motivations to limit fertility; and (3) the availability of means by which fertility can be regulated.

6. Higher fertility will persist if these preconditions have not been met, as evidenced by the situations in Jordan, India, and Mexico.

7. The need for children as sources of labor and security against risk, as well as the desire for sons, are some of the structural features of a society that may contribute to high fertility.

8. In almost all of the higher fertility nations in the world today there are genuine stirrings of a fertility decline, as high-fertility norms and behavior give way to low-fertility preferences.

9. Lower fertility is often associated with increasing income and with higher levels of education.

10. The fertility transition from high to low levels is well illustrated by England, Japan, the United States, and Canada.

11. The Baby Boom is one of the most famous demographic events in United States history (and a Baby Boom was part of the demographic landscape of England and Canada, as well), but it was really a readily explainable detour on a long road of declining fertility punctuated by the Baby Bust and the baby boomlet.

12. In the world today, a woman gives birth to a child more than four times each second. (We've got to find this woman and stop her!)

Suggested Readings

1. Richard A. Easterlin and Eileen M. Crimmins, 1985. The Fertility Revolution: A Supply–demand Analysis (Chicago: The University of Chicago Press).

   Although many prior and subsequent authors have elaborated on the supply–demand framework, this is Easterlin's own statement of the perspective that is often associated with his name.

2. John Cleland and Christopher Wilson, 1987. "Demand Theories of the Fertility Transition: an Iconoclastic View," Population Studies 41: 5–30.

   This article helped to shape the debate regarding the merits of the innovation-diffusion perspective compared to the supply–demand framework.

3. Warren C. Robinson, 1997, "The Economic Theory of Fertility Over Three Decades," Population Studies 51:63–74.

   A well-documented and very readable review and critique of the economic theories of fertility, with a discussion by an economist of the limitations of economics in explaining demographic behavior.

4. Karen Oppenheim Mason, 1997, "Explaining Fertility Transitions," Demography 34:443–454.

   A review and synthesis of the various and sometimes competing theories of the fertility transition, with an emphasis on the distinction between prenatal fertility limitation and postnatal population controls.

5. John B. Casterline, Ronald D. Lee, and Karen A. Foote (editors), 1996, Fertility in the United States: New Patterns, New Theories, Supplement to Volume 22 of Population and Development Review (New York: Population Council).

An authoritative volume on recent trends in the United States including an excellent overview chapter accompanied by chapters that attempt to explain what's going on.

🌐 Websites of Interest

Remember that websites are not as permanent as books and journals, so I cannot guarantee that each of the following websites still exists at the moment that your are reading this:

1. **http://www.census.gov/ipc/www/idbacc.html**

   It is time again for you to visit the Census Bureau's International Database, this time for the purpose of selecting data on fertility (select "Age-specific fertility rates and selected derived measures") for the country(s) and date(s) that interest you.

2. **http://unfpa.org/SWP/SWPMAIN.HTM**

   The UNFPA is the United Nations Population Fund—the population "outreach" arm of the UN. They publish an informative annual report and in 1997 the report was "The Right to Choose: Reproductive Rights and Reproductive Health," which gets to the heart of the three preconditions for a fertility decline.

3. **http://www.pcbs.org/english/sel_stat.htm**

   The Palestinian population has one of the highest fertility rates in the world, so this website of the Palestinian National Authority's Palestinian Central Bureau of Statistics is of considerable interest. The site includes results from recent demographic surveys and has several tables showing fertility differentials by selected population characteristics.

4. **http://www.hist.umn.edu/~rmccaa/**

   Professor Robert McCaa at the University of Minnesota has a rich and varied website devoted to the population history of Mexico, including informative slides on the recent decline in fertility.

5. **http://www.mofa.go.jp/j_info/japan/socsec/**

   At this website of Japan's Ministry of Foreign Affairs are speeches given overseas by influential Japanese scholars. Two speeches, in particular, will provide you with interesting insights into the causes and consequences of the fertility decline in Japan: **.../sodei.html** takes you to a speech in 1997 by Takako Sodei, Professor of Gerontology at Ochanomizu University, and **.../ogawa.html** takes you to a speech by Naohira Ogawa, Professor of Economics at Nihon University.

CHAPTER 7
Migration

"The sole cause of man's unhappiness," quipped Pascal in the 17th century, "is that he does not know to stay quietly in his room." If this is so, unhappiness is enjoying unprecedented popularity as people are choosing to leave their rooms, so to speak, in record numbers. Sometimes they are fleeing from unhappiness; sometimes they are producing it. Because migration brings together people who have probably grown up with quite different views of the world, ways of approaching life, attitudes, and behavior patterns, it contributes to many of the tensions that confront the world, leading Kingsley Davis to comment that "so dubious are the advantages of immigration that one wonders why the governments of industrial nations favor it" (Davis 1974:105). The popular literature reflects this ambivalence. Comments such as "Once, migration caused statues to be erected and poems to be written. . . . There is no monument, however, to the new immigrants" (Breslin 1982; quoted by Sheppard 1982:72); "Emigration is an unnatural act between consenting adults . . . an act of desperation, endured by immigrants and hosts alike without gratitude or sympathy, a placebo, not a cure" (Cornelison 1980; quoted by Strouse 1980:99); "Boat people arouse Japan's Xenophobia" (Rubinfein 1989); "Immigrants to Europe from the third world face racial animosity" (Horwitz and Forman 1990); "Refugees from Bosnia find scant welcome in western countries" (Milbank 1994); or "So, does America want them or not?" (Economist 1997) reflect the concern surrounding the influx of strangers into our midst.

Even if a country tries to slam its doors to immigrants, however, will that stop people from migrating? Nearly 90 million people are being added to the world's population each year. What are they to do? As it becomes ever harder for a person to find a niche in the world economy, a would-be worker is often compelled to move. An old Mexican saying goes, "Don't ask God to give it to you, ask Him to put you where it is." "Where it is" for many in Mexico is the United States. Though migrants come from all over to the United States, Mexico sends the greatest number, accounting for 15 percent of legal migrants and 85 percent of illegal or undocumented migrants. Today's pilgrims are from places such as Jalisco, Sinaloa, and Michoacán. They look to their northern neighbor for economic advancement in life and, although entrants from Mexico often cross the border with every intention of returning home, many never do. Similar tales are told of Algerians migrating to France; Moroccans to Spain, and Pakistanis to the United Kingdom.

This vast transnational migration is an integral part of the demographic transition, as was discussed earlier with reference to the perspective of demographic change and response. Population growth changes the ratio of people to resources and this forces some kind of local adjustment. One such adjustment is to move somewhere else—to "get out of town" or to "head west, young person." Humans have been migrating throughout history (or else we would not be found in every nook and cranny of the globe), but the advent of relatively inexpensive and quick ground and air transportation has given migration a new dimension. It is easier than ever for people to move, and when they do, they are less likely to assume that they will stay in the new location forever. The globalization of the world's labor force, brought about by advances in transportation and communication (in concert with the demise of centrally planned economies), has turned some immigrants into a new type of nomad (Kraly and Warren 1992). As people move in and around, they contribute to population growth and change, both in the short run and in the long run; and migration—when and where it occurs—is a population process of considerable importance.

Migration can profoundly alter a community or an entire country within a short time. Although it is one of the three population processes (along with fertility and mortality), it is different in many respects beyond the obvious. To begin with, migration is very hard to measure, and consequently we know less about it than we do about mortality and fertility. That means, of course, that we understand even less about the complexities of why people migrate than we do about why they have babies (although that is a tough problem also) and why they die. Furthermore, despite the fact that there is much we do not know about who migrates and why, migration has been a subject of far more government control than has either fertility or mortality.

I begin this chapter with some comments on the definition and measurement of migration. Then I move on to a discussion of some of the explanations that have been offered for why people voluntarily move, followed by a review of some of the major consequences of migration (since the potentially dramatic results of migration are leading factors in the many attempts by governments to control the movement of people). I also examine migration that occurs forcibly, including slavery and refugee movements. That takes us to an examination of the actual patterns of migration: Where do people move? We look first at global patterns of migration, then focus on North America, looking especially at patterns of migration within the United States, and migration in and out of the country. Who moves into and out of the United States? How many people move about within North America, and where do they move?

Defining Migration

Migration is defined as any permanent change in residence. It involves the "detachment from the organization of activities at one place and the movement of the total round of activities to another" (Goldscheider 1971:64). Thus the most important aspect of migration is that it is spatial by definition. You cannot be a migrant unless you "leave your room." However, just because you leave your room, you are not necessarily a migrant. You may be a traveler or perhaps a daily commuter from your home to work. These activities represent mobility, but not migration. You might be a temporary resident elsewhere (such as a construction worker on a job away from home for a few weeks or even months), or a seasonal worker (returning regularly to a permanent home), or a **sojourner** (typically an international migrant seeking temporary paid employment in another country; see, for example, Estrella 1992). Again, such people are mobile, but they are not migrants because they have not changed their residence permanently. Of course, even when you change your permanent residence, if your new home is only a short distance away and you do not have to alter your round of activities (you still go to the same school, have the same job, shop at the same stores), then you are a mover (and maybe even a shaker), but not a migrant. All migrants are movers, but not all movers are migrants.

Less clear conceptually are transients and nomads. Technically, you could say that because they constantly are changing their residence and round of activities, then they should be thought of as migrants. However, the lack of a permanent residence creates a problem in defining them as migrants. Most demographers deal with this by simply ignoring transients and nomads, and I do much the same, although there is a discussion of the homeless in Chapter 11.

Anyone who moves permanently (which the United Nations defines as spending at least 1 year in the new locale) to another geographic region of the same country, and all who move permanently to another country, can be defined unambiguously as migrants. Migrants usually are categorized for research purposes according to whether they crossed political boundaries, and if so, what kind of boundary (county line, state line, international border), and also according to the points of origin and destination. The major distinction, however, is simply between **internal** and **international migration.** Internal migration involves a permanent change of residence within national boundaries. With reference to your area of origin (the place you left behind), you are an **out-migrant,** whereas you become an **in-migrant** with respect to your destination.

If you move from one country to another, you become an international migrant—an **emigrant** in terms of the area of origin and an **immigrant** in terms of the area of destination. The distinction between internal and international is important because the latter is usually more difficult to accomplish than the former, meaning that the motivation to move may have to be much stronger. In addition, the cultural impact of international migration is typically greater than that involved in internal migration. Crossing an international border is far more likely to involve a change of language, customs, and politics—in general, a change of lifestyle and world view—than is a move within a country. Because commuters and sojourners also may cross international boundaries, the United Nations has tried to tighten the definition of an international migrant by developing the concept of a **long-term immigrant,** which includes all persons who arrive in a country during a year and whose length of stay in the country of arrival is more than 1 year (Kraly and Warren 1992).

International migration can be further differentiated between **legal immigrants, illegal** (or undocumented) **immigrants, refugees,** and **asylees.** Legal immigrants are those who have legal and political permission to make the move they undertake, whereas illegal or undocumented migrants do not. A refugee is defined by the United Nations (and by most countries of the world) as "any person who is outside his or her country of nationality and is unable or unwilling to return to that country because of persecution or a well-founded fear of persecution. Claims of persecution may be based on race, religion, nationality, membership in a particular social group, or public opinion" (U.S. Immigration and Naturalization Service 1997:72). An asylee is a refugee—with a geographic twist. He or she already is in the country to which they are applying for admission, whereas a refugee is outside the country at the time of application.

You can see that the definition of migration is confounded by the fact that migration is an activity (changing residence) carried out by people (the migrants) under varying legal and sociopolitical circumstances. If we have this much trouble defining migration, you can be sure that it is hard to measure.

Measuring Migration

The difficulty of measuring migration is apparent when we examine its complexity. To begin with, there are no biological imperatives or restrictions to deal with, as there are with mortality and fertility. In theory, at least, no one would ever have to migrate. If migration does occur, it may or may not recur, and if it does recur, it may be a return to the original point or another move to a new destination. In addition, to tangle the situation further, migration may involve more than a single individual—a family or even an entire village may migrate together. A ghost town, it has been suggested, does not necessarily signal the end of a community, only its relocation. People may move short or long distances, and they may or may not cross political boundaries (such as between states or between countries). In other words, the sheer act of migration is an amazingly difficult phenomenon to measure. Indeed, even the definition of migration is subject to some negotiation about what is meant by "permanent" and "residence." Those kinds of ambiguities obviously do not exist with births and deaths (except in very esoteric circumstances).

Most of the information we have about migration into and within the United States, Canada, or Mexico is collected by asking people where they lived at a certain previous time. For example, the 1990 U.S. census on April 1, 1990, asked where people had lived on April 1, 1985 (similar questions have been asked on previous censuses and similar questions are asked in both Canada and Mexico). Thus from those data we can tell that 47 percent of the population aged 5 or older in 1990 had lived in some other house in 1985; 21 percent were migrants, as defined by crossing county lines, and 2 percent were immigrants from another country.

However, we have no idea how many times, or to what places, people may have migrated between those two dates. Additionally in the United States, the Current Population Survey each March asks a sample of Americans where they were living on March 1 the year before—again, we have the same problem of not knowing what happened in the interim. Every other year the Census Bureau also conducts the American Housing Survey, which tracks changes in the nation's housing stock and thereby generates data on residential mobility.

In the United States, the Immigration and Naturalization Service (INS) keeps track of legal immigrants but has essentially no record of people who emigrate. The same can be said for Canada, where Citizenship and Immigration Canada records the arrival of immigrants, but has no data on people who leave the country. This stands in contrast to at least a few European nations, which, as I mentioned in Chapter 2, maintain population registers and thus have a pretty accurate fix on the extent of both internal and international migration. Most countries, however, have little available information, and we either do not know what is happening or we have to rely on sample surveys or other indirect evidence (such as the number of foreign-born people counted in a census) to infer patterns of migration.

When data are available, migration is measured with rates that are similar to those we construct for fertility and mortality. Gross or total out-migration represents all people who leave a particular region during a given time period (usually a year), and so the **gross rate of out-migration** relates those people to the total midyear population in the region (and then we multiply by 1,000):

$$\text{Gross rate of out-migration} = \frac{\text{Total out-migrants}}{\text{Total midyear population}} \times 1{,}000$$

Similarly, the **gross rate of in-migration** is the ratio of all people who moved into the region during a given time period to the total midyear population in that region:

$$\text{Gross rate of in-migration} = \frac{\text{Total in-migrants}}{\text{Total midyear population}} \times 1{,}000$$

As was true for other crude rates (especially the crude birth rate), the gross rate of in-migration is a little misleading because the midyear population refers to the people living in the area of destination, which is not the group of people at risk of moving in (indeed, they are precisely the people who are *not* at risk of moving in because they are already there). Nonetheless, the in-migration rate does give us a sense of the impact that in-migration has on the region in question, and so it is useful for that reason alone. It is also useful because we use it to calculate the net migration rate.

The difference between those who move in and those who move out is called **net migration.** If these numbers are the same, then the net migration rate is zero, even if there has been a lot of migration activity. If there are more in-migrants than out-migrants, the rate is positive; and if the out-migrants exceed the in-migrants, the rate is negative. The **crude net migration rate** is the net number of migrants in a year per 1,000 people in a population. The crude net migration rate (CNMR) thus is calculated as follows:

$$\text{CNMR} = \frac{\text{Total in-migrants} - \text{Total out-migrants}}{\text{Total midyear population}} \times 1{,}000$$

The total volume of migration also may be of interest to us (because that can have a substantial impact on a community even if the net rate is low), and we can measure this as the **total migration rate:**

$$\text{Total migration rate} = \frac{\text{Total in-migrants} + \text{Total out-migrants}}{\text{Total midyear population}} \times 1{,}000$$

By comparing the total migration rate with the net migration rate, we gain a sense of the turnover of people that migration is generating. We could find a very low net migration rate, for example, and still have had a high volume of people moving in and out. Thus the **migration turnover rate** is the ratio of the total migration rate to the crude net migration rate:

$$\text{Migration turnover rate} = \frac{\text{Total migration rate}}{\text{Crude net migration rate}} \times 1{,}000$$

The flip side of the turnover rate (no pun intended) is called the rate of **migration effectiveness** (Gober 1993; Plane and Rogerson 1994). It measures how "effective" the total volume of migration is in redistributing the population. For example, if there were a total of 10 migrants in a region in a year and all 10 were in-migrants, the "effectiveness" of migration would be 10/10, or 100 percent; whereas if 4 were in-migrants and 6 were out-migrants, the effectiveness would be much lower: $(4 - 6)/10$, or -20 percent. In general, the rate of effectiveness is as follows:

$$\text{Migration effectiveness (E)} = \frac{\text{Crude net migration rate}}{\text{Total migration rate}} \times 100$$

Because we often do not have complete sets of data on the number of in- and out-migrants, we can "back into" the migration rate by solving the demographic equation (see Chapter 1) for migration. This is known as the **components of change method of estimating migration.** The demographic equation states that population growth between two dates is a result of the addition of births, the subtraction of deaths, and the net effect of migration (the number of in-migrants minus the number of out-migrants). If we know the amount of population growth between two dates, and we know also the number of births and deaths, then by subtraction we can estimate the amount of net migration. Let me give you an example. Based on the 1980 census of the United States, we can estimate that on April 1, 1980, there were 226,505,000 residents in the country. Between that date and April 1, 1990, there were 37,447,000 births and 20,626,300 deaths in the country. Thus on April 1, 1990, we should have expected to find 243,325,700 residents if no migration had occurred. However, the 1990 census suggests that there were, in fact, 248,710,000 people. That difference of 5,384,300 people we estimate to be the result of migration (note that a small fraction of the difference could also be the result of differences in coverage error between the two censuses, as discussed in Chapter 2).

We can also calculate intercensal net migration rates for each age group and gender by combining census data with life-table probabilities of survival—a procedure called the **forward survival method of migration estimation.** For example, in 1980 in the United States there were 20,317,510 males aged 20 to 29. Life-table values (see the Appendix) suggest that 98.15 percent of those men (or 19,941,636) should still have been alive at ages 30 to 39 in 1990. Yet, the 1990 census counted 21,332,000 men in that age group, or 1,390,364 more than expected. We assume, then, that those "extra" people were migrants.

There are, by the way, no universally agreed-upon measures of migration that summarize the overall levels in the same way that the total fertility rate summarizes fertility and life expectancy captures a population's experience with mortality. However, one way of measuring the contribution that migration makes to population growth is to calculate the ratio of migration to natural increase. Thus the **migration ratio** is:

$$\text{Migration ratio} = \frac{\text{Net number of migrants}}{\text{Number of Births} - \text{Number of Deaths}} \times 1,000$$

For example, again using the data for the United States from 1980 to 1990, we can calculate the migration ratio as (5,384,300 net migrants) divided by (37,447,000 births minus 20,626,300 deaths) times 1,000 equals 320 migrants added to the American population during the decade of the 1980s for each 1,000 people added through natural increase. Looked at another way, migrants accounted for 24 percent of the total population growth in the United States during that period of time.

Now, having worn you out trying to measure the nearly unmeasurable, let us move on to yet another difficult (but frequently more interesting) task: explaining why people migrate.

Why Do People Choose to Migrate?

In the premodern world, rates of migration typically were fairly low, just as birth and death rates were generally high. The demographic transition helped to unleash migration, and it is reasonable to suggest that a **migration transition** has occurred in concert with the fertility and mortality transitions discussed in earlier chapters (Zelinsky 1971). The theory of demographic change and response (see Chapter 3) suggested that migration is a ready adaptation that humans (or other animals, for that matter) can make to the pressure on local resources generated by population increase. However, people do not tend to move someplace at random—they tend to go where they believe opportunity exists. Because the demographic transition occurred historically in the context of economic development, which involves the centralization of economic functions in cities, migrants have been drawn to cities, and urbanization is an important part of the migration transition, as discussed more fully in Chapter 11. Yet the mere existence of a migration transition does not explain who moves, and when, why, and where they go. We need to dig deeper for those explanations.

The Push–Pull Theory

Over time, the most frequently heard explanation for migration has been the so-called **push–pull theory,** which says that some people move because they are pushed out of their former location, whereas others move because they have been pulled or attracted to someplace else. This idea was first put forward by Ravenstein (1889), who analyzed migration in England using data from the 1881 census of England and Wales. He concluded that pull factors were more important than push factors: "Bad or oppressive laws, heavy taxation, an unattractive climate, uncongenial social surroundings, and even compulsion (slave trade, transportation), all have produced and are still producing currents of migration, but none of these currents can compare in volume with that which arises from the desire inherent in most men to 'better' themselves in material respects." Thus Ravenstein is saying that it is the desire to get ahead more than the desire to escape an unpleasant situation that is most responsible for the voluntary migration of people, at least in late-19th-century England. This theme should sound familiar to you. Is it not the same point made by Davis (1963) in discussing personal motivation for having small families (see Chapter 6)? Remember, Davis argued that it is the pursuit of pleasure or the fear of so-cial slippage, not the desire to escape from poverty, that motivates people to limit their fertility.

In everyday language, we could label the factors that might push a person to mi-grate as stress or strain. However, it is probably rare for people to respond to stress by voluntarily migrating unless they feel that there is some reasonably attractive al-ternative, which we could call a pull factor. The social science model conjures up an image of the decision maker computing a calculated cost-benefit analysis of the sit-uation. The potential migrant weighs the push and pull factors and moves if the ben-efits of doing so exceed the costs (Kosinski and Prothero 1975; Massey 1990; Stone 1975). For example, if you lost your job, it could benefit you to move if there are no other jobs available where you live now, unemployment compensation and wel-fare benefits have expired, and there is a possibility of a job at another location. Or, to be more sanguine about your employability, the process may start, for example, when you are offered an excellent executive spot in a large firm in another city. Will the added income and prestige exceed the costs of uprooting the family and leaving the familiar house, community, and friends behind? In truth, whether or not you migrate will likely depend on a more complicated set of circumstances than this simple example might suggest. The decision to move usually occurs over a fairly long period of time, proceeding from a desire to move, to the expectation of mov-ing, to the actual fact of migrating (Rossi 1955). In Rossi's longitudinal sample of families in the 1950s, half of those interviewed expressed a desire to move, but only about 20 percent of them actually did so. Sell and DeJong (1983) produced a set of longitudinal data for the 1970s that reinforces Rossi's findings—migration rates re-flect a whole spectrum of attitudes, ranging from people who are "entrenched non-movers" (who have no desire to move and no expectation of moving, and who do not migrate) to "consistent decision-maker movers" (who desire to, expect to, and do migrate).

Between the desire to move and the actual decision to do so there also may be **intervening obstacles** (Lee 1966). The distance of the expected destination, the cost of getting there, poor health, and other such factors may inhibit migration. These obstacles are hard to predict on any wide scale, however, and so we tend to ignore them and concentrate our attention on explaining the desire to move. Economic variables dominate most explanations of why people migrate. Analysis of data from the University of Michigan's Panel Study of Income Dynamics underscores the im-portance of job-related factors in influencing migration decisions. People who are unemployed and those who are dissatisfied with their present jobs are more likely than others to say they are planning a move, and to actually follow through with those plans (DaVanzo 1976).

Migration associated with career advancement, as happens so often in the mil-itary, in academics, and in large companies, illustrates the hypothesis that migration decisions "arise from a system of strategies adopted by the individual in the course of passing through the life cycle" (Stone 1975:97). If it is assumed that individuals spend much of their lifetimes pursuing various goals, then migration may be seen as a possible means—an **implementing strategy**—whereby a goal (such as more educa-tion, a better job, a nicer house, a more pleasant environment, and so on) might be attained. Although this is not a startling new hypothesis (it is little more than a modern restatement of Ravenstein's 19th-century conclusions), it is nonetheless a very reasonable one. Indeed, Lee (1966) has observed that two of the more enduring generalizations that can be made about migration are:

1. Migration is selective (that is, not everyone migrates, only a selected portion of the population).
2. The heightened propensity to migrate at certain stages of the life cycle is important in the selection of migrants.

One particular stage of life disproportionately associated with migration is that of reaching maturity. This is the age at which the demand or desire for obtaining more education tends to peak, along with the process of finding a job or a career, and getting married.

Migration Selectivity

Selectivity by Age   In virtually every human society young adults are far more likely to migrate than people at any other age. The data in Figure 7.1 show the age pattern of intercounty migrants in the United States between 1995 and 1996, using data from the Current Population Survey. As you can see, young adults were much more mobile than people of other ages, and although these data are for the United States for 1995–96, the same pattern has existed in the United States in the past and holds true in other countries as well (Long 1988).

The young adult ages, 20 to 29, are clearly those at which migration predominates. One in seven American females aged 20 to 24 was an intercounty migrant during the year, whereas migration peaked a little later for males—25 to 29—at which age one in seven was a migrant during the year from 1995–96. Migration rates thus peak in the 20s, and from there the percentage of people who migrate drops off steeply. At ages younger than 20, children typically are just following their parents around, so it is not surprising that the younger children (who have the youngest parents) move more than older children. We can see then that age is an important determinant of migration because it is related to life-cycle changes that affect most humans in most societies.

Selectivity by Life Cycle and Education   Most human groups expect that young adults will leave their parents' home, establish an independent household, marry, have children, and/or pursue an employment career. Some people choose to ignore one or more of these expectations, of course, but they exist nonetheless as behavioral guidelines. They also influence migration because each of these several phases of young adulthood may precipitate a move. For example, the divorced, separated, and widowed have the highest migration rates at every age (Long 1988), and longitudinal data from the Census Bureau's Survey of Income and Program Participation (SIPP) have shown that people whose marital status changed between two consecutive interviews were far more likely to have moved than those whose marital status remained unaltered (DeAre 1990).

Figure 7.1   Young Adults Are Most Likely to be Migrants

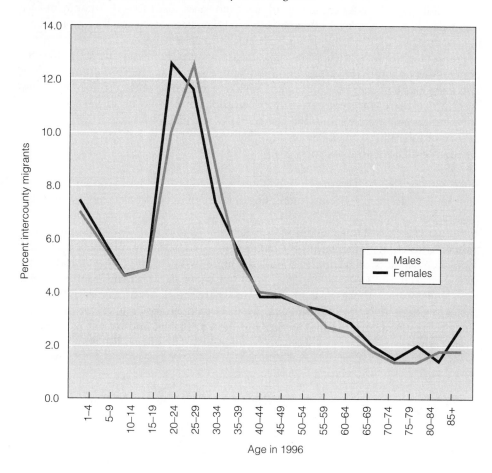

Source: Adapted from Kristin A. Hansen, 1997, "Geographical Mobility: March 1995 to March 1996," U.S. Bureau of the Census, Current Population Reports, P20-497, Table 2.

The incidence of migration also varies according to the number and ages of children. Among young couples, the smaller the family and the younger the children, the greater the probability of migration. For example, among married couples with the householder aged 25 to 34 in 1996, 12 percent of those with no children had migrated from one county to another between 1995 and 1996, dropping to 8 percent for couples with at least one child. Furthermore, the likelihood of migration was greater if the oldest child was under 6 years old; once a child is old enough to start school, the temptation to move seems to go down (Hansen 1997). Migration, in its turn, may temporarily disrupt family building activity. The period just before migration, as people plan for their move, has been found to be a time of lower than expected fertility, although it may be followed by a "catch-up" time after migration is completed (Bean and Swicegood 1985; Goldstein and Goldstein 1981).

The levels of migration also tend to go up as occupational levels rise, as income levels go up, and as educational attainment increases. Because the attainment of a particular educational level often sets up a whole chain of events leading to a certain occupation and income, it is a particularly crucial aspect of the life cycle. Immigrants to both the United States and Canada tend to be better educated than the populations in both the areas of origin and destination. Within countries it is also true that the better educated are more likely to be migrants. You can see in Figure 7.2 that there is a clear pattern for migration rates to go up as educational attainment goes up—a person in the United States with a college degree had twice the chance of migrating across county lines between 1995 and 1996 as a person with less than a high school education.

Selectivity by Gender   In the United States women have virtually the same rates of migration as do men, reflecting increasing gender equality. Women have similar opportunities for education as do men, as well as an increasing ability to postpone or avoid marriage and family-building. In so-called "traditional" societies, the role of women is to be at home caring for children and other family members and under such circumstances migration is dominated more by males than by females. Therefore it should not surprise you that men are most likely to outnumber women among migrants in those areas of the world where the status of women is lowest—Africa, western Asia, and South Asia, whereas women are as likely or even more likely to be migrants in Latin America and the Caribbean, and in east Asia (except China, where males are more likely to be migrants) (Chant and Radcliffe 1992).

Figure 7.2   The Better Educated You Are, the More Likely You Are to Migrate

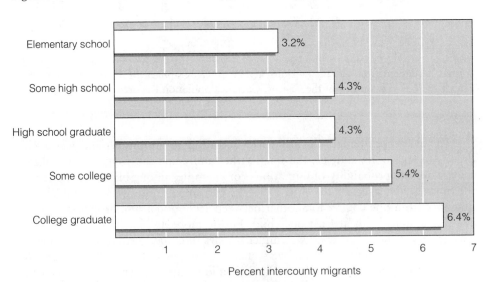

Source: Adapted from Kristin A. Hansen, 1997, "Geographical Mobility: March 1995 to March 1996," U.S. Bureau of the Census, Current Population Reports, P20-497, Table 5.

It is still true even in the United States that marriage and child-rearing responsibilities constrain the migration of married women (Enchautegui 1996), but especially for single women migration can be seen as a form of emancipation from these traditional roles, and global patterns of migration show a rise in the percentage of all migrants who are women (Campani 1995). Whereas it was once true that men were "pioneer" migrants who were later joined by their wives, there is now a world market for female labor that encourages migration and may serve to improve the status of women.

Conceptualizing the Migration Process

So far we have seen that migration may be a response to a number of factors (push and pull), but that young adults—both male and female—disproportionately are the ones responding. Agreement is nearly universal on these ideas, but "the devil is in the details." There are so many different aspects to the process of migration that no one yet has produced an all-encompassing theory of migration. Figure 7.3 provides an overview of the major aspects of the migration process that require explanation, adapted from a conceptual model devised by DeJong and Fawcett (1981). These are analogous in certain respects to the three preconditions for a fertility decline. The migration process has three major stages including: (1) the propensity to migrate in general; (2) the motivation to migrate to a specific location; and (3) the actual decision to migrate.

Figure 7.3   A Conceptual Model of Migration Decision Making

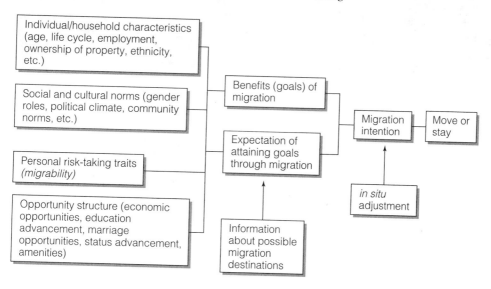

Source: Adapted from G. DeJong and J. Fawcett, 1981, "Motivations for migration: An assessment and a value-expectancy research model," in G. DeJong and R. Gardner (eds.), Migration Decision Making (New York: Pergamon Press), Figure 2.2.

The migration process begins with individuals and household members in the context of a given culture and society. The decision about who will migrate, when, and to where may often be part of a household strategy for improving the group's quality of life—consistent with the perspective of demographic change and response, as you recall from Chapter 3. Furthermore, the household decision is not made in a vacuum; it is influenced by the sociocultural environment in which the household members live. Individual and household characteristics are important because of the selectivity of migrants—households with no young adults are less likely to contemplate migration. Social and cultural norms are important because they provide the context in which people might think consciously of migration as a necessary or desirable thing to do. As interprovincial migration has recently increased in mainland China, women are far less likely to migrate than men, probably because of the already discussed gender discrimination in China (Li and Li 1995). Social norms can play a role in discouraging migration by emphasizing the importance of place and community. However, political and economic instability may cause people to rethink their commitment to an area. Violence in rural Guatemala, for example, has been shown to increase the likelihood that people will migrate toward the cities (Morrison 1993).

Personal traits are important because some people are greater risk takers than others. Rajulton (1991) has suggested that people can be scaled along a dimension of *migrability*, reflecting the probability that, all things being equal, they would risk a migration. In the United States, the "average" person will move 11 times in his or her lifetime (Hansen 1994), but that figure hides a lot of variability. Some people account for a disproportionate amount of migration by migrating frequently, while others never move. Part of the explanation for the propensity to move may be cultural, of course. Long (1991) has examined rates of residential mobility (which are highly correlated to migration rates) for a number of developed nations. He found that the "overseas European" nations, including the United States, Canada, Australia, and New Zealand—populated by migrants who supplanted the indigenous population—are the countries with the highest rates of mobility.

Demographic characteristics combine with societal and cultural norms about migration to shape the values that people hold with respect to migration. Such values or goals represent clusters of motivations to move, including the desires for wealth, status, comfort (better living or working conditions), stimulation (including entertainment and recreation), autonomy (personal freedom), affiliation (joining family or friends), and morality (especially religious beliefs). At the same time, personal traits (such as being a risk-taking person) combine with the opportunity structure for migration to affect one's value system with respect to migration. All of these personal and social environmental factors combine to affect a person's expectation of actually achieving the goals they have in mind that might be facilitated by migration. The amount of information a person has about the comparative advantages of moving further contributes to the expectation of attaining migration values or goals. Goals and expectations jointly influence the intentions that a person has toward migration. This represents the results of a cost-benefit analysis that a person (probably in concert with other household members) has made about migration. If the benefits appear to outweigh the costs, then the person may decide to migrate.

Given the intention to move, a person may discover that, by making adjustments in his or her current situation, personal or family goals can be achieved without having to move. "Such adjustments might include a change in occupation, alterations to the physical structure of the house, a change in daily and friendship patterns, or lifestyle changes" (DeJong and Fawcett 1981:56). Finally, the intention to move (or to stay) leads ultimately to the act of moving (or staying) itself, although unanticipated events may affect that decision.

Of the factors laid out in Figure 7.3, the most important elements in explaining migration in the modern world appear to be: (1) the creation of new opportunity structures for migration, which raise the benefit of migrating (pull factors) partly by undermining existing local relationships between people and resources (push factors); while (2) cheaper and quicker transportation and communication increase the information that people have about a potential new location (lowering the risk of migration by closing the gap between the anticipated benefits and the perceived likelihood of attaining those goals; and (3) making it easier to migrate and to return home if things do not work out (also lowering the risk of migration by making the decision to migrate more reversible).

The decision about whether you are going to move cannot be divorced from where you might be going. Data from the American Housing Surveys suggest that nearly one in four internal migrants in the United States is involved in a job transfer, implying that the choice of destination may have been in someone else's hands. An additional one in ten migrants was moving closer to relatives, and obviously their location fixes the migrant's destination.

Step migration and **chain migration** are two migration strategies that have stood out over time helping to determine where migrants go. Step migration is a process whereby migrants attempt to reduce the risk of their decision by sort of inching away from home. The rural resident may go to a nearby city, and from there to a larger city, and perhaps eventually to a huge megalopolis. Chain migration reduces risk because it involves migrants in an established flow from a common origin to a predetermined destination where earlier migrants have already scoped out the situation and laid the groundwork for the new arrivals. This is a pattern that especially characterizes migration from Mexico to the United States and it reminds us that the choice of where to move is a large component of the decision to migrate internationally.

Differences between Internal and International Migration

Internal migration typically is "free" in the sense that people are choosing to migrate or not, often basing that decision on economic factors, as I have discussed. This is not to say that, within a country, people are never forced to move. Witness the migration of 250,000 Egyptians who were forced to relocate so that the Aswan Dam could be built, or the massive transmigration that Indonesia periodically attempts—moving people from the crowded island of Java to other, less populous islands. But such internal migration, though forced, is usually planned. People's needs are anticipated in advance and, presumably, the migration is expected to improve the lives of the people involved.

Migration across international boundaries is sometimes free, but it usually means that a person has met fairly stringent entrance requirements, is entering illegally, or is being granted refugee status, fleeing from a political, social, or military conflict. Each of these instances is apt to be more stressful than internal migration, and on top of that is heaped the burden of accommodating to a new culture and often a new language, being dominated perhaps by a different religion, being provided different types and levels of government services, and adjusting to different sets of social expectations and obligations.

Referring back to Figure 7.3, we can make the general statement that internal migration is more strongly influenced by individual characteristics of people, whereas international migrants are more apt to be influenced by the social and political climate and by the opportunity structure (especially the lack of barriers to migration). The kinds of migration goals that internal migrants have are also likely to differ somewhat from those of international migrants. Because international migration is a more drastic change in life than internal migration, and because it has important implications for social, political, and economic policy, it has received considerable attention in the literature. In particular, Massey and his associates (1993; 1994; 1997) have reviewed and evaluated various theories that try to explain contemporary patterns of international migration.

Theories of International Migration

The major theories that exist to help explain various aspects of international migration, as outlined by Massey and his associates (1993; 1994), include those that focus on the initiation of migration patterns: (1) neoclassical economics; (2) the new household economics of migration; (3) dual labor market theory; and (4) world systems theory. Then there are three perspectives that help to explain the perpetuation of migration, once started: (1) network theory; (2) institutional theory; and (3) cumulative causation.

The Neoclassical Economic Approach   By applying the classic supply and demand paradigm to migration, this theory argues that migration is a process of labor adjustment caused by geographic differences in the supply of and demand for labor. Countries with a growing economy and a scarce labor force have higher wages than a region with a less developed economy and a larger labor force. The differential in wages causes people to move from the lower wage to the higher wage region. This continues until the gap in wages is reduced merely to the costs of migration (both monetary and psychosocial). At the individual level, migration is viewed as an investment in human capital. People choose to migrate to places where the greatest opportunities exist. This may not be where the wages are currently the highest, but rather where the individual migrant believes that, in the long run, his or her skills will earn the greatest income. These skills include education, experience, training, and language capabilities.

This approach has been used to explain internal as well as international migration. It is also the principle that underlies Ravenstein's conceptualization of push factors (especially low wages in the region of origin) and pull factors (especially high wages in the destination region).

The New Household Economics of Migration   The neoclassical approach assumed that the individual was the appropriate unit of analysis, but the new household economics of migration approach argues that decisions about migration are often made in the context of what is best for an entire family or household. This approach accepts the idea that people act collectively not only to maximize their expected income, but also to minimize risk. Thus migration is not just a way to get rid of people; it is also a way to diversify the family's sources of income. Migrating members of the household have their journey subsidized and then remit portions of their earnings back home. This cushions households against the risk inherent in societies with weak institutions. If there is no unemployment insurance, no welfare, no bank from which to borrow money or even to invest money safely, then the remittances from migrant family members can be cornerstones of a household's economic well-being.

Dual Labor Market Theory   This theory offers a reason for the creation of opportunities for migration. It suggests that in developed regions of the world there are essentially two kinds of job markets—the primary sector, which employs well-educated people, pays them well, and offers them security and benefits; and the secondary labor market, characterized by low wages, unstable working conditions, and lack of reasonable prospects for advancement. It is easy enough to recruit people into the primary sector, but the secondary sector is not so attractive. Historically, women, teenagers, and racial and ethnic minorities were recruited into these jobs, but in the past few decades women and racial and ethnic minority groups have succeeded in moving increasingly into the primary sector, at the same time that the low birth rate has diminished the supply of teenagers available to work. Yet the lower echelon of jobs still needs to be filled, and so immigrants from developing countries are recruited—either actively (as in the case of agricultural workers) or passively (the diffusion of information that such jobs are available).

World Systems Theory   This theory offers a different perspective on the emerging opportunity structure for migration in the contemporary world. The argument is that since the 16th century (and as part of the Industrial Revolution in Europe) the world market has been developing and expanding into a set of core nations (those with capital and other forms of material wealth) and a set of peripheral countries (in essence, the rest of the world) that have become dependent on the core, as the core countries have entered the peripheral countries in search of land, raw materials, labor, and new consumer markets.

According to world systems theory, migration is a natural outgrowth of disruptions and dislocations that inevitably occur in the process of capitalist development. As capitalism has expanded outward from its core in Western Europe, North America, Oceania, and Japan, ever-larger portions of the globe and growing shares of the human population have been incorporated into the world market economy. As land, raw material, and labor within peripheral regions come under the influence and control of markets, migration flows are inevitably generated (Massey, et al. 1993:445).

Migration flows do not tend to be random, however. In particular, peripheral countries are most likely to send migrants (including refugees and asylees) to those core nations with which they have had greatest contact, whether this contact be economic, political, or military (Rumbaut 1991).

Network Theory   Once migration has begun, it may well take on a life of its own, quite separate from the forces that got it going in the first place. Network theory argues that migrants establish interpersonal ties that "connect migrants, former migrants, and nonmigrants in origin and destination areas through ties of kinship, friendship, and shared community origin. They increase the likelihood of international movement because they lower the costs and risks of movement and increase the expected net returns to migration" (Massey, et al. 1993:449). Once started, migration sustains itself through the process of diffusion until everyone who wishes to migrate can do so. In developing countries, such migration eventually may become a rite of passage into adulthood for community members, having little to do with economic supply and demand.

Institutional Theory   Once started, migration also may be perpetuated by institutions that develop precisely to facilitate (and profit from) the continued flow of immigrants. These organizations may provide a range of services, from humanitarian protection of exploited persons to more illicit operations such as smuggling people across borders and providing counterfeit documents, and might include more benign services such as arranging for lodging or credit in the receiving country. These organizations help perpetuate migration in the face of government attempts to limit the flow of migrants.

Cumulative Causation   This perspective recognizes that each act of migration changes the likelihood of subsequent decisions about migration because migration has an impact on the social environments in both the sending and receiving regions. In the sending countries, the sending of remittances increases the income levels of migrants' families relative to others in the community, and in this way may contribute to an increase in the motivation of other households to send migrants. Migrants themselves may become part of a culture of migration and be more likely to move again, increasing the overall volume of migration. In the receiving country, the entry of immigrants into certain occupational sectors may label them as "immigrant" jobs, which reinforces the demand for immigrants to fill those jobs continually.

Which Theories Are Best?

Massey and his associates (1994) attempted to evaluate the adequacy of each of the previous theories in explaining contemporary patterns of international migration. Their conclusion was that each of the theories is supported in some way or another by the available evidence and, in particular, none of the theories is specifically refuted. This serves only to underscore the point made at the beginning of the chapter that migration is a very complex process. No single theory seems able to capture all of its nuances, but all of the previous perspectives add something to our understanding of migration. Recognizing now that the reasons for migrating are numerous and complex, we also must bear in mind that when people migrate the impact is felt deeply at both the individual and the societal levels.

Consequences of Migration

The process of migration has both individual and group consequences. For the individual, migration may produce anxiety and stress as a new social environment has to be negotiated. For communities, especially in the receiving region, the result may be **xenophobia,** the fear and mistrust of strangers, and this may lead to discrimination and even acts of violence against immigrants.

One of the ways in which migrants cope with a new environment is to seek out others who share their cultural and geographic backgrounds. This is often aided or even forced by the existence of enclaves or ghettos of recent and former migrants from the same or similar donor areas. In fact, the development of an enclave may facilitate migration, since a potential migrant need not be too fearful of the unknown. The host area has guides to the new environment in former migrants who have made the adaptation and stand ready to aid in the social adjustment and integration of new migrants. "For subsequent generations, preservation of the ethnic community, even if more dispersed, can also have significant advantages. Among the entrepreneurially inclined, ethnic ties translate into access to working capital, protected markets, and pools of labor" (Portes and Rumbaut 1990:54).

Although finding people of similar background may ease the burden of coping for a new migrant, there is some evidence to suggest that the long-run social consequences of "flocking together" (especially among relatives) will be a retardation of the migrant's adjustment to and assimilation into the new setting. John F. Kennedy's comment that "the way out of the ghetto lies not with muscle, but with the mastery of English," is more than a facile phrase. It points to a key to educational success and labor force entry (Rumbaut 1991). Similarly, a study of Korean immigrants in Chicago showed that ethnic ties were important in getting a business started, but long-term success required catering to other groups in other neighborhoods (Yoon 1990).

On arrival in the host area, an immigrant may go through a brief period of euphoria and hopefulness—a sort of honeymoon. However, that may be followed by a period of shock and depression, especially for refugees. Rumbaut (1995) quotes a Cambodian refugee to the United States he interviewed:

> I was feeling great the first few months. But then, after that, I started to face all kinds of worries and sadness. I started to see the real thing of the United States, and I missed home more and more. I missed everything about our country: people, family, relatives and friends, way of life, everything. Then, my spirit started to go down; I lost sleep; my physical health weakened; and there started the stressful and depressing times. By now [almost 3 years after arrival] I feel kind of better, a lot better! Knowing my sons are in school as their father would have wanted [she was widowed], and doing well, makes me feel more secure [Rumbaut 1995:260].

Immigrants undergo a process of **adaptation** or adjustment to the new environment, in which they adjust to the new physical and social environment and learn how best to negotiate everyday life. Some immigrants never go beyond this, but most proceed to some level of **acculturation,** in which they adopt the host language, bring their diet more in line with the host culture, listen to the music and read the newspapers, magazines, and books of the host culture, and make friends outside of their immigrant group. This may be more likely to happen if the immigrant has children, because they often are exposed to the new culture more intensively than are adults. Language use frequently is employed as an indicator of acculturation. Many migrants never go beyond this, but some migrants (and especially their children raised in the host culture) **assimilate,** in which they take on not just the outer trappings of the host culture, but also assume the behaviors and attitudes of members of the host culture (Alba and Nee 1997; Rumbaut 1997). Intermarriage with a member of the host society often is used as an index of assimilation.

These individual adjustments to a receiving society assume an open society and assume that immigrants are considered on an individual basis. In fact, nations rarely are open with regard to immigrants, and because immigration occurs regularly in clumps (with groups of refugees, or new guest workers arriving nearly en masse), immigrants often are treated categorically. Although assimilation is one model by which a society might incorporate immigrants into its midst, there are at least three other types of incorporation: **integration** (mutual accommodation); **exclusion** (in which immigrants are kept separate from most members of the host society and are maintained in separate enclaves or ghettos); and **multiculturalism** (in which immigrants retain their ethnic communities but share the same legal rights as other members of the host society) (Zlotnick 1994).

Multiculturalism, in particular, is enhanced by a new class of migrant—the **transnational migrant**—who sets roots in the host society while still maintaining strong linkages to the donor society (Schiller, Basch, and Blanc 1995). The leading edge of this phenomenon is what Findlay (1995) has called "skilled transients"—relatively skilled workers moving internationally on assignment and, in the process, having an impact on the area of destination while always intending to return to the area of origin. In parts of sub-Saharan Africa, a type of transnationalism has been institutionalized into the social structure among migratory laborers (Guilmoto 1998), and there is evidence that elsewhere in the world less skilled workers are adopting this strategy of living dual lives—working in one environment, but maintaining familial ties in another (Smith and Guarnizo 1998).

Although the consequences of migration for the individual are of considerable interest (especially to the one uprooted), a more pervasive aspect of the social consequences of migration is the impact on the demographic composition and social structure of both the donor and host areas. The demographic composition is influenced by the selective nature of migration, particularly the selectivity by age. The **donor area** typically loses people from its young adult population, those people then being added to the **host area.** Further, because it is at those ages that the bulk of reproduction occurs, the host area has its level of natural increase augmented at the expense of the donor area. This natural-increase effect of migration is further enhanced by the relatively low probability of death of young adults compared with the higher probability in the older portion of a population.

The selective nature of migration, when combined with its high volume, such as in the United States and Canada, helps to alter the patterns of social relationships and social organization. The extended kinship relations are weakened, although not destroyed, and local economic, political, and educational institutions have to adjust to shifts in the number of people serviced by each.

In summary, migration has the greatest short-run impact on society of any of the three demographic processes. It is a selective process that always requires changes and adjustments on the part of the individual migrant. More important, when migration occurs with any appreciable volume, it may have a significant impact on the social, cultural, and economic structure of both donor and host regions. Because of their potential impact, patterns of migration are harbingers of social change in a society.

People Who Are Forced to Migrate

Slavery

There can be no doubt that the most hideous of migratory movements are those endured by slaves. Slavery has existed within various human societies for millennia, and even in the 1990s it was reported that slave-trading was still taking place in the Sudan, in sub-Saharan Africa (Davis 1998). McDaniel (1995) has summarized the early historical situation as follows:

> The international slave trade in Africans began with the Arab conquests in northern and eastern Africa and the Mediterranean coast in the seventh century. From the seventh to the eleventh century, Arabs and Africans brought large numbers of European slaves into the North African ports of Tangier, Algiers, Tunis, Tripoli, and Fez. In fact, most of the slaves traded throughout the Mediterranean before the fall of Constantinople were European. Between the thirteenth and fifteenth centuries Africans, along with Turks, Russians, Bulgarians, and Greeks, were slaves on the plantations of Cyprus [p. 11].

However, the most massive migration of slaves was that of the Atlantic Slave Trade which transported an estimated 11 million African slaves to the western hemisphere between the end of the 15th century and the middle of the 19th century (Thomas 1997). The slaves came largely from the west coast of sub-Saharan Africa, from countries that now comprise Senegal, Sierra Leone, Ivory Coast, Dahomey, Benin, Cameroon, Gabon, Nigeria, and the Congo. "The preponderance of Africans who were sold into slavery were taken by force. Some were taken directly by Arab or European slave traders, but most were sold into slavery by the elite Africans who had captured them in warfare or who were holding them either for their own use as slaves or to be traded as slaves later" (McDaniel 1995:14).

The destinations were largely the sugar and coffee plantations of the Caribbean and Brazil, but hundreds of thousands were also sold in the United States to serve as laborers on cotton and tobacco plantations. The slave traders themselves were initially Portuguese and Spanish, but the French, Dutch, and especially the British were active later on. It was eventually the British, however, who pushed for a worldwide abolition of slavery, and it was finally outlawed in North America in the latter half of the 19th century.

Refugees

There were more than 14 million refugees in the world in 1997, according to data compiled by the United Nations High Commissioner for Refugees (United Nations High Commissioner for Refugees 1997) and the U.S. Committee for Refugees, as can be seen in Table 7.1. Refugees from Palestine represent the largest group, although the majority of these individuals have been born into refugee status and so the United Nations does not always include them in their statistics. Jordan is the only one of the Middle Eastern countries that typically has allowed any Palestinian to become a citizen of the country of asylum, and so Palestinians have remained as refugees for nearly a half century. Afghanistan has produced the largest group of "current" refugees (all of them within the last two decades), with nearly 2.7 million. During the 1980s, virtually one third of the Afghani population—6 million people— became refugees as the Soviet Union became embroiled in the civil war there. The majority of these refugees went to neighboring Pakistan and Iran, although about two thirds have since been repatriated.

The breakup of the former Yugoslavia has produced more than 1 million refugees to date. Germany has taken more refugees from Yugoslavia than has any other country, but every country in Europe is housing a group of Yugoslav refugees. More than 30 years of rule by Saddam Hussein, including wars with Iran and the U.S.-led forces in 1992 had produced more than 600,000 Iraqi refugees as of 1997. The remainder of the large refugee groups in the world originated in countries of sub-Saharan Africa, as can be seen in Table 7.1. In 1993 a bloody civil

Table 7.1 World Refugees by Country of Origin and Current Country of Asylum (1997)

Country of Origin	Number	Country of Asylum	Number
Palestine	3,112,000	Iran	2,014,000
Afghanistan	2,675,000	Jordan	1,359,000
Bosnia & Herzegovina	700,000	Germany	1,266,000
Iraq	630,000	Pakistan	1,200,000
Liberia	480,000	United States	597,000
Somalia	452,000	F.R. Yugoslavia	548,000
Sudan	407,000	Congo	450,000
Eritrea	328,000	Sudan	405,000
Angola	300,000	Guinea	401,000
Sierra Leone	280,000	Ethiopia	354,000
Rwanda	255,000	Côte d'Ivoire	210,000
All others	4,881,000	All others	5,696,000
	14,500,000		14,500,000

Note: Most Palestinian refugees fall under the mandate of the United Nations Relief and Works Agency for Palestine Refugees in the Near East (UNRWA), and are thus not included in the UNHCR data, whereas the USCR includes them as refugees in their data tables.

Sources: United Nations High Commissioner for Refugees, 1997, "UNHCR at a Glance," Refugees Magazine, No. 109; U.S. Committee for Refugees (USCR), 1998, **http://www.refugees.org.**

war in Burundi sent nearly one half million of its citizens to neighboring Rwanda, which returned the favor with a civil war of its own late in 1994. Three years later, in 1997, more than a quarter of a million of the survivors were still listed as refugees by the United Nations.

The three major solutions to the problem of refugee populations include: (1) repatriation; (2) resettlement in the country to which they initially fled; and (3) resettlement in a third country. As an illustration, each of these solutions has been applied to Vietnamese refugees since the fall of the South Vietnamese government in 1975. Hong Kong, before rejoining the People's Republic of China, attempted to repatriate several thousand Vietnamese refugees, although Vietnam resisted the plan. At the same time, more than 200,000 Vietnamese have been resettled in China, where they had fled directly upon leaving Indochina. Hundreds of thousands of Vietnamese have also been resettled in the United States and France (with smaller numbers in other nations) after first spending time (sometimes years) in refugee camps in places like Thailand, the Philippines, and Guam. Many Vietnamese refugees are still literally warehoused in refugee camps in Asia, hoping that someone will come to their rescue. Hong Kong's repatriation efforts may seem especially heartless because that area's own success was built on the backs of an earlier generation of refugees from the communist revolution in China. To some, refugees represent a pool of potential human capital; to others, a trough of potential trouble.

Where Do People Migrate?

Global Patterns of Migration

The massive waves of international migration that characterized the 19th and early 20th centuries have already been described in Chapter 1, and the main flows are shown in Figure 7.4. They represented primarily the voluntary movement of people out of Europe into the "new" worlds of North and South America and Oceania. Restrictive immigration laws throughout the world (not just in the United States) and the worldwide economic depression between World Wars I and II severely limited international migration in the 1920s and 1930s. However, World War II unleashed a new cycle of European and Asian migration—this time a forced push of people out of war-torn countries as boundaries were realigned and ethnic groups were transferred between countries (United Nations 1979). Shortly thereafter, the 1947 partition of the Indian subcontinent into India and Pakistan led to the transfer of more than 15 million people—Muslims into Pakistan and Hindus into India. Meanwhile, in the Middle East, the partitioning of Palestine to create a new state of Israel produced 700,000 Palestinian out-migrants and an influx of a large proportion of the Middle Eastern Jewish population into that area. Substantial migration into Israel from Europe, the Soviet Union, and other areas continued well into the 1960s (Wolffsohn 1987). In the 1980s, the flow of migrants into Israel began to dry up, replaced by a small but steady stream of out-migrants, but this trend was quickly turned around in the early 1990s following the Soviet Union's decision to allow Soviet Jews to emigrate. Those choosing to leave headed primarily for Israel and the United States.

Figure 7.4   Main Currents of Intercontinental Migration (1500–1950)

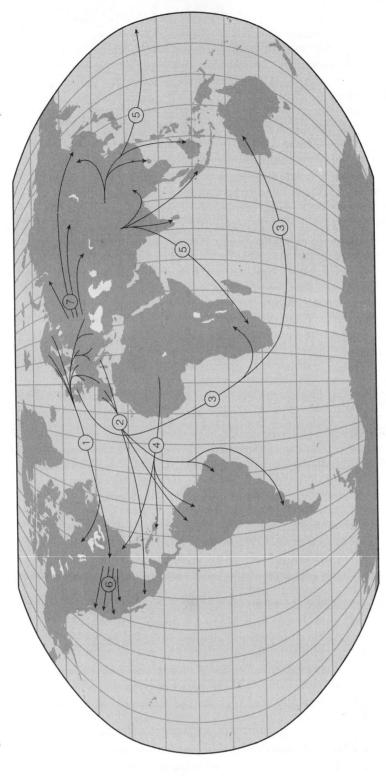

Note: Major currents of intercontinental migration have been (1) from all parts of Europe to North America; (2) from Latin countries of Europe to Middle and South America; (3) from Great Britain to Africa and Australia; (4) from Africa to America; (5) from China and India abroad (partly intercontinental, partly intracontinental). Important currents of internal migration have been (6) the westward movement in the United States and (7) the eastward movement in Russia. Before the end of World War II, the main currents of migration were out of the more densely settled regions in Europe and Asia and into North and South America and Oceania. Major shifts since 1950 include a net flow of people back into Europe and a net flow out of many Latin American countries.

Source: W. S. and E. S. Woytinsky, 1953, World Population Production (New York: The Twentieth Century Fund), p. 68. c 1953 by The Twentieth Century Fund, New York. Reprinted with permission.

254

Another unexpected political event in eastern Europe in the late 1980s and early 1990s was the collapse of the Berlin Wall and the amazingly rapid reunification of Germany. Stimulated by Gorbachev's policy of openness in the Soviet Union (and by the former USSR's economic inability to continue subsidizing other communist nations), the reunification of Germany was both the cause and effect of migration from East Germany into West Germany (Heilig, Büttner, and Lutz 1990). Between 1950 and 1988 more than 3 million East Germans had fled to the West, but most of those had done so before the Berlin Wall went up in 1961. Then, in 1989, East Germany relaxed its visa policies, allowing East Germans to visit West Germany, and Hungary relaxed the patrol of its border with Austria, allowing vacationing East Germans to escape to the West. Within weeks, migration from east to west was transformed from a trickle into a flash flood.

This east–west migration in Europe is in many respects a continuation of a pattern that has evolved over centuries. Between 1850 and 1913 (the start of World War I) more than 40 million Europeans moved from east to west to pop- ulate North America (Hatton and Williamson 1994), but as that was occurring, Polish, Slavic, and Ukrainian workers were migrating west to Germany and France, and Italians were migrating west to France to find work. The Cold War, which cut off much of that flow, was simply a temporary aberration in a long- term trend.

To this trend has been added a mixture of other patterns: (1) south-north mi- gration (particularly of migrant laborers from developing countries of the "south" to developed countries of the "north"); (2) a flow of migrant laborers from some of the poorer developing countries to some of the "emerging" economies, especially in south and southeast Asia; (3) a flow of workers in the Persian Gulf region from the non–oil-producing to the oil-producing nations; and (4) a flow of refugees, especially in Africa and western Asia. Figure 7.5 shows the countries of the world that sent or received the greatest absolute number of net international migrants (including refugees) in 1990–95, according to World Bank estimates. You can see that the United States is the top receiving country, followed by Germany and Canada. Yemen is on the list because the two Yemens were united in 1990, creating a pattern of return migration, combined with the estimated impact of the Gulf War (which also is estimated to have affected the number of migrants coming from Jordan or going to Kuwait). At the other extreme, China is estimated to lose the greatest number of people each year (contributing to a phenomenon known as the Chinese Diaspora), but that is just a tiny fraction of its total population. Mexico is second on the list, followed by India.

The global picture of sending and receiving countries is shown in Figure 7.6. The countries *receiving* an estimated 20,000 or more people during 1990–95 are shown as the darkest shade, and those countries *sending* an estimated 20,000 or more are shown as the lightest shade. In between are those countries with little net migration in absolute terms. The principal receiving countries include the United States, Canada, Australia, most of the nations of western Europe and Scandinavia, as well as Russia (which has been receiving immigrants—especially repatriated Rus- sians—from its former republics). The sending countries tend to be in Asia and Latin America (especially Mexico and Central America).

Figure 7.5 The United States Takes in More Immigrants (Net) Than Any Other Country, and China Sends More Emigrants Than Any Other Country (Estimates for 1990–95)

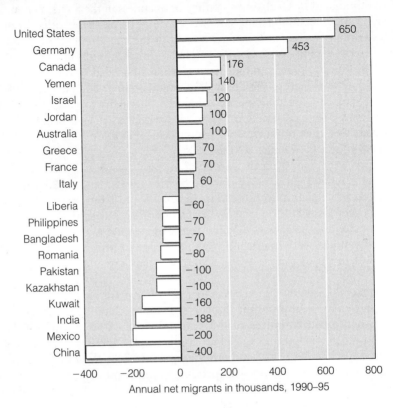

Source: Adapted by the author from World Bank estimates.

Labor Migration

Improved communication and transportation technology have greatly facilitated a time-honored way of solving short-term labor shortages—the importation of workers from elsewhere. Rapid population growth in less developed nations has put incredible pressure on resources in those nations, while declining rates of population growth in the more developed nations has, in many instances, heightened the demand for lower-cost workers from the Third World. In 1989 there had been 1.7 million guest workers in West Germany, largely from Turkey, Yugoslavia, Italy, and Greece. East Germany had nearly 100,000 guest workers, the majority from Vietnam (Heilig, Büttner, and Lutz 1990). England, France, and the Netherlands in particular also have received large numbers of immigrants from their former colonies. Of demographic importance for the future is the fact that immigrant groups from developing countries tend to have higher fertility levels than prevail in their European host nations. Thus the next generation throughout much of Europe will have a disproportionate share of children whose parents were born outside of Europe. The European Community has been trying to close its doors to immigrants. Europe has decided that it will open its

Figure 7.6   Net Receiving and Sending Countries in the World, 1990–95

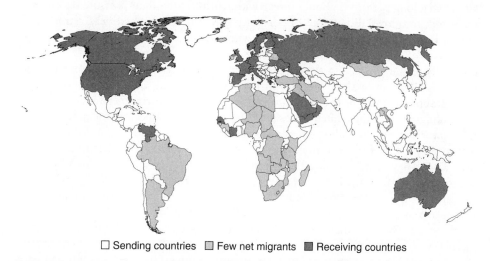

☐ Sending countries   ☐ Few net migrants   ■ Receiving countries

Source: Adapted by the author from World Bank data.

borders to "internal" migration—allowing labor migration between its 12 member nations, rather than permit people to enter Europe from outside those nations.

Meanwhile, guest worker programs have flourished in the Middle East and Africa (see, for example, Gould and Findlay 1994). The Arab oil-producing nations became centers of rapid immigration in the late 1970s and early 1980s as they benefited economically from the rise in oil prices. Libya and Saudi Arabia, in particular, became attractive destinations for Egyptian and Jordanian migrants, along with substantial numbers of Indians and Pakistanis.

Guest-worker programs have also been historically common in sub-Saharan Africa, but the movement toward decolonization and the creation of independent states has altered some of those practices as governments attempt to control migration. In general, migration in Africa has involved males, who may be gone from their families for as long as 2 years while working in a resource-rich neighboring country.

Hidden from the view of most eyes is the tremendous economic benefit that third-world sending countries receive when their citizens go off to work in first-world nations. Most such workers send part of their pay back home, thereby raising the standard of living of the family members who stayed behind. As I have already mentioned, remittances are an important part of the family economy of many households in developing countries (Stark and Lucas 1988).

Migration into North America

Prior to World War I, there were few restrictions on migration into the United States and Canada, so the number of immigrants was determined more by the desire of people to move than anything else. Particularly important as a stimulus to migration, of course, was the drop in the death rate in Europe during the 19th century, which launched a long period of population growth, with its attendant pressure on Europe's economic resources. Economic opportunities in America looked very attractive to young Europeans who were competing with increasing numbers of young people for jobs. Voluntary migration from Europe to the temperate zones of the world—especially to the United States—represents one of the significant movements of people across international boundaries in history. The social, cultural, economic, and demographic impacts of this migration have been truly enormous (Davis 1988).

Immigration to the United States and Canada reached a peak in the first decade of this century, when 1.6 million entered Canada and nearly 9 million entered the United States, accounting for more than one in ten of all Americans at that time (see Figure 7.7). "They came thinking the streets were paved with gold, but found that the streets weren't paved at all and that they were expected to do the paving" (Leroux 1984).

Immigration to the United States in most of the 19th century was dominated by people arriving from northern and western Europe (beginning with England, Ireland, Scotland, and Sweden, then stretching to the south and east to draw immigrants from Germany especially) as seen in Figure 7.7. There was a lull during the American Civil War, but after the war the pace of immigration resumed. By the late 19th and early 20th centuries, the immigrants from northern and western Europe were being augmented by people from southern and eastern Europe (Spain and Italy, then stretching farther east to Poland and Russia). This represents one of the most massive population shifts in history and virtually all of the theories of international migration discussed earlier have something to offer by way of explanation (Hatton and Williamson 1994; Massey, et al. 1994; Moch 1992). Compared to the United States, European wages were low and unemployment rates high. Capital markets were beginning to have a disruptive effect in some of the less developed areas of southern and eastern Europe. Eastern Europe was also undergoing tremendous social and political instability, and the Russian pogrom against Jews caused many people to flee the region.

Nor was all of this emigration from Europe aimed at North America. Millions of Italians and Spaniards, as well as Austrians, Germans, and other Europeans settled throughout Latin America during this period. Most notable among the destinations were Brazil and Argentina (Sanchez-Albornoz 1988). The migration peaked at about the time of World War I, and Europe never since has experienced emigration of this magnitude, partly because wages rose in Europe, helping to keep people there (and encouraging some return migration from the Americas), compounded by the Great Depression of the 1930s (which also encouraged return migration to one's "roots"). In the latter part of the 19th century, the percentage of the population enumerated in the U.S. census as being foreign born hovered near 14 percent, as seen in Figure 7.7. Never since has it reached that level.

The Great Depression, coupled with more restrictive immigration laws (discussed in Chapter 14), followed by World War II, effectively slowed migration to the United States and Canada to a trickle during the 1930s and 1940s. In the 1950s, there was a brief post-World War II upsurge in migration from northern and western Europe, but as seen in Figure 7.7, the 1950s represented a transition time to a new set of "origins and destinies" for immigrants to North America (Rumbaut 1994). Since the 1960s, European immigrants have been replaced almost totally by those from Latin America (especially Mexico) and Asia (especially the Philippines and southeast Asia—Vietnam, Cambodia, Laos).

Figure 7.7 The Geographic Origin of Immigrants to the United States Has Changed Dramatically over the Decades

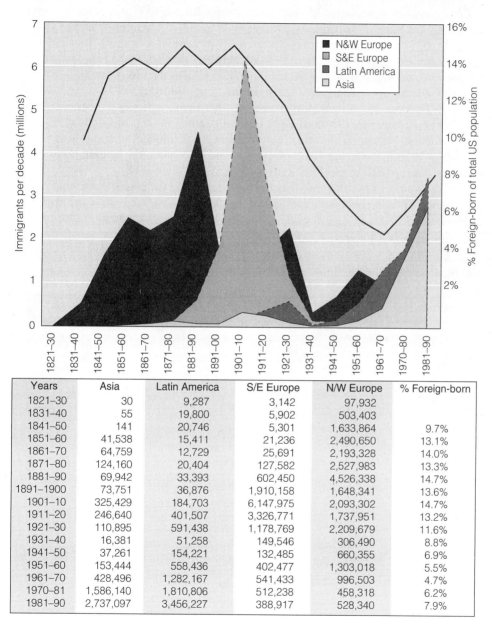

Years	Asia	Latin America	S/E Europe	N/W Europe	% Foreign-born
1821–30	30	9,287	3,142	97,932	
1831–40	55	19,800	5,902	503,403	
1841–50	141	20,746	5,301	1,633,864	9.7%
1851–60	41,538	15,411	21,236	2,490,650	13.1%
1861–70	64,759	12,729	25,691	2,193,328	14.0%
1871–80	124,160	20,404	127,582	2,527,983	13.3%
1881–90	69,942	33,393	602,450	4,526,338	14.7%
1891–1900	73,751	36,876	1,910,158	1,648,341	13.6%
1901–10	325,429	184,703	6,147,975	2,093,302	14.7%
1911–20	246,640	401,507	3,326,771	1,737,951	13.2%
1921–30	110,895	591,438	1,178,769	2,209,679	11.6%
1931–40	16,381	51,258	149,546	306,490	8.8%
1941–50	37,261	154,221	132,485	660,355	6.9%
1951–60	153,444	558,436	402,477	1,303,018	5.5%
1961–70	428,496	1,282,167	541,433	996,503	4.7%
1970–81	1,586,140	1,810,806	512,238	458,318	6.2%
1981–90	2,737,097	3,456,227	388,917	528,340	7.9%

Sources: Data for 1820 through 1890 are from U.S. Bureau of the Census, 1976, *Historical Statistics of the United States from Colonial Times, Centennial Edition, Part 1* (Washington, D.C.: Government Printing Office), Tables A1-5 and A6-8; and from U.S. Immigration and Naturalization Service, 1993, *1992 Statistical Yearbook of the Immigration and Naturalization Service* (Washington, D.C.: Government Printing Office), Table 2; data for 1900 through 1990 are from R. Rumbaut, 1994, "Origins and destinies: Immigration to the United States since World War II," *Sociological Forum* 9(4), Table 1.

Canada's immigration experience was similar, but not identical to the pattern in the United States. Although Canada has historically had a high level of immigration, it also experienced considerable emigration (people leaving to return home, or entering the United States from Canada) until after World War II, as shown in Figure 7.8. Since the 1950s immigration has increased to its present level of more than 200,000 per year, but emigration has slowed considerably and this has pushed the level of net migration to unprecedented high levels. The 1996 Census data for Canada show that 1 in 10 Canadians speaks a language other than English or French at home, and the number of people reporting a mother tongue other than English or French increased by 15 per cent between 1991 and 1996 to total nearly 5 million of Canada's 30 million residents. Chinese is the most common language spoken at home other than English or French, replacing German, Italian, and Ukrainian, the leading nonofficial languages as reported in the 1971 census (Migration News 1998).

As the number of immigrants has fluctuated, so have the places from which they have come. This diversity is illustrated by the data in Table 7.2, where information is summarized for the major geographic regions from which immigrants to the

Figure 7.8   Fluctuations in the Net Number of Immigrants to Canada

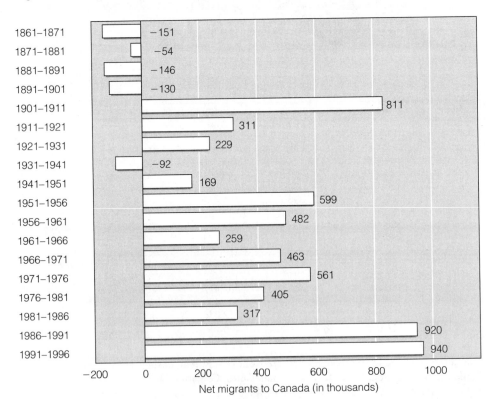

Source: Statistics Canada, 1998, "Population and Growth Components," http://www. statcan.ca/english/Pgdb/People/Population/demo03.htm

Table 7.2 The Immigrant Population in the United States Is a Diverse Group

Region/Country of Birth	Foreign-Born Persons in 1990	Immigrated to United States in the 1980s (%)	Naturalized U.S. Citizen (%)	Median Age (Years)	Female (%)	College Graduates (%)	Speaks English Not Well or Not at All (%)	Poverty Rate (%)	Children Born per Woman (35–44)	Children <18 with 2 Parents (%)
Europe and Canada	5,095,233	20	63	53	57	19	9	9	1.8	86
Italy	580,595	5	76	59	52	9	16	8	2.1	85
Germany	711,929	11	72	53	65	19	2	8	1.8	75
United Kingdom	640,145	25	50	50	60	23	0	7	1.8	85
Latin America	8,416,924	50	27	33	48	9	40	24	2.7	69
Cuba	736,971	26	51	49	52	16	40	15	1.8	72
El Salvador	485,433	76	15	29	46	5	49	25	2.7	61
Mexico	4,298,014	50	23	30	45	4	49	30	3.3	73
Asia	4,979,037	57	41	35	51	38	22	16	2.0	83
China	529,837	55	44	45	51	31	44	16	1.8	87
Korea	568,397	55	41	35	57	34	30	16	1.8	87
Philippines	912,674	51	54	39	50	43	34	6	2.1	89
Vietnam	543,262	64	43	30	47	16	31	26	2.5	73
Africa	363,819	61	34	34	41	47	5	16	2.2	75
All foreign-born	19,767,316	44	41	37	51	20	26	18	2.3	74
U.S.-born										
White, non-Hispanic				33	51	20	1	13	1.9	73
African-American				35	51	22	1	9	1.8	80
Mexican-origin				28	53	11	1	30	2.2	37
Asian-origin				18	50	9	10	24	2.5	69
				15	50	36	6	10	1.5	86

Source: Adapted from R. G. Rumbaut, 1994, "Origins and destinies: Immigration to the United States since World War II," *Sociological Forum* 9(4), Tables II, III, and IV.

261

United States have come, along with examples of specific countries in each region. More than 5 million persons were enumerated in the 1990 census as having been born in either Europe or Canada. Only 20 percent of them had migrated as recently as the 1980s, and nearly two thirds were naturalized citizens. The median age for this group was 53 (compared to 33 for the U.S.-born population in 1990), and 19 percent were college graduates (nearly identical to the U.S.-born population). In general, European and Canadian immigrants represent an older version of the demographic profile of the U.S.-born population. The more recent immigrants, especially from Latin America and Asia, are less similar to the U.S.-born population (as well as differing from each other).

Migration from Latin America to the United States   Although only one person is recorded as having migrated from Mexico to the United States in 1820, the number of Mexican and other Latin American immigrants has increased so tremendously since then that the legal and illegal immigrants from Mexico probably account for one in four of all foreigners now moving into the United States. Between 1820 and 1995 more than 5 million Mexicans migrated legally into the United States, with more than one half of those people arriving since 1980 (U.S. Immigration and Naturalization Service 1997). Migration between Mexico and the United States is now the largest sustained flow of migrant workers in the contemporary world (Massey, et al. 1994). That increase coincided, of course, with the beginnings of the current rapid rate of population increase in Mexico. As Mexico's population has grown, so has the difficulty of finding adequate employment for the burgeoning number of young adults. That "push," accompanied by the "pull" of available jobs with higher wages in the United States, has stimulated a tremendous migration stream. The flow in this stream can build quickly as the number of social ties between sending and receiving areas grows, "creating a social network that progressively reduces the cost of international movement. . . . The range of social contacts in the network expands with the entry of each migrant, thus encouraging still more migration and ultimately leading to the emergence of international migration as a mass phenomenon" (Massey, et al. 1987:5). Data from Massey's Mexican Migration Project suggest that migration streams are much easier to start than to stop, because migration cumulatively begets more migration as community members in the sending area derive a real benefit (**human capital**) from migration and as expanded networks (**social capital**) make it increasingly easier to migrate (Massey and Espinosa 1997). In this context, human capital refers to the benefits derived by the migrants who have been to the United States, whereas social capital refers to the links with friends and relatives already abroad.

Immigrants from Latin America are arriving from countries in which the mandatory level of education typically is 6 years of schooling, and they represent a relatively poorly educated, lower-skilled group. The principal exception to this generalization has been Cuban immigrants, but recent immigrants from Cuba are less well educated and have fewer skills than the group of exiles who fled Castro's takeover of the island initially. Table 7.2 shows that, of foreign-born persons enumerated in the 1990 census, 50 percent of Mexican migrants and 76 percent of Salvadoran migrants entered the United States during the 1980s. They were the youngest of the immigrant groups, and males outnumbered females. Among Mexican immigrants, only 4 percent had a college degree and nearly one half did not speak English well or did not speak English at all. They also tended to be the least well-off economically, as evidenced by the 30 percent poverty rate. Consistent with lower levels of education and income, Mexican immigrant women aged 35 to 44 had given birth to 3.3 children, compared to the average of 1.9 for all U.S.-born women, and 2.5 for U.S.-born Mexican-origin women. Still, this was lower than the 4.3 children born to the average Mexican woman that age, according to the 1987 Mexican Fertility Survey.

Migration from Asia to the United States Asian immigration to the United States began in the middle of the 19th century, but a series of increasingly restrictive immigration laws (which I will discuss in Chapter 14) kept the number relatively low until the late 1960s, when the laws were substantially liberalized. In the entire decade of 1841–50, scarcely more than 100 Asians immigrated (see Figure 7.7), but in the following decade the demand for labor on the West Coast produced a 10-year total of 41,000 Asian immigrants, virtually all of whom were from China. Between 1871 and 1880 the level of immigration from China climbed to 124,000, and that volume led to an unfortunate series of restrictive laws limiting the ability of Chinese to enter the country. As migration from China abated, migration from Japan increased; in the decade 1901–10, only 21,000 Chinese immigrated, whereas 130,000 came from Japan.

MIGRATION FROM MEXICO AND THE IMPACT OF THE NORTH AMERICAN FREE TRADE AGREEMENT

Mexico and the United States are unique in the fact that nowhere else on earth does a highly developed nation share a long (2000 mile) border with a less developed nation. Given the geographic proximity and the long-standing economic disparities between the two countries, it would be truly remarkable if there were not a good deal of interaction and migration taking place.

Migration from Mexico to the United States began in earnest early in this century as a reaction to the Mexican Revolution that started in 1910 ending in the creation of the modern United Mexican States (the official name of the Republic of Mexico). As seen in Figure 7.7, the migration northward from Latin America (largely Mexico) began to increase in the 1920s and 1930s. That flow was then halted by a combination of the

Great Depression, which raised unemployment levels in the United States, and the concomitant discrimination against immigrants that surfaced during this period, leading to massive deportations of many immigrants (legal and otherwise), including thousands from Mexico.

Labor shortages in agriculture during World War II, however, led to a renewed invitation for Mexican workers to migrate to the United States. In 1942, the United States signed a treaty with the government of Mexico to create a system of contract labor whereby Mexican laborers ("braceros") would enter the United States for a specified period of time to work. After the war ended, the bracero program remained in place, but it was not until the early 1950s that the number of Mexican contract workers began to increase noticeably (Garcia y Griego, Weeks, and

(Continued)

MIGRATION FROM MEXICO AND THE IMPACT OF THE NORTH AMERICAN FREE TRADE AGREEMENT—CONT'D

Ham-Chande 1990). By the mid-1950s, the contract workers had been joined by many undocumented immigrants from Mexico, and the United States reacted in 1954 by deporting more than 1 million Mexicans (some later found to be U.S. citizens) in what was called "Operation Wetback."

The bracero program was ended formally in 1964, and in 1965 the new immigration act, which ended the national origins quota system, also put a numerical limit on the number of legal immigrants to the United States from countries in the Western Hemisphere. Neither action noticeably slowed the migration from Mexico, which was by then well entrenched because there was constant demand for immigrant labor. However, they did jointly conspire to increase the number of those immigrants who were classified as illegal or undocumented (U.S. Immigration and Naturalization Service 1991).

Migration from Mexico to the United States has persisted for virtually all of the reasons discussed in this chapter concerning the various theories of international migration. The drop in the death rates in Mexico that began in the late 1930s, but accelerated after World War II, produced the equivalent of a baby boom in that country and, despite rapid economic growth, Mexico has had difficulty creating enough jobs to go around. An obvious economic strategy for many households was to help a household member get to the United States to work and send money back home. Massey and his associates estimate that in many villages in west central Mexico nearly all males have migrated at least once to the United States by age 30 (Massey and Espinosa 1997). Many do remit money home, but do not themselves return, preferring instead to remain in the United States. Although the principal jobs available to immigrants from Mexico are low-wage, low-skill jobs (as predicted by the segmented dual labor market theory), experience and English language skills offer opportunities for better paying jobs. Furthermore, as the number of Mexicans living in the United States has increased and the network of immigrant families has widened, the costs of arriving in a new country have dropped substantially, making it less

risky for villagers from Mexico to "try out" the U.S. labor market.

Meanwhile, back in Mexico, the ending of the bracero program had created problems for Mexico, because there was a labor pool of un- or underemployed persons along the northern border of Mexico waiting for the chance to return to work in the United States. In 1965 the government of Mexico decided to tap into this pool of labor by creating the Border Industrialization Program. The aim was to employ people in assembly plants (which came to be known as *maquiladoras*) in which parts were shipped to Mexico by multinational corporations, assembled in Mexico, and then reexported to another country for sale. Taxes were levied only on the value added to the product during the time that it was in Mexico—meaning primarily labor, which is much cheaper in Mexico than in the United States or Japan (which were the two countries that took greatest advantage of the Border Industrialization Program).

By the late 1990s nearly 1 million Mexican workers were employed in the maquiladoras, especially in the northern border cities of Ciudad Juarez (the twin city of El Paso, Texas), Tijuana (the twin city of San Diego, California), and Matamoros (the twin city of Brownsville, Texas) and these assembly plants have been an important profit center for Mexico. The Mexican government had suggested in the 1960s that the Border Industrialization Program would create a wall of jobs that would limit migration to the United States; however, the economic crisis that followed the devaluation of the peso in 1982, coupled with continued job opportunities in the United States, helped to generate more, not less, migration, once again including large numbers of undocumented migrants. The U.S. government responded with the Immigration Reform and Control Act (IRCA) of 1986 (discussed in more detail in Chapter 14), which attempted to limit undocumented migration by penalizing U.S. employers of such persons. However, it also permitted the legalization of about 1.7 million people who had entered the United States previously without documents. Nearly 70 percent of these persons were from Mexico, and one half lived in California, primarily in Los Angeles (U.S. Immigration and Naturalization Service 1992).

Shortly after IRCA was passed, the number of apprehensions of illegal border crossers went down, signaling that the law was discouraging such migration. However, subsequent research has suggested that this was only temporary, and that illegal migration soon climbed back to higher levels (Bean, Vernez, and Keely 1989; Donato, Durand, and Massey 1992). It appears that the purchase of counterfeit documents allowed many illegal immigrants to be hired, effectively circumventing the control mechanism built into IRCA (Lowell and Jing 1994).

Although the Border Industrialization Program did not create enough jobs to reduce migration from Mexico to the United States significantly, it did provide much of the inspiration for the North American Free Trade Agreement (NAFTA). When NAFTA was signed into law in January 1993, it committed the United States, Mexico, and Canada to economic cooperation designed to lower trade barriers and increase the flow of capital, goods, and services across the borders between the three countries. Would that create more or less migration to the United States from Mexico?

If we accept the idea of cumulative causation, discussed in this chapter as a way of explaining why migration persists even after the reasons for its initiation may have disappeared, then we might conclude that NAFTA will have little effect on legal migration from Mexico to the United States (see, for example, Massey and Espinosa 1997). The pattern of legal migration from Mexico is now occurring for reasons that transcend economic improvement in Mexico and is unlikely to be affected dramatically by events in either Mexico or the United States. The migration that is most apt to be affected is illegal or undocumented migration. What are the possible scenarios?

Some of the provisions of NAFTA encouraged Mexico to reclaim land from peasant farmers in order to increase agricultural production. This was seen by some people as a stimulus to landless Mexican peasants to migrate to the United States (Cornelius and Martin 1993). Others looked at NAFTA and projected a long-term increase in economic well being in Mexico that will serve as a potential deterrent to emigration from Mexico (Acevedo and Espenshade 1992).

At this point, it is still too early to tell what the long-run impact of NAFTA will be in terms of the permanent migration of undocumented immigrants from Mexico to the United States. The evidence seems clear that Mexicans continue to migrate at least temporarily to work (Bustamante 1997), but the available data also suggest that most Mexican workers who come to the United States return to Mexico. Although the U.S. Immigration and Naturalization Service reports more than 1 million apprehensions per year of people trying illegally to enter the United States, the U.S. Commission on Immigration Reform concluded in 1997 that only about 100,000 undocumented immigrants per year from Mexico were remaining as permanent residents in the United States (Loaeza Tovar and Martin 1997).

A variety of factors having little direct relationship to NAFTA may be operating to slow down the rate of permanent migration. A higher minimum wage and an increased supply of low-skill workers being dropped off the welfare rolls may lower the demand in the United States for cheap immigrant labor. At the same time, the declining fertility rate in Mexico is starting to pay off with a drop in the number of new entrants into the labor market, thus lowering the supply of potential immigrants (Loaeza Tovar and Martin 1997). However, we do have evidence that economic crises in Mexico still have the impact of sending undocumented migrants to the United States. Late in 1994, the Mexican government abruptly devalued the peso yet again, sending the Mexican economy into a tailspin and causing many Mexican families to pack the bags of at least one member of the household and send him or her north. Despite the efforts of the U.S. Border Patrol to stop such people at the border with "Operation Hold the Line" in El Paso and "Operation Gatekeeper" in San Diego, it appears that if someone from Mexico wants to get to the United States to find work, they will (Espenshade 1994).

The National Origins Quota system, which went into effect in the 1920s (see Chapter 14) dramatically reduced migration from all parts of Asia. Thus of the 528,000 immigrants to the United States between 1931 and 1940, only 16,000 were from Asia. However, following the 1965 revisions of the immigration laws in the United States, the picture changed. Since 1965 Asians have accounted for more than four in ten new arrivals into the United States, and in the period from 1991–95 the list was led by the Philippines, and followed by China (including the People's Republic of China, Taiwan, and Hong Kong), India, and Vietnam (U.S. Immigration and Naturalization Service 1997).

Immigration from the Philippines began in the 1920s, when free migration was permitted between the two countries after the United States gained control of the country from Spain as a result of the Spanish-American War in 1898. By 1930 45,000 Filipinos lived in the United States, almost all of them males. The majority of these men came without families, and either returned home or stayed and married non-Filipino women. Although migration was fairly heavy in the 1920s, it was stopped almost entirely during the Depression and World War II by discriminatory, anti-Filipino legislation that stemmed from displeasure at the extent to which Filipino men dated and married white women (Kitano 1997). After World War II, however, that sentiment changed somewhat as a result of Americans' wartime experiences as allies and defenders of the Philippines, and there was an influx of Filipino veterans after the war. But the biggest boon to Filipino immigration was the change in the law in the 1960s that gave highest immigration preference to family members. Americans of Filipino origin began to sponsor the immigration of their relatives (including especially parents and siblings). Thus since 1980 more than three fourths of the 40,000 to 60,000 immigrants each year from the Philippines were relatives of U.S. citizens (U.S. Immigration and Naturalization Service 1997).

Although migration from Asia clearly increased during the 1960s, the volume was swollen by the large number of Indochinese refugees who were resettled in the United States as a result of the war in Vietnam. Since the end of American involvement in the military conflicts of Southeast Asia in 1975, nearly 900,000 refugees have fled Indochina for the United States. About two thirds of this population is Vietnamese (including ethnic Chinese), with the numbers of Laotians, Cambodians, and Hmong being nearly equal. California has attracted at least one third of the refugees, and eight other states (Texas, Washington, Minnesota, Pennsylvania, Illinois, Oregon, Virginia, and New York) account for another third. The ethnic diversity of these new immigrants is matched by their socioeconomic and educational differences. Many of those who arrived in the first wave of immigration (1975–78) were highly skilled, educated, urban residents who had strong ties to the American government and Western culture. Those who arrived in the second wave were more rural in origin and had fewer occupational and educational skills. In addition, whereas many in the first group of refugees spent little time in Asian refugee camps, the majority of post-1978 refugees spent many months and even years in profound hardship, both in their escape from Vietnam, Laos, and Cambodia, and in their tenure at the various camps in Thailand, Hong Kong, Malaysia, Indonesia, and the Philippines.

Vietnamese refugees are twice as likely as Mexican immigrants to have become U.S. citizens, probably because most of the Vietnamese do not expect to return to their homeland, whereas a large portion of Mexican immigrants expect to return (even if they eventually do not). This may change over time as a result of the law that went into effect in Mexico in 1998 allowing Mexicans to retain their Mexican nationality even if they become a citizen of another country such as the United States. Under the Mexican Nationality Act, people who lost their Mexican citizenship when they became naturalized U.S. citizens have until 2003 to go to a Mexican consulate' to re-acquire their Mexican nationality and receive a Mexican passport. This is not the same as dual citizenship, however, because dual nationals are not allowed to run for public office or vote in Mexico or serve in the armed forces. However, the dual nationality status means that Mexican nationals who are U.S. citizens will have the right to buy and sell land in Mexico free of the restrictions imposed on foreigners and to receive better treatment under investment and inheritance laws in Mexico, to attend public schools and universities as Mexicans, and to access other Mexican government services and jobs.

The sociodemographic factor that most distinguishes Asian from Latin American immigrants is the level of education. Overall, 38 percent of Asian immigrants were college graduates in 1990 (the figure for immigrants from India was 65 percent), compared with an average of 9 percent for Latin American immigrants (see Table 7.2). These percentages are then related to levels of English fluency, poverty, and other measures of integration into the U.S. economy.

Migration out of the United States

The United States Immigration and Naturalization Service keeps track only of expatriates—those who renounce American citizenship—and these number no more than a few thousand people annually. However, a larger number of people migrate out of the country each year, and although no records are kept for such individuals, it is estimated that 160,000 people emigrate each year from the United States (Warren and Kraly 1985). Estimates for both the United States and Canada suggest that in the 1990s there were about 5 immigrants for each emigrant (Statistics Canada 1998; U.S. Immigration and Naturalization Service 1997).

Using data from the Current Population Surveys and other sources, Woodrow-Lafield (1996) has estimated that there are approximately 220,000 emigrants from the United States each year, of whom about 133,000 (83 percent) are foreign-born persons (probably returning to their country of origin). We know, for example, that in 1996 372,000 people were living abroad and receiving American social security benefits. Canada had the greatest number of such people (84,000), followed by Mexico (52,000) and Italy (36,000), and then by the United Kingdom (23,000), Germany (22,000), Greece (19,000), and the Philippines (19,000). About 90 percent of them were not born in the United States, and most (95 percent) appear to have migrated to the United States at a younger age, worked to build up a social security account, and then retired to their country of origin as older migrants (Kraly 1982).

The figure of 220,000 emigrants annually from the United States is assumed to be essentially constant, and is factored into population estimates that the Census Bureau makes each year of the U.S. population (Day 1996). However, emigration almost certainly has varied in its tempo over time, and Passel and Edmonston (1994) have compared immigration data with the census counts of the foreign-born population to estimate the extent of return migration. Their data suggest that immigrants to the United States (like those to Canada) in the first part of this century were far more likely to return home than are more recent immigrants. For example, in the period between 1900 and 1930 18.1 million immigrants entered the United States, whereas Passel and Edmonston estimate that 7.9 million people emigrated from the United States during that time. Thus the ratio of immigrants to emigrants was 2.3 to 1. However, between 1960 and 1990 there were 21.4 million immigrants and an estimated 3.7 million emigrants, for a ratio of 5.8 to 1—more than twice the level of earlier in the century.

Migration within North America

The United States is a nation on the move, and it always has been. The Census Bureau has estimated that 43 million Americans (16 percent of the population) aged 1 and older in 1996 were living in a different house than in the year before. Some of these people had undoubtedly moved more than once during that period, so that represents a probable minimum of mobility. Of those movers, 14 million crossed county lines and would be considered migrants, not just movers. An additional 1.4 million persons had moved in from outside the United States during the previous year; including U.S. citizens returning from abroad (about one third of that total) and immigrants (Hansen 1997).

Americans, though certainly very mobile, are not totally unique in that respect. In an international comparison of residential mobility data, Long (1988) found that Australians and Canadians have rates that are virtually identical to Americans. On the other hand, residents in those three countries are much more likely to migrate than are the British or Japanese. Long (1991) argues that the United States, Australia, and Canada have high rates of geographic mobility partly because all three are nations of immigrants. Migration is thus not a new or innovative idea, but persists from generation to generation. As a result, we are more likely to turn to migration as a life strategy than are people in countries where migration is less common. At the same time, we can note that the number of migrants has kept pace with the population in the United States, but the actual rates of migration (the percentage of people moving each year) have remained remarkably steady since the end of World War II and, if anything, they have declined, not risen (Gober 1994; Hansen 1997).

Intimately bound up with the reasons for moving and the number of people who migrate is, as mentioned previously, the question of where people go. Because migration is so often associated with life-cycle stages and represents attempts to improve the quality of life, migrants naturally tend to go where they perceive opportunities to be greatest. Economic motives dictate that migrants go where business is good and leave behind those places where it is not so good, as discussed earlier in the chapter.

Within the United States, there were several decades when migration was in the direction of the industrializing centers in the northeastern and north central states and to the rich farmland and industry in the midwestern states. The strongest of these movements was the one westward (Shryock 1964). At first this meant that the mountain valley areas west of the Atlantic seacoast were migration destinations; then the plains states were settled; and, especially since the end of World War II, the Pacific Coast states have been popular destinations.

Until about 1950 migrants had also been heading out of the southern states and into the northeastern and north central states. This generally represented rural-to-urban migration out of the economically depressed South into the industrialized cities of the North. In the 1950s this pattern of net out-migration from the South reversed itself and the northeastern and north central states found themselves increasingly to be migration origins rather than destinations.

In the 1950s, migrants began heading not only west, but south as well. This pattern continued into the 1990s, as can be seen in Table 7.3. The northeastern states (the "Rust Belt" or "Snow Belt") have had more out-migrants than in-migrants, and people have been moving west and south to the "Sun Belt." In the 1960s and 1970s, the movement still was more strongly to the west than to the south, but in the 1980s and the 1990s, the south was gaining more through net migration than were the western states.

In the early to mid-1990s, the nearly unthinkable pattern unfolded of net out-migration from California, at least in terms of internal migrants. As a result of the defense cutbacks associated with the end of the Cold War, on top of a general economic recession, the California economy slowed dramatically in the early 1990s, and in 1993 demographers estimated that 252,000 more persons left California for other states than moved to California from elsewhere in the United States (California Department of Finance 1994). Most of the people leaving California were going to the nearby states of Arizona, Nevada, Utah, Idaho, and Washington (Gober 1994). In spite of the net domestic out-migration, California's population was continuing to increase numerically, not only through an excess of births over deaths, but also through the continuation of net international migration from Latin American and Asian

Table 7.3 Americans in the Mid-1990s Were Heading South Rather Than West

Destination in 1996:	Origin in 1995:				
	Northeast	Midwest	South	West	Totals
Northeast	−234	118	252	71	441
Midwest	127	67	480	235	842
South	366	449	151	469	1,284
West	182	208	401	16	791
Totals	675	775	1,133	775	3,358

Note: Between 1995 and 1996 there were 3,358,000 Americans who moved between the four major geographic regions of the country. This table shows the number of migrants (in thousands) according to their origin (where they lived in 1995) and their destination (where they lived in 1996). The diagonal (highlighted) shows the net number of migrants for that region. Thus there were 675,000 people who left the northeast between 1995 and 1996 and 441,000 who moved into the northeast during that year, for a negative net migration of −234,000. The South gained more from this migration than did any other region.

Source: Adapted from Kristin A. Hansen, 1997, "Geographical Mobility: March 1995 to March 1996," U.S. Bureau of the Census, Current Population Reports, P20-497, Table 20.

countries. By 1998 the flow of migrants had turned around again and was heading back to California, as domestic in-migrants to the state exceeded the out-migrants by 21,000 (California Department of Finance 1998). Of course, international immigrants continued to flow into California at the pace of nearly 250,000 per year.

The influx of international migrants has confounded the pattern of internal migration to a certain extent. In the first place, immigrants tend to congregate in a relatively few states (especially California, Texas, New York, and Florida), and within those states they tend to congregate in specific metropolitan areas, such as Los Angeles and Houston (Frey 1996). Furthermore, when they move internally in the United States, they are more apt to move in order to be close to relatives and others of the same nationality than they are to move for purely economic reasons (Kritz and Nogle 1994).

Internal migration patterns among immigrants represent part of an increasingly complex picture of overall migration patterns in the United States, as Frey (1995) has noted:

> The spatial demographic shifts that characterize the 1980s and 1990s are a far cry from the 1970s days of Snowbelt urban declines, Texas oil booms, and California dreaming. Nor is there much talk of a back-to-nature rural renaissance. The new, post-1980 urban revival is an uneven one—rewarding corporate nodes, information centers, and other tie-ins to the global economy. Areas specializing in high-tech manufacturing, recreation, or retirement have also grown. And while these kinds of areas can be found in most parts of the country, they are now especially prominent in newly developing regions—the South Atlantic coastal states, and states around California [p. 333].

Frey elaborates five trends that he believes capture the complexity of contemporary patterns of population movement in the United States: (1) an uneven urban revival—a select few metropolitan areas (those with diverse economies that can withstand industrial restructuring) are gaining migrants at the expense of others; (2) regional racial division—the influx of immigrants from Asia and Latin America has diversified dramatically the racial and ethnic composition of the major receiving states (California, Texas, and New York), and has separated those states from the rest of the country, which remains predominantly non-Hispanic white; (3) regional divisions by skill level and poverty—the "hourglass" economy that has come to characterize the highly industrialized nations has widened the income gap between college graduates and those with less education, and the geographic redistribution of knowledge-based industries creates in its wake a migration of those with higher education; (4) Baby Boom and elderly realignments—the early baby boomers fueled the movement west and south, whereas the later baby boomers are having to look elsewhere for opportunities; meantime, the elderly continue to move to the Sun Belt; and (5) suburban dominance and city isolation—suburban areas "captured the bulk of employment and residential growth in 1980s. The modal commuter now both lives and works in the suburbs" (Frey 1995:275).

The current trend, then, is a multiplicity of migration trends, reminding us once again of the complexity of migration. Economic motives may dominate individual decisions to migrate, but even in the United States, the world-system affects where economic opportunities are likely to be, and who is likely to be attracted to them.

Summary and Conclusion

Migration is any permanent change of residence. It is the most complex of the three population processes because we have to account for the wide variety in the number of times people may move, the vast array of places migrants may go, and the incredible diversity of reasons there may be for who goes where, and when. For decades, migration theory advanced little beyond the basic idea of push and pull factors operating in the context of migration selectivity. More recently, conceptual models have developed that are very reminiscent of the explanations of the fertility transition. To begin with, we need a model of how the migration decision is arrived at (not unlike the first precondition for a fertility decline—the acceptance of the idea that you are empowered to act). Then, most importantly for the study of migration, we need to understand what might motivate a person to migrate. A variety of theoretical perspectives have been offered, including the neoclassical approach, the new home economics of migration approach, the theory of the dual labor market, world-systems theory. Finally, the means available to migrate is explored at least partly by network theory, institutional explanations, and the concept of cumulative causation.

Migration has dynamic consequences for the migrants themselves, for the areas from which they came, and for the areas to which they go. Some of these consequences, especially for the areas of origin and destination, are fairly predictable if we know the characteristics of the migrants. If in-migrants are well-educated young adults, for example, they will be looking for well-paying jobs, they may add to the economic prosperity of an area, and they will probably be establishing families, which will further add to the area's population and increase the demand for services.

Throughout the world, population growth has induced an increase in the volume of migration, both legal and illegal. The United States passed legislation designed to limit the entry of illegal immigrants in 1986, whereas India had approached the issue more straightforwardly in 1983 by constructing a barbed-wire fence along its border with Bangladesh, designed to keep illegal migrants out of India. "Temporary" labor migration has also increased throughout the world as jobs have become available in developed societies for workers from third-world countries. Understandably, workers are often reluctant to leave the higher-income countries, even when the economies in those places slow down and pressure builds for foreigners to go home. Such people are only a few steps away from the unhappily large fraction of the world's migrants who are refugees, seeking asylum in other countries after being forced out of their own. Of the more than 14 million refugees scattered throughout the globe in the 1990s, disproportionate shares are found in western Asia and sub-Saharan Africa.

Although it is not always apparent, the quality of our everyday life is greatly affected by the process of migration, for even if we ourselves never move, we will spend a good part of our lifetime adjusting to people who have migrated into our lives and to the loss of people who have moved away. Each new person coming into our life greatly expands our social capital by increasing the potential size of our social network, especially since many people who move away do so physically but not symbolically; that is, we remain in communication.

The next chapter examines the way in which migration, fertility, and mortality operate to shape the age and sex structure of a population—a structure that affects the lives of each of us by defining how many people of different ages and sex we must deal with.

Main Points

1. Migration is the process of changing residence and of moving your whole round of social activities from one place to another.

2. International migrants move between countries, whereas internal migrants do their moving within national boundaries.

3. Explanations of why people move typically begin with the push–pull theory, first formulated in the late 19th century.

4. Migration is selective and is associated especially with different stages in the life cycle, giving rise to the idea that migration is an implementing strategy—a means to a desired end.

5. Young adults are geographically more mobile than people of other ages.

6. The more highly educated you are, the more likely you are to migrate.

7. Major theories offered to explain international migration include neoclassical economics, the new household economics of migration, the dual labor market theory, world systems theory, network theory, institutional theory, and cumulative causation.

8. Migration forces adjustment to a new environment on the part of the migrant, and forces a societal response to the immigrant on the part of the receiving society.

9. Major concepts in the adjustment of immigrants include adaptation, acculturation, assimilation, integration, incorporation, exclusion, multiculturalism, and transnationalism.

10. There are more than 14 million refugees in the world, concentrated in Africa and western Asia.

11. Global patterns of migration include the east–west movement of Europeans, the north–south movement of labor, and the circulation of labor from less developed to emerging economies.

12. Labor migration accounts for a large part of the overall volume of global migration.

13. In the 19th and early 20th centuries, Europeans were the principal migrants to the United States and Canada; in the 1930s and 1940s more migrants left than came; and since the 1960s migration largely has been from Latin America and Asia.

14. An estimated 220,000 people leave the United States each year, probably to return home to their place of origin.

15. Ross Baker once suggested that the "First Law of Demographic Directionality" is that a body that has headed west remains at west, but that is no longer so true in the United States.

Suggested Readings

1. Everett Lee, 1966, "A theory of migration," Demography 3:47–57.

   This widely quoted article summarizes and updates the famous work of Ravenstein, who published a still influential treatise on migration in 1889.

2. Douglas Massey, Joaquín Arango, Graeme Hugo, Ali Kouaouci, Adela Pellegrino, and J. Edward Taylor, 1993, "Theories of international migration: A review and appraisal," Population and Development Review 19(3):431–466; and 1994, "An evaluation of international migration theory: The North American case," Population and Development Review 20(4):699–752.

   These two articles, representing a matched set, will have a long-lasting influence on the field of international migration analysis because they bring together a long needed review of theoretical perspectives.

3. Timothy J. Hatton and Jeffrey G. Williamson, 1994, "What drove the mass migrations from Europe in the late nineteenth century?" Population and Development Review 20(3):503–531.

   The modern world was shaped importantly by the migration of Europeans to the Americas in the late 19th century, and this article reviews the evidence and the theories that can help explain what happened and why.

4. James P. Smith and Barry (eds.), 1998, The Immigration Debate: Studies on the Economic, Demographic, and Fiscal Effects of Immigration (Washington, D.C.: National Academy Press).

   In 1990 the U. S. Congress created a bipartisan commission to examine immigration reform and that commission in turn asked the National Research Council to generate expert opinion on immigration issues. The result was this edited volume of papers covering both research and policy issues related to migration to the United States.

5. Robin Cohen (ed.), 1995, The Cambridge Survey of World Migration (Cambridge: Cambridge University Press).

   This massive volume of essays is almost encyclopedic in its coverage of the historic, geographic, and social dimensions of human migration.

Websites of Interest

Remember that websites are not as permanent as books and journals, so I cannot guarantee that each of the following websites still exists at the moment that you are reading this:

1. http://migration.ucdavis.edu

   This website was developed by Philip Martin at the University of California, Davis, and is a tremendous resource especially because it includes full issues of the newsletter Migration News, in which newspaper, magazine, and journal articles and books about migration are summarized and commented upon.

2. **http://www.ins.usdoj.gov**

   This is the website of the Immigration and Naturalization Service (INS) which is part of the Department of Justice (DOJ) within the U.S. government (GOV). The site includes information about and for immigrants and refugees.

3. **http://cicnet.ci.gc.ca**

   In Canada the immigration agency is known as Citizenship and Immigration Canada (CIC) and their website includes a section for potential immigrants called "Coming to Canada" as well as research and publication information.

4. **http://www.refugees.org**

   The United States Committee for Refugees (USCR) is a private, nonprofit organization working to help refugees throughout the world. This site includes information about refugees and asylees for almost every country of the world. Because the estimation of refugees is as much an art as it is a science, you should also visit the site of the United Nations High Commissioner for Refugees (UNHCR) at **http://www.unhcr.ch** for comparisons.

5. **http://www.ercomer.org**

   The European Research Centre on Migration and Ethnic Relations (ERCOMER) is located at Utrecht University in the Netherlands and is devoted to comparative migration research with a focus on Europe.

Part Three
Population Structure
and Characteristics

Thus far I have concentrated on the demographic perspective and on the three population processes: mortality, fertility, and migration. I have emphasized how social forces shape the trends and levels of each process, and, to a lesser extent, I have analyzed the social impact of changes in each of these processes. With this background, it is time now to turn to a more detailed analysis of other factors that are intimately intertwined with population processes. These factors are commonly called population characteristics and they include the distribution of a population by age, sex (and its social component, gender), race, family and household structure, education, occupation, income, and residence in urban or rural areas.

I have divided these characteristics into four chapters to discuss them in detail. Chapter 8 deals with age and sex in general terms, whereas in Chapter 9, I go into greater detail about population aging—the demographic phenomenon that has beset the more economically advanced nations. In Chapter 10, I review the sociodemographic characteristics that tend to determine a person's chances in life in modern society, including family and household structure, education, labor force participation, occupation, income, and race and ethnicity. All of these characteristics of people and households are brought into play when we contrast life in urban places with that in rural areas, and that is the subject of Chapter 11.

Throughout the next four chapters the analysis is two-pronged. We look at the way in which population processes influence characteristics and the way in which those characteristics in turn affect population processes. By understanding these interrelationships—these feedback systems—you enhance your own demographic perspective.

Part Three
Reparation Administration
and Character Effects

CHAPTER 8
Age and Sex Structure

You can't see a population grow in the same way that you can watch a crowd fill up a football stadium. If you leave a place for a few years and then come back, the change may be apparent to you—something akin to time-lapse photography. For the most part, however, we observe demographic changes by seeing their effect on the age and sex structure of an area—on the number of people of each age and sex. For example, recent changes in fertility influence the number of children in elementary school, whereas recent migration or prior fertility levels may affect the number of new apartments and houses being built (or vacated) to accommodate young families. In general, it is the interaction of fertility, mortality, and migration that produces the age/sex structure, which can be viewed as a key to the life of a social group—a record of past history and a hint of the future.

Population processes not only produce the age/sex structure but are, in turn, affected by it, another example of the complexity of the world when seen through your demographic "eye." It would not be exaggerating too much to say that changes in the age/sex structure affect virtually all social institutions and represent a major force in social change. In this chapter I escort you through that complexity by first defining age and sex structures and examining how we measure and use them. Then I look at the impact of each of the population processes on the age/sex structure, and finally I examine the potential contributions that changing age/sex structures make to social change.

What Is an Age/Sex Structure?

Strictly speaking, a structure is something that is built or constructed. In social science it refers more broadly to a pattern of interrelationships between parts of a society. An **age/sex structure** actually combines both definitions, since it represents the number of people of a given age and sex in society and is built from the input of births at age zero and deaths and migration at every age.

Age and sex influence the working of society in important ways because society assigns social roles and frequently organizes people into groups on the basis of their age and gender. Note that I use the term *sex* when referring to the strictly biological differential between males and females, reserving the term *gender* for the social aspects of behavior. Young people are treated differently from old people, and different kinds of behavior are expected of each. Women are treated differently from men in most human societies and different kinds of behavior are often expected from each. Regardless of your ideological position as to the rightness or wrongness of these distinctions, they do exist to some degree in every nation. Further, at very young and very old ages, people are more dependent on others for survival, and so the proportions of people at these ages will influence how society works.

Measuring the Age Structure

A population is considered old or young depending on the proportion of people at different ages. In general, a population with more than about 35 percent of its people under age 15 is "young," and a population with more than about 10 percent of its people aged 65 or older can be considered "old." Further, as the proportion of young people increases relative to the total, we speak of the population as growing younger. Conversely, an aging population is one in which the proportion of older people is increasing relative to the total. We can graphically or statistically quantify the age structure in three major ways. These include constructing a **population pyramid,** calculating the **average age of a population,** and calculating the **dependency ratio.**

Population Pyramids   A population pyramid (or **age/sex pyramid**) is a graphic representation of the distribution of a population by age and sex. It is called a pyramid because the "classic" picture is of a high fertility, high mortality society (which characterized most of the world until only several decades ago) with a broad base built of numerous births, rapidly tapering to the top (the older ages) because of high death rates in combination with the high birth rate. Nigeria's age/sex structure as of the year 2000 reflects the classic look of the population pyramids, as you can see in Figure 8.1. Until very recently Mexico also looked like that, but the decline in fertility since the 1970s has begun to narrow the base of the pyramid rather noticeably. Developed countries such as the United States and Canada have age/sex distributions that are more rectangular or barrel-shaped (see Figure 8.1), but we still call the graph a population pyramid. The shaded areas between ages 35 and 54 in the pyramids for both the United States and Canada indicate the approximate ages of the baby boomers in both countries as of the year 2000. Later in the chapter I return to a more detailed look at interpreting population pyramids.

Although a picture may be worth a thousand words, there are times when we like to summarize an age/sex structure in only a few short words or, even better, in a few numbers. The average age and dependency ratio are two measures to help us do just that.

Average Age and Dependency Ratio   The average age in a population is generally calculated as the median, which measures the age above which is found half of the population and below which is the other half. In Figure 8.1, the population pyramids of Nigeria, Mexico, Canada, and the United States reflect median ages of 17.4, 23.3, 36.9, and 35.7, respectively. Thus the obvious differences in the shapes of the age distributions for less developed and more developed countries are reflected in the clear differences in median ages.

An index commonly used to measure the social and economic impact of different age structures is the dependency ratio—the ratio of the dependent-age population (the young and the old) to the working-age population. The higher this ratio is, the more people each potential worker is having to support; conversely, the lower it is, the fewer people there are dependent on each worker. For example: suppose that a population of 100 people had 45 members under age 15, 3 people 65 or older, and the rest of approximately economically active ages (15 to 64). This is the situation in Nigeria, one of the highest fertility nations in the world. Forty-eight percent of the population is of dependent age (0 to 14 and 65+) compared with the remaining 52 percent of working age. Thus the dependency ratio is 48/52 or 0.92, which means that there are 0.92 dependents per working-age person—nearly one dependent for each person of working age, which is a fairly heavy load, especially as in most societies you will not find everyone of working age actually working.

Figure 8.1   Age Pyramids Display the Age/Sex Structure of a Population

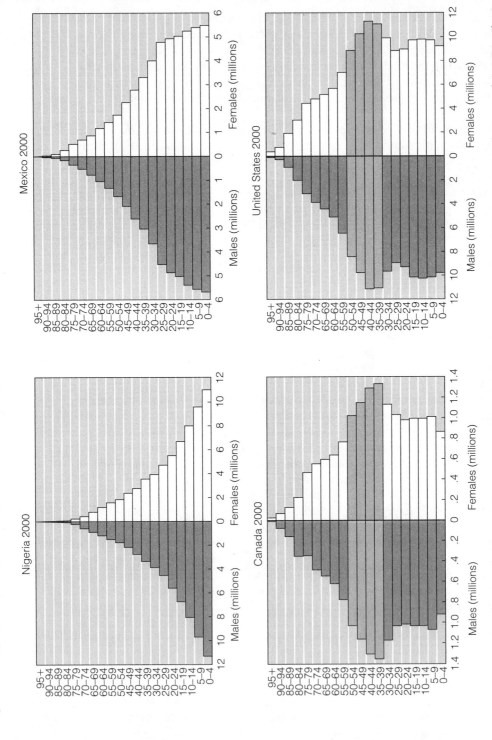

Sources: Data for Nigeria, Mexico, and Canada are medium-variant projections from United Nations, 1997, The Sex and Age Distribution of the World Populations, The 1996 Revision (New York: United Nations); Data for the United States are from J. C. Day, 1996, "Population Projections of the United States, by Age, Sex, Race, and Hispanic Origin: 1995 to 2050," U.S. Bureau of the Census, Current Population Reports, P25-1130, Table 2.

We can compare this dependency ratio of 0.92 with that for a population of 100 people in which 19 are under 15, 13 are 65 or over, and the rest (68) are of working age. This is the situation in Canada and it produces a dependency ratio of 32/68 or 0.47, which means that one person of working age would be supporting about one half as many dependents as are supported in Nigeria. When you consider that it is more difficult for a parent (usually the mother) to be as economically productive when she has to tend to the needs of children, you can appreciate that the actual difference in dependency between Nigeria and Canada is even greater than these numbers suggest, a theme returned to in Chapter 12.

The dependency ratio does not capture all the intricacies of the age structure, but it is a useful indicator of the burden (or lack thereof) that some age structures place on a population. For individuals with large families, the impact of a youthful age structure, for example, will be immediately apparent. But even for the childless or those with only a few children, the effect may be higher taxes to pay for schools, health facilities, and subsidized housing. For those in business (whether government or private), an age structure that includes numerous dependents may mean that workers are able to save less, having to spend it on families, while government taxes must go toward subsidizing food, housing, and education rather than to financing industry or economic infrastructure, such as roads, railways, and power and communications systems.

Measuring the Sex Structure

It is a common assumption that there are the same numbers of males and females at each age—actually, this is rarely the case. Migration, mortality, and fertility operate differently to create inequalities in the ratio of males to females (known as the **sex ratio**). For example, in some instances females are more likely to migrate (and thus to be added to or subtracted from an age/sex structure), and in other situations males are more likely to be the migrants and thus also produce inequalities in the age/sex structure.

Mortality creates sex inequalities because at every age males have higher death rates than females, as I discussed in Chapter 4. As mortality has declined, women have benefited disproportionately, and Western nations have thus become increasingly characterized by having substantially more older females than males. Several of these patterns are observable in Figure 8.2 where I have plotted for you the sex ratios at each age group for the four countries (Nigeria, Mexico, Canada, and the United States) whose population pyramids are shown in Figure 8.1. All four countries show the general pattern that the sex ratio declines with increasing age, but it is more noticeable for the lower mortality countries of Canada and the United States than it is for Mexico and Nigeria. At the younger ages you can see that the sex ratio in Canada is the highest of the four countries and this is almost certainly due to the impact of international immigrants, who are somewhat more likely to be males than females. The flip side is that the sex ratio is lower than might be expected in those ages in Mexico, where the male population is disproportionately affected by international emigration.

Fertility has the most predictable impact on the sex ratio because in virtually every known human society more boys are born than girls. This is perhaps a biological

Figure 8.2   Sex Ratios by Age

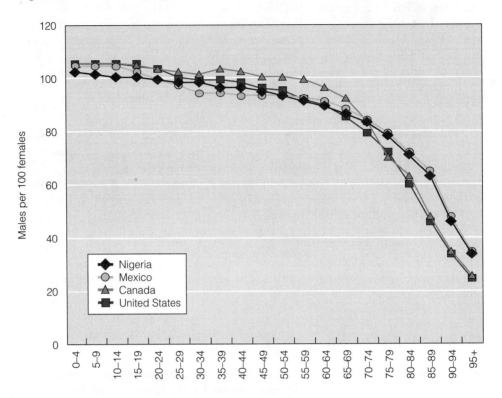

Sources: Data for Nigeria, Mexico, and Canada are medium-variant projections from United Nations, 1997, The Sex and Age Distribution of the World Populations, The 1996 Revision (New York: United Nations); Data for the United States are from J. C. Day, 1996, "Population Projections of the United States, by Age, Sex, Race, and Hispanic Origin: 1995 to 2050," U.S. Bureau of the Census, Current Population Reports, P25-1130, Table 2.

adaptation to compensate partially for the higher male death rates. In most countries, including the United States, there are normally 105 boys born for every 100 girls, with the ratio of boys to girls being slightly higher for U.S. whites than African-Americans. Over time the sex ratio at birth for African-Americans has gradually increased in the United States, which is probably associated with improvements in health.

Let me also note that couples in most modern societies have at least some ability to choose the sex of their offspring. The gender of the fetus can usually be identified through amniocentesis, and selective abortion can then be utilized to select for a girl or a boy. Less drastic means that are in the experimental stage include separating the X- and Y-bearing sperm and subsequently impregnating a woman through artificial insemination; and timing coitus through the ovulatory cycle so that it occurs either early or late, to enhance the probability of conceiving a male, or in the middle of the cycle, to increase the likelihood of conceiving a female. A more drastic approach that is, of course, already available is infanticide. Westoff and Rindfuss

(1974) believe that if these methods ever enjoy widespread acceptance, there would be a short-run rise in the sex ratio at birth, since a preference for sons as first children (and for more total sons than daughters) is fairly common throughout the world, as discussed in Chapter 6. However, Westoff and Rindfuss also conclude that after an initial transition period, the sex ratio at birth would probably revert to the natural level of about 105 males per 100 females, because the disadvantage of too many or too few of either sex would be controlled by a shift to the other sex.

Ultimately, the effect of sex selection would likely be a return to the state of affairs exemplified by the current situation in the United States, where the excess of males over females prevails throughout the younger ages until higher male mortality takes its toll. In the United States, for example, there are more males than females at every age younger than 25 (see Figure 8.1), whereas females outnumber males at every age from 25 on. Now that you have a feel for the essential ingredients of an age/sex structure, let us examine how those age/sex structures come about as a consequence of the three population processes.

Impact of Population Processes on the Age/Sex Structure

Each of the three population processes—migration, mortality, and fertility—makes its own imprint on the age/sex structure, and I list them in that order for a reason. Migration has the most dramatic, short-term impact on the distribution of people by age and sex, but over the long run its influence is negligible. Mortality can have both short-run and long-run effects on the age/sex structure, but in neither case is the impact very dramatic. Finally, fertility has relatively little short-term effect, but in the long run it is by far the most important of the three population processes in influencing the shape of the age pyramid.

The Impact of Migration

A population experiencing net in- or out-migration (and virtually all populations except the world as a whole do experience one or the other) will almost certainly have its age/sex structure altered as a consequence. Since immigration has been especially important in the United States, it provides a good beginning for our analysis.

Impact of Immigration to the United States   We can assess the potential impact of international migrants into the United States by looking at the age/sex distribution of legal immigrants into the United States in 1995 as measured by the U.S. Immigration and Naturalization Service (1997). You can see in Figure 8.3 that men outnumber women slightly up to age 20, and then there are more female than male immigrants from that point to the oldest ages. However, the most significant part of the picture is that both males and females exhibit the pattern of migration by age that I discussed in Chapter 7. It is true that there are migrants at virtually every age, but migrants are far more likely to be young adults than they are to be people of any other age.

Figure 8.3   The Age Structure of Immigrants to the United States

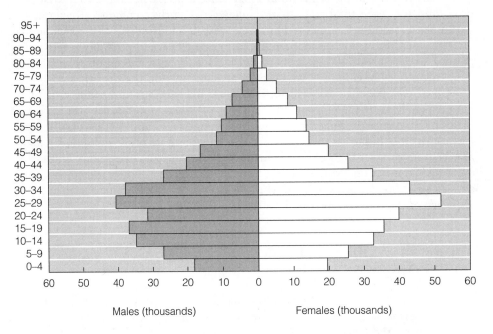

Males (thousands)   Females (thousands)

Source: Adapted from U.S. Immigration and Naturalization Service, 1997, 1995 Statistical Yearbook of the Immigration and Naturalization Service (Washington, DC: Government Printing Office), Table 12.

In the long run, the impact of migration is felt indirectly through its influence on reproduction, because immigrants typically are of prime reproductive ages. The 820,000 net migrants into the United States each year represent substantially less than one percent of the total U.S. population. Even in Canada, where immigrants represent a larger fraction of the population, the 250,000 immigrants per year still represent less than one percent of the total Canadian population. Yet, in both countries, if you consider that all North Americans except Native Americans historically are descendants of fairly recent migrants it becomes apparent that the indirect effect of immigration has been enormous over the long run.

In the short run, however, migration affects those local areas where migrants show up or depart more readily than it does national populations. This can be exemplified by looking at the differential impact of migration within a city.

Impact of Internal Migration within a City   The impact of migration on the age structure in a city is particularly noticeable when that area contains a social institution, such as a military base, college, or retirement community, which attracts a particular age group or sex. Examples of these situations are shown in Figure 8.4. Each of these age and sex distributions is drawn from areas of San Diego as measured by the 1990 census. The first example is the population in the Montezuma area near San Diego State University, where you can see the heavy bulges in the late teens and

Figure 8.4   Age Pyramids in Different Areas of San Diego

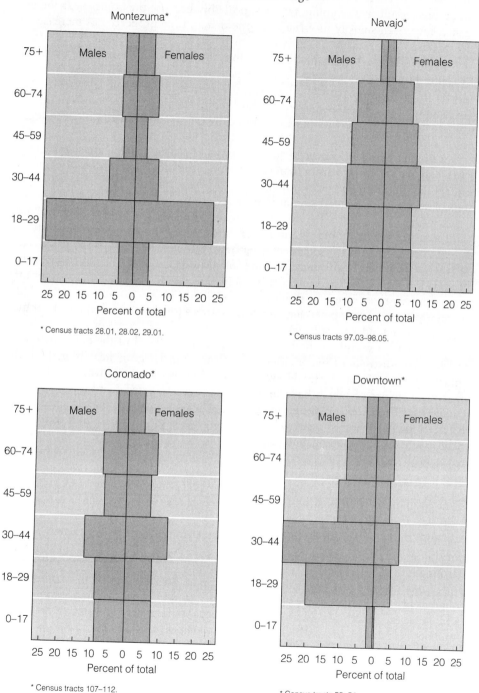

Montezuma*

* Census tracts 28.01, 28.02, 29.01.

Navajo*

* Census tracts 97.03–98.05.

Coronado*

* Census tracts 107–112.

Downtown*

* Census tracts 53, 54.

Note: The age/sex structures of a city vary considerably, depending on the nature of social and economic activity in the neighborhood. See text for full discussion.

Source: U.S. Bureau of the Census, 1990 Census of Population and Housing.

early twenties. There are relatively few married couples with young children. This is not an area conducive to young families, probably because the homes near the university have risen sharply in value since the demand for property has increased as the university's enrollment has grown to one of the largest in the state of California. The typical young adult couple with children thus cannot afford to move into the area, and homeowners tend to be older residents, as can be seen by the bulges at ages 60 to 74. This age pattern is by no means unique, however. Most American college communities are characterized by this kind of bipolar distribution—young students living side by side with (although typically not in contact with) older residents (often retirees).

Young families are located in adjacent parts of the city, which are slightly more suburban and of lower density than the area near the university. You can see in Figure 8.4 that in the Navajo area, for example, there are relatively few college-age persons. The area is composed primarily of older adults and their children.

The third example is the age and sex distribution for the city of Coronado, California, an island suburb of San Diego. It is the city that serves the North Island Naval Air Station, but it experienced substantial gentrification in the 1970s and 1980s (see Chapter 11 for more on the gentrification process). Thus Coronado has become too expensive for younger enlisted people, but it has a sizable group of officers (who are more likely to show up in the 30 to 44 age group) and other baby boomer yuppies. At the older ages are Navy retirees (or their widows) and wealthier, recently arrived older persons moving into expensive beachfront condominiums.

The fourth example is the most bizarre, yet it is typical of the downtown central business districts of many cities. Downtown San Diego is heavily male in the ages up to retirement. Virtually everyone residing in this area lives in a hotel or apartment, many of which have been deliberately converted to residences serving the elderly. The excess of young males is a reflection partly of the presence of the Navy, of sailors living in nearby, but off-base, housing and partly of a more transient population, including homeless persons. At the middle ages, the excess of males probably also reflects a population of derelicts and other transients who move into the area for lack of anywhere else to go. At the retirement ages, the greater proportion of females is a consequence of the greater longevity of women, who have been attracted to the downtown area by the construction of new, high-security, high-rise, low-rent apartments.

These graphs show how the age and sex distributions in an area reflect the history of that area and reveal important social differences. In each case, the major determinant of the age/sex structure was migration, either in or out, and this illustrates the dramatic effect that migration may have on an area. Migration in two instances acted to produce bulges in the young adult ages, and in another instance to produce a bulge at the older ages. In one case, excessive male migration affected the structure of the population, while an excess of females was apparent in two of the examples. Migration, then, can affect either sex and can affect different ages to varying degrees. Of course, migration was not the only influence. Note that the abundance or lack of children in one area or another reflects high or low levels of fertility as well.

It is generally true that the more precisely you define a geographic area, the more likely it is that the age/sex structure will have been affected by migration, and the more likely it is that the area's "personality" will be affected by (and, of course, reciprocally influence) the age/sex structure. The age/sex structure for the state of California is very similar to that for the United States. As you narrow in on San Diego, local variations show up from the influence of the Navy (primarily) and the universities (there are three universities within the city limits). As you close in on specific sections of the city, as in Figure 8.4, the variations grow even wider. The character of each neighborhood begins to emerge, shaped by and helping to shape the age/sex distribution, which in turn affects and is affected by migration. Of course, the effects of the other two demographic processes, mortality and fertility, always go along with the effects of migration.

The Impact of Mortality

Long-Term Impact of Mortality Changes   Mortality is similar to migration in that it affects all ages and both sexes. But the age and sex pattern of dying is more predictable from one society to the next, as I discussed in Chapter 4. In virtually all societies, the very youngest and the very oldest ages are most susceptible to death, and in modern societies (where maternal mortality is fairly low), males are more likely than females to die at any given age, a trend that accelerates with age.

Not only is there a fairly consistent pattern of mortality by age and sex, but when mortality levels change, all ages tend to be affected, even though some are affected more than others. Thus improved health conditions in a society lower death rates at all ages, although the rates will be reduced proportionately more at the youngest and oldest ages. Similarly, if a famine or epidemic of disease strikes, death rates will rise at all ages, although again, the youngest and oldest will be most affected.

The upshot of these facts is that severe alterations in the level of mortality in a society have far less dramatic consequences for the age and sex structure than does migration or, as we shall see, fertility. Over the long run, changes in mortality by themselves do not appreciably affect the age and sex structure of a society. However, in populations with medium to high mortality, the impact of a decline in mortality is to make the population slightly younger—a seemingly paradoxical result, since it seems as though lower mortality should age the population by allowing people to live longer. This more expected pattern does occur when mortality is lower, but in higher mortality populations the disproportionately larger decline in infant mortality as overall levels of mortality decline tends to have the opposite effect. This impact of a mortality decline on the age/sex structure can be very noticeable in the short run if it is not accompanied by a decline in fertility.

Short-Term Impact of Mortality Changes   In the short run, a decrease in mortality levels in all but low mortality societies can substantially increase the number of young people, and one of the best studies of this effect is by Arriaga (1970) in his analysis of Latin American countries. Arriaga examined data from 11 countries for which information was available on the mortality decline from 1930 to the 1960s. He discovered that "of the 27 million people alive in all the eleven countries in the 1960s who would not have been alive if there had not been a mortality decline since the 1930s, 16 million—59 percent—were under 15" (1970:103). In relative terms, a lowering of mortality in Latin America noticeably raised the proportion of people at the young ages, slightly elevated the proportion at old ages, and lowered the proportion at the middle ages (14 to 64). However, in absolute terms, the number of people at all ages increased. Declining mortality had an impact similar to that of a rise in fertility, while also making a contribution to higher fertility.

The appearance of higher fertility is, of course, produced by the greater proportion of children surviving through each age of childhood. It is as though women were bearing more children, thereby broadening the base of the age structure. The actual contribution to higher fertility is generated by the higher probabilities of women (and their spouses) surviving through the reproductive ages, since under conditions of high mortality, a certain percentage of women will die before giving birth to as many children as they might have. When death rates go down, a higher percentage of women live to give birth to more children, assuming that social changes are not producing motivations for limiting fertility. The effect on fertility of changing mortality has been studied by Ridley and her associates (1967), among others, and they were able to demonstrate that improved chances for survival increase the number of children ever born per woman and also increase the net reproduction rate.

Note that the only time that a change in mortality generates a change in the age/sex distribution is when the mortality shifts are different at different ages. If there is a change in the probability of survival from one age to the next that is exactly equal for all ages and for both sexes, then the age/sex structure will remain unchanged. On the other hand, in a low mortality society such as the United States, where nearly three fourths of all deaths occur at ages 65 or older, a drop in mortality will age the population largely because it is very hard to improve on death rates at the younger ages.

The Impact of Fertility

Both migration and mortality can affect all ages and differentially affect each sex. The impact of fertility is not quite the same. Fertility obviously adds people only at age zero to begin with, but that effect stays with the population age after age. Thus if the birth rate were to drop suddenly in one year (as it did, for example, in Japan in 1966; see Chapter 6), then as those people get older, there will still always be fewer of them than there are people just older and (at least in the Japanese case) just younger. If fertility goes up, then there will be more people in each younger age group. Both of these situations—rising and declining fertility—have had strong influences on the age/sex structure of the United States, as I will discuss in a moment.

In general, the impact of fertility levels is so important that with exactly the same level of mortality, just altering the level of fertility can produce age/sex structures that run the gamut from those that might characterize primitive to highly developed populations. For example, let us suppose that we are looking at two countries with high female life expectancies of 74 years (such as the United Arab Emirates and Bulgaria). However, one country (United Arab Emirates) still has high fertility—a total fertility rate (TFR) of 4.1 in 1995, whereas the other has very low fertility (a TFR of only 1.4 in 1995). The respective age distributions of these two nations are already very different, because the United Arab Emirates has had high fertility for a long time, whereas Bulgaria has had low fertility for some time now. In the United Arab Emirates, 32 percent of the population was under age 15 in 1995, compared with only 19 percent of the Bulgarian population. However, if each of these countries maintained these same levels of fertility and mortality for several decades, particularly in the absence of any migration, the differences would be even greater. In the long run, the United Arab Emirates would have nearly 37 percent of its population under age 15 and only 6 percent aged 65 or older (despite the very low mortality). By contrast, only 14 percent of the Bulgarian population would be under 15 in the long run, and fully 24 percent would be aged 65 and older. The average age of a resident of the Emirates would be only 27, whereas the average Bulgarian would be 45. The Emirates would be supporting a population with a higher proportion of younger people and a dependency ratio of 0.76, whereas the Bulgarian population would be supporting a much larger older population, but the dependency ratio would be quite a bit lower at 0.61.

Having discussed the way in which the age/sex structure is built on the foundation of migration, mortality, and fertility rates, it is worth spending a short while reminding you of the opposite side of that coin: How does the age structure influence rates of population growth?

Impact of Age Structure on Population Processes

The actual rate of population growth is of course determined by the combination of fertility, mortality, and migration, and the rate of each of these is influenced by the age structure—a fact alluded to in earlier chapters, but worth repeating. A bulge in the age structure at the young adult (reproductive) ages will tend to raise the crude birth rate by producing a large number of children relative to the total population. As we shall see later in this chapter, even in otherwise low fertility societies such as the United States and Canada, a large population of young women produces a substantial number of births. At the other extreme, a population with a relatively small proportion of young people, but with a high proportion of older people, will have a substantial number of deaths each year, even if life expectancy is high, just because there are so many people moving into those higher-risk years. This will generate a crude death rate higher than might otherwise be expected.

A young age structure produced by high fertility may also encourage a relatively large number of out-migrants. Each year such a population would produce a greater number of young adults, who are at greatest risk of migrating. Such migration, if it occurs, may well contribute to the sorts of massive changes almost invariably created by population growth. The question of social changes wrought by population processes leads us to examine the more general, and generally more interesting question: How do age/sex structures influence what a society is like?

The Dynamics of Age/Sex Structures

It may seem mundane, if not downright boring, on the surface, but the age/sex structure of a population is actually one of the most potent forces of social change known to us. It is through this mechanism that all demographic changes are translated into a force with which we must cope. A high birth rate does not simply mean more people: It means that 6 years from now there will be more kids entering school than before; that 18 years from now there will be more new job hopefuls and college freshmen than before. An influx of young adult refugees this year means a larger than average number of older people 30 or 40 years from now (and it may mean an immediate sudden rise in the number of births, with all the attendant consequences). We study these phenomena by doing **population modeling.** This involves setting up hypothetical demographic situations—by constructing stable and stationary populations and projecting populations into the future (or into the past), and doing so with specific models in mind. Then we look at those hypothetical demographic scenarios and ask what are the likely social, economic, and political ramifications of each. Let me go through the process to show you what I mean.

Stable and Stationary Populations

The long-run influences of mortality and fertility are best expressed by formal demographic models called stable and stationary populations. A **stable population** is one in which neither the age-specific birth rates nor the age-specific death rates have changed for a long time. Thus, a stable population is stable in the sense that the percentages of people at each age and sex do not change over time. However, a stable population could be growing at a constant rate (that is, the birth rate is higher than the death rate), it could be declining at a constant rate (the birth rate is lower than the death rate), or it could be unchanging (the birth rate equals the death rate). If the latter case prevails, we call it a **stationary population.** Thus, a stationary population is a special case of a stable population—all stationary populations are stable, but not all stable populations are stationary. The life table (discussed in Chapter 4) is one type of stationary population model.

For analytical purposes, a stable population is usually assumed to be closed to migration. Since 1760, when Leonhard Euler first devised the idea of a stable population, demographers have used the concept to explore the exact influence of differing levels of mortality and fertility on the age/sex structure. Such analyses are possible using a stable population model because it smooths out the dents and bumps in the age structure created by migration and by shifts in the death rate or the birth rate. Thus, if demographers were forced to study only real populations, we would be unable to ferret out all the kinds of relationships I discussed in previous sections. Demographers look at real populations and then apply stable population models to the real setting to understand the underlying demographic processes that influence the structure of a population by age and sex.

In Figure 8.5 I have employed stable population models to show how different fertility levels can affect the shape of the age/sex structure if everything else is held constant. Figure 8.5 assumes (as do most stable population models) that no migration is occurring. Then it assumes that mortality is constant, with a life expectancy of 71 years. The high fertility level is equivalent to a total fertility rate of 7.1 (similar to Somalia in the mid-1990s). All other things being equal, a high fertility, low mortality society will have a very youthful age distribution. Indeed, the average age

Figure 8.5 Different Levels of Fertility Have Dramatically Different Effects on the Age/Sex Distribution

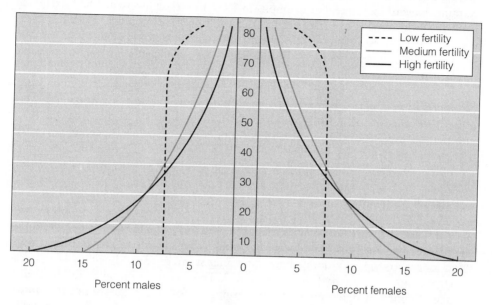

Note: In this graph you can see the large impact that different levels of fertility have on the age/sex structure of a population when mortality and migration are held constant. In all three cases the assumption is that there is no migration and that mortality is low. When fertility is also low, the age pyramid looks like a barrel, whereas the pyramidal shape is associated with higher levels of fertility.

in the population is 20.7 years, with 48 percent of the population being under age 15. The median fertility level is equivalent to a total fertility rate of 4.3 children per woman (similar to the Philippines in the mid-1990s). This level of fertility still produces a youthful age structure, with an average age of 26.1 years and with 38 percent of the population under age 15. At the low fertility level, we are talking about a total fertility rate of 2.1, which is exact replacement. This population, which is close to the stable population model for the United States, has an average age of 37.9 years, and only 21 percent of the population is under age 15.

Stable populations are also invaluable aids in estimating demographic measures. Let us suppose, for example, that we knew what the mortality and fertility rates were for a population, but no census had been conducted so we were unsure about the age and sex structure of the population. As I mentioned in Chapter 2, it is not uncommon for less developed nations to have such incomplete information. We do not have to give up or sit on our thumbs waiting for the United Nations to finance a census. We can use a **model stable population**—the age/sex structure implied by an unchanging set of mortality and fertility rates (Coale and Demeny 1983). A country that experiences little migration and has had relatively little change in mortality and fertility will have an age and sex structure similar to that of a stable population. Thus if we know (or can reasonably estimate) two important facts, such as age-specific death rates (so that we can calculate life expectancy) and the rate of population increase, we can determine what the age and sex structure would almost certainly have to look like. For example, in 1991 the results were just being released from the 1987 census in the eastern African nation of Zimbabwe. With a life expectancy in 1987 estimated to be 57 years, an annual rate of growth of 3.5 percent per year, and only a very modest decline in fertility in recent years, it was possible to use the stable population models to infer that the median age in Zimbabwe should have been 16.3, with 47 percent of the population under 15, and only 3 percent aged 65 or older. Subsequently, Census data showed that these inferences were quite close to reality. The 1987 census revealed that the actual median age was 16.7, and that 45 percent of the population was under age 15, while 3 percent was aged 65 or older (United Nations 1991).

Population Momentum

Youthful age structures carry with them the potential for substantial **momentum of population growth**. This is a concept not unlike that idea that a heavy freight train takes longer to stop than does a light commuter train, or that a Boeing 747 requires a longer runway for takeoff than does a 737. The amount of momentum built into an age structure is determined by answering the following question: How much larger would the population eventually be, compared to now, if replacement fertility were instituted in the population right now? Put another way, how much larger would be the stable population based on replacement-level fertility compared to the current population? (Keyfitz 1971; Kim and Schoen 1997). You can probably see intuitively that if a population has a large fraction of women in the reproductive years, then before the population stops growing they will contribute a lot of additional babies to the future population even at replacement level, whereas an older population, with fewer women in the reproductive ages, will contribute fewer babies and thus will stop growing with fewer people added. Preston and Guillot (1997) have shown, for example, that the populations of Africa and Asia would grow by a factor of 1.56 (they would be 56 percent larger in size when they finally stopped growing) if replacement level fertility had been introduced in 1997. By contrast, the momentum factor for North America was calculated to be 1.10, indicating that the less youthful age structure of North America would not grow by very much more even if replacement level fertility had been achieved in 1997.

Population models are extremely useful in understanding the dynamics of population processes, and thus they are themselves aids in the development of demographic theory. However, we are usually interested in what we anticipate will happen in the real world, so we use our theories to project the population into the future to assess what we can most likely expect.

Population Projections

A **population projection** is the calculation of the number of persons we can expect to be alive at a future date given the number now alive and given reasonable assumptions about age-specific mortality and fertility rates (Keyfitz 1968). In many respects, population projections are the single most useful set of tools available to us for demographic analysis. By enabling us to see what the future size and composition of the population might be under varying assumptions about the trends in mortality and fertility, we can intelligently evaluate what the likely course of events will be many years from now. Also, by projecting the population forward through time from some point in history, we are able to determine the sources of change in the population over time. A word of caution, however, is in order. Population projections always are based on a conditional future—this is what will happen if a certain set of conditions are met. Demographic theory is not yet sophisticated enough to be able to predict future shifts in demographic processes, especially fertility. Thus we must distinguish projections from forecasts. As Nathan Keyfitz has observed, "Forecasts of weather and earthquakes, where the next few hours are the subject of interest, and of unemployment, where the next year or two is what counts, are difficult enough. Population forecasts, where one peers a generation or two ahead, are even more difficult" (1982:746).

There are several ways in which a demographer might project the population. These include: (1) extrapolation methods, (2) the components of growth method, and (3) the cohort component method. In addition, some of these methods can be used to project backward, not just into the future.

Extrapolation   The easiest way to project a population is to extrapolate past trends into the future. This can be done either using a linear (straight line) extrapolation, or a logarithmic (curved line) method. Both methods assume that we have total population counts or estimates at two different dates. For example, in 1980, there were 226,542,199 people enumerated in the U.S. Census, and in 1990 the census counted 248,709,873. We can calculate the average annual **rate of linear growth** (r_{lin}) between 1980 and 1990 using the following formula:

$$r_{lin} = \left[\frac{\text{Population at Time 2} - \text{Population at Time 1}}{\text{Population at Time 1}} \right] / n$$

In this example, the population at Time 1 is the census count in 1980; the population at Time 2 is the census count in 1990, and n is the number of years (10) between the two censuses. You can plug in the above numbers to see that the average annual linear rate of growth turns out to be .0098 (or 9.8 per thousand per year). Now, we use that rate of growth to extrapolate the population forward, for example, from our **base year** of 1990 to a **target year** of 2000, using the following formula:

$$\text{Population in target year} = \text{Population in base year} \times [1 + (r_{lin} \times n)]$$

In this formula, r_{lin} is the average annual linear rate of growth just calculated (.0098) and n is the number of years (in this case 10) between the base year (1990) and the target year (2000). So, plugging in the numbers from above, the projected population in the year 2000 is

$$[248,709,873 \times (1.098)] = 273,083,441$$

You may recall from Chapter 1 that populations typically are thought to grow exponentially, not in a straight line fashion, and this hearkens back to the formula that we used in that chapter to derive an estimate of the doubling time of the population. In that problem, our concern was estimating the likely time that it would take a population to double in size when we knew the rate of population growth. By algebraically rearranging that equation, we can produce a formula that expresses the logarithmic growth of a population, assuming a constant rate of growth. This formula is as follows:

$$\text{Population at Time 2} = \text{Population at Time 1} \times e^{rn}$$

In this case, r represents the geometric or exponential rate of increase and is calculated with the following formula:

Geometric (exponential) average annual rate of population growth (r_{exp})

$$= [ln\ (\text{Population at Time 2} / \text{Population at Time 1})] / n$$

The term ln represents the natural logarithm of the ratio of the population at time 2 to the population at time 1. It is one of the function buttons on most hand-held calculators. Once again, n is the time between censuses. So, to calculate the exponential average annual rate of population growth between 1980 and 1990, first find that the ratio of the population at those two dates (248,709,873/226,542,199) is 1.0979, then find that the natural logarithm of that number is .0934, which then is divided by 10 to find that the rate of increase is .0093 (or 9.3 per thousand per year). Next, plug this rate of growth (.0093) back into the formula for exponential or logarithmic growth (above) in order to project the population forward from 1990 to 2000. The answer is:

$$(248,709,873 \times 1.0975) = 272,959,086$$

This is similar to, but a little less than, the projection using the linear extrapolation method.

Notice that these extrapolation methods of population projection do not take into account births, deaths, or migration. If we have such information available, we can project a population using the **components of growth** method.

Components of Growth   This is an adaptation of the demographic balancing equation mentioned earlier in the book. The population of the United States in the year 2000 will be equal to the population in 1990 plus all the births between 1990 and 2000, less the deaths, plus the net migration between those two dates. We already know what some of these numbers are, of course, but we have to estimate data for the most recent years. Using data from the vital statistics (Ventura, Martin, Curtin, and Matthews 1997) and estimates made by the U.S. Bureau of the Census (Day 1996), we can project that between 1990 and 2000 there will be a total of 39,810,000 babies born in the United States. During this same time, there will be a projected 22,813,000 deaths, and a projected net immigration of 8,200,000. So, starting with the 1990 base year population of 248,709,873, then adding births, subtracting deaths, and adding net migrants, we arrive at a projection of the target year population in 2000 of 273,906,873. This is close to, but slightly higher than either of the projections based on extrapolating from total census counts in 1980 and 1990.

None of the previously discussed methods provides a way of actually estimating the number of future births, deaths, and migrants, and none of them takes into account the age/sex structure of a society, which is an extremely important ingredient in the impact that population growth has on a society. What is needed is a more sophisticated approach to population projections, which can be found in the cohort component method.

Cohort Component Method   To make a population projection using the **cohort component method,** we begin with an age/sex distribution (in absolute frequencies, not percentages) for a specific base year. Usually a base year is a year for which we have the most complete and accurate data—typically a census year. Besides age/sex distributions, you need to have base-year age-specific mortality rates (that is, a base-year life table); base-year age-specific fertility rates; and, if possible, age-specific rates of in- and out-migration. Data are usually arranged in 5-year intervals, such as ages 0 to 4, 5 to 9, 10 to 14, and so on, which facilitates projecting a population forward in time in 5-year intervals. For example, if we are making projections from a base year of 1990 to a target year of 2010, we would make intermediate projections for 1995, 2000, and 2005.

With the base-year data in hand and a target year in mind, we must next make some assumptions about the future course of each component of population growth between the base year and the target year. Will mortality continue to drop? If so, which age will be most affected and how big will the changes be? Will fertility decline, remain stable, or possibly rise at some ages while dropping in others? If there is an expected change, how big will it be? Can we expect rates of in- and out-migration to change? Note that if our population is an entire country, our concern will be with international migration only, whereas if we are projecting the population of an area such as a state, county, or city, we will have to consider both internal and international migration. However, since adequate data on migration are often not available, it is occasionally ignored in population projections.

The actual process of projecting a population involves several steps and is carried out for each interval (usually 5 years, remember) between the base and target years. First, the age-specific mortality data are applied to each 5-year age group in the base-year population to estimate the number of survivors in that age range 5 years into the future. Thus since there were 9,262,000 females aged 20 to 24 in the United States in 1990 and the probability of a female surviving from age 20 to 24 to age 25 to 29 (derived from the life table; see the Appendix) is 0.997, then in 1995 there should have been 9,234,000 women aged 25 to 29. This process of "surviving" a population forward through time is carried out for all age groups in the base-year population. The probabilities of migration (if such data are available) are applied in the same way as are mortality data.

In projecting a population forward at 5-year intervals, the task of fertility estimation is to (1) calculate the number of children likely to be born during the 5-year intervals, and (2) calculate how many of those born will also die during those intervals. The number to be born is estimated by multiplying the appropriate age-specific fertility rate by the number of women in each of the childbearing ages. Then we add up the total number of children and apply to that number the probability of survival from birth to the end of the 5-year interval. Experience suggests that fertility behavior often changes more rapidly (both up and down) than demographers may expect, so population projectionists hedge their bets by producing a range of estimates from high to low, with a middle or medium projection that incorporates what the demographer thinks is the most likely scenario. The highest estimate reflects the demographer's estimate of the highest fertility trend possible in the future, along with the highest decline likely in mortality, and the maximum net immigration likely. Conversely, the lowest projection incorporates the most rapid decline in fertility, the least rapid drop in mortality, and the lowest level of net immigration probable.

Differences in fertility usually account for the biggest chunk of variation between high and low projections. For example, in the projection of the population of Mexico made by the U.S. Bureau of the Census in the late 1970s (U.S. Bureau of the Census 1979), the high projection for 1990 was 102,349,000, whereas the low projection was 88,103,000. Virtually all the difference is found in the ages 0 to 14. The 1990 census in Mexico, however, counted "only" 86 million people. Part of the difference was almost certainly due to an acknowledged undercount in the Mexican census, but most of the difference was a result of the success of the campaign in that country to lower the birth rate.

Calculating a cohort component projection is somewhat complicated and other sources contain more details (Murdock and Ellis 1991; Smith 1992). Using these methods, the U.S. Bureau of the Census completed a series of population projections with 1995 as the base year and 2050 as the ultimate target year, using varying assumptions in their projections about the future course of fertility, mortality, and migration (Day 1996). Intermediate calculations were done for 5-year intervals, and it turned out that, for the year 2000, the highest projected population was 278,129,000, whereas the lowest projection for that year is 271,237,000. The middle (most likely) scenario produced a projected population of 274,634,000—a bit higher than the number generated by the various other methods of population projection that I have reviewed here for you.

Backward or Inverse Projection   Population data are projected into the future in order to estimate what could happen down the road. Similar methods can be used to work backward to try to understand what happened in the past. The basic idea is to begin with census data that provide a reasonably accurate age/sex distribution for a given year. Then, making various assumptions about the historical trends in fertility, mortality, and migration, you work back through time to "project" what earlier populations must have been like in terms of the number of people by age and sex. Wrigley and Schofield (1981) used this method to work backward from the 1871 census of England and Wales to reconstruct that region's population history. Whitmore (1992) used a complex backward projection model to show how it was possible for European contact to have led to a 90 percent depopulation of the indigenous peoples living in the Basin of Mexico at the time that Cortes arrived. Details of the methods are contained in other sources (see, for example, Oeppen 1993; Smith 1992).

Cohort component methods of population projection give clues about the future (and even about the past) by letting us view alternative age/sex distributions. From this demographic information we draw inferences about the impact on society by calling on our knowledge of how prior changes in the age structure have altered the fabric of social life. These kinds of interrelationships are well described by the perspectives of age stratification and cohort flow.

Age Stratification and Cohort Flow

The perspective of **age stratification** and **cohort flow** was first put forward as a cohesive package by Matilda White Riley, Marilyn Johnson, and Anne Foner; it has been expanded and detailed since then especially by Riley (1976a; Riley 1979) and Foner (1975). The notion of age status is not new; Kingsley Davis noted in 1949 that "all societies recognize age as a basis of status, but some of them emphasize it more than others" (Davis 1949:104). Likewise, the importance of cohorts in analyzing social change is not unique to Riley, Johnson, and Foner; Norman Ryder suggested in 1964 that "social change occurs to the extent that successive cohorts do something other than merely repeat the patterns of behavior of their predecessors" (Ryder 1964:461). What this perspective does is to integrate these concepts of age status and cohorts in a way that helps to explain social change, and thus helps us to understand how and why the status of older people has been shifting over time.

The age-stratification theory begins with the proposition that age is a basis of social differentiation in a manner analogous to stratification by social class (Foner 1975). The term *stratification* implies a set of inequalities, and in this case it refers to the fact that societies distribute resources unequally by age. These resources include not only economic goods but also such crucial intangibles as social approval, acceptance, and respect. This theory is not a mere description of status, however; it introduces a dynamic element by recognizing that aging is a process of social mobility. Foner (1975) notes that "as the individual ages, he too moves within a social hierarchy. He goes from one set of age-related social roles to another and at each level receives greater or lesser rewards than before" (1975:156). Contrasted to other forms of social mobility, however, which may rely on merit, luck, or accident of birth, social mobility in the age hierarchy is "inevitable, universal and unidirectional in that the individual can never grow younger" (1975:156).

Age strata, though identifiable, are not viewed as fixed and unchanging. The assumption is that the number of age strata, and the prestige and power associated with each, are influenced by the needs of society and by characteristics of people at each age (their numbers and sociodemographic characteristics). European society of a few hundred years ago seems to have been characterized by three age strata—infancy, adulthood, and old age (Aries 1962); and power (highest status) seems to have been concentrated in the hands of older people (Simmons 1960). Modern western societies appear to have at least seven strata—infancy, childhood, adolescence, young adulthood, middle age, young old, and old old, with power typically concentrated in the hands of the middle-aged and the young old.

As we age from birth to death we are allocated to **social statuses** and **social roles** considered appropriate to our age. Thus children and adolescents are currently allocated to appropriate educational statuses; adults to appropriate positions of power and prestige; and the elderly to positions of retirement. We all learn the roles that society deems appropriate to our age, and we reward each other for fulfilling those roles and tend to cast disapproval on those who do not fulfill the societal expectation. But neither the allocation process nor the overall **socialization** process is static (as I discuss later in the chapter). They are in constant flux as changing cohorts alter social conditions and as social conditions, in turn, alter the characteristics of cohorts. This leads to the concept of cohort flow.

A cohort, in this case, refers to a group of people born during the same time period. As Riley (1976b) points out:

> Each cohort starts out with a given size which, save for additions from immigration, is the maximum size it can ever attain. Over the life course of the cohort, some portion of its members survive, while others move away or die until the entire cohort is destroyed. Each cohort starts out also with a given composition; it consists of members born with certain characteristics and dispositions. Over the life course of the individual, some of these characteristics are relatively stable (a person's sex, color, genetic makeup, country of birth, or—at entry into adulthood in our society—the level of educational attainment are unlikely to change). . . . When successive cohorts are compared, they resemble each other in certain respects, but differ markedly in other respects: in initial size and composition, in age-specific patterns of survival (or longevity), and in the period of history covered by their respective life span [pp 194–195].

At any given moment, a cross section of all cohorts defines the current age strata in a society (see Figure 8.6). As cohorts flow through time, their respective sizes and characteristics may alter the allocation of status and thus the socialization into various age-related roles. Additionally, some characteristics of cohorts may change in response to changing social and economic conditions (such as wars, famines, and economic prosperity), and those changing conditions will influence the formation of new cohorts. This continual feedback between the dynamics of successive cohorts and the dynamics of other changes in society produces a constant shifting in the status and meaning attached to each age stratum.

The older population of the United States has been transformed by the changing characteristics of the new cohorts moving into old age. Cohorts of different sizes and backgrounds, and with different facets of social history having affected them, shape old age somewhat differently. For example, there were 10.3 million Americans aged 65 to 69 in 1990 (representing the birth cohorts of 1921 through 1925). Between 1921 and 1925, there were about 15 million Americans born, and, based on the known death rates for people born during those years, we should have expected to find only about 9 million of them still alive in 1990. Where did the others come from? Europe, mainly, but also Asia, Latin America, and, to a much lesser extent, Africa. Many were recent arrivals, but most had come as younger people and had grown old in the United States. The members of this birth cohort were youngsters right after World War I and then were hit hard by the Depression that descended in the early 1930s. Many of the men found themselves fighting in World War II, and the entire cohort was entangled in the war one way or another. Life was not completely bleak for this cohort, however. After the war its members provided leaders who forged the incredible rise in national wealth and personal income in the United States, and they became eligible for social security benefits at a time when those benefits hit record high levels. These people represent, in fact, the best-educated and highest-income group of older people so far in the United States, a topic to which I return in the next chapter.

One of the more widely discussed analyses of the impact of changing cohort size on a population is the "Easterlin hypothesis," which I discussed in previous chapters. Yet, you need not accept all of Easterlin's assumptions about the effect of the relative size of cohorts to know that the U.S. Baby Boom generation has, through sheer cohort size, vastly influenced the social world of the United States, and boomers in Canada have had a similarly dramatic impact on that nation. One example of this from the United States is the relationship between the age structure and the crime rate (see the Essay in this chapter).

Figure 8.6   Cohort Flow and Age Strata Are Closely Intertwined

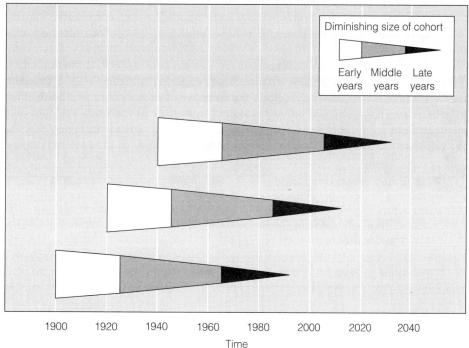

Note: Cohorts flow through time, gradually diminishing in size and impinging on society as they move through time. At any given date the number and characteristics of people in each life stage will influence the system of age strata.
Source: Adapted from M. Riley, M. Johnson, and A. Foner, Aging and Society: A Sociology of Age Stratification, Vol. III (New York: Russell Sage Foundation, c 1972), p. 10. Used by permission.

Another issue that researchers have examined has to do with the economic well-being of different cohorts. Neoclassical economic theory would suggest that, all things being equal, a larger cohort would increase the supply of labor relative to demand and thus depress wages, leaving the Baby Boom generation less well off than its parents. Has this happened?

Easterlin and his colleagues (1993) examined this issue and concluded that baby boomers are not less well off (at least economically) than their parents. The reason: Things were not equal. Early on, baby boomers anticipated and responded to the threat of depressed wages that were implied by their larger numbers. They adjusted. They altered their life cycle decisions dramatically by delaying marriage, postponing children in marriage, combining the income of spouses, and combining mother's work with childbearing. These are now familiar attributes of the Baby Boom (Russell 1992), and they represent the underlying causes of much of the social change that the Baby Boom has produced in American society. In the simplest formulation, these changes can be seen as the consequences of the way in which baby boomers have reacted to the age distribution pressures that have confronted them all of their lives.

The downside of the Baby Boom adjustments may be that the penalties attached to a large cohort have just been postponed to an older age. Easterlin and associates (1993) also note that baby boomers have not provided so well for their retirement years. They have neither the financial nor the familial resources for retirement that their parents have.

Another example of the influence of the age structure on social phenomena is the change in the American suicide rate. This is another variation on the Easterlin theme that changing cohort size alters the fortunes of each cohort. Besides leading to delayed marriage, delayed parenthood, and slower social and economic mobility, changing cohort size may have contributed to a rise in the suicide rate, just as it did the crime rate. The Easterlin approach suggests that:

> When relative cohort size (the ratio of younger to older males) increases, it is expected to lead to an increase in suicide rates for young males (as their labor market position relative to their income aspirations deteriorates) and a decrease in suicide rates for older males (as their labor market position relative to their income aspiration improves) [Ahlburg and Schapiro 1984:99].

The economic effect that is so crucial to the Easterlin perspective may be augmented by a psychosocial effect:

> The large baby boom cohort, because of imperfect socialization arising from the rising child/adult ratio, was "saddled with expectations and beliefs which were less realistic, less suited to particular individual lives, less flexible in the face of adversity" (Carlson 1980). Such imperfect socialization may be expected to result in an increase in the suicide rate [Ahlburg and Schapiro 1984:100].

Ahlburg and Schapiro show that the data from the United States do, in fact, support the idea that between 1948 and 1976 the suicide rate went up among younger people as (1) the size of the younger cohorts increased relative to the older, and (2) the ratio of children to adults increased. Thus the age structure itself may help to account for at least part of the dramatic rise in teenage suicides during the late 1960s and 1970s, and forecasts a rise in suicide rates for middle-aged and older males as the Baby Boom cohort flows through time.

Changes in economic well-being and in suicide rates occasioned by the changing age distribution of the United States represent only two of numerous ways in which we can "read" an age structure and, in essence, try to tell a population's fortune. Let us do this in more detail for the United States.

Reading the American Age Structure

Changes Induced by Fertility Fluctuations   We can trace the history of fertility (and recount the economic situation of the time) in the age pyramid for the United States (Figure 8.7). Actually, above age 55 or 60 the effect of accelerating death rates obscures changes that might have occurred, but for slightly younger

Figure 8.7   Age Pyramid for the United States (1990)

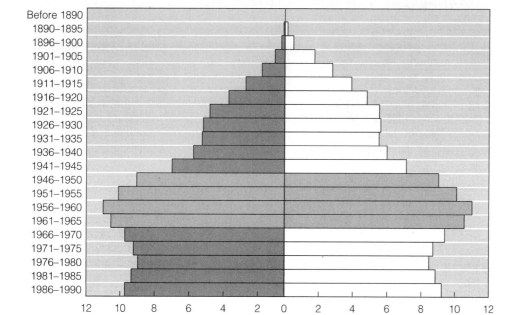

Males (millions)   Females (millions)

Note: The population pyramid for the United States in 1990 shows the impact of low birth rates during the Depression (births in 1931–35), the higher birth rate of the Baby Boom generation (births in 1946–65—the shaded area), the decline in fertility in the late 1960s and 1970s, and the slight resurgence in fertility in the 1980s.

Source: Adapted from U.S. Bureau of the Census, 1990 Census of Population and Housing.

CRIME AND THE AGE STRUCTURE

There were 48 percent more arrests in the United States in 1980 than in 1970, according to the FBI (data from U.S. Department of Justice 1971; 1981). Had society fallen apart? Had the nation lapsed into a police state? Neither one. What happened was that the Baby Boom generation was moving through the late teen and early adult years, when crime rates are highest. They swelled those ages and crime went up at least partly because there were more people "at risk" of committing a crime. At these younger ages, a person's fancy seems to turn not just to love, but disproportionately to crime as well. In most societies crime is the province of the young, and it stands to reason that a youthful age structure will produce higher overall crime rates (that is, crude crime rates) than will an older age structure (Cohen and Land 1987; Steffensmeier and Harer 1987).

Calculating crime rates by age is a bit tricky for large geographic areas such as the United States, because not all regions of the country report crimes, nor do those reporting necessarily use the same criteria. Furthermore, you know the age of an alleged criminal only after you have caught him or her, so areas that are more efficient at catching criminals will have different data from those that are less efficient. Nonetheless, these limitations have not deterred me from comparing crime data over time in the United States.

Between 1970 and 1980 the arrest rate in the United States increased from 33 arrests per 1000 people to 43 per 1000. Nearly half of that increase was due to the fact that there were more young adults in the high-risk ages. This is very similar to the findings of the 1967 National Crime Commission, which concluded that 40 to 50 percent of the increase in the U.S. crime rate between 1960 and 1965 was due to the changing age structure—the early impact of the Baby Boom.

It is possible that police were more likely to arrest people in 1980 than in 1970 because more money had been pumped into law enforcement in response to the crime siege brought on by the baby boomers. Another explanation for rising crime rates is that the number of children per adult rose during the Baby Boom, potentially lowering the amount of social control exercised over each child, thus leading to more deviant behavior that resulted in arrest. That, of course, would represent a cohort phenomenon resulting from fertility's impact on the age structure rather than a direct result of the age structure itself. Nonetheless, there is a fair amount of evidence to suggest that declining social control over teenagers (whether a function of the age structure or not) has played a role in rising crime in the United States (Osgood, et al. 1996; Sampson and Laub 1990).

A less obvious, but potentially very important, aspect of the age structure's influence on crime is the fact that the age structure of the population of potential victims may also affect the opportunities for crime and thus the crime rate itself. In the United States in the 1970s and 1980s, the proportion of the population that was fairly young, at work during the day, and affluent enough to own easily stolen and fenced goods rose dramatically, and thus the opportunity for crime rose. Cohen and Felson (1979) argue that this substantial increase in the opportunity to commit crime was, in fact, the most important explanation of the rise in the crime rate. If so, the age structure effect was less direct, but no less influential.

In the accompanying graph, you can see that in both 1979 and 1989 the arrest rates in the United States were distinctly highest in the 15 to 29 age group. You may have personally been victimized by the increase in rates among all young adults in that 10-year interval. Consistent with Cohen and Felson's argument, the rates by age increased at these younger ages in step with the rise in the potential opportunity for crime. Gender equality has probably also played a role. In 1970, only 6 percent of those arrested were female, whereas by 1989 the figure had jumped to 18 percent.

Despite the obvious rise in arrest rates for younger people between 1979 and 1989, you can see at the right side of the graph that the overall arrest rate scarcely changed. How could that be? The baby boomers, who still make up a disproportionate share of the population, had mercifully moved on beyond the high-crime ages, leaving crime to the younger, smaller cohort of the baby busters.

One of the more interesting aspects of the debate over whether demography or law enforcement is responsible for the changing crime rate is that the relative importance of the two explana-tions shifted in less than a decade. In the mid-1970s people were only beginning to accept the idea that crime bore any relation to the changing age structure. By the mid-1980s that influence was so widely accepted that law enforcement personnel were clamoring to remind us that demography does not explain everything. As usual, of course, the truth seems to lie somewhere between those two extremes (Cohen and Land 1987; Gartner 1990).

Arrest Rates by Age, 1979 and 1989

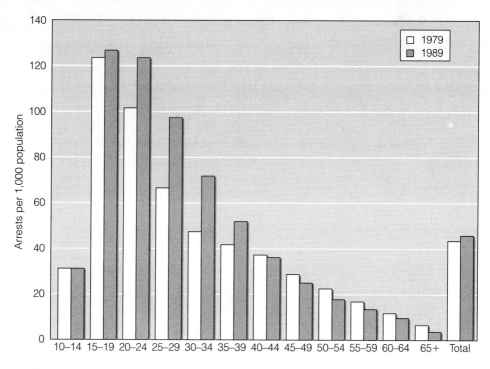

Sources: Rates calculated by the author from (1) arrest data from U.S. Department of Justice, 1980, Crime in the United States, 1979 (Washington, D.C.: Government Printing Office); 1990, Crime in the United States, 1989 (Washington, D.C.: Government Printing Office); and (2) population data from the U.S. Bureau of the Census, Current Population Reports, Series P-25, for the appropriate years.

people, you can see the effect the Depression had on the birth rate. For example, people who were between 55 and 60 years old in 1990 were born between 1931 and 1935, which was the bottom of the Depression. In those years the birth rate was very low (below replacement level, as I discussed in Chapter 6), and you can see further that fertility had been declining since 1921–25. Had you looked only at the age distribution for males, you might well have jumped to the conclusion that the dent at those ages was a result of World War II casualties, but a glance at the female side would cause you to reject that idea, since females were not engaged in actual combat.

In the late 1930s, the birth rate picked up a little and, of course, the Baby Boom occurred in the 1940s and 1950s as the economy recovered after World War II. You can spot the Baby Boom generation as the birth cohorts of approximately 1946 through 1965—those people who were 25 to 44 years old in 1990 (the shaded ages in Figure 8.7). Between 1960 and 1980 the birth rate declined, with each year generally seeing fewer people added than the year before. Thus in 1990 there were fewer people between ages 10 and 15 than there were in any 5-year age group up to age 55. That "birth dearth" or "Baby Bust" will be a feature of the age structure until the middle of the twenty-first century and will be a long-lived reminder of the inflation, tight job market, energy shortage, and female liberation from the late 1960s through the 1970s. The birth rate rebounded slightly in the 1980s (the "Baby Boomlet"), and this is reflected in the slight bulge at the base of the population pyramid for 1990.

In the United States in 1990, 22 percent of the population was under age 15, so it was not exactly young, while more than 12 percent was 65 and older, putting it in the category of "old" (remember my discussion earlier in the chapter). In Figure 8.7 notice that the Baby Boom bulge is centered around the late twenties and early forties, with fewer people at the younger ages as a result of the decline in fertility in the 1960s and 1970s. It was actually that decline in fertility that has aged the American population proportionately, although of course the decline in mortality throughout the twentieth century has meant that more and more people are able to survive to old age.

The distortions you see in the age structure have some interesting implications for the future, and in the following section I discuss them in connection with two different sets of projections about the future course of birth rates. One projection assumes that birth rates were manipulated to achieve **zero population growth** (ZPG) in the 1990s, whereas the other assumes that fertility will remain unchanged from 1990 levels, thus eventually (rather than abruptly) leading to ZPG.

ZPG—Now or Later

The public worry over population growth in the United States quieted considerably in the 1970s after the commotion created in the 1960s. Interestingly enough, at the time of greatest public concern, about 1966–70, the U.S. birth rate had already begun to drop, and the concern was in many ways a belated reaction to the Baby Boom (as well as a corollary to other social concerns especially prevalent in the 1960s). However, the decline of interest in population issues in the United States in the 1970s was premature, since the population is certain to continue its increase well into the 21st century. Much of this growth is being fueled by immigrants, who became a focus of public concern in population issues in the 1990s. Another source of population growth, of course, is through the birth rate, and there has been considerable public debate over the issues of abortion and federally funded family planning programs—both of which have at least some influence on national fertility levels, as you recall from Chapters 5 and 6.

At current levels of fertility and with relatively unchanging levels of migration and slowly improving levels of mortality, the U.S. Census Bureau expects continued population growth beyond the middle of the next century. Their middle series projects a total U.S. population of 394 million in the year 2050—131 million more than there were in 1990, which amounts to a 50 percent increase in population size in that 60-year period. This would be equivalent to adding the entire population of Japan to the United States over the period from 1990 to 2050. There are those who have argued that the country cannot, and what's more should not, tolerate additional people and that ZPG should be achieved immediately. The principal objection to population growth is that Americans consume a radically disproportionate share of the world's resources and adding to the U.S. population exacerbates the world's environmental problems and aggravates the distorted distribution of income in the world.

What are the consequences of these two different growth trajectories—of ZPG now or later? The contrast in the age/sex structure represents the single most important source of difference, and regardless of the path of future growth, age/sex structure changes will play a crucial role in the future development of the United States.

To bring these issues into sharper focus, I have made two projections for the U.S. population from 1990 to 2070. The first one (*ZPG now*) assumes that from 1990 on, the American population would not grow in size. The second (*ZPG later*) is based on the assumption that ZPG would come later as a result of a continuation of 1990 fertility levels, which were, as you know by now, below replacement level.

For the sake of simplicity, I have assumed in both sets of projections that mortality will remain at the levels prevailing in 1990, and that the number of net migrants per year by age and sex will remain at the 1990 level. If mortality were to continue to decline dramatically from the 1990 levels, then my projections would be too low, because greater proportions of people would survive to each successive age. However, if significant restrictions were placed on legal immigration into the United States, the number of immigrants could drop from the 1990 levels. On the other hand, there have been suggestions that immigration levels should be allowed to slide up, rather than down, because that would allow immigrants to pay the Social Security taxes for the Baby Boom generation when the boomers retire. Although there are many more immigrants to than emigrants from the United States, keep in mind that emigrants may be a bit older than immigrants, so that could have at least a small impact on the age structure (Bouvier, Poston, and Zhai 1997). As you look at these alternative projections, you will do well to keep in mind the old Chinese proverb: "Prediction is very difficult—especially with regard to the future."

1990   Let me put the projections into perspective by briefly reviewing the situation of 1990, adding to the comments that I made earlier about reading the age structure. The declining birth rate that followed the Baby Boom forced adjustments in several segments of American society. During the early years of the Baby Boom there were too few classrooms and teachers, so new schools were built and new programs were developed to train teachers. But the drop in birth rates resulted in a bumper crop of unemployed teachers and, furthermore, companies that had made big profits on babies were forced to rethink their markets. The Baby Boom children have grown up and have been having fewer children than their parents had. (This is the real story of how baby shampoo became a beauty shampoo for adults, as I discuss further in Chapter 15.) As the Baby Boom children hit the labor market, their annually increasing numbers put a severe strain on the ability of the economy to provide jobs, so it should be no surprise that during the 1970s unemployment rates were persistently high and that in the 1980s women were in the labor force in record proportions, helping to produce an adequate family income.

The children of the Baby Bust (born from 1965–77) were just emerging into young adulthood in 1990 and were labeled "Generation X"—the generation forgotten by America because it grew up in the shadow of the Baby Boom, coping with the detritus of the boomer's lifestyle choices. However, in the later middle ages (50 to 64) we find cohorts of people who were among the most advantaged in American history. They were born during the Depression (and thus, you recall, there are not many of them) and they were too young for World War II, although some were involved in the Korean War. They were advantaged primarily in an economic sense, since they entered the labor market during a time of relative economic expansion. Because there were fewer of them than of older cohorts, they were absorbed into the labor force more quickly and had less competition for career advancement. Overall, the rise in their standard of living from childhood to mid-adulthood was probably greater than for any other generation of Americans.

At the oldest ages, at and above the age of retirement, the number of people was larger than at any time in history. The problems created by the Baby Boom generation, though, tended to steal the spotlight from the older generation, which struggled to gain a major voice in American policy making. Their increasing numbers have given rise to new construction of retirement communities, an expansion of employment opportunities servicing the elderly, and fears that the social security system will be bankrupted, topics to which I return in the next chapter.

Would the age/sex structure be more or less problematic in 2010 if ZPG had been achieved in 1990? Let us see.

2010   Zero population growth in the absence of migration means that the number of births in any given year just equals the number of people dying. But because I am taking migration into account in these projections of the United States (where immigration vastly exceeds emigration), there would have to be fewer births than deaths to maintain ZPG. This is reflected in Figure 8.8, where you can see that in 2010 the age structure would be heavily weighted toward the middle adult population, at the expense of the young, in order to achieve zero population growth. Indeed, in this model there are virtually twice as many people aged 45 to 54 as there are aged 0 to 9—nearly a complete turnabout from the early Baby Boom years.

To have achieved ZPG in the 1990s and maintained it until 2010 would have required a substantial reduction in fertility beyond the already low levels. Consequences of that rapidly falling fertility include the obvious fact that, after the bottom fell out of the higher education market and the number of new families formed dropped dramatically by the early 2000s, there would be only a slight rebounding by 2010. Since the number of large or growing families would be severely cut back, the housing construction industry would also be affected. However, there might be an increased demand for specialized housing and services for the elderly, whose numbers will, in all events, increase. The smaller numbers of people in the young-adult age cohorts, added to the prospect of even smaller cohorts younger than they, would provide the young adults with a relative advantage in entering the labor market. Rapid upward mobility would probably be rather difficult for those people, however, because the Baby Boom generation would still be glutting the market. In addition, the smaller family size might well add even greater proportions of women to the labor market, increasing competition for jobs and promotions. The problem of creating jobs and career opportunities might be compounded by the difficulty of expanding the economy in the face of no population growth. Americans (especially in the middle class) have generally taken progress for granted and tend to feel deprived when their standard of living is not materially improving, but it is possible that with ZPG, the quality of life would have to be measured in other than material terms.

It is probable, of course, that the future of American society will be a quest for nonmaterial progress whether we have ZPG now or later. The increasing worldwide demand for resources is placing ever-increasing constraints on the ability of Americans actually to improve their standard of living. Indeed, there are some people who argue that material progress in the United States is really a sideways movement, not a true increase in the quality of life. The replacement of homemade cookies with snack foods and the popularity of fast-food restaurants and electronic games, for instance, represent changes in lifestyle but not necessarily improvements. A rise in the standard of living is perhaps better measured in the more standard terms of improved health, more comfortable housing, more and better education, and greater artistic achievement and appreciation. The future will in all probability see at least a measure of attention being focused increasingly on these aspects of the quality of existence rather than on simply diversifying consumer goods and services.

It should be obvious to you that in 2010 the only difference between the ZPG-now and ZPG-later positions would be the number of people at ages under 20. You can see from Figure 8.8 that with 1990 levels of fertility in the ZPG-later projection, there was a slight rise in the number of births between 1990 and 1995, resulting in a slight rise in the number of people aged 15 to 19 in 2010. This is a result of the last of the Baby Boom women in 1990 moving through their reproductive ages; even at a rate of almost exactly two children per woman, there were still so many women

Figure 8.8   ZPG-Now Age Pyramids Are More Distorted for the Future than those for ZPG-Later

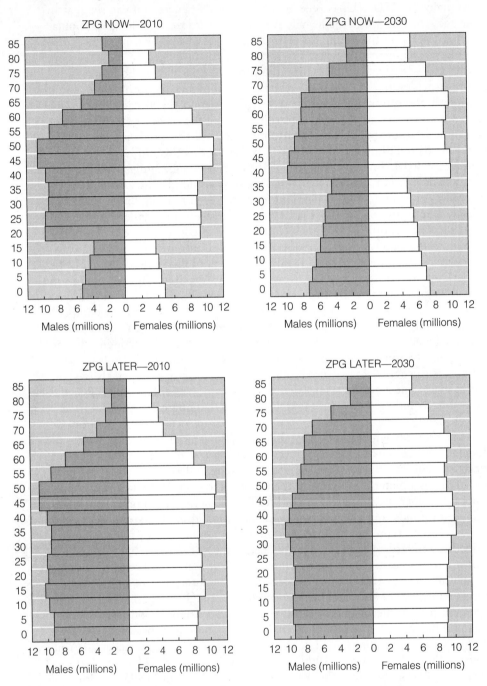

Note: Projections over time of the U.S. population from 1990 to 2070 (assuming constant numbers of migrants and constant mortality rates) show that the ZPG-later path would provide a smooth transition to nongrowth, whereas the ZPG-now path could create a new cycle of problems similar to the baby boom.
Source: Calculations made by the author.

Figure 8.8 (continued)

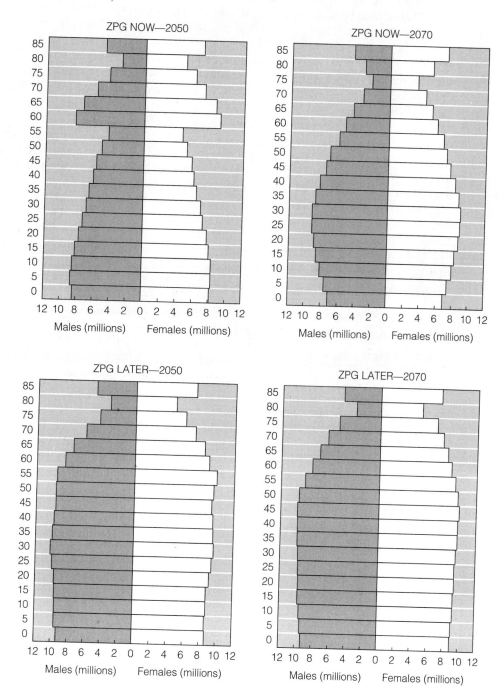

of reproductive age in 1990 that their fertility increased the number of annual births. This would be offset in 1995–2010 by the relatively smaller number of young women (born in the 1960s and 1970s) moving through their reproductive years. Overall, however, the number of births in the ZPG-later projection is fairly stable and the social, political, and economic dislocations would probably be minimal. The price for the stability, of course, is a population that would still be growing. Between 1990 and 2010 the population of the United States would increase from 248 million to 280 million.

2030   By the year 2030, the number of old people will be rising each year and contributing to the possibility of rising fertility. In the ZPG-now projection the age structure broadens out slightly at the bottom as a new Baby Boom begins to replace the old boom generation, which by 2030 has reached retirement age.

The economic and social makeup of society would be considerably different than it is now, being dominated by older middle-aged people but in the midst of a transition to a renewed emphasis on families and youth. The couples who had only one child to achieve ZPG could be thinking about a greater number of grandchildren. The economy then would have to be moving in two directions at once—accommodating both the increasing demands of children and the increasing needs of the elderly. It is conceivable that this would be a fairly rigid society in terms of social behavior, dominated by the older middle-aged adults. Indeed, from 1990 to 2030 the pattern is one of a lessening likelihood of innovative behavior on the part of the young as a consequence of the rise in the ratio of adults to children and the probable resulting increase in social control. On the other hand, the Baby Boom generation has been noted for its high degree of flexibility and innovativeness (see Jones 1981; Russell 1992), and it is reasonable to suppose that such behavior will be carried into older age (Weeks 1984).

By the year 2030 the population in the ZPG-later projection would have grown to about 305 million people, a 23 percent increase over the 1990 population but only 9 percent larger than the 2010 population. As you can see in Figure 8.8, the age structure suggests a population dominated numerically by people in their older middle ages. But unlike the ZPG-now population, the ZPG-later population in 2030 would have a fairly even distribution by age. The average age of the population would become steadily older, but the process is a gradual one, unencumbered by the fluctuations in the youthful population, which would be required at this date in the ZPG-now projection.

2050 and 2070   In the year 2050 ZPG-now would produce an age structure from ages 0 through 55 that looks very much like an underdeveloped country (see Figure 8.8). Each year for five decades the number of babies born would have been greater to offset the increasing number of deaths generated by the burgeoning population of the elderly. The number of young children (aged 0 to 4) would now exceed the number alive in 1990, adding to the dependency burden made awesome by the size of the retirement-age population. Everyone in the Baby Boom generation would be either dead or retired in 2050, but retirees could represent an enormous social and political force in American society. Their power in society would be augmented by the fact that the retirement age would have to be virtually eliminated or else jobs could go begging in society (and recall that immigration has been added to the picture).

Although dominated by the elderly, the American society of 2050 would have to start looking forward to the future (say, about 2070), when younger adults would again be a major influence in the economy, politics, and social fabric of the country, as they were in 1990. In fact, the "middle series" population projections made by the U.S. Census Bureau, which represent what they believe is the most likely demographic scenario for the future, suggest that by the year 2050 there will be considerably more teenagers in the United States than there were in 1990 (Day 1996).

In 2070 the United States would be dealing with a new Baby Boom generation of young adults, but a new Baby Bust phenomenon at the very youngest ages, and the ferris-wheel cycle of fertility would start all over again. That cycle would continue for as long as U.S. policy was oriented toward maintaining ZPG. By now, you may well have supposed that the massive shifting every decade or two in the numbers of people at each age might well have led to an abandonment of the ZPG policy in the interest of economic, political, and social stability. For example, the economic impact of the fluctuating age structure over time is evident in the dependency ratio. In Figure 8.8 you can see that with ZPG now, in the 40 years between 2010 and 2050 the dependency ratio would more than double, but in another 20 years (to 2070) the ratio would have dropped to nearly the 1990 level. The consequence would be a series of booms and busts in schools, retirement homes, job creation for youth, the demand for durable goods, and so forth.

As the ZPG-later projection progresses through time, the age/sex structure continues to smooth out to a population with a decreasing emphasis on children and youth and an increasing predominance of middle-aged and elderly persons. The majority of the population is of working age, and this could be more economically advantageous than the situation of ZPG now. We can conclude that over time the abrupt changes in fertility levels required just to bring a low fertility society such as the United States to ZPG now would produce lifelong distortions in the age structure, which could create constant problems of adjustment for society. This is the dilemma that American society has had to face with the Baby Boom generation. At first there were the problems of schooling and jobs; now there are problems of economic consumption (finding affordable housing, getting promotions, and so on); and society is approaching the problems of retirement and providing pension and medical benefits.

Summary and Conclusion

The age/sex structure of a society is a subtle, commonly overlooked aspect of the social structure. The number of people at each age and of each sex is an important factor in how a society is organized and how it operates. The age/sex structure is determined completely by the interaction of the three demographic processes. Migration can have a sizable impact, because migrants tend to be concentrated in particular age groups and, in addition, migration is sometimes selective for one sex or the other. Mortality has the smallest short-run impact on the age/sex distribution, but when mortality declines suddenly (as in the less developed nations), the impact is to make the population more youthful. At the same time, a decline in mortality influences the sex structure at the older ages by producing increasingly greater numbers of females than males.

Changes in fertility generally produce the biggest changes in a society's age structure; a decline in fertility ages the population, just as a rise makes it more youthful. A rise in fertility also tends to produce a greater number of males than females, since more boys are born than girls. For example, males in the world outnumber females, because continued high levels of reproduction have generated a youthful world age structure in which the excess of male babies at birth results in the overall majority of males.

We study the effects of population processes on the age/sex structure by calculating stable and stationary population models, which are types of population projections—one of the most useful methodological tools created in the field of population studies. Looking backward through time, we can see how different birth cohorts have flowed through time, being influenced differentially at each age by social events, and in turn have shaped events by their passage through the age structure. Using population projections to look ahead, we can chart the potential course of change implied by different age/sex structures. The analysis of two alternative future patterns of population change in the United States (ZPG-now and ZPG-later) illustrates the impact that changing fertility could have on the age/sex structure and, by implication, on the overall structure of society. It is reasonable to suppose that distortions in the age structure could lead to changes in economic organization, political dominance, and social stability. In all of the developed societies in the world today, fertility has been low for long enough that population aging has become a major societal concern—not because we are afraid of old people, but rather because the demands on societal resources are very different for an older than for a younger population. In the next chapter I discuss population aging in more detail.

Main Points

1. An age/sex structure represents the number of people of a given age and sex in society and is built from the input of births at age zero and from deaths and migration at every age.

2. A young population is one with a high proportion of young people, whereas an old population is one with a high proportion of elderly people.

3. The age structure is typically graphed as a population pyramid and measured by the median age or the dependency ratio.

4. More male babies than females are generally born, but the age structure is further influenced by the fact that at almost every age more males than females die.

5. Migration can have a very dramatic short-run impact on the age/sex structure of society, especially in local areas.

6. Migration most drastically affects the number of young adults.

7. Mortality has very little long-run impact on the age structure, but in the short run a decline in mortality typically makes the population younger in medium to high mortality societies, and a little older in lower mortality societies.

8. Fertility is the most important determinant of the shape of the age/sex structure.

9. High fertility produces a young age structure, whereas low fertility produces an older age structure.

10. The age/sex structure is a powerful stimulant to social change.

11. The influences of fertility and mortality on the age/sex structure are usually understood by applying stable population models to real populations.

12. Population projections are developed from applying the age/sex distribution for a base year to sets of age-specific mortality, fertility, and migration rates for the interval between the base year and the target year.

13. The concepts of age stratification and cohort flow place age structures in their proper sociohistorical context.

14. In the United States, even with current low levels of fertility, a policy of zero population growth right now would produce rather severe dislocations in social organization as a result of changes in the age structure that would take place if the population suddenly stopped growing.

Suggested Readings

1. Jeanne Ridley, Mindel Sheps, J. Lingner, and Jane Menken, 1967, "The effects of changing mortality on natality," Milbank Memorial Fund Quarterly 55(1):77–97.

   The title of this article pretty well sums up the topic under discussion, but it fails to tell you that this was one of the first important statements of the link between mortality and fertility.

2. United Nations, 1997, The Sex and Age Distribution of the World Populations: The 1996 Revisions (New York: United Nations).

   Every several years the Population Division of the United Nations updates its projections of every country of the world by age and sex. This volume includes estimates of the population by age and sex for the years 1950 through 1990 (in five-year increments) and low, medium, and high projections by age and sex for the years from 1995 through 2050. Because they are probably not yet available on the internet, this publication is the only way to access this vast reservoir of information.

3. Carl Haub, 1987, "Understanding population projections," Population Bulletin 42(4).

   This is a well-written summary of the major types of population projections and their interpretations.

4. Cheryl Russell, 1993, The Master Trend: How the Baby Generation is Remaking America (New York: Plenum Press).

   The author shares her insight as a boomer and boomer-watcher to assess the role that the Baby Boom generation has played in the wide swath of social change that has characterized the United States over the past few decades.

5. Ronald Lee, W. Brian Arthur, and Gerry Rodgers (eds.), 1988, Economics of Changing Age Distributions in Developed Countries (Oxford: Clarendon Press).

   Nearly every country of the world has had to cope with the impact of changing age distributions. The chapters in this book give you insight into how Israel, Japan, and Great Britain have dealt with the effects of baby booms and busts.

✿ Websites of Interest

Remember that websites are not as permanent as books and journals, so I cannot guarantee that each of the following websites still exists at the moment that you are reading this:

1. **http://www.census.gov/ipc/www/idbpyr.html**

   At this website the U.S. Bureau of the Census has a program to draw an age/sex pyramid "on the fly" at your request for any country in the world, and they also have a dynamic pyramid which graphically changes the age/sex structure over time for any given country.

2. **http://www.census.gov/ipc/www/idbacc.html**

   If you want the actual numbers of people by age and sex for a specific country for which a pyramid was drawn above, you must go to this address to access the information.

3. **http://www.census.gov/population/www/projections/popproj.html**

   The U.S. Bureau of the Census allows you to freely download the file containing the full set of documentation and tables for their detailed projections of the United States by age sex and race/ethnicity from 1995–2050. More limited data are available for Canada (age and sex for 1996–2106) at **www.statcan.ca:80/english/Pgdb/People/Population/demo23a.htm**; and data for Mexico by age and sex for 1990–1995 are available at: **www.inegi.gob.mx/homeing/estadistica/sociodem/poblacion/pob-5.html**

4. **http://members.aol.com/genxcoal/genxcoal.htm**

   Are you tired of reading about the impact of baby boomers on American society? People of the Baby Bust, known as Generation Xers, have created a literature of their own which can be accessed at this website.

5. **http://www.rand.org/publications/CF/CF124/CF124.chap6.html**

   In 1995 the RAND Corporation in Santa Monica held a conference dealing with population issues in Russia. One of the contributions was this paper by Sergei A. Vassin titled "The Determinants and Implications of an Aging Population in Russia." It includes text and graphics covering changes in the age and sex structure of the Russian population in the past, with projections to the year 2015.

CHAPTER 9
Population Aging

Most old people are pretty much alike—they have no interest in sex, they are miserable most of the time, they cannot work as effectively as younger people, and they have incomes below the poverty level. Now, if you believe all that, I have some swampland in Florida you might want to buy! The truth is that older people are not all pretty much alike. On the contrary, exposure to several decades of history and countless different patterns of social interaction tends to make older people even less alike than people at younger ages. Thus we find a wide range of interest in sex among older people, a wide range of life satisfaction (from miserable to extremely happy), a wide range of worker effectiveness, and a wide range of income. This means that you stereotype older people at your own risk; indeed, at an increasing risk, because the older population of the world is growing at a rapid rate, especially in the industrialized nations.

The decline in mortality throughout the world has even broader implications than more babies surviving to adulthood; it means more adults surviving to old age as well. As fertility rates also decline, then the *proportion* of the population that is older will also be increasing, not just the *number* of elderly, as you will recall from the previous chapter. Migration, especially urbanization, also influences the proportion of the population that is older because younger people tend to migrate more than do older people. Because older people now represent the fastest growing segment of the world's population, it is of some importance to understand the causes and consequences of the growth in their number and the increase in their percent in the total population. I begin by defining what we mean by the "older" population—both in biological and in social scientific terms. Then I examine the growth and distribution of the older population, looking at the influence of mortality, fertility, and migration on these trends. Next, I review the basic sociodemographic characteristics of the elderly to see how life course differed in important respects from the younger population. In the process of seeing what older people are like, I weave in a discussion of how the aging process will likely influence the future course of society.

What Is Old?

Age as we usually think of it is a social construct—something we talk about, define, and redefine on the basis of social categories, not purely biological ones. A good way to visualize this concept is to contemplate Satchel Paige's famous question: "How old would you be, if you didn't know how old you was?" which illustrates the point that age takes on its meaning from our interaction with other people in the social world. If we are defined by others as being old, we may be treated like an old person regardless of our own feelings about whether or not we are, in fact, old. We humans depend heavily upon visual clues to assess the age of other people and we learn quickly that there are certain kinds of outward physical changes that are typically associated with aging—the graying of hair, the wrinkling of the skin, the decline of muscle tone, and changing shape caused by the redistribution of fat. These are taken as signs of physical decline and it is fair to say that most of us are, at best, ambivalent about the aging process—an attitude that is itself probably as old as human society (Warren 1998).

Especially among women, the physical symptoms of aging may be associated with the end of reproduction. Indeed, humans are among the few species that have a substantial post-reproductive existence (Albert and Cattell 1994). Still, these changes do not follow a rigid time schedule, and some can be successfully hidden or disguised (think of dyed hair, hair implants, face lifts, and liposuction). In general, then, there is no inherent chronological threshold to old age; however, in the United States and much of the world, old age has come to be defined as beginning at age 65. The number 65 assumes its almost mystical quality in the United States because it is the age at which important government-funded benefits such as social security and Medicare have been fully available.

In 1935, when the present social security benefits were designed in the United States, the eligibility age was set at 65, "more because of custom than deliberate design. That age had become the normal retirement age under the few American pension plans then in existence and under the social insurance system in Germany" (Viscusi 1979:96). The Older Americans Act passed by the U.S. Congress in 1965 did provide benefits in some programs for people aged 60 and over, but age 65 is a milestone firmly entrenched in the Western world, and so I too will adopt that definition in this chapter. However, remember that the number is arbitrary, and people obviously will not fit neatly into any senior citizen mold on reaching their 65th birthday. In fact, most people in low mortality societies do not think of themselves as being old until well beyond that age, whereas most people in high mortality societies probably would have thought of themselves as old well before age 65.

Our demographic interest in the older population comes from the fact that as the number and proportion of older people increase (as the *population* ages), changes are wrought in the organization of society. These changes are the result of the process of *individual* aging—people change biologically with age and societies react differently to older than to younger people, producing the social changes that we see accompanying population aging. I have outlined these concepts in Figure 9.1, where I point out that the exact nature of the social changes will depend partly on the social context, including aspects of society such as cohort-specific historical

Figure 9.1   Population Aging and Individual Aging Combine to Produce Change in Society

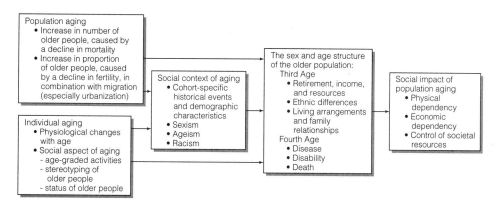

events that have affected different cohorts through the life course, but also societal levels of sexism, ageism, and racism. The impact of population aging will also differ according to the proportion of people who are young-old (the "third" age of life—in which the impact of aging on society is more social), compared with the proportion who are old-old (the "fourth" age of life—in which society must cope more with the individual biological aspects of aging). My goal in this chapter is to discuss the concepts shown in Figure 9.1, working from left to right in that figure. Let us begin by examining the dimensions of population aging.

Population Aging

How Many Older People Are There?

As of the year 2000 there will be 416 million people in the world aged 65 and older, according to United Nations estimates (United Nations 1997). Indeed, if they all lived together under one flag, they would represent the third largest nation in the world. As a fraction of the total world population, the elderly account for 7 percent, but bear in mind that this percentage varies considerably from one part of the world to another. For example, in the year 2000 only 19 percent of the total population of the world will be living in more developed nations, yet 41 percent of the world's population aged 65 or older will be there, accounting for 14 percent of the population of developing countries. Still, that means that more than one half of all people in the world aged 65 and older will be living in the developing countries, even though representing only 5 percent of the population of developing nations. You can see these data in Table 9.1.

The data in Table 9.1 show that between the years 2000 and 2010 the average annual rate of growth of the population aged 65 and older in the world is projected to be 1.99 percent, compared with 1.23 percent per year for the total population. That decade, however, represents the lull before the storm because the huge batch of babies born after World War II will move into the older ages in the following decade, between 2010 and 2020. During that decade, the population aged 65 and older will be increasing by 3.02 percent per year, while the total world population is projected to be growing by only 1.07 percent. Furthermore, ever since 1970 the older population has been growing more quickly in less developed countries than in the more developed countries, even though the percent of the population that is 65 and older is still lower in those parts of the world. You can see in Table 9.1 that in 1970 the older population was almost evenly divided between more and less developed nations, but by the year 2020 we can expect there to be more than twice as many older people in the less developed nations as in the more developed.

If we compare the United States and Canada with Mexico, we can see that, although the number of older Mexicans currently is only about one tenth the number of older people in the United States, the rate of growth of the older Mexican population has been much higher than in the United States since 1970. Even between 2010 and 2020, when the U.S. and Canadian Baby Boom generations will be ballooning the older ages (a much-dreaded event in the United

States, as I discuss in the accompanying essay in this chapter), the population of older Mexicans will actually be growing much more quickly.

Not only has the number of older people been increasing everywhere in the world, but the percentage of the population that is 65 and older has also increased, as you can see in Figure 9.2, which graphs those data from Table 9.1. Near the middle in that graph are the bars that show the data for the world as a whole. In 1980 there was a clear increase in the percent aged 65 and older and it is projected to climb in the foreseeable future. In the back rows of the graph are the more developed countries generally, along with data for the United States and Canada more specifically. Since the end of World War II, but beginning before that, there has been an uninterrupted increase in the fraction of the population that is aged 65 and older. However, in the less developed nations, typified in Figure 9.2 by Mexico, the percent aged 65 and older actually went down between 1950 and 1970 before starting a sustained rise in 1980. In the period right after World War II, the death rate was declining but the birth rate was not and, as you remember from Chapter 4, the declines in the death rate tend to favor the young, so that had the early effect of increasing the youthful population at the proportionate expense of the elderly. As birth rates have begun to decline in many less developed nations, including Mexico, that has led to an increase in the fraction of the population that is older. I will return to that theme in a minute.

It is worth reminding you that the most populous countries of the world with respect to total numbers are also those with the greatest number of older people, almost regardless of the percent of the population that is 65 and older. In Table 9.2 you can see that China has the greatest number of older people as of the year 2000, with more than 87 million, followed in order by India and the United States—the same countries that are the top three in terms of overall population size. Below that, however, all of the remaining top ten are developed nations except for Indonesia, which is the fourth most populous nation, but is ninth in terms of the older population. The bottom panel of Table 9.2 shows the top ten countries with respect to the percentage of the population that is 65 and older and you can see that Italy heads this list with more than 18 percent of the total population. Every one of the countries on this list except Japan is in Europe, which, not coincidentally, is where the demographic transition occurred first and where fertility and mortality levels are now lowest. Let us examine the components of the mortality, fertility, and migration transitions to see how they contribute to the pattern of population aging.

The Effect of Declining Mortality

As death rates drop and life expectancy increases, the older population increases partly because life expectancy goes up somewhat at the older ages and so people live longer when old. However, the most important impact on aging is a result of the fact that lower mortality increases the probability of surviving to old age. These relationships are shown for you in Table 9.3, using data for the United States. In 1900 the life expectancy in the United States was lower than in almost every country in the world today. At birth a male could expect to live an average of 46.3 years, compared to 48.3 for females. These life expectancies were about the same as England

Table 9.1 The Number of Older People Has Been Growing More Rapidly Than the Rest of the Population All Over the World

Region	1950	1960	1970	1980	1990	2000	2010	2020	2030
WORLD									
Total population	2,523,878	3,026,541	3,701,909	4,447,374	5,282,306	6,091,351	6,890,775	7,671,924	8,371,602
Population aged 65+	130,670	161,084	202,185	262,878	325,774	415,704	507,221	686,093	934,336
%65+	5.18%	5.32%	5.46%	5.91%	6.17%	6.82%	7.36%	8.94%	11.16%
r(total)		1.82%	2.01%	1.83%	1.72%	1.43%	1.23%	1.07%	0.87%
r(65+)		2.09%	2.27%	2.63%	2.15%	2.44%	1.99%	3.02%	3.09%
MORE DEVELOPED									
Total population	812,687	915,841	1,007,667	1,082,859	1,148,119	1,186,990	1,206,375	1,218,526	1,212,147
Population aged 65+	64,052	78,315	99,646	125,988	143,005	168,457	186,670	224,773	259,435
%65+	7.88%	8.55%	9.89%	11.63%	12.46%	14.19%	15.47%	18.45%	21.40%
r(total)		1.19%	0.96%	0.72%	0.59%	0.33%	0.16%	0.10%	-0.05%
r(65+)		2.01%	2.41%	2.35%	1.27%	1.64%	1.03%	1.86%	1.43%
LESS DEVELOPED									
Total population	1,711,191	2,110,700	2,694,242	3,364,515	4,134,187	4,904,360	5,684,400	6,453,398	7,159,455
Population aged 65+	66,618	82,769	102,539	136,890	182,066	247,247	320,550	461,321	669,901
%65+	3.89%	3.92%	3.81%	4.07%	4.40%	5.04%	5.64%	7.15%	9.36%
r(total)		2.10%	2.44%	2.22%	2.06%	1.71%	1.48%	1.27%	1.04%
r(65+)		2.17%	2.14%	2.89%	2.85%	3.06%	2.60%	3.64%	3.73%

UNITED STATES

Total population	157,813	186,158	210,111	230,406	254,106	277,825	298,885	322,280	337,277
Population aged 65+	13,043	17,101	21,666	25,793	31,477	34,529	38,822	52,439	67,610
%65+	8.26%	9.19%	10.31%	11.19%	12.39%	12.43%	12.99%	16.27%	20.05%
r(total)		1.65%	1.21%	0.92%	0.98%	0.89%	0.73%	0.75%	0.45%
r(65+)		2.71%	2.37%	1.74%	1.99%	0.93%	1.17%	3.01%	2.54%

CANADA

Total population	13,737	17,909	21,324	24,593	27,791	30,679	33,010	35,338	36,633
Population aged 65+	1,054	1,343	1,676	2,309	3,118	3,866	4,634	6,406	8,383
%65+	7.67%	7.50%	7.86%	9.39%	11.22%	12.60%	14.04%	18.13%	22.88%
r(total)		2.65%	1.75%	1.43%	1.22%	0.99%	0.73%	0.68%	0.36%
r(65+)		2.42%	2.22%	3.20%	3.00%	2.15%	1.81%	3.24%	2.69%

MEXICO

Total population	27,737	36,945	50,596	67,570	83,226	98,881	112,891	124,976	136,240
Population aged 65+	1,232	1,698	2,156	2,564	3,295	4,671	6,686	9,866	14,674
%65+	4.44%	4.60%	4.26%	3.79%	3.96%	4.72%	5.92%	7.89%	10.77%
r(total)		2.87%	3.14%	2.89%	2.08%	1.72%	1.33%	1.02%	0.86%
r(65+)		3.21%	2.39%	1.73%	2.51%	3.49%	3.59%	3.89%	3.97%

Note: r(total) = average annual rate of growth of the total population. r(65+) = average annual rate of growth of the older population.

Source: United Nations, 1997, The Sex and Age Distribution of the World Populations: The 1996 Revisions (New York: United Nations); Population figures are in thousands.

Figure 9.2 The Percent of the Population that Is 65 and Older Is Increasing All Over the Globe

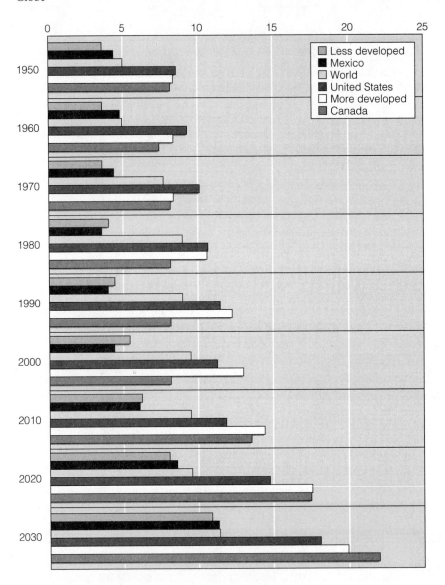

Sources: Adapted from United Nations, 1997, The Sex and Age Distribution of the World Populations: The 1996 Revision (New York: United Nations), data for 2000 through 2030 refer to the medium variant projections; note that the United Nations projections are not necessarily identical to projections made for the respective countries by the U.S. Bureau of the Census, Statistics Canada, or INEGI.

Table 9.2   The Top Ten Countries in Terms of the Number of People Aged 65 and Older, and in Terms of the Percentage Aged 65 and Older as of the Year 2000

Top Ten Countries in Terms of Number of People Aged 65 and Older			
Rank	Country	Number 65+	Percent 65+
1	China	87,457,789	6.96%
2	India	47,522,109	4.67%
3	United States	33,128,874	12.05%
4	Japan	21,467,582	16.98%
5	Russia	18,352,063	12.58%
6	Germany	13,518,659	16.47%
7	Italy	10,313,388	18.19%
8	France	9,548,868	16.15%
9	Indonesia	9,370,691	4.27%
10	United Kingdom	9,314,387	16.08%

Top Ten Countries in Terms of Percent Aged 65 and Older			
Rank	Country	Number 65+	Percent 65+
1	Italy	10,313,388	18.19%
2	Sweden	1,539,592	17.22%
3	Greece	1,845,550	17.17%
4	Belgium	1,742,306	17.11%
5	Japan	21,467,582	16.98%
6	Spain	6,604,910	16.85%
7	Bulgaria	1,348,358	16.53%
8	Germany	13,518,659	16.47%
9	France	9,548,868	16.15%
10	United Kingdom	9,314,387	16.08%

Source: Adapted from U.S. Bureau of the Census International Database

at the time, but lower than most Scandinavian nations. At this level of mortality, a male baby had only a 36 percent chance of surviving from birth to age 65, and a female baby had a 41 percent chance of living that long. For those who reached 65, the pattern of mortality in 1900 produced a life expectancy at age 65 of 11.5 additional years for males, and 12.2 additional years for females. Scarcely one fourth of males aged 65 and only about one third of females aged 65 could expect to still be alive at age 80.

Significant improvements in mortality occurred in the United States and other developed countries between 1900 and 1950, as you can see in Table 9.3, and the improvements have continued, albeit at a slower pace, since then. By 1995, females in the United States had a life expectancy at birth of 78.9 years—30.6 years more

Table 9.3 Declining Mortality Increases the Likelihood
of Survival to and within Old Age, United States 1900–95

Year	Life Expectancy at Birth:		Proportion Surviving from Birth to Age 65:		Life Expectancy at Age 65:		Proportion Alive at Age 65 who Survive to Age 80:	
	Males	Females	Males	Females	Males	Females	Males	Females
1900	46.3	48.3	0.36	0.41	11.5	12.2	0.27	0.33
1950	65.6	71.1	0.66	0.82	12.8	15.0	0.38	0.47
1995	72.5	78.9	0.75	0.86	15.6	18.9	0.53	0.67
1900–95	+26.2	+30.6	+108%	+110%	+4.1	+6.7	+96%	+103%

Source: National Center for Health Statistics, 1997, Health, United States, 1996–97 (Hyattsville, MD: Public Health Service), Table 29, and author's own calculations of proportions surviving based on the given life expectancies.

than at the turn of this century. This has more than doubled the proportion of female babies surviving from birth to age 65 and has been a major reason for the increase in the number of older persons in the United States. It is now commonplace for people to reach old age and we take it nearly for granted, as I discussed in Chapter 4. On reaching age 65, females can now expect to live an additional 18.9 years on average—an improvement of 6.7 years more beyond age 65 than was true in 1900. This translates into a doubling of the proportion of women age 65 who will still be alive at age 80. Indeed, in 1995 in the United States, two thirds of the women alive at age 65 will still be alive at age 80. The situation is not quite so favorable for males, but it has been improving for them, too.

The Effect of Declining Fertility

Although declining mortality will always lead to an increase in the number of older people, the percent age 65 and older will only increase noticeably if fertility declines, as it always has historically in the context of the demographic transition. We can use the stable population models that I discussed in Chapter 8 to illustrate this. In Table 9.4 I have calculated the percent of the population that would be aged 65 and older if a population maintained over time the various combinations of mortality and fertility that I have shown in the table. For example, a country whose life expectancy was only 30 years would have 3.9 percent of the population aged 65 and older if the total fertility rate (TFR) were 5 children per woman, and it would drop to 2.8 percent if the TFR went up to 6 (note that at a TFR of 4 or below, the population would be depopulating, so the percent aged 65 and older would be temporarily high, until everybody died off).

Let us stay focused for the moment on the total fertility rate of 5 children. You can see that as life expectancy increases from 30 years to 60 years, the percent aged 65 and older actually goes down slightly. Remember that this is exactly what we ob-

Table 9.4   The Percent of the Population that Is 65
and Older is Determined More by Fertility Than by Mortality

Life Expectancy at Birth	Total Fertility Rate (TFR)				
	2	3	4	5	6
30	a	a	a	3.9	2.8
40	a	a	5.6	3.8	2.7
50	a	8.8	5.5	3.7	2.6
60	15.0	8.8	5.4	3.6	2.5
70	16.5	9.2	5.7	3.7	2.6
75	18.0	9.9	6.1	4.0	2.8

[a] No calculation made because this represents a situation of depopulation.

Source: Data are based on Coale-Demeny "West" Stable Population Models.

served for Mexico in Table 9.1. Mortality was declining in Mexico between 1950 and 1970, but fertility had not begun to decline and so the older population was declining as a percent of the total population. However, you can also see in Table 9.4 that, as mortality continues to decline, eventually it gets low enough that the percent representing the elderly begins to increase even if fertility does not change. Of course, a population with a life expectancy of 75 years and a total fertility rate of 5 children would be doubling in size every 20 years, and the number of older people eventually would be very large, even if they represented only 4 percent of the total.

If you look at any given level of life expectancy, you can see clearly that at each lower level of fertility, the percent of the population that is elderly is higher. Now, finally, if you go down the diagonal of this table from the top right (low life expectancy and high fertility) to the bottom left (high life expectancy and low fertility), you can trace the typical path of the percent elderly as a country passes through the demographic transition.

The Effect of Migration

The impact of migration on the growth of the older population is, as always, more complex than the effect of either mortality or fertility. Younger people are more likely to migrate than the elderly, but there is no biological limitation to migration and so we find that, although in general migration tends to leave the elderly behind, there are other times when the elderly move disproportionately to specific areas (such as Florida in the United States) in search of "amenity-rich communities with sunnier, warmer and recreationally more enjoyable environments" (Rogers 1992:3). Haas and Serow (1993) have pointed out that this is a type of pull migration in which people use their information sources to decide where to migrate, in a manner similar to patterns of international migration, especially from Mexico to the United States, which I discussed in Chapter 7. And, in a fashion that is also similar to Mexicans migrating to the United States, many older migrants are seasonal—perhaps we could call them "trans-communal"—having incorporated the movement between two communities (such as one in the Sun Belt and another in the Snow Belt) as a regular part of their life (Hogan and Steinnes 1998).

Migration affects population aging most by what older people do not do—they do not migrate very much. Throughout the world, older people tend to age in place. As a consequence, the process of urbanization, which occurs in concert with the demographic transition, typically leaves the elderly abandoned in the countryside by their children who migrate to the cities to work. In the developed nations, those children who migrated to the cities will age in place in the city and may find themselves abandoned by their own children, who are more likely to move out to the suburbs. These patterns affect the social relations of older persons, as I discuss later. For now, it is enough to remember that outmigration tends to increase the percent of the population that is elderly in a region, whereas in-migration has the opposite effect, except in those relatively rare cases where the in-migrants are themselves older persons.

Individual Aging

Population aging would not be of any particular consequence if there were no physical changes taking place in the aging body to which people and society in general can or should respond. So, to understand population aging, we must also understand what those physical changes are and how people react to them.

Individual aging has both a biological and a social component. The biological component refers to those physical changes that are the defining aspect of the aging process—the changes that tell us and others that we are, in fact, aging. The social component refers to the way that social systems react to these physical symptoms of growing older.

Biological Aspects of Aging

Could you live forever if you were able to avoid fatal accidents and fatal communicable diseases, and if you had as healthy a lifestyle as possible? The answer seems to be no. Biologists (Carnes and Olshansky 1993) suggest to us that as we move past the reproductive years (past our biological "usefulness"), we undergo a set of concurrent processes know as **senescence**: a decline in physical viability accompanied by a rise in vulnerability to disease (Spence 1989). Some important generalizations about physical aging are: (1) the physiological changes that take place tend to be steady but gradual through adulthood into old age; (2) the more complex the bodily function, the more rapid is its rate of decline; (3) individuals age at different rates, and different tissues and systems within one person also may age at different rates; (4) aging brings with it a lowered ability to respond to stress; and (5) aging brings a diminished ability to resist disease (Weg 1975).

There are several theories in vogue as to why people become susceptible to disease and death as age increases. These can be roughly divided into theories of "wear and tear" and "planned obsolescence." Wear and tear is one of the most popularly appealing theories of aging and likens humans to machines that eventually wear out due to the stresses and strains of constant use. But which biological mechanisms might actually account for the wearing out? One possibility is that "as cells continue to function, random error may occur somewhere in the process of the synthesis of new proteins" (Rockstein and Sussman 1979:42). Protein synthesis involves a long and complex series of events, beginning with the DNA in the nucleus and ending with the production of new proteins. At several steps in this delicate process, it seems possible that molecular errors can occur that lead to irreversible damage (and thus aging) of a cell. Of special concern is the possibility that errors may occur in the body's immune system, so that the body begins to attack its own normal cells rather than just foreign invaders. This process is called autoimmunity. Alternatively, the immune system may lose its ability to attack the outside cells, leading to the functional equivalent of AIDS, in which the body no longer can fight off disease.

The planned obsolescence theories revolve around the idea that each of us has a built-in biological time clock that ticks for a predetermined length of time and then is still. It essentially proposes that you will die "when your number is up," because each cell in your body will regenerate only a certain number of times and no more (Hayflick 1979). Which of these theories makes most sense? Good question. If you had a solid answer, you could bottle it and retire. Short of that, let me remind you that current evidence points to two basic conclusions: (1) aging is much more complex than we have previously assumed, and each of the theories may fill in part of the puzzle; and (2) we have not yet discovered the basic, underlying mechanism of aging that (if it exists) would explain everything. The programmed time clock theory could explain why animal species each have a different lifespan, whereas the various aspects of wear and tear seem better able to explain why members of the same species show so much variability in the actual aging process. Olshansky and Carnes (1994) have offered their opinion that "Although there is probably not a genetic program for death, the basic biology of our species, shaped by the forces of evolution acting on us since our inception, places inherent limits on human longevity" (p. 76).

As we discuss human aging it is important not to confuse senescence, the biological process of aging, with **senility.** Senility (or senile dementia, as it is known formally) has often been used as a derogatory term applied to forgetfulness or seemingly erratic behavior on the part of older people. We now know, however, that senility is not a normal part of aging. Instead it represents the consequences of several different organic brain disorders, especially **Alzheimer's disease.** In recent years a great deal of federally funded research has been aimed at the identification of exact disease mechanisms so that a treatment or cure can be found.

The physical changes that actually take place as we age are often exaggerated in our minds when we are younger. The gradualness with which changes occur usually provides ample time to learn to cope with them, and actual changes may well occur later or be different from what we suspected when we were younger. For example, one of the most blatant stereotypes of aging has to do with sexual functioning in old age. As a matter of fact, sexual activity tends to remain fairly strong well into old age, at least until death removes a person's partner (Atchely 1996). The frequency and enjoyment of sexual relations in old age seem to be more closely related to individual preference than to age itself. Thus people who enjoyed sex and were sexually active when younger will likely remain that way when older. Though your physical capacity to engage in intercourse may diminish with age, your level of sexual activity will likely be influenced more by social factors than by biology.

Social Aspects of Aging

The social world of the elderly is different from that of younger people because virtually every society has some built-in system of age stratification—the assignment of social roles and social status on the basis of age. As people age, their sets of social obligations and expectations change, and certain kinds of behavior are deemed appropriate for some ages but not for others. I discussed this phenomenon in general terms in the previous chapter, but here I will highlight the importance of age stratification for the older population.

One aspect of age stratification that crops up especially in Western societies is the stereotyping of older persons. Older people are often seen as being politically and socially more conservative than younger people, intolerant of others (especially younger people), and more prone to despondency and impairment. So pervasive are they that older people themselves have these stereotypes of other older persons (Hummert, et al. 1995). Stereotypes are most likely to arise when groups of people do not share a great deal of their life in common, and age stratification encourages this generational separation in modern societies.

Age stratification and generational separation have also been implicated as determinants of the status of older people in society. Indicators of status such as access to economic resources, policy-making influence, and breadth of social relations all suggest that the status of the elderly in Western nations underwent a decline for several decades until recently, when it seems to have risen once again. How and why have these changes occurred? The most widely accepted explanation of these events is based on the joint processes of the demographic transition and modernization.

Modernization and the Status of the Elderly   Most historians accept the idea that the status of older persons is relatively high in premodern, agrarian societies, even though there may be considerable variety in exactly how high that status is (see, for example, Kertzer 1995). However, things began to change as societies modernized and people began to live longer. The flip side of the increasing survivability of people to old age is the increasing likelihood that a young or middle-aged adult will have surviving parents. For example, under a constant mortality regime of 40 years of life expectancy, a person aged 30 has less than one chance in four of having both parents still alive, and there is about one chance in two that one parent will be still alive. A high level of mortality like that increases the odds that a younger person will be able to inherit the family farm or business, or will be able to move into some other position in society being vacated by the relatively early deaths of other people. However, at a life expectancy of 70 years, nearly two out of every three adults aged 30 can expect to have both parents still alive, and nearly 8 in 10 can expect to have at least one parent still alive. This pattern essentially clogs up familial and societal mobility, and makes the younger population somewhat more redundant. The problem is alleviated as the birth rate drops, but in the middle of the demographic transition, when mortality is declining more rapidly than fertility, the pressure on population has historically produced a migration out of the less developed, more agricultural areas and into urban areas, or into foreign countries, where younger people may have economic opportunities that would not otherwise have been available to them.

These ideas are summarized in Kingsley Davis's theory of demographic change and response (discussed in Chapter 3), although his focus is not on the older population. The conceptual tie to the elderly has been provided by Donald Cowgill (1979), and his framework is displayed in Figure 9.3. In Cowgill's model, four factors involved in modernization—health technology, economic technology, urbanization, and education—combine to lower the status of the elderly.

The health technology accompanying modernization increases longevity, as you well know. That greater longevity in turn creates intergenerational competition for jobs, because people are no longer so likely to die and create room for the young to enter and advance within the labor market. This competition has led to the phenomenon we call retirement. Work itself is a valued activity in society, since work produces income; because income is a major source of status, retirement leads inexorably to decreased status.

The economic technology that accompanies modernization is closely associated with the creation of new, especially urban, occupations. Because the young are more

Figure 9.3   How Is the Status of the Elderly Influenced by Modernization?

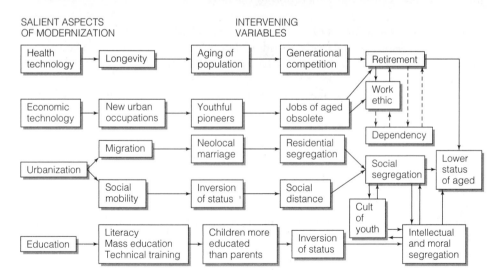

Note: Donald Cowgill's framework is an attempt to explain the lower status of the elderly in modernized societies as the result of the combined influence of changes in health technology, economic technology, urbanization, and education.
Source: Donald Cowgill, 1979, "Aging and modernization: A revision of the theory," in J. Gubrium (ed.), *Late Life: Communities and Environmental Policy* (Springfield, Illinois: Charles C. Thomas, Publisher), Figure 5. Reprinted with permission.

likely to migrate to cities than are the old, and because urban occupations tend to pay more than those in rural areas (as I will discuss in Chapter 11), the result is "an inversion of status, with children achieving higher status than their parents instead of merely moving up to the status of their parents as in most pre-modern societies" (Cowgill 1979:59).

The third factor that Cowgill mentions as leading to a decline in the status of the elderly is urbanization. In addition to its relationship to mobility and the inversion of status, as I just mentioned, urbanization produces a segregation of residence of the young and the old. In turn, that produces a greater social distance between generations, relegating aging parents to a more peripheral role in the lives of children. The greater physical distance between generations thus accelerates the already inverted social distance and further depresses the status of the elderly.

The process of modernization is closely bound up with an increase in formal education. In preliterate societies education is based especially on the recitation of life's experiences by the older people. The school of hard knocks prevails, and experience is the best teacher—a feature of life that enhances the status of the elderly. Modernization, though, tends to bring with it institutionalized, formal, mass education, and the content is more technical than experiential. The main targets of such educational programs are always the young, and this leads to the young being better educated than their parents. This further exacerbates the status of the older person because it leads to a greater intellectual and moral (or value) distance between the generations. Education brings with it (indeed is part of) changes in the value system of society, and the constant upward spiral of education has led to shifting values and intellectual levels from one generation to the next.

Rising Status in Later Stages of Development   Cowgill's theory of the loss of status for the elderly leaves them largely abandoned in the countryside (this reached a peak in the United States before World War II). However, eventually the countryside empties out—at least that is what has happened in the now developed countries—and the vast majority of the population becomes urban (this has happened in the United States and other industrialized nations in the past few decades). The people who abandoned their parents by moving to the city have now grown old in the city, and this has reduced the likelihood that they too will be abandoned by their children. More importantly, growing old in the city has given them broader access to the social, economic, and political resources, and makes it less likely that they will be economically deprived and abandoned, even if their children do move somewhere else. Indeed, the end of urbanization and a slowdown in the rate of growth of development affords the elderly the opportunity to regain control of resources, which implies a rise in their status.

The older households in the United States are those in which net worth is the greatest. More specifically, we can see who controls the United States by looking at the Forbes Magazine list of 400 richest Americans to see what their ages are. Figure 9.4 shows the data for 1994, but the pattern has been the same for a number of years now. It is very clear that the elderly are far more likely to be very wealthy than are younger people. Of the wealthiest 400 (actually there were 406 on the list because of ties), 206, or virtually one half, were ages 65 or older, whereas the elderly comprise only about 12 percent of the population, as you will recall from

Figure 9.4   The Wealthiest 400 People in the United States Are Disproportionately Aged 65 or Older

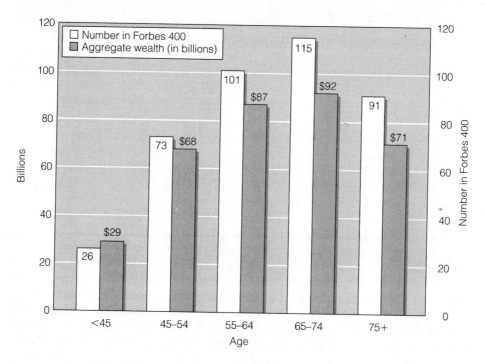

Source: Adapted from data in Forbes Magazine, October 17, 1994, pp: 326–340.

Table 9.1. The peak age for wealth is 65 to 74, and there are not only more wealthy people at that age, they also command the greatest amount of wealth as a group. However, per person wealth is highest for the youngest age group (who average $1.1 billion), and it goes down with each older group (people aged 75 and older average "only" $776 million).

The wealthiest person in the United States in 1994 was 38-year-old William Gates, the founder of Microsoft—a nearly $10 billion fortune that he and associates built. However, nearly six in ten of the richest people in the under-45 age group inherited their wealth, which underscores the resources of the elderly—it is they who pass the wealth on. The improvement of fortunes for older people relative to the middle-aged population in the United States has been abetted by the lack of growth in real incomes among younger adults over the past three decades (as I will discuss in the next chapter), which has allowed the elderly (who experienced rapid increases in income when they were working) to increase the economic distance between themselves and the middle-aged. As I discussed in the previous chapter, the cohort of people now aged 65 and older in the United States may be better off than any that came before or will come after.

As the United States and other industrialized countries move into the final stages of the demographic transition, the data suggest that the elderly have regained much, if not all, of the status that they may have lost during the initial phases of modernization. At the same time, less developed nations are now moving into those middle stages of the demographic transition, wherein a reduction in status of the elderly seems almost inevitable.

The Declining Status of the Elderly in Developing Countries   It is consistent with Cowgill's modernization theory that as less developed societies go through the process of modernization the status of their older population should drop, much as it did in the now developed regions of the world. So it is that in country after country the complaint is being heard that the elderly, increasing numerically with each new cohort, are being left behind by the younger generation and forgotten by the rest of society. As we move into the 21st century, nearly two out of every three older people in the world will be in the less developed nations, and their plight will be a significant factor in world affairs.

An anecdotal, but perhaps significant, bit of evidence about the status of the elderly in developing countries emerged in the People's Republic of China in 1997 when the government passed a law protecting the rights and interests of the elderly. In a nation that made famous the concept of *filial piety*—respect of children for the elder parents—a law has now been required to "forbid discrimination against the aged by insulting, mistreating, or forsaking them" and the law calls for appropriate measures to be taken against anyone committing such abuse (Global Aging Report 1997). Bolivia also put in place in 1997 a law that guarantees older people legal and police protection, among other things. The necessity for such guarantees in Latin America is illustrated by a study in Costa Rica indicating that older family members were more likely to be abused if they were members of a larger, rather than a smaller household—there was no safety in numbers. A report from Ethiopia indicated that the widespread poverty in the capital city of Addis Ababa was particularly brutal for older people, some of whom were living in cardboard boxes with a plastic cover to keep out the rain (Global Aging Report 1997). From Kenya has come the opinion that "due to modern changes, the traditional family commitment of caring for the aged is slowly dying" (Oriol 1982:34–35). In another African nation, Zimbabwe, the elderly have been described as being at the edge of destitution (Adamchak, et al. 1991; Hampson 1983) and the conclusion drawn that "within a society that values and respects the elderly, a significant proportion feel marginalised and bereft of support" (Hampson 1989:208).

As modernization occurs, the process of urbanization, combined with education and a redefinition of the labor force, pushes the elderly aside. At the same time the older people themselves begin to redefine their own expectations, anticipating less from their children in the way of old-age support (Coombs and Sun 1981). This is the crucial moment of change in the obligations and expectations that the old and the young have toward each other that may be the tipping point in the generational flow of wealth, in the perceived value of children, and ultimately in the birth rate. Over the long run, of course, we would expect the status of the elderly to rise again in the less developed nations, as their achievement of higher standards of living and lower birth rates elevates the demographic and social profiles of the older population, as has happened in the more developed nations of the world.

The Social Context of Aging and the Life Course

The fall and rise in the status of the elderly reminds us that the impact of aging on society is defined in large part by the social context in which it takes place. As historical forces shape and influence the status of the elderly in a society, those same forces are shaping the character of the older people themselves. This is the basis of the concept of cohort flow, which I discussed in Chapter 8. Thus in an era of rapid social change, such as we have experienced throughout the last century, we should expect that each new generation of people moving into the old ages will be substantially different from the previous one. This has tremendous implications for society. Younger people may be more likely to engage in innovative behavior than are older people, but those innovations are taken with them into old age, transforming every age group as they proceed through the life course. Some of the innovations are responses to cohort-specific historical events.

Researchers at the University of Michigan have found that people remember as most important in their lives those events that took place in late adolescence and early adulthood and they carry those "generational memories" (and their responses to them) with them as they age (Schuman and Scott 1990). This is not to be confused, however, with the idea that somehow older people *dwell* on the past. Older people remember the past and are influenced by it, but in the Berkeley Older Generation Study the period of life mentioned most often by older people as bringing them the most satisfaction was "right now" (Field 1997).

It is these cohort-specific reactions to innovation and change that probably represent the major force of social change over time. Some events, such as current economic and political crises, may be reacted to in very similar ways by people of all ages, but deeper social issues, such as sexism, racism, and even ageism, are more likely to be challenged and changed by younger people than by the elderly. However, those attitudes do not melt away with age and so we can expect that higher levels of tolerance among younger people today will evolve into greater tolerance at older ages in the future.

Similarly, the high level of political advocacy of youth in the 1960s will almost certainly be translated into greater political advocacy among the elderly when those people reach old age. Already people aged 65 and older in the United States are the most likely to be registered to vote and most likely to actually vote in national elections (U.S. Bureau of the Census 1997b). As the current generation of younger people grows older, we should find the older population becoming more tolerant of a wide range of lifestyles and more understanding of the problems and issues facing older people.

Another important change that is occurring within the older population all over the world is that it is becoming increasingly better educated. Of course, if we look at a cross section of society, we see that older people have lower levels of education than do younger adults. For example, in 1996, only 59 percent of people aged 75 and older in the United States were high school graduates, compared with 87 percent of those aged 25 to 44 (Day and Curry 1997). Similarly, only 13 percent of the elderly (75 and older) were college graduates, compared with 27 percent of those aged 25 to 44. Despite this cross-sectional, or period, difference, the cohort of people now in the older ages is the best educated ever to have occupied those ages. In fact, in just the

20-year span between 1970 and 1990, the proportion of people aged 65 and older in the United States who were high school graduates rose from 28 percent to 55 percent (U.S. Bureau of the Census 1992). This rise was almost entirely a result of the higher educational levels of the people moving into the older years, not the recent, but still sparse, attempts to bring older people back into the educational system.

On average, older people are not less well educated because they are less competent, less able to achieve higher levels of education, or even less ambitious. Rather, it is because the world has been experiencing an upward spiral of literacy and education for the last few hundred years, with the result that each generation tends to be better educated than the previous one. For example, in Italy in 1881, less than 20 percent of the women aged 71 and older were literate, compared with almost 40 percent literacy among women aged 21 to 30. Comparisons were similar for men, and they were also similar for other European nations in the 19th century (Cipolla 1969).

Because educational programs will almost certainly continue to be oriented largely to younger people, we can assess at least the minimum educational attainment of older people in the future by looking at today's younger cohorts. For example, the fact that 87 percent of people aged 25 to 44 in 1996 had a high school diploma foreshadows the year 2046, in which we can expect that nearly 90 percent of the population aged 75 and older in the United States will be high school graduates—a substantial increase from the 59 percent in 1996.

People carry a lot of personal and social baggage with them into the older years, but improved mortality at the older ages means that people who are lucky enough to reach that point in life have more years than ever before in human history in which to enjoy, or at least cope with, the process of aging. This means that old age is more complicated than it used to be and symptomatic of those complications is the increasing diversity by sex and age.

The Sex and Age Structure of the Older Population

In the past, and still today in many less developed societies, old age was implicitly that age at which a person could no longer make a full economic contribution to the household economy due to one or more disabilities that would eventually (and sooner rather than later in most cases) lead to death. The same wealth in the modern world that has lowered death rates has also made it possible to separate the decline in economic productivity from a decline in physical functioning. This has allowed the institution of retirement to arise, creating what Laslett (1991) has called the "Third Age," distinguishing it from the later phase of aging—disability and death—which he has termed the "Fourth Age." Although I organize the discussion of sociodemographic characteristics of the elderly around these two broad age categories, we first must examine the increasing feminization of old age.

The Feminization of Old Age

Women live longer than men in almost every human society and so they disproportionately populate the older ages. As recently as 1930 in the United States, there were equal numbers of males and females at ages 65 and older (partly a result of the earlier influx of immigrant males) (Siegel 1993; Taueber 1993), but by 1990 there were only 70 males for every 100 females at the older ages. In Table 9.5 you can see the ratio of males per 100 females for several countries, with data broken down by the younger-old (65 to 74) and the older-old (75 and older).

In the United States the ratio of males to females aged 65 to 74 declined steeply between 1950 and 1970, leveled off, and is projected to continue a slow rise from 1980 through 2030 as men begin to close the gender gap in life expectancy. Both the United Nations and the U.S. Census Bureau projections assume that life expectancy gains will be greater for males than females (a reversal of historical trends), and that will help to push up the ratio of men to women as time goes by. At ages 75 and older, the higher mortality of males really has taken its toll: in 1950, there were 83 males per 100 females at that age in the United States, but it declined to only 54 males per 100 females by the 1990 census. That means that at ages 75 and older, two thirds of the people alive in the United States are women. Again, the projections

Table 9.5   In Most Countries, Women Outnumber Men at the Older Ages

Country	Males per 100 Females								
	1950	1960	1970	1980	1990	2000	2010	2020	2030
United States									
65–74	0.90	0.84	0.77	0.77	0.81	0.84	0.86	0.87	0.89
75+	0.81	0.72	0.62	0.54	0.54	0.58	0.59	0.61	0.64
Canada									
65–74	1.10	0.97	0.88	0.84	0.81	0.87	0.89	0.89	0.92
75+	0.94	0.90	0.75	0.63	0.60	0.61	0.63	0.65	0.68
Japan									
65–74	0.78	0.86	0.85	0.78	0.73	0.86	0.87	0.90	0.91
75+	0.59	0.59	0.64	0.65	0.60	0.56	0.63	0.65	0.67
Mexico									
65–74	0.85	0.85	0.85	0.85	0.85	0.86	0.86	0.84	0.82
75+	0.79	0.79	0.80	0.78	0.76	0.75	0.74	0.73	0.71
Pakistan									
65–74	1.42	1.16	1.25	1.16	1.06	0.96	0.96	0.99	0.98
75+	1.38	1.33	1.10	1.18	1.13	1.02	0.92	0.87	0.88

Source: United Nations, 1997, The Sex and Age Distribution of the World Populations: The 1996 Revisions (New York: United Nations).

imply that this trend will be reversed over the next few decades, although women will continue to outnumber men by a large margin at the oldest ages for the foreseeable future.

The general pattern in the sex ratio at the older ages has been similar in Canada and the United States, but the actual level of the sex ratio is consistently higher (albeit not by much) in Canada. This is probably due to the joint effects of immigration (immigrants account for a higher fraction of the Canadian than of the U.S. population), and the fact that the gender difference in life expectancy has been slightly less in Canada than in the United States.

In Japan, both males and females live longer than in the United States and the difference between them in life expectancy is also somewhat less. As a result, the ratio of males to females in Japan has been slightly higher than in the United States since 1970, and is estimated to be nearly identical to the sex ratio for Canada from about the year 2000 on. Between 1950 and 1970, the ratio was affected by the high mortality experienced by males in World War II.

In Mexico, the feminization of old age is clearly underway (Ham-Chande 1993), especially in urban areas where death rates are lowest. You can see that the projections for Mexico suggest that by the year 2020, the ratio of males to females at ages 65 to 74 in Mexico will actually be lower than in the United States.

The data for Pakistan tell a different story. Pakistan is one of several countries in the world where there are still more men than women at the older ages. This is a result of the lower status of women which increases their mortality rates relative to men over the entire life course. In 1950 in Pakistan, there were 142 men aged 65 to 74 for every 100 women that age, and the ratio was nearly that high at ages 75 and older. Nonetheless, you can see that progress is being made to bring about gender equality, and the United Nations projections suggest that by the year 2010 there will be slightly fewer males than females at the older ages.

Even though Pakistan and several other countries may not yet have a female majority at the older ages, they are nonetheless moving in the direction of an increasing feminization of the elderly. This has a wide range of social ramifications for the economic well-being of older persons. Let us examine these in more detail.

The Third Age (The Young-Old)

You have probably grown up expecting, and probably looking forward to, a time interval between the end of your work career and death—a time of leisurely retirement. This is a very recent invention. In the earlier part of this century, the older population in the United States, for example, was concentrated in the ages 65 to 74, as you can see in Figure 9.5. High mortality meant that people at that age were pretty close to death, and most worked for as long as they were physically able. As mortality has declined, a greater fraction of people have survived into the older ages, thus creating a new period between the traditional entrance into old age (approximately age 65) and the time when death begins to stalk us. We have taken advantage of this in society and have been able to act on the old adage that "youth is wasted on the young" by giving ourselves a youth-like care-

Figure 9.5 The Percentage of the Older Population that Is Old-Old Has Been Increasing Over Time, United States, 1900–2020

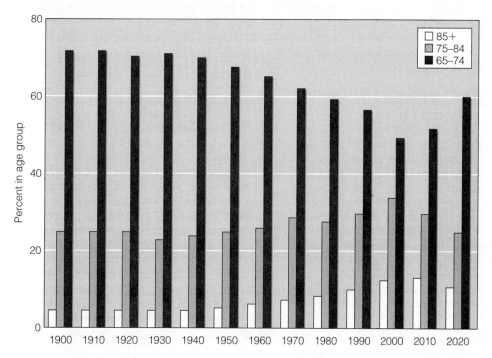

Note: The proportion of the older population that is aged 65 to 74 has declined steadily over time, as the older population has itself aged; this process will be temporarily reversed early in the 21st century as the Baby Boom generation enters old age.
Source: Adapted from C. Taeuber, 1993, "Sixty-five plus in America," U.S. Bureau of the Census, Current Population Reports, Special Studies, Series P23-178RV, Table 2-1.

free period toward the end of life. This is the so-called Third Age, a time when we are still healthy enough to engage in all of the normal activities of daily life, but are able to be free of regular economic activity.

Retirement, Income, and Resources In industrialized societies, old age is now stereotypically a time of retirement from labor force activity. Indeed, when the Social Security Act was passed in the United States in the middle of the Great Depression in the 1930s, it was designed quite literally to encourage people to leave the labor force. At the time the idea was to remove older workers from the work force to replace them with younger workers and thus lower the rate of unemployment among younger people. The arbitrarily chosen age of 65 became etched in stone as the age of retirement (Segerberg 1974). Most companies and government entities alike turned age 65 into a mandatory age of retirement. More than half of all workers since the early 1960s have avoided mandatory retirement (which is now 70, rather than 65) by retiring early. In 1956 for women, and in 1961 for men,

Congress allowed reduced social security pensions to be available at age 62, and this has been a popular option. It appears that the availability of an adequate retirement income is a strong inducement for people to leave the labor force. In 1890, 75 percent of all men remained in the labor force past age 64, whereas by 1993 only 16 percent were still employed at that age (Hobbs and Damon 1996; Viscusi

THE AMERICAN BABY BOOM AND SOCIAL SECURITY

Back in 1935, when President Roosevelt's Committee on Economic Security was putting the finishing touches on the social security legislation, two committee members met to discuss the projections that had been made for social security expenditures for 1935 through 1980. Treasury Secretary Henry Morgenthau, Jr., and Harry Hopkins, head of the Federal Emergency Relief Administration, were aware of possible problems ahead, as is evidenced by their comments at the meeting (quoted in Graebner 1980:256):

> *Hopkins:* Well, there are going to be twice as many old people thirty years from now, Henry, than there are now.
> *Morgenthau:* Well, I've gotten a very good analysis of this thing . . . and I want to show them [other members of the committee] the bad curves.
> *Hopkins:* That old age thing is a bad curve.

That bad curve referred to the ratio of workers to retirees, which, though quite favorable in the early years of social security, could be foreseen to worsen over the years as the small birth cohorts of the early 1930s tried to support the numerically larger older cohorts. Despite the fact that reference is often made to the term "trust fund," most of you are probably aware that social security systems in most countries, including the United States, were never designed to have the government actually deposit money in an account with your name on it and have the money accrue principal and interest until you retire and start withdrawing your pension. Rather, almost every system is "pay as you go"—current benefits are paid from current revenue. Morgenthau's data suggested that by 1980, social security expenditures would have risen to one billion dollars (in 1935 currency) (Graebner 1980). This curve

looked bad then, but in truth it turned out to be even sharper than expected: life expectancy has increased; the unexpected Baby Boom has injected a large cohort that will ultimately have to be dealt with in retirement (but that in the meantime has helped to delay the funding crisis because of its members' payroll contributions); and Congress has regularly expanded social security coverage and raised benefits.

The demographic impact on the social security system was felt keenly through the 1980s as the older population grew more rapidly than the number of younger workers. By 1990, for example, there were 10 percent more people aged 60 to 69 (people moving into retirement) in the United States than in 1980, yet there were 3 percent fewer people aged 20 to 29 (people moving into the labor force). These changes, of course, had been projected for some time, and in the mid-1970s (and again in the 1980s) Congress made adjustments to increase payroll taxes and cut back on the annual allowable increase in social security payments. These measures (along with a little borrowing from the disability and Medicare trust funds) allowed the system to survive the 1980s and early 1990s. The late 1990s has seen a hiatus in the Social Security crisis because of the slowdown in the increase of new retirees as the Depression-era cohorts reach old age. For example, between 1990 and 2000, the number of people aged 60 to 69 in the United States declined by 6 percent. This has eased the pressure of expenditures while revenues rose.

Part of the problem as we look ahead to the huge influx of Baby Boom retirees is that when the baby boomers were younger Congress felt generous about retirement benefits because the Baby Boom cohort was supplying an influx of new workers to pay taxes, and inflation was showering

1979). The employment rate of older women has also declined, although not so much. In 1960, 11 percent of older women were active in the economy, compared with 8 percent in 1993, but that was actually higher than a decade earlier. As the labor force participation rates increase at younger ages, the fraction of older women who are in the labor force may continue to rise.

social security with unexpected revenue. In 1972, Congress boosted retirement benefits by 20 percent and built in an automatic adjustment to keep benefits increasing each year along with inflation.

Back in the early 1980s, Robert Myers, former chief actuary of the Social Security Administration worried that Congress would use the growing surplus of the 1990s to increase benefits, or to lower taxes, or to pay off the national debt. Such a course of action could be disastrous, he felt, because around 2010 the Baby Boom generation will really crunch the pension system. Between the years 2000 and 2010, there will be nearly a 50 percent increase in the population aged 60 to 69—an unprecedented rise in the number of people who might be retiring—and this increase will continue until 2030. The Trustees of the Social Security Administration reported in 1998 that Social Security expenses will exceed income by the year 2013. In theory, things will be all right, however, until 2032 because the government will be able to draw down the surplus that was built up prior to the Baby Boom began retiring. The only problem with the surplus is that it was invested in government bonds which will have to redeemed when the time comes, and that will have some impact on the overall economy. In the 1960s there were nearly 4 workers for every Social Security retiree, but by 2030 that will have dropped to only 2. The burden on the younger generation will obviously be intense.

As a result, there may be considerable pressure on the elderly to be more self-sufficient—not only to work longer (retire later) but also to become involved in mutual self-help organizations that could relieve some of the burden on public agencies. It is ironic indeed that the social security system, which was designed in large part to encourage older people to leave the work force, may in the future be bailed out because people can stay in the labor force longer. Congress has already thought of this, of course, and in the United States persons born after 1959 will have to wait until age 67 before receiving full Social Security benefits. They will still be able to retire as early as age 62, but only at a lower benefit level than currently prevails. Another hedge, less predictable, of course, is for the economy to be growing fairly rapidly in the next century as the Baby Boom generation reaches retirement age. The kind of structural mobility (when the economic situation of everyone is improving) that has typically accompanied rapid economic growth in the United States would permit the transfer of money from the younger to the older generation to be a little less painful.

The experience of countries such as the United States has pushed less developed countries to think of different ways of attempting to finance retirement for their own aging populations. The model popularized in the 1990s was the "Chile" model, crafted for that country by economists trained at the University of Chicago. The concept is that people must save for their own retirement, but they must be forced to do so by the government (or else the temptation is to spend the money on other things), while at the same time having a reasonably low level of risk of losing their money. In Chile (and now many other countries as well), workers are required to put a certain fraction of their earnings into a governmentally regulated (but privately managed) set of mutual funds. The savings provide a pool of investment funds that is supposed to help the national economy develop, thus ensuring that the workers will have a nice pension benefit when they retire. The programs are so far too new for us to evaluate how successful they will be.

The average age at retirement in the United States for both men and women dropped from 66 years in the mid-1950s to just below 63 by the mid-1970s, but since then it has leveled off (Gendell and Siegel 1996). Increasingly people are choosing more complex routes to retirement than in the past, partly by necessity and partly by choice (Henretta 1997). The necessity came from the popularity of corporate "downsizing" in the 1980s and 90s which forced many workers into retirement earlier than anticipated and so they officially retired from one job, but maintained a presence in the labor force nevertheless. The choice comes from people who prefer a "blurred transition" to retirement rather than a "crisp exit" (Mutchler, et al. 1997). A blurred transition involves, for example, taking retirement benefits from a previous full-time job, but continuing to work part-time, or even full-time but at a different job for less money. Indeed, it seems that money is often the reason for a person not wanting to retire fully.

Needless to say, when people drop out of the labor force their income is likely to decline. This is certainly true for older people. The minimum social security benefit is below the poverty level, although in 1974 the establishment of Supplemental Security Income (SSI) for older persons meant that most old people in the United States are now guaranteed an income no lower than the poverty level. Indeed, between 1959 and 1989, the percentage of people aged 65 and older with poverty-level incomes in the United States dropped from 35 percent to 12 percent (U.S. Bureau of the Census 1995), and it had declined further still to just under 11 percent by 1996 (Lamison-White 1997).

How much money do older Americans have to live on? In 1996, the median income of households in which the head of the household was 65 years or older was $19,449 per year, about one half the $40,941 average for households whose heads are less than 65, and well below the median of $50,472 earned by householders aged 45 to 54 (U.S. Bureau of the Census 1997a). The difference stems in large part from the fact that older people are not in the labor force. Note that in 1996 males aged 65 and older who were still year-round full-time workers had an average income of $42,836 per year—higher than the average for those under 65 ($33,321) and, in fact, higher than any younger age group. By contrast, the average income from Social Security benefits in that year among older persons was $8,329.

In the United States more than 90 percent of all households headed by someone aged 65 or older receives a Social Security check (Grad 1996), and almost all families are able to combine Social Security with other sources of income such as a pension, dividends or interest income, property income, earnings from a job (including self-employment), or public assistance (principally SSI). In-kind benefits such as food stamps and Medicare are available to shore up impoverished elders, and for the better off, security in old age may be bolstered by homeownership and other assets. In fact, in 1993, the age at which median net worth peaked was 65 to 69. Yet even people 75 and older had higher net worth than those aged 45 to 54.

How can a person most likely arrive at an advantageous position in old age? The answer is to be in an advantageous position when you are young and to plan well for your future. Educational attainment, for example, is closely related to income in old age. In 1996 in the United States, females aged 65 and older with a college education had a median annual income of $35,956, compared to only $11,194 for women that age with just a high school diploma. Thus college was worth nearly

$25,000 a year in additional income in old age. For men, the incomes were higher than for women, but the absolute difference was essentially the same. Males aged 65 and older with a college degree had a mean income of $42,056 in 1996, compared to $17,993 for men with only a high school diploma (U.S. Bureau of the Census 1997a). The lesson: a college degree can make a huge difference in your income in old age.

For at least one segment of the American population, however, a decent living in old age has been made more difficult because opportunities to do well in younger years were not always available. These are the minority elderly.

Ethnic Differences   Persons who are both old and members of an ethnic minority group are said to be in double jeopardy. As minority group members, they spent their lives dealing with prejudice and discrimination; as older persons, they also face the prejudice and discrimination that may befall a person just because of being old. In fact, higher mortality rates at younger ages among African-Americans and Hispanics means that the probability of reaching old age is less for these groups than for white non-Hispanics or Asians, for whom life expectancy is higher. Indeed, in 1990 the higher mortality and higher fertility among African-Americans than whites meant that only 8.2 percent of the African-American population in the United States was aged 65 and older, compared with 13.4 percent of the white population. Among Hispanics, the combination of higher mortality, higher fertility, and the recency of immigration, meant that only 5.1 percent of the Hispanic population in 1990 was 65 or older. Once they do reach old age, minority elders are less likely than their counterparts in the majority to be doing well economically.

The economic plight of older minority group members can be appreciated by noting the percentage below the poverty level. In 1996, when 9.4 percent of whites aged 65 and over lived below the poverty line, 25.3 percent of African-American elders lived at that level, as did 24.4 percent of older Hispanics (Lamison-White 1997).

In terms of income, the patterns for older people by race and ethnicity are similar to those for younger people, but the levels are lower. From the Current Population Survey in the United States, we have data for the two most populous ethnic minority groups—African-Americans and Hispanics. In 1996 the average (median) family income for Hispanics with a householder aged 65 and over was $14,006, compared with $14,019 for older African-American households. As a comparison, white householders aged 65 or older had a median household income of $19,977. We might say then that in the United States there is a $6,000 per year penalty among older householders for not being white.

Lower incomes in old age are largely a consequence of lower incomes at the younger ages, and the effects cumulate over a lifetime, producing a considerable inequality in health (Schoenbaum and Waidmann 1997) and wealth (monetary assets) in old age (Smith 1997). Housing is an important asset and minority group members are less likely be well-housed at younger ages so it is no surprise that they have the least adequate housing among older persons. Housing inadequacy is especially noticeable among Blacks living in the south (Markham and Gilderbloom 1998).

Living Arrangements and Family Relationships   In the past, and still today in many less developed countries, the higher status of the elderly was tied partly to the fact that as old age approached, they were situated in their own housing unit. Even if they lived with their children, it was likely that the children (typically a son with his wife and children) were actually living in the parental home, rather than the other way around (Kertzer 1995). The concept of filial piety, of respect for one's parents, has been a traditional value in most cultures, encouraging children to take care of their parents when the need arises. Of course, in high mortality societies, the probability that your parents would survive to old age (and the probability that you would survive to help them) was low enough so that relatively few people ever had to make good on that concept, as I mentioned previously.

In the modern world, society after society has bemoaned the fact that the multigenerational family has been a victim of "the movement toward smaller families, the expansion of the female labor market, the geographic mobility of villagers, and the tendency of the young toward more individualistic life styles" (Sung 1995). That description was applied to Korea, but it is echoed in Brazil (Ramos 1993) and many other places. Older people are no longer assured that they will live out their days nestled in the bosom of their family. To be sure, not all older people necessarily want to live with their children, especially if they are forced to be dependent on the children. A worldwide phenomenon has been emerging of older people wanting "intimacy at a distance." In Malaysia, for example, recent survey results show that the more economically advantaged older people are, the less likely they are to live with adult children. Those who co-reside with children do so out of necessity, not necessarily because they prefer that arrangement (DaVanzo and Chan 1994). Diversity in living arrangements is as much a part of the lives of older people as it is among the young, as I will discuss in the next chapter.

In developed societies, the diversity of living arrangements among the elderly is compounded by the patterns of marriage, divorce, and remarriage in combination with the differences in mortality between males and females. The unbalanced sex ratio at the older ages in most societies signals a change in marital status, which in turn means a change in living arrangements for many people as they grow older. According to the 1990 census, in the United States 70 percent of all women aged 35 to 39 were married and living with their spouse; by ages 65 to 74, the percentage had dropped to 51; and by ages 75 and older, only 24 percent were married and living with a spouse. Indeed, at ages 75 and older, two thirds of all American women are widows, and the United States tends to mirror much of the world in these terms. Males, of course, are less likely to experience a change in marital status as they grow older, because they are more likely to be outlived by their wives and more likely than older women to remarry.

Does a change in marital status affect living arrangements? The answer is yes, and it means especially that more women wind up living alone. In fact, the proportion of older people living alone in the United States has been constantly increasing for the past several decades (Choi 1991). This is in contrast to the fairly common notion that old people are packed off to the "home." Only about 5 percent of older people in the United States live in group quarters designed for the elderly (Taueber 1993). It is also in contrast to the pattern earlier in this century, when widowhood in old age was more apt to mean that a woman would be living with one or more of her children (Harevan and Uhlenberg 1995).

Thus for women in the United States, old age more likely means living alone than being institutionalized or living in a group environment with other seniors. In 1996, 31 percent of all women aged 65 to 74 were living alone, and at ages 75 and older 53 percent were living alone (Saluter and Lugaila 1998). For men, living alone is less common. At ages 65 to 74, 15 percent of men were living alone, whereas at ages 75 and above, 21 percent were living alone. In American and European societies, older people are much more separate from their children than in Asian societies such as Japan (Kinsella 1995). In Japan it has long been the norm for older parents to live with a child (usually the eldest son). Will the Japanese become westernized and alter the pattern of elders living with their children? Maybe, but it is likely to be a slow process, since the respect and dignity afforded older people in Japan are not likely to be eroded in a single generation. The proportion of older parents living with children in Japan is decreasing, especially in urban areas and among the highly educated (as we might expect from Cowgill's theory), but the 1995 Census of Japan still found that only 6 percent of males and 16 percent of females aged 65 and older were living alone (Statistics Bureau of Japan 1998)—far lower numbers than the ones that I noted above for the United States.

Similar trends have been experienced by ethnic minority groups in the United States. For example, there is a strong tradition in the Mexican-origin community in the United States that family members should care for older parents, but the reality is that in the United States, as in Mexico, the demographic and social trends are changing the patterns of intergenerational relations (Dietz 1995). Older Mexican-Americans are increasingly living independently of their children, although they, like African-Americans, are still less likely than non-Hispanic whites to be homeowners. Data from the 1993 American Housing Survey show that among Hispanics, 64 percent of households headed by a person aged 65 or older were owner-occupied, similar to the 63 percent for African-Americans. These numbers are in contrast to the 79 percent ownership rate among older non-Hispanic white households (U.S. Bureau of the Census 1995).

The Fourth Age (The Old-Old)

Definitions of the old-old population vary (Laslett 1995), although Suzman and associates (Suzman, Willis, and Manton 1992) have suggested that being 85 or older is a reasonable criterion for the *oldest*-old. The distinguishing characteristic of this stage of life is an increasing susceptibility to senescence—increases in the incidence of chronic disease and associated disabilities, and, of course, death. In the United States, the pattern that has emerged in these later years of life is for women to have long-term, chronic disabling diseases, whereas men tend to develop relatively short-term fatal diseases (Taeuber 1993).

One of the greatest fears that many younger people have about growing old is that they will be shuffled off to a nursing home, where they will be neglected and die. Data from the U.S. National Health Interview Survey indicate that only 1 percent of people aged 65 to 69 had used a nursing home during the previous year, and even at ages 75 to 79 only 6 percent had required nursing home care. However, if you survive to age 90 or older, the chances increase dramatically that you will spend time in a nursing home—46 percent for women and 31 percent for men (Feinleib, Cunningham, and Short 1994).

Many of the functions of nursing homes are gradually being replaced by home health care and hospice care, but the move to a long-term care facility is usually occasioned by disabilities that limit a person in his or her ability to carry out the activities of daily living (ADLs, as they are usually called). I mentioned earlier that this is often associated with the progression of Alzheimer's disease. At ages 65 to 69 most males and females in the United States report their health to be good or excellent and they have no difficulties with the activities of daily living. However, by ages 85 and older, nearly one third of men and one half of women report at least one difficulty with ADLs (Mermelstein, et al. 1993). Different kinds of disabilities in old age require very different responses on the part of health care and social service providers, and the challenge of any aging society will be to adequately assess and plan for those needs.

As the older population continues to grow, demand will also grow for health care, long-term care facilities, assisted transportation, subsidized housing, day-care services, respite care services for home care providers, prepared meals, personal care, and housekeeping assistance, to name just a few of the items required by increasingly dependent adults. Because a large portion of the cost of such services for the elderly is borne by the public in all industrialized societies, it is clear that the aging of the population, and within that group the aging of the older population, will generate continued demands for public funds in virtually all of the low mortality, low fertility developed nations. Who will pay? Will young people pay higher taxes in order to meet the needs of their parents and grandparents? Will immigrants from the developing countries bail out the older generation of people in the developed nations? Or will some of the needs of older people be ignored? Stay tuned.

Summary and Conclusion

The older population is the most rapidly growing segment of the population almost everywhere in the world. This is a result of declining mortality. In addition, as fertility drops, the percentage of the population that is elderly also begins to rise.

Changes occur in many aspects of social organization as a population ages, because the aging process brings with it numerous changes in individuals themselves, both biological and social. The biological changes are related to the gradual decline in physical functioning and the concomitant rise in susceptibility to disease. Social changes are related especially to the system of age stratification, which in modern industrialized societies has relegated the elderly to a lower status than was formerly true or than is true in less developed, agrarian societies. The downgrading of status seems to be a combined result of greater longevity, which has led to retirement (which has lower status than work); economic technology, which renders the skills of the elderly obsolete; urbanization, which segregates the generations and reinforces the inversion of status between generations engendered by the upward spiral of technology; and improvements in education, which means that children tend to be better educated than their parents. Related to these historical changes is the fact that each cohort is unique in its social and historical experiences as it moves through the life cycle. This introduces a dynamic element into the process of aging, signaling future changes in the demographic characteristics and lifestyles of the elderly.

The older population in the United States is characterized by an unbalanced sex ratio, because male mortality is higher than female mortality. This means that as women age they are increasingly susceptible to widowhood and the prospect of living alone. Old age also brings with it a fairly dramatic drop in income as people are encouraged or forced to leave the labor force. Minority elders are in double jeopardy because they, as younger people, were less likely to have had access to the higher educational and occupational levels that are associated with high incomes among the elderly.

The future prospects for the elderly differ according to the level of development of a society. In industrialized societies, the options available to older people are increasing. These include the ability to work longer, the growth in lifelong learning opportunities—college courses designed for people of all ages, and travel discounts for seniors that may open up opportunities previously closed. These trends are likely to continue almost unabated well into the next century as a result of the previously mentioned cohort flow. For example, the greater participation of women in the labor force and the resulting increase in female financial independence will likely mean a greater sense of personal freedom for these women as they grow older. Indeed, Easterlin and associates (1993) have concluded that the baby boomers are likely to be in good economic shape when they reach retirement, because of the economic and demographic adjustments that they have made so far in their lives, including delayed marriage, lower fertility, and high labor force participation of women. Life in old age also seems to be improving from a health perspective and data from the 1986 Longitudinal Study of Aging in the United States imply that, as healthier cohorts move into the older ages, the overall health status of the elderly should improve (Rogers, Rogers, and Belanger 1990).

In developing societies, the situation seems to be moving in the opposite direction for older people. The demographic transition is widening the gap between the older and younger generations in some of the basic sociodemographic characteristics including education, occupation, and income, and, in the process, pushing the older population to the margins of society. Although I have discussed these sociodemographic characteristics in this chapter as they relate to the older population, it is time to expand that analysis to all ages and I do so in the next chapter.

Main Points

1. The population aged 65 and older in the world numbers more than 400 million and represents the fastest growing segment of the world's population.

2. The growth of the older population is actually more rapid in the less developed than in the more developed countries.

3. The cause of the increase in the number of older people is the long-term drop in death rates throughout the world.

4. The percentage of the population that is 65 and older is greatest in the more developed countries, because its increase is largely dependent on a decline in the birth rate.

5. Most people age in place, being left behind by younger people who migrate. However, the migration of older people can be an important source of demographic change in selected areas, especially those attractive to retirees.

6. Biological aging refers to a decline in physical abilities accompanied by an increased risk of illness.

7. Socially, aging carries with it a changing set of obligations and role expectations.

8. Increased longevity, higher levels of education, urbanization, and economic technology may all have conspired to alter the status of older persons in industrialized nations.

9. The status of older persons in developed nations currently is on the rise, moving in the opposite direction of the situation in less developed nations.

10. The upward spiral of literacy and education over the last few hundred years has meant that each generation of older people tends to be better educated than the previous generation.

11. In most societies females outnumber males at the older ages. The exceptions are nations in which the social status of women tends to be clearly inferior to that of men.

12. The age structure of the elderly can usefully be divided between the "Third Age" (when social forces are most important), and the "Fourth Age" (characterized by disease, disability, and death).

13. The older ages are typically characterized by an increase in widowhood and an increase in the proportion of people (especially women) who live alone.

14. In industrialized nations old age is characterized by retirement from full-time employment and by a subsequent drop in income.

15. Minority elders are said to be in double jeopardy, discriminated against both for their minority status and for their age.

16. Despite the fears of many younger people about their own aging, most older people live independently and only require nursing home care at very advanced ages, typically as a result of the onset of cognitive dysfunction, such as Alzheimer's disease.

Suggested Readings

1. Donald Cowgill, 1986, Aging Around the World (Belmont, CA: Wadsworth Publishing).
   Cowgill popularized the modernization theory of aging, which has a heavy demographic emphasis, and in this work he puts the theory into global perspective.

2. Peter Laslett, 1991, A Fresh Map of Life: The Emergence of the Third Age (Cambridge, MA: Harvard University Press).
   A world famous British demographer, best known for studies of historical demography, has himself reached the later decades of life and has produced a very scholarly and readable analysis of the demographic aspects of aging.

3. Jacob S. Siegel (with contributing authors Murray Gendell and Sally L. Hoover), 1993, A Generation of Change: A Profile of America's Older Population (New York: Russell Sage Foundation).

This nearly encyclopedic work covers all important demographic aspects of the aging population in the United States.

4. David Kertzer and Peter Laslett (eds.), 1995, Aging in the Past: Demography, Society, and Old Age (Berkeley and Los Angeles: University of California Press).

Although this is nominally a demographic history of aging, most of the chapters are highly relevant to the current understanding of aging in the world, and the book as a whole provides a great deal of insight into the demographic aspects of aging in society.

5. Jay Sokolovsky (ed.), 1997, The Cultural Context of Aging: Worldwide Perspectives, Second Edition (Westport, CT: Bergin & Garvey).

This book pulls together a wide range of authors bringing a global (and largely anthropological) perspective to the study of aging and will serve to round out your perspective on the aging process.

🌐 Websites of Interest

Remember that websites are not as permanent as books and journals, so I cannot guarantee that each of the following websites still exists at the moment that you are reading this:

1. **http://www.census.gov/population/www/socdemo/age.html#elderly**

The U.S. Bureau of the Census regularly updates its analyses of the demographics of aging in the United States and their current offerings should be available at this website.

2. **http://www.ssa.gov/OACT/TRSUM/trsummary.html**

The Social Security Administration (SSA) is the U.S. agency charged with administering the Social Security system, including pension benefits. The demographic changes taking place in American society have been of concern to them for a long time and are fully addressed in this report of the Trustees of the SSA.

3. **http://www.nih.gov/nia/bsr/bsrdda.htm**

The National Institute on Aging (NIA) supports much of the research taking place on the older population of the United States, and the demographic research is coordinated by the Office of Demography of Aging, within the NIA's Behavioral and Social Research (BSR) Program. Their web site includes linkages to other aging-related websites that may interest you.

4. **http://www.umich.edu/~hrswww**

One of the projects funded jointly by the National Institute on Aging, the Social Security Administration and other agencies is the Asset and Health Dynamics Among the Oldest Old (AHEAD) project at the University of Michigan. This is a longitudinal study of a representative sample of U.S. residents aged 70 and older, and the website offers details and results about the study, as well as links to other projects.

5. **www-nais.ccm.emr.ca/schoolnet/issues/agepop/eaging.html**

This is a portion of a bigger educational project put together by Natural Resources Canada and it has a very good description of the older Canadian population along with thematic maps showing the geographic distribution of the elderly in Canada.

CHAPTER 10
Family Demography and Life Chances

Households used to be created by marriage and dissolved by death—in between there were children. Throughout the world this pattern has been transformed by what some have called the "second demographic transition." Households no longer depend on marriage for their creation, nor do they depend on death to dissolve them, and children are encountered in a wide array of household and living arrangements. "The family is in crisis, as witnessed by increasing instability of unions, fluidity of the 'marital' home and the economic stress experienced by women and children of disrupted marriages" (Makinwa-Adebusoye 1994:48). Although it certainly sounds like the subject matter of books and talk shows in the United States, the author of that quote is referring to sub-Saharan Africa. "Marriage is becoming rarer and the age at first marriage is increasing; the number of couples who cohabit before and without marrying is rapidly increasing; and so as a consequence, are births out of wedlock" (Blossfeld and de Rose 1992:73). That, too, could be a description of the United States, but in this instance the authors are talking about Italy.

All over the world changes in household formation and living arrangements are being discussed and debated. Curiously, however, these debates rarely include a review of the underlying demographic changes that helped to spawn this massive social shift, and so my purpose in this chapter is to rectify that deficiency for you. I begin the chapter with a discussion of exactly how the structure of households and living arrangements have, in fact, changed over time—how big is this transition in the United States and elsewhere?

Then I turn to the specific demographic changes that have contributed to the increasing diversity in household structure. The key element is the changing set of life chances that people are experiencing in the United States and all over the world—changes in the **population (or demographic) characteristics** that influence how your life will turn out. These include especially education, labor force participation, occupation, and income, which in turn affect gender roles and marital status. All of these things have influenced the changing family and household structure, although differently for some cultural groups than others. Indeed, race and ethnicity, along with religion, mediate the impact of life chances in every human society. Although population characteristics affect family and household structure, your own life chances are influenced in their turn by the kind of family and household in which you grow up, and I conclude the chapter by reviewing some of the major implications of household and living arrangement changes for the future, especially the impact on children.

Defining Family Demography and Life Chances

Have you ever given serious thought to the reasons why you may be seeking, or have achieved, a college education? How will your life be different because you are educated? What does it matter if you are African-American in the United States, Arab in Israel, Indian in Malaysia? Each of these different aspects of who you are will affect your life chances—your probability of having a high-prestige job, lots of money, a stable marriage or not marrying at all, and a small family or no family at all. These differences in life chances, of course, are not necessarily a reflection of your worth as an individual, but they are reflections of the social and economic makeup of society—indicators of the demographic characteristics that help to define what a society and its members are like.

We are born with certain **ascribed characteristics,** such as gender and race and ethnicity, over which we have essentially no control (except in extreme cases). These characteristics affect life chances in very important ways because virtually every society uses such identifiable human attributes to the advantage of some people and the disadvantage of others. Religion is not exactly an ascribed characteristic, but worldwide it is typically a function of race or ethnicity and, as with other ascribed characteristics, it is often a focal point for prejudice and discrimination, which influence life chances.

Life chances are also directly related to **achieved characteristics,** those sociodemographic characteristics, such as education, occupation, labor force participation, income, and marital status, over which we do exercise some degree of control. For example, the better educated you are, the higher is your occupational status apt to be, and thus the higher your level of income will likely be. Indeed, income is a crass, but widely accepted, index of how your life is turning out. Ascribed characteristics have their impact on your life chances primarily by affecting your access to achieved characteristics, which then become major ingredients of social status—education, occupation, and income. Population characteristics affect your own demographic behavior, especially fertility and family-building, although they also influence mortality and migration, as I have already discussed in Chapters 4 and 7, respectively. Demographic behavior, in its turn, affects life chances through its ability to facilitate or retard your access to opportunities for higher education, a higher-status occupation, or a better-paying job. All of these aspects of population characteristics and their influence on life chances converge to affect the kind of family we choose to create and the type of household we form, and I have illustrated these relationships for you in Figure 10.1.

It is now a given that a major metamorphosis has been taking place in the structure of families and households in much of the world. To study this change we must,

Figure 10.1 Population Characteristics Influence Family and Household Structure and Jointly They Affect and Are Affected by Population Processes

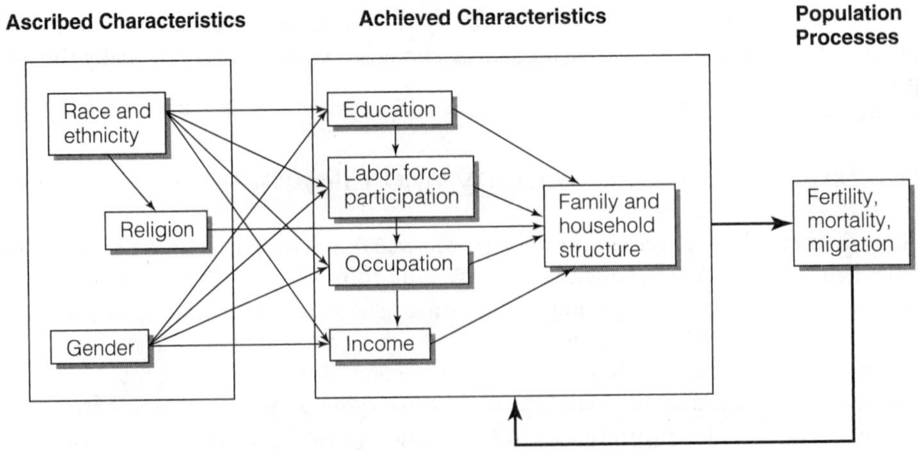

Note: See text for discussion

of course, define some terms. In a general sense, a **family** is any group of people who are related to one another by marriage, birth, or adoption. The nature of the family, then, is that it is a *kinship* unit. We usually make a distinction between the **nuclear family** (at least one parent and their/his/her children) and the **extended family**, which can extend to other generations (add in grandparents and maybe even great-grandparents) and can also extend laterally to other people within each generation (brothers and sisters of the parents, aunts and uncles, and so forth). Now, the question of interest to us is where do these people live? People who share a **housing unit** are said to have formed a **household**. A housing unit is the physical space used as a separate living quarters for people. It may be a house, an apartment, a mobile home or trailer, or even a single room or group of rooms. The household is thus a *residential unit* and you can see that a **family household** is a housing unit occupied by people who are all related to one another, whereas a **nonfamily household** is a housing unit shared by people who are not related to one another.

The Transformation of Families and Households

The "traditional" family household of a married couple and their children is no longer the statistical norm in North America and in many other parts of the world, even if it remains the ideal type of household in the minds of many people. Families headed by females, especially with no husband present, are increasingly common, as are "nontraditional" households inhabited by unmarried people (including never-married, divorced, widowed, and cohabiting couples). You are almost certainly aware of these societal shifts through personal experience or the mass media. What may be less obvious is that these changes are closely linked to demographic trends.

Household Composition

The total number of households in the United States increased from 63 million in 1970 to 96 million in the mid-1990s, but within that increase was a dramatic change in the composition of the American household, as Figure 10.2 illustrates. In 1970 the classic "married with children" households accounted for 40 percent of all households in the United States. Married couples without children (either before building a family or after the kids were grown) accounted for another 30 percent. The "other" families include male- and, disproportionately, female-headed households in which other family members (usually children) are living with the householder. In Figure 10.2 you can see that the light shading represents all family households (a household in which the householder is living with one or more persons related to her or him by birth or marriage). In 1970 they had comprised 81 percent of all households—a drop from 90 percent in 1940, when the Census Bureau first began to compile these data (Rawlings 1994). By the mid-1990s it had dropped even further to 71 percent, and by the mid-1990s scarcely one in four households included a married couple with children.

As married couples with children have become less common, they have been replaced most often by female-headed, mother-only families, and by nonfamily households, including people who live alone, and nonfamily coresidents (friends living together, a single householder who rents out rooms, cohabiting couples, etc.). The phrase that perhaps best describes the changes in household composition as shown in Figure 10.2 is increased diversity. There is no single model of family and household structure that can now characterize life in the United States. Although the United States may be leading the pack in this regard, Canada and most European nations have also experienced a decline in the relative importance of households composed of a married couple with children, along with an increase in female-headed households with children present (McLanahan and Casper 1995).

Female-Headed Households

Between 1970 and the mid-1990s the number of female-headed families with children (mother-only families) went up from 2.8 million to 7.2 million in the United States, whereas the number of married-couple-with-children families went down

Figure 10.2   Households Have Become Considerably More Diverse in the United States

Source: S. Rawlings, 1994, "Household and family characteristics: March 1993," U.S. Bureau of the Census, Current Population Reports, Series P20-477, Table A.

from 25.5 million to 24.7 million. In 1970 female-headed households accounted for 10 percent of families with children, but by the mid-1990s that had jumped to 22 percent (see Figure 10.3). Even those numbers hide the true scope of the transformation, because a married-couple family in the mid-1990s was more likely than in 1970 to be a recombined family, involving previously married spouses and children from other unions. At the end of this chapter, I discuss the social impact of these transformations, but my goal at present is to describe the changes.

Widely discussed in public debate is the fact that substantial racial/ethnic differences exist with respect to female-headed households, especially among families with children, as shown in Figure 10.3. Although the rise in mother-only families has been experienced by whites, African-Americans, and Hispanics, in 1970 African-American families were already more apt to be headed by a female than were white or Hispanic households in the mid-1990s. By the mid-1990s nearly 60 percent of all African-American families with children were mother-only households. Ruggles (1994) has found that since at least 1880, African-American children have been far more likely to reside with only their mothers (or with neither parent) than white children. Nonetheless, the data in Figure 10.3 show that the percentage of African-American mother-only families has increased considerably just in the past few decades.

Figure 10.3   Racial/Ethnic Differences in the Percentage of Family Households with Children that are Mother-Only Families, United States

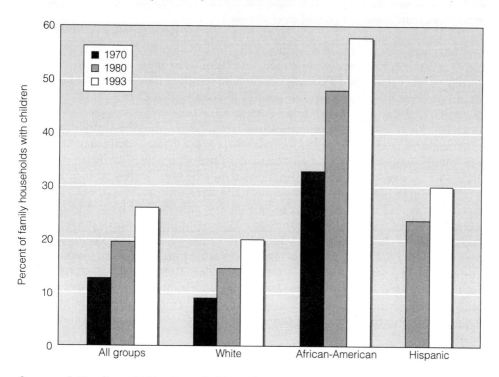

Source: S. Rawlings, 1994, "Household and family characteristics: March 1993," U.S. Bureau of the Census, Current Population Reports, Series P20-477, Table F.

Other "Nontraditional" Households

In the mid-1990s three out of ten households in the United States were nonfamily households, as you can see looking back at Figure 10.2. This is part of the trend away from what is often thought of as the traditional family, enshrined by old TV sitcoms—a family in which a married couple live together with their children and the husband works full-time while the wife cares for the children and attends to domestic chores (Hernandez 1993). In fact, this type of "Leave It to Beaver" family is a relatively new phenomenon historically—a product of the demographic transition. High mortality alone (but especially when combined with high fertility and an agrarian economic environment) prevented this type of household from being the norm for most of human history. Let me explain.

Human beings are by nature social animals. We prefer to live with others and nearly all human economic activities are based on cooperation and collaboration with other humans. Our identity as individuals is paradoxically dependent on our interaction with other people. We only know who we are by measuring the reaction of other people to us, and we depend on others to teach us how to behave and how to negotiate the physical and social worlds. Furthermore, humans are completely dependent creatures at birth, and every known society has organized itself into social units (households/families) to ensure the survival of children and the reproduction of society. In high-mortality societies, the rules about who can and should be a part of that social unit must be a bit flexible because death can take a mother, father, or other important household member at almost any time.

I referred in Chapter 4 to the Nuatl-speaking Mexican families in Morelos in the 16th century, who lived in a "demographic hell," where high mortality produced high rates of orphan- and widowhood. In response to these "vagaries of severe mortality," they developed a complex household structure that was "extremely fluid and in constant flux. Headship and household composition shifted rapidly because marriages and death occurred at what must have been a dizzying pace" (McCaa 1994:10). Zhao (1994) has used genealogies from China from the 13th through the 19th centuries to show that high mortality kept most Chinese from actually living in a multigenerational family at any particular time, although as in Mexico the high death rate produced a complex form of the family because of shifting membership.

The bottom line here is that what we think of as the "traditional" family is, in fact, not very traditional at all. It depends on low mortality, which is a historically recent phenomenon, in combination with a fairly young structure, characterized by young adults with their children. This combination of demographic processes is found largely in the middle phase of the demographic transition, when mortality has dropped, but fertility is still above replacement level. Over time, as mortality remains low and fertility drops to low levels, the population ages, as I discussed in the previous two chapters, and older married couples are left without children any longer in the household, and then the women are left without their husbands in the household.

Given the fact that household structure has historically responded to demographic conditions, it should be no surprise to you that in the post-World War II era, when demographic conditions have been undergoing tremendous change all over the globe, we should see a rise in what we think of as "nontraditional" households.

Let us now examine some of the direct proximate causes of the transformation in household structure that societies have been experiencing.

Proximate Determinants of Changing Household Structure

The increasing diversity in household structure is a result of several interdependent trends taking place in society. At the younger ages this is especially due to a delay in marriage, accompanied by young people leaving their parent's home (which has increased the incidence of cohabitation), and an increase in divorce (which also contributes to cohabitation); whereas at the older ages, the greater survivability of women over men has increased the incidence of widowhood and that has an obvious impact on family and household structure.

Delayed Marriage Accompanied by Leaving the Parental Nest

In general, in the United States the social penalties for early marriage have gradually eased as the economic well-being of the population has increased, divorce laws (and pressures against divorce) have eased, fertility control has increased, and social sanctions against premarital sexual activity have rather dramatically been lowered. In 1890, more than one third of all women aged 14 and older (34 percent) and close to one half of all men (44 percent) were single. Between 1890 and 1960, being single became progressively less common as women, and especially men, married at earlier ages. Only since the 1960s has there been a resurgence of delayed marriage. By 1996, the average (median) age at marriage for females had risen to 24.8, the highest level in U.S. history, while for males it had increased to 27.1, also higher than the level of 1890 (Saluter and Lugaila 1998). In fact, in 1996, 38 percent of all American women aged 25 to 29 were still single, compared with only 11 percent in 1970, and 52 percent of all men of that age had not yet married, a huge increase from only 19 percent in 1970. Changes in the popularity of early marriage have been roughly similar for both African-Americans and whites, although African-Americans have been more likely than whites to delay marriage or remain single (McLanahan and Casper 1995).

Since the 1960s the slowdown of economic growth, accompanied by the stiff competition for jobs brought about by the Baby Boom children growing up, has made it more advantageous for couples to postpone marriage to take maximum advantage of educational and career opportunities. And, very importantly, the contraceptive revolution, especially the birth control pill, has allowed people to disconnect marriage from sexual intercourse. Methods of birth control have been known about and used for a long time (and you can refer to Chapter 5 if you need a review). However, the failure rate of all of those pre-pill methods was higher than the pill, and a couple engaging in intercourse ran a clear risk of an unintended pregnancy. Prior to 1973 and the legalization of abortion in the United States, an American woman could end an unintended pregnancy only by flying to a country such as Sweden, where abortion was legal, or by seeking an illegal (and often dangerous) abortion.

In more traditional societies (including the United States and Canada until the 1960s), an unintended pregnancy was most apt to lead to marriage, although a woman might also bear the child quietly and give it up for adoption. Illegitimacy was widely stigmatized and having a child out of wedlock was the course of last resort. Marriage was the only genuinely acceptable route to regular sexual activity, and only married couples were routinely granted access to available methods of birth control.

In the late 19th century, the older age at marriage already alluded to in North America and Europe had been accomplished by a delay in the onset of regular sexual activity—the Malthusian approach to life. Intercourse was delayed until marriage, and in this way nuptiality was the main determinant of the birth rate: early marriage meant a higher birth rate, and delayed marriage meant a lower one (Wrigley and Schofield 1981). A variety of social and economic conditions might discourage an early marriage. The societal expectation that a man should be able to provide economic support for his wife and children tended to delay marriage until those expectations could be met. Under conditions of rising material expectations, as was the case in the late nineteenth century, marriage had to be delayed even a bit longer than in previous generations because the bar had been raised higher than before. Delayed marriage typically meant that young people stayed with their parents in order to save enough money to get ahead financially and thus be able to afford marriage. Staying with parents also minimized the opportunities for younger people to be able to engage in premarital sexual intercourse, which might lead to an unintended pregnancy and destroy plans for the future. In those days, delayed marriage did not lead younger people to leave home and set up their own household independently of their parents prior to marriage.

In the early post-World War II period, economic robustness allowed a young person to leave the parental home at an earlier age without an economic penalty and since the risk of pregnancy meant that intercourse was still tied closely to marriage, the age at marriage reached historic lows in the United States and Europe. In discussing the situation in Germany, Blossfeld and de Rose (1992) argue that "until the late 1960s, the opportunity for children to leave their parental home had increased remarkably because of the improvement of economic conditions. But the social norm that they had to be married if they wanted to live together with a partner of the opposite sex was still valid, so that age at entry into marriage was decreasing until the end of the 1960s" (p. 75). Similar arguments could be made for the United States and for other European countries.

Disconnecting sex from marriage has thus been accompanied by a dramatic rise in the fraction of young people who leave their parental home before marriage. In 1940, 82 percent of unmarried males and females aged 18 to 24 lived with their parents (Goldscheider and Goldscheider 1993); by the mid-1990s that had declined to 50 percent (Saluter 1993). "By the 1980s, leaving home before marriage had evidently become institutionalized in the United States" (Goldscheider and Goldscheider 1993:34). Especially telling are the data shown in Figure 10.4, drawn from an analysis of the National Survey of Families and Households (Goldscheider and Goldscheider 1994). Until the early 1960s the majority of women who left home after age 18 were doing so to get married. That pattern changed in the 1960s, and by the late 1980s only a third of women were leaving home to marry. Men, however, have been leaving home for

Figure 10.4 Women Are Increasingly Leaving Home for Reasons Other than to Get Married

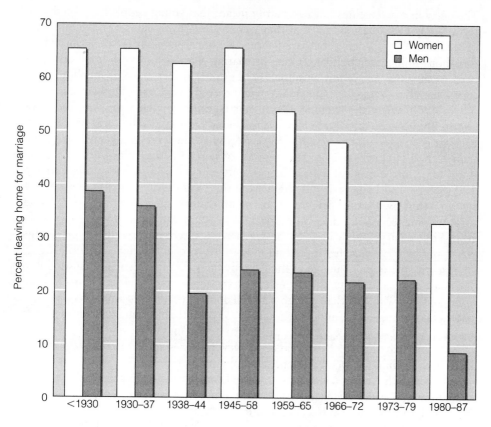

Source: Adapted from F. Goldscheider, and C. Goldscheider, 1994, "Leaving and returning home in 20th century America," Population Bulletin 48(4); Figures 4 and 5.

reasons other than marriage for a long time. So, the recent delay in marriage and nest-leaving behavior has been more significant as a force of change for women than for men.

You may have asked yourself how the data on increasing independence of young people can square with the wide array of stories about these same people returning home to live with their parents. The answer is that as the fraction of young people who leave home to live independently increases, the absolute volume of people available to move back in with parents increases. Children graduating from college and those returning from the military are especially likely to return home for a short while, whereas those who left home to marry are the least likely to return home (Goldscheider and Goldscheider 1994). Nonetheless, because children do not necessarily return for very long, a snapshot photo of society (such as the census or a survey) does not capture many young people who have returned home, even though a fairly high percentage may do so at some time or another.

Cohabitation

The delay in marriage has not necessarily meant that young people have been avoiding a family-like situation, nor that they have necessarily avoided having children out of wedlock. When leaving the parental home, young people may set up an independent household either by living alone or sharing a household with nonfamily members, or they may move into nonhousehold group quarters such as a college dormitory. From these vantage points, many then proceed to **cohabitation** before marriage.

In 1970, when the average age at marriage for women was 20.3, there were about 500,000 cohabiting couples in the United States, and the ratio of cohabiting to married couples was 1 to 100. By 1996, when the average age at marriage for women had climbed to 24.8, the number of cohabiting couples had increased to 4 million and the ratio of cohabiting to married couples also had jumped to 7 per 100 (Saluter and Lugaila 1998). Data from the 1992–94 National Surveys of Family Growth show that 49 percent of American women aged 25 to 39 had cohabited at some time, especially prior to marriage (Bumpass and Sweet 1995). Survey data from France suggest that in 1965 only 8 percent of couples cohabited before marriage, but by 1995 that figure had jumped to 90 percent (Toulemon 1997).

Despite the popular image in the United States that this trend was started in the 1970s by college students, the data show that cohabitation is more common among young people with less than a high school education than it is among college graduates (Thornton, Axinn, and Teachman 1995). As cohabitation has increased in frequency, American college students appear to be the imitators, not the perpetrators, of the trend (Cherlin 1992). In France, however, it does appear that college students have been leading the trend (Villeneuve-Gokalp 1991).

Although the rise in cohabitation has led commentators to predict the demise of the family in Western civilization, the reality appears to be that most of the increase in cohabitation in the United States and in other countries has been a prelude to marriage, rather than a substitute for marriage (Cherlin 1992; Leridon 1990). Only among the lower social classes has cohabitation remained a viable permanent option (Finnäs 1995).

The delay in marriage, accompanied as it has been by high rates of premarital sexual activity, is also associated with an increasing fraction of births out of wedlock—an event that immediately transforms a woman living alone, or an unmarried couple living together, from a nonfamily to a family household. Between 1970 and 1995, the birth rate for unmarried women in the United States nearly doubled, but there have been important differences within the population. For whites, the rate more than doubled, whereas for blacks the rate actually declined (although it remained more than twice the rate for whites), and most of the increase for unmarried white women occurred during the 1980s (Ventura, et al 1997). During this time the birth rate for married women had been declining and the proportion of young women who were not married had been going up, so the fraction of all births that were out of wedlock has been increasing more rapidly than would be indicated by just looking at the birth rate for unmarried women. In 1970 only 11 percent of all babies born in the United States were illegitimate, but that had jumped to 32 percent by 1995, ranging from 21 percent of babies born to non-Hispanic white mothers, to 38 percent of babies to Mexican-origin women, to 70 percent of babies to African-American mothers.

The desire to have children may determine the timing of formal marriage for many cohabiting couples, although in some instances the birth of the child may precede the marriage. Nonetheless, cohabiting couples seem to have much lower levels of fertility than married couples and, once married, couples who had cohabited prior to marriage appear to follow essentially the same pattern of family building as couples who had not cohabited (Manning 1995).

Increase in Divorce

Not only has marriage been increasingly pushed to a later age, but once accomplished, marriages also are more likely to end in divorce than at any previous time in history (see Preston and MacDonald 1979; Schoen and Weinick 1993). This trend reflects many things, including the loosening hold of men over women and the longer lives we are leading, both of which may produce greater conflict within marriage. In 1857 in the United States, there was only a 27 percent chance that a husband aged 25 and a wife aged 22 would both still be alive when the wife reached 65, but for couples marrying in the 1990s, the chances have more than doubled to 60 percent. Conversely, only about 5 percent of marriages contracted in 1867 ended in divorce (Ruggles 1997), whereas it has been estimated that more than half of the marriages contracted in 1967 will end in divorce (Martin and Bumpass 1989; Preston and MacDonald 1979).

The United States is certainly not unique in experiencing an increase in divorce probabilities. Goode (1993) compiled data for Europe showing that throughout Europe the percentage of marriages that will end in divorce virtually doubled between 1970 and the mid-1980s. For example, in Germany in 1970 it is estimated that 16 percent of marriages would end in divorce, increasing to 30 percent in 1985. In France, the increase went from 12 to 31 percent during that same period of time. Australia has experienced similar trends (Bracher, et al. 1993).

Given the previous discussion of cohabitation, it is of some interest to note that, contrary to popular belief in the value of "trial" marriage, cohabitation before marriage appears to be one of the factors that increases the odds that a marriage will end in divorce, at least in North America. Several studies have suggested that cohabitation is selective of people who are mistrustful of marriage and, once married, they may thus be more prone to end a marriage by divorce (Axinn and Thornton 1992; Hall and Zhao 1995).

Many marriages that in earlier days would have been dissolved by death are now dissolved by divorce. This seems apparent from the fact that the annual combined rate of marital dissolution from both the death of one spouse and divorce remained remarkably constant for more than a century. In Figure 10.5 you can see that the overall rate of marital dissolution was essentially unchanged between 1860 and 1970. As widowhood declined, divorce rose proportionately. Only with the rapid increase in divorce during the 1970s did that pattern begin to diverge. By 1985, the number of existing marriages dissolved by death had dropped to an all-time low (not unexpectedly, of course, since death rates were at an all-time low), and for the first time in history, divorces in the 1980s accounted for more than half of all dissolved marriages. The increased frequency of divorce shortened the average length of marriages in the United States from 36.4 years in 1960–66 to 25.2 years in 1972–76 (Goldman 1984). So dramatic was the rise in divorce in the 1970s that in the mid-1960s the elimination of divorce would have added an additional 6.7 years to the average marriage, whereas by the mid-1970s its elimination would have added 17.2 years.

Divorce is not the only way in which a marriage may be disrupted without requiring that one spouse die. African-Americans and Hispanics, for example, are more likely than whites to split up without a formal divorce. They show up in statistics among the people who are married but whose spouse is absent. In 1996, such people accounted for 3 percent of ever-married whites compared with 8 percent of African-Americans and 7 percent of Hispanics. One result is that marital disruption among African-Americans is almost twice as high as you would guess from looking only at divorce statistics. A second result is that, although divorce rates are lower among Hispanics than whites, the overall level of marital disruption is higher among Hispanics, because they are more than twice as likely to leave their spouse without getting a divorce (Saluter and Lugaila 1998).

The lower propensity of African-Americans to marry has been seen as both a cause and a consequence of the seeming disintegration of contemporary African-American family life (see, for example, Staples 1985), and some observers have attributed it to the disproportionately high fraction of young African-American males

Figure 10.5 The Annual Rate of Marital Dissolution Changed Little in the United States until the 1970s

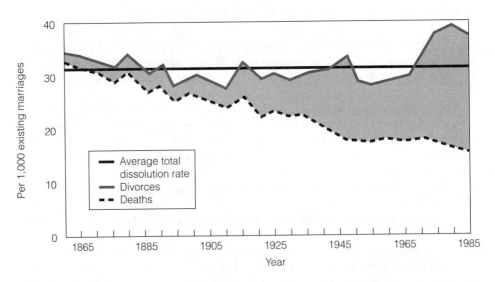

Note: Although the divorce rate has been rising for a long time, it had been just offsetting the drop in marital dissolutions from death until the 1970s. Thus, each year from the 1860s to the late 1960s, the number of total dissolutions per 1,000 existing marriages has remained remarkably stable. The total dissolution rate climbed in the 1970s, but dropped again in more recent years. The line represents the average total dissolution rate for the period 1860–1985.
Source: Adapted from K. Davis, 1972, "The American family in relation to demographic change," in C. Westoff and R. Parke, Jr. (eds.), U.S. Commission on Population Growth and the American Future, Vol. 1, Demographic and Social Aspects of Population Growth (Washington, D.C.: Government Printing Office), Table 8, updated to 1985 by the author.

who are in prison. Lichter and his associates (1992) analyzed data from the National Longitudinal Survey of Youth and concluded that too few men accounts for some but not all of the difference between whites and blacks in the transition to marriage, a finding mirrored in other studies (see, for example, Schoen and Kluegel 1988). The difference in marriage patterns has existed for many decades (Ruggles 1994), for reasons that are not well understood, and neither do we adequately understand why the gap has widened in the past few decades.

Widowhood

As death has receded to the older ages, the incidence of widowhood has steadily been pushed to the older years as well. Divorce is a more important cause of not being married than is widowhood up to age 65, beyond which widowhood increases geometrically because of the higher death rate of men, undoubtedly compounded by the tendency of divorced women to change their status to widow upon the death of a former husband. As is true with so many social facts in the United States, African-American women are disadvantaged relative to whites in terms of marital status. At every age, blacks are more likely to be either divorced or widowed.

The Combination of these Determinants

Table 10.1 summarizes the basic dimensions of the changing patterns of marriage and divorce in the United States that have contributed to the increasing diversity in

Table 10.1   Basic Dimensions of the Changing
Patterns of Marriage and Divorce in the United States

Category	Males		Females	
	1970	1988	1970	1988
Life expectancy at birth	67.2	70.7	74.8	78.5
Percent ever marrying of those surviving to age 15	96.0	83.5	96.5	87.9
Average age at first marriage	23.4	27.5	21.8	25.1
Percent of marriages ending in divorce	37.3	42.7	35.7	43.2
Percent of marriages ending in widowhood	18.7	18.3	45.3	39.3
Percent of marriages ending in own death	44.0	39.0	19.0	17.6
Percent of widowed persons remarrying	27.0	17.3	10.3	6.3
Percent of divorced persons remarrying	86.0	78.3	80.2	72.3
Average duration of a marriage	26.5	24.5	26.8	24.8
Percent of life spent never married	35.0	46.2	30.6	39.1
Percent of life spent currently married	58.4	44.8	50.5	41.2
Percent of life spent widowed	3.1	2.8	12.6	10.0
Percent of life spent divorced	3.6	6.2	6.3	9.6

Source: R. Schoen and R. M. Weinick, 1993, "The slowing metabolism of marriage: figures from 1988 U.S. Marital Status Life Tables," Demography, 30(4):737–746, Table 1.

HAS THE SEX RATIO AFFECTED HOUSEHOLD FORMATION?

In this chapter I point out that the rise in the status of women in the United States and other industrialized nations has been aided by a series of long-term demographic events, but that a catalytic force such as the women's movement was necessary to unleash the power built into demographic forces. Ironically, many analysts (for example, Davis and van den Oever 1982; Heer and Grossbard-Schechtman 1981) have argued that the women's movement itself was triggered by one of the major demographic phenomena of the Baby Boom—the marriage squeeze. The marriage squeeze is "an evocative though imprecise term used to describe the effects of an imbalance between the number of males and females in the prime marriage ages" (Schoen 1983:61). It is a by-product of the combination of different-sized cohorts and the fact that women do not usually marry men their own age.

You may recall from Chapter 8 that, although more boy babies are born than girl babies, higher male mortality evens the sex ratio by early adulthood. Thus a young woman expecting to marry someone her own age would have virtually a 100 percent chance of finding someone. However, in nearly every human society, women typically marry someone older than themselves. Now, if there are fewer older males, due to the fact that the number of births had been lower prior to the woman's year of birth, her probability drops of finding someone of the usual marriageable age. Since men her own age may not be expecting to marry for another 2 to 3 years, she too may be forced to postpone marriage.

American and Canadian women, on average, marry a man who is 2 years older, and over time the vast majority of American women have preferred marriage (at least at some point in life) to remaining single. Assuming that people will continue to want to marry, the imbalance in the number of males or females can lead to a delay in marriage for the sex that has the excess numbers. This is exactly what has happened to women of the Baby Boom. In the accompanying table you can see that in 1950 there were 98 men aged 22 to 26 for every 100 women 2 years younger (aged 20 to 24). Thus there was near equality in the number of males and females of typical marriageable ages. If we assume that women preferred men 3 years older

than themselves (the age gap in 1950 was 3 years and it has since dropped to 2), the ratio was even closer to equity. There were 99 men aged 23 to 27 in 1950 per 100 women aged 20 to 24. But that ratio dropped precipitously through the years when the pre-Baby Boom and Baby Boom cohorts were reaching marriage age, because in nearly every year from 1937 through 1957 there were more people born in the United States than in the previous year. Women looking at the age group 2 to 3 years older kept seeing fewer men than women of their own age. The result was that many women were forced to postpone marriage, a phenomenon I have already discussed in this chapter, and which also has been well documented for Canada (Foot 1996).

By 1980 relative equity had been achieved again in the sex ratio at the usual marriage ages, but by that time some remarkable changes had already taken place in American gender roles. "The Women's Liberation Movement may be interpreted, on the one hand, as a collective means by which women helped themselves to reorient each other to the new lower compensation for the traditional female role and, on the other, as the means by which the increasing number of women outside of the traditional wife-mother role sought to combat the discrimination meted out to women in the job world" (Heer and Grossbard-Schechtman 1981:49). Either interpretation can be viewed as women's responses to the demographic forces at work in the marriage market (Grossbard-Schechtman 1985; 1993). In the late 1980s and early 1990s North American *males* were experiencing a marriage squeeze as the declining number of births through the 1960s and early 1970s produced a dearth of women of the usual marriage age (see the accompanying table). This led to a better bargaining position for women in the marriage market, and almost certainly helped in some measure to improve gender equality.

Guttentag and Secord (1983) used the concept of sex ratio to develop their theory of gender relations, based on Becker's new household economic theory. In general, their hypothesis is that a shortage of women relative to men encourages domesticity, marital stability, and a maintenance of male dominance over women. Conversely, a low sex ratio (more women than men) encourages women to find alternatives to traditional role models, since there are too few men of marriageable age. Guttentag and Secord used 19th-century America as an example. On the frontier the sex ratio was high and female roles were typically very traditional, with an emphasis on family building and family life. At the same time, the New England states had a much lower sex ratio and it was in that part of the country, perhaps not coincidentally, that the suffrage movement began.

African-Americans have for several decades had a sex ratio lower than that of whites because of higher male mortality from birth on. The sex ratio has been further lowered because of the higher proportion of African-American than white males in prison. Guttentag and Secord argue that this low sex ratio has led to greater independence among African-American women and to lower marital stability as well, because a relative surplus of women allows men to be choosier and thus more assertive.

As always in the social sciences, however, cause and effect is maddeningly hard to pin down. In those nations of the world where females are most oppressed, especially in the Middle East and Africa, sex ratios tend to be among the highest in the world, precisely because the low status of women leads to higher mortality among women than among men. So far, the high sex ratio has not helped to elevate the status of women in such cultures. Help may be coming, however. Schoen (1983) noted that a number of developing countries increasingly are experiencing marriage squeezes as a result of the declines in infant mortality since the end of World War II that have pro-

duced a demographic effect very similar to the American Baby Boom. The relative scarcity of husbands for women in those countries may help lead to the delays in marriage that I have identified as being so crucial for women in avoiding the oppression of early motherhood and the maintenance of male domination.

As Schoen predicted, marriage squeezes have been increasingly sighted in developing countries. China is experiencing this problem (Tien, et al. 1992), as is India, where the marriage squeeze has been implicated as a factor forcing families to use ever more of their limited resources for a dowry in order to marry their daughters in an increasingly competitive marriage market (Rao 1993). In sub-Saharan Africa the marriage squeeze may have increased the number of polygamous marriages (running counter to the idea that marriage squeezes necessarily work to improve the status of women by delaying marriage), and in Brazil the rise in cohabitation has been linked to the marriage squeeze in that country (Greene and Rao 1995). Goodkind (1997) has even discovered what he calls a "double marriage squeeze" for Vietnamese. Within Vietnam the combination of population change, war, and excess male migration produced a shortage of men and thus a marriage squeeze for females, leading to a rise in the age at marriage and a decrease in the marriage rate. At the same time, the male migrants who left the country have been facing an overseas marriage squeeze because there are not enough overseas Vietnamese women to go around, and so the men too have had to delay or forego marriage.

The Sex Ratio of the Baby Boom Produced a Dearth of Potential Husbands

| | Year | | | | |
Cohort	1950	1960	1970	1980	1990
Men aged 22–26 / Women aged 20–24	98	84	87	98	108
Men aged 23–27 / Women aged 20–24	99	86	85	96	111

Sources: U.S. Bureau of the Census, 1983, 1980 Census of Population, General Population Characteristics (Washington, D.C.: Government Printing Office), Table 2; and U.S. Bureau of the Census, 1982, "Projections of the population of the United States: 1982 to 2050 (Advance Report)," Current Population Reports, Series P-25, No. 922, Table 2.

household structure. Using life table methodology, Schoen and Weinick (1993) estimated the changing probabilities of marital events between 1970 and 1988. The data show, for example, that between 1970 and 1988, the proportion of women surviving to age 15 who can expect ever to marry declined from 96.5 percent to 87.9 percent. At the same time, the average age at marriage was increasing, the percentage of marriages ending in divorce was increasing, and the percentage of marriages ending in widowhood was declining. Furthermore, as life expectancy increases while the average duration of a marriage shortens, and the percentage of divorcing people remarrying goes down, the percentage of a person's life spent being currently married has declined from 50.5 percent in 1970 to 41.2 percent in 1988.

Now having described the principal features of the transformation of families and households in industrialized societies, let us see how it is that they have been influenced by changing life chances.

Changing Life Chances and the Transformation of Families and Households

The leading explanations for the shift in household structure in Western nations combine elements of the demographic transition perspective with the life course perspective. The demographic transition perspective relates changing demographic conditions to the rise in the status of women, which has altered life chances by offering women greater economic and social freedom to choose their own pattern of living. This rise in status is exemplified especially by the increase in female labor force participation, which affords women the option of economic independence. As these changes have unfolded, the differing life chances of women have contributed to the transformation of families and households. To understand how these changes arose, however, it helps to review the demographic conditions that facilitated the domination of men over women, and the demographic conditions that have helped to loosen that domination.

Demographic Conditions Facilitating Male Domination

By now you are familiar with the fact that pronatalist pressures are strong in societies characterized by high mortality and high fertility, especially agricultural societies. In those areas, several children must be born just to ensure that enough will survive to replace the adult membership. Thus one component of the social status of women is that with a regime of high mortality, women are busy with pregnancy, nursing, and child care, and men, who are biologically removed from the first two of these activities, are able to manipulate and exploit women by tying the status of women to their performance in reproduction and the rearing of children.

Furthermore, high mortality means that childbearing must begin at an early age, because the risk of death even as an adult may be high enough that those younger, prime reproductive years cannot afford to be "wasted" on activities other than family building. In a premodern society with a life expectancy of about 30 years, fully one third of women aged 20 die before reaching age 45, making it imperative that childbearing begin as soon as possible.

Women who marry young and begin having children may be "twice-cursed"—having more years to be burdened with children and also being in a more vulnerable position to be dominated by a husband. Men need not marry as young as women since they are not the childbearers, and they also remain fecund longer. The older and more socially experienced a husband is relative to his wife, the easier it may be for him to dominate her; it is no coincidence that in the predominantly Muslim countries of the Middle East, where women are probably less free than anywhere in the world, men are about 8 years older than their wives on average. By contrast, in the United States since 1970 the age gap at first marriage has been 2 years (Clarke 1995).

Among the uglier aspects of the way in which men have exercised dominance over women is the practice of female genital mutilation (FGM), sometimes known as female circumcision, but involving practices that are technially *clitoridectomy* and *infibulation*. Clitoridectomy accounts for about 80 percent of such mutilation practices (Chelala 1998) and involves the total removal of the clitoris, whereas infibulation involves cutting the clitoris, the labia minora, and adjacent parts of the labia majora and then stitching up the two sides of the vulva. These are useless and dangerous practices to which an estimated 2 million girls and women in at least 28 countries in northern and sub-Saharan Africa and parts of Asia are subjected annually. As was true for footbinding in China, the purpose is to "control access to females and ensure female chastity and fidelity" (Mackie 1996).

The migration of refugees from African countries such as Somalia, where the practice is common, to North America and Europe helped to ignite worldwide knowledge of and concern about female genital mutilation and there are now international movements in place to bring pressure on governments to make it illegal. A significant event occurred in 1997 when the highest court in Egypt upheld the government's ban on FGM. There is evidence as well that educated women are less willing to submit to FGM, so the trend in Africa toward increasing female education bodes well for the eventual elimination of this form of male domination (The Economist 1996).

Demographic Factors Facilitating Higher Status for Women

Three demographic processes—a decline in mortality, a drop in fertility, and increasing urbanization—have importantly influenced the ability of women to expand their social roles and improve their life chances. A major factor influencing the rise in the status of women has been the more general liberation of humans from early death. The decline in mortality does not mean that pressures to have children have evaporated. That is far from the case, but there is a greater chance that the pressures will be less; indeed, remaining single and/or childless is more acceptable for a woman now than at any time in history.

In North America, Europe, and much of Latin America and Asia, most of a woman's adult lifetime is now spent doing something besides bearing and raising children, because she is having fewer children than in previous generations and she is also living longer. An average American woman bearing two children in her late 20s and early 30s would, at most, spend about 20 years bearing and rearing them.

That many years is far fewer, of course, than she will actually have of relative (indeed increasing) independence from child-rearing obligations, since if her two children are spaced 2 years apart and the first child is born when she is 30, then by age 38 her youngest child will be in school all day, and she will still have 43 more years of expected life. Is it any wonder, then, that women have searched for alternatives to family building and have been recruited for their labor by the formal economic sector? In a sociocultural setting in which the reproduction of children consumes a great deal of societal energy, the domestic labor of women is integral to the functioning of the economy. With a slackening in demand for that type of activity, it is only natural that a woman's time and energy would be employed elsewhere, and elsewhere is increasingly likely to be in a city.

In contrast to rural places, cities provide occupational pursuits for both women and men that encourage a delay in marriage (thus potentially lowering fertility) and lead to a smaller desired number of children within marriage. Furthermore, since urbanization involves migration from rural to urban areas, this has meant that women, as they migrated, were removed from the promarital and pronatalist pressures that may have existed in their parents' homes. Thus migration may have led to a greater ability to respond independently to the social environment of urban areas, which tends to devalue children. Young adults are especially prone to migration and every adult who moves may well be leaving a mother behind. That means that the migrant will have more time on her hands to look for alternatives and to question the social norms that prescribe greater submissiveness, a lower status, and fewer out-of-the-home opportunities for women than for men.

I should note, however, that historically the process of urbanization in the Western world initially probably led to an increase in the dependency of women before influencing liberation (Nielsen 1978). Urbanization is typically associated with a transfer of the workplace from the home to an outside location—a severing of the household economy and the establishment of what Kingsley Davis (1984) has called the "breadwinner system," in which a member of the family (usually the male) leaves home each day to earn income to be shared with other family members. In premodern societies, women generally made a substantial contribution to the family economy through agricultural work and the marketing of produce (Boserup 1970), but the city changed all that. Men were expected to be breadwinners (a task that women had previously shared), while women were charged with domestic responsibility (tasks that men had previously shared). From our vantage point in history, the breadwinner system seems "traditional," but from a longer historical view, it is really an anomaly:

> The breadwinner system develops slowly in the early phase [of development], characterizing the burgeoning but small middle class rather than the peasantry or the proletariat. Then, after reaching a climax in which virtually no married women are employed, the arrangement declines as more and more wives enter white-collar employment in offices, schools, hospitals, stores, and government agencies. In the United States the heyday of the breadwinner system was from about 1860 to 1920 [Davis 1984b:404].

As the life expectancy of the urban woman increased and as her childbearing activity declined, the lack of alternative activities was bound to create pressures for change.

In Figure 10.6 I have diagrammed the major paths by which mortality, fertility, and urbanization influence the status of women and lead to more egalitarian gender roles. Increased longevity lessens the pressure for high fertility and lessens the pressure to marry early. These changes permit a woman greater freedom for alternative activities before marrying and having children, as well as providing more years of life beyond childbearing. Women are left to search for the alternatives and society is offered a "new" resource—nondomestic female labor. Urbanization reinforces the low fertility norms while freeing women from the constraints of traditional views held by agrarian family members. The urban economic environment typically offers greater access to education and to jobs outside the home, which combine to increase the social and economic independence of women.

Again, I emphasize that these demographic conditions are necessary, but not sufficient, to initiate the current rise in the status of women in industrialized societies. What is also required is some change in circumstance to act as a catalyst for the underlying demographic factors. This is where the life course perspective is drawn upon, because women who have grown up in a different demographic and social milieu, and thus see the world differently than did earlier generations of

Figure 10.6   Demographic Components of the Changing Status of Women

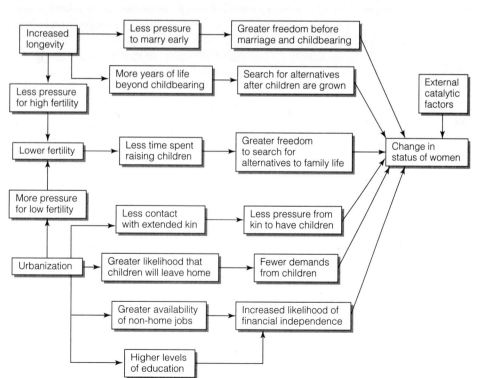

Note: Increased longevity, lower fertility, and urbanization are demographic processes that have contributed to the rise in the status of women in developed nations.

women, have the potential to generate change in society. The women's movement in North America, for example, seems to have provided that catalytic force, operating especially through the mechanism of increasing the job opportunities for women. Now, how do these changes translate into delayed marriage, increasing marital dissolution, and a restructuring of families and households? One part of the explanation is based on the way in which the changing characteristics of people (and the households in which they grow up) affect life. Nothing is more important than education.

Education

Becoming educated is probably the most dramatic and significant change that you can introduce into your life. It is the locomotive that drives much of the economic development throughout the world, and it is a vehicle for personal success used by generation after generation of people in the highly developed nations of the world. However, the relative recency with which advanced education has taken root in American society can be seen in Figure 10.7. In 1940, only one out of four Americans aged 25 or older had graduated from high school, and less than one in ten was a college graduate. A historically short five decades later, three of every four people had high school diplomas and one in five had graduated from college. Even higher proportions of people are educated at the younger adult ages, where recent gains are primarily registered. In 1970, 16 percent of Americans aged 25 to 34 had completed 4 or more years of college, but by 1990 an incredible 70 percent of all people of that age had been to college and 24 percent had completed at least 4 years.

Of special importance is the fact that women have been increasing their level of educational attainment relative to men. Although gender equality was reached several decades ago with respect to graduating from high school, only since the 1970s has the college gap been wiped out. In 1970 women were just as likely to graduate from high school as men, but only 17 percent of female high school graduates aged 25 to 34 had gone on to complete college, compared with 26 percent of men. By 1996, a slightly higher percentage of women than men had completed high school in the United States, and a slightly higher fraction of those female high school graduates had become college graduates than was true for men. The United States mirrors Canada in the trend for women to surpass the educational achievements of men. Data from the 1996 Census of Canada show that 20 percent of women aged 20 to 44 held a university degree, compared with 19 percent of men that age.

The world as a whole has been experiencing an increasing equalization of education among males and females—an important component in raising the global status of women. The data in Figure 10.8 show that between 1970 and 1990, the ratio of females per 100 males attending secondary school increased worldwide from 68 to 76. In some areas of the world, such as North America (not shown in the graph) and Europe, women have already achieved parity with men, or are even more likely than men to be enrolled in secondary school (that is especially the pattern in Central and Eastern Europe). In sub-Saharan Africa and south Asia, where the status of women has been notably low, there are nonetheless notable improvements in the ratio of girls to boys attending school.

Figure 10.7   Educational Attainment Has Increased Significantly Over Time in the United States (1940 to 1990)

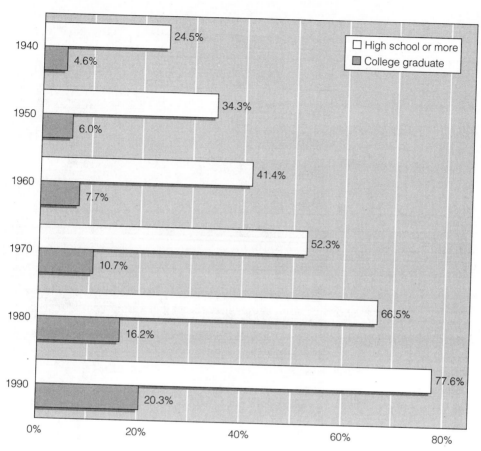

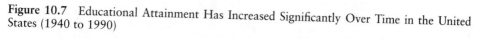

Sources: U.S. Bureau of the Census, 1983, 1980 Census of Population, Vol. One, General Social and Economic Characteristics, U.S. Summary (Washington, D.C.: Government Printing Office), Table 83; U.S. Bureau of the Census, 1991, Current Population Reports, Series P-20, No. 449, Table 1.

Our interest in education lies especially in the fact that in altering your world view, education tends to influence nearly every aspect of your demographic behavior and outcomes, as I have discussed in varying degrees in the previous chapters. Data from censuses and from sources such as the Demographic and Health Surveys show that nearly anywhere you go in the world, the more educated a woman is, the fewer children she will have had. Not that education is inherently antinatalist; rather, it opens up new vistas—new opportunities and alternative approaches to life, other than simply building a family—and in so doing, it delays the onset of childbearing, which is a crucial factor in setting the tone for subsequent fertility (Marini 1984; Rindfuss, Morgan, and Offutt 1996). Among cohorts of women in the United States there is, in fact, a striking relationship between higher education

Figure 10.8   Women Are Improving Their Educational Status Relative to Men All Over the Globe

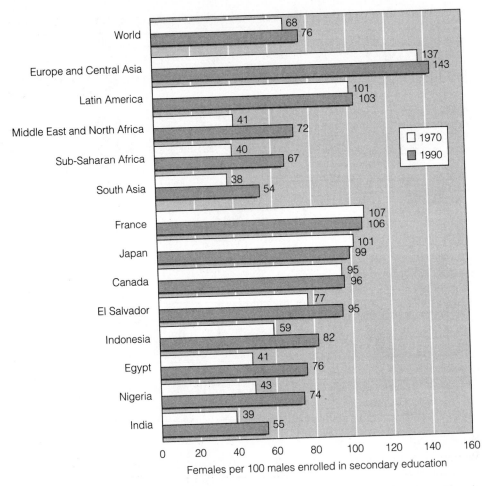

Females per 100 males enrolled in secondary education

Source: Adapted from World Bank, 1997, World Development Indicators (accessed on the internet in 1998).

and childlessness (Bloom and Trussell 1984). Even in the United States, where fertility levels are among the lowest in the world, there is a significant difference in childbearing behavior between the two extremes of educational attainment. Among women aged 25 to 34, when interviewed in 1995 by the U.S. Bureau of the Census, 57 percent of college graduates were still childless, compared with only 15 percent childlessness among women that age who lacked a high school diploma. Looking ahead to the rest of their lives, college graduates expect to have fewer children than those who had not finished high school and, once born, it also turns out that the children of better-educated women have a greater chance of survival through infancy and childhood (see, for example, Bourne and Walker 1991; Cramer 1987).

The fact that education alters the way you view the world also has implications for the marriage market in the United States. For most of American history, a major concern in choosing a marriage partner was to pick someone who shared your religious background (social scientists call this "religious homogamy"). Over time, however, the salience of religion has given way to "educational homogamy"—people want to marry someone with similar amounts of education, and so education has been replacing religion as an important factor in spouse selection (Kalmijn 1991; Mare 1991).

The greater proportion of people going to college has altered the life-styles of many young Americans. It has been accompanied by delayed marriage, delayed and a diminished amount of childbearing, and, consequently, higher per-person income among young adult householders. Young people today have more personal disposable income than almost any previous generation. In 1996, women consistently earned less than men (which I will discuss later in the chapter), but for both men and women, the data in Table 10.2 show that those with a post-graduate degree were earning more than twice as much per year as were those who were only high school graduates. This has become known as the "college premium," and its existence has almost certainly contributed to increased education (investment in human capital) and increased economic productivity (Phillips 1996). For the investment in education to pay off for people, households, and society more generally, it must lead to higher levels of labor force participation and more productive occupations. Let us examine these aspects of your life chances.

Labor Force Participation

As education increases, so does the chance of being in the labor force. Among both males and females in the United States in 1996, the higher the level of education attainment among people aged 25 and older, the higher the percentage of people who were currently in the labor force. For example, 85 percent of male college graduates

Table 10.2   Better Educated Workers Have Higher Incomes

Educational Attainment	Annual Median Income		Ratio of Female to Male Income
	Males	Females	
Less than high school	$20,464	$14,345	0.70
High school graduate	28,542	19,649	0.69
Some college	32,684	23,776	0.73
College graduate	42,602	30,798	0.72
Post-graduate degree	58,388	40,193	0.69
All workers	$32,858	$23,619	0.72

Note: Data are for year-round full-time workers aged 25 and older in the United States in 1996.
Source: Jennifer C. Day and Andrea Curry, 1997, "Educational attainment in the United States: March 1996 (Update)," *Current Population Reports*, P20-493: Table 9.

were in the labor force, compared with 76 percent of males with only a high school diploma. The differences are even greater for females: 75 percent of college graduates were working, compared with only 56 percent of women with only a high school diploma. Women are less likely than men to be in the labor force at any given level of education, but it is nonetheless true that the pattern over time has been for women to be working more, and men to be working less. Prior to the 1970s, for example, women who worked typically did so only before they married or became pregnant. Thus the labor force participation rates peaked in the early 20s and decline after that. That is still a common pattern in many developing countries, but it no longer characterizes women in the United States, Canada, and most of Europe. Labor force participation rates by age are now very similar for males and females.

Keep in mind as we talk about labor force participation rates that most countries include unemployed persons as being in the labor force. Thus if you are looking for work, even though not actually working or even if you have never before held a job, you are considered to be in the labor force. Unemployment rates are strongly related to age—the older the age, the lower the rate. At the younger ages there are considerable numbers of people looking for work even if they haven't found it yet, whereas at the older ages people are more likely to give up on employment and seek a retirement pension as soon as it is available, if they experience difficulty finding a job. Women also tend to have lower unemployment rates than do men.

By far the biggest gain in employment over the past few decades has been the movement of Baby Boom women into the labor market. They literally burst their way into the work force in the 1970s, and by 1996, 77 percent of women aged 35 to 44 (the boomers who entered the labor force ages in the 1970s) had a job. The biggest increase in labor force participation brought about by the baby boomers was to open up the U.S. labor market to married women. Table 10.3 shows that in 1960, just before the baby boomers come of age, labor force participation rates among married people in the United States were well below those for single women at every age from 20 through 64. By 1970, as the first wave of baby boomers had reached adulthood, the rates for married women were clearly on the rise, and by 1995 they had nearly reached those of single women.

Working, as I have mentioned before, cuts down on fertility under normal circumstances. It is certainly no coincidence that the birth rate in the United States began to drop at about the same time that the labor force participation rates for married women began to rise. For example, in 1995, among women aged 25 to 34, those in the labor force had given birth to 1.1 children, compared with 1.9 children born to women not in the labor force. Some of that difference is due to the fact that working women, even those currently married, are more likely to be childless than women who are not in the labor force (Bachu 1997). Women who do work outside the home tend to be optimistic about their future plans for having children, but even if we combine children already born with those that women expect in the future, working women still have an average of half a child less (Bachu 1993).

The relationship between labor force participation of women and their fertility has always been clouded by a chicken-and-egg argument over which causes which—does low fertility produce greater opportunities for work, or does working cause a woman to reduce her family building activity? The evidence suggests that, in truth,

Table 10.3 Married Women Have Dramatically Increased Their Labor Force Participation in the United States

Marital Status and Year:	Female Labor Force Participation Rate:			
	20–24	25–34	35–44	45–64
Single				
1960	77.2	83.4	82.9	79.8
1970	73.0	81.4	78.6	73.0
1980	75.2	83.3	76.9	65.6
1990	74.7	81.2	81.0	66.1
1995	72.9	80.2	79.5	67.3
Married:				
1960	31.7	28.8	37.2	36.0
1970	47.9	38.8	46.8	44.0
1980	61.4	58.8	61.8	46.9
1990	66.5	69.8	74.0	56.5
1995	64.7	72.0	75.7	62.7

Source: U.S. Bureau of the Census, 1996, The American Almanac 1996–1997: Statistical Abstract of the United States (Austin, TX: Reference Press), Table 624.

both factors are involved (Cramer 1980; Groat, et al. 1982). Traditionally, women who worked after marriage were disproportionately those who were subfecund. Thus it was low fertility among such women that may have encouraged them to remain in or go back into the labor force. More recently, however, it is clear that a high proportion of women work before marriage, remain working after marriage, and adjust their fertility downward to accommodate their working. This is true not just in North America, but in other parts of the world as well.

In Mexico an increase in labor force participation has accompanied a decline in fertility. In 1979, when the total fertility rate in Mexico was 5.0, only 21.5 percent of women were in the labor force; by the mid-1990s the total fertility had dropped to 3.4 and the female labor force participation rate had climbed to 31.4 percent (Fleck and Sorrentino 1994). It is unlikely that one directly caused the other; rather, each change helped stimulate the other. An initial decline in fertility, for example, may lead to a small increase in the percentage of women in the labor force, which may encourage other women to limit fertility, which may promote other opportunities in the labor force. Each event cumulatively causes the other, and when two things occur together, it is difficult to sort out cause and effect.

In a similar vein, a study of two seemingly different villages in India revealed that in the village where women were employed rolling cigarettes for a local contractor, the employment gave women, very serendipitously, substantial autonomy and increased their use of contraception. In another village without employment opportunities, the researchers expected to find that contraceptive use had remained low. Yet, in that village the time spent not working had been spent instead on additional education for women and that, it seems, had increased contraceptive utilization (Dharmalingam and Morgan 1996). There are multiple and cumulative paths to low fertility and to improved social situations for women.

A clear concomitant (if not cause) of the rise in female labor force participation in the United States has been a marked increase in women's wages. In fact, in the United States between 1947 and 1977, "female wage rates appear to be the dominant factor in explaining variations in fertility and female labor force participation . . . with increases in female earning leading to both depressed fertility and increased labor force participation of women" (Devaney 1983:147). Another study using data from the National Longitudinal Survey of Young Women has shown that the higher the wages on her job, the less likely is a woman to quit because of a pregnancy (Felmlee 1984). Using data from the 1970, 1980, and 1990 censuses, McLanahan and Casper (1995) have estimated that 70 percent of the decline in marriage among white women in the United States between 1970 and 1990 is accounted for by the increase in women's earning power, whereas 8 percent of the change can be attributed to a decline in the earning power of men.

New Household Economics and Female Labor Force Activity

We have already discussed the new household economics as an approach to explaining why fertility is kept low in developed societies and why households might encourage family members to migrate. Now, we call on it again to explain why the rise in the status of women and increased female labor force participation might generate the household transformations we have been reviewing in this chapter. The basic idea is that "the rises in women's employment opportunities and earning power have reduced the benefits of marriage and made divorce and single life more attractive. While marriage still offers women the benefits associated with sharing income and household costs with spouses, for some women these benefits do not outweigh other costs, whatever these may be" (McLanahan and Casper 1995:33).

In most social systems, people who can take care of themselves and have enough money to be self-reliant have higher status and greater freedom than those who are economically dependent on others. Further, a pecking order tends to exist among those who are economically independent, with higher incomes being associated with higher status than are low incomes. Being independent, though, is definitely the starting point, and an increasing number of women are arriving at that point.

Although mortality and fertility have been declining since the 19th century in the United States and urbanization has been occurring throughout that time, it was during World War II that the particular combination of demographic and economic circumstances arose to provide the leading edge of a shift toward equality of the sexes. The demand for armaments and other goods of war in the early 1940s came at the same time that men were moving out of civilian jobs into the military, and there was an increasing demand for civilian labor of almost every type. Earlier in American history, the demand for labor would have been met by foreign workers migrating into the country, but the Immigration Act passed in the 1920s (see Chapter 14) had set up national quotas that severely limited immigration. The only quotas large enough to have made a difference were those for immigrants from countries also involved in the war and thus not a potential source of labor.

With neither males nor immigrants to meet the labor demand, women were called into the labor force. Indeed, not just women per se, but more significantly, married women, and even more specifically, married women with children. Single women had been consistently employable and employed since at least the beginning of the century, as each year 45 to 50 percent of them had been economically active, as I already pointed out earlier. But in the early 1940s there were not enough young single women to meet labor needs, partly because the improved economy was also making it easier for young couples to get married and start a family. It was older women, past their childbearing years, who were particularly responsive to making up the deficit in the labor force (Oppenheimer 1967).

These were the women who broke the new ground in female employment in America, with the biggest increase in labor force participation between 1940 and 1950 coming from women aged 45 to 54, and because more than 92 percent of those women were married, this obviously represented a break with the past. Who were these women? They were the mothers of the Depression, mothers who had sacrificed the larger families that they wanted (see Chapter 6) in order to scrape by during one of America's worst economic crises. They were women who had smaller families than their mothers and thus were more easily able to participate in the labor force. However, the ideal family size remained more than three children, and the improved economy permitted the low fertility of the 1930s to give way to higher levels in the 1940s and 1950s. Women with the small families from the Depression actually opened the door to employment for married women, but younger women were not ready to respond to those opportunities in the 1940s and 1950s. Indeed, between 1940 and 1950 the labor force activity rates of women aged 25 to 34 actually declined. Those women were busy marrying and having children rather than becoming wage earners. Between 1940 and 1950 the percentage of women aged 20 to 24 who were married increased from 53 percent to 68 percent. In 1940 the total fertility rate in the United States was 2.3 children, and 10 years later it had gone up to 3.2—an increase of nearly one child per woman—and that was only the middle of the rise. By 1960 the total fertility rate had soared to 3.7 children per woman. The postwar Baby Boom, though, was an aberration in American history, a large bump on a long road toward low fertility, and in the late 1960s the number of babies women were having resumed its decline.

Between 1950 and the mid-1990s there was a substantial increase in the number and proportion of American women who were in the labor force and earning independent incomes. In 1950, for example, there were 29 female, year-round, full-time workers for every 100 males in that category; by the mid-1990s, there were 69 females working full-time, year-round per 100 male workers. This increase in labor force activity was accomplished especially by younger women. Each cohort of women born before the Depression began adulthood with approximately one third of its members in the workplace. But the younger the cohort, the more rapidly did labor force involvement increase, especially beyond ages 25 to 34, when the children tend to be in school. The Depression and World War II cohort started out at higher levels of labor force participation than had previous groups, but the rapid increase in labor force participation coincided with the unprecedented increase in working experienced by the Baby Boom cohorts.

Getting a job is one thing, but the kind of job you get—your occupation—depends heavily on education, and also is influenced by factors such as gender and race/ethnicity.

Occupation

Occupation is without question one of the most defining aspects of a person's social identity in industrialized society. It is a clue to education, income, and residence—in general, a clue to life-style and an indicator of social status, pointing to a person's position in the social hierarchy. From a social point of view, occupation is so important that it is often the first (and occasionally the only) question that a stranger may ask about you. It provides information about what kind of behavior can be expected from you as well as how others will be expected to behave toward you. Although such a comment may offend you if you believe that "people are people," it is nonetheless true that there is no society in which all people are actually treated equally.

Since there are literally thousands of different occupations in every country, we need a way of fitting occupations into a few slots. The U.S. Census Bureau has devised such a classification scheme to divide occupations into several mutually exclusive categories, and in Table 10.4 I have listed the occupational distribution of employed males and females in the United States as measured in the Current Population Survey in 1996. These occupational categories were developed originally for the 1980 census. Prior to that time, the broad categories had included terms such as "white collar," "blue collar," "service," and "farm." In their turn, the categories shown in Table 10.4 have been modified somewhat for the Census 2000, as I mentioned in Chapter 2. Statistics Canada developed a slightly different classification scheme for occupations in Canada and the bottom panel of Table 10.4 shows you the distribution of data from the 1996 Census of Canada.

A striking feature of occupational distributions in both the United States and Canada is the clear difference between males and females. You may already have appreciated this difference from your own observation of human society, but the numbers bring the point home (Anker 1998). In the United States, you can see that some occupations such as "precision production, craft, and repair" and "transportation and material handling" are almost entirely populated by men, whereas "administrative support, including clerical" is especially likely to involve women. The names of the categories are different for the Canadian data, but the gender differences are similar.

One of the distinguishing characteristics of people holding the higher status occupational positions in North American society (indeed, everywhere in the world) is that they tend to be better educated than those people holding the jobs at the other end of the occupational ladder. The professional specialty occupations in the United States, for example, are almost by definition jobs that require a college education, and nearly three fourths of both men and women in this category hold at least a bachelor's degree, as you can see in Table 10.5. In the executive, administrative, and managerial occupations a majority of men hold college degrees, whereas a smaller fraction of women do. Fewer than one in ten of the workers in the less prestigious

Table 10.4 Occupational Distributions Are Different for Males and Females

United States, 1996

	Males	Females
Executive, administrative, and managerial	16.3%	14.6%
Professional specialty	14.0%	19.3%
Technicians and related support	2.7%	3.9%
Sales	11.2%	11.0%
Administrative support, including clerical	5.7%	25.1%
Private household	0.0%	1.1%
Other service	8.5%	14.4%
Farming, forestry, and fishing	3.6%	1.0%
Precision production, craft, and repair	19.1%	2.1%
Machine operators, assemblers, and inspectors	7.4%	5.3%
Transportation and material moving	7.3%	0.9%
Handlers, equipment cleaners, helpers, and laborers	4.4%	1.4%
Number of people	55,035,000	47,857,000

Canada, 1996

	Males	Females
Management occupations	11.3%	6.2%
Business, finance, and administrative occupations	9.9%	29.8%
Natural and applied sciences and related occupations	7.5%	1.9%
Health occupations	2.0%	8.7%
Occupations in social science, education, government service, and religion	5.1%	8.9%
Occupations in art, culture, recreation, and sports	2.3%	3.2%
Sales and service occupations	20.7%	32.3%
Trades, transport and equipment operators, and related occupations	24.4%	1.9%
Occupations unique to primary industry	6.9%	2.2%
Occupations unique to processing, manufacturing, and utilities	9.9%	4.9%
Number of people	7,768,490	6,549,050

Sources: Data for the United States refer to people aged 25–64 and are from Jennifer C. Day and Andrea Curry, 1997, "Educational attainment in the United States: March 1996 (Update)," Current Population Reports, P20-493: Table 7; data for Canada refer to people aged 15 and older and are from the 1996 Census of Canada (**http://www.statcan.ca:80/english/census96/mar17/occupa/table2/t2p00.htm**).

"precision production, craft, and repair," "machine operators, assemblers, and inspectors," "transportation and material moving," and "handlers, equipment cleaners, helpers, and laborers" categories had completed college. Farming, forestry, and fishing, on the other hand, have increasingly become driven by technology, rather than just representing the muscular domination of nature, and you can see that a

Table 10.5 Higher Educational Attainment Goes Hand-in-Hand with Higher Occupational Status in the United States

| | Percent with college degree: | | | |
| | Ages 25–64 | | Ages 25–34 | |
	Males	Females	Males	Females
Executive, administrative, and managerial	54.2%	41.2%	58.2%	48.2%
Professional specialty	78.7%	73.5%	78.3%	75.6%
Technicians and related support	37.0%	28.3%	36.8%	30.7%
Sales	37.8%	19.9%	37.9%	26.8%
Administrative support, including clerical	23.2%	13.5%	24.0%	16.4%
Private household	n/a	6.3%	n/a	10.1%
Other service	11.8%	5.8%	11.6%	7.2%
Farming, forestry, and fishing	9.4%	12.2%	6.1%	12.2%
Precision production, craft, and repair	6.9%	9.2%	6.5%	11.9%
Machine operators, assemblers, and inspectors	4.6%	4.0%	5.2%	4.7%
Transportation and material moving	5.6%	3.9%	5.1%	2.4%
Handlers, equipment cleaners, helpers, and laborers	4.5%	8.3%	3.8%	8.8%

Sources: Data refer to people aged 25–64 in 1996 and are from Jennifer C. Day and Andrea Curry, 1997, "Educational attainment in the United States: March 1996 (Update)," Current Population Reports, P20-493: Table 7.

fairly high fraction of workers in those occupations have college degrees. Women are less likely to be as well educated as men at almost every occupational level, but these patterns are changing as women catch up with men educationally, evidenced by the fact that the educational difference by gender is less at the younger ages of 25 to 34 than it is for all of the ages 25 to 64.

People holding the higher status occupations are more likely to think of themselves as having a career as opposed to just a job, and they are apt to derive more intrinsic satisfaction from their work. The odds are also in their favor that they earn more money—a key to economic and social independence, especially for women.

Income

It has been estimated that shortly before his death in 1976, John Paul Getty, the oil magnate, had a daily income of $300,000. "Assuming a normal working day, this means that he earned $75,000 by the time the doughnuts arrived for his morning coffee break" (Dalphin 1981:1). Even by 1996 less than one in six American households made $75,000 in an entire year. The very rich tend to be that way because of birth. Of course every generation produces its share of self-made people like Bill Gates of Microsoft, but his children and grandchildren will inherit their wealth, rather than having to produce it for themselves. Wealth and its attendant high income are essentially ascribed characteristics for those born into families that own huge homes, large amounts of real estate, and tremendous interests in stocks and bonds or other business assets. Indeed, for such people a good education has historically been as much a result of high income as it was a resource by which to improve their social standing. Huntington Hartford, heir to the A & P fortune, is said to have commented that although he was just a "C" student at St. Paul's (a posh prep school), he was automatically accepted at Harvard (Dalphin 1981). But, for the rest of us, income is at least partially a consequence of the way in which we have parlayed a good education into a good job.

Occupation may be the primary clue that people have about our social standing, yet our well-being is thought by most people to be a product of our level of income. Little has changed since the late 1970s when Coleman and Rainwater concluded that "money, far more than anything else, is what Americans associate with the idea of social class" (Coleman and Rainwater 1978:29). It is not just having the money; rather it is how you spend it that signals to others where you stand in society. The principal indicators of having money include the kind of house in which you live, the way your home is furnished, the car or cars you drive, the clothes you wear, and the vacations and recreations (the "toys") that you can afford.

It is no mystery that there is an uneven distribution of income in American society. In 1996 in the United States, the richest 20 percent of families earned 49 percent of the total income of the nation, while the poorest 20 percent earned less than 4 percent (U.S. Bureau of the Census 1997). This was a continued deterioration of the distribution that had prevailed in the late 1960s, when the top 20 percent commanded 43 percent, while the bottom 20 percent shared more than 4 percent of the nation's income. These changes have produced the "hourglass" economy, which bulges with households at the high and low ends of incomes and is squeezed in the middle. Three broad explanations have been offered for the increasing inequality: (1) public policy changes, such as tax "reforms" that benefit some groups more than others; (2) labor market changes (occurring throughout the world, not just in the United States), such as an increasing mismatch between jobs and the skills of the labor force, or a polarization of jobs into those that require high skills and those that require few skills, with little in between; and (3) changes in demographic structure, such as the increasing fraction of households headed by females. All three of these changes probably have played a role in the increasing income inequality in the United States (Danziger and Gottschalk 1993; Morris, Bernhardt, and Handcock 1994), but let us focus for the moment on the third explanation, relating to demographic change.

Disparities by Race/Ethnicity   Family income rose steadily for American families during the 1950s and 1960s, leveled off in the 1970s and early 1980s, and then rose, albeit unevenly, between the mid-1980s and mid-1990s. These patterns can be seen in Figure 10.9. In 1950, a white non-Hispanic family had an income equivalent to $21,295 at 1996 prices, and that had more than doubled by 1996 to $47,023. African-American families also have experienced more than a doubling of income since the end of World War II. In 1950, the average African-American family had an income of $10,768 (in 1996 dollars) and that had risen to $26,522 in 1996, although, as for whites, most of those gains were made before the 1970s. In relative terms, income for African-American families has grown slightly faster than that for

Figure 10.9 The Gap in Family Income of Whites and African-Americans Widened between 1947 and 1996

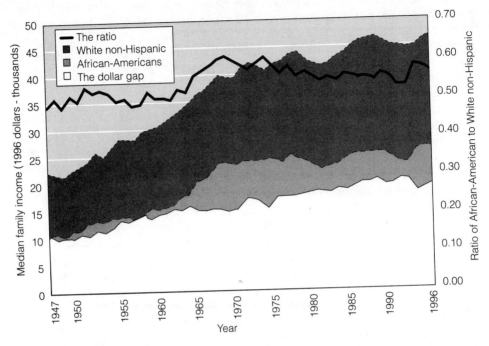

Sources: Adapted from U.S. Bureau of the Census, 1992, Current Population Reports, P60-180, Table B-7; 1995, P60-188, Table A; and 1997, P60-197, Table 1. Data are in constant 1996 dollars.

white non-Hispanic families. Thus "the ratio" in Figure 10.9 shows that in 1950 African-American family income was exactly half that of whites (a ratio of 0.50). By 1996 that ratio had crept up to 0.56. Yet, despite more than doubling their family income, African-Americans actually were losing ground to white families in absolute terms. In 1950, the "dollar gap" (in 1996 prices) between African-American and white families was $11,011, but by 1996 it had increased to $20,501.

Thus African-Americans have been in the peculiar position of having their incomes rise faster in percentage terms than non-Hispanic whites, but in dollar terms (the actual money available to spend on family members) they were falling further behind. This is one of the paradoxes that results from **structural mobility**—that situation in which an entire society is experiencing an upward mobility. That is the only time in which one group can improve itself socially or economically without forcing an absolute sacrifice from another group. When structural mobility slowed down in the 1970s, then the gains of African-Americans also hit the brakes, because at that point any rise in income would had to have been the result of a deliberate policy of income redistribution among ethnic groups.

I should add a word of caution about the comparison of family incomes of whites and African-Americans in the United States. African-Americans have fewer earners per family than do whites (1.4 for African-Americans compared with 1.6 for whites in 1996), and thus have less family income on that account alone. Furthermore, 47 percent of African-American families are headed by a female, compared with 14 percent of white families (U.S. Bureau of the Census 1997). Since females tend to earn less than males, that too lowers African-American family income relative to whites. We can control for these factors by looking separately at the incomes of only those people who are year-round, full-time earners.

If we were to assume that every family in the United States had a male and a female year-round, full-time worker, each earning the median income for his or her gender and race, then we would find that in 1955 (the first year for which such data are available) white families would have had an average family income of $42,400 (in 1996 dollars), compared with $24,365 for African-American families. Thus under these controlled circumstances, African-American families would have had $18,035 less money to spend each year—less than two thirds of the income of white families. By 1996, however, the situation had definitely improved for everybody. White two-earner families would have been making an average of $57,126, compared with $47,877 for African-American families. That amounts to a difference of $9,249, with African-American family income at a level that was 84 percent that of whites. Between 1955 and 1996, the average income of these hypothetical white couples would have increased 35 percent, while the increase for the hypothetical African-American couples during this time amounted to 96 percent! Furthermore, if all other things had been equal, the actual dollar income gap between African-Americans and whites would have been cut nearly in half.

Things rarely are equal, however, and so we find that the comparison between actual families is not as rosy as in our controlled case. In the controlled case, the income gap was due almost entirely to the fact that black male full-time year-round workers earn consistently less money than do white males. But in the real world, the disadvantage of African-Americans relative to whites in terms of family income is greatly exacerbated by the high rates of nonmarriage and marital disruption among African-Americans.

Disparities by Gender   Let us consider in more detail the issue of income disparities by gender. In 1996 the average female wage earner, working year-round, full-time, earned only 72 percent of the income garnered by males (as you will recall from Table 10.2). This was an improvement from 1977, however, when pay for women was 58 percent of that for men. Some of this difference is due to the fact that women are likely to have been in the labor force less time than men and in their current job for a shorter period of time, and that women are more likely than men to delay the completion of their education (which delays their reaching the better-paid occupational levels). Yet, the evidence suggests that the gender income gap persists even after taking all of these factors into account (Wellington 1994).

Why is there a differential in income by gender? The answer is that there is discrimination against women, in terms of what kinds of jobs they are hired for and what pay they receive (Marini and Fan 1997). This is true almost wherever you go in the world. It has been quipped that "Japan is the land of the rising sun, but only the son rises." In the German Democratic Republic right up to the period before the Berlin Wall collapsed, communism was supposed to guarantee gender equality. Yet, women with the same education as men, working at the same jobs as men, were receiving less pay than men (Sörenson and Trappe 1995). One of the few countries where true gender equality appears to have been reached, at least outside of the home, is in Sweden (Casper, McLanahan, and Garfinkel 1994).

In the United States the data clearly seem to suggest that the gap in status between men and women has closed most rapidly for younger people (Bianchi 1995), and part of the explanation for this is the fact that women's wages have been rising more quickly than those for younger men (Bianchi 1995; Oppenheimer 1994). The U.S. economy, for example, had to swallow a Big Gulp to find employment for the baby boomers, and although the economy did eventually absorb these people, it did so without offering much improvement in earnings over previous generations. Economic improvement for households has required that two-earner households become the norm, and this has certainly contributed to the delay in marriage and the rise in divorce.

Poverty   If you have several children the odds increase that your income will be below average, and if, on top of that, you are a divorced mother, the chance skyrockets that you will be living below the poverty level. In 1996, one in every three (33 percent) families headed by a woman (with no husband present) was below the poverty level, but nearly two in three of such families (59 percent) with children under 6 were below the poverty level (Lamison-White 1997).

To imagine the struggle it is to manage successfully in the United States on so little money, it is necessary only to review the definition of the poverty level. The poverty index was devised initially in 1964 by Mollie Orshansky of the U.S. Social Security Administration. It was a measure of need based on the finding of a 1955 Department of Agriculture study showing that approximately one third of a poor family's income is spent on food, and on a 1961 Department of Agriculture estimate of the cost of an "economy food plan"—a plan defined as a minimally nutritious diet for emergency or temporary use (Orshansky 1969). By calculating the cost of an economy food plan and multiplying it by 3, the poverty level was born. It has been revised along the way, but the idea has remained the same and since 1964 it has been raised at the same rate as the consumer price index.

In 1996, the poverty threshold for a single person under the age of 65 was $8,163 (the equivalent of earning $3.70 an hour if you were a year-round, full-time worker). A single parent with two children under the age of 18 could be earning $12,641 (the equivalent of $5.72 per hour for a year-round, full-time worker) and still be right at the poverty level threshold. You can compare these numbers with the national minimum wage in the United States, which in 1996 was $4.75 an hour—above the poverty level for a single person, but not by much. Between 1960 and 1973 the percentage of Americans living below the poverty level was cut in half, from 22 to 11 percent. However, since that time the poverty level has leveled off and has been hovering just under 15 percent (14 percent in 1996). The seeming lack of improvement in poverty levels has been questioned by some as being an artifact of how poverty is measured. For example, Brown (1994) has argued that the poor in the United States were actually better off in 1988 than in 1973 (despite the poverty level data suggesting otherwise) because of an increase in unreported income. She drew this conclusion by looking not at the reports of what poor families reported as income, but rather what they reported as their expenditures (derived from the Consumer Expenditure Surveys conducted regularly by the U.S. Bureau of the Census in collaboration with the Bureau of Labor Statistics). Poor families in 1988 were reporting expenditures that reflected a higher level of living than had been true in 1973.

Canada has adopted a similar strategy for measuring the lower end of the income scale, but the Canadian government has tried to avoid controversy by not officially defining a poverty threshold. Rather, Statistics Canada produces what are labeled "low income cut-offs." As of 1996, 18 percent of Canadian families lived below that cut-off, and 21 percent of Canadian children under age 18 were living in low-income families.

On the basis of global comparisons, it might be argued that very few people in North America are poor in absolute terms—it is the relative deprivation that is socially and morally degrading. A widely used index of poverty in India, for example, is the number of meals per day that a person can afford. If you can afford at least two meals a day, then you are above the poverty line, and by that definition only about 20 percent of Indians were below the poverty level in 1996 (Jordan 1996), but the overwhelming majority of Indians live far below the U.S. poverty line.

The International Labour Office (ILO, an agency within the United Nations) has offered worldwide comparisons of poverty by defining poverty as an annual income of less than US $370 (in constant 1985 dollars; that would translate to 17 cents an hour or less than $1 per day income for a year-round full-time worker). Using this definition, the ILO has estimated that 43 percent of the population of South Asia was living below the poverty line in 1993, as was 39 percent of the population of sub-Saharan Africa, 26 percent in East Asia, and 24 percent in Latin America and the Caribbean (International Labour Office 1997). On the basis of this same definition of poverty, the World Bank estimates that as of 1993 there were 1.3 billion people in the world (approximately 22 percent) living in poverty (World Bank 1996b). Very few of these people are in North America or Europe.

In rural Kenya, the World Bank has estimated that half of the population was living in poverty in 1992, and the poorest households were those headed by women with no male support (World Bank 1996a). This is a familiar pattern throughout the world and has generated the idea of the "feminization of poverty." It is likely that more than half of the poor in the world are females, aggravated by situations in which women are heads of households in societies in which they are generally not able to earn as much money as men. If we assume that people living in female-headed households are at greatest risk of being below the poverty line, then Marcoux (1998) estimates that the regions of the world most vulnerable to the feminization of poverty are the Caribbean, Latin America, East Asia, and sub-Saharan Africa.

Wealth

Poverty implies not only the lack of adequate income from any and all sources, but also the lack of any other assets on which a person might draw for sustenance. As people obtain and build assets, they create wealth. An asset is something that retains value or has the potential to increase in value over time. For most people, a home is the most important asset that will be acquired in a lifetime, but assets can also include personal property such as jewelry or other collectibles, stock in companies or mutual funds, savings accounts in banks, or ownership in a business venture. Typically, wealth is measured as net worth—the difference between the value of assets and the money owed on those assets. If your only asset is the house you just bought for $150,000, you are in the process of building wealth, but your net worth may be close to zero because you still may owe as much on the mortgage as the house is worth. There are three basic ways to generate wealth: (1) inherit assets from your parents or other relatives (the easiest way); (2) save part of your income in order to purchase assets (the hardest way); and (3) borrow money in order to purchase assets (the riskiest way).

The only sources of data on the wealth of Americans are household surveys, which always present problems of accurate reporting. The most representative source of information almost certainly is a set of questions asked periodically in the Survey of Income and Program Participation (SIPP) conducted by the U.S. Bureau of the Census. These data show that net worth in 1993 averaged $37,587 per household—a drop from the $43,999 (in 1993 dollars) found in the 1988 survey. The drop is attributed by some to the fact that Baby Boomers tend to spend what they earn (and sometimes more, by running up credit cards), rather than saving money and building assets. However, the SIPP data show that part of the problem is the increasing difficulty that people have in affording a home (Eller 1994; 1995).

In general, wealth increases with income and with age. If the two are put together, then it is correct to guess that older people with the highest incomes are the ones who have had the greatest opportunity to acquire assets and pay off their debts (as I mentioned in Chapter 9). Figure 10.10 puts this information into graphic perspective. At one end of the wealth continuum is the household headed by someone under age 35 with income in the lowest fifth of income levels. Such households (of which there were more than 5 million in 1993) have an average (median) net worth of $478. The other end of the continuum is the household headed by someone 75 or older still earning at the highest level. Such households (and there were nearly a half a million of them in 1993) average almost half a million dollars in net worth.

Marriage also is an important ingredient in accumulating wealth (as long as the couple stays married) and thus it can be found that the highest level of net worth occurs among older (65 and older) married-couple households, whose net worth in 1993 of $129,790 was more than twice the net worth of unmarried males ($60,741) or unmarried females ($57,679) of that age. At the younger ages (under 35), married-couple households averaged $12,941 in net worth, compared with only $1,342 for female-headed households.

Race and Ethnicity

To be a member of a subordinate racial, ethnic, or religious group in any society is to be at jeopardy of impaired life chances. African-Americans, Hispanics, Asians, and American Indians in the United States are well aware of this, as are indigenous people in Mexico, Tamils in Sri Lanka, Muslims in Israel or in India, Indians in

Figure 10.10   Median Net Worth Increases with Income and Age (United States, 1993)

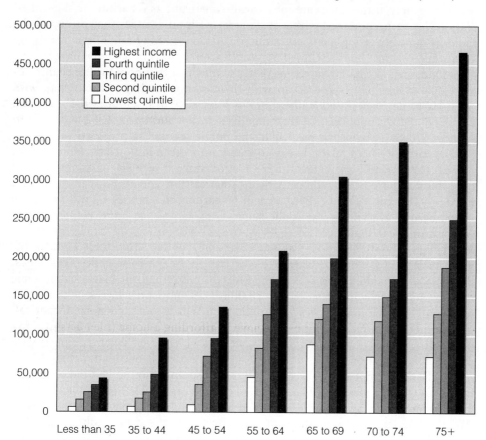

Note: People in each successively higher income group (grouped into quintiles of income) have higher net worth, and at each age net worth is higher.

Source: Adapted from Eller, T. J., and Fraser Wallace, 1995, "Asset Ownership of House-holds: 1993," U.S. Bureau of the Census, Current Population Reports, Series P70-47; Table D.

Malaysia, and virtually any foreigner in Japan or China. In this section I will sketch the impact on life chances of being a member of a minority racial/ethnic group in the United States, then I will briefly review patterns of discrimination in Canada and Mexico, as well as in other parts of the world.

Race and Ethnicity in the United States   In the United States, African-Americans represent the largest minority group, with a population of nearly 30 million (12 per-cent of the total population), according to the 1990 census. Being of African origin in the United States is associated with higher probabilities of death, lower levels of education, lower levels of occupational status, lower incomes, and higher levels of marital disruption than for the white population.

Earlier in this chapter I noted that in recent decades income levels have risen proportionately at the same rate for African-Americans as for whites in the United States. This fact, combined with a wide array of legal changes that protected the rights of minority group members, induced Wilson (1978) to talk of the "declining significance of race" in the United States: "Race relations in America have undergone fundamental changes in recent years, so much so that now the life chances of individual blacks have more to do with their economic class position than with their day-to-day encounters with whites" (Wilson 1978:1). Hout (1984) has used data from the Occupational Changes in a Generation Surveys of 1962 and 1973 to support Wilson's claims. In particular, the public sector has provided substantial new opportunities for black males to move into the middle class. Nonetheless, Bianchi (1981) and Farley (1984) have cautioned us to be wary of overestimating the gains that have been made by African-Americans. If you are African-American, there is still a good chance that you will be earning less money than if you were white, even if everything else is the same. For example, in 1996 income for an African-American male aged 35 to 44 who was a year-round full-time worker and had a high school education was more than $8,100 less than for a comparable white male; and the difference for females was $4,300. Among college graduates, the income for white males was about $15,000 more than for black males, while the advantage for white females was $8,000 (Day and Curry 1997). Why the difference? The obvious reason is discrimination, which looms as a significant (although difficult to measure) factor, accounting for the fact that members of a minority group may not earn as much money as whites even when equipped with comparable educational skills and working at comparable occupational levels.

A classic attempt to measure the effect of discrimination (albeit indirectly and inferentially) was made by Hauser and Featherman (1974). They noted that in 1962, white males aged 35 to 44 earned $3,755 more per year on average than African-American males of the same age. By 1972, the income differential had dropped to $3,195. They then controlled for factors such as family background (father's education, father's occupation, and farm background); the number of siblings (too large a family might have cut down opportunity); education of the respondent; and occupational status, to see whether differences in those social structural factors could explain all the differences in income. They could not. In 1962 and 1972, nearly 40 percent of the income differential was left unexplained, and the researchers speculated that discrimination might be the missing factor.

The lower levels of educational attainment, occupational status, income, and marital stability for African-Americans than for whites in the United States have sometimes been attributed to the fact that African-Americans come from larger families on average than whites and thus are launched with fewer resources per person. Although the study by Hauser and Featherman (1974) just discussed did not lend much credence to that idea, it is nonetheless true that African-Americans tend to have more children, even though the difference is no longer very large. Over time the trend in fertility among African-Americans has tended to follow that for the rest of the population, but at a higher level (O'Hare, et al. 1991). African-American fertility declined during the Depression, but did not drop as low as for whites; it rose during the Baby Boom, and to a higher level than for whites, it declined during the Baby Bust, but not as much as for whites, and it has risen in the more recent baby boomlet. Fertility preferences among younger women have now nearly converged for these two groups in the population. In particular, the fertility behavior of college-educated African-American and white females is virtually indistinguishable.

McDaniel (1996) has suggested that the different life chances of whites and blacks in American society are due to *racial stratification,* which he defines as a socially constructed system that characterizes one or more groups as being distinctly different. Your membership in a group defined as different from the others then creates, in essence, a different social world for you than for those who are in other groups, and this affects your behavior and your life chances in society, because there is no genuine societal expectation that you will be assimilated into the rest of the society. He offers an illustration:

> The ability of a group to be assimilated depends on whether it is considered an ethnic or racial group . . . For example, immigrants from Nigeria and Ghana assimilate into the African American race, and immigrants from Sweden and Ireland assimilate into the European American race . . . Ethnic groups such as those from Europe and segments of the Hispanic population show definite signs of assimilability; however, the African-descendent population continues to be blocked from assimilation within the United States [McDaniel 1996:139].

People of Hispanic origin (largely of Latin American origin) comprise the second largest minority group in the United States, with more than 22 million counted in the 1990 census. "Hispanic" is a broad term, encompassing people of widely differing cultural backgrounds, including Mexican (13.5 million), Puerto Rican (2.7 million), Cuban (1 million), and a variety of others (5.1 million). According to a 1978 directive of the U.S. Office of Management and Budget, the term "Hispanic" describes a person of Mexican, Puerto Rican, Cuban, Central or South American, or other Spanish culture or origin, regardless of race. The Hispanic population in the United States is the fifth largest population of Spanish origin in the world, behind Mexico, Spain, Argentina, and Colombia. In general, Hispanics have infant mortality rates lower than those of whites but they fall in between whites and African-Americans with respect to income, as can be seen in Table 10.6. On the other hand, to show you how untidy the social world can be, average educational levels are lower than for African-Americans or whites. Most of the educational disadvantage, and its subsequent deleterious effect on occupational status among Hispanics, appears to be due to language barriers (Stolzenberg 1990).

Since people of Mexican origin represent 60 percent of the Hispanic population in the United States, their levels of fertility have a substantial influence on the overall fertility pattern of Hispanics. You might expect to find that Mexican-Americans, having their origin fairly recently in a high-fertility society, would have birth rates noticeably higher than those for the population as a whole. In general, you would not be disappointed; the data do indeed reveal that pattern, although birth rates have been declining for the Hispanic population as they have for all groups in the United States (as well as in Mexico).

The 7 million people who comprise the Asian and Pacific Islander population in the United States represent an even more culturally heterogeneous group than the Hispanics. According to the 1990 census, Asians of Chinese origin were most

Table 10.6 Life Chances Differ Across Racial and Ethnic Groups in the United States

Population Characteristics[1]	Whites	African-Americans	American Indians	Asians and Pacific Islanders	Hispanics
Education					
Persons 25+ who are high school graduates (%)	77.9	63.1	65.5	77.5	49.8
Persons 25+ who are college graduates (%)	21.6	11.4	9.3	36.6	9.2
Labor Force					
Females 16+ in the labor force (%)	56.4	59.5	55.1	60.1	56.0
Males 16+ unemployed (%)	5.3	13.7	15.4	5.1	9.8
Mean Household Income	$40,308	$25,872	$26,206	$46,695	$30,301
Households with less than $25,000 (%)	39.2	59.3	59.5	33.8	51.5
Households with $100,000 or more (%)	4.8	1.3	1.4	7.5	2.0
Marital and Family Status					
Households headed by married couples with children under 18 (%)	26.4	18.8	28.6	40.3	37.5
Households headed by females with children under 18 (%)	4.5	19.1	12.9	4.7	11.6
Ratio of married-couple/ female-headed households	5.9	0.9	2.2	8.6	3.2
Population Processes[2]					
Total fertility rate per year	1,840	2,450	2,760	2,480	2,900
Female life expectancy at birth (years)	80.0	74.5	81.6	86.2	82.8
Net number of international migrants per year	9,867	21,494	73	177,924	140,642

[1] Population characteristics data refer to 1990 and are from the 1990 Census of Population and Housing Summary Tape File 3C.

[2] Population processes data refer to 1993 and are from J. C. Day, 1993, "Population projections of the United States, by age, sex, race, and Hispanic origin: 1993 to 2050," U.S. Bureau of the Census, Current Population Reports, P25-1104.

numerous, followed by Filipino, Indochinese, Japanese, Asian Indian, and Korean. From the standpoint of sociodemographic characteristics, Asians and Pacific Islanders represent a widely diverse set of ethnic groups, yet overall they are the most advantaged of the broadly defined racial/ethnic groups in the United States, as can be seen in Table 10.6. Data from the 1990 census reveal a mean family income of $46,695, which exceeds the level of all other groups. Furthermore, one out of every three Asian-Americans aged 25 and older is a college graduate, nearly double the percentage for any other group. Unemployment rates are lower, and female participation rates are higher than for any other group. The heterogeneity of this population is reflected in the fact that the percentage of people below the poverty level is higher than for whites. Indochinese refugees and some of the Pacific Islander groups, in particular, are less likely than the Chinese, Japanese, and Koreans to be sharing in the higher levels of education and income.

American Indians are in the unenviably unique position of being the first and yet the most disadvantaged of the major racial/ethnic groups in the United States. At the time of Columbus's arrival in the Caribbean in 1492, the North American native population is estimated to have been between 2 and 5 million (Snipp 1989), as I mentioned in Chapter 1. Contact with Europeans quickly decimated the native population, however, partly through warfare and partly by starvation caused by massive population dislocation, but largely through disease. The Europeans carried diseases never before seen in the western hemisphere, and the lack of resistance on the part of the local population had disastrous results. So devastating were these effects that they have been likened to a holocaust (Thornton 1987).

The 1990 census count of 2 million American Indians (including Eskimo and Aleut tribes) suggests that finally, 500 years after Columbus's arrival, the American Indian population has rebounded at least to the lower likely number of such persons living on the continent at the time of his arrival. On average, the educational and income levels of American Indians are approximately the same as for African-Americans and Hispanics, while the rate of unemployment is the highest. At the same time, life expectancy for females is lower than for any group except African-Americans, and the total fertility rate is higher than for any group except Hispanics.

Ethnicity in Canada and Mexico   The United States does not have a corner on the racial and ethnic minority market, nor are demographic differences by race and ethnicity peculiar to the United States. In Canada, ethnicity is measured as much by language as any other characteristic, leading the Francophone (French-speaking) population in the eastern edge of the country (Quebec) to attempt to secede from the Anglophone (English-speaking) remainder of the country. Demographics played a role in defeating the referendum on separation held in 1995, however, because the traditionally Catholic Francophone population now has very low levels of fertility—probably the lowest anywhere in North America. French Canadians are not replacing themselves and non-Francophones (especially recent immigrants) generally do not support separation, leading Canadian demographers correctly to predict that separation would not be approved by the voters (Kaplan 1994; Samuel 1994). Still, the controversy over language in

Canada underscores the power of society to turn any population characteristic into a sign of difference, from which prejudice and discrimination often follow.

Language is also an issue in Mexico, where the lowest stratum of society tends to be occupied by those who speak an indigenous language (linguistically related to Aztec and Mayan languages) rather than Spanish. The Indian population of Mexico is concentrated in the south especially. The state of Chiapas, scene of Indian-speaking people challenging the authority of the central government in the 1990s, is the area in Mexico where the people are the poorest and fertility is highest. Language minorities thus represent both the geographic and demographic extremes in North America, from Quebec in the northeast with very low fertility to Chiapas in the south with very high fertility.

Ethnicity in Other Countries   Goldscheider (1996) has examined the demographics of ethnic pluralism in Israel and found that the Muslim Arab population in that country has been growing at twice the rate of the Jewish population and in 1982 represented 13 percent of the total, up from 8 percent in 1961; by 1998 nearly one in five residents of Israel was Arab (Israel Central Bureau of Statistics 1998), of whom about 75 percent are Muslim (the remainder are Christian and Druze). Like minority group members in the United States, the Muslim minority in Israel is less educated, holds lower-status jobs on average, and has a lower life expectancy and higher fertility levels: "In most cases, Jews in Israel believe that the State should prefer Jews to Arabs in all arenas of public policy. Only a small proportion of Jews subscribe to the notion of 'equal opportunity' for all groups in areas of higher education, jobs, and financial support from the government" (Goldscheider 1989:23).

Fertility has declined among Arabs in Israel since 1960, but the drop has been modest, and since Jewish fertility has also declined, the gap in fertility levels remains fairly wide. In 1996 the total fertility rate for the Jewish population was 2.6 children per woman, compared with 4.6 for the Muslim population (Israel Central Bureau of Statistics 1998). However, high fertility in Israel is not simply a function of being Arab, because the Christian Arab minority (representing 14 percent of the total Arab population) has the lowest total fertility rate (just under the Jewish average of 2.6 children per woman) of any group in Israel. The contrasts between Muslims, Jews, and Christians takes us next into the realm of religion.

Religion

Virtually everyone is born into some kind of religious context, which is why I have likened religion to an ascribed characteristic. Yet, people can willingly change their religious preference during their lifetime, and so it is akin to an achieved status. Despite the appearance of choice, however, most people do not alter religious affiliation and so it is a nearly permanent feature of their social world. Like race and ethnicity, religion sets people apart from one another and is a common source of intergroup conflict throughout the world (Choucri 1984). Because it is an often-discussed sociodemographic characteristic, religion has regularly come under the demographer's microscope, with particular attention being paid to its potential influence on fertility.

America's history of **religious pluralism,** in which a wide variety of religious preferences have existed side by side, perhaps sensitized American demographers to the role of religion in influencing people's lives. A good deal of attention has been focused on the comparison between Protestants and Catholics. During the past century at least, Catholics have tended to want and to have more children than have Protestants in the United States, and internationally it has been true that predominantly Protestant areas (such as the United States and northern Europe) experienced low fertility sooner than did predominantly Catholic areas (such as southern and eastern Europe). More recently, however, data from the 1988 National Fertility Study in the United States have suggested that the long-time differences between Protestant and Catholic fertility levels have nearly disappeared (Goldscheider and Mosher 1991). This appears to be closely related to the substantial drop in church attendance over time, especially among Catholic women (Williams and Zimmer 1990).

Is religion less important now than it used to be? Are there other factors besides religion that explain the seeming differences sometimes noted between religious groups? Obviously, the relationship between religion and fertility is not a simple one, and, in fact, there are three major themes that run through the literature: (1) religion plays its most important role in the middle stage of the demographic transition; (2) religiosity may be more important than actual religious belief; and (3) religion and ethnicity are inextricably bound up with each other. Let us examine each of these ideas.

In a study of religious differentials in Lebanese fertility, Chamie (1981) concluded that a major effect of religion may be to retard the adoption of more modern, lower-fertility attitudes during the transitional phase of the demographic revolution. Adherents to religious beliefs that have been traditionally associated with high fertility will be slower to give ground than will people whose religious beliefs are more flexible with respect to fertility. In the United States, Jews have generally had lower fertility levels than the rest of the population. Trends in Jewish fertility have followed the American pattern (a decline in the Depression, a rise with the Baby Boom, a drop with the Baby Bust), but at a consistently lower level. "Widespread secularization processes, upward social mobility, a value system emphasizing individual achievement, and awareness of minority status have all been indicated as factors that are both typical of American Jews and conducive to low fertility" (DellaPergola 1980:261). Indeed, it is not just American Jews whose fertility is low. DellaPergola (1980) points out that Jewish communities in central and western Europe have also been characterized by low fertility since as early as the second half of the 19th century, largely because contraception is readily accepted into the Jewish normative system (at least among non-Orthodox Jews). Leasure (1982) also found that, in the United States, fertility declined earliest in those areas dominated by more secularized religious groups. People who are more traditional in their religious beliefs tend to be less educated, have less income, and are thus more prone to higher fertility.

The former Yugoslavia is the site of centuries-old ethnic battles, with religion as one of the important ethnic identifiers. Courbage (1992) has shown that, until recently, the Muslims in that region of Europe (who are predominantly Slavic in ethnic origin, not Arab) had considerably higher fertility levels than did Christians.

However, by the time of the 1981 census in Yugoslovia an interesting pattern had emerged, in which people who identified themselves as "Muslim nationals" (the population that now comprises the Muslim part of Bosnia) had lowered their fertility to "European" levels, similar to the the Slavic Christian population, whereas those who identified themselves as "Albanians" were Muslims whose fertility levels remained closer to Middle Eastern levels, rather than being at European levels.

In looking at religious differentials in fertility it may be at least as important to examine the religiosity of people as to know their specific religious beliefs. It is likely, in fact, that fertility differences by religion will always exist in any society to some extent just because some groups inspire greater religious fervor than others and, in almost all instances, religious zealotry is associated with a desire for larger than average families because there is usually a desire to maintain or return to more traditional value systems in which large families are the norm, and this may lead religious fundamentalism to have a negative influence on educational attainment among its adherents (Darnell and Sherkat 1997). In many respects, this is the flip side of saying that secularism causes a decline in fertility. To eschew education and secularism generally means to maintain high fertility.

For the United States I have already discussed the high fertility of the Hutterites (see Chapter 5). Another, larger religious group whose higher than average level of religiosity seems to contribute to higher than average fertility is the Mormon Church. Mormons dominate the social, political, and demographic fabric of the state of Utah, explaining the fact that Utah's birth rate is the highest of any state—twice the national average—and Provo, Utah (the site of Brigham Young University, the principal Mormon institution of higher education) has the highest birth rate of any city in the United States.

The Social Impact of Demographic Shifts in Household Composition

In the United States, the changes taking place in the family structure are occurring mainly within the Baby Boom cohort, but similar changes have taken place earlier in Europe (especially Sweden and Denmark), where the baby boom was much smaller. Westoff has suggested that the institutions of marriage and the family show signs of change because

> The economic transformation of society has been accomplished by a decline in traditional and religious authority, the diffusion of an ethos of rationality and individualism, the universal education of both sexes, the increasing equality of women, the increasing survival of children and the emergence of a consumer-oriented culture that is increasingly aimed at maximizing personal gratification [1978:53].

It has been argued that many of these cultural changes have followed, rather than preceded, the changes in household structure. They may not have initiated the trends, but they have reinforced the transformations and ensured their spread within each country and from one country to another. Global changes are, however, occurring unevenly.

In predominantly Muslim nations, the gender gap between men and women is especially wide (Ahmed 1992), strengthened over the centuries by the Koran, which states that "man has authority over women because Allah made the one superior to the other" (quoted by Pool 1970). Throughout the Third World, development is bringing about an increase in the breadwinner system (remember the earlier comments by Davis [1984]) and is increasing the economic and political dependence of women. Charlton (1984) contends that in India the early stages of modernization were accompanied by a declining status of women. This is a story confirmed by the Report of the National Committee on the Status of Women in India, which concluded that "to our dismay, we found that such disabilities [centers of resistance to change in the status of women] have sometimes been aggravated by the process of development itself" (National Committee on the Status of Women 1975:2).

Worldwide economic changes have emphasized the value of nondomestic labor and this has marginalized the economic contribution that women make to the family economy. It has also aggravated the problems associated with cultural practices such as the dowry that Indian families are expected to pay to their new in-laws on the marriage of each daughter. Because traditionally in India girls do not receive an inheritance on the death of parents, the dowry is in essence an "upfront" payment of a potential inheritance. However, from a practical point of view, it means that any investment in a girl will accrue to her husband's family, and so parents are not necessarily motivated to do much for daughters. Indeed, it has been suggested that child labor is used in India as a way for girls to defray the costs of the dowry (The Economist 1994). A solution to such a problem of "cultural entrapments embroidered by female submissiveness" (Mhloyi 1994) was signed into law in the Indian state of Maharashtra in 1994 when the legislators of that state passed laws that granted women the same rights of inheritance as men, and they also reserved up to 30 percent of government jobs for women.

The cultural model prevalent for the past two to three decades in industrialized nations has been that self-fulfillment and individual autonomy are the most important values in life and serve to justify scrapping a marriage (Schiffren 1995). If women are approaching the level of economic independence previously reserved for men, perhaps the value of marriage has been permanently eroded, and marriage will (or has) become only one option among many from which people might reasonably choose.

One of the complaints about marriage often registered by women is that the move toward gender equality in the division of labor in the formal marketplace has not necessarily been translated into equity in the division of labor within the household. Women are able to operate in society independently of a husband or other male patriarch or protector, but they may not have the same ability to have an equal relationship at home with a husband. Data from the National Survey of Families and Households show that married couples are the least egalitarian of all households with respect to the division of domestic chores. Marriage is the household setting in which women do the greatest amount of domestic work (regardless of their own labor force status), while men do the least (South and Spitze 1994).

So, does marriage matter? An increasing body of evidence suggests that marriage matters very much even in a modern, industrialized society—it enhances household income and wealth and promotes the well-being of spouses and children, while even adding to sexual activity in the bargain. Waite (1995) has reviewed the literature and analyzed appropriate data sets in order to draw the following conclusions about the benefits of marriage: (1) married couples have higher household income than the unmarried; (2) married couples save more of their income than do the unmarried; and therefore (3) the married have more wealth than the unmarried; (4) married men and women live longer than the unmarried, and engage in fewer high-risk behaviors; (5) children in a marriage are better off financially than those in a one-parent family; (6) children in a marriage are less likely to drop out of school, less likely to have a teenage pregnancy, and less likely to be "idle" (out of both school and work) as a young adult than are children in a one-parent family; and (7) married couples have sex more often and derive greater satisfaction from it than do the unmarried.

The social impact of marriage derives from these personal benefits. Perhaps most compelling is the fact that the family remains the primary social unit in which society is reproduced—in which children are taught the rights and reciprocal obligations of membership in human society. The evidence seems to suggest that this is accomplished most efficiently in a household/family unit that includes both parents of the children in question. The evidence is persuasive that children derive few, if any, positive benefits from growing up without a father and, indeed, tend to suffer both short-term and long-term ill effects of being fatherless, and the same is probably true for motherless families, although we have fewer studies of such family settings (Furstenberg and Cherlin 1991; McLanahan and Sandefur 1994). "The data clearly indicate that a healthy two-parent family optimizes both the economic well-being and the physical and mental health of children" (Angel and Angel 1993:199). If this is correct, then the implications are that diversity of household composition is problematic only if children are involved. The negative effects of "nontraditional" families seem to fall disproportionately on children, and as these consequences become more widely understood in society, it is possible that new attitudes may lead to new behavior (and, of course, vice versa). Part of the complication is that the traditional family was not necessarily a bed of roses, either, and it is probably harder on children to be in an intact family where the parents are constantly battling each other than to be in some other type of household situation where there is less conflict (Furstenberg 1998).

It appears that the diversification of households in industrialized societies has leveled off. There has been much less change in the late 1980s and the 1990s than there was in the 1970s and early 1980s. Waite and her associates found that the increasing independence of women with respect to living arrangements helped to shape a change in attitudes among women toward the division of labor within the family as well as in society generally (Waite, Goldscheider, and Witsberger 1986). As men also adopt those attitudes, their satisfaction with a marriage appears to increase (Amato and Booth 1995). Thus we may venture to guess, with (Goldscheider and Waite 1991) that "Couples with modern attitudes toward both women's work outside and inside the family will create more egalitarian families in which children can grow up and later emulate" (p. 141).

Summary and Conclusion

The past few decades have witnessed a fundamental shift in household structure in the United States and other industrialized nations. Married couple households have become less common, being replaced especially by female-headed households and also by nonfamily households. This greater diversity in household structure is a direct result of the trends in marriage and divorce. Marriage has been increasingly delayed (although most still do eventually marry), but people are leaving the parental nest to live independently by themselves, with friends, or in a cohabiting relationship prior to marriage. Once married, there is an increased tendency to dissolve the marriage. Accompanying these trends has been an increase in the proportion of children born out of wedlock, contributing to the increased percentage of children who are living with only one parent.

Less directly, but no less importantly, the transformation of family and household structure has been a result of changing population characteristics, especially the improvement in the life chances of women. Women have become less dependent on men as they have begun to live longer and spend more of their lives without children in an urban environment, where there are alternatives to childbearing and family life. Throughout the world women are closing the education gap between themselves and men and, especially in industrialized nations, have joined the labor force and are moving up the occupational ladder. These new opportunities to be more fully engaged in all aspects of social, economic, and political life have been simultaneously the cause and consequence of declining fertility and improved life expectancy. They have enabled women to delay marriage while becoming educated and establishing a career, choose marriage or not (most do), choose children or not (most choose two), and if married, to choose to stay so (only about half choose to). Therein lie the principal explanations for the increased diversity of families and households.

Not all people have equal access to societal resources such as advanced education, a well-paying job, and other assets with which to build wealth. In the United States, African-Americans, Hispanics, and American Indians are less likely than others to be highly educated, and this may contribute to their relative social and economic disadvantage in American society. On the other hand, Asian-Americans tend to have higher levels of education than other groups, which may help account for their higher levels of income (and higher life expectancies, as well). In most countries of the world we find one or more groups that, for reasons of discrimination beyond their control, are disadvantaged relative to the dominant group.

The impact of the trends toward greater family and household diversity (as well as reduced life chances) falls disproportionately on children, for whom growing up in other than a two-parent family lowers the household income and increases the odds of health and social problems in childhood and young adulthood. However, there is another group that also will bear the brunt of dissolved marriages. Today's divorced women could become the biggest group of elderly poor in the future. This is a trend that we will have to watch over time, but it will be easier to track than in previous generations because most households with older people in the industrialized nations are now located in urban areas where they may be more visible than the elderly poor in rural areas of less developed nations. This transition from a predominantly rural to a predominantly urban society is an important feature of the trends inherent in the demographic transition, and I turn to that topic in the next chapter.

Main Points

1. Married couple households are declining as a fraction of all households, being replaced by female-headed households and by nonfamily households.

2. The direct causes of these changes in household composition are a delay in marriage, a rise in the propensity to divorce, and to a lesser extent widowhood in the older population.

3. Delayed marriage has been accompanied by an increase in leaving the parental home to live independently before marriage; this increasingly leads to cohabitation prior to marriage.

4. Declining mortality has helped to contribute to a freeing of women from childbearing responsibility and has provided more years of pre-reproductive and post-reproductive time.

5. Urbanization has helped to improve the status of women by providing opportunities for nonfamilial activities.

6. The transformation of families and households has been an accompaniment to improved life chances for women, including higher levels of education, labor force participation, occupation, and income.

7. Average educational attainment has increased substantially over time in most countries, and especially in industrialized nations women have been rapidly closing the gender gap in education.

8. In the United States during World War II, a combination of demand for labor and too few traditional labor force entrants created an opening for married women to move into jobs previously denied them.

9. Since 1940 the rates of labor force participation have risen for women, especially married women, while declining for men.

10. Women represent nearly 50 percent of all workers, but they are still concentrated disproportionately in administrative support, sales, and service occupations.

11. The better educated you are, the more money you can expect to earn in your lifetime.

12. Americans of almost all statuses are wealthier in real absolute terms now than they were in the 1940s, but there have been only minor changes in the relative status of most groups.

13. Even at comparable occupational levels, with similar educational backgrounds, African-Americans on average earn less money in the United States than whites.

14. Even at comparable occupational levels, with similar educational backgrounds, women on average earn less money than men.

15. Race may be just "a pigment of your imagination," but African-Americans, Hispanics, and American Indians tend to be disadvantaged compared with whites in American society, whereas Asian-Americans tend to be better off in socioeconomic terms.

16. The diversity of households seems to have plateaued, and the future of married couple families may depend on the willingness of men to adopt more egalitarian attitudes toward domestic roles within the household.

17. Demographers can prove that the average person in Miami, Florida, is born Cuban and dies Jewish.

Suggested Readings

1. Reynolds Farley, 1996, The New American Reality: Who We Are, How We Got There, Where We Are Going (New York: Russell Sage Foundation).

   In order to expedite the social and economic analysis of data from the 1990 census in the United States, the Russell Sage Foundation helped to coordinate the publication of a three-volume series in which various experts provide chapters analyzing the census data on the population characteristics that are the focus of their research. This is the third volume, the preceding volumes having been edited by Farley—State of the Union: America in the 1990s, Volume One: Economic Trends, and Volume Two: Social Trends

2. Jorge del Pinal and Audrey Singer, 1997, "Generation of Diversity: Latinos in the United States," Population Bulletin, 52(3).

   Latinos (or Hispanics) represent the fastest growing ethnic group in the United States and early in the 21st century will become the majority in California and the largest minority group in the United States. This Bulletin spells out the demographic causes and consequences of this part of the cultural mosaic in the United States.

3. Andrew J. Cherlin, 1992, Marriage, Divorce, Remarriage, Revised and Enlarged Edition (Cambridge: Harvard University Press).

   An excellent summary of the demographic trends in marriage, divorce, and remarriage and as assessment of the causes underlying the trends.

4. Nancy E. Riley, 1997, "Gender, Power, and Population Change," Population Bulletin 52(1).

   Underlying the changing family and household patterns throughout the world are changing power relationships between men and women. Riley discusses these trends and relates them directly to issues of population growth.

5. Frances K. Goldsheider and Linda J. Waite, 1991, New Families, No Families? The Transformation of the American Home (Berkeley and Los Angeles: The University of California Press).

   The culmination of years of collaborative data analysis, this book examines the impact on the family of the division of labor in the formal economy and the division of labor within the household.

🏴 Websites of Interest

Remember that websites are not as permanent as books and journals, so I cannot guarantee that each of the following websites still exists at the moment that you are reading this:

1. **http://www.slip.net/~ccf/**

   In the late 1990s a battle broke out in academia between those who were generally more politically progressive in their views on family and household diversity and those who were more conservative. This is the website for the Council on Contemporary Families (CCF), the group that is more progressive; whereas: **http://www.kidscampaigns.org/** is the website for the KidsCampaigns, which can be characterized as being more politically moderate with respect to its view on family and household structure.

2. **http://www.umich.edu/~psid/**

   The Panel Study of Income Dynamics (PSID) is a longitudinal study emphasizing the dynamic (changing) aspects of demographic and economic behavior in American society. It began in 1968 and is housed at the University of Michigan.

3. **http://www.irc.essex.ac.uk**

   The Economic and Social Research Council (ESRC) Research Centre on Micro-Social Change is the home of the British Household Panel Study, a longitudinal study of British households begun in 1989.

4. **http://stats.bls.gov/ocohome.htm**

   Do you want to know for yourself which occupations seem to have the greatest growth potential? The U.S. Bureau of Labor Statistics regularly updates their Occupational Outlook Handbook and you can access it at this site.

5. **http://www.hrdc-drhc.gc.ca/**

   Human Resources Development Canada (HRDC) is the government agency in Canada that deals with issues of occupational and employment development, and one tool they use is a survey of recent college graduates to see what they are up to. The results are located at this website.

CHAPTER 11
The Urban Transition

The world is rapidly becoming urban. Consider that as recently as 1850 only 2 percent of the entire population of the world lived in cities of 100,000 or more people. By 1900 that figure had edged up to 6 percent, and it had risen to 16 percent by 1950 (Davis 1972). So for most of human history almost no one had lived in the city—cities were small islands in a sea of rurality. But within just one century, from 1850 to 1950, cities had jumped out to grab 1 in 6 human beings, and a short 50 years after that, in the year 2000, virtually 1 in every 2 people is estimated to be living in cities of 100,000 or more, and nearly 2 in 3 will be living in places that are labeled as urban (United Nations 1996).

> The present historical epoch, then, is marked by population *redistribution* as well as by population *increase*. The consequences of this redistribution—this "urban transition" from a predominantly rural, agricultural world to a predominantly urban, nonagricultural world—are likely to be of the same order of magnitude as those of the more widely-heralded increase in world population [Firebaugh 1979: 199]

The majority of Americans live in—indeed were born in—cities, and most people of the Western world share that urban experience. Some of us take the city for granted, some curse it, some find its attractions irresistible, but no one denies that urban life is the center of Western industrial civilization. Cities, of course, are nothing new, and their influence on society is not a uniquely modern feature of life; however, the widespread emergence of urban life—the explosive growth of the urban population—is very much a recent feature of human existence.

This urban transition is one of the most significant demographic movements in world history partly because it is intimately tied to the population explosion itself. It is not too much of an exaggeration to say that world population growth is *occurring* in the countryside, but *showing up* in the cities. What are the demographic components of urbanization, and what are the demographic consequences for society of an ever-increasing concentration of people in urban areas? These are the questions I will focus on in this chapter. Clearly the city is implicated in a wide range of problems, issues, and triumphs in all societies, but my intention here is not to review life in the city (which could, and has, filled volumes). Rather, I want to provide you with a demographic perspective on **urbanization** (a word that I will use interchangeably with the term **urban transition**). To accomplish that, I begin with an overview of urbanization, putting it into broad historical perspective. Then I turn to an analysis of the demographic components of urbanization, looking especially at the way in which migration, mortality, and fertility interact to produce the process of urbanization. In this context, we see how the urban transition fits into the overall pattern of the demographic transition and how the rise of cities as the center of social and economic life has created new worldwide interlinkages among urban areas. Finally, I examine some of the spatial processes taking place in urban areas that are both causes and consequences of many of the demographic trends that have been discussed in the previous chapters. These include trends in metropolitanization, suburbanization, patterns of residential segregation within urban areas, and the response of humans to being crowded into urban places.

What Is Urban?

An **urban** place can be defined as a spatial concentration of people whose lives are organized around nonagricultural activities; the essential characteristic here is that urban means nonagricultural. A farming village of 5,000 people should not be called urban, whereas a tourist spa or an artist colony of 2,500 people may well be correctly designated an urban place. You can appreciate, then, that "urban" is a fairly complex concept. It is a function of (1) sheer population size, (2) space (land area), (3) the ratio of population to space (density or concentration), and (4) economic and social organization.

The definitions of urban used in most demographic research unfortunately rarely encompass all the above ingredients. Due to limitations in available data and sometimes simply for expediency, researchers (and government bureaucrats as well) typically define urban places on the basis of population size alone. Thus all places with a population of 2,000, 5,000, 10,000, or more (the lower limit varies) might be considered urban for research purposes. Of course you should recognize that an arbitrary cutoff disguises a lot of variation in human behavior. Although the difference between **rural** and urban areas may at first appear to be a dichotomy, it is really a continuum in which we might find an aboriginal hunter-gatherer near one end and an apartment dweller in Manhattan near the other. In between will be varying shades of difference. Indeed, the next time you drive from the city to the country (or the other way around), you might ask yourself where you would arbitrarily make a dividing line between the two. In the United States in the 19th and early 20th centuries, rural turned into urban when you reached streets laid out in a grid. Today, such clearly defined transitions are rare and, besides, even living in a rural area in most industrialized societies does not preclude your participation in urban life. The flexibility of the automobile combined with the power of telecommunications put most people in touch with as much of urban life (and rural life, what is left of it) as they might want. In the most remote areas of developing countries, radio and satellite-relayed television broadcasts can make rural villagers knowledgeable about urban life, even if they have never seen it in person (Critchfield 1994).

An essential ingredient of being urban is economic and social life organized around nonagricultural activities. Thus there is an explicit recognition that urban people order their lives differently than rural people do; they perceive the world differently and behave differently, a topic I will return to later. Now, however, let me discuss the demographic aspects of the process whereby a society is transformed from rural to urban—the urban transition, or the process of urbanization.

An Overview of Urbanization

Urbanization refers to the change in the proportion of a population living in urban places; it is a relative measure ranging from 0 percent, if a population is entirely agricultural, to 100 percent, if a population is entirely urban. Figure 11.1 maps the countries of the world in the mid-1990s according to the percent of the population living in urban places. These data are based on each individual country's own

Figure 11.1 Europe and the Americas Have the Highest Percent Urban Population, Whereas Africa and Asia Have Lower Percentages of Urban Population

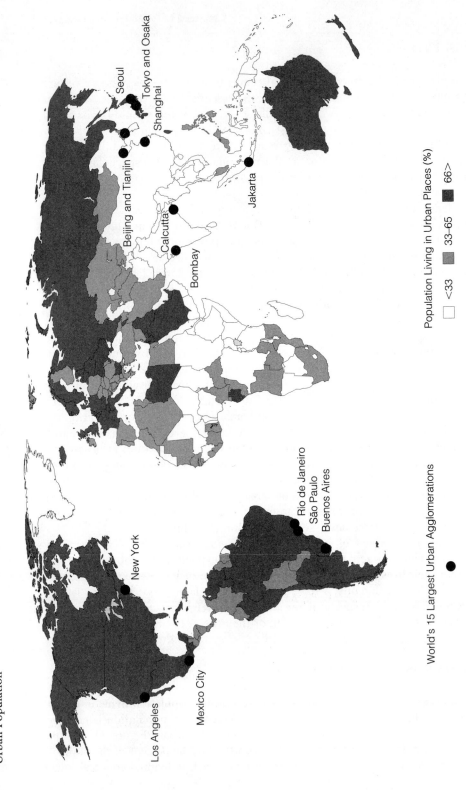

World's 15 Largest Urban Agglomerations ●

Population Living in Urban Places (%)

□ <33   ▨ 33–65   ■ 66>

Source: Adapted by the author from United Nations, 1994, Urban Agglomerations 1994 (New York: United Nations Population Division), and from Population Reference Bureau, 1995 World Population Data Sheet (Washington, D.C.: Population Reference Bureau).

definition of "urban," but typically it refers to places with 2,000 people or more. A little more than one fourth of the countries of the world (28 percent) have less than 33 percent of the population living in urban places. These include the two most populous nations, China and India, as well as the rest of the Indian subcontinent, most of southeast Asia, and nearly half of the sub-Saharan nations. More than one third of the world's nations (38 percent) have between 33 and 65 percent of the population living in urban places. In the Americas, these include countries in Central and South America with large indigenous populations. Sub-Saharan Africa is sprinkled with countries in this range, and most of the former Soviet republics in Central Asia are in this category. Some European countries are also in this range, as you can see from the map, including Spain and Portugal, as well as the region from Greece northward through the Balkan states up into Finland. The remaining third of the world's nations have 66 percent or more of the population residing in urban places. This includes most countries in North and South America, as well as most European nations, along with Japan and Korea in Asia, and Australia and New Zealand in Oceania. In essence, the highest percentage of the urban population tends to be found among European and "overseas" European nations.

Cities before Urbanization

How did we get to this situation in which the world is shifting from rural to urban places? Urbanization is a phenomenon as recent as, and part of, the demographic transition. The earliest cities were not very large, because most of them were not demographically self-sustaining. The ancient city of Babylon might have had 50,000 people, Athens possibly 80,000, and Rome as many as 500,000; but they represented a tiny fragment of the total population. They were symbols of civilization, visible centers that were written about, discussed by travelers, and densely enough settled to be later dug up by archaeologists. Our view of ancient history is colored by the fact that our knowledge of societal detail is limited primarily to the cities, although we can be sure that most people actually lived in the countryside.

Early cities had to be constantly replenished by migrants from the hinterlands, because they had higher death rates and lower birth rates than did the countryside, usually resulting in an annual excess of deaths over births. The self-sustaining character of modern urban areas began with the transformation of economies based on agriculture (produced in the country) to those based on manufactured goods (produced in the city). Control of the economy made it far easier for cities to dominate the rural areas politically and thus ensure their own continued existence in economic terms. A crucial transition in this process came between about 1500 and 1800 with the European discovery of "new" lands, the rise of mercantilistic states (that is, based on goods rather than landholdings), and the inception of the Industrial Revolution. These events were inextricably intertwined, and they added up to a diversity of trade that gave a powerful stimulus to the European economy. This was a period of building a base for subsequent industrialization, but it was still a preindustrial and largely preurban era. During this time, for example, cities in England were growing at only a slightly higher rate than the total population, and thus the proportion of urban population was rising only very slowly. Between 1600 and

1800 London grew from about 200,000 people to slightly less than 1 million (Wrigley 1987)—an average rate of growth considerably less than 1 percent per year; also during this span of 200 years, London's population increased from 2 percent of the total population of England to 10 percent—significant, but not necessarily remarkable, especially considering that in 1800 London was the largest city in Europe. In 1801 only 18 percent of the population in England lived in cities of 30,000 people or more, and nearly two thirds of those urban residents were concentrated in London. Thus on the eve of the Industrial Revolution, Europe (like the rest of the world) was predominantly agrarian.

Neither England nor any other country was at that time urbanizing with any speed, because industry had not yet grown sufficiently to demand a sizable urban population and because cities could not yet sustain their populations through natural increase. Not until the 19th century did urbanization take off, with a timing closely tied to industrialization and the decline in mortality that triggered population growth.

Cities and Development

As economic development occurred, cities grew because they were economically efficient places. For example, commercial centers bring together in one place the buyers and sellers of goods and services. Likewise, industrial centers bring together raw materials, laborers, and the financial capital necessary for the profitable production of goods. They are efficient politically because they centralize power and thus make more efficient the administrative activities of the power base that supports them. In sum, cities perform most functions of society more efficiently than is possible when people are spatially spread out. Mumford says it well: "There is indeed no single urban activity that has not been performed successfully in isolated units in the open country. But there is one function that the city alone can perform, namely the synthesis and synergy of the many separate parts by continually bringing them together in a common meeting place where direct fact-to-face intercourse is possible. The office of the city, then, is to increase the variety, the velocity, the extent, and the continuity of human intercourse" (Mumford 1968:447).

Cities are efficient partly because they reduce costs by congregating together both producers and consumers of a variety of goods and services. By reducing costs, urban places increase the benefits accruing to industry—meaning, naturally, higher profits. Those profits translate into higher standards of living, and that is why cities have flourished in the modern world. They are part and parcel of the modern rise of capitalism, and the patterns of urbanization in different parts of the world reflect where a country entered the modern economic system.

Early competitive, laissez-faire capitalism characterized the economies of Europe and North America from the late 18th through the mid-19th centuries and cities were still largely commercial in nature, although in the 19th century there was a clear transition to the industrial economy. Urban factory jobs were the classic magnets sucking young people out of the countryside in the 19th century. This happened earliest in England, and in Figure 11.2 you can see the rapid rise in urbanization in the United Kingdom in response to early industrialization.

Figure 11.2   Industrialized Nations Are Highly Urbanized

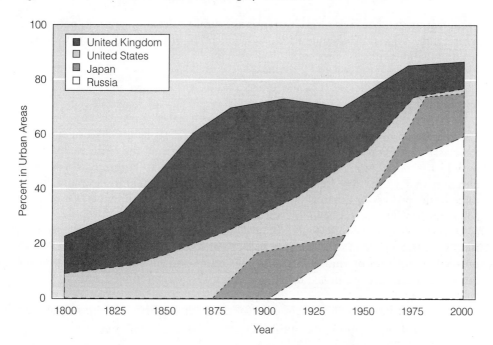

Note: Until the post-World War II period, the process of urbanization was closely associated with industrialization. England and Wales (comprising most of the United Kingdom) began to industrialize before the United States did, and so their urban fraction increased sooner. Japan began to industrialize and urbanize in the late nineteenth century, whereas those two processes were delayed in Russia until after the revolution.

Sources: Kingsley Davis, 1965, "The urbanization of the human population," Scientific American, 213(3):47, copyright 1965 by Scientific American, Inc., all rights reserved; updated by the author from United Nations, 1991, World Urbanization Prospects 1990 (New York: United Nations), Tables A.2 and A.4.

Japan and Russia entered the industrial world later than the United Kingdom or the United States, and you can see in Figure 11.2 that their pattern of urbanization was therefore delayed. Once started, however, urbanization proceeded quickly in these countries.

In the contemporary post-industrial world characterized by what is often called "advanced capitalism," the function of cities is changing again. In the developed, already urbanized part of the world, cities are losing their industrial base and are increasingly service centers to economic activities occurring in the hinterlands of the same country, or in another country altogether. In less developed countries, commercial and industrial activities combine with historically unprecedented rates of city growth to generate patterns of urbanization somewhat different from that which occurred in the now developed nations. We return to a description of the current situation later in the chapter. Now, however, it is time to ask how the shift of people from rural to urban places actually occurs.

Demographic Components of Urbanization

Urbanization can occur as a result of internal rural-to-urban migration, natural increase, international urban migration, reclassification of places from rural to urban, or combinations of these processes.

Internal Rural-to-Urban Migration

The migration of people within a country from rural to urban places represents the classic definition of urbanization because it is intuitively the most obvious way by which a population can be shifted from countryside to curbside. There is no question that in the developed countries, rural-to-urban migration was a major force in the process of urbanization. Over time the agricultural population of these countries has tended to decline in absolute numbers, as well as in relative terms, even in the face of overall population growth. In less developed countries, though, rural-to-urban migration is occurring in large absolute terms but without a consequent depopulation of rural areas. The reason, of course, is the difference in the rates of natural increase in less developed countries compared with rates in the developed nations.

Had it not been for migration, cities of the 19th century and before could not have grown in population size. In fact, in the absence of migration, the excess of deaths over births would actually have produced deurbanization. Of course, migration did occur, because economic development created a demand for an urban population that has been largely met by migrants from rural areas. Industrial cities drew the largest crowds, but commercial cities, even in nonindustrial countries, also generated a demand for jobs and created opportunities for people to move from agrarian to urban areas. The cities of most previously colonized countries bear witness to this fact. For example, migration accounted for 75 to 100 percent of the total growth of 19th-century cities in Latin America (Weller, Macisco, and Martine 1971). The growth of cities in southern Asia and western Africa was also stimulated by the commercial contacts of an expanding European economy. Naturally, in the highly industrialized, highly urbanized countries the agricultural population is so small that cities (also nations) now depend on the natural increase of urban areas rather than migration for population growth.

Natural Increase

The underlying source of urbanization throughout the world is the growth of the rural population. The decline in death rates in rural places, without a commensurate drop in the birth rate, leads to genuine **overpopulation** in rural areas (too many people for the available number of jobs) and causes people to seek employment elsewhere (the now-familiar tale of demographic change and response). If there were no opportunities for rural-to-urban migration, then the result might simply be that the death rate would eventually rise again in rural areas to achieve a balance between population and resources (the "Malthusian" solution). However, opportunities typically have existed elsewhere in urban places precisely because the innovations that led to a drop in the rural death rate have originated in the cities—the site of technological and material progress and the source of economic development.

The speed of urbanization—the number of years it takes to go from a low percent urban to a high percent—depends partly on the difference in the rates of natural increase between urban and rural areas. In turn, the rate of natural increase is dependent on trends in both mortality and fertility, and these patterns have changed dramatically over time.

Urbanization and Mortality   Kingsley Davis (1973) has estimated that in the city of Stockholm, Sweden, in 1861–70, the average life expectancy at birth was only 28 years; for the country as a whole at that time, the life expectancy was 45 years. I discussed in Chapter 4 that the ability to resist death has been passed to the rest of the world by the industrialized nations, and the diffusion of death control has usually started in the cities and spread from there to the countryside. Thus there has been a crucial reversal in the urban-rural difference in mortality. When the now industrialized nations were urbanizing, death rates were higher in the city than in the countryside (see, for example, Williams and Galley 1995) and this helped keep the rate of natural increase in the city lower than in rural areas. In turn, that meant that rural-to-urban migration was a more important factor influencing the percent urban in a country.

For the past several decades, however, death rates have been lower in the city than in the country in nearly every part of the world. As a consequence, the process of urbanization in the less developed countries is taking place in the context of historically high rates of urban natural increase. Furthermore, when mortality declines as a response to economic development, there are also structural changes that tend to reduce fertility; but when death control is introduced independently of economic development, mortality and fertility declines lose their common source, and mortality decreases whereas fertility may not. This results in fertility levels being higher today in the less developed countries (urban and rural places alike) than they were at a comparable stage of mortality decline in the currently advanced countries.

Urbanization and Fertility   It is almost an axiom in population studies that urban fertility levels are lower than rural levels; it is also true, of course, that fertility is higher in less developed than in developed nations. Putting these two generalizations together, you can conclude that urban fertility in less developed nations will be lower than rural fertility but still higher than the urban fertility of cities in the industrialized nations. High fertility persists in the cities of less developed nations partly because the urban environment, as bad as it may seem, is less hostile to reproduction than it used to be. Less developed nations often have systems of public welfare, subsidized housing, free education, and accessible maternal and child health clinics. Nonetheless, the lower fertility that almost always prevails in urban places deserves closer scrutiny.

We can usually anticipate that people residing in urban areas will have fairly distinctive ways of behaving compared with rural dwellers. So important and obvious are these differences demographically that urban and rural differentials in fertility are among the most well documented in the literature of demographic research. John Graunt, the 17th-century English demographer whose name I first mentioned in Chapter 2, concluded that London marriages were less fruitful than those in the country because of "the intemperance in feeding, and especially the Adulteries and Fornications, supposed more frequent in London than elsewhere . . . and . . . the minds of men in London are more thoughtful and full of business than in the Country" (quoted by Eversley 1959:38).

In rural areas large families may be useful (for the labor power), but even if they are not, a family can "take care of" too many members by encouraging migration to the city. Once in the city, people have to cope more immediately with the problems that large families might create and, besides, the city offers many more alternatives to family life than do rural areas. In recent decades the once wide divergence in urban and rural fertility levels has narrowed as rural fertility has decreased relative to urban levels, reflecting the growing dependence of rural places on urban production and an urban lifestyle.

In 1940 in the United States there were still substantial differences in the number of children women had according to where they lived. Rural farm women, for example, at every age over 19 had at least twice as many children as urban women. In the early 1990s, fewer than 2 percent of women in the United States of reproductive age were living on farms, and their fertility was only slightly higher than the other 98 percent of the population (Bachu 1993). By the mid-1990s, farm residence was not even reported in the data for the United States, although women who worked in occupations listed as "farming, forestry, and fishing" did have higher fertility levels than any other group of employed women (Bachu 1997). Still, the differences were small; so small that Long and Nucci (1996) were willing to conclude that "the traditional urban-rural fertility differential in the United States has ended, at least temporarily" (p. 19).

Migration is also related to fertility, since migrants tend to be young adults of reproductive ages. Furthermore, migrants from rural areas typically wind up having levels of fertility lower than those in the rural areas from which they left but still higher than levels in the urban areas to which they have moved (Goldstein and Goldstein 1981; Ritchey and Stokes 1972; Zarate and de Zarate 1975). To some extent, of course, the fertility impact of the migrants will depend on whether males or females (or neither) predominate in the migration stream.

Sex Ratios in Cities   Females tend to be more mobile in North and South America and Europe, whereas in Africa and Asia more men than women migrate from rural to urban places, as I mentioned in Chapter 7. The differences in the sex ratios of migrants seem to be determined largely by the employment opportunities for women. The pattern of agricultural labor in Europe, North America, and South America has been for males to do most of the outside-the-home work, relegating women to domestic work for the most part; whereas in Africa and Asia (including Arab countries and India), the role of women in regular agricultural work (especially commercial enterprises associated with agricultural produce bazaars, and so on) has been more prominent. Thus it is at least a reasonable thesis that as an economy develops and urban opportunities arise, females will be more responsive than males to these opportunities if they are less actively involved in the agrarian labor force. In countries of Europe, North America, and South America, where women have been less active in agriculture, the sex ratios in cities is more feminine than in rural areas. For example, in Romania in 1992, the urban sex ratio at ages 20 to 24 (when people are especially likely to migrate) was 92 males per 100 females, whereas in rural areas at that age it was 120.

In African and Asian countries, urban sex ratios tend to be higher (that is, more masculine) than rural sex ratios. In China in 1990 the urban sex ratio at ages 20 to 24 was 120, compared with the rural sex ratio of 102. Nonetheless, Asian women have been heading for the cities in increasingly greater proportions, stimulated by the new availability of jobs in electronics, textiles, and other industries in which there has been a demand for cheap labor (Paul 1994; Smith, Khoo, and Go 1983), and so we find that in 1990 in the Philippines the urban sex ratio at ages 20 to 24 was 97 males per 100 females compared with the rural sex ratio of 107.

International Urbanward Migration

International migration also operates to increase the level of urbanization, because most international migrants move to cities in the host area regardless of where they lived in the donor area. From the standpoint of the host area, then, the impact of international migration is naturally to add to the urban population without adding significantly to the rural population, thereby shifting a greater proportion of the total to urban places. Certainly, most immigrants to the United States wind up as urban residents in big cities or their suburbs (U.S. Bureau of the Census 1993).

In spite of the massive increase in international migration in the past few decades and the fact that most international migrants live in urban areas, the actual impact on the process of urbanization throughout the globe has been less dramatic than it might seem. The reason for this is that many international migrants are headed toward cities in countries that are already highly urbanized (Berry 1993).

Reclassification

It is also possible for "in-place" urbanization to occur. This happens when the absolute size of a place grows so large, whether by migration, natural increase, or both, that it reaches or exceeds the minimum size criterion used to distinguish urban from rural places. In the United States, any incorporated town that exceeds 2,500 is considered urban because, back in the 1920s when the current definition was devised, any place that large seemed urban to the people living in the surrounding countryside. So in 1980, the tiny village of Troy, North Carolina became an urban area, to be included in the same category as New York City. A more restrictive definition is that of an "urbanized area," which the Census Bureau defines as any city and its surrounding urban fringe (suburbs) with a total population of 50,000, as long as the surrounding territory has a density of at least 1,000 persons per square mile. The 1990 census identified 396 urbanized areas in the United States, and one of the new ones was Watsonville, California, whose rural roots are obvious in its designation as the "artichoke capital of the world."

Note that reclassification is more of an administrative phenomenon than anything else and is based on a unidimensional (size only) definition of urban places, rather than also incorporating any concept of economic and social activity. Of course, it is quite probable that as a place grows in absolute size it will at the same time diversify economically and socially, probably away from agricultural activities into more urban enterprises. This tends to be part of the social change that occurs everywhere in response to an increase in population size; an agricultural population can become quickly redundant and the lure of urban activities (such as industry, commerce, and services) may be strong under those conditions.

Another administrative trick that can lead to rapid city growth is annexation. City boundaries are almost constantly changing, and the effect is often to bring into the city limits people who might otherwise be classified as rural. Urban growth rates can thus be misleading. For example, "The city of Houston grew 29 percent during the 1970s—one of the most rapidly growing large cities in the country. But the city also annexed a quarter of a million people. Without the annexation, the city would have grown only modestly" (Miller 1984:31).

Metropolitanization and Agglomeration

In countries such as the United States, cities have grown so large and their influence has extended so far that a distinction is often made between metropolitan and nonmetropolitan counties, rather than the more vague distinction between urban and rural. In 1949 the U.S. Census Bureau developed the concept of standard metropolitan area (SMA). An SMA consists of a county with a core city of at least 50,000 people and with a population density of at least 1,000 people per square mile. The concept proved useful and was subsequently revised to be called the standard metropolitan statistical area (SMSA). In 1983 the U.S. Office of Management and Budget (OMB) ordered a new revision, and the United States is now divided into MSAs (metropolitan statistical areas), CMSAs (consolidated metropolitan statistical areas), and PMSAs (primary metropolitan statistical areas). An MSA is a county that has a core city with at least 50,000 people. The MSA then includes that population and the surrounding population, which itself may or may not be urban in character. The definition follows the concept of the core-periphery model of urbanization, which I will discuss later in the chapter.

As of the mid-1990s there were 255 "stand-alone" MSAs and 18 CMSAs (each with several MSAs of its own) in the United States, encompassing within their boundaries 80 percent of the total U.S. population. An example of a stand-alone MSA would be the Raleigh-Durham-Chapel Hill MSA in North Carolina, which was tagged by *Money* magazine in the 1990s as the best community in America in which to live. Although the MSA comprises the three places in its title, along with several smaller suburban cities and areas, it is not part of any larger unit. By contrast, the nation's most populous MSA, New York City (which, incidentally, was 123rd on the list of the best places to live), is only one of 15 MSAs that make up the "New York-Northern New Jersey-Long Island, NY-NJ-CT-PA" CMSA. Each one of those constituent MSAs is thus known as a PMSA (I'm not making this up). OMB has ordered that the MSA definitions be revised once again for use with the Census 2000 data, but the revisions had not been finalized as this book went to press.

In Canada the definition of metropolitan is similar, although not identical, to that in the United States The Census Metropolitan Area (CMA) has a core urban area of at least 100,000 and includes the adjacent urban and rural areas, which have a high degree of economic and social integration with that urban area. In Mexico the government agencies have not defined metropolitan areas quite as precisely as in Canada and the United States.

Another level of aggregation is the **urban agglomeration,** a term used largely by the United Nations, but consistent with the more popular term "**megalopolis.**" The UN defines an urban agglomeration as "the population contained within the contours of contiguous territory inhabited at urban levels of residential density without regard to administrative boundaries" (United Nations 1994). The concept is similar to the U.S. Bureau of the Census's definition of a CMSA and to Statistics Canada's definition of a Census Metropolitan Area. In Table 11.1, I have listed for you the largest metropolitan areas in the United States, Canada, and Mexico according to the United Nations' definition of urban agglomeration, with the comparison for the United States and Canada of the census-derived definitions for those two countries (comparable census definitions are not available for Mexico). In general, the United Nations definitions are more limited geographically than those of the U.S. census, but are nearly identical to those of the Canadian definitions. By UN definitions Mexico City is just slightly more populous than the New York metropolitan area, although the U.S. census definition of the CMSA of New York includes 3.5 million more people than the UN definition. Toronto is Canada's most populous metropolitan area and its size would make it the fourth largest metro area in the United States and the second largest in Mexico.

As a further "refinement," the United Nations refers to any urban agglomeration with more than 8 million people as a **mega-city.** By this definition there were 24 mega-cities in the world in 1996. Figure 11.3 shows the top 15 mega-cities in the world as of 1996. You can locate them geographically by looking back at Figure 11.1. Tokyo is the largest urban agglomeration with 27.2 million people, followed by Mexico City, São Paulo, and New York City. Just a few years ago, Mexico City was the largest and many people projected it to continue growing almost forever. However, the serious environmental problems in the Valley of Mexico created by population growth convinced the Mexican government to undertake a concerted effort to move industry out of the area and to divert migrants to other metropolitan areas. The effort clearly paid off, because the 1990 census counted fewer people than anticipated and on that basis the United Nations demographers revised their projections of population growth in Mexico City. Still, the 16.9 million people living in Mexico City in 1996 was a huge increase from the 3.1 million in 1950, and the United Nations projects that the population of Mexico City will top 19 million by the year 2015.

The bottom panel of Figure 11.3 shows the ten largest urban agglomerations in 1950. In that year only three of the top cities were in developing countries—Shanghai, China; Buenos Aires, Argentina; and Calcutta, India. By 1996 only three of the ten largest cities were in developed nations—Tokyo, New York, and Los Angeles. This is because the cities of the less developed nations have simply skyrocketed in their population size.

Table 11.1   The Largest Metropolitan Areas in the United States, Canada, and Mexico

Country	Metropolitan Area	Population (in millions) in 1996 according to:	
		United Nations Urban Agglomeration	Census Definition
USA			
	New York	16.4	19.9
	Los Angeles	12.6	15.5
	Chicago	6.9	8.6
	Philadelphia	4.3	6.0
	San Francisco	3.9	6.6
	Dallas	3.7	4.6
	Detroit	3.7	5.3
	Houston	3.2	4.3
	Boston	2.9	5.6
	San Diego	2.8	2.7
	Atlanta	2.5	3.5
	Phoenix	2.4	2.7
	Minneapolis	2.3	2.8
	Miami	2.1	3.5
	Baltimore	2.0	7.2
	Saint Louis	2.0	2.5
	Seattle	2.0	3.3
	Tampa	1.9	2.2
	Cleveland	1.7	2.9
	Norfolk	1.7	1.5
	Pittsburgh	1.7	2.4
	Denver	1.6	2.3
	San Jose	1.6	(included in San Francisco)
	Riverside–San Bernardino	1.5	(included in Los Angeles)
	Fort Lauderdale	1.4	(included in Miami)
	Kansas City	1.4	1.7
	Cincinnati	1.3	1.9
	Milwaukee	1.3	1.6
	Portland	1.3	2.1
	Sacramento	1.3	1.6
	San Antonio	1.2	1.5
	New Orleans	1.1	1.3
	Orlando	1.1	1.4
	Columbus	1.0	1.4
Canada			
	Toronto	4.4	4.4
	Montreal	3.3	3.4
	Vancouver	1.9	1.9
	Ottawa	1.0	1.0
Mexico			
	Mexico City	16.9	
	Guadalajara	3.5	
	Monterrey	3.1	
	Puebla	1.8	

Sources: United Nations, 1996, Urban Agglomerations 1996 (New York: United Nations); U.S. Bureau of the Census, 1997, Population Estimates Released on the Internet; Statistics Canada, 1998, Population of Census Metropolitan Areas.

Figure 11.3 The World's Largest Urban Agglomerations Changed Dramatically between 1950 and 1996

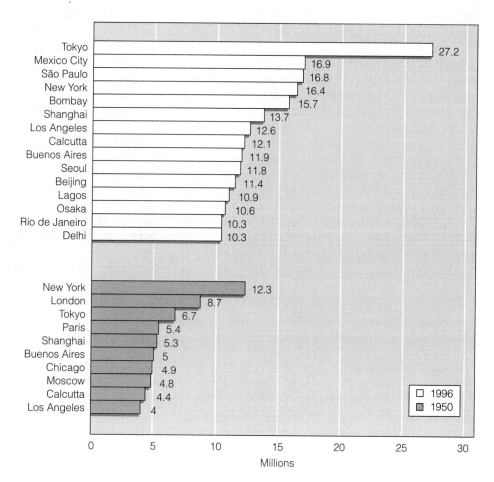

Source: United Nations, 1996, Urban Agglomerations 1996 (New York: United Nations Population Division). Population figures are in millions.

The rapid growth of cities in the context of rapid population growth in developing countries deserves more scrutiny on our part because it represents a potent source of social change with which societies must cope. To understand it, however, we must first examine some major theoretical perspectives about the place of cities in the global economic system.

Theories of the Urban Transition

Two major theories exist to explain the process of urbanization: (1) modernization theory; and (2) core-periphery models associated with dependency or world systems theory. You have read of these theories before in previous chapters.

Modernization Theory

I have already described the major elements of modernization theory as they relate to urbanization because the theory of the demographic transition is derived from the modernization theory. The basic thesis is that economic development is built on the efficiencies of cities. Cities are the engines of growth because of the concentration of capital (to build industry), labor (to perform the industrial tasks), and the financial, governmental, and administrative services necessary to manufacture, distribute, sell, and regulate the goods that comprise the essence of economic development. Once economic growth begins, the principle of cumulative causation (another familiar concept by now) kicks in to promote further local development, often at the expense of other regions.

Modernization theory tends to focus on individual nations, examining the role that cities play in the development process in that particular country. The fact that cities are crucial to modernization has led many developing nations such as Chile, Kenya, Malaysia, and Mexico to establish industrial cities precisely to mimic the process of modernization that occurred in the more developed nations (Potter 1992). The results have been mixed at best.

Modernization theory offers a generally good explanation of what happened in the now developed countries, especially in the period up to World War II, as individual nation building was taking place, but transportation and communication still limited the amount of direct interaction between countries. Most of the still-quoted theories of modernization were developed in the 1950s (such as Hirschman 1958; Myrdal 1957), at a time when the "Third World" had not yet been defined as such, and when urbanization in developing countries was still in its early stages. As urbanization proceeded in developing nations it became obvious that the process of modernization did not describe adequately what was happening.

Theories Built on the Core-Periphery Model

Empirical studies of city systems within countries have shown that a common pattern of cities by size is expressed by the rank-size rule. As set out by Zipf (1949), this says that the population size of a given city (P_i) within a country will be approximately equal to the population of the largest city (P_1) divided by the rank of city i by population size (R_i), or:

$$\text{rank-size rule: } P_i = P_1/R_i$$

So, if the largest city in a country has a population of 5 million, then the second largest city should have a population size of approximately 5/2 or 2.5 million, whereas the third largest city should have a population of 5/3 or 1.7 million people.

Many countries generally follow a pattern similar to the rank-size rule, but there are important exceptions. For example, in Mexico, the largest city (Mexico City) has a population of 16.9 million people, so the rank-size rule would predict a population of 16.9/2, or 8.5 million, for the second largest city (Guadalajara) and a population of 16.9/3, or 5.6 million, for the third largest city (Monterrey). However, the actual sizes of Guadalajara and Monterrey in 1996, as you can see in Table 11.1, were only 3.5 million and 3.1 million, respectively. What happened? Mexico is one of a number of countries characterized by a **primate city**—a disproportionately large leading city holding a central place in the economy of the country.

The importance of these empirical generalizations is that they led to the realization that most countries had a somewhat predictable system of cities that might be amenable to a consistent theoretical interpretation. An important set of such theoretical perspectives was the core-periphery model put forth by Friedmann in 1966 (Potter 1992). Prior to economic development, a series of independent cities may exist in a region, but development tends to begin in, and will be concentrated in, one major site (the primate city). This is especially apt to happen in less developed countries that have a history of colonial domination. Over time, the development process diffuses to other cities, but this happens unequally because the primate city (the core) controls the resources, and the smaller cities (the periphery) are dependent on the larger city. It was only a small step, then, to apply these ideas of the core and periphery to a world system of countries and cities.

The world systems theory is based on the notion that inequality is part of the global economic structure, and has been since at least the rise of capitalism 500 years ago (Wallerstein 1974). "Core" countries are defined as the highly developed nations that dominate the world economy and were the former colonizers of much of the rest of the world. The "periphery" is composed of those countries that, in order to become a part of the global economic system (the alternative to which is to remain isolated and undeveloped), have been forced to be dependent on the core nations, which control the basic resources required for development and a higher standard of living. Because countries tend to be dominated by cities, the world systems theory predicts that cities of core countries will control global resources, operating especially through multinational corporations headquartered in core cities (Chase-Dunn 1989; Chase-Dunn and Hall 1997), and cities in peripheral countries will be dependent on the core cities for their own growth and development (and, of course, that will filter down from the primate city to the other cities in a nation's own city system).

Fitting the Theories to the Real World

As is so often the case, no single theory seems capable of explaining all of the complexity in the process of urbanization. Several tests of modernization and world systems theories have suggested that variations of both contribute to our understanding of the real world experience of urbanization (see, for example, Bradshaw 1987; London 1987; Firebaugh 1994). Keep this in mind as I illustrate urbanization in developing countries with examples from Mexico and China.

An Illustration from Mexico

The impact of population processes on urbanization is illustrated by what has happened to people in one small village in Mexico—Tzintzuntzan, situated midway between Guadalajara and Mexico City in the state of Michoacán. Historically, the site was the capital of the Tarascan empire (Brandes 1990), but today it is a village of artisans, farmers, merchants, and teachers. For nearly 400 years the population of Tzintzuntzan stayed right at about 1,000 people (Foster 1967). In the mid-1940s, as Foster began studying the village, the population size was starting to climb slowly, because death rates had started to decline in the late 1930s at about the time a government project gave the village electricity, running water, and a hard-surfaced highway connecting it to the outside world (Kemper and Foster 1975). In 1940 the population was 1,077 and the death rate was about 30 per 1,000, while the birth rate of 47 per 1,000 was leading to a rate of natural increase of 17 per 1,000. For some time there had been small-scale, local out-migration from the village to keep its population in balance with the limited local resources, but by 1950 the death rate was down to 17 per 1,000 and the birth rate had risen. Better medical care had reduced the incidence of miscarriage and stillbirth, and in 1950 the village had 1,336 people (Foster 1967). By 1970 the population had reached about 2,200 (twice the 1940 size); however, were it not for out-migration draining away virtually all the natural increase of Tzintzuntzan, the population would again have doubled in about 20 years (Kemper and Foster 1975).

What, you ask, does growth in a small Mexican village have to do with urbanization? The answer, of course, is that the migrants headed for the cities. Seventy percent of those people who leave Tzintzuntzan go to urban places, with Mexico City (230 miles away) having been the most popular destination (Kemper and Foster 1975). One of the easiest demographic responses that people can make to population pressure is migration, and in Mexico, as in most countries of the world, the city has been the receiving ground. Furthermore, the demographic characteristics of those who go to the city are what you would expect; they tend to be younger, slightly better educated, of higher occupational status, and more innovative than nonmigrants (Kemper 1977).

For Tzintzuntzeños, migration to Mexico City has raised the standard of living of migrant families, altered the world view of both adults and their children toward greater independence and achievement, and, indirectly, "urbanized" the village they left behind. This last effect is due to the fact that having friends and relatives in Mexico City is one factor that leads the villagers to be aware of their participation in a wider world. This makes it easier for each generation to make the move to Mexico City, because they know what to expect when they arrive and they know people who can help them.

Tzintzuntzeños have also been attracted to the United States, initially recruited through the Bracero program (which I mentioned in Chapter 7), and the same patterns of mutual assistance have encouraged the flow of money and ideas from cities in the United States to this small village in the interior of Mexico (Kemper 1991). Migrants have brought back to the village many of the accoutrements of urban life, from new stoves and sewing machines to stereos and television sets, in effect urbanizing what was once a remote village. This in-place urbanization ultimately has had an effect on fertility. With rising education and incomes, birth rates appear finally to be dropping here as they are throughout Mexico. In the 1990 census, the village was enumerated at about 3,000 people, which was only three times the 1940 population, compared with the tenfold increase in the size of Mexico City during that half-century interval.

From a theoretical perspective we can see that modernization is the key to the transformation of lives of Tzintzuntzeños. However, the process of modernization was not endogenous—it sought the villagers out, rather than the other way around. In a very literal sense the modernization of the village and its inhabitants was dependent on what was happening elsewhere. Government leaders in Mexico City (the core) made the decision to provide rural areas (the periphery) with health care, electricity, and paved highways. The rest, as they say, is history, because few villagers, when given the choice, turn down the opportunity for a higher standard of living (Critchfield 1994).

An Illustration from China

China is a very interesting case of urbanization because it is one of the few countries in the world where the government fairly successfully "kept them down on the farm." The Chinese Communist Party officially adopted an anti-urban policy when it came to power in 1949, believing that cities were a negative "Western" influence, and Chinese government policies in the 1960s and 1970s were designed to counteract the process of urbanization that has occurred in most of the rest of the world. These policies attempted to "promote wider income distribution, reduce regional inequalities, and create a more balanced urban hierarchy, which would lead to a greater decentralization of economic activities. In doing so, the intention was to slow population growth in the largest cities, while allowing continued increases in medium-sized and smaller urban centers" (Goldstein 1988 as quoted in Bradshaw and Fraser 1989:989). As wonderful as that may sound, the basic policy is a rigid system of household registration called *hukou,* which created a type of occupational apartheid in China. "Anyone in a rural county is automatically registered as a farmer, anyone in a city as a non-farmer; and the distinction is near rigid. A city-dwelling woman (though not a man) who marries a farmer loses the right to urban life" (The Economist 1998:42).

Government policies in China thus prevented the high rate of natural increase in the rural areas from spilling over disproportionately into migration to urban areas. This was mitigated partly by the governmental emphasis on heavy industry, often located in rural areas, rather than on the light and service industries that typify newer cities in the world today (Hsu 1994). Of course, that does not mean that the urban population was not growing. Quite the contrary. Between 1953 and 1990, 326 new cities were created in China (Hsu 1994), and urban growth occurred primarily in small to medium-sized places (Han and Wong 1994), just as the government had planned. Also according to plan, urbanization was slow by world standards—scarcely more than one in four Chinese (29 percent) lived in urban places in the mid-1990s.

In the 1980s the Chinese government relaxed some of its restrictions, permitting rural residents to become temporary, "guest" workers in urban areas, although without the right to become permanent residents. In 1993, it was suggested that "freed from the strictures of a command economy, residents in places like Humen discovered they could make more money in light industry, property speculation or running sing-along bars than they could on their farms" (Brauchli 1993:A12). It has been estimated that a "floating population" of as many as 80 to 110 million Chinese are working in cities even though they technically belong in the countryside (Chang 1996), and the government has been trying to deal with this by offering legal urban residency and a job to rural migrants who agree to stay in the smaller "buffer" cities rather than seek work in one of China's mega-cities (Johnson 1997).

Impact of Urbanization on the Human Condition

It is not an exaggeration to suggest that urbanization is a revolutionary shift in society. In 1964, Reissman described it this way:

> Urbanization is social change on a vast scale. It means deep and irrevocable changes that alter all sectors of a society. In our own history [the United States] the shift from an agricultural to an industrial society has altered every aspect of social life . . . the whole institutional structure was affected as a consequence of our urban development. Apparently, the process is irreversible once begun. The impetus of urbanization upon society is such that society gives way to urban institutions, urban values, and urban demands [Reissman 1964:154].

The benefits of cities, of course, are what make them attractive, and they at least partially explain the massive transformation of countries like the United States and Canada from predominantly rural to primarily urban nations within a few generations. The negative impact of urbanization on the human condition represents the set of unintended consequences that may prevent the city from being as attractive as it might otherwise be.

The efficiencies of the city have generally been translated into higher incomes for city dwellers than for farmers. In fact, it even tends to be true that the larger the city, the higher the wages. Wage differentials undoubtedly have been and continue to be prime motivations for individuals to move to cities and stay there. That is not to say that people necessarily prefer to live in cities; the reverse may actually be true. Throughout American history the sins and foibles of urban life have been decried, and the city is often compared unfavorably with a pastoral existence (see Fischer 1988). Of course, people recognize the advantages of the city, and Americans prefer to be near a city though not right in it. Fuguitt and Zuiches have pointed out that public opinion polls since 1948 have shown that a vast majority of Americans indicate a preference for living in rural areas or small cities and towns. When, however, they asked for the first time a survey question about the desire to be near a large city, those rural preferences became more specific. Their data showed that of all the people who said they preferred living in rural areas or in small cities, 61 percent also wanted to be within 30 miles of a central city (Fuguitt and Zuiches 1975). A replication of the survey in 1988 revealed a remarkable consistency over time in those residential preferences (Fuguitt and Brown 1990). In general, Americans like it both ways. They aspire to the freedom of space in the country but also prefer the economic and social advantages of the city. The compromise, of course, is the suburb, and it turns out that the average American is already living in this preferred location, leading Frey (1995) to declare that "America is in the suburbs. The suburbs are America" (p. 314).

Suburbanization in the United States

A century has passed since Adna Weber (1899) noted that American cities were beginning to suburbanize—to grow in the outlying rings of the city. It was not until the 1920s, however, that suburbanization really took off. Hawley (1972) has noted that between 1900 and 1920 people were still concentrating in the centers of cities, but after 1920 the suburbs began regularly to grow in population at a faster pace than the central cities. Two factors related to suburbanization are the desire of Americans to live in the less crowded environment of the outlying areas and their ability to do so—a result of increasing wealth and the availability of buses and especially automobiles (Tobin 1976). Such transportation has added an element of geographic flexibility not possible when suburbanites depended on fixed-rail trolleys to transport them between home in the suburbs and work in the central city.

From the 1920s through the 1960s the process of suburbanization continued almost unabated in the United States (as indeed in most cities of the world). Admittedly, the process was hurried along by automobile manufacturers and tire companies that bought local trolley lines in order to dismantle and replace them with gasoline powered buses (Kunstler 1993). Nonetheless, the advantages of the automobile are numerous and it was inevitable that cars would influence the shape of urban areas. By 1990 in the United States, 73 percent of all workers got to work by driving alone in their automobile (car, truck, or van) and an additional 13 percent carpooled to work in an automobile. According to data from the 1990 census, more people walk to work (6 percent) in the United States than use public transportation (5 percent). As you can see in Table 11.2, the vast majority of Americans in every part of the country get to work in an automobile. Not surprisingly, Detroit—the home of automobile manufacturing—has one of the highest percentages of commuters going to work in cars, but somewhat surprisingly, the figure for Detroit is noticeably higher even than for the Los Angeles metropolitan area—which is a legend in its own time for its love of automobiles. The use of public transportation is highest in the older cities of the northeast, such as New York and Boston.

The fact that America is now predominantly suburban adds new complexity to the metropolitan structure. Several trends are worth commenting on: (1) there has been a decided westward tilt to urbanization in the United States, which has facilitated suburbanization through the creation of new places; (2) many of those new places are **edge cities** within the suburbs, replacing the functions of the old central city; and (3) older parts of cities have been **gentrified.** Let me examine each of these in a bit more detail, remembering that although my comments are directed primarily at the United States, many of these same trends are being seen in other developed nations as well.

Table 11.2 The Majority of Commuters in
the United States Use Automobiles to Get to Work

Region of the United States	Drive Alone (%)	Carpool (%)	Use Public Transportation (%)	Average One-Way Commute Time (min.)
Entire United States	73.2	13.4	5.3	22.4
Northeast	66.5	11.6	12.8	24.5
New York City CMSA	53.6	10.3	26.6	30.6
Midwest	76.4	11.6	3.5	20.7
Detroit CMSA	82.7	10.1	2.4	23.4
South	75.7	15.1	2.6	22.0
Raleigh-Durham-Chapel Hill MSA	78.0	13.3	2.0	19.7
West	72.1	14.3	4.1	22.7
Los Angeles CMSA	72.3	15.5	4.6	26.4

Source: 1990 Census of Population and Housing, Summary Tape File 3C.

The western United States, the land of open spaces, has become the most highly metropolitanized area of the country (Abbott 1993). A higher fraction of residents lives in metropolitan areas in the west than in any other part of the country. This has happened because the flow of migration in the United States has been consistently westward, especially since the end of World War II, as you will recall from Chapter 7, and migration in the modern world is almost always toward or between urban places. People and jobs have been moving west (as well as south). Kasarda (1995) has noted that as recently as 1960, 25 percent of all Fortune 500 firms were headquartered in New York City, but by 1990 that had declined to 8 percent. Companies have shifted their operations to Sunbelt cities, but not necessarily the central parts of those cities. The suburbs are increasingly the sites of company headquarters, congregating near major highways and regional airports. It is perhaps a sign of the times that the richest person in the United States (William Gates) runs a company (Microsoft) that is located in the suburbs (Redmond, Washington) of a western city (Seattle).

Increasing suburbanization has meant greater metropolitan complexity, as new areas spring up on the edges of cities, competing with one another for jobs and amenities (Frey 1995; Hughes 1993). Garreau (1991) has coined the term "edge city" to describe the suburban entities that are emerging in the rings and beltways of metropolitan areas and are replicating, if not replacing, the functions of the older central cities. Some of the edge cities are actually within the same city limits as the central city, but are distinct from it. Furthermore, larger metropolitan areas may have several edge cities, each with its own pattern of dominance over specific economic functions (such as high technology or financial services) in conjunction with a full range of retail shops and dining and entertainment establishments. Muller (1997) argues, in fact, that it is precisely in these edge cities that the globalization of American cities is taking place.

The growth of edge cities and the increasing economic and social complexity of the suburbs helps to explain the shift in commuting patterns in the United States. In essence, the flexibility of the automobile allows people to live and work almost anywhere else within the same general area. Thus data from the 1990 census show that the number of commuters going from one suburban area to another far exceeds the number of commuters going from the suburbs to the central city (Kasarda 1995). Those cars on the freeway in the morning are not all headed downtown—they are headed every which way.

These trends in metropolitan complexity and diversity have tended to leave the central cities with a daytime population of "suits" who commute downtown to work at the various service companies (especially government administration and financial services industries) that have remained in the central city. At the same time, shopping centers, corporate headquarters, many new high-technology industries, and traffic gridlock have all relocated to the suburbs, leaving some central cities to be little more than residential areas for low-income people. At the same time, older suburban areas have been facing problems that used to be associated with the inner city. "The aging of suburbia is causing radical changes in the character and politics of countless bedroom communities. It is leading to school closings, property-tax revolts, and demands for new or expanded housing, transportation and recreation services for the elderly" (Lublin 1984:1).

The Baby Boom generation in the United States grew up in the suburbs to a greater extent than any previous cohort, but as they reached an age to buy homes, baby boomers found themselves caught in the midst of spiraling housing costs in the suburbs. One alternative was to head even farther out of town, to what sometimes are called the *exurbs*—the suburbs of the suburbs. Another innovative response to higher housing prices has been the purchase and rehabilitation of abandoned homes in older sections of central cities—especially in the older cities in the eastern half of the nation. Younger people, typically without children, have been buying homes that were sometimes mere shells and were thus sold very cheaply, and then renovating them and moving in. Because these innovative renovators tended to be white and upwardly mobile, they have been likened to the gentry moving back into the city, and thus the term *gentrification* is applied. Yet, despite the radical transformation of a few neighborhoods in places like New York City, Chicago, Baltimore, and Washington, D.C., as well as Paris and London (Carpenter and Lees 1995), gentrification has never involved enough people to counter the continuing movement of people out of the central cities and into the suburbs (Frey 1990).

Residential Segregation

Although suburbia has become a legendary part of American society, suburbanization was disproportionately engaged in by whites until the 1970s. For example, in 1970 in 15 large areas studied by Farley (1976), 58 percent of the whites lived in the suburbs compared with 17 percent of the nonwhites. Since the 1930s the proportion of whites living in central cities has declined steadily and the proportion of African-Americans has risen steeply (Schnore, Andre, and Sharp 1976); the African-

American population underwent a very rapid urbanization at the same time that whites were suburbanizing.

During the period 1910–30, there was a substantial movement of African-Americans out of the South destined for the cities of the North and West. The urban population of African-Americans grew by more than 3 percent per year during that 20-year period, whereas the rural population declined not only relatively but in absolute terms as well. The reasons for migration out of rural areas were primarily economic, with the decline in the world demand for southern agricultural products providing the push out of the South. But there were concurrent pull factors as well in the form of demands in northern and western cities for labor, which could be met cheaply by African-Americans moving from the South (Farley 1970). During the Depression there was a slowdown in the urbanization of African-Americans, but by the beginning of World War II half the nation's African-Americans lived in cities, reaching that level of urbanization 30 years later than whites had. After World War II the urbanization and rural depopulation of African-Americans resumed at an even higher level than after World War I, and by 1960 the African-American population was 58 percent urban in the South and 95 percent urban in the North and West. Urbanization was associated not only with the economic recovery after the war but also with the severely restricted international migration (see Chapter 14), which, until the law was changed in 1965, meant that foreigners no longer were entering the labor force to take newly created jobs, thus providing a market for African-American labor. The consequence of the urbanization of African-Americans, their urban rates of natural increase, and the relatively higher rates of suburbanization of whites has been the segregation of African-American and white populations within metropolitan areas.

The segregation of people into different neighborhoods on the basis of different social characteristics (such as ethnicity, occupation, or income) is a fairly common feature of human society (Farley 1976; Zlaff 1973). However, in the United States residential segregation by race is much more intense than segregation by any other measurable category. For example, Farley (1976) has demonstrated that in both predominantly white and predominantly African-American areas there is a fair amount of residential segregation by education, occupation, and income—whether you look at central cities or suburbs. Residential segregation of blacks in the United States has been called an "American apartheid system" (Massey and Denton 1993), and the maintenance of this pattern until fairly recently has been explained by Farley and Frey (1994) as being due especially to the following factors: (1) mortgage lending policies were discriminatory; (2) African-Americans who sought housing in white areas faced intimidation and violence similar to that occurring during World War I; (3) after World War II, suburbs developed strategies for keeping African-Americans out; and (4) federally sponsored public housing encouraged segregation in many cities.

Of course, the United States in the 1990s is not the same as it was earlier. For one thing, the 1968 Fair Housing Act now has had more than three decades to work, and the 1965 changes in the Immigration Act (see Chapter 7) have also diversified the ethnic structure of the country. The data suggest that Asians and Hispanics have a greater propensity or ability to suburbanize than do African-Americans (Denton and Massey 1991; Logan and Alba 1993; Logan, 1996), and that the slower rate of suburbanization of African-Americans continues to be the result of discrimination on the part of whites (consistent with the concept of racial stratification that I discussed in Chapter 10). It appears that whites generally do not object to a few African-American neighbors of the same income and education, but larger numbers may be seen as threatening (Clark 1991).

From the standpoint of demographic characteristics, the suburbs are composed especially of higher income married couple families (Alba and Logan 1991)—a pattern that disproportionately would exclude African-Americans. But demographic components of suburbanization do not explain residential segregation; they merely point to its existence. The explanations are essentially social in nature, and one of the prevailing ones is based on the idea that "status rankings are operationalized in society through the imposition of social distance" (Berry, et al. 1976:249). In race relations, the social status of African-Americans has been historically lower than that of whites. That status ranking used to be maintained symbolically by such devices as uniforms, separate facilities, and so forth, which were obvious enough to allow social distance even though African-Americans and whites lived in close proximity to each other. However, as African-Americans left the South and moved into industrial urban settings, many of those negative status symbols were also left behind. As a result, spatial segregation serves as a means of maintaining social distance "where 'etiquette'—the recognition of social distance symbols—breaks down" (Berry et al. 1976:249). Thus as African-Americans have improved in education, income, and occupational status, whites have maintained social distance by means of residential segregation facilitated by suburbanization.

Massey (1996) has suggested that the spatial isolation of poverty and especially of low-income blacks portends future community instability and violence in America. However, there is some evidence that a trend toward desegregation does exist and it appears to be related to the westerly drift of the population and the increasing suburbanization taking place especially in the south and west (Frey and Farley 1996). Comparing data from the 1980 and 1990 censuses, Farley and Frey (1994) have concluded that the largest decreases in segregation scores by residential area have occurred in the more recently built suburbs of southern and western metropolitan areas. You can see, for example, in Figure 11.4 that the most segregated metropolitan areas in 1990 tended to be in the Rustbelt cities of the industrial north, whereas the lowest scores tended to be in smaller metropolitan areas of the south and the west.

European cities are also characterized by a certain amount of residential segregation, largely with respect to the ethnic minority groups that have comprised the guest-worker populations. However, in many parts of Europe a large segment of the housing market for working class families is subsidized and controlled by the government, and this has limited the scope of residential segregation taking place (Huttman, Blauw, and Saltman 1991).

The Impact of Urban Crowding

For centuries the crowding of people into cities was doubtless harmful to existence. Packing people together into unsanitary houses in dirty cities raised death rates.

Figure 11.4   Residential Segregation of African-Americans in the United States Is Greatest in Older Cities in the North, and Lowest in Cities in the South and West

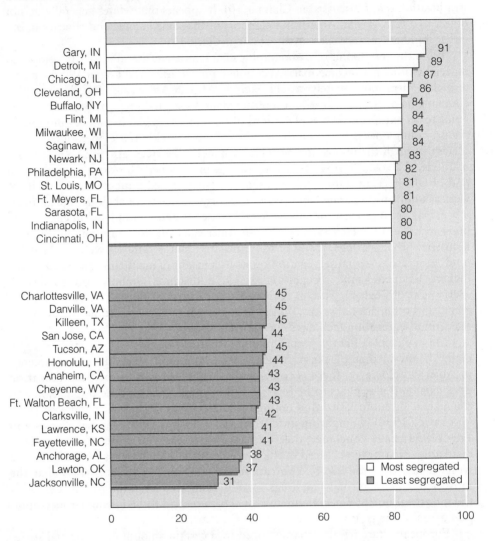

Note: Segregation score is measured as the percent of African-Americans in the metropolitan area who would have to move to another census block in order to achieve residential balance by race throughout the metropolitan area.

Source: Adapted from R. Farley and W. H. Frey, 1994, "Changes in the segregation of whites from blacks during the 1980s: Small steps toward a more integrated society," *American Sociological Review* 59:23–45; Table 1.

Furthermore, as is so often the case, as cities grew to unprecedented sizes in 19th-century Europe, death struck unevenly within the population. Mortality went down faster for the better off, leaving the slums as the places where lower-income people were crowded into areas "with their sickening odor of disease, vice and crime" (Weber 1899:414).

When early students of the effects of urbanization such as Weber and Bertillon discussed crowding and overcrowding, they had in mind a relatively simple concept of **density**—the number of people per room, or per block, or per square mile. Thus Weber quotes the 1891 census of England, "regarding as overcrowded all the 'ordinary tenements that had more than two occupants to a room, bedrooms and sitting rooms included'" (1899:416). The prescription for the ill effects (literally) of overcrowding was fairly straightforward as far as Weber was concerned. "The requirement of a definite amount of air space to each occupant of a room will prevent some of the worst evils of overcrowding; plenty of water, good paving, drainage, etc. will render the sanitary conditions good."

Crime and vice are also often believed to be linked to urban life and, as a matter of fact, crime rates are higher in cities than in the countryside, at least in the United States. But what is it about crowding that might lead to differences in social behavior between urban and rural people? To examine that question, you have to ask more specifically what crowding is.

The simplest definition of crowding is essentially demographic and refers to density—the ratio of people to physical space. As more and more people occupy a given area, the density increases and it becomes therefore relatively more crowded. Under these conditions, what changes in behavior can you expect? In a 1905 essay, Georg Simmel suggested that the result of crowding was an "intensification of nervous stimulation" (1905:48), which produced stress and, in turn, was adapted to by people reacting with their heads rather than their hearts. "This means that urban dwellers tend to become intellectual, rational, calculating, and emotionally distant from one another" (Fischer 1976:30). Here were the early murmurs of the **urbanism** concept—that the crowding of people into cities changes behavior—a concept often expressed with negative overtones.

Perhaps the most famous expression of the negative consequences of the city is Louis Wirth's paper "Urbanism as a Way of Life" (1938), in which he argued that urbanism will result in isolation and the disorganization of social life. Density, Wirth argued, encourages impersonality and leads to people exploiting one another. For two decades there was little questioning of Wirth's thesis and, as Hawley has put it, "in one short paper, Wirth determined the interpretation of density for an entire generation of social scientists" (1972:524). The idea that increased population density had harmful side effects lay idle for a while, but it was revived with considerable enthusiasm in the 1960s following a report by Calhoun on the behavior of rats under crowded conditions.

Crowding among Rats   Although he initiated his studies of crowding among rats in 1947, it was not until 1958 that Calhoun began his most famous experiments (published in 1962). In a barn in Rockville, Maryland, he designed a series of experiments in which rat populations could build up freely under conditions that would permit detailed observations without humans influencing the behavior of the rats in relation to one another.

Calhoun built four pens, each with all the accoutrements for normal rat life and divided by electrified partitions. Initially eight infant rats were placed in each pen, and when they reached maturity Calhoun installed ramps between each pen. At that point the experiment took its own course in terms of the effects of population growth in a limited area. Normally, rats have a fairly simple form of social

HOMELESSNESS IN URBAN AMERICA

Homelessness has become an increasing public concern in all North American cities. "Street people" have been the subject of intense media coverage, Congressional hearings, immense speculation, and to a somewhat lesser degree, scientific inquiry. Public support has been mobilized to serve the needs of the homeless primarily on the basis of two assumptions: (1) the number of homeless is very large and increasing at a rapid rate; and (2) today's street people are qualitatively different from the traditional skid row bums; included now are substantial fractions of families with children and others who are not disengaged from society, but who simply have no place to live because they have been left out of the housing market. These are assumptions in search of verification, and thus far the available research findings have sharpened our focus without fully clarifying the picture.

There can be little doubt that homelessness is a serious problem for at least some urban Americans. But how many homeless people are there in the United States? Many were asking that question in the late 1980s, and so the U.S. Congress mandated the Census Bureau to implement a plan to count the homeless as part of the 1990 census enumeration. The evening and early morning hours of March 20–21, 1990 were designated as S-Night (referring to "streets and shelters"). Trained census workers across the country "fanned out to canvass city streets, freeway overpasses, and (from the outside) abandoned buildings—anywhere homeless people were likely to take shelter" (Haupt 1990:3). However, enumerators were specifically instructed to avoid dangerous places, and no attempt was made to track down the many nearly invisible nooks and crannies of a city and its environs where people might hide in the dark.

Advocates of the homeless had encouraged the Census Bureau to count people at the soup kitchens, because people will quite literally crawl out of the woodwork for a free meal. The Bureau rejected that idea on the grounds that not all soup kitchen attendees are homeless, and it was worried about double-counting (Wilkens and Fried 1991). When found, a homeless person was asked the same questions as were others who were enumerated (see Chapter 2), except that, obviously, no information was gathered about housing. As in the general count, every sixth homeless person was administered the long form of the census questionnaire.

The Census Bureau has reported that this procedure generated a national count of 228,621 homeless persons, including 178,828 people in shelters and 49,793 on the street. This number was lower than almost anybody's precensus estimates, which had ranged from about 250,000 to a high of as many as 3 million (Haupt 1990). From the outset the Bureau had publicly warned of the possible shortcomings of its enumeration process, but the low count only served to fan the flames of debate. Do the people counted represent only the tip of the iceberg, and if so, how big is the iceberg? Or, does the count represent the best possible estimate and one that may be close to the actual number, even if admittedly too low?

Of interest in this debate is the finding that virtually all independent systematic counts of homeless persons in specific cities have produced fewer homeless people than expected (Wright 1989). One of the best examples of such a study is a series of counts carried out in Nashville, Tennessee. Between 1983 and 1988 the Nashville Coalition for the Homeless conducted biannual surveys of the homeless population in that city to determine the size, growth rate, and composition of the homeless population. They developed a set of procedures that served as a model for the Census Bureau's 1990 enumeration and found that, despite the public perception of an increase in the number of homeless in Nashville, the actual count had remained quite stable at around 700 (Lee 1989). The Nashville results showing the number of homeless to be fewer than expected and to be stable over time echoed similar findings in an earlier research effort in Chicago (Rossi 1989), as in several other locales.

Advocates for the homeless have frequently decried these results, complaining about the inability of researchers to find the homeless and arguing that many homeless persons specifically avoid being counted (Wilkens and Fried 1991). Yet one important reason for the disagreement about how many homeless persons there are may lie in the lack of consensus about the definition of the very term itself. "There is a very fine line between being precariously housed and being literally homeless; and between short-term (episodic) homelessness and long-term (chronic) homelessness. These distinctions make for some slippery definitional problems" (Knox 1994:315).

Who are the homeless? Experts disagree about the details, but do agree that the profile includes many people who are not the quintessential vagrants. The homeless are the poorest of the poor. They are predominantly young to middle-aged single, white males, but African-Americans and Hispanics are overrepresented. Family units among the homeless are largely composed of women with young children, and this group seems to have experienced the largest increase, even though still representing a small fraction of all homeless.

The characteristics of the homeless are closely entwined with the causes of homelessness. Of course, the basic problem for the very poor is usually lack of income caused by unemployment, often exacerbated by mental illness, substance abuse (including alcohol as well as drugs), and a disengagement from social networks where information and assistance might be more readily found. But these problems do not lead inexorably to lack of shelter. At least three major trends have contributed to the current problem of homelessness in urban America. One trend has been the decline in outpatient care and community resources for mentally ill persons (Jencks 1994). It has been argued that this is the largest, single reason for the "new" homeless (see, for example, Torrey 1988 and Isaac and Armat 1990), whereas others are persuaded that such people represent only a small fraction of the homeless (see, for example, Momeni 1989). At

least some of this difference in perspective may be caused by a lack of consistency in defining mental illness (Filer 1990).

A second trend has been the increased availability of crack—a more affordable form of cocaine, and the increase in homelessness is at least coincident with a rise in crack addiction (Jencks 1994). The third trend contributing to homelessness is the increasingly scarce supply of low-income housing. This problem has two parts: (1) the slowdown in new construction of low-income housing in the 1980s; and (2) the demolition of many single-room occupancy hotels (SROs) and other marginal living quarters as part of central-city revitalization or redevelopment projects throughout the country. The closing of places where the very poor had been living put them out on the streets, where the lack of alternative housing has kept them, heightening their visibility and contributing to public awareness of their plight. The issue of housing is important to the distinction between the *houseless* and the *homeless*. Passaro (1996) points out that women, especially those with children, are fairly quickly rehoused when they become homeless. The vast majority of the persistently homeless, on the other hand, are unattached men who are left to fend for themselves.

Prior to the 1980s, Americans were most likely to imagine homelessness as something that happens in the slums of Calcutta or other third-world cities. But any visitor to or resident of North American cities now takes for granted that it does occur in North America, and European cities have also witnessed some increase in homelessness over the past several years, although the incidence from one country to the next seems to depend on how universal is the access to public welfare (Hope and Young 1990). Thus the size and characteristics of the homeless population will likely vary according to the extent to which a society views homelessness as a social problem rather than just a punishment for personal failure and can accompany that social concern with economic resources to provide housing for everyone who needs it.

organization, characterized by groups of 10 to 12 hierarchically ranked rats defending their common territory. There is usually one male dominating the group, and status is indicated by the amount of territory open to an individual.

As Calhoun's rat population grew from the original 32 to 60, one dominant male took over each of the two end pens and established harems of 8 to 10 females. The remaining rats were congregated in the middle two pens, where problems developed over congestion at the feeding hoppers. As the population grew from 60 to 80, behavior patterns developed into what Calhoun called a "behavioral sink"— gross distortions of behavior resulting from animal crowding. Behavior remained fairly normal in the two end pens, where each dominant male defended his territory by sleeping at the end of the ramp, but in the middle two pens there were severe changes in sexual, nesting, and territorial behavior. Some of the males became sexually passive; others became sexually hyperactive, chasing females unmercifully; and still another group of males was observed mounting other males as well as females. Females became disorganized in their nesting habits, building very poor nests, getting litters mixed up, and losing track of their young. Infant mortality rose significantly. Finally, males appeared to alter their concept of territoriality. With no space to defend, the males in the middle two pens substituted time for territory, and three times a day the males fought at the eating bin.

Calhoun's study can be summarized by noting that among his rats, crowding (an increase in the number of rats within a fixed amount of space) led to the disruption of important social functions and to social disorganization. Related to these changes in social behavior were signs of physiological stress, such as changes in the hormonal system that made it difficult for females to bring pregnancies to term and care for the young. Other studies have shown that not only rats, but monkeys, hares, shrews, fish, elephants, and house mice also tend to respond to higher density by reducing their fertility (Galle, Gove, and McPherson 1972). Now the important question must be raised: Does the behavior of rats and other animals signal an analogous response to crowding on the part of humans?

Does It Apply to Humans?   Although the severe distortions of behavior that Calhoun witnessed among rats have never been replicated among humans, research has suggested that at the macro (group) level there may be some fairly predictable consequences of increasing population density (mainly as a result of increases in population size). For example, Mayhew and Levinger note that violent interaction can be expected to increase as population size increases; "the opportunity structure for murder, robbery, and aggravated assault increases at an increasing rate with aggregate size" (Mayhew and Levinger 1976:98). There are more people with whom to have conflict, and an increasingly small proportion of people over whom we exercise direct social control (which would lessen the likelihood of conflict leading to violence). Increasing size leads to greater superficiality and to more transitory human interaction, that is, greater anonymity. Mayhew and Levinger point out that "since humans are by nature finite organisms with a finite amount of time to devote to the total stream of incoming signals, it is necessarily the case that the average amount of time they can devote to the increasing volume of contacts . . . is a decreasing function of aggregate size. This will occur by chance alone" (1976:100).

Because no person has the time to develop deeply personal relationships (primary relations) with more than a few people, the more people there are entering a person's life, the smaller the proportion that one can deal with in depth. This leads to the appearance that people in cities are more estranged from one another than in rural settings, but Fischer (1981) offers evidence that in all settings people are distrustful of strangers and we just encounter more of them in the city. This may lead to personal stress as people try to sort out the vast array of human contacts since the more people there are, the greater is the variety both of expectations that others have of you and of obligations that you have toward others. The problems of not enough time to go around and of contradictory expectations lead to "role strain"—a perceived difficulty in fulfilling role obligations. Most of these problems of size seem to arise naturally in large metropolitan areas and, indeed, the problems are most intense there.

But do these stresses of living in a city and/or living in a crowded household have any important demographic consequences? Social research commonly produces mixed results, of course, and the study of crowding is no exception. Thus Choldin (1978) reviewed the literature on density and its effects and concluded that "when social structural differences among neighborhoods are considered (held constant), population density appears to make a trivial difference in predicting pathology rates" (1978:109). On the other hand, a study of rural villages in India has suggested that high density does appear to dampen fertility levels (Firebaugh 1982)—a finding consistent with studies among other animals, but a study in Chicago suggested that the effects of density on fertility are trivial (Loftin and Ward 1983). An extensive survey of the effects of crowding in Bangkok, which is far more crowded than any North American city, suggested that crowding, although not necessarily a good thing, had little effect on sexual behavior or on health in general (Edwards, et al. 1994). On a more positive note, data from the General Social Survey of the National Opinion Research Center have been used to suggest that moving to a big city increases your tolerance for other human beings, rather than the other way around (Wilson 1991).

The jury is therefore still out on the impact of crowding on human behavior. To the extent that urban crowding makes a difference in life, however, its effects are most likely to be felt in third-world cities, where space and resources are at a premium.

Third-World Cities

"Poor countries' cities are bursting at the seams, yet rural migrants are coming in faster than ever; a social and environmental meltdown is waiting to happen" (The Economist 1996:44). This is how one writer described the rationale for the United Nations Conference on Human Settlements (Habitat II) held in Istanbul in 1996. The opportunities that third-world cities offer to rural peasants in less developed nations may seem meager to those of us raised in a highly developed society. Most third-world cities have long since outgrown their infrastructure and, as a result, drinkable water may be scarce, sewage probably is not properly disposed of, housing is hard to find, transportation is inadequate, and electricity may be only sporadically available. This is not a pretty picture, but it is still an improvement on the average rural village! Thus "despite these problems, the flood of migrants to the cities continues apace. Why? The answer lies in the natural population increase in rural areas, limited rural economic development, and the decision-making calculus of urban migrants. . . . What this all means, of course, is that the primary cause of what some have termed 'overurbanization' (more urban residents than the economies of cities can sustain) is increasingly severe 'overruralization' (more rural residents than the economies of rural areas can sustain)" (Dogan and Kasarda 1988:19).

The most obvious symbols of crowding and overurbanization in third-world cities are the squatter settlements that line the outer ring of almost every large city in developing countries. They may be called *colonias* in Mexico or *favelas* in Brazil, but the meaning is the same—slums in which people live in small dirt-floored huts with no running water, electricity, or sewage disposal systems. This is typically a re-sult of cities growing faster than governments can afford to build an urban infra-structure. In other places, such as Mumbai (formerly Bombay), the government may also be to blame for the inner-city slums, but for a different reason. It has been re-ported that "a decent-sized three-bedroom flat in southern Bombay costs $2 million. Meanwhile, two thirds of the city's 10 million residents live either in one-room shacks or on the pavement" (The Economist 1995). Government regulations con-trolling rent and land use have contributed to this situation, and of course this has encouraged population growth in the suburbs without relieving the inner city crowding.

An evaluation of the quality of life in the 100 largest cities of the world re-vealed, not surprisingly, that metropolitan areas of developing countries are gener-ally less livable than the cities of Europe, North America, and Oceania. The Popu-lation Crisis Committee/Population Action International (1990) assembled data on crime, health, income, education, infrastructure, traffic, and related indexes of the quality of urban life to produce an "urban living standards score." Of 21 cities ranked as "very good," the list was led by Melbourne, Montreal, and Seattle-Tacoma, and only one third-world city, the city-state of Singapore, was included. Only 3 of the 23 cities categorized as "good" were in developing countries. On the other hand, 25 of 26 cities rated as "fair," and all 28 cities listed as "poor" were in the third world. The two lowest rated of the 100 largest cities were both in sub-Saharan African—Kinshasa, the capital of Congo (formerly Zaire), and Lagos, the capital of Nigeria. Indeed, the Second African Population Conference in 1984 con-cluded that African cities are "plagued by the adverse effects of rapid growth—ur-ban sprawl, unemployment, delinquency, inadequate social services, traffic conges-tion, and poor housing" (quoted by Goliber 1989:17). A major reason for this situation is that urban areas in Africa have consistently been growing at more rapid rates than urban areas anywhere else in the world.

As dismal as urban life is in developing countries, it is a step up from rural areas. The United Nations estimates that in the 1990s, 96 percent of urban residents in developing countries had access to health care, compared with only 76 percent of rural residents; 87 percent of urban residents had access to water; only 60 percent in rural areas; and 72 percent of urban residents had access to sanitation services compared with only 20 percent living in the countryside (United Nations Development Programme 1997). Cities are where economic development is occurring in the world, and it is where infrastructure and housing will continue to be built. The demand for housing itself is a function of the relationship between the population of young adults wanting to form their own family household and the number of people who are dying and thus presumably freeing up existing housing. The high rates of growth and attendant youthful age structures of the developing countries tell us that the worldwide demand for housing will come predominantly in the cities of the third world. Along with that housing will have to come infrastructure improvements—water, sewer, electricity, roads, and public transportation that make urban life possible. Between the beginning and the middle of the 21st century, the world will likely experience the need to shelter an additional 2 billion families—the majority of them in third-world cities (Weeks and Fuller 1996). This is a daunting task considering that the volume of housing needed is unprecedented in world history and, of course, we have not come even close to adequately sheltering the current generation of humans.

Summary and Conclusion

The urban transition describes the process whereby a society shifts from being largely bound to the country to being bound to the city. It is a process that historically has been the close companion of economic development, which, of itself, suggests the close theoretical connection of urbanization with demographic processes. Although rural-to-urban migration is a major aspect of urbanization, mortality and fertility are importantly associated as well, as both causes and consequences. Population pressures created by declining mortality in rural areas, combined with the economic opportunities offered by cities, have been historically linked to urbanization. On the other hand, mortality now tends typically to be lower in cities than in rural areas, which permits higher rates of urban natural increase than in the past.

The development of the industrialized countries is replete with examples of how urban life helps to generate or ignite the first two of Ansley Coale's three preconditions for a fertility decline—the acceptance of calculated choice as an element in personal family size decisions and the perception of advantages to small families (see Chapter 6). In cities of the less developed nations, urban fertility, which is lower than rural fertility, is still typically much higher than in the cities of the developed world. As a result, cities in third-world nations are the most rapidly growing places on earth. It seems that the most readily identifiable beneficial consequences of living in urban places are those associated with higher standards of living compared with rural areas. The detrimental consequences that lead to lower quality of life in some of the world's largest cities seem primarily to be the result of the economic inability of third-world cities to adequately absorb the tremendous numbers of newcomers to the city.

My purpose in this chapter has been to analyze the demographic components of urbanization—one of the world's abiding social issues. In the next chapter I will continue in this vein by applying the demographic perspective to the debate over economic development.

Main Points

1. The world is rapidly approaching a situation in which one of every two people lives in an urban area.

2. Urbanization refers to the increase in the proportion of people living in urban places.

3. One of the most striking features of urbanization is its recency in world history. Highly urban nations like England and the United States were almost entirely agricultural at the beginning of the 19th century.

4. Cities are related to the development process because they are centers of economic efficiency.

5. Urbanization can occur as a result of internal rural-to-urban migration, higher rates of natural increase in urban than in rural areas, international urban migration, and reclassification of places from rural to urban.

6. Until the 20th century, death rates in cities were so high and fertility was low enough that cities could not have grown had it not been for the migration of people from the countryside.

7. In virtually every society, fertility levels are lower in cities than in rural areas.

8. Despite lower fertility in cities, cities in less developed nations almost always have higher fertility levels than cities in developed nations.

9. Eighty percent of the U.S. population resides within the boundaries of the nation's metropolitan areas.

10. Tokyo is the largest urban agglomeration in the world, followed in size by Mexico City, São Paulo, and New York City.

11. Urbanization in developed countries may be best explained by the modernization theory, whereas elements of world-systems theory seem to help explain urbanization in less developed nations.

12. Urbanization in the United States has now turned into suburbanization, and most Americans live in suburbs.

13. The suburbanization process has heightened residential segregation in the United States as it has, to a lesser extent, in other developed nations.

14. Population growth in cities has given rise to fears about potential harmful effects of crowding. These fears seem most capable of expression in third-world cities.

15. Homelessness is a serious problem in urban America, although the number of homeless people counted in the 1990 census was well below the expected number.

Suggested Readings

1. Claude Fischer, 1988, The Urban Experience, Second edition. (San Diego, CA: Harcourt Brace Jovanovich).

   This is one of the most insightful and well-written books ever published on the subject of how humans live in cities.

2. Richard T. LeGates and Frederic Stout, editors, 1996, The City Reader (New York: Routledge).

   This book of readings provides a wide range of expert articles on urbanization and the structure of cities—a good companion to the Fischer book (above).

3. Richard Critchfield, 1994, The Villagers: Changed Values, Altered Lives: The Closing of the Urban-Rural Gap (New York: Anchor Books).

   The author spent his life studying villagers (both urban and rural) and this book, published just before his death, is a culmination of a life of thinking about the common thread of global villagers.

4. John D. Kasarda, 1995, "Industrial restructuring and the changing location of jobs," Chapter 5 in Reynolds Farley (ed.), State of the Union: American in the 1990s, Volume One: Economic Trends (New York: Russell Sage Foundation).

   A concise and fact-packed analysis of why metropolitan structure is changing in the United States—follow the jobs and the answers are probably not far behind.

5. Curtis Roseman, Günther Thieme, and Hans Dieter Laux, editors, 1996, EthniCity: Geographic Perspectives on Ethnic Change in Modern Cities (Lanham, MD: Rowman and Littlefield Publishers).

   Each chapter in this book takes you to a different city in the developed world to examine the impact on that city of the increasing ethnic diversity created largely by urbanward transnational migrants.

🌐 Websites of Interest

Remember that websites are not as permanent as books and journals, so I cannot guarantee that each of the following websites still exists at the moment that you are reading this:

1. http://www.un.org/Conferences/habitat/

   This is the official website for the United Nations Conference on Human Settlements (Habitat II) held in Istanbul in 1996.

2. http://www.undp.org/popin/wdtrends/urb/urb.htm

   You can check on the latest updates of population estimates of the world's cities at this website of the United Nations Development Programme.

3. http://lanic.utexas.edu/la/Mexico/water/book.html

   No city can survive without water, and no city in the world has a bigger water problem than Mexico City. At this website you will find the text of the book "Mexico City's Water Supply: Improving the Outlook for Sustainability" published in 1995 by the National Research Council.

4. http://130.166.124.2/library.html

   Census data from 1990 for several major cities in the United States are mapped for you at the census tract level in this Digital Atlas of New York City, a website developed by

William Bowen at California State University, Northridge. Included are New York City, Boston, Washington, DC, and Los Angeles.

5. **http://edcwwww2.cr.usgs.gov/umap/umap.html**

   You can watch the urban growth of the San Francisco Bay Area and the Washington DC area unfold before your eyes in these temporal maps developed by Keith Clarke at the University of California, Santa Barbara in a project with the U.S. Geological Survey. You have to download some internet software to watch the "movie" but it is worth it.

Part Four
Population, Development, and the Environment

Do the "boomsters" or the "doomsters" have the best handle on the ability of developing nations to improve their well-being in the face of continued population growth? Is it possible to keep feeding billions of people each year, and if so, to do it without completely degrading the environment? These are among the most important questions confronting us in the world and they require a demographic perspective for a good understanding on our part. The desire of less developed nations to improve their levels of living has generated massive foreign debt for many countries, often (if not always) compounded by high rates of population growth. Are people a drag on development, or a blessing in disguise? These issues are taken up in Chapter 12. More people certainly mean greater competition for the earth's scarce resources, and one of the most important global issues confronting us is whether environmental degradation can be halted in the face of the massive new additions of human beings to the planet each year, all of whom not only need to be fed, but need a healthier diet as well. This is the substance of Chapter 13.

CHAPTER 12
Population Growth and Development

Is the control of population growth a necessary precursor to economic development? A lot of people think so, but not everyone agrees and the debate over this question gets to the heart of policy planning throughout the world. At stake, quite literally, is the level of living and overall well-being for you and me and, especially, people living in less developed nations. The debate has three sides to it, and this chapter is devoted to examining those three sides, then seeing how they might be reconciled into a comprehensible perspective on population growth and economic development.

In the United States, the debate is often seen as juxtaposing the views of the **doomsters** with those of the **boomsters**. Doomsters are the neo-Malthusians, exemplified most famously by Paul Ehrlich, who has argued for decades that continued population growth will lead to certain economic and environmental collapse in a worldwide tragedy of the commons. For Ehrlich the policy choices are clear—population control must be a part of any development strategy, or that strategy will fail. The boomsters have been most famously influenced by the late Julian Simon, who argued for decades that population growth stimulates development, rather than slowing it down. For Simon, the policy choices are also clear—development strategies should not deliberately slow down population growth because such growth is both a cause and a symptom of economic development. Outside of the United States a third perspective is often put forth: a **neo-Marxist** view that population growth has nothing to do with economic development at all. Economic development, where it lags, is held back by the injustice of the world system that creates dependency by the periphery on the dominant core countries. Each of these perspectives carries with it a very different set of policy prescriptions of population and economic planning, so it matters which one is right. In this chapter and the next, I discuss the evidence for which perspective may be correct. I will not leave you in suspense on this—all three perspectives have merit. I discuss why that is true and how the three can be reconciled to produce a well-grounded perspective on the relationship between population growth and economic development. First, though, we need some definitions.

What Is Economic Development?

The most common definition of **economic development** is that it represents a growth in average income, usually defined as per capita (per person) income. A closely related idea is that economic development is occurring when the output per worker is increasing; since more output should lead to higher incomes, you can appreciate that they are really two sides of the same coin. Of course, if you are holding down two jobs this year just to keep afloat financially, you know that producing more will not necessarily improve your economic situation. Rather, it may only keep it from getting worse. Thus a more meaningful definition of economic development refers to a rise in real income—an increase in the amount of goods and services that you can actually buy.

An important aspect of development is that it is concerned with improving the welfare of human beings. It includes more than just increased productivity; it includes the resulting rise in the ability of people to consume (either buy or have available to them) the things they need to improve their level of living. Included in the list of improvements might be higher income, stable employment, more education, better health, consumption of more and healthier food, better housing; and increased public services such as water, sewerage, power, transportation, entertainment, and police and fire protection. Naturally, these improvements in human welfare help to increase economic productivity because the relationship is synergistic.

Economic Growth and Economic Development

Gross National Product

Economic growth refers to an increase in the total amount of productivity or income in a nation (or whatever your geographic unit of analysis might be) without regard to the total number of people, whereas economic development relates that amount of income to the number of people. But how do we measure that income? The most commonly used index of a nation's income is the **gross national product (GNP),** which the World Bank defines (somewhat obscurely) as "the sum of value added by all resident producers plus any taxes (less subsidies) that are not included in the valuation of output plus net receipts of primary income (employee compensation and property income) from nonresident sources" (World Bank 1998). Basically, if you add up the value of all of the paid work that goes on in a country, and then add in the money received from other countries, you have the measure of gross national product. If you exclude the money from abroad and just include the income generated within a country's own geographic boundaries, you have **gross domestic product (GDP).**

Gross national product is the most widely used measure of economic well-being in the world, but it is important for you to keep in mind the things that GNP does *not* measure: (1) it does not take into account the depletion and degradation of natural resources (a topic we return to in the next chapter); (2) it does not make any deduction for depreciation of "man-made" assets such as infrastructure; (3) it does not measure the value of unpaid domestic labor such as that generated especially by women in developing nations; and (4) it does not account adequately for regional or national differences in purchasing power.

This latter limitation is one that the World Bank has been particularly interested in. Although GNP figures are usually expressed in terms of U.S. dollars, a dollar may go farther in Zimbabwe than it will in England, even when the exchange rates have been taken into account. The United Nations and the World Bank have sponsored a number of household expenditure surveys in developing countries to try to estimate the actual differences in the standard of living, in order to produce more meaningful income comparisons. The wealthier nations have also been encouraged to conduct such surveys, along the lines of the U.S. Bureau of Labor Statistics' Consumer Expenditure Survey.

Purchasing Power Parity

The product of these efforts is a measure called **purchasing power parity** (PPP) and is defined as "the number of units of a country's currency required to buy the same amount of goods and services in the domestic market as one dollar would buy in the United States" (World Bank 1994:246). One way of expressing this concept is through the use of what *The Economist* calls its "Big Mac Index" (1997). McDonalds sells its hamburgers in more than 100 countries and in each country the sandwich must conform to the same exact standards of ingredients and preparation. If the Big Mac costs $2.42 in the United States, then it should cost the same in real terms anywhere else in the world. So, if you go to Chile and discover that you're paying 1,200 pesos for a Big Mac, that should tell you that there are 1200/2.42 = 496 Chilean pesos per U.S. dollar. In fact, that was pretty close to the official exchange rate in the late 1990s and presumably you could use that ratio to convert all prices in Chile into their U.S. dollar equivalent, so that international comparisons can be made. For the moment, the biggest drawback in using PPP is that the consumer expenditure data required by the PPP calculations (beyond the Big Mac) are not as widely available as is the information that goes into the calculation of the GNP. I use both measures in this chapter.

In the mid-1990s the gross national product in the United States was 7.4 trillion dollars (World Bank 1998). That was by far the highest figure in the world. In fact, the GNP in the United States in the mid-1990s was greater than the combined GNP of all 88 countries that the World Bank defines as low- and middle-income. Figure 12.1 shows the GNP figures for the ten largest economies of the world ex-

Figure 12.1   The Ten Largest Economies of the World as Measured by Gross National Product (GNP) and Purchasing Power Parity (PPP)

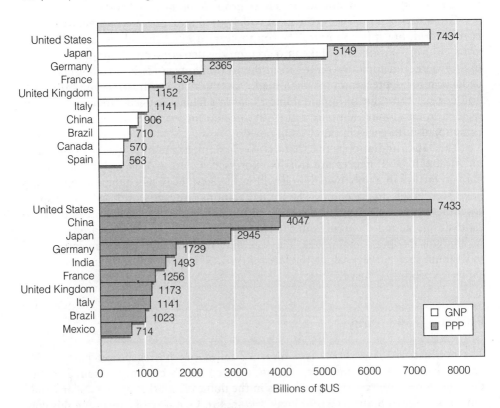

Source: World Bank, 1998, World Development Indicators 1998, Table 1.1

pressed both in GNP and in PPP. You can see that, in terms of total gross national product, after the United States come Japan, Germany, France, the United Kingdom, Italy, China, Brazil, Canada, and Spain. However, if we value goods and services not by official exchange rates, but by purchasing power, then we see in the bottom panel of Figure 12.1 that the order shifts a bit. Measured in terms of PPP, China has the second largest economy after the United States, followed then by Japan, Germany, India (which is not among the top ten defined by GNP), France, the United Kingdom, Italy, Brazil, and Mexico (which essentially replaces Canada on the list).

If we divide the total national income by the number of people, we obtain a measure of per capita income, which gives a sense of the relative well-being of people in one country compared with another. In Table 12.1 I have listed the per person income for selected countries according to both per person GNP and per person purchasing power parity. Even a glance at the table shows that when we take purchasing power into account, the per person income figures do not look quite so extreme. For example, Switzerland had a higher per capita GNP than the United States, but when you take into account the actual cost of living in Switzerland compared with that of the United States by computing per person PPP, you can see the average Swiss is not quite as well off as the average American. At the other extreme, Mozambique has the lowest per person GNP in the world at a mere $60 per person per year, but the purchasing power comparison raises that to the equivalent of what $500 would buy in the United States. Still, even taking relative purchasing power into account, a person in Mozambique has only 2 percent of the level of living as that of the average American.

China may have one of the largest economies in the world (see Figure 12.1), but its huge population means that the per person income (measured in purchasing power parity—see Table 12.1) is only a tiny fraction of that in the United States. On a per capita basis, the Chinese have only 12 percent of the income of the average American—or, put another way, the average American is 8 times better off than the average person in China, and 17 times better off than the average person in India.

So far I have discussed economic development in terms of average income, but those averages often hide inequalities and disparities in the distribution of income. It may happen that the per capita increase in productivity profits only a few people rather than the entire population, and, in fact, some economists argue that a concentration of income is the only way that enough money can be saved for further investment and further economic growth. Kuznets (1965), for example, suggested that income inequality characterizes the early phases of economic development when capital formation is so crucial; only later is it possible to spread the income around. A loose analogy might be made to a family that wants to "develop economically" by purchasing a home. Assuming that a substantial amount of cash must be saved for the down payment and closing costs, the members of the family may well have to do without things they would like to consume, because all the extra money is being accumulated for the house purchase. Only when the house has been bought can the sacrifice cease and the family income be spread around more among its members. The analogy is not perfect, but it illustrates a point well known to early industrial entrepreneurs—a delay of gratification is required if income is going to be reinvested for further growth.

Table 12.1 The Top Ten and Bottom Ten Countries and Other Selected Countries in Terms of Per Person GNP and Per Person PPP

Country	Per Person GNP ($US)	Ratio to United States	Per Person PPP ($US)	Ratio to United States
Top Ten Countries				
Switzerland	44,350	1.58	26,340	0.94
Japan	40,940	1.46	23,420	0.84
Norway	34,510	1.23	23,220	0.83
Denmark	32,100	1.15	22,120	0.79
Singapore	30,550	1.09	26,910	0.96
Germany	28,870	1.03	21,110	0.75
Austria	28,110	1.00	21,650	0.77
United States	28,020	1.00	28,020	1.00
Belgium	26,440	0.94	22,390	0.80
France	26,270	0.94	21,510	0.77
Other Selected Countries				
United Kingdom	19,600	0.70	19,890	0.71
Canada	19,020	0.68	21,380	0.76
Mexico	3,670	0.13	7,660	0.27
Russia	2,410	0.09	4,190	0.15
Indonesia	1,080	0.04	3,310	0.12
China	750	0.03	3,330	0.12
India	380	0.01	1,580	0.06
Bottom Ten Countries				
Niger	200	0.01	920	0.03
Sierra Leone	200	0.01	510	0.02
Rwanda	190	0.01	630	0.02
Malawi	180	0.01	690	0.02
Burundi	170	0.01	590	0.02
Tanzania	170	0.01	N/A	N/A
Chad	160	0.01	880	0.03
Congo, Dem. Rep.	130	0.00	790	0.03
Ethiopia	100	0.00	500	0.02
Mozambique	80	0.00	500	0.02

GNP = gross national product; PPP = purchasing power parity; N/A = no data available.

Source: World Bank, 1998, World Development Indicators 1998, Table 1.1.

Data on income distribution are not available for all countries, but the evidence from World Bank statistics (World Bank 1994) tends to bear out Kuznet's hypothesis. If we look at the percentage share of the total income or consumption in a nation that is accounted for by the wealthiest 20 percent of the population, we find that it is higher in the middle income countries (51 percent) than in the low income countries (48 percent) and is lowest of all in the high income countries (40 percent). Thus income inequality is highest in those nations that are emerging into higher overall income levels, whereas it is lowest in the nations that have already achieved high income levels.

National Wealth

Where does the income come from that goes into these measures? An important part of it comes from the transformation of natural resources into things that are more useful to us—converting a tree into a house and furniture; converting the "fruit" of cotton plants into shirts and dresses; converting minerals found in rocks into the steel body of an automobile; transforming a hidden pool of underground oil into fuel used by machines. Then, these products must be packaged, delivered, sold, and people have to coordinate all of that, and make sure that the infrastructure exists to do everything that needs to be done. We can divide the resources that go into producing income into two broad categories—natural resources (what is given to us on the planet), and human resources (how clever and successful we are in making something of those natural resources). Together, this combination of resources can be thought of as the **wealth** of a nation (Dixon and Hamilton 1996). Measuring these things is not easy, as you can imagine, although researchers at the World Bank have been working on it (applying their human resources to this issue, as it were) and their preliminary results suggest, for example, that Canada has a higher per person level of wealth than the United States, based on the natural resources in Canada, and that the value of human resources in the United States exceeds the natural resources wealth of the country (World Bank 1995). More importantly, their analysis suggests that natural capital is distributed fairly evenly around the inhabitable portions of the globe, so the variable factors in global economic well-being tend to be the level of human resources in one area compared with another, and the number of humans in one area compared with another amongst whom these resources are shared. Either way, demography seems to be playing a role.

The Debate—Statistical Bases

There is a nearly (but not completely) indisputable statistical association between economic development and population growth; that is, when one changes, the other also tends to change. As you no doubt already know, though, two things may be related to each other without one causing the other. Furthermore, the patterns of cause and effect can conceivably change over time. Does population growth promote economic development? Are population growth and economic development only coincidentally associated with each other? Or is population growth a hindrance to economic development? That is the debate.

The problem is that the data presently available lend themselves to a variety of interpretations. If we look at the global pattern of per person income (measured with per capita GNP, because per capita PPP is not universally available), you can see in Figure 12.2 that the poorest countries are in sub-Saharan Africa and Asia, whereas the richest countries are the European and "overseas" European countries (the United States, Canada, Australia, and New Zealand) as well as Japan. The rest of the world generally falls in between, with the Latin American and middle eastern oil-producing nations being better off than eastern Europe and Western Asia. The data for Figure 12.2 are based on the World Bank's ranking of countries based on levels of per capita GNP. Low income countries, for example, are those whose per capita GNP in the mid-1990s was less than $676, whereas incomes of $676 through $2,695 defined the lower-middle income group, incomes of $2,696 through $8,355 defined the upper-middle income group, and incomes over $8,355 were defined as high income.

There is a high, although not perfect, correlation between being a high income country by World Bank standards and being a "Core" country as defined by world systems analysts. "Core nations are linked to other core countries and to semiperipheral and peripheral nations; peripheral nations are linked to core nations but not to each other, and semiperipheral nations are linked to core countries and other semiperipheral nations, but are only weakly linked to peripheral nations" (Bollen and Appold 1993:286–87). The upper-middle income countries tend to be the "semiperipheral" nations and the lower-middle and low income nations tend to be the "peripheral" countries.

The geographic pattern shown in Figure 12.2 is similar to maps showing the components of the demographic transition. Those places in the world where incomes are highest also tend to have the lowest levels of fertility, the lowest mortality levels, the lowest overall rates of population growth, and the highest levels of urbanization. The picture is fairly compelling, buttressed by the data in Figure 12.3. In that graph I have compiled data for the 118 countries for which World Bank estimates exist of the per person income measured in terms of PPP. The income data refer to the mid-1990s. The population growth data represent the average annual rate of growth during the previous 12-year period. You can see that in general, countries in which average income levels are low tend to be clustered around the high end of population growth rates. I have partitioned the graph into four parts. The dashed horizontal line represents the average per person income for all countries ($6,337) and the dashed vertical line represents the average rate of population growth between the early 1980s and the mid-1990s (1.88 percent). Thus Jordan is a nation that is below the world average in terms of income, but well above with respect to population growth. Joining Jordan in that quadrant are countries like India and most of the non-oil-producing nations of Africa, Asia, and Latin America.

At the other extreme are the wealthier countries with low rates of population growth, exemplified by the United States, Japan, Switzerland, and the other countries of western and northern Europe. The countries that do not fit the expected pattern include the oil-producing Middle Eastern nations such as Saudi Arabia and the United Arab Emirates, where fertility rates are only now beginning to decline, following the rise in oil-based wealth. Also off the track are those countries that have not been able to translate low rates of population growth into higher

Figure 12.2 The Highest Per Capita Incomes Are Found in the "North"

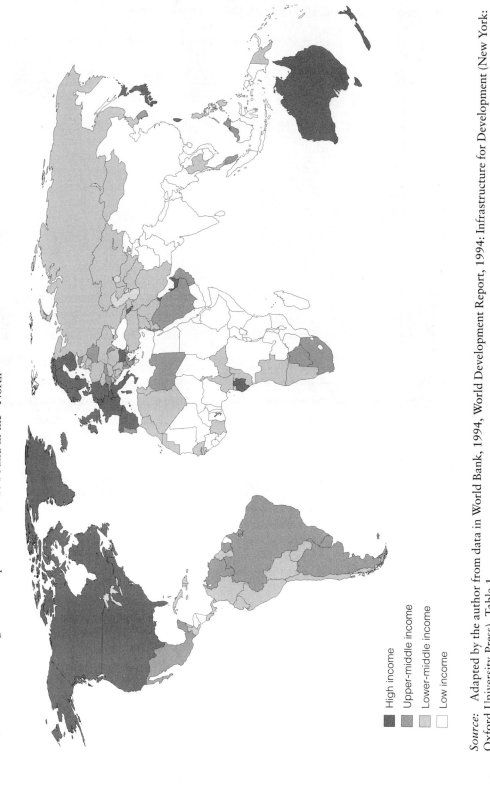

High income

Upper-middle income

Lower-middle income

Low income

Source: Adapted by the author from data in World Bank, 1994, World Development Report, 1994: Infrastructure for Development (New York: Oxford University Press), Table 1.

Figure 12.3 Countries with High Per Capita Income Tend to Have Low Rates of Population Growth

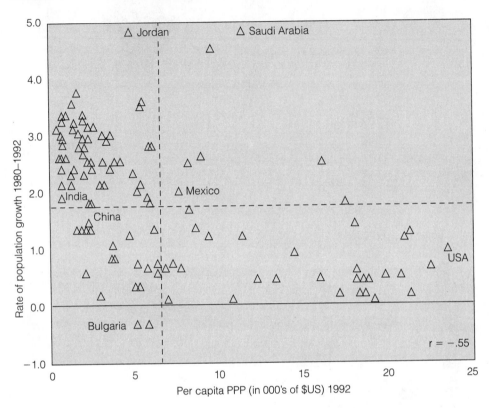

Note: In the 1990s there was a clear tendency for countries with high per capita gross national product to have had low rates of population growth during the previous 12 years. The vertical dotted line represents the average income for all countries, and the horizontal dotted line indicates the average rate of population growth for all countries. Income is measured in terms of purchasing power parity (PPP) per person.

Source: Adapted from data in World Bank, 1994, World Development Report, 1994 (New York: Oxford University Press), Tables 26 and 30.

standards of living. Virtually all of these countries, like Bulgaria, are either just emerging from decades of state socialism or are still influenced by central planning, such as the People's Republic of China—even though some changes are under way in that nation.

Clearly, a low rate of population growth is no assurance of a high income, but that point is obvious because for most of human history the rate of population growth was low and so was the overall standard of living. The data in Figure 12.3 also show that, at least in the short run, countries with sufficient resources (especially oil) can achieve high levels of income even in the face of rapid population growth. Nonetheless, the data also exhibit a reasonably strong negative relationship as evidenced by the correlation coefficient (r) of −.55. Mexico, for example, has been been sliding down the slope toward a combination of lower rates of population growth and higher incomes in spite of the occasional setbacks experienced by the Mexican economy. Furthermore, a country like China now appears to be taking economic advantage of the slowdown in population growth that it has been experiencing over the past two decades (see the essay that accompanies this chapter).

Despite the intuitive appeal of the negative association shown in Figure 12.3 between population growth and levels of national income, it is not a perfect association, and it does not allow us to establish a direct cause-and-effect relationship. A few decades ago it was widely accepted that rapid population growth posed a serious threat to successful economic development. However, that perception has been eroded by studies showing the relationship to be very complex, as well as by a general neglect of research effort in this area (Demeny 1994). Menken (1994:60) has summarized the major findings of studies of the relationship between population growth and economic development as follows:

1. Population growth has important effects on economic development, but they are, in some cases, not as great as was thought in the past.

2. Feedback mechanisms are important—societies adapt; if this is not taken into account, we can vastly overestimate the impact of population growth in our forecasting.

3. Population growth has not prevented economic growth, although it may have reduced it.

4. The effects of population growth depend on the situation of individual countries—for example, concern about self-sufficiency in food production is quite different for Kuwait, which had oil to trade in international markets, than for Bangladesh, where trade is far less likely to make up for internal shortages.

5. No singly directed policy can address all the ills of societies. Reducing population growth through declining fertility is not a panacea; it will not, in and of itself, make a poor country rich. Rather, rapid population growth exacerbates problems, but it is not a root cause of many of the deterrents to economic development.

6. Reducing population growth acts both directly on some aspects of economic development, and indirectly by "buying time" for root causes of developmental deterrents to be addressed.

7. Effects also go in the opposite direction: Economic change can affect population growth and change.

Of the studies reviewed by Menken, the most influential was probably the report by the Working Group on Population Growth and Economic Development of the National Research Council (1986). This study concluded, in particular, that changes over time in the rate of population growth are only vaguely related to changes over time in income levels. They attributed this ambiguity to the inadequate research base that currently exists, but at the same time they concluded that rapid population growth may be somewhat detrimental to economic development, at least in the poorest nations. Critics have argued that the National Research Council report ignored or misinterpreted information that could have led them more convincingly to conclude that population growth is bad for economic development (Daly 1986; McNicoll 1995; Potter 1986). Others have viewed the inconclusiveness as evidence that no relationship exists between population growth and economic development (Simon 1989).

A newer study, not yet available to Menken as she was summarizing the previous literature in the field, suggests that the lack of association that was found in the National Research Council study may have been due to the fact that it, like other studies, had not yet had access to data for the 1980s. In an analysis prepared for the World Bank, Kelley and Schmidt (1994) conclude that "a statistically significant and quantitatively important negative impact of population growth on the rate of per capita output growth appears to have emerged in the 1980s. Whether this represents a new trend, or whether the 1990s will witness a return to the pattern of the previous two decades of no observed correlation, is uncertain" (p. 82). The continued uncertainty about the relationship between population growth and economic development means, of course, that the relationship can be interpreted in several different ways, depending on your ideological predilections.

The Debate—Ideological Bases

The debate over population growth and economic development historically has been three-cornered. In the first corner, arguing that population growth stimulates development, you will typically find **nationalists**—people seeking freedom for their country from economic and political exploitation by more powerful nations. A frequent corollary of nationalism is the idea that more people will bring more productivity and greater power. Another form of nationalism is that which appeared as the official United States position at the 1984 International Population Conference in Mexico City—that in any free market system population growth stimulates demand and thus helps the national economy.

In the second corner you will find Marxists, neo-Marxists, and others arguing that social and economic injustice result simultaneously from the lack (or slowness) of economic development and the (erroneous) belief that there is a population problem. The Marxist position maintains that no cause-and-effect relationship exists between population growth and economic development—that poverty, hunger, and other social welfare problems associated with lack of economic development are a result of unjust social and economic institutions, not population growth.

Finally, in the third corner you find those who have historically antagonized the others, namely the neo-Malthusians. They are, of course, latter-day advocates of the thesis that population growth, unless checked, will wipe out economic gain. The difference between Malthus and the neo-Malthusians is that Malthus was opposed to birth control, as you will recall from Chapter 3, whereas neo-Malthusians are strong advocates of birth control as a preventive check to population growth. Let me examine these three positions in more detail.

Is Population Growth a Stimulus to Economic Development?

An early proponent of the idea that population growth is the trigger of economic development is the Danish economist, Ester Boserup. In a set of extremely influential writings (see, especially, 1965; 1981), she advanced the idea that, in the long run, a growing population is more likely than either a nongrowing or a declining population to lead to economic development. The history of Europe shows that the Industrial Revolution and the increase in agricultural production were accompanied almost universally by population growth. Boserup's argument (also advanced by Clark 1967) is based on the thesis that population growth is the motivating force that brings about the clearing of uncultivated land, the draining of swamps, and the development of new crops, fertilizers, and irrigation techniques, all of which are linked to revolutions in agriculture. The kernel of the argument has been well stated by British agricultural economist, Colin Clark:

> [Population growth] is the only force powerful enough to make such communities change their methods, and in the long run transforms them into much more advanced and productive societies. The world has immense physical resources for agriculture and mineral production still unused. In industrial communities, the beneficial economic effects of large and expanding markets are abundantly clear. The principal problems created by population growth are not those of poverty, but of exceptionally rapid increase of wealth in certain favoured regions of growing population, their attraction of further population by migration, and the unmanageable spread of their cities [Clark, 1967:preface].

This same line of reasoning is part of a strategy of development forwarded by Hirschman (1958), who has argued as follows: (1) An increase in population size will lower a population's standard of living unless people reorganize their lives to increase production; (2) It is "a fundamental psychological postulate" that people will resist a lowering in their standard of living; and (3) ". . . the activity undertaken by the community in resisting a decline in its standard of living causes an increase in its ability to control its environment and to organize itself for development. As a result, the community will now be able to exploit the opportunities for economic growth that existed previously but were left unutilized" (1958:177).

The thesis that population growth is beneficial to economic development does have some foundation in fact. In Europe and the United States development may well have been stimulated by population increase. Indeed, some historians regard preindustrial declines in death rates in Europe, associated partly with the disappearance of the plague (perhaps also with the introduction of the potato) as the spark that set off the Industrial Revolution. The reasoning goes that the lowered death rates created a rise in the rate of population growth, which then created a demand for more resources (Clark 1967, has given a review). An analogous example of population growth influencing development is the case of the American railroad, which opened up the frontier and hastened resource development in the United States. Fishlow (1965) has demonstrated that the railroad (which helped to accelerate the economic development of the western states) was actually following people westward rather than the other way around.

Although history may show that population growth was good for development in the now highly industrialized nations, statistics also reveal very important differences between the European-American experience and that of modern, less developed nations. The less developed countries today are not, in general, retracing the steps of the currently developed nations. For example, less developed nations are building from a base of much lower levels of living than those that prevailed in either Europe or the United States in the early phases of economic development (Kuznets 1965; Boserup 1981). Furthermore, although the rate of economic growth in many underdeveloped countries has recently been higher than at comparable periods in the history of the developed nations, population growth is also significantly higher. They have much higher rates of population growth than European or American countries ever had, with the possible exception of the colonial period of American history. In fact, the rates of population growth in the underdeveloped world are virtually unparalleled in human history.

It appears that population growth may have helped to stimulate economic growth in the developed countries "by forcing men out of their natural torpor and inducing innovation and technical change, or by speeding up the replacement of the labour force with better educated labour" (Ohlin 1976:9). The less developed nations of today, however, do not seem to require any kind of internal stimulation to be innovative. They can see in the world around them the fruits of economic development, and quite naturally they want to share in as many of those goodies as possible—a situation often referred to as "the revolution of rising expectations." People in less developed nations today know what economic development is, and by studying the history of the highly industrialized nations they can see at least how it used to be achievable. If it was ever true that more people meant a greater chance of producing the genius that will solve the world's problems, it is a difficult argument to sustain today. As Nathan Keyfitz has pointed out, "the England that produced Shakespeare and shortly after that Newton held in all 5 million people, and probably not more than one million of these could read or write. . . . The thought that with more people there will be more talent for politics, for administration, for enterprise, for technological advance, is best dismissed. . . . For the most part, innovation comes from those who are comfortably located and have plenty of resources at their disposal" (quoted in United Nations Population Fund 1987:16). In any event, it seems unlikely that a spark such as population growth is necessary any longer, although there really is little solid evidence one way or another.

Controversy is always fed by contradictory evidence, and in the 1980s Julian Simon popularized his thesis that a growing human population is the "ultimate resource" in the search for economic improvement. Eschewing the Malthusian idea that resources are finite, Simon suggests that resources are limited only by our ability to invent them and that, in essence, such inventiveness increases in proportion to the number of brains trying to solve problems. Coal replaced wood as a source of energy only to be replaced by oil, which may ultimately be replaced by solar energy—if we can figure out how to do it properly. From Simon's vantage point, innovation goes hand in hand with population growth, although he is quick to point out that moderate, rather than fast (or very slow) population growth is most conducive to an improvement in human welfare (Simon 1981).

Simon, however, made a crucial assumption: To be beneficial, population growth must occur in an environment in which people are free to be expressive and creative. To him, that meant a free market or capitalist system. Ironically, the idea that the important element in economic development is the type of economic system that prevails draws him into the same perspective on population growth and economic development as that shared by Marxists and neo-Marxists.

Is Population Growth Unrelated to Economic Development?

In his later writings, Simon moved toward the position that population growth is far less important an issue in economic development than is the marketplace itself. He suggested that "the key factor in a country's economic development is its economic and political system. . . . Misplaced attention to population growth has resulted in disastrously unsound economic advice being given to developing nations" (Simon 1992:xiii). With respect to the relationship between population growth and economic development, this is not far from the usual Marxist view that population problems will disappear when other problems are solved, and that economic development can occur readily in a socialist society. Marx (and Engels) believed that each country at each historical period has its own law of population, and that economic development is related to the political–economic structure of society, not at all to population growth. Indeed, Marx seemed to be arguing that whether or not population grew as a nation advanced economically was due to the nature of social organization. In an exploitive capitalist society, the government would encourage population increase to keep wages low, whereas in a socialist state there would be no such encouragement. Socialists argue that every member of society is born with the means to provide his or her own subsistence; thus economic development should proportionately benefit every person. The only reason why it might not is if society is organized to exploit the workers by letting capitalists take large profits, thereby depriving laborers of the full share of their earnings.

It is increasingly difficult to find economists who believe that socialism is a more successful route to economic development than is capitalism, but neo-Marxists share many ideas in common with the world systems approach to understanding the global economic situation. The world system, it is argued, works much as Marx described capitalism—except that the scope is global, not country-specific. The developed nations of the West "are charged with buying raw materials cheap from developing countries and selling manufactured goods dear, thus putting developing countries permanently in the role of debtors and dependents" (Walsh 1974:1144). If the economic power of the developed nations could be reduced and that of developing nations enhanced, the boost to development in those nations would dissipate problems such as hunger and poverty that are currently believed to be a result of too many people. At such time the population problem will disappear because, it is argued, it is not really a problem after all. When all other social problems (primarily economic in origin) are taken care of, people will deal easily with any population problem if, indeed, one occurs. This was obviously the attitude of Friedrich Engels, who wrote in a letter in 1881, "If at some stage communist society finds itself obligated to regulate the production of human beings . . . , it will be precisely this and this society alone, which can carry this out without difficulty" (quoted in Hansen 1970:47).

Marxist supporters of this position have some evidence to which they can appeal. In Russia in the 1920s after the Communist revolution, Lenin repealed anti-abortion laws and abolished the restrictions on divorce in order to free women; the result was a fairly rapid decline in the birth rate (although it turned out to be too rapid for the government's taste and in the 1930s abortions were again made illegal). On the other hand, the Cuban response to a Marxist government was exactly the opposite. Shortly after the Cuban revolution in 1959, the crude birth rate soared from 27 births per 1,000 population in 1958 to 37 per 1,000 in 1962. A Cuban demographer, Juan Perez de la Riva, has explained that after the revolution, rural unemployment disappeared, new opportunities arose in towns, and an exuberant optimism led to a lowering of the age at marriage and an abandonment of family planning (Stycos 1971)—an ironically "Malthusian" response to a Marxist reorganization of society. The birth rate in Cuba later did reestablish its prerevolutionary decline, facilitated by the easing of restrictions on abortion and the increasing availability of contraceptives (Hollerbach 1980). The underlying causes of the renewed decline in fertility were similarly non-Marxist in nature, at least to begin with: (1) increasing modernization, especially of the rural population, and (2) an economy that deteriorated after the initial flush of revolutionary success (Diaz-Briquets and Perez 1981). However, low fertility has been maintained over time by the tried-and-true factors of improved female education and increasing participation of women in the labor force (Catasús Cervera and Fraga 1996).

In sum, the evidence from countries such as Russia, Cuba, and, indeed, China (see Chapter 14 for more on China) suggests that a revolution may alter the demographic picture of a nation, but the relationship to economic development is somewhat cloudy. Indeed, the previously Marxist, centrally planned economies of eastern Europe and Russia are those in which low rates of population growth were not translated into commensurately high levels of living. On the contrary, there is widespread speculation that low birth rates in those countries had been a response to the economic limits placed on the family—especially scarce housing and limited consumer goods, although as in Cuba there had been an emphasis on female education and labor force participation.

The idea that population growth and economic development may be only tenuously related to one another is also reflected in empirical work that seems to support a non-Marxist, but still "neutral," view of the relationship between population growth and per capita income. Using data for developing societies for the period 1965–84, Bloom and Freeman (1986) concluded that despite rapid population growth, the labor markets in most developing countries were able to absorb large population increases at the same time that per worker incomes were rising and productivity was increasing. In other words, just as Davis had pointed out in the theory of demographic change and response, a society's initial response to rapid population growth is to work harder to support its new members. But can that be sustained? Preston (1986) has argued that it could be in those areas that have sufficient natural resources and, more importantly, are making increasingly efficient use of the major societal resource—human capital. This means not simply more people (as Simon seemed to infer) but a better-educated and better-managed labor force, combined with improved methods of communication and transportation (the economic infrastructure).

Small, oil-rich nations have been able quickly to increase their per person wealth through the sale of a highly valued resource, and they have done so without much concern over their high rates of population growth. At the same time, there are other areas of the world in which problems are so deeply rooted and resources are so scarce that every additional human being will likely aggravate the economic condition of the society. In Bangladesh, it appears that real agricultural wages in the 1970s were actually below the level of the 1830s, and much of this decline occurred after population growth accelerated in the 1950s (Preston 1986). For such a country, it is fairly easy to make the case that population growth is probably detrimental to economic development.

Is Population Growth Detrimental to Economic Development?

In the industrialized world it is popular to support the neo-Malthusian position that economic development is hindered by rapid population growth. In its basic form, it is a simple proposition. Regardless of the reason for an economy starting to grow, that growth will not be translated into development unless the population is growing slower than the economy. An analogy can be made to business. A storekeeper will make a profit only if expenses (the overhead) add up to less than gross sales. For an economy, the addition of people involves expenses (a demographic overhead) in terms of feeding, clothing, sheltering, and providing education and other goods and services, and if the demographic overhead equals or exceeds national income, then no improvement (that is, no profit) will have occurred in the overall standard of living. If the overhead exceeds income, then, just like a family, disaster can be averted for awhile by borrowing the money, but eventually that money has to be repaid and if the loan has been used simply to pay the overhead, rather than being invested in human capital, it is unlikely that the money will ever be there to pay the loan when it comes due. So, the loan is extended or refinanced, and disaster is averted for just a little while longer . . .

Let me illustrate the point further with a few numbers. Between 1980 and the mid-1990s, the population of Nigeria increased from 71 million to 102 million people, representing an average annual rate of population growth of 3 percent per year. The economy was also growing, thanks to substantial oil resources. In fact, at a rate of 2.3 percent per year, Nigeria's economy was growing at almost exactly the same clip as were most European nations during the 1980–92 period (World Bank 1994). However, that was not fast enough to keep up with population growth in Nigeria and, as a consequence, the per capita gross national product in Nigeria in the mid-1990s was only two thirds what it had been in 1980. Despite a growing economy, the average person was worse off. It is easy to ignore the impact of population growth when you are dealing with a country such as Nigeria where an oppressive statist government has skimmed off much of the country's wealth for the benefit of a small elite group. This is a blatant example of what Leibenstein (1957) politely called "organizational inefficiencies"—the existence of corruption and bribery in an environment without a stable banking or insurance system. In such a context it is easy to see why people would pay little attention to population growth, but we can be sure, nonetheless, that population growth has been compounding an already bad situation.

No country in the world is more aware of its demography than China, and it was taking advantage of its population slowdown during the same period of time that Nigeria was squandering its wealth. Between 1980 and the mid-1990s, China's population increased from 980 million to 1.2 billion, representing an average annual rate of increase of 1.4 percent—less than half the rate in Nigeria, and considerably lower than the rate had been in China in the previous decade. Economic liberalization allowed the economy to boom along at an estimated rate of 9.1 percent per year—a rate that was exceeded in Asia only by South Korea. The rapid rate of economic growth combined with population slowing meant that the per person gross national product in China was increasing by an impressive 7.6 percent per year, although of course the average income was still very low by world standards (see Table 12.1). These numbers have been very attractive to foreign investors and corporations from all over the world have been scrambling to gain a foothold in the Chinese economy to take advantage of the increasing well-being of the average Chinese consumer. This probably will have the cumulative effect of encouraging both continued economic growth and continued low fertility in China.

Although I hand-picked the above two examples to make a point, the situation seems simple enough. If populations are growing more slowly, then economic development can take place more easily, and so the neo-Malthusian concludes that population growth is detrimental to economic development. This position is so prevalent in Western society that I will discuss it in more detail.

Economic Development as a Source of Population Change   So far I have been looking at only one side of the relationship between population growth and economic development, that is, the consequences for economic change when population changes. The other side of the coin is what happens to demographic dynamics when the economy changes. Even if we accept Boserup's contention that it was population growth that spurred the Industrial Revolution we know that hidden within that economic improvement were the changes that produced a long-term drop in both mortality and fertility and unleashed the rural population and sent it scampering to the cities.

The idea that economic development was a stimulus first to a rise in the rate of population growth (through its effect on lowering mortality) and then to a slowing in its growth rate (through its effect on lowering the birth rate) underlies the theory of the demographic transition and rebuts the Malthusian argument that population growth inevitably leads just to more people living in misery.

These relationships were clearly outlined in 1958 by Ansley Coale and Edgar Hoover in a study that is unprecedented in its impact on theory and research on population growth and economic development. They note that economic development led to a mortality decline in the developed countries, and that it was also the economic development of those countries that has led to mortality declines in the rest of the world. The important point here is that the demographic transition theory suggests that the same economic development that lowered death rates will have within it the motivation for couples to lower birth rates. Yet, because the death rates in the less developed nations dropped as a result of someone else's economic development, why should we expect a rise in motivation to limit fertility without similar intervention? This question, more than any other, has plagued policy-makers, as I will discuss in more detail in Chapter 14. If a decline in the death rate is induced from the outside, should outsiders encourage fertility limitation on the expectation that the population growth from declining mortality in the absence of fertility decline will be harmful to everybody's health; do we do all we can to encourage economic development in those countries so that fertility can respond in its own way, or do we turn our heads and hope that disaster doesn't strike?

Impact of Population Growth Rates on Economic Development   It seems intuitively obvious to neo-Malthusians (and a lot of other people who do not think of themselves as neo-Malthusians) that population growth can make a difference in how many resources are consumed in the world—how much we have to pay for things like food and gasoline, and how much elbow room we have in the world. Population growth anywhere in the world may well threaten the quality of our own existence, as well as inhibit the improvement of life in those countries struggling to develop economically under the burden of daily increasing numbers of people. Actually, at least three different aspects of population change affect the course of economic development—rates of growth, population size, and the age structure.

The starting point of economic development is the investment of **capital**. Capital represents a stock of goods used for the production of other things rather than for immediate enjoyment. Although capital may be money spent on heavy machinery or on an assembly line or on infrastructure such as highways and telecommunication, it can better be thought of as anything we invest today to yield income tomorrow (Spengler 1974). This means not only equipment and construction, but also investments in human capital—education, health, and, in general, the accumulation and application of knowledge. For an economy to grow, the level of capital investment must grow. Clearly, the higher the rate of population growth, the higher the rate of investment must be; this is what Leibenstein (1957) called the "population hurdle." If a population is growing so fast that it overreaches the rate of investment, then it will be stuck in a vicious Malthusian cycle of poverty; the economic growth will have been enough to feed more mouths but not enough to escape from poverty.

The problem is complicated by the fact that in today's rapidly growing populations poverty is already rampant, as I already mentioned in Chapter 10, potentially impeding the ability of a nation to save enough money for the investment required to push its economy into rapid growth (Mason 1988). Furthermore, most of the less developed nations in the world today have histories of being colonized and dependent on other countries for their economic and political fortunes. Often this has meant that "not only were economic problems neglected but the native leadership was trained in political conflict rather than economic statesmanship" (Kuznets 1965:182). Heavy reliance is therefore placed on foreign capital—money earned and saved by slower growing, richer nations. In those countries, of course, the initial capital investment required for development was much less in relative terms than it is today. There are several reasons for this, including the facts that the developed nations started out with considerably lower rates of population growth and they did not have to jump into a well-advanced world economic system that requires high levels of technology to compete.

The less developed nations face a different set of world circumstances in trying to improve economically than did the developed nations. Many of these circumstances are probably (although not necessarily) hurdles. One example is energy—where will it come from? Vast amounts of energy, of course, are required for agriculture, manufacturing, transportation, and daily living. Countries that are hardest hit are the less developed nations that have few energy resources themselves. Only if oil-producing nations invest their profits in those countries with few resources can the less developed nations hope to keep their economies growing, and regardless of the economic order prevailing in a nation or in the world, a rapidly growing population will make it harder for an economy to find its way than would a slowly growing population. This hearkens back to the earlier discussions about the dilution effect of too many children in a family. If you as a parent have only two children and are able to save enough money from your earnings so that they can both get a college degree without going into debt for their education, then their income upon graduation will be immediately available to them to improve their standard of living, rather than having to maintain a lower standard of living for some period of time while their loans are repaid. What happens if you have a third child, but aren't able to save any more money? Do all three of your children have to borrow at least some of their college expenses, thus implicitly lowering their future standard of living? Or, do you make choices amongst the three, and devote all of your resources on only one or two of the children, leading to heightened inequality within the family? What if you have four children and now you can't save anything, so none of them are likely to be able to attend college without substantial outside support, meaning that if they do get a college education it will be even that much longer into the future before they will be able to profit from it? Those are the kinds of demographically induced choices being faced by developing countries right now.

Impact of Population Size on Economic Development   As a population grows larger, the ability to garner resources for development may grow progressively smaller. This is true for individual nations just as it is true for the entire world. Although we can conjure up images of standing room only as the point at which all economic activity most certainly would have to stop, in reality the limit is far less than that. But how much less? This is a question that is still puzzling but has been the object of a good deal of scrutiny, as researchers have tried to define an **optimum population size** for the earth or for a particular country. In trying to determine an optimum size, we ask how large a population can be before the level of living begins to decline.

It is widely recognized that there are economies of scale associated with size; that is, too few people may retard economic development as surely as too many people might. The world is much better off economically with nearly 6 billion people than it was with 1 billion. General Motors can produce a car far more cheaply than you or I could build one, precisely because they sell so many cars that they can afford the expensive assembly plants that reduce production costs per car. Granting that larger is sometimes more economical, a population may grow too large to be efficient or so large that, at a given level of living, it will exhaust resources. When it reaches that point, it is said to have exceeded its carrying capacity—that size of population that could theoretically be maintained indefinitely at a given level of living.

The carrying capacity will vary according to which level of living you might choose for the world's population. The lower that level, the greater the number of people that can be indefinitely sustained. On the other hand, if the desired level of living is too high, you may well exceed the carrying capacity and start draining resources at a rate that will lead to their exhaustion. Once you have done that, you reduce the long-run carrying capacity. For example, if you and everyone else in the world were content to live at the level of the typical South Asian peasant, then the number of humans that the world could carry would be considerably larger than if everyone were trying to live like the board of directors of General Motors. Indeed, it is improbable that the world has enough resources for 6 billion people to ever approach the level of living of a successful business executive.

One of the most elaborate and well-known empirical investigations of an optimum population size for the world is the Club of Rome study, Limits to Growth (Meadows 1974; Meadows, et al. 1972). Their study addressed the question of what size of population will enable the earth to maximize the socioeconomic well-being of its citizens. After building a computer model simulating various paths of population growth and capital investment in resource development, this team of social scientists came to the conclusion that the world's population is so large and is consuming resources at such a prodigious rate that by the year 2100 resources will be exhausted, the world economy will collapse, and the world's population size will plummet. After introducing their most optimistic assumptions into the model, the Meadows' team describes the potential result in the following way:

> Resources are fully exploited, and 75 percent of those used are recycled. Pollution generation is reduced to one-fourth of its 1970 value. Land yields are doubled, and effective methods of birth control are made available to the world population. The result is a temporary achievement of a constant population with a world average income per capita that reaches nearly the present U.S. level. Finally, though, industrial growth is halted, and the death rate rises as resources are depleted, pollution accumulates, and food production declines [1972:147].

This was the gloomiest forecast of the impact of population size on economic development since the publication of Ehrlich's Population Bomb (1968), in which worldwide famine and war were seen as almost inevitable results of continued increases in the world's population. It was another variation on the Malthusian theme that the growth of population tends to outstrip resources, and, taken at face value, it could be viewed as so discouraging that there would be no point in worrying any longer; the population that we have at hand is already too large and has too much momentum for continued growth to permit further sustained improvements in the human condition. Of course, as the authors of Limits to Growth freely acknowledge, their models do not replicate the complexities of the real world, nor are they attempting to predict the future. Nonetheless, the study demonstrates the possibility that for the world as a whole, the optimum population is probably no larger than the present level, and there are groups who advocate a lowering of world population size.

The implications of this position are rather striking. Meadows and associates discuss the need for "dynamic equilibrium" in which population and capital remain constant, while other "desirable and satisfying activities of man—education, art, music, religion, basic scientific research, athletics, and social institutions . . ." also flourish (1972:180). In 1977, President Carter directed that a study be conducted "of the probable changes in the world's population, natural resources, and environment through the end of the century" (Council on Environmental Quality 1980). The result was the "Global 2000" report, released in 1980, and containing equally dreary conclusions:

> If present trends continue, the world in 2000 will be more crowded, more polluted, less stable ecologically, and more vulnerable to disruption than the world we live in now. . . . Barring revolutionary advances in technology, life for most people on earth will be more precarious in 2000 than it is now. . . . At present and projected growth rates, the world's population would reach 10 billion by 2030 and would approach 30 billion by the end of the twenty-first century. These levels correspond closely to estimates by the U.S. National Academy of Sciences of the maximum carrying capacity of the entire earth [Council on Environmental Quality 1980:1–3].

You tell me. Is the world in 2000 "more crowded, more polluted, less stable ecologically, and more vulnerable to disruption" than the world in 1980? The answer has to be yes, although warnings such as these have helped to shape policies designed to mitigate the worst effects of population growth and environmental degradation. Still, there is little comfort here for countries not yet fully developed, since for them, the implication is that they should cease growing demographically and hope for a redistribution of income from the wealthier nations.

Before you take the gloomiest of these forebodings too seriously, let me offer a comment on the idea of a population decline in the wake of economic collapse. As van de Walle (1975) has noted, the Limits to Growth model assumes historical reversibility—that mortality rates could rise in the future just as they declined in the past. This is an unwarranted assumption, according to van de Walle, because our knowledge of nutrition would not be lost, nor would we lose our ability to reorganize life around different standards of living that could maintain health despite a reduced supply of food. Let me remind you as well of Simon's argument that specific resources, such as oil, may be finite, but in broad, generic terms, energy as a resource may be infinite—we just need to be clever enough to unlock the mysteries of the universe.

The Malthusian specter of sheer numbers of people exhausting available resources is rather overwhelming, and as a result it disguises other, more subtle negative consequences that population growth has for economic development—consequences that are far more certain to be problems than are worldwide famine, war, or economic collapse. The consequences to which I refer are those associated with the age structure of rapidly growing populations.

Impact of Age Structure on Economic Development   A rapidly growing population has a young age structure. As you will remember from Chapter 8, this means that a relatively high proportion of the population is found in the young ages. Two important economic consequences of this youthfulness are that the age structure affects the level of dependency, and it puts severe strains on the economy to generate savings for the investment needed for industry and to create the jobs sought by an ever-increasing number of new entrants into the labor force.

Dependency   A major theme of the Coale and Hoover (1958) study of economic development (which I mentioned earlier in the chapter) is that a high rate of population growth leads to a situation in which the ratio of workers (people of working age) to dependents (people either too young or too old to work) is much lower than if a population is growing slowly. This means that in a rapidly growing society each worker will have to produce more goods (that is, work harder) just to maintain the same level of living for each person as in a more slowly growing society. I have made this point before: the parent of six children will have to earn more money than the parent of three just to keep his or her family living at the same level as the smaller family. But it goes deeper than that. A nation depends at least partially on savings from within its population to generate investment capital with which to expand the economy, regardless of the kind of political system that exists. With a very young age structure, money gets siphoned off into taking care of more people (buying more food, and so on) rather than into savings per se (Kelley 1973).

One demographic part of the "Asian Economic Miracle" (the rapid economic development in east Asia) has been the fact that declining fertility in that region has been associated with an increase in savings (which generates capital for investment in an economy) (Higgins and Williamson 1997; Mason 1997). An analysis of data in Thailand has shown that even in rural populations in a poor country, couples with fewer children are better able to accumulate wealth than are couples with large families (Havanon, Knodel, and Sittitrai 1992). As Mason (1988) has pointed out, a very old age structure may also be conducive to low levels of saving, since in the retirement ages people may be taking money out rather than putting it in.

Entry into the Labor Force   In a growing population the number of prospective entrants into the labor force is also growing every year, as each group of young people matures to an economically active age. If economic development is to occur, the number of new jobs must at least keep pace with the number of people looking for them. The expansion of jobs is, of course, related to economic growth, which in turn relies on investment, which may be harder to generate with a young age structure:

> In countries like Pakistan and Mexico, for example, the work force will grow at about 3% a year [between 1985 and 2000]. In contrast, growth rates in the United States, Canada, and Spain will be closer to 1% a year, Japan's work force will grow just 0.5%, and Germany's work force (including the Eastern sector) will actually decline [Johnston 1991:116].

When the rate of labor force growth is slow, the new job entrants can simply step into the places vacated by people dying or retiring. As the rate of population growth increases and the age pyramid spreads out at the bottom, the ratio of new job seekers to those leaving the labor force goes up rapidly. You can get a sense of this by looking at Table 12.2, where I have calculated the ratio of males aged 15 to

AGE STRUCTURE AND ECONOMIC DEVELOPMENT IN ASIA

In this chapter I have argued that the youthful age structure of a growing population may be an impediment to economic development since it carries with it a high demographic overhead, which drains resources away from development. However, the other side of the coin, emphasized by Crenshaw, Ameen, and Christenson (1997) is that there is a delayed age structure reaction to a *decline* in fertility that produces a favorable swelling of the age structure—a burgeoning adult population less fettered by large families. This produces a potential "demographic windfall," which has been noticeable in the 1990s in Asia, and has contributed to the economic success of the so-called "Asian Tigers" by freeing up incomes to be saved, rather than consumed right now, and that savings is what generates the capital necessary for investment (Higgins and Williamson 1997). Let me point out the situation in both Indonesia and China, then comment on whether such a windfall seems likely to improve the future economic scenario in Mexico.

After World War II and their revolution of the 1950s, the Indonesian death rate went down (while the birth rate possibly rose), leading to a very young population; by 1960 40 percent of all Indonesians were under age 15, whereas 24 percent were in the young adult ages of 20 to 34, as you can see from the accompanying table. By 1970, the age structure was even younger because the birth rate had not yet begun to fall, and the population aged 0 to 14 was growing at a rate of 2.7 percent per year—well above the 1.3 percent annual increase in the 20 to 34 year age group. Until the mid-1960s, when the Sukarno government was replaced, Indonesia's population policy was officially pronatalist. However, the Suharto government inaugurated a family planning program on the main island of Java in 1969, from where it has spread to the other islands (Warwick 1987), probably stimulated by the national process of economic development built on the back of the country's substantial oil reserves (Gertler and Molyneaux 1994). Between 1965 and 1998 the total fertility rate dropped in Indonesia from 5.8 to 2.6 (U.S. Bureau of the Census 1998).

The demographic consequence of the decline in fertility can be seen in the changing age distribution for 1980 and 1990. By 1980 the 20 to 34 age group was growing faster than the 0 to 14 group and by 1990, not only was that gap even larger, but the percentage of the population that was 20 to 34 was going up, whereas the percentage at ages 0 to 14 was declining. This will have peaked by the year 2000 and then the demographic windfall will be over. Has this shock wave of troops into the labor force been beneficial? The answer seems to be yes. You can see in the table that the most demographically beneficial period was from approximately 1990 to the year 2000 and the average annual increase in gross national product was highest for that period. On the other hand, the demographics will not be so beneficial in the first decade of the 21st century, so even if there had not been an economic crisis in late 1997 and early 1998, the economy would have slowed down from the influence of the age structure, if all else had remained the same.

The data for China show a pattern similar to that in Indonesia, except that the rate of fertility decline was steeper than in Indonesia, so the overall rate of population growth was lower than in Indonesia, leading to a larger economic benefit from the changes in the age structure than occurred in Indonesia. You can see in the table that in 1970, as fertility was beginning to decline in China, the rate of growth in the 0 to 14 age group was 2.5 percent per year, compared with 1.5 percent for the working age population of 20 to 34. By 1980, just after the imposition of the one-child family policy in China, the rate of growth in the 20 to 34 group was already much higher than at the youngest ages, and by 1990 there was a negative rate of growth in the youngest ages. This helps to explain, then, why the rate of per capita GNP was higher than in Indonesia, and why that rise in per capita GNP occurred earlier in Indonesia than in China.

Asia is not unique in having the potential to benefit from the changing age structure. In the case of Mexico, which I discuss in some detail in this chapter, you can see that when fertility began to decline in the 1970s, the fraction of the population in the 0 to 14 age group was already quite a bit higher than in either Indonesia or China, so Mexico had more demographic ground to make up. By 1990 the rate of growth of the age group 20 to 34 was clearly more rapid than the 0 to 14 age group, and that situation continued through the 1990s, almost certainly contributing to Mexico's ability to rebound from the economic crises that beset the economy in the 1980s. You can also, see, however, that the overall rate of population increase in 2000 was higher than in either Indonesia or China and, in fact, in the year 2000 the population aged 0 to 14 is still slightly more numerous than the population aged 20 to 34. These differences contribute to Mexico's inability to make as much hay from its demographic windfall as have Indonesia and China—it hasn't experienced as big a windfall.

Lest you think that these relationships are only observable in developing countries, I offer you data for Canada, which experienced a Baby Boom during the 1950s and 1960s, followed by a sustained decline in fertility since then. As the Canadian baby boomers moved into the labor market in the 1970s and 1980s, without having very many children themselves, the per person GNP rose at a rate of 3.3 percent per year. However, as fertility has remained low and the population has continued to age (the post-windfall period, we might call it), there has been a decided slowdown in the rate of GNP increase per year.

Age Structure and Economic Indicators for Selected Countries

		1960	1970	1980	1990	2000	2010
Indonesia	Percent age 0–14	40.2%	42.3%	41.0%	35.7%	30.8%	26.8%
	Percent age 20–34	24.4%	22.1%	22.5%	25.5%	27.2%	26.0%
	Growth rate for all ages	1.9%	2.2%	2.3%	1.9%	1.5%	1.2%
	Growth rate for ages 0–14	2.2%	2.7%	2.0%	0.5%	0.0%	−0.2%
	Growth rate for ages 20–34	2.4%	1.3%	2.5%	3.1%	2.1%	0.7%
	Per capita GNP growth rate			5.2%	4.2%	5.8%	
China	Percent age 0–14	38.9%	39.7%	35.5%	27.5%	24.9%	20.3%
	Percent age 20–34	22.2%	20.4%	24.3%	27.9%	27.2%	22.8%
	Growth rate for all ages	1.7%	2.3%	1.8%	1.5%	1.0%	0.7%
	Growth rate for ages 0–14	3.2%	2.5%	0.7%	−1.1%	0.0%	−1.3%
	Growth rate for ages 20–34	1.1%	1.5%	3.6%	2.8%	0.7%	−1.1%
	Per capita GNP growth rate			4.1%	8.2%	8.9%	
Mexico	Percent age 0–14	45.0%	46.5%	45.1%	38.6%	33.1%	28.4%
	Percent age 20–34	20.7%	20.5%	21.8%	24.6%	27.2%	25.4%
	Growth rate for all ages	2.9%	3.1%	2.9%	2.1%	1.7%	1.3%
	Growth rate for ages 0–14	3.6%	3.5%	2.6%	0.5%	0.2%	−0.2%
	Growth rate for ages 20–34	2.8%	3.1%	3.5%	3.3%	2.7%	0.6%
	Per capita GNP growth rate			3.6%	−0.1%	4.7%	
Canada	Percent age 0–14	33.5%	30.2%	22.7%	20.7%	19.3%	17.1%
	Percent age 20–34	20.8%	21.8%	27.2%	26.3%	20.8%	19.7%
	Growth rate for all ages	2.7%	1.7%	1.4%	1.2%	1.0%	0.7%
	Growth rate for ages 0–14	3.9%	0.7%	−1.4%	0.3%	0.3%	−0.5%
	Growth rate for ages 20–34	1.3%	2.2%	3.6%	0.9%	−1.3%	0.2%
	Per capita GNP growth rate			3.3%	1.4%	0.5%	

Sources: Age data were calculated from United Nations, 1997, The Sex and Age Distribution of the World Populations: The 1996 Revision (New York: The United Nations); data for GNP per capita were derived from United Nations Development Programme, 1997, Human Development Report 1997 (New York: Oxford University Press). Note that the GNP figures listed under 1980 are for the period 1965–80; the figures listed under 1990 are for 1980–93, and the figures listed under the year 2000 are for 1995–96.

19 (an age to be entering the labor force) to males aged 60 to 64 (people at an age to be leaving the labor force) for several countries, using projections for the year 2000. Because all of these people are already alive, the projection is a very good estimate of what the actual situation will be in the year 2000. You can see, for example, that in Nigeria in 2000 there will be 7,224,000 males aged 15 to 19, but only 972,000 males aged 60 to 64. Thus the ratio of 7.43 males aged 15 to 19 to every male aged 60 to 64 tells us that the Nigerian economy will have to produce more than six jobs for young men entering the work force for every one job being vacated by an older man. At the other extreme, Germany will actually have fewer young men than older men, creating continued pressure for guest workers to enter the country. Partly because of immigration, the United States will have almost twice as many young men aged 15 to 19 as there are men aged 60 to 64, so the economy will have to generate nearly one additional job for every one job being vacated by older men. Mexico, however, will have to create more than four new jobs by the year 2000 for every one being left behind by retirement. Can the Mexican economy work that much harder than the United States economy? Perhaps—but, if not, the pressure on Mexicans to migrate to the United States will continue.

The data in Table 12.2 tell only part of the story because they are restricted to males. In less developing nations it is still true that the formal paid labor force is dominated by males (Bloom and Brender 1993). However, efforts are being pushed globally to bring women into the labor force in greater proportions. In order for this to happen, the economies of the developing nations will either have to grow even that much faster, or else women will have to replace men in the labor force, rather than simply join them.

Table 12.2   A Young Age Structure Forces an Economy to Create Many New Jobs

Country	Males Aged 15–19 in 2000 (000s)	Males Aged 60–64 in 2000 (000s)	Ratio of Males Aged 15–19 to Males Aged 60–64 in 2000
Nigeria	7,224	972	7.43 : 1.0
Bangladesh	8,059	1,258	6.41 : 1.0
Jordan	258	42	6.14 : 1.0
Mexico	5,168	952	5.43 : 1.0
Egypt	3,559	682	5.22 : 1.0
Indonesia	11,506	2,666	4.32 : 1.0
India	52,862	13,700	3.86 : 1.0
China	46,996	21,010	2.24 : 1.0
United States	9,540	5,082	1.88 : 1.0
Switzerland	208	183	1.14 : 1.0
Japan	3,804	3,705	1.03 : 1.0
Germany	2,315	2,715	0.85 : 1.0

Source: Adapted from World Bank, 1994, World Population Projections 1994–95, database prepared by the International Economics Department of the World Bank.

Mexico as a Case Study

In their classic study of population growth and economic development, Coale and Hoover (1958) looked briefly at the situation in Mexico, although their major effort was devoted to an examination of India. Unlike India, which was the first country in the world to institutionalize family planning programs as a national government policy, Mexico was until the mid-1970s pronatalist in its official policy. Since Mexico had one of the highest rates of natural increase in the world until the late 1970s, it is of considerable interest to ask if population growth has been a stimulus or a hindrance to development. And what has been the government's position with respect to population programs?

The Analysis by Coale and Hoover

When Coale and Hoover began their analysis in 1958, the Mexican population was already growing very rapidly as a result of declining mortality and sustained high fertility. Indeed, Mexico's population growth dates back to at least 1930, when mortality began to decline. By 1955 the life expectancy at birth (for both sexes) was about 53 years. This was equivalent to a crude death rate of 14 per 1,000 population, and the rate of natural increase was 30 per 1,000, or 3 percent per year—well above the world average. Coale and Hoover made projections of future population size in Mexico, assuming that mortality would decline steadily so that by 1985 the life expectancy at birth would be 70 years. They made three different projections, based on three alternate assumptions about future trends in fertility: (1) it would remain unchanged; (2) it would decline by 50 percent between 1955 and 1980; and (3) it would decline by 50 percent between 1965 and 1980. They then examined the differential impact on future economic growth and development.

In 1955 Mexico had a population of about 31 million, and the projected population by 1970 was 50 million; only if fertility had dropped dramatically from 1955 through 1970 would the population have been less than that. Coale and Hoover argued that the higher the rate of population growth and the larger the population, the more difficult economic development would be. With sustained high growth, (1) Mexico would have potential difficulty in maintaining its agricultural self-sufficiency; (2) exports would have to be curtailed; (3) the import of consumer goods would rise at the expense of capital goods; and (4) foreign investment would also decline as the high rates of population growth made the future of economic growth more uncertain. In general, high rates of population growth mean more consumption, less investment, and ultimately a lower level of per capita income. The lower the fertility rate, the faster per person output will rise, simply because more money can be used for development of the economy rather than for maintenance of the population. Coale and Hoover saw no reason to expect a decline in fertility in Mexico, at least not as a consequence of government action. However, they did see some possibility, since the Mexican population was more than one third urban, and urban fertility is lower than rural fertility. In general, their analysis suggested that population growth would be a substantial deterrent to economic development in Mexico. Has it been?

What Has Happened?

In the 1970 Mexican census the population was listed as about 51 million, 1 million more even than the maximum estimate by Coale and Hoover based on the 1955 population. Mortality declined slightly faster than they had anticipated; their estimate suggested that the life expectancy at birth for Mexican males in 1970 would be 61 years, and the actual figure was 64 years. They anticipated the continued high fertility, and their model had suggested an average annual rate of population growth of 3.4 percent per year by 1970, compared with the observed rate of 3.3 percent. Out-migration to the United States probably kept the growth rate lower than it might otherwise have been. The 1990 census of Mexico counted 86 million, with a growth rate of about 2.4 percent per year, and the mini-census in 1995 estimated the population to be 91 million, with the growth rate down to 2.1 percent per year, and more than 70 percent of the population living in cities.

Fertility has declined rather dramatically in Mexico over the years, as I mentioned in Chapter 6. In 1970 the total fertility rate was 6.8 children per woman, but by 1998 it was estimated to have dropped to 2.9—a decline of four children per woman in scarcely more than one generation (U.S. Bureau of the Census 1998). But you will recall that the population was continuing to grow even as the fertility declined because of the momentum built into the youthful age structure. What has been the relationship of population growth to economic development?

Between 1970 and 1980, the total GNP in Mexico was increasing by a rather phenomenal 6.3 percent per year, while the United States economy was expanding by only 2.8 percent per year (Figure 12.4). Yet during this same time, the population in Mexico was increasing at a rate of 2.9 percent per year, compared to only 1.1 percent in the United States. The net result was that per person GNP was increasing in both countries at almost exactly the same rate (2.2 percent per year in Mexico; 2.1 percent in the United States). During this period it was true in Mexico that the economy and the population were both growing quickly, but population increase was eating up (quite literally) the lion's share of economic growth.

The rapid economic growth of the 1970s was not sustainable in Mexico, especially because it was based heavily on borrowing against future oil sales. The price of oil fell and the Mexican economy stumbled badly in 1982, and the country spent the rest of the 1980s recovering. Thus the average annual rate of growth in the GNP in Mexico between 1980 and the early 1990s was only 1.5 percent, compared to 2.7 percent in the United States (see the bottom panel of Figure 12.4). Population growth did begin to slow down noticeably in Mexico during the 1980s, down to an average of just over 2 percent per year, but that was still twice the rate of population growth that the United States was experiencing (1 percent per year). The combination of economic growth and population growth meant that between 1980 and the early 1990s, the per capita GNP in Mexico was actually declining rather than going up. The average Mexican was worse off in the early 1990s than in 1980. In the United States, however, slower population growth meant that most of the growth in the economy was translated into higher standards of living.

Figure 12.4   Population Growth Has Not Aided Mexico's Push for Economic Development

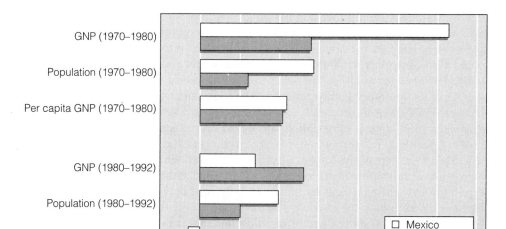

Source: Adapted from data in World Bank, 1994, World Development Report, 1994 (New York: Oxford University Press), Tables 1, 3, and 26.

In 1960 the dependency ratio (the ratio of people aged 0 to 14 and 65 and older to people aged 15 to 64) was 0.90 in Mexico, and by 1979 it had increased to 0.94. Thus the dependency burden increased in Mexico even while per person income was rising by 55 percent, demonstrating that a rising dependency ratio did not foreclose the possibility of economic development. But Coale and Hoover had also foreseen that possibility, suggesting that "Mexico may be able to achieve rising economic welfare for 30 years or even for a half-century without major declines in fertility" (Coale and Hoover 1958:335). However, they argued that the relative deterioration of the economy (which is at least aggravated by high fertility) would eventually catch up with Mexico. In the early 1970s Isbister (1973) concluded that a decline in fertility in Mexico would almost certainly stimulate savings and thus the economy, but continued high fertility would tend to undermine both. Coale himself reevaluated the situation in Mexico in the late 1970s and again in the 1980s and drew essentially the same conclusions (1978; 1986). By 1990 the dependency ratio had declined to .70, entirely a result of the dramatic drop in fertility. Mexico has thus experienced a multiphasic response to population growth. The first response was for the economy to expand to meet the needs of the multiplying numbers of people. This was accompanied by the second response—migration from rural to urban areas (especially to Mexico City, as I discussed in Chapter 11), and migration out of the country (especially to the United States, as I discussed in Chapter 7). The third response has been for the birth rate to drop, a phenomenon that has been aided by public policy.

Population Policy Implications

It is by no means certain that population growth has held back economic development in Mexico, but the suspicion is certainly there (Urquidi 1992). The Mexican government has had the same suspicion, and in 1973 it took an initial step toward population control. In 1970 President Luis Echeverria had campaigned on a pronatalist platform, but in 1973 that policy was reversed and the government launched a program of voluntary family planning. The Mexican constitution was modified in January 1974 to guarantee every couple the right to plan their family freely, and since then birth control information and services have been made available in hundreds of health clinics throughout Mexico. The result has been a surprisingly swift reaction on the part of the Mexican population. Faced with inflation, recession, and high levels of unemployment, Mexicans have turned rather dramatically to contraception to ease their burdens by limiting family size. Be mindful, however, that the battle is far from over for Mexico. The birth rate is still almost one child above the replacement level, and even with a rapid decline in fertility, Mexico's population is projected still to be growing by the year 2050, at which date the United Nations projects a total population of 154 million (United Nations 1997)—more than 50 percent larger than in 1995. Needless to say, the economy of Mexico will be fully exercised by that time.

Can the Three Positions Be Reconciled?

The fact that all three positions in the debate over population growth and economic development tend to be strongly held suggests that there are compelling elements of theoretical and empirical validity to each position. Blanchet (1991) has suggested that the reason for this may be that the relationship between the two factors cannot be described by a simple, straight line. In the early stages of economic development, population growth may simultaneously stimulate and be spurred by economic development, as I suggested was implied by the demographic transition theory and by Davis's theory of demographic change and response (see Chapter 3). Sustaining that economic momentum in the short run may have little to do with population growth. It is more dependent upon things like political stability, organizational efficiency, and other cultural and economic factors. However, to reach high levels of income in the long run may require that the rate of population growth slow substantially. Thus the relationship between population growth and economic development changes in rough accordance with the stage of the demographic transition through which a country is passing. The single most important demographic element is probably the age structure, rather than the total population size. Thus what matters is how the numbers of people at different ages are increasing. Crenshaw and his associates (Crenshaw, Ameen, and Christenson 1997) have examined the pattern of age-specific growth rates and economic development for the period 1965 to 1990 and have concluded that an increase in the child population (the impact of declining mortality in a high fertility society) does indeed hinder economic progress. On the other hand, an increase in the adult population relative to other ages (the delayed effect of a decline in fertility) fosters economic development, producing what they call a "demographic windfall effect whereby the demographic transition allows a massive, one-time boost in economic development as rapid labor force growth occurs in the absence of burgeoning youth dependency" (p. 974). I illustrate this idea with examples in the essay that accompanies this chapter.

The relationship between population growth and development is further complicated by the increasing globalization of the world economy. Most analyses of population growth and economic development deal with countries as separate and distinct entities. While that may have characterized the past century, it is becoming less relevant. We are in the midst of a massive intermingling of national economies and work forces (see, for example, Johnston 1991). World systems theorists argue, of course, that these relationships become understandable if we know a nation's place in the global system. Core countries are assumed to dictate economic terms to the rest of the world, with peripheral nations being dependent on decisions made by the core and semiperipheral nations. The latter group of countries are those with economies large enough to have some clout, but not powerful enough to be more than intermediaries between core and peripheral nations. Where a nation happens to be in the international pecking order may then determine how much economic development is going to take place, regardless of the rate of population growth.

In Figure 12.5 I have tried to bring these ideas together in a diagram showing the change over time in the relationship between population rates of growth and levels of economic development, taking into account whether economic development occurred early (the core countries) or later (the peripheral and semiperipheral

Figure 12.5   Reconciling the Various Perspectives on the Relationship between Population Growth and Economic Development

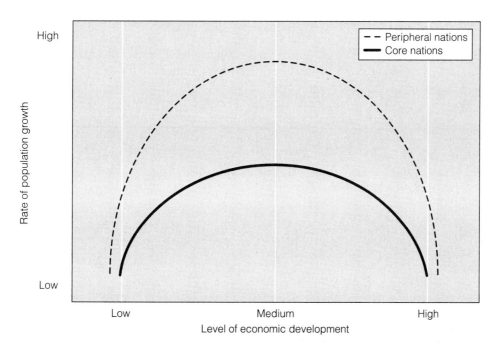

countries). In the lower left corner of the graph is the starting point of low levels of economic development and low rates of population growth—the situation that characterized most of the world until the onset of the Industrial Revolution. The trajectory over time for the now developed (core) nations is shown by the solid line. Early gains in income were accomplished in the context of increasing rates of population growth, which leads to a positive relationship between population growth and economic development in the early stages of the development process.

The situation of the developing or peripheral nations is shown by the dotted line. These countries have experienced more rapid rates of population growth in the early stages of economic development than did the now developed countries, but in all events there may still be a positive relationship between population growth and increasing levels of economic development at the earliest stages of development, largely due to the fact that the population growth is induced by a decline in mortality generated by the intervention of outside countries into the public health systems of developing nations, and the developed nations then respond to the growing population by funneling economic aid to those same countries, and by encouraging those countries to "develop" their economies by tapping into their sources of wealth. This positive relationship can be maintained only for a short while, however, because too rapid a rate of growth in the young, dependent population can severely undermine these development efforts.

As we approach the middle levels of economic development, the rate of population growth slows down. Indeed, we might even speculate that a nation will not reach the middle levels of economic development unless its rate of population growth begins to slow. At that point, as population growth flattens, you can see that a snapshot of just the middle of the graph in Figure 12.5 would make it appear as though there were no relationship at all between population growth and economic development. But, if we return to the idea that it is the rate of growth at specific ages that matters, not the overall rate of population growth, then we can understand that the leveling off of the rate of growth may mean that the rate of increase is declining at the youngest ages, but increasing at the adult ages—a positive situation as detailed by Crenshaw and his associates (Crenshaw, Ameen, and Christenson 1997). Again, however, this point will occur for peripheral nations after a higher rate of population growth has been reached than that experienced by the core nations.

Finally, a country reaches the higher levels of economic development only as (or because) its rate of population growth slows down considerably. At this point, at the later stages of economic development, the relationship between population growth and economic development becomes negative—a country will be able to achieve (and presumably sustain) a higher level of economic development only if its rate of population growth is approaching zero.

Figure 12.5 incorporates all three sides of the debate about population growth and economic development, taking into account the complexity of a world system in constant transition. Early in the process of economic development population growth may well be a stimulus to economic growth, although the situation will be a bit different for the core nations (those that developed first) than for the peripheral nations (those whose development is now occurring). The notion that there is no relationship captures what happens at the pivotal middle of the transition to higher levels of economic development. Paradoxically, this may be where the rate of population growth (at the adult ages) is actually most strongly related to economic development through its "windfall" effect. The demographic opportunity won't necessarily be converted to economic improvement unless other important non-demographic factors are in place (Cincotta and Engelman 1997). Individual freedom, property rights, free markets, and democratic governments may well be the keys to whether a country will continue to develop and will achieve high levels of economic development, or will falter at this point and stagnate economically. However, to reach a high level of economic development it seems reasonable to suggest that a country must bring its rate of population growth down to a very low level, and that must happen more quickly for the peripheral nations than occurred in the history of the core nations, because the less developed nations have experienced economic development in the context of much more rapid rates of population growth than did the now developed countries.

Summary and Conclusion

If people did not think that population growth was related in some way to economic development there would be far less interest in population issues than exists today in the world. Although most people seem to share the view that too much population growth is not good, it has been frustratingly difficult to prove a cause-and-effect relationship. Into this void of conclusions have leapt three competing perspectives. The doomsters are neo-Malthusians, who argue that population growth must slow down or the economic well-being of the planet will deteriorate over time. Boomsters are those who believe that population growth is a good thing—it stimulates development and is a sign of well-being, rather than being a threat to well-being. A third perspective is shared by Marxists, neo-Marxists, and adherents (whether Marxist or not) to the world systems perspective. This position in the debate argues that population growth is irrelevant to economic development. It is the relationships among national economies, and a nation's place in the global economic system, that determine the pace and pattern of economic development.

Looking at a country like Mexico, we can see that the beginning stages of economic development were accompanied by high rates of population growth. Neither the rapid economic nor demographic growth could be sustained, however, and the Mexican population is slowing its rate of growth and is attempting to move forward from a disastrous economic decade during the 1980s. Those economic problems were not directly caused by rapid population growth, but the government's desire to push the economy faster than it could be pushed almost certainly was an indirect response to the relentless pounding of feet on the pavement as people searched for work. Population growth creates long-term pressures on societal resources that must be dealt with.

In the final analysis, all three of the basic perspectives of the relationship between population growth and economic development probably have merit—it is just that each is describing a different part of a complicated process, one that is unfolding differently for today's less developed nations than it did historically for the now developed nations. A major difference is the much more rapid rate of population growth now than in the history of the wealthier nations. This creates more opportunities for commerce than might otherwise exist (more feet to be shod, more food to be processed, more people to buy TVs and CD players), but it also means that we have been consuming resources at a historically unprecedented rate. Early capitalists were accustomed to saving in order to build up capital (hence the name), which was crucial as the building block for future success. However, rapid population growth has meant that we have been dipping into our capital—not just the financial capital, but the resource base of the planet. We have been using up our environmental resources and changing the very nature of human life in the process. That part of development—the relationship between population and the environment—is the topic we turn to next.

Main Points

1. Economic development represents a growth in average income—a rise in the material well-being of people in a society.

2. Economic growth often occurs in conjunction with population growth, but economic development may be hampered by a high rate of population growth.

3. Data for the world indicate that high levels of average income typically are associated with low rates of population growth, whereas high rates of population growth generally are accompanied by low levels of average income.

4. The worldwide debate over the relationship between population growth and economic development has three main ideological sides, which can roughly be characterized as doomsters (anti-growth), boomsters (pro-growth), and an amalgam of Marxists, neo-Marxists, and world system analysts (growth neutral).

5. Boserup and, more recently, Simon have each argued that population growth serves as a stimulus for technological development and economic advancement.

6. Those concerned about international economic relationships argue that population growth is irrelevant to the process of economic development, and that no meaningful relationship exists.

7. Neo-Malthusians argue that population growth is detrimental to economic development.

8. It may be that high rates of population growth are detrimental to economic development, whereas low rates of growth are either not harmful or may in some cases actually be a stimulus to development.

9. Ultimately, continued population growth would lead to a population too large for the world's resources.

10. High rates of population growth lead to a young age structure, which may create demographic overhead costs that are too high for a developing economy.

11. Mexico is a country in which early population growth was accompanied by economic development; but more recently the sustained increase in size has created an economic strain, leading the government to advocate family planning.

12. As fertility declines, the burgeoning young adult population may produce a "demographic windfall" that aids in the transformation to higher levels of living.

13. In the end, a case can be made that all three seemingly different perspectives on the relationship between population growth and economic development are really just describing different phases in a complicated process.

Suggested Readings

1. Ansley Coale and Edgar Hoover, 1958, Population Growth and Economic Development in Low-Income Countries (Princeton, NJ: Princeton University Press).

   This book is such a classic in the area of the demography of economic development that it is the basis of comparison for evaluating virtually all theories and research findings.

2. Ester Boserup, 1981, Population and Technological Change: A Study of Long-Term Trends (Chicago: University of Chicago Press).

   Boserup's work has been extremely influential in the field of population growth and economic development, and this volume will show you why.

3. National Research Council, Committee on Population, 1986, Population Growth and Economic Development: Policy Questions (Washington, D.C.: National Academy Press).

   A highly respected group of economic demographers evaluated a wide range of research data and produced this assessment of the current status of our knowledge about the relationship between population growth and economic development; their conclusion that little relationship could be found generated a storm of controversy.

4. Robert Cassen and contributors, 1994, Population and Development: Old Debates, New Conclusions (New Brunswick: Transaction Publishers).

   In anticipation of the 1994 United Nations Conference on Population and Development, the Overseas Development Council asked a group of experts to contribute chapters to this volume on the debate concerning population growth and economic development, hoping to update and clarify some of the conclusions of the National Research Council study.

5. Richard P. Cincotta and Robert Engleman, 1997, Economics and Rapid Change: The Influence of Population Growth (Washington, DC: Population Action International).

   This is a fairly short, but very accessible overview of the relationship between population growth and economic development, also written to "correct" the conclusions of the National Research Council study. The publisher might be characterized as a "doomster" organization, but the essay is even-handed.

🌐 Websites of Interest

Remember that websites are not as permanent as books and journals, so I cannot guarantee that each of the following websites still exists at the moment that you are reading this:

1. http://www.pbs.org/thinktank/archive/transcripts/transcript.112.html

   With the death of Julian Simon in 1998, Ben Wattenberg, author of "The Birth Dearth," became the standard-bearer of the "boomsters" in the United States. In 1994 he interviewed Julian Simon, along with demographers Samuel Preston and John Bongaarts on his PBS TV show "Think Tank." This is the transcript from that show.

2. http://www.populationaction.org/why_pop/birthdearthfs.htm

   Population Action International is a nonprofit group promoting a slow-down in world-wide rates of population growth. They have responded to Ben Wattenberg's boomster arguments with this internet document detailing the consequences of continued population growth in the world.

3. http://connectory.sdsu.edu

   Most communities in the world are interested in economic development at the local level—the improvement of the local economy for the benefit of local residents, and this website shows you how this process can be facilitated using the resources of the internet.

4. http://www.undp.org/undp/hdro/

   The United Nations Development Programme publishes an annual Human Development Report, which includes analysis and statistics that relate directly to the relationship between population and economic development, and summaries can be obtained from this website.

5. http://www.worldbank.org/wdi/wdi/wdi.htm

   Over the years the World Bank has replaced the United Nations as the best single source for comparative world data that relate to topics such as population growth and economic development. This website summarizes their most recent World Development Indicators (WDI), and contains several tables of data that can be viewed on-line or downloaded and analyzed on your own computer.

CHAPTER 13
Population Growth, Food, and the Environment

It is elementary, my dear Watsons: Humans cannot survive without food and water. Those favored few of us in the world who can rely on water from the tap and groceries from the supermarket deal with this principle pretty much on a theoretical level. We know intellectually that some areas of the world have regularly been faced with the prospect of famine and drought. We also know that Malthus was already stewing 200 years ago about population growth outstripping the food supply. Although it is certainly a shame that all people cannot find a seat on the gravy train, the fact is that Malthus was wrong. Right? After all, food production has actually outpaced population growth over the past 200 years. It is a fact. The same boomsters mentioned in the last chapter as believing that population growth stimulates economic development also believe that the food record speaks for itself—we can grow it as we need it (in this context the boomsters are known as *cornucopians*).

However, look a little closer—the picture is less rosy, even for us. The clues increasingly are pointing to the grim reality that we will all be paying a very heavy price for coaxing ever higher yields from our increasingly overburdened planet. The plot of our mystery has taken a turn. Maybe Malthus was right. Although the formula for ultimate disaster was more complicated than he knew, critical resources such as land and water are finite. At some point, we may exhaust the earth's capacity to produce—then everybody loses.

The Food and Agricultural Organization (FAO) of the United Nations estimates the number of people in the world with inadequate access to food to be 841 million (Food and Agricultural Organization 1996)—almost the same number of people as inhabited the entire globe when Malthus first wrote his Essay on Population in 1798. During the next minute, as you read this page, 20 children under the age of 5 years will die of diseases related to malnutrition, although their places will be more than taken by the 275 babies who will be born during the same minute. The good news is that the number of undernourished humans in the 1990s is less than in the 1970s, although not by much. The uncertainty is whether we can maintain a growth in the food supply that not only provides food for the more than 80 million additional people on the planet every year, but also improves everybody's diet in the bargain? We do not really know. Do we have enough fresh water to support all of those people? We do not know that either.

An important reason for our uncertainty about the answers to those two questions is the growing certainty that the effort to grow more food and raise the standard of living around the world is doing great damage to the environment. A growing body of evidence suggests that if we stay on our current course of environmental degradation, we will permanently lower the sustainable level of living on the planet. We are in the midst of what Paul Harrison (1992) has called "the pollution crisis," and we will either work our way through this on a global scale or be faced with a major eco-catastrophe that could greatly diminish the quality of life for all of us. The World Bank, reporting on the situation in Africa, put it this way:

> With their economies largely linked to agricultural production, most West African countries must battle simultaneously to alleviate widespread poverty, ensure food security and achieve environmentally sustainable development. This has to be accomplished against a background of high illiteracy rates, rapidly growing populations, low and erratic rainfall, inherently infertile soils, and development strategies which have had a strong urban

bias. Under such conditions, traditional production systems are unable to sustain the population. Without significant change, land degradation will accelerate and the natural resource base on which agriculture production depends will continue to decline [World Bank 1998].

Yet despite the high level of poverty and undernourishment and the threat to the environment in places like sub-Saharan Africa, it is in the wealthy industrialized countries of North America, Europe, and East Asia that the biggest threats to our planet exist. In this chapter I try to uncover some of the clues to solving this unsettling dilemma. Some may lie in our history. How did we get to this crisis and what can we learn from the past?

The Relationship between Agricultural and Industrial Revolutions

It was roughly 10,000 years ago that humans first began to domesticate plants and animals, thereby making it possible to grow food and settle down in permanent villages. The domestication of plants, of course, hinged on the use of tools to work the ground near the settlement site, and the invention of those tools and their application to farming can be traced to many different areas of the world. Some of the earliest known sites are in the Dead Sea region of the Middle East (Cipolla 1965), where the Agricultural Revolution apparently took place around 8000 B.C. From the eastern end of the Mediterranean, the agricultural innovations spread slowly west through Europe (being picked up in the British Isles around 3000 B.C.) and east through Asia.

Plants and animals were also domesticated in the Western Hemisphere several thousand years ago (Harlan 1976), resulting in an increase in the amount of food that could be produced per person. As you know by now, increased food production was associated with population growth (possibly as a cause, possibly as a consequence). Overall, the Agricultural Revolution "created an economy which, by . . . giving men a more reliable supply of food, permitted them to multiply to a hitherto unknown degree" (Sanchez-Albornoz 1974:24). The classic Malthusian view, of course, is that the cultivation of land was the cause of population increase by lowering mortality and possibly raising fertility. However, the Boserupian view is that independent increases in population size among hunter-gatherers, perhaps through a long-run excess of births over deaths, led to a need for more innovative ways of obtaining food, and so of necessity the revolution in agriculture gradually occurred. Seen from this perspective, the Agricultural Revolution was the result of a "resource crisis," in which population growth, slow though it may have been 10,000 years ago, generated more people than could be fed just by hunting and gathering. The crisis led to a revolution in human control over the environment—humans began to deliberately produce food, rather than just take what nature provided. In turn, this had the cumulative effect of sustaining the slow but steady growth in population in most areas of the world for several thousand years preceding the Industrial Revolution.

Industrialization requires a massive increase in energy use. Economic development, as discussed in the previous chapter, is defined as an increase in the per person amount of resources transformed for human consumption. If everyone is consuming more, it is because production per person has increased, and that comes about by the application of nonhuman energy to tasks that were previously done less efficiently by humans, or not done at all because humans could not do them. Wood has served as the major source of energy for most of human history, but in Europe, population growth and the beginnings of industrialization led inexorably to deforestation, producing an "energy crisis" (Harrison 1992). That crisis forced a new way of thinking about energy sources—a new way of controlling the environment—and the result was the Industrial Revolution.

Keep in mind that the Agricultural and Industrial Revolutions are linked together. The Industrial Revolution of the 19th century was preceded by, indeed made possible by, important changes in agriculture that significantly improved output (Clough 1968). Through most of human history, including in Malthus's day, increases in the food supply depended largely upon **extensification of agriculture**—putting more land under production. However, since we have essentially run out of new land that can be farmed (as I discuss later in the chapter) the modern rise in agricultural output has come about through the **intensification of agriculture**—getting more out of the land than you used to. In Europe and North America those factors helping to increase agricultural productivity in a relatively short time included (1) mechanization of cultivating and harvesting processes, (2) increased use of fertilizers and irrigation, and (3) reorganization of land holdings (Poleman 1975; Walsh 1975).

The Industrial Revolution generated a host of mechanical devices, especially mechanical reapers, to greatly speed up harvesting. Drawn first by horses or oxen, reapers were pulled later by an even more efficient energy converter—the tractor. Like most early engines of industrialization, these tractors were driven by steam. Their thirst for fuel was quenched by wood as long as it lasted, but the use of coal became necessary as a result of deforestation. The idea behind the steam engine, by the way, has been around for a long time, just waiting for the right moment to be adapted to something dramatically useful. In the ancient world "Greek mechanics invented amusing steam-operated automata but never developed the steam engine; the crankshaft and connecting rod were not invented until the middle ages, and without a crankshaft it is impossible to transform longitudinal into circular motion" (Veyne 1987:137). Overall, the mechanization of agriculture vastly increased the number of acres that one or a few people could farm, and also increased the amount of land that could be devoted to more than one crop per growing season, since land could be cultivated and harvested so much more easily.

Although mechanization was certainly a prime mover of increased productivity in agriculture, especially in North America where land was plentiful in relation to people, it is not an absolute requirement. In North America, where population density was low and labor was scarce, the increase in energy needed to intensify agriculture came from mechanical devices. In Japan, on the other hand, where even at the beginning of industrialization labor was at a surplus, the initial increase in energy came from people—people working harder and more efficiently on the land (Gordon 1975).

One method of intensifying agriculture is to **multiple crop**, that is, to grow more than one crop per year on the same plot of ground. Multiple cropping, as well as making more intensive use of single-cropped land by using it every year (rather than letting it lie fallow), has been greatly aided by increased use of fertilizers and more extensive irrigation of areas previously too dry to cultivate.

Many agricultural innovations have also been made possible by reorganizing agricultural land and developing better policies for the use of land (Dyson 1996). Collecting farms into large units and using meadows and pastures for cultivation rather than extensive grazing have increased production, particularly in the United States and Europe, since large farms introduce economies of scale that permit investment in expensive tractors, harvesters, fertilizers, irrigation systems, and the like. In the United States this is a process that has a long history and is still continuing; for example, between 1950 and 1982 the number of small farms (less than 180 acres) in the United States decreased from 4,120 to 1,138, a 72 percent decline (U.S. Bureau of the Census 1996). The 1992 Census of Agriculture in the United States revealed that the total number of farms dropped below 2 million (to 1.9 million) for the first time since 1850, after reaching a peak of 6.8 million in 1935. This does not mean that the number of acres under cultivation has declined much (it has changed very little), but rather that there is a trend toward large commercial farms and away from small family farms.

Although it may be intuitively obvious, it bears repeating that industrial expansion cannot occur unless agricultural production increases proportionately. Industrialization is typically associated with the migration of people out of rural and into urban areas, naturally resulting in a shift of workers out of agriculture and into industry. Therefore, those workers left behind must be able to produce more—enough for themselves and also for the nonagricultural sector of the population. Thus you can see that the Industrial Revolution would have been impossible if agricultural production had not increased. Adam Smith, the classical economist, once remarked that "when by the improvement and cultivation of land . . . the labor of half the society becomes sufficient to provide food for the whole, the other half . . . can be employed . . . in satisfying the other wants and fancies of mankind" (quoted by Nicholls 1970:296).

In the modern world those wants and fancies are usually related to economic development, and so we are reminded again that the world goal is not just to feed more people, but rather it is to allow an increasingly small fraction of people to grow enough food to feed more people better than they were being fed before. This is a tall order, and before we get to it, we need to keep in mind the basic earthly ingredients at our disposal for such a task.

Environmental Concepts and Definitions

The world inhabited by humans is known to scientists as the **biosphere**—the zone of Earth in which life is found (Miller 1998). As mentioned in Chapter 1, the biosphere consists of three major parts: (1) the lower part of the **atmosphere** (known as the **troposphere**—the first 11 miles or so of the atmosphere above the surface of the earth); (2) the **hydrosphere** (most surface water and groundwater); and (3) the **lithosphere** (the upper part of the earth's crust containing the soils, minerals, and fuels that plants and animals require for life). Within the biosphere are **ecosystems** representing communities of species interacting with one another and with the inanimate world. All of the world's ecosystems then represent the **ecosphere,** which is the living portion of the biosphere.

All living organisms in the biosphere require three basic things: (1) resources (food, water, and energy); (2) space to live; and (3) space to "dump waste." The carrying capacity of the biosphere, or of any ecosystem within the biosphere is, as mentioned in the previous chapter, the number of organisms that can be sustained indefinitely—the number for whom there are renewable resources, sufficient space to live, and sufficient space to get rid of waste products (all forms of life generate waste products). If the population exceeds an ecosystem's carrying capacity, we have a situation of **overshoot** or **overpopulation.**

Overshooting the Carrying Capacity?

In animal populations, overshoot occurs with a certain regularity in some ecosystems, and the consequence is a die-back of animals to a level that is consistent with resources, or at the extreme, a complete die-off in that area. A good rain one winter may produce an abundance of food for one species, creating an abundance of food for its predators, and so on. In classic Malthusian fashion, each well-fed species breeds beyond the region's carrying capacity, and when normal rainfall returns the following season there is not enough food to go around, and the death rate goes up from one end of the food chain to the other. Biologists have been documenting such stories for a long time.

Premodern humans were susceptible to the same phenomena. The apocryphal story is told of the goat that destroyed a civilization. A civilization existed that depended heavily on goats for meat and milk. "The goat population thrived, vegetation disappeared, erosion destroyed the arable land, sedimentation clogged what once had been a highly efficient irrigation system. The final result was no water to drink or food to eat. It did not happen overnight, but gradually the people had to leave to survive and the civilization perished" (Freeman 1992:3). In general form, this is apparently what happened to the great Mesopotamian civilizations of Sumeria and Babylon that flourished in Western Asia nearly 9,000 years ago. The region at the time was covered with productive forests and grasslands, but each generation over time made greater and greater modifications to the environment—deforesting the area and building great irrigation canals. Around 1900 B.C. it appears that the population peaked at a level that greatly exceeded the ecosystem's carrying capacity (Simmons 1993). A combination of environmental degradation, climate change, drought, and a series of invading armies led to a long-term decline in population in the region (Miller 1998), and the area became the barren desert that today makes up parts of Iran and Iraq. In more recent history, the Mayan civilization in Central America reached a peak of population size about the year A.D. 800, but the civilization collapsed as the population overshot the region's agricultural capacities, perhaps aggravated by a severe drought (Hodell, Curtis, and Brenner 1995).

These are only a few among many stories of premodern humans exceeding the carrying capacity of a region. Human life has survived these catastrophes, but human civilizations have not. Carrying capacity is, to be sure, a moving target. We know that the carrying capacity of the earth is greater than Malthus thought because we have discovered that certain kinds of technological and organizational improvements can improve the productivity of the land. Cornucopians (the boomsters in farmer attire) assume that human ingenuity will permit a continued expansion of carrying capacity up to the point at which the world's population stops growing of its own accord which, according to the United Nations projections as of 1998 could be in the year 2200 with a population of 11 billion.

However, the stories of Mesopotamia and of the Mayans remind us that factors generally beyond our control such as the weather can also *reduce* the carrying capacity of a region—and perhaps of the entire planet. Furthermore, as I discuss later in the chapter, we as humans have been in the process of trashing our environment to the point that we might ourselves be lowering the carrying capacity of the earth, rather than increasing it. This phenomenon is less visible at the local level than it used to be because, especially in the highly industrialized nations, we use up resources that are very distant from where we live, and we try to dump our pollution as far away from us as we can. I discuss this more in the essay that accompanies this chapter, dealing with the concept of the **ecological footprint.**

Let us assume that for all of human history up to the beginning of the 19th century the carrying capacity of the globe was essentially fixed (even if it fluctuated over time from region to region) and was greater than the existing global population, but that the gap between population and sustainable resources had been narrowing—the perception of which spurred Malthus to write his Essay on Population. The Industrial Revolution was associated with increasing population growth, of course, but also with innovations in agriculture which allowed food production to stay ahead of that population growth, as I discussed earlier, and at least some of these innovations in growing food are certainly sustainable, so it is reasonable to assume (even if cannot be proven) that the carrying capacity is greater today than it has ever been in history. Thus the assumption is that since the dawn of the Industrial Revolution both population size and the global carrying capacity have increased. But the problem is that we don't know whether or not we have now already exceeded the carrying capacity. If so, then we are in a period of global overshoot and will face a catastrophe down the road (the global collapse as modeled by Meadows, et al., and discussed in the previous chapter). There is strong evidence that we have, indeed, exceeded the global carrying capacity for sustaining more than about two billion people at the current North American standard of living (Cohen 1995; Pimentel, et al. 1997). That implies that we are doomed to global inequality with respect to the consumption of resources, unless we in the highly developed world are willing (or are forced) to dramatically lower our standard of living. But, even if we assume that we have not yet exceeded the global carrying capacity, how much additional room do we have before we do? Are we headed in the same direction as Mesopotamia and the Mayan Civilization or can we feed billions more without a significant overshoot of our carrying capacity?

How Can Food Production Be Increased?

There are two aspects to feeding the world's growing population. The first is the technical problem of growing enough food, and the second is the organizational problem of getting it to the people who need it. I will focus on the technical issue, although later in the chapter I will comment on how the food is distributed. As I have already mentioned, food production can be increased though either *extensification* or *intensification* of agriculture. Let us examine each approach.

HOW BIG IS YOUR ECOLOGICAL FOOTPRINT?

Demographers are sometimes at a loss to explain why the relationship between population growth and change does not show up more clearly in the statistics of world development. It is intuitively obvious that more people consuming resources and leaving behind the detritus of the industrial world is detrimental to the long-term health of the planet. But it is maddeningly difficult to show that population growth in Mexico, for example, is more or less damaging to the earth than population growth in Indonesia. One of the problems is that those of us in the highly industrialized countries don't always make the biggest mess where we live—we are able to get someone else somewhere else to make our mess for us, which means that they (usually in less developed nations) in turn are having an environmental impact that really should not be directly attributed to them. The best way to visualize this is with the concept of the **ecological footprint.**

An ecological footprint has been defined as the land and water area that is required to support indefinitely the material standard of living of a given human population, using prevailing technology (Rees 1996b; Rees and Wackernagel 1994; Wackernagel and Rees 1996). For most of human history this was easy to figure out. If you farmed 10 acres and all of your needs were met from the resources within those 10 acres and all of your waste products were deposited within that acreage, then you did not influence life outside of that zone in any demonstrable way. But urbanization changed all that because, as I noted in Chapter 11, an urban population requires that someone else in the countryside grow their food, cut their wood, gather their sources of energy, and stow their trash. Thus it is not easy to tell how big an impact an urban resident has on the resources of the earth since the resources are drawn from multiple sources and the waste is spread out in multiple directions. Urban areas are

not sustainable on their own; they must borrow their carrying capacity from elsewhere, and this is why it has been so important historically for cities to establish political and economic dominance over the countryside. Who are the city residents borrowing from, and how much are they borrowing? Those are questions that William Rees of the University of British Columbia sought to answer. He devised a set of calculations to estimate the total land and water required to generate the materials used by humans (including the materials for food, shelter, consumer goods, energy, and so forth) and the land and water needed by each human to deposit the waste products generated by consumption of those materials. The benchmark for the late 1990s was that 2.0 hectares of land were available to each human being for their use.

The wealth of cities virtually ensures that they will exceed the average footprint. Rees calculated that the 472,000 residents of his home city of Vancouver, British Columbia generated an average ecological footprint of more than four hectares per person which meant that the city had a footprint of more than two million hectares. Vancouver itself comprises only 11,400 hectares, so Rees points out that "the ecological locations of cities no longer coincide with their locations on the map" (Rees 1996a:2). A city like Tokyo, for example, has an ecological footprint that would cover a large section of southeast Asia if it were aggregated all in one place.

The same analysis can be applied to countries. Many wealthy countries exceed their own carrying capacity by borrowing the ecological resources from other parts of the globe. In the long run, sustainability means that those countries that are running an ecological deficit (we might call them the "exploiters") must be offset by those who have an ecological surplus. The numbers in the accompanying table summarize data for several countries of the world. It probably will not sur-

Increasing Farmland

Water covers about 71 percent of the earth's surface, leaving the remaining 29 percent for us to scratch out our respective livings. Only 11 percent of the world's land surface is readily suitable for crop production, and an additional 26 percent is

prise you to learn that the average resident of the United States has the largest ecological footprint in the world, at 10.3 hectares per person. The vastness of its size and resources offers the average American an ecological capacity of 6.7 hectares per person—well above the world average of 2.0. Yet, Americans still exceed that capacity by 3.6 hectares per person. Canada, on the other hand, has an estimated available ecological capacity of 9.6 hectares, and Canadians use 7.7 hectares each, leaving surplus ecological capacity for the moment.

The wealth of most European countries causes them to exceed their capacity (although France is near a point of equilibrium), but even Mexico is estimated to be exceeding its ecological capacity as of 1997, and so were China and India. In fact, the bottom line of the table shows that, by these calculations, the average person in the world in the late 1990s required the constant production of 2.8 hectares in order to maintain their standard of living, whereas only 2.0 were calculated to be available on a sustainable basis. If these data represent the real impact of humans on the environment, then we have already overshot our carrying capacity, like people who continue to charge things on their credit card without knowing if or how they will pay back that indebtedness.

Ecological Footprints of Nations

Country	World Rank in Size of Ecological Footprint	Ecological Footprint in Hectares per Capita	Available Ecological Capacity	Ecological Deficit (if Negative) in Hectares per Capita
United States	1	10.3	6.7	−3.6
Australia	2	9.0	14.0	5.0
Canada	3	7.7	9.6	1.9
New Zealand	4	7.6	20.4	12.8
Russia	7	6.0	3.7	−2.3
United Kingdom	14	5.2	1.7	−3.5
France	23	4.1	4.2	0.1
Mexico	36	2.6	1.4	−1.2
Indonesia	45	1.4	2.6	1.2
China	46	1.2	0.8	−0.4
India	50	0.8	0.5	−0.3
WORLD		2.8	2.0	−0.8

Source: Mathis Wackernagel, et al., 1998, Ecological Footprints of Nations: How Much Nature Do They Use?—How Much Nature Do They Have? (Veracruz, Mexico: Centro de Estudios para la Sustentabilidad), Table 1; data refer to 1997.

devoted to permanent pasture (World Resources Institute 1996). Forests and woodlands cover about 32 percent of the land surface, and the remaining 31 percent is too hot or too cold for any of those things, or is used for other purposes (such as cities and highways). Most of the land that could be fairly readily cultivated is already cultivated; the rest is covered by ice, is too dry or too wet, is too hot or too cold, is too steep, or has soils that are unsuitable for growing crops. Of course, as Lester Brown, President of Worldwatch Institute, once commented, "If you are willing to pay the price, you can farm the slope of Mt. Everest" (Newsweek 1974:62).

In 1860 there were an estimated 572 million hectares of land in the world cleared for agricultural use (Revelle 1984). As the populations of Europe and North America expanded in the late 19th century, the amount of farmland in these regions virtually doubled. More recently, the population pressure in developing countries has been accompanied by an expansion of farmland in those parts of the world, often generated by slash and burn techniques in relatively fragile ecosystems. All of this adds up to a total of nearly 1.5 billion hectares of farmland in the world today—nearly a tripling since 1860. This seems to be the real limit of decent quality farmland, and you can see in Figure 13.1 that the amount of land in agricultural production did not change much in the world between 1960 and the mid-1990s, according to estimates made by the United Nations Food and Agricultural Organization.

Figure 13.1 World Food Production Has Kept Pace with Population Growth through the Intensification of Agriculture

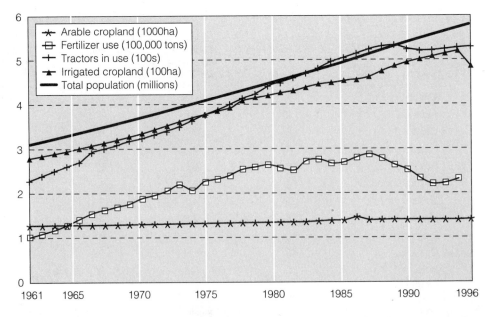

Source: Adapted from data in the United Nations Food and Agricultural Organization Internet Database (accessed in 1998).

The Cost of Extensification In reaching the limits of readily cultivable land, we have been encroaching on land that supports habitat for plants and animals that we really cannot afford to do without. We are only now beginning to discover how dependent we are on biological diversity on the planet (Miller 1998), and we threaten our environment as we search for more land on which to grow food. If that list of problems is not dismal enough, consider that the amount of good farmland is actually shrinking. In some parts of the world this is a result of soil erosion or desertification, whereas in many other places it is a consequence of urban sprawl. The sites of most major cities are in abundant agricultural regions that can provide fresh food daily to the city populations; only recently have transportation and refrigeration lessened (but not eliminated) that need. As cities have grown in size, nearby agricultural land increasingly has been graded and paved for higher-profit residential or business uses. Only about one percent of the land surface of the earth is devoted to urban use (World Resources Institute 1996), which may sound like a small amount, but it looms larger next to the 11 percent that is devoted to crops, because of the fact that so much cropland is in the vicinity of cities. The use of good farmland in the United States for nonagricultural purposes has become enough of an issue that in 1979 two federal agencies (the Department of Agriculture and the Council on Environmental Quality) commissioned a joint 18-month study to investigate the problem. The conclusion was that the United States was losing 3 million acres per year (an area roughly the size of Connecticut) of actual or potential cropland through conversion to nonfarm use.

> The allure of farm land to the developer is clear: it's generally flat and well-drained, thus good for building. It's probably outside town limits and therefore taxed at cheaper county rates. And the land is probably family-owned, an important factor with farming costs going up, children moving to the city and nonfarming family members pressuring for a quick sale [Carlson 1981:25].

Such conversion is extremely difficult to reverse and, as the Assistant Secretary of Agriculture of the United States said in 1980, "When farmland goes, food goes. Asphalt is the land's last crop" (The Environmental Fund 1981:1). You will not be surprised to discover that the government study had only a limited impact. Gardner (1996) estimates that between 1982 and 1992, the United States was still losing two New York City's worth of cropland each year (and, yes, that is a little less than the size of Connecticut, but not by much). The problem is not peculiar to the United States. China's economic expansion has devoured a great deal of prime farm land, although the government has been bringing more marginal land into production to try to counteract the loss (Gardner 1996).

Soil erosion is a major problem throughout the world, as more topsoil is washed or blown away each year, a result of overplowing, overgrazing, and deforestation, all of which leaves the ground unprotected from rain and wind (Brown 1998). In the world as a whole, it appears that we are losing soil to erosion faster than nature can rebuild the supply. Thus we are losing ground (quite literally) at the same time that few places in the world still await the plow. Some portions of sub-Saharan Africa and areas in the interior of South America still have land that could be cultivated, but they could hardly be considered choice since they are on hillsides and other places where it is more expensive and more environmentally degrading to farm. Already there is evidence that the disruption of rain forests in Africa, Asia, and Latin America is leading to the degradation of fragile ecosystems and is threatening the planet's biodiversity (Tuxill and Bright 1998).

Farming the Ocean   It has also been suggested that a viable source of "land" is the sea—mariculture, or the "blue revolution," as some have called it. Farming the sea includes both fishing and harvesting kelp and algae for human consumption, but the expense of growing kelp and other plants is again so great that it does not appear to be an economically viable alternative to cultivating land. Fish, of course, are an excellent source of protein, and fish by-products include a fairly cheap form of fertilizer, making them an attractive resource. Between 1950 and the mid-1990s the annual fish harvest (largely from the oceans) increased from 21 to 75 million tons (Brown 1981; World Resources Institute 1996). But the supply of fish that seemed almost inexhaustible in the 1970s is now at or near its peak. Aquaculture, or "fish farms," now accounts for about 1 in 7 fish consumed in the world. "By the late eighties, however, fishers had tapped all the high-volume, low-value fisheries, and were pushing the limits of a number of the tuna and squid fisheries. Although some of the high-value and other minor fisheries continued to expand, major declines in other fisheries were off-setting these modest gains" (Weber 1994). It appears that we have reached the level of the ocean's sustainable fish catch, and given that the human population is increasing, it is reasonable to conclude that seafood prices will continue to rise and per person consumption will decline (Brown 1995).

It seems very doubtful that either the extension of agricultural land or farming the ocean will produce the amount of food needed by the world in the next century, given current projections of world population. The data in Figure 13.1 showed clearly that population has been increasing whereas farmland has not, and the numbers behind those trends are shown in Table 13.1. In 1983 there were 0.31 hectares (0.77 acres) of farmland under production for each person alive in the world. By 1993 that had dropped to 0.26 (0.64 acres). The output per acre of land under production is going to have to continue its increase if we are to feed billions more.

Increasing Per Acre Yield

There are several different ways that output from the land can be increased, and often methods must be combined if substantial success is to be realized. Those methods include plant breeding, increased irrigation, and increased use of pesticides and fertilizers. In combination, they add up to the **Green Revolution.**

The Green Revolution   The Green Revolution began quietly in the 1940s in Mexico at the Rockefeller Foundation's International Maize and Wheat Improvement Center. The goal was to provide a means to increase grain production, and under the direction of Norman Borlaug, **new high-yield varieties (HYV)** of wheat were developed that are known as dwarf types, because they have shorter stems that produce more stalks than do most traditional varieties. In the mid-1960s these varieties of wheat were introduced into a number of countries, notably India and

Table 13.1　There is Wide Global Variability in the Inputs to Food Production

	Cropland (Hectares) per Person		Irrigated Land as a Percent of Cropland	Fertilizer Use (Kilograms per Hectare of Cropland)
	1983	1993	1991–93	1993
World	0.31	0.26	17	83
Africa	0.34	0.27	7	21
Europe	0.20	0.19	12	116
North & South America	0.71	0.61	11	95
South America	0.41	0.33	9	59
Asia	0.16	0.14	34	118
Oceania	2.09	1.86	5	41
Egypt	0.05	0.05	100	357
Nigeria	0.39	0.31	3	16
Rwanda	0.19	0.15	0	2
France	0.35	0.34	8	237
UK	0.12	0.11	2	338
Ukraine	0.70	0.67	8	39
Canada	1.82	1.58	2	60
USA	0.81	0.73	11	108
Mexico	0.34	0.27	24	71
Argentina	0.92	0.81	6	11
Chile	0.37	0.31	30	58
India	0.23	0.19	28	73
China	0.10	0.08	52	261
Indonesia	0.16	0.16	15	85
Japan	0.04	0.04	63	407
Australia	2.97	2.64	4	32

Source: World Resources Institute, 1997, World Resources 1996–97: A Guide to the Global Environment (New York: Oxford University Press); Table 10.2.

Pakistan, with spectacular early success—a result that had been anticipated after what the researchers saw in Mexico (Chandler 1971). In 1954 the best wheat yields in Mexico had been about 3 metric tons per hectare, but the introduction of the HYV wheat (now used in almost all of Mexico's wheat land) raised yields to 6 or even 8 tons if crops were carefully managed. A major difference was that the more traditional varieties were too tall and tended to lodge (fall over) prior to harvest, thus raising the loss per acre, whereas the dwarf varieties (being shorter) prevented lodging. This is critical, because lodging can be devastating; it destroys some ears of grain and damages others. Furthermore, resistance to lodging makes possible the heavy fertilization and irrigation necessary for high yields.

The Green Revolution (a term coined by the U.S. Agency for International Development back in the 1960s) was not restricted to high-yield wheat and maize, and in 1962 the Ford Foundation began to research rice breeding at the International Rice Research Institute in the Philippines. In a few short years a high-yield variety of dwarf rice had been developed that, like HYV wheat, dramatically raised per acre yields. Rice production was increased in India and Pakistan, as well as in the Philippines, Indonesia, South Vietnam, and several other less developed countries. China and India have both embraced the Green Revolution as a means for ensuring the food security of their people. *Food security*, by the way, is a United Nations term meaning that people have physical and economic access to the basic food they need in order to work and function normally; that is, the food is there, and they can afford to buy it.

There is, however, a catch in the Green Revolution—success requires more than simply planting a new type of seed. These plants require fertilizers, pesticides, and irrigation in rather large amounts, a problem compounded by the fact that fertilizer and pesticides are normally petroleum-based and the irrigation system requires fuel for pumping. These are expensive items and usually demand that large amounts of adjacent land be devoted to the same crops and the same methods of farming, which in its turn often means the use of tractors and other farm machinery in place of the less efficient human labor. This is the true meaning of the revolution.

Plant Breeding   The plants involved in the Green Revolution have been principally wheat, maize, and rice, but there is the potential, of course, for breeding HYV soybeans, peanuts, and other high-protein plants. Since HYV grains (wheat, maize, and rice) have already found wide acceptance throughout the world, though, you probably should not look to an overwhelming increase in HYV planting as a solution; it is not a solution for poor farmers because they simply cannot afford it. One possible avenue of genetic research is to raise the nutrient levels of crops now being cultivated. Remember that HYV wheat and rice are about equal nutritionally to the wheat and rice they replace, so if foods can be developed that offer more nutrition per plant, then we will have improved the ability to reduce malnutrition even without increasing per acre yields.

An example is the development of synthetic species such as triticale, which is derived from a cross of wheat and rye and is very hardy and high yield, although so far it has been mainly used as forage for feed animals in Europe rather than for people in less developed nations. Another candidate is the winged bean (or goa bean), sometimes known as "a supermarket on a stalk" because the plant combines the desirable nutritional characteristics of the green bean, garden pea, spinach, mushroom, soybean, bean sprout, and potato—all on one plant that is almost entirely edible, save the stalk. It is becoming a staple in poorer regions of Africa and South Asia because it grows quickly, is disease resistant, and is high in protein (Herbst 1995).

At least as important as the nutritional aspect of plant breeding is the development of disease and pest resistance. The rapid change in pest populations requires constant surveillance and alteration of seed strains. Insects are very much our competitors for the world food supply, and it has been estimated that pests of all kinds have the potential to wipe out as much as a third of all crops in the world each year (Wittwer 1977). The pesky devils are a problem both before and after the crops are harvested. In addition, other major obstacles to the increase in yield per acre even with (or especially with) HYV seeds include the availability of water, fertilizers, and pesticides (in order of usual importance).

Water   The high-yield seeds generally require substantial amounts of water to be successful, and irrigation is the only way to ensure that they get it (nature is a bit too fickle). This is because they, like all crops, grow best with a controlled supply of water and also because irrigation can increase the opportunities for multiple cropping. It has been estimated that only 17 percent of the world's cropland was irrigated as of the mid-1990s (see Table 13.1), but that accounted for 40 percent of all the world's food (Postel 1996). Irrigation, of course, requires a water source (typically a reservoir created by damming a river), an initial capital investment to dig canals and install pipes, and energy to drive the pumps. Each of these elements is in increasingly short supply. "There is widespread agreement that the future supply of water for agriculture represents a much more significant constraint to raising food production than do any likely foreseeable difficulties relating to soil or land" (Dyson 1996:149).

The problem is that almost all of the choice dam sites have already been exploited, and most within the historically short time span since the end of World War II. To give you some sense of the magnitude of the water issue, it takes about half a million gallons of water to grow an acre of rice, and irrigated agriculture accounts for about 70 percent of the water consumed worldwide (Falkenmark and Widstrand 1992b). The tremendous expense of providing irrigation—estimated to be more than $4,000 per acre in Africa—(World Bank 1984) imposes serious limits to any sizable future increase in the amount of land being irrigated in developing nations. You can see in Figure 13.1 that the upward trend of irrigated land matched the trend of population growth in the years between 1961 and 1995, although there was a drop-off in 1996. Table 13.1 shows that although only 17 percent of the world's cropland is irrigated (the remainder is fed by rain), there is considerable variability in the world. As a country, Egypt leads the world in having virtually all of its cropland irrigated, but it is in Asia that we find the highest overall levels of irrigation. Japan, for example, uses irrigation on 63 percent of its cropland, and China has 52 percent of its cropland under irrigation. Europe can get by with very little irrigation because the growing season is associated with considerable rainfall. That isn't usually the case in sub-Saharan Africa, however, where nonetheless only a small fraction of cropland is irrigated.

Fertilizers and Pesticides   In order to maximize yields, plants must be fed (fertilized) and protected (sprayed with pesticides). These are key ingredients in the success of the Green Revolution. When fertilizer use goes up, an increase in food production generally follows, unless the pests get there first. In Figure 13.1 you can see that world fertilizer use increased steadily between 1961 and 1989, and so did the world's grain yield per acre. Since 1989, the use of fertilizer has dropped off, largely as a result of the collapse of the Soviet Union, and the decline in international subsidies of fertilizer purchases by developing nations. In the early 1980s it was still true that two thirds of all fertilizers were being used in Europe and the United States, but in 1992 the use of fertilizers in developing countries exceeded that in the more developed countries for the first time ever (Brown, Kane, and Roodman 1994). China has dramatically increased its fertilizer use in recent years, and by 1993 was using as much fertilizer per hectare of arable land as were European nations, as you can see in Table 13.1.

Pesticide use is decidedly a two-edged sword. Although heavy use of pesticides initially killed insects and increased per acre yield, pesticides can also kill beneficial predators of insects and diseases that feed off of the crops. Pesticide production has increased steadily worldwide, but its use has become more judicious—too much pesticide in the short term actually lowers crop productivity in the long term (Weber, 1994b).

Incentives for Increasing Yields   The Green Revolution, to be effective in all parts of the world, would require major changes in the way social life is organized in rural areas, not just a change in the plants grown or the fertilizers used. This is because the Green Revolution is based on Western (especially American and Canadian) methods of farming, in which the emphasis is on using expensive supplies and equipment and on the high-risk, high-profit principle of economies of scale—plant one crop in high volume and do it well.

If you are a traditional, subsistence farmer, that kind of agriculture will not necessarily appeal to you. You will be accustomed to growing a variety of crops, not only to round out your diet but also to lower your risk of failure, since even in poor seasons there is a good chance that something can be salvaged. You may also prefer to save money on new seeds (the HYV seeds are not cheap) by holding back some of your crop from this year to use as next year's seed—a practice of subsistence farmers for thousands of years. Of course, in minimizing your risks you minimize your profit (that is, the ability to grow a big surplus) as well. Let me add that the major risk lies in the vagaries of nature—particularly bad weather (which I discuss later in the chapter—an early frost can wipe out a crop and so can heavy rains), but also new kinds of pests and new kinds of diseases. Furthermore, although HYV plants are bred to resist the most important diseases, and pesticides are effective against most pests, the life cycle of immunity is only about 5 years. After that time, new forms of pests and diseases have had a chance to establish themselves and can wipe out a crop. Obviously, if all your eggs are in one basket, so to speak, a crop failure can be economically disastrous. This kind of risk taking is commonplace in Western societies but requires a new way of thinking about the world for subsistence farmers in less developed nations.

Another important danger inherent in planting one crop or very few crops is nutritional deficiency. As I already noted, the HYVs are not much different in nutritional value from the earlier varieties of wheat and rice, and they may wind up replacing more nutritious crops such as beans, peas, and lentils. Poor nutrition can retard physical and mental growth, creating health and social problems for a society, which compound the difficulties of trying to increase agricultural productivity and economic development. Thus the kinds of changes that Western farmers might make to increase the food supply could (indeed, already have in some cases) lower the quality of existence in some areas of the less developed world. Subsistence farmers, who proliferate in less developed nations, tend to operate by minimizing risks, which generally also means that they are not producing even near maximum capacity. What will motivate them to increase production? The answer may be land reform, because a fairly high probability of increased profit could be an important motivation for greater output, and one way to achieve this is to put the ownership of the land into the hands of the people who are doing the work.

Another potential method for increasing incentives is to subsidize the farmer's risk taking. This can be accomplished by government subsidies in the form of credits and price supports. Without institutional support (from governments or large cooperatives) it is unlikely that yields per acre can be systematically increased. The flip side of this coin is that agricultural production is enhanced when governments do not implicitly discourage it. In many less developed nations the governments' need for cash has led to policies that encourage the growing of exportable, nonfood cash crops at the expense of basic foodstuffs (Hendry 1988; Stonich 1989). The result is that farmers have a greater risk of losing money if they grow the usual food crops. Increasingly, governments have become aware of these problems and efforts to alter policy have resulted in increased food production.

Reducing Waste   A subtle, but effective, way of getting more out of each acre of food production is to waste less. Governments that help farmers use water and fertilizer more efficiently will clearly have more of those resources to spread around. But a great deal of waste occurs once the food is grown, both in its storage (where it may become spoiled or eaten by other creatures) and in the hands of consumers. There are at least two important aspects to consumer waste: (1) overconsumption by wealthy nations; and (2) throwing less food away.

With respect to consumption it is certainly true that Americans, for example, could eat less beef and still be well nourished, and there is increasing pressure in that direction. It takes several pounds of grain to produce one pound of red meat, and there are other, more efficient ways to get protein (such as soybeans, peanuts, peas, and beans). The amino acids found in meat, but not in the vegetable proteins, may possibly be obtained from such things as lysine-enriched corn (Brown, et al. 1991). Cutting back on animal protein could then free up the production of grain for human rather than animal consumption.

The shift to a diet high in fat, cholesterol and sugar (and often accompanied by a sedentary lifestyle) has been described by Popkin (1993) as a *nutritional transition* that has accompanied the epidemiological transition in western nations, and is increasingly seen in developing countries as they change their diet, at least partly in response to the demands of the Green Revolution. Of course, most Americans do not welcome a suggestion that they should eat less meat, since eating beef, especially, is as much a part of American culture as not eating meat is in India. Thus in the United States there is resistance to the idea of meatless meals just as in India many people resist killing cows, monkeys, and even rats in the belief that all living things are sacred.

The issue of rats is central to a more invidious kind of waste—the destruction of food after it is produced but before it gets to the consumer. In India this happens at least partly because rats get the food before humans do. It was estimated that in the 1970s that it would have taken a train almost 3,000 miles long to haul the grain eaten by rats in a single year in India (Ehrlich and Ehrlich 1972). Official estimates suggest that the rats (which outnumber humans by 8 to 1) continue annually to consume at least 15 percent of India's agricultural output (United Nations Population Fund 1992). In China, the government reportedly has tried to keep its rat population under control with a rat poison that unfortunately seems to kill off the rat's natural enemies—cats and weasels—faster than it exterminates the rodents. The rat eats the poison, the cat eats the rat, and both die. Rats breed faster than cats or weasels, however, so the rat population is still growing, while cats are disappearing.

Keeping Pace with Population Growth

The Green Revolution has clearly brought some breathing room for feeding the world's population. Food production per capita was higher in the mid-1990s than it had been twenty years earlier and in general the world is a better fed place than it used to be, with disproportionate gains in food production per capita being registered in the developing countries, as you can see in Figure 13.2. There is a fair amount of regional variability, however. The biggest gains in food production have been in Asia, whereas per capita food production has dropped over time in much of sub-Saharan Africa and in Mexico and Central America, as well (World Resources Institute 1996).

About half of all food eaten in the world is based on grain (wheat, corn, and rice, for example), so if you know who has or does not have grain, you have a pretty good handle on the food situation. In Figure 13.3 I provide some examples of the major world exporters and importers of grain. Most countries of the world outside of North America and Europe depend at least in part on imported grains to feed their populations, whereas the major exporters of grain have been for a long time now the United States, Canada, Argentina, and Australia. For example, although Americans represent 5 percent of the world's total population, in 1994 American

Figure 13.2   Population and Food Production Growth

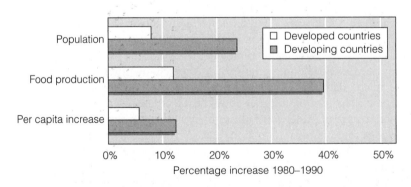

Note: Data from the Food and Agriculture Organization of the United Nations show that food production stayed ahead of population growth from 1980 to 1990, even in developing nations.

Source: Adapted from data in Food and Agriculture Organization of the United Nations, 1991, Production Yearbook, 1990 (Rome: FAO), Tables 3 and 4.

farmers produced 12 percent of the world's wheat and 45 percent of the world's corn (U.S. Bureau of the Census 1996).

Although rich and poor countries alike fall into the importer category, important demographic differences exist between them. The richer nations are growing only very slowly in population size, thus their demand for food is increasing more because of a desire to improve (or at least change) their diet than because of more mouths to feed. The developing nations have a younger population with a great need for a nutritious, higher calorie diet in order to maximize the health and productivity of the labor force. Yet many developing countries are struggling just to

Figure 13.3   The World Remains Dependent on North American Grain

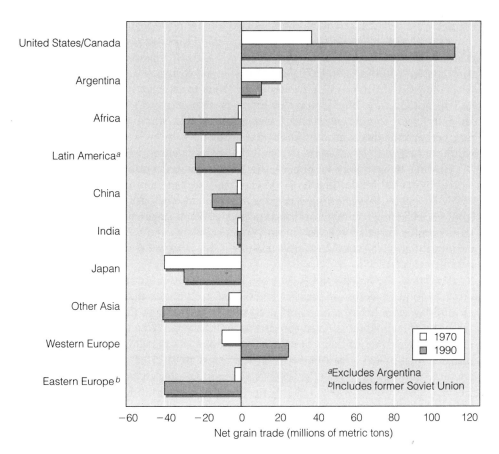

Note: In the two decades between 1970 and 1990, the world continued to be dependent on North America for its grain. The United States, Canada, and Argentina are the major net exporters of grain. In the 1970s the former Soviet Union dramatically increased its importation of North American grain in order to feed more cattle to increase the availability of meat to the Soviet population. India and China have hovered near grain self-sufficiency.

Sources: Food and Agriculture Organization of the United Nations, 1973, Trade Yearbook 1972 (Rome: FAO), Table 42; 1991, Trade Yearbook 1990 (Rome: FAO), Table 38.

maintain their current caloric intake per person, much less improve on it. Will it be possible, then, not only to feed the growing billions, but also to feed them better?

How Many People Can Be Fed?

We are approaching the limits of exploitable land and water, but per acre yield can still be increased, and we can reduce waste. Could this combination produce enough food to meet the needs of the 10 to 12 billion people that we anticipate will inhabit the planet in the next century? We know how dangerous it is to try to predict the future, but the value of trying to do so is that we may be able to invent a future that is more to our liking. In 1968 Paul Ehrlich wrote in The Population Bomb that the world population situation "boils down to a few elementary facts. There is not enough food today. How much there will be tomorrow is open to debate. If the optimists are correct, today's level of misery will be perpetuated for perhaps two decades into the future. If the pessimists are correct, massive famines will occur soon, possibly in the early 1970's, certainly by the early 1980's" (Ehrlich 1968:44). Over the years many people have derided Ehrlich for being so wrong, but of course he had noted in the final chapter of his book that this is a situation in which the "penalty" for being wrong is that fewer people will be starving than expected, and that perhaps the dire warnings about the problems of population growth and food will have helped to spur action to avoid that consequence.

Bearing these things in mind, others step forward periodically to assess the world's potential for feeding itself. Vaclav Smil, a Canadian geographer, did so in 1994. He begins by reviewing estimates made during the past 100 years that range from Ehrlich's low estimate of 2 billion people being sustainable from the world's food supply (Ehrlich 1968) to Simon's conclusion that the food supply has no upper limit (Simon 1981). Most other estimates are between 6 billion and 40 billion (see also Cohen 1995).

Smil then reminds us that no reasonable calculation of the earth's capacity for growing food generates an estimate even close to the idea that the nearly 6 billion alive today would have a diet similar to that of the average American (Smil 1994). We have, for all intents and purposes, exceeded the carrying capacity with respect to the average American diet. Smil goes on to point out that this is not as big a problem as it might seem, however, because Americans are overfed and very wasteful. To be sure, throughout the world there is tremendous slack ("recoverable inefficiencies") in the way in which food is produced and eaten. Smil calculates that by improving agricultural practices, reducing waste, and promoting a healthier diet (limiting fat intake to 30 percent of total energy, especially by reducing meat intake), the world in 1990 could have had a 60 percent gain in the efficiency of food production without putting a single additional acre under production. This would have fed an additional 3.1 billion in 1990, raising the supportable total in 1990 to 8.4 rather than the actual 5.3 billion.

Supportability in 1990 does not necessarily imply sustainability down the road. Smil introduces a series of conservative assumptions to suggest how a population of 10 to 11 billion could be supported during the next century, even without assuming some kind of magical technological fix. Applying the same logic of reducing inefficiencies, he calculates that the biggest gains (47 percent) could come from increasing the per acre yield, followed by a continued extension of cultivated land (33 percent), cultivating idle land, employing high-efficiency irrigation, reducing beef production, irrigating some crops with salt water, and farming the sea. Underlying these estimates are assumptions that all populations will have a healthier (not necessarily a higher calorie) diet, and that the food will be grown where it can be and distributed to where the demand is. These are huge assumptions and tell us only that in the best of circumstances it might be possible to feed a much larger population than we currently have.

The assumptions of changing diets and food distribution take us from the realm of technology to social organization and culture. People are remarkably adaptable if they want to be, but "culture" often intervenes to prevent the "rational" response to situations. For example, the idea that nations need to be self-sufficient with respect to food (which underlies the idea of food security) may wind up wasting a nation's resources that could be used to produce something else that can be sold in order to buy food (Davis 1991). Most of us living in developed nations are not personally self-sufficient with respect to either food or water. We rely on the good faith of strangers to provide us with what we need because we are willing to pay for it. The billions of people to be sharing the planet with us in the next century will be properly fed only if the entire world adopts that same trading principle. Trade and a surplus for aid are necessary antidotes to the maldistribution of the physical and social resources required for growing food. Important among the physical resources is the weather, on which farmers still depend the world over.

Famine and the Weather

Drought and famine have hit Africa with frightening regularity over the centuries and the most recent bout began in the late 1960s and early 1970s in the area just south of the Sahara Desert, in the countries of Chad and Niger. In northern Chad, an area "seared by eight years of drought, one group of parents implored a United Nations relief official not to send drugs when a diphtheria epidemic broke out. It was better, they explained, for their children to die straight off than to suffer further from hunger—or to grow up with their minds stunted and their spirit crippled by lingering malnutrition" (Newsweek 1974).

In the mid-1980s to the early 1990s the drought (and resulting famine) spread to many other African nations (among which Ethiopia and Somalia cornered the greatest amount of attention) and relief aid was mobilized on a massive scale. We are usually prone to take weather for granted and neglect its impact on agricultural development, but research has increasingly called attention to weather as a factor affecting the relationship between population growth and the food supply. In a review of evidence linking weather to population changes in the preindustrial world, Galloway concluded that "an important driving force behind long term fluctuations in populations may very well be long term variations in climate and its effects on agricultural yields and vital rates" (Galloway 1984:27). Historical data for China and Europe suggest that, much as you would expect, bad weather contributes to poor harvests, higher mortality, and slower population growth, whereas good weather has the opposite effect.

A team of anthropologists working in Niger found, however, that the demographic impact of the drought was not quite what they expected. Faulkingham and Thorbahn (1975) discovered that, in a village considered by them to be representative of the region, the population actually grew by 11 percent during the drought period. Despite the drought and near famine conditions, the death rate was not abnormally high for the region, and women were bearing an average of more than six children each. In the African sahel south of the Sahara, drought occurs with devastating predictability, and for the thousands of years that people have lived there, they have learned to combat drought and high mortality by having large families (see Chapter 6 on the need to replenish society). In the years just before the drought of the 1970s, the death rate had been declining, especially as a result of the better care women had been giving their children—an influence of a maternity hospital in the vicinity of the village. The impact of the drought was to slow down the decline in the death rate, while it had no apparent effect on the birth rate.

Drought is only one of several adverse weather conditions that can lead to a famine—defined as "food shortage accompanied by a significant increase in deaths" (Dyson 1991:5). Famine periodically hits South Asia either as a result of a drought caused by the lack of monsoon rains or by flooding caused by over-heavy rains. Either extreme can devastate crops and lead to an increase in the death rate. As in Africa, the high death rates are typically compensated for by high birth rates, and the famine-struck regions of the world continue to increase rapidly in population size. That maintains the pressure on the agricultural sector to improve productivity all it can. Still, we always have to come back to the fact that in the long run the only solution is to halt population growth; at some point the finite limit to resources will close the gate on population growth.

One factor already at work to limit future food production, which will also endanger human health and thus produce a deteriorating human condition, is the degradation of the environment especially related to the attempt to be ever more productive agriculturally as well as industrially, and to produce ever more energy to mobilize that productivity. It is ironic that some of the techniques that have seemed to offer the greatest hope for increasing the supply of food and improving the level of living may be changing the very ecosystem upon which food production depends.

The Degradation of the Environment

The environmental issues that confront the world today deal largely with the side-effects of trying to feed and otherwise raise the standard of living of an immense number of human beings. In coping with an ever-increasing number of humans, we are damaging the lithosphere, the hydrosphere, and the atmosphere. This is because the reality of the environment is that everything is connected:

> The web of life is seamless, and the consequences of disruption to one part of the ecosystem ripple throughout the whole; soil erosion in the Himalayas contributes to massive flooding in Bangladesh; the deforestation of the Amazon may alter the atmospheric balance over the whole globe; and chemicals and gases produced in the richer industrialized countries are destroying the ozone layer that protects everyone, rich and poor alike [Seager 1990].

Damage to the Lithosphere

We survive on the thin crust of the earth's surface. Actually, we live on only 29 percent of the surface. The rest is covered with water, especially the oceans which we tend to treat as open sewers, but we also exploit resources that are in the ground under that water. It is obviously on the land surface of the earth where most things that we humans are interested in grow, and the damage we do to this part of the environment has the potential to lower the ability of plants and animals to survive. We have been busy doing damage such as: (1) soil erosion; (2) soil degradation from excess salts and water; (3) desertification; (4) deforestation; (5) loss of biodiversity; (6) strip mining for energy resources; and (7) dumping of hazardous waste.

Almost every step of improving agricultural productivity has its environmental costs—from irrigation to the use of fertilizers and pesticides to the creation of energy sources and the production of machinery. "In spite of the increasing pace of world industrialization and urbanization, it is ploughing and pastoralism which are responsible for many of our most serious environmental problems and which are still causing some of our most widespread changes in the landscape" (Goudie 1990:323).

Degrading the Agricultural Environment   If not carefully managed, farming can lead to an actual destruction of the land. For example, improper irrigation is one of several causes of soil erosion, to which valuable farm land is lost every year. In the United States alone it is estimated that during the past 200 years, at least one third of the topsoil on croplands has been lost, ruining as much as 100 million acres of cultivated land (Brown 1998; Pimentel, et al. 1976). Even if cropland is not ruined, its productivity is lowered by erosion, because few good chemical additives exist that can adequately replace the nutrition of natural topsoil. Unfortunately, the push for greater yield per acre may lead a farmer to achieve short-term rises in productivity without concern for the longer-term ability of the land to remain productive:

> In many human cultures, agriculture is practiced as an extractive industry and soils continue to be degraded throughout the world. Continuation of the observed rate of soil degradation from 1945 to 1990 suggests an effective half-life of the vegetated soils of the earth of about 182 years. Such conversion of land to agricultural purposes alters the entire ecosystem, and the resulting impact on soil structure and fertility, quality and quantity of both surface and groundwater and the biodiversity of both terrestrial and aquatic communities diminishes both present and future productivity [Vanderpool 1995].

Crop rotation and the application of livestock manure help to reduce soil erosion, but in some parts of the world the land is robbed even of cow dung by the need of growing populations for something to burn as fuel for cooking and staying warm. The eroded soil has to go somewhere, of course, and its usual destinations are river beds and lake bottoms, where it often causes secondary problems by choking reservoirs (Seitz 1995). Desertification and deforestation are ecological disasters associated with the pressure of population growth on the environment. The southern portion of the Sahara desert has been growing in size as overgrazing (complicated by drought) has denuded wide swaths of land (Goliber 1985). It has been estimated that 70,000 square kilometers of farmland are abandoned each year in the world because the ground is worn out (United Nations Population Fund 1991). In the Philippines, the growing population is pushing rural villagers further up the hillsides in search of farmland, which leads to increasing deforestation which then hastens soil erosion.

Deforestation   At the dawn of human civilization, forests covered about half of the earth's land surface (excepting Greenland and Antarctica). Only about half of that forest is left and "each year another 16 million hectares of forest disappear as land is cleared by timber operations or converted to other uses, such as cattle ranches, plantations, or small farms" (Abramovitz 1998). Nearly 80 percent of that deforestation has been attributed to the impact of population growth, even if indirectly (United Nations Population Fund 1991). In Brazil, population pressure has led people to head for the Amazon Basin in search of land. The land they find is covered with a rain forest, which they have been cutting down at a prodigious rate. Deforestation then contributes to soil erosion by removing physical barriers that prevent wind and water from carrying the topsoil away.

Forests are also susceptible to the effects of air pollution, which can damage the vegetation and lessen the plant's resistance to disease. In their turn, fewer trees and less healthy trees may alter the climate because the forests play a key role in the **hydrologic cycle** as well as in the carbon cycle. In the hydrologic cycle water is being continuously converted from one status to another as it rotates from the ocean, the air, the land, through living organisms, and then back to the ocean. Solar energy causes evaporation of water from the oceans and from land, and it condenses into liquid as clouds, from whence comes the rain, sleet, and snow to return the water to the ground. Trees are important in this cycle both directly because water transpires through the plants and is evaporated into the air, and indirectly because the trees slow down the runoff and heighten the local land's absorption of the water. More than half of the moisture in the air above a forest comes from the forest itself (Miller 1998), so when the forest is gone, the local climate will become drier. These changes can mean that an area once covered by lush and biologically diverse tropical forest can be converted into a sparse grassland or even a desert.

The *carbon cycle* is that process through which carbons, central to life on the planet, are exchanged between living organisms and inanimate matter. Plants play an important role in this cycle through photosynthesis and the forests are sometimes called the earth's "lungs." Deforestation thus has the effect of reducing the planet's lung capacity, so to speak, and that contributes to global warming because it increases the amount of greenhouse gases which, in the right number, otherwise keep us at just the right temperature for normal existence.

Damage to the Atmosphere

The atmosphere is the mixture of gases surrounding the planet, and it is a layered affair (each layer being a "sphere"). As I mentioned earlier in the chapter, we spend our life in the troposphere, that part of the atmosphere near the surface where all the weather takes place. But other layers are of importance as well, such as the ozone in the stratosphere which protects us from the ultraviolet radiation from the sun. Most famous of the gases are the **greenhouse gases** (mainly carbon dioxide and water, but also ozone, methane, nitrous oxide, and chloroflourocarbons) which allow light and infrared radiation from the sun to pass through the troposphere and warm the earth's surface, from which it then rises back into the troposphere. Some of it just escapes back into space, but some of this heat is trapped by the greenhouse gases and this has the effect of warming the air, which radiates the heat back to the earth (Miller 1998). The greenhouse effect is a good thing, because without it the average temperature on the planet would be zero degrees Fahrenheit (-18 °C) and life would not exist in its present form, but too many greenhouse gases have the effect of **global warming**—an increase in the global temperature.

Global warming has the potential to change climatic zones, warm up and expand the oceans, and melt ice caps. The result would be a rise in average sea level, inundating coastal areas (where a disproportionate share of humans live), and a shift in the zones of the world where agriculture is most productive. The evidence is almost overwhelming that we have been adding to the greenhouse gases and that human activity is contributing to a rise in global temperature (Gelbspan 1997). This has happened as a polluting side-effect of trying to support more humans, and to do so at a higher standard of living.

Population growth, the intensification of agriculture, and the overall increase in the standard of living of human beings have been made possible by substantial increases in the amount of energy that we use. Holdren (1990) has estimated that in 1890, when the world's population was 1.5 billion, the annual world energy use was 1.0 terawatts. (A terawatt is equal to five billion barrels of oil.) One hundred years later, in 1990, when the world's population was at 5.3 billion, total world energy use had rocketed to 13.7 terawatts. This is an important number because "energy supply accounts for a major share of human impact on the global environment" (Holdren 1990:159).

The byproducts of our energy use (especially carbon dioxide and methane) wind up disproportionately in the atmosphere and contribute to global warming. As you can see from Figure 13.4, we know who is pumping carbon dioxide (CO_2) into the atmosphere and the enemy is us. The United States leads the list of CO_2 producers, both in absolute terms and in terms of per person emissions. Canada is ninth on the list in absolute emissions but is third in terms of per person emissions. Mexico is also on the top 15 list in terms of total output of carbon dioxide emissions, but its volume is only a tiny fraction of the United States and its per capita use is one of the lowest on the list. You may already have noticed that this list (Figure 13.4) is similar to the list of the largest economies of the world juxtaposed with the countries by per person income, which I showed you in Chapter 12 (Figure 12.1 and Table 12.1). In general, you could say that those with the biggest incomes make the biggest mess.

Other gases that we send into the environment—especially chloroflourocarbons—have the potential to thin the ozone layer protecting us from deadly ultraviolet light. These "holes" in the ozone layer, which have been documented especially in the southern hemisphere, can damage crops and livestock and, of course, humans as well.

Although the switch from wood to coal for creating steam may have helped to save forests, the by-product of burning coal is "acid rain"—sulfur particles trapped in the air, which then cause damage by killing plants, undermining animal habitat,

Figure 13.4   Countries with the Highest Industrial Emissions of Carbon Dioxide

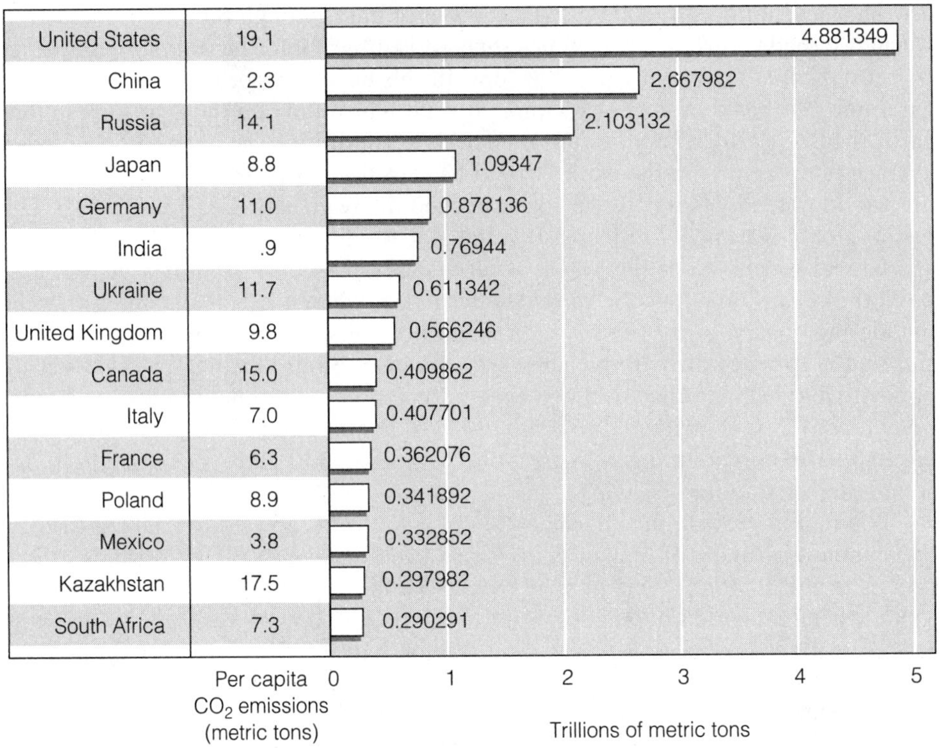

Country	Per capita	Trillions of metric tons
United States	19.1	4.881349
China	2.3	2.667982
Russia	14.1	2.103132
Japan	8.8	1.09347
Germany	11.0	0.878136
India	.9	0.76944
Ukraine	11.7	0.611342
United Kingdom	9.8	0.566246
Canada	15.0	0.409862
Italy	7.0	0.407701
France	6.3	0.362076
Poland	8.9	0.341892
Mexico	3.8	0.332852
Kazakhstan	17.5	0.297982
South Africa	7.3	0.290291

Per capita CO$_2$ emissions (metric tons)   0   1   2   3   4   5

Trillions of metric tons

Source: World Resource Institute, 1997, World Resources 1996–97 (New York: Oxford University Press); Table 14.1 and Data Table 14.1

and eroding human-built structures, especially those made of marble and limestone. Photochemical smog produced by automobile and industrial emissions creates a wide band of air pollution that is known to be harmful to humans, other animals, and to plants as well. In a variation on the theme "what goes up, must come down," the gases and particles that we pump into the atmosphere come back to haunt us in myriad ways, none of them beneficial.

Damage to the Hydrosphere

Water is an amazing liquid. It covers 71 percent of the earth's surface, including almost all of the southern hemisphere and nearly half of the northern hemisphere. You are full of it—about 65 percent of your weight is water. Despite all of that water, only a small fraction—3 percent—of it is the fresh water that humans and plants need. Furthermore, most of that 3 percent is water that is locked up as ice in the poles and glaciers or in extremely deep groundwater. Only about 0.003 percent of the total volume of water on the planet is fresh water readily available to us in lakes, soil moisture, exploitable groundwater, atmospheric water vapor, and streams (Falkenmark and Widstrand 1992a). Although fixed in amount, the water supply is constantly renewed in the hydrologic cycle of evaporation, condensation, and precipitation. The principle issues with respect to water have to do with its management (distributing it to where it is needed), purity from disease (in order to be drinkable), and pollution.

It has been estimated that in 1850 the freshwater resources in the world were equivalent to 33,000 cubic meters per person per year (United Nations Population Fund 1991) but by 1995 that had shrunk to less than 7,200 (World Resources Institute 1997). Within the ecosphere, salt water is converted to fresh water through the hydrologic cycle, but it is very expensive to mimic nature. In fact, it has been joked that the two most difficult things to get out of water are politics and salt. Most desalination plants are based on a process of distillation that imitates the water cycle by heating water to produce vapor, which is then condensed to produce fresh, potable (drinkable) water. The problem is that it is very costly to heat the water and as a result desalinated water is typically several times more expensive than drinkable local water. Reverse osmosis as a desalination process may hold some promise, but it seems unlikely that anything but naturally generated fresh water will be able to supply human needs for the foreseeable future, and we will have to survive by using that resource more efficiently than in the past (Postel 1993).

All over the globe more people are competing for water even as water consumption per person has been on the rise, and, all the while, we have been sending pollutants into the water, degrading the already limited supply. Some of the pollution goes directly into the water, and some goes into the ground where it seeps into the water supply or into the air where it then falls on us as acid rain.

A good example of bad practices is that of the Reserve Mining Company, of Silver Bay, Minnesota, north of Duluth. The company was established in the early 1950s to extract magnetite or iron oxide from taconite, a low-grade iron ore. From 1955 to 1974 the company daily dumped thousands of tons of taconite tailings into Lake Superior, the earth's largest body of fresh water, which, because of its natural purity, was used for drinking water without filtration by Duluth and other cities around the lake. However, in 1972 it was discovered that the drinking water in Duluth contained asbestos fibers, which were traceable to the taconite dumpings. Asbestos is now known to be a carcinogen, and nearly 50 percent of all asbestos workers die of cancer (compared with 18 percent for the general population in the United States). The State of Minnesota sued the plant and after a lengthy trial, the company agreed to build an inland storage basin. However, by that time the demand for steel had dropped and the company closed down in the mid-1980s.

Unfortunately, this is just one of thousands of similar examples of how we as humans have messed up our most important renewable, but fixed resource. We know as well that polluted water can alter marine life, killing fish and other sources of marine food. Ironically, one of the sources of water pollution is from the chemicals that we add to the soil to improve agricultural productivity, and this is aggravated by the use of irrigation which increases the amount of water that is exposed to the chemicals. Irrigation requires dams, of course, and there has been a worldwide movement to stop the construction of dams as we learn more about the ecological damage caused upstream, downstream, and on the cropland itself by dams and the irrigation water; not to mention the millions of people who have been displaced around the world because their home was going to be under water in the reservoir behind the dam. I mentioned earlier in the chapter that most of the choice dam sites have already been taken, but not all. You will recall from Table 13.1 that Asia has a higher percentage of its land under irrigation than any other area of the world, and China, in particular, is adding to the total. As of the late 1990s, China was in the midst of building three major dams, the largest and most famous of which is the Three Gorges Dam being built on the Yangtze River. When finished in 2009 it will supply irrigation water, generate hydroelectric power (estimated at more than ten percent of China's total), and will have forced the migration of more than one million people (The Economist 1997).

Direct Health Hazards in Food Production

Much of the success in increasing food production during and after World War II has been achieved by applying chemical pesticides. No matter how much water and fertilizer you apply to a crop, it will be for naught if pests damage or destroy the crops either before or after harvest. DDT was the first widely used pesticide, but the buildup of its toxicity throughout the food chain, culminating in increased likelihood of death for humans, was most poignantly publicized by Rachel Carson's book, The Silent Spring, which first appeared in 1962. Since then DDT has been banned in the United States and is no longer widely used in the world. However, many chemical pesticides that have replaced DDT have also come under attack. Of course, chemicals not only are in the plants and the soil but are also washed into rivers and underground streams and eaten by other animals that humans may in turn eat, and at each stage the buildup becomes more highly concentrated. The chemical industry has really been around only since the early 1940s (Carson, 1962), and the substantial rise in chemical pesticides has obviously been even more recent than that. Because the effect on your health may be delayed for 20 to 30 years or even longer, we have only recently moved into an era when pesticide use may have a discernible effect on mortality (and, of course, there are many who point out that the cancer rate is rising—although the cause-and-effect relationship is not firmly established). In the meantime, seed breeders have continued to look for organic sources of pesticides, since the problem of controlling pests remains a major issue in food storage and production.

Yet other potential dangers are involved in the production of some foods, such as additives. Especially in developed nations, chemical substances are added to food to protect its nutritional value, to lengthen its shelf life (preservatives), and to change or enhance flavors and colors. Additives in some cases greatly aid the process of feeding people by keeping food from spoiling and helping to preserve its value. This has aided in the mass distribution of food and has made it possible for people to live considerable distances from food sources. The use of preservatives is one means of deterring the spoilage of food by microorganisms, and worldwide food shortages could be at least partially alleviated by their more widespread use. For example, the World Health Organization estimates that about 20 percent of the world's food supply is lost to microorganism spoilage. Nonetheless, preservatives have increasingly been attacked as potential cancer agents. Sodium nitrate, used for centuries in curing meat to prevent botulism, is now suspected of being a carcinogen, at least in some dietary combinations. In 1984 the U.S. Food and Drug Administration approved the use of low doses of irradiation as a form of food preservation and data suggest that it is about as safe as food prepared in a microwave oven. Irradiated foods have been available in Europe for some time, but the first plant in the United States did not open until 1991 when a firm in Florida began irradiating fruits and vegetables (Ingersoll 1991).

Assessing the Role of Population in Environmental Degradation

"The real world is inconveniently complicated. It is a vast system in which the immense complexity of human society interacts with the even greater complexity of the natural world. The character of that interaction varies from one place to another. And in all places, it changes over time" (Harrison 1992:236). The role of population in environmental degradation thus differs from place to place, from time to time, and depends on what type of degradation we are discussing.

In general terms, however, environmental degradation can be seen as the combined result of population growth, the growth in production (transformation of products of the natural environment for human use) that we call economic development, and the technology applied to that transformation process. Ehrlich and Ehrlich (1990) have summarized this relationship in their impact equation:

$$\text{Impact (I)} = \text{Population (P)} \times \text{Affluence (A)} \times \text{Technology (T)}$$

Impact refers to the amount of a particular kind of environmental degradation; *population* refers to the absolute size of the population; *affluence* refers to per person income; and *technology* refers to the environmentally damaging properties of the particular techniques by which goods are produced (measured per unit of a good produced). "Technology is double-edged. An increase in the technical armoury sometimes increases environmental impact, sometimes decreases it. When throwaway cans replaced reusable bottles, technology change increased environmental impact. When fuel efficiency in cars was increased, impact was reduced" (Harrison 1992:237).

Barry Commoner (1972; 1994) has proposed a slight variation on the Ehrlich and Ehrlich formula that better allows the researcher to assess the relative contributions that population growth, affluence, and technology might have on environmental degradation by examining specific types of pollutants:

$$\text{Pollution} = \text{Population} \times (\text{good/population}) \times (\text{pollution/good})$$

If we want to measure the relative impact of population on the pollution generated by motor vehicle traffic (the "good"), we measure the (good/population) as being equal to the number of vehicle miles per person; whereas the (pollution/good) is measured as carbon monoxide emissions per vehicle mile driven. For example, between 1970 and 1987 carbon monoxide emissions declined in the United States by 42 percent, due to a combination of increased regulation and higher fuel costs that spurred automobile manufacturers to lower emissions levels and improve the fuel efficiency of cars. Commoner (1994) has shown that technology change (a 66 percent reduction in pollution per automobile mile driven) lowered the overall pollution from carbon monoxide, despite the fact that population in the United States increased 19 percent during that time and the number of miles driven per person increased by 45 percent. Population and affluence were pushing pollution upward, but that was counteracted by technology change. That is a hopeful trend.

Less hopeful are similar analyses carried out by Harrison (1992) showing that despite improving technology, the combination of population growth and increased consumption per person has driven up the level of environmental degradation in a number of key areas. Table 13.2 summarizes some of Harrison's results. For example, you can see that between 1961 and 1985 there was a 15 percent increase in the amount of acreage under production in developing countries. Each additional acre of land used for cultivation increases the environmental degradation. Because technological change was permitting an increase in yield per acre, technology was work-

Table 13.2   Estimating the Impact of Population on Environmental Degradation

Type of Environmental Degradation	Overall Change (%)	Change Due to		
		Population (%)	Consumption (%)	Technology (%)
Arable land growth 1961–85 (contributes to deforestation, loss of wetlands, species loss)				
Developing countries	15	+72	+28	−100
Developed countries	3	+46	+54	−100
Growth in livestock numbers (cattle, sheep, and goats) 1961–85 (contributes to erosion, deforestation, methane emissions)				
Developing countries	36	+69	+31	−100
Developed countries	8	+59	+41	−100
Growth in fertilizer use 1961–85 (contributes to water pollution, global warming, micronutrient depletion)				
Developing countries	1,568	+22	+8	+70
Developed countries	208	+21	+18	+61

Source: Adapted from Paul Harrison, 1992, The Third Revolution: Environment, Population and a Sustainable World (London: I. B. Tauris & Co., Ltd), page 243.

ing to lower the impact of land use. Most of the increase in arable land use (72 percent) was attributable to population increase, whereas 28 percent was attributable to an increase in food production per person.

Two things are important to keep in mind as you examine Table 13.2. Notice first that environmental degradation has been increasing faster in developing than in the developed countries. Although those of us living in developed nations still consume a vastly disproportionate share of the earth's resources and thus contribute disproportionately to the pollution crisis, the rates of population growth and economic development in developing countries mean that the global impact is shifting increasingly in that direction. For example, the data in Figure 13.4 show that China and India are among the biggest CO_2 polluters in the world, despite very low per person rates in those two relatively poor countries. Secondly, notice that technological improvements are already operating to dampen the environmentally degrading impact of consumption, but population growth has been exerting continual upward pressure on degradation.

Commoner (1994) has argued that of the three components of the impact equation, production technology has the single most important impact on the environment. You can see the dilemma here: just to maintain the current impact on the environment, technology must completely counteract the impact of population growth and increasing affluence. Much of the affluence in the developed nations has come at the expense of the rest of the world—we have used resources without paying for them because the cost of goods that we purchased did not typically include the environmental costs associated with their production and consumption (see, for example, Brown and Mitchell 1998; Turner, Pearce, and Bateman 1993). We cannot continue indefinitely to draw down the "capital" of nature to supplement our income. The price of goods increasingly will have to include some measure of the cost of dealing with the environmental impact of making that product (the pollution from the manufacturing process) and the cost of getting rid of the product when it is used up (the pollution from waste). Measuring the cost of goods in this way may slow down the rate of economic development, measured in a purely economic way, but it should increase the overall human well-being by balancing economic growth with its environmental impact. Nowhere in this set of equations can it be concluded, however, that increased population is beneficial. Population growth is something that must be coped with at the same time that we continue to try to slow it down, because "rational people do not pursue collective doom; they organize to avoid it" (Stephen Sandford, quoted by Harrison, 1992:264).

Summary and Conclusion

The world's rapidly growing population naturally requires an equally rapid increase in food production. Since the world is almost out of land that can be readily cultivated, increases in yield per acre seem to offer the principal hope for the future. Indeed, that is what the Green Revolution has been all about—combining plant genetics with pesticides, fertilizer, irrigation, crop rotation, land reorganization, and multiple cropping to get more food out of each acre. At current levels of technology it may be reasonable to suppose that the world's population could be fed for many years to come if food can be properly distributed, if farmers in less developed nations are able to reach their potential for production, and if environmental degradation does not intervene to limit productivity. Whether that happens is more a political, social, and economic question than anything else. What does seem clear is that it is almost inconceivable that all the people of the world will ever be able to eat as Americans currently do. We have almost certainly exceeded the carrying capacity for that level of living.

A major constraint in providing food for the growing billions of newcomers to the planet is that in the process of providing for our sustenance we are degrading the environment—perhaps irreparably. Paul Harrison (1992:270) has summarized the situation very succinctly:

> On the eve of the Third Millennium, we are in the embrace of an environmental crisis that is coiling around more and more regions and ecosystems. Accelerating deforestation in the South, forest death in the North, red tides, the ozone hole, the threat of global warming: all have arrived over the space of a mere fifteen years. Underlying them is the long attrition of biological diversity, and the progressive degradation of land.
>
> These problems arose when perhaps no more than 1.4 billion people were consuming at levels of at least moderate material affluence.
>
> Yet ahead lie four decades of the fastest growth in human numbers in history.

Demographers, of course, cannot provide solutions to the problems of feeding the population or of halting environmental degradation, but they have wrestled mightily with the task of slowing down the rate of population growth, as I discuss in the next chapter.

Main Points

1. The Agricultural Revolution provided the first opportunity for an increase in the number of humans, involving as it did the domestication of plants and animals.

2. Population growth during the Industrial Revolution also was associated with increased agricultural output.

3. Food production will obviously have to continue to increase in order to feed the inevitably larger world population.

4. Food production can be increased by increasing farmland or per acre yield. The latter can be accomplished by continued plant breeding and increased use of irrigation, fertilizers, and pesticides.

5. The Green Revolution is a recent phenomenon involving an increase in grain yields through the development of new strains of plants.

6. The high-yield varieties of wheat and rice (and later maize) that formed the basis of the Green Revolution produced early spectacular success, but more recently the gains have been slower because of the increased costs of pesticides, fertilizers, and irrigation schemes.

7. In India it is reported that rats eat stored grain and are, in turn, eaten by snakes. However, the food chain tends to stop there, since the snakes in India are rarely eaten by humans.

8. It is estimated that the world could not grow enough food for the entire population to eat an average American diet.

9. It is perhaps within the realm of possibility that 11 to 12 billion people could be sustained even at current levels of agricultural technology if the food production and distribution worked far more efficiently and if environmental degradation did not intervene to lower agricultural productivity.

10. A still unpredictable element in the ability to grow food and sustain life in any given location is the weather.

11. In trying to feed ever more people and attempting to elevate world standards of living, we have managed to do a great deal of damage to the environment.

12. We have degraded the lithosphere especially through soil erosion, desertification, deforestation, and the dumping of waste on the land.

13. We have degraded the hydrosphere especially through depletion of aquifers and pollution of groundwater.

14. We have degraded the atmosphere especially by sending gases into the air that generate photochemical smog, acid rain, ozone holes, and global warming.

15. The immediate or proximate causes of environmental degradation are population growth, economic development, and technology. The relative contribution of each varies from place to place, from time to time, and according to the specific type of degradation.

Suggested Readings

1. World Resources Institute, 1996, World Resources: A Guide to the Global Environment 1996–97 (New York: Oxford University Press).

   This report is part of a series and by the time you read this, the newer ones in the series may be available. The World Resources Institute, a nonprofit organization in Washington, DC, collaborated on this volume with the United Nations Development Programme and the World Bank, and the first six chapters are devoted to topics relating to the Habitat II Conference in Istanbul (referred to in Chapter 11).

2. Lester Brown, 1995, Who Will Feed China?: Wake-up Call for a Small Planet (New York: W. W. Norton & Company).

   China is just barely self-sufficient with respect to food. What would happen if China had to import large amounts of food? In answering this question, Brown explores most of the issues discussed in this chapter.

3. Paul Harrison, 1992, The Third Revolution: Environment, Population and a Sustainable World (London: I. B. Tauris & Co.).

   This is very well-written and well-reasoned examination of the history of the impact of humans on the environment. From the numerous references made in this chapter, you know that I hold this book in high esteem.

4. Lourdes Arizpe, M. Priscilla Stone, and David C. Major, 1994, Population & Environment: Rethinking the Debate (Boulder, CO: Westview Press).

Based on papers prepared for a special conference in Mexico City in 1992, the editors produced a volume to serve as input for the 1992 Earth Summit as well as the 1994 Population Conference.

5. United Nations, 1994, Population, Environment and Development: Proceedings of the United Nations Expert Group Meeting on Population, Environment and Development (New York: United Nations).

This volume is also based on a conference—in this instance one of the six expert group meetings organized by the United Nations Population Division in preparation for the 1994 Conference on Population and Development.

Websites of Interest

Remember that websites are not as permanent as books and journals, so I cannot guarantee that each of the following websites still exists at the moment that you are reading this:

1. **http://www.fao.org**

The Food and Agricultural Organization (FAO) of the United Nations is the principal source of data on land use and agricultural production. The data from their annual Production Yearbook are available on-line at this website in an interactive mode that lets you choose the information you want and the format you want it in. Search the site carefully because there are a lot of other things there.

2. **http://www.sarep.ucdavis.edu**

The Sustainable Agricultural Research and Education Program (SAREP) at the University of California, Davis is a world-renown organization devoted to researching new ways to engage in sustainable agriculture and then disseminating that information to those who can make the best use of it.

3. **http://www.den.doi.gov/wwprac/reports/west.htm**

There are few places in the world where population growth has impacted water resources more than in the southwestern states of the United States (and northern Mexico). In 1995 President Clinton appointed the Western Water Policy Review Advisory Commission (WWPRAC) to study the allocation and use of water in the western states. The reports from the Commission, several of which discuss the population–water interaction, are available at this website.

4. **http://www.umass.edu/newsoffice/press/98/0422cli.html**

In 1998 a very important study on global climate change was published in Nature by two scientists at the University of Massachusetts and one at the University of Arizona. By using tree-ring data they were able to estimate that 1990, 1995, and 1997 were the three warmest years on the planet anytime during the last 600 years. Visit the paper and the authors at this website.

5. **http://www.ucsusa.org/resources/index.html**

The Union of Concerned Scientists (UCS) is an organization devoted to presenting what they believe is the most objective and trustworthy information about global environmental issues. There are many good resources here, but click first on "Global Resources" and then on "Population Growth" and go from there.

Part Five
Using the Demographic Perspective

One of my major purposes in this book has been to provide you with a demographic perspective that will help you understand the past, cope with the present, and contemplate the future. Population growth has been, is, and will continue to be an integral part of your world, and the more you know about it (realizing the myriad ways in which it can influence social change), the better you can handle it. Most people, at least in the developed nations, feel that indeed we should do more than just cope; we should actively work toward reducing population growth. To do so demands that you develop a population policy by putting your demographic perspective into action to influence the future course of events.

Those of you who decide to have no children or only a few, as well as those who decide to have several, are putting a personal population policy of sorts into practice. However, in the next chapter I will be discussing the large picture—the population policies (both explicit and implicit) of governments that affect where and how populations grow. What can be done to influence or directly induce a change in the rate of growth or the distribution of a population? What has been done in the past? Those are two of the questions that I will be examining in Chapter 14, and I hope that uppermost in your mind as you read is the question: What would I do?

Formulating and implementing population policies, however, are not the only (nor necessarily even the most frequent) uses to which demographic perspectives are put. People engaged in private business, in public social planning, and in the broader realm of politics have increasingly been using demographics to understand who their clients, constituents, or prospective customers are, where they live, and how they might be reached. These applications of population perspectives and methods may do more to educate you about the issues of population growth and change than all the other awareness campaigns put together, because they help make the direct connection between your life and the seemingly anonymous demographic events swirling around you. In Chapter 15 (the last chapter) I show you how businesses and other organizations apply demography to help illuminate and solve their problems.

CHAPTER 14
Population Policy

What can you do about it if you think the population is growing too rapidly or too slowly? In this chapter your demographic perspective will be put to work trying to find answers by examining how others have tried to influence demographic events. This is an important use to which a demographic perspective can be put—employing your understanding of the causes and consequences of population growth to improve the human condition, including your own.

I will begin with a discussion of what a policy is, so that you will appreciate the fact that the complexity of policy making leads to an almost inevitable lack of agreement about what should be done. The disagreement has generated several different types of policies designed to pursue the basic goals of either retarding growth, promoting growth, or maintaining growth. As the chapter proceeds, I will introduce you to some of these different policy orientations and also to specific proposals that have been or could be implemented to achieve the desired goals. In particular, I will review the outcome of the 1994 United Nations International Conference on Population and Development (the Cairo Conference). First, though, what is a policy?

What Is a Policy?

A policy is a formalized set of procedures designed to guide behavior. Its purpose is either to maintain consistency in behavior or to alter behavior, in order to achieve a specified goal. **Population policy** represents a strategy for achieving a particular pattern of population change. The strategy may consist of only one specific component—a single-purpose goal—such as to lower the crude birth rate by 10 points during a 5-year period. Or it may be multifaceted, such as an attempt to improve the reproductive health of women. Naturally, in both cases the objective requires a policy only if there is some indication that the goal may not be achieved unless a policy is implemented. Note that in the foregoing situations I am referring to a direct population policy, one aimed specifically at altering demographic behavior. There are also indirect population policies, which are not necessarily designed to influence population changes but wind up doing so anyway. I will return later in the chapter to a brief discussion of indirect policies and will focus here only on the direct ones.

I have outlined the basic elements of analyzing policy formulations in Figure 14.1. Your first step is to assess carefully the current demographic situation, a technique that I have tried to illustrate in virtually every chapter of this book. This is obviously a crucial task, since you have to know where you are now if you expect to chart a course for the future. Assuming that you can accurately measure (or even carefully estimate) the present situation, your next step is to analyze what the future would bring if society were left to its own devices.

Assessing the Future

We humans have been preoccupied with looking into the future for centuries, since knowledge of what is coming gives us the power to prepare for it or possibly to change it. Predicting or forecasting the future, though, is an almost impossible task

Figure 14.1   Formulating a Population Policy

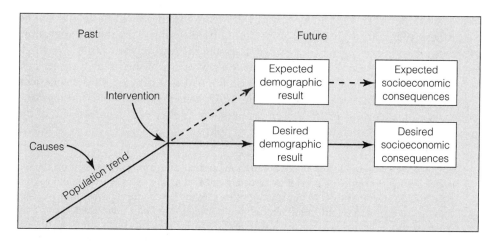

Note: The formulation of a population policy requires an assessment of current trends (requiring an understanding of the causes of population change) and an evaluation of the expected consequences of current trends. If the expected consequences differ from the desired result, then a policy may be implemented to alter the course of demographic events.
Source: Kingsley Davis, 1975, "Demographic Reality and Policy in Nepal's Future," Workshop Conference on Population, Family Planning, and Development in Nepal, University of California, Berkeley, p. 2. Used by permission.

except in very general terms, because of the amazing complexity of the social and physical world. As a result, demographers rarely stretch their necks out to predict the future but rely instead on projections, which are statements about what would happen under a given, specified set of conditions. Indeed, throughout the book I have made frequent references to projections (see especially Chapter 8). For example, when I say that in the late 1990s the world's population was growing so fast that it would double in size in 49 years, that is a projection. In other words, if neither birth rates nor death rates changed from their late 1990s levels, twice as many people would be living in the world in the middle of the 21st century as were alive at the end of the 20th. I am not saying that there necessarily will be that many people; there might be fewer (if fertility continues to decline or if mortality rises) or there might be more (if mortality declines further without a compensatory drop in fertility); I am only indicating the consequences of current rates.

Projections also permit you to ask questions about other possible paths into the future if conditions change. In the previous two chapters I emphasized the precarious relationship that exists in the world between population and resources. Whether the population is expected to grow by a lot or by a little makes a very big difference in the future impact of humans on the environment. Thus by looking at an array of alternatives, you can lay the groundwork for anticipating various courses of events.

It is at this stage, in deciding what the future might look like, that most attempts at policy making (whether in the area of population or anything else) break down.

Because most of us have different insights and perceptions of the world (even though differences may be small), it is sometimes impossible to reach a consensus about anything much beyond how many 19-year-olds there will be in the United States next year, and even that is complicated by differing estimates of migration. However, if you try to avoid the issue of assessing the future, you void the possibility of implementing a policy because you won't have established a case for it. This is what happened to the U.S. Commission on Population Growth and the American Future, a body legislated into existence in 1969, which presented its findings and policy recommendations in 1972. The commission sidestepped an assessment of the probable future and thereby lost most of its political clout:

> For our part, it is enough to make population, and all that it means, explicit on the national agenda, to signal its impact on our national life, to sort out the issues, and to propose how to start toward a better state of affairs. By its very nature, population is a continuing concern and should receive continuing attentions. *Later generations, and later commissions will be able to see the right path further into the future.* In any case, no generation needs to know the ultimate goal or the final means, only the direction in which they will be found [Commission on Population Growth and the American Future 1972:8]. [Emphasis is mine.]

If a policy is to be implemented with the expectation of producing predictable results (and controllable side effects), you must get a bead on the most likely future. If you do, then not only is your policy well grounded, but you are in a position to decide later if your assessment was wrong, and then either abandon your policy as incorrect or alter it in accordance with your revised estimate of where your society is going.

Establishing a Goal

Once you have an idea of what the future may be, or at least a range of reasonable alternatives, you are in a position to compare that with what you aspire to in demographic and social terms. Establishing a goal is not an easy task, and it grows more difficult as the number of people involved in setting the goals grows. As a result, goals are usually general and idealistic in nature; those related to population issues might include improving the standard of living, reducing economic inequalities, promoting gender equality, eliminating hunger and racial/ethnic tension, reducing environmental degradation, preserving international peace, and increasing personal freedom. For example, in Cairo in 1994, delegates to the United Nations International Conference on Population and Development agreed to a set of 15 principles embracing the concept that population-related goals and policies are integral parts of cultural, economic, and social development, all of which are aimed at improving the quality of life of all people. At the end of the chapter I will discuss this conference in more detail.

The demographic future is assessed primarily with an eye toward determining whether projected demographic trends will enhance or detract from the ability to achieve other broad goals. In other words, population control is rarely an end in itself but rather an "implementing strategy" that helps to achieve other goals. This is analogous to my comment in Chapter 6, that, for individuals, having children is typically a means to other ends, rather than being intrinsically a desired goal. Thus when you look at the future course of demographic events, you might ask whether projected population growth will undermine the ability of an economy to develop. Will projected shifts in the age/sex distribution affect the ability of an economy to provide jobs, thereby leading to lower incomes or a greater welfare burden? Will projected growth and urbanization in one ethnic group or another lead to greater intergroup tension and hostility? Will the projected growth of the population lead to a catastrophic economic–demographic collapse that will drastically restructure world politics? Will declining fertility lead to a huge unmanageable population of older people?

Because population policies are only a means to one or more of these other ends, it is easy for population policies to be "hijacked" by the proponents or opponents of these other goals (Basu 1997), but whatever your goal, if it is discrepant with the projected future, you can use demographic knowledge to propose specific policies to avert unhappy consequences. Your work does not end there, of course, because once your policies are implemented, you have to continually evaluate them to see whether they are accomplishing what you had hoped and to make sure they are not producing undesirable side effects.

Who Needs a Population Policy?

The process of policy formulation outlined in Figure 14.1 assumes a country or a group of people oriented toward the future and anticipating change. If a population or its government leadership is "traditional" and expects tomorrow to be just like today, policies aimed at altering behavior are not likely to be adopted. If demographic policies exist at all in such a country, they may have as a goal the maintenance of the status quo and are probably coercively pronatalist; that is, they might forbid divorce or abortion, impede the progress of women, and so forth. In general, they would probably discourage the kind of innovative behavior that might lead to demographic change. To be sure, that type of attitude may even be associated with poor perceptions of reality which, in their turn, help to preserve the status quo. Randall (1996) provides an interesting example of a traditionally nomadic population in Mali in sub-Saharan Africa. Their fertility levels are lower than those of sedentary farmers in the area in which they reside, and demographic analysis has revealed that this is due to the influence of monogamous marriages in a society in which husbands are substantially older than wives, and where high adult mortality produces a marriage squeeze that reduces the exposure of women to reproduction. However, the local perception is that women have a very high rate of sub-fecundity and that this is a problem that needs to be solved. Any policy implemented on the basis of that perception would likely have very different consequences than a policy based on demographic reality.

It is also easy to misperceive the important role that demography plays in the modern world. It is, as Westoff (1997) has noted, a large problem with low visibility. As you can appreciate from having read this book, there are few corners of the earth and few aspects of human existence that are not influenced by demography, but the demographic roots of social and technological change are not always recognized for what they are, and so their influence may be denigrated or ignored (Smail 1997).

Who Has a Population Policy?

No country in the world today can completely ignore the issue of population policy. This is partly because population growth is a world renown issue, even if its visibility may be lower than it should be, and partly because the United Nations Population Division regularly queries each country of the world about its policy position on population growth. The first survey was conducted in 1974, and it has been repeated every few years since. The number of governments saying that their policy is to slow down the rate of population growth has been steadily growing over time. Looking specifically at government attitudes in the 1990s toward the birth rate, Figure 14.2 shows you the worldwide distribution of countries according to whether their government thinks the birth rate is too high, too low, or satisfactory. Governments that think the birth rate is too high are likely to have a policy to try to lower it. Fortunately, only a few of the governments saying that the birth rate is too low are trying actively to raise it.

Of the 180 countries plotted in Figure 14.2, 44 percent have governments that think the birth rate is too high, whereas 12 percent think it is too low, and the remainder are satisfied with the current situation. Within that latter group most are countries that already have achieved low fertility (such as the United States, Canada, and most European nations), but there are also some countries in Africa and Asia where the government is satisfied with a high birth rate. The government of Somalia, for example, says that it is satisfied with the current birth rate—which translates into an average of 7.0 children per woman. Saudi Arabia's government is similarly satisfied with its birth rate which produces an average of 6.4 children per woman, and the Laotian government seems unconcerned about its birth rate which generates an average of 5.9 children per woman (Population Reference Bureau 1998).

If a government is so inclined, there are several different basic policy orientations that can be adopted to either close the gap or maintain the fit between goals and projections. These orientations include (1) retarding growth, (2) promoting growth, and (3) maintaining growth. Each policy orientation offers a wide variety of specific means that can be implemented to achieve the desired kind of demographic future. The most controversial and widespread of those three policy orientations is retarding growth; I will discuss it first and in greater depth than the others.

Retarding Growth

The major policies advocated for (and in) most areas of the world are those that will slow down population growth. In previous chapters I have indicated that since no

Figure 14.2 Governments' View of Their Country's Birth Rate

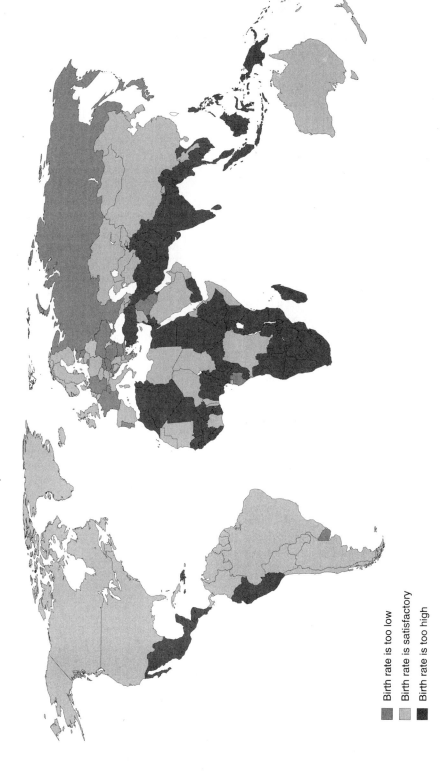

Birth rate is too low

Birth rate is satisfactory

Birth rate is too high

Source: Adapted by the author from data from United Nations Population Division 1992 questionnaire, courtesy of Population Reference Bureau, Inc., Washington, D.C.

WHAT IS THE POPULATION POLICY OF THE UNITED STATES?

In response to the United Nations surveys on population policies, the United States has consistently indicated that the government is satisfied with current growth rates and fertility rates and has no plans for intervention. From that fact might flow the facile conclusion that there is no population policy in the United States. On the contrary, the United States has an implicit domestic population policy and an explicit international population policy.

Domestic Policy   Implicitly, the United States supports the ability of Americans to maintain low fertility through the availability of contraceptives, voluntary sterilization, and abortion. The government also encourages legal immigration while discouraging illegal immigration. Explicitly, the closest that the country has come to a population policy was under President Nixon, who in 1969 proposed a commission (which Congress thereafter legislated into existence) to examine the growth of population in the United States, to assess the impact of that growth on the American future, and "to make recommendations on how the nation can best cope with that impact" (Commission on Population Growth and the American Future 1972:preface). Before producing its findings in 1972, the commission spent 3 years gathering information, listening to the public and to experts, and deliberating. Its essential conclusion was that "no substantial benefits would result from continued growth of the nation's population" (Commission, 1972:1), and its principal recommendation was that the United States pursue a policy of population stabilization (that is, of ZPG later—see Chapter 8). But how should that be accomplished?

The commission made a number of specific recommendations that would help to stabilize population while also maintaining freedom of individual choice. I have listed three of the most important recommendations here and have indicated what their fate was as of the late 1990s.

1. States ought to ratify the Equal Rights Amendment to the Constitution so that women would have the same legal rights as males, thereby potentially expanding their ability to be financially and socially independent. The ERA was initially passed by Congress in 1972 and required approval by 38 states before it could become part of the U.S. Constitution. After 10 years without ratification, the ERA expired in 1982, only to be reintroduced into the House and then defeated there in 1983. President Reagan argued that a constitutional amendment was not needed to fight sexual bias, and a majority of congressional members apparently agreed.

2. Abortion should be legalized and performed on request by duly licensed physicians under conditions of medical safety. This change in American law did take place in January 1973 when a U.S. Supreme Court decision (Roe v. Wade) essentially legalized abortion on demand during the first 3 months of pregnancy. In 1977, however, Congress passed a law forbidding the use of federal funds for abortion, whereas in 1983 the Supreme Court invalidated a series of city and state laws that had been designed to make abortions more difficult to obtain. The battle between "pro-choice" and "pro-life" groups has since 1984 escalated with the bombing of several abortion clinics in the United States and the murder of a physician who performed abortions. In 1986, the U.S. Supreme Court struck down a Pennsylvania law that had restricted access to abortion and, in doing so, reaffirmed the Court's 1973 decision. Yet, in 1989, the Court upheld a number of Missouri state law prohibitions on abortion, signaling a shift in the other direction. Roe v. Wade was put to the test again in 1992 when the Supreme Court upheld another Pennsylvania statute that added hurdles to the process of obtaining an abortion in that State. However, the ruling narrowly applied only to that case, and signaled

that the court did not yet have the critical fifth vote needed to overturn Roe v. Wade. Abortion remains legal in the United States despite the efforts of anti-abortion groups to limit access to such services.

3. Congress should immediately consider the serious situation of illegal immigration and pass legislation to impose civil and criminal sanctions on employers of illegal border-crossers or aliens in an immigration status in which employment is not authorized. Legislation that would accomplish this was introduced into Congress in the early 1980s and, although an early version was defeated in Congress in 1984, a modified version was passed in 1986. Known as the Immigration Reform and Control Act of 1986, the bill provided for amnesty to long-term undocumented workers, along with employer sanctions for hiring undocumented workers.

International Policy   The United States has played a key role in explicitly encouraging governments of developing nations to slow down the rate of population growth, and it has provided a great deal of money for the establishment and maintenance of family planning programs all over the world. Donaldson (1990) has chronicled the growing American consciousness from the 1940s through the 1960s that rapid population growth in less developed nations might not be good either for those nations specifically or for the United States more generally. When President Kennedy established the Agency for International Development (AID) within the Department of State in 1961, there was an early recognition that population was an important factor in development. By 1967 that idea bore fruit when Congress earmarked AID funds for the purpose of providing population assistance along with economic assistance. For the next 18 years the United States played a crucial part in helping to slow down the rate of population growth in the world—an effort that may have reduced the world population by nearly a half-billion people compared with the

growth that would have occurred in the absence of organized family planning efforts in developing countries (Bongaarts, Mauldin, and Phillips 1990).

Things changed in 1985, however, as a result of Reagan administration policies that were carried over into the Bush administration. The opposition of both administrations to abortion created controversy in both the implicit domestic policy, because of the push to reverse Roe v. Wade, and in the international policy, because in 1985 the opposition to abortions in China was used as a pretext for cutting off all United States government funding for the United Nations Population Fund (see Crane and Finkle 1989 for a full discussion). The UNFPA has provided long-term support to the Chinese government for family planning activities, but despite protestations that the UNFPA has no role in the allegedly coercive abortions in China, the United States did not restore UNFPA funding until after Clinton was elected President in 1992.

The hiatus in U.S. funding for the UNFPA did not mean the end of population assistance, of course. Indeed, AID has remained the single, largest source of international support for family planning programs. Such assistance is typically provided through other organizations working in various parts of the world. AID also funds important programs in training and research activities. Furthermore, other countries, especially Germany, Japan, and Norway, have picked up some of the slack left by the United States.

A majority of Americans are in favor of providing aid to help slow the growth of population in developing nations, according to results from the Gallup Poll (Newport and Saad 1992). However, people tend to be fairly ignorant of the causes of the rapid population growth. Lack of sex education and lack of access to birth control are the most commonly given reasons by Americans for rapid population growth. Ignorance of the problem may also help explain why, after Republicans gained control of Congress in the mid-1990s, a movement began in Congress to cut back on funding of population programs in developing countries.

nation's population can grow indefinitely, a policy to slow growth may produce more desirable results than letting demographic, social, and economic events take their respective courses. This seems to be particularly true if economic development is also a desired goal. Growth can be retarded by attempting to manipulate one or more of the three population processes of mortality, migration, and fertility.

Influencing Mortality

Although no nation advocates it, it is possible for population growth to be slowed by returning to former high levels of mortality. It has been argued, at least facetiously, that since the dramatic drop in mortality in less developed countries precipitated current world concern about population growth, if fertility levels do not come down, mortality levels should be "allowed" to rise. The "lifeboat ethic" and "triage" are two orientations that would retard growth by selectively raising death rates.

The lifeboat ethic is based on the premise that since a lifeboat holds only so many people and any more than that will cause the whole boat to sink, only those with a reasonable chance to survive (those with low fertility) should be allowed into the lifeboat. Withholding food and medical supplies could drastically raise the death rates in less developed nations and thus provide a longer voyage for those wealthier nations already riding in the lifeboat. A closely related doctrine is that of *triage*, which is the French word for sorting or picking, and refers to an army hospital practice of sorting the wounded into three groups—those who are in sufficiently good shape that they can survive without immediate treatment, those who will survive if they are treated without delay, and those "basket cases" who will die regardless of what treatment might be applied. As with the lifeboat ethic, it translates into selectivity in providing food and economic aid should the day come when supplies of each are far less than demand. It means sending aid only to those countries that show promise of being able to bring their rates of population growth under control and abandoning those nations that are not likely to improve.

Most people probably share the opinion that raising mortality is better grist for science fiction than for population policy. However, there is another aspect of mortality control that is often an integral part of policies aimed at controlling growth—lowering mortality. Despite a reduction in mortality in most countries of the world, death rates are still commonly higher in rapidly growing, less developed nations than in the more highly advanced countries. Thus most governments find it difficult to commit time and money to any policy aimed at reducing growth without also paying attention to lowering mortality. Access to health care is routinely viewed as a fundamental human right, but at the same time lowering mortality is a policy that exacerbates the problems of growth. Although there is no necessary link between lowering fertility and lowering mortality, the two programs are increasingly linked administratively because of the emphasis on the total spectrum of reproductive health from improved health for mothers and their children, and increased choice with respect to methods for fertility limitation (and, in some cases, fertility enhancement).

I discussed in Chapter 6 the intuitively appealing idea that fertility will not drop until parents are sure that their children will survive, so the beginning point of any policy to lower fertility should be to lower infant and childhood mortality. If women actually do bear more children than they want just to overcome high mortality, then it is reasonable to propose that lowering infant mortality will lower the need for children. This would hopefully be translated into a demand for birth control. Although that argument certainly makes sense in general, there is very little consistent evidence to support it at the individual family level (Ahn and Shariff 1994; Preston 1978).

The difficulty with lowering infant mortality as a means of ultimately lowering fertility is that it assumes that people are having more children than they actually want simply to overcome the effect of high mortality. I discussed the dubiousness of that proposition in Chapter 6. The data suggest a more complex relationship. Life expectancy always improves in the context of broader social and economic changes that affect much more than just the survivability of children. All of these changes, including the drop in infant mortality rates, influence the attitudes of couples regarding their family size (see Ross and Frankenberg 1993).

The consequence is that a further decline in infant mortality could have the unintended (but certainly predictable) effect of raising, not lowering, the rate of population growth. Indeed, this is what happened in one of India's largest states between 1951 and 1975 when infant and childhood mortality declined noticeably. Even with an official family planning program, the fertility levels showed little sign of change (Srinivasan, Reddy, and Raju 1978). Data for Guatemala suggest that it may take at least two generations for the reduction in infant and child mortality to influence the level of fertility (Pebley, Delgado, and Brinemann 1979). In fact, it may be that the relationship works the other way around—that the best way to lower infant mortality is to develop a comprehensive family planning program. Couples throughout the world have known for a long time that the health of both mothers and infants is improved by greater spacing between births (Gadalla 1978), and research in the Gambia has shown that women there are not shy about using modern contraceptive techniques to regulate the spacing of children in order to achieve higher rather than lower fertility (Bledsoe, et al. 1994).

It is probably best to assume that reducing mortality, especially infant and childhood mortality, is always good in its own right, and that in so doing the resulting higher rate of population growth will have to be adjusted for in some way or another. Perhaps migration can be manipulated.

Influencing Migration

Migration is the most easily controlled of the three population processes, at least in theory. You cannot legislate against death (except for laws prohibiting homicide or suicide), and few have dared try to legislate directly against babies. But you can set up legal and even physical barriers to migration. In practice, of course, controlling migration is difficult if people are highly motivated to move. Nonetheless, local and national governments often try. In the following sections I will first look at efforts to control international migration, and then will examine policies directed at internal migration.

International Migration   Immigration may be the sincerest form of flattery, but few countries encourage it. Migration between countries is fraught with the potential for conflict among people of different cultural backgrounds and, as technology has made transportation and thus migration easier, policies to deal with international migration have come into sharper focus. Nowhere is this more true than in the United States and Canada, where international migration has been a way of life for more than two centuries and toward which a large fraction of the annual volume of the world's international migrants still heads. I will focus on policies in the United States but, as I point out in this discussion, the two countries have tended to implement and change their policies in tandem with, and sometimes in reaction to, each other.

Historical Background in the United States   Prior to World War I there were few restrictions on migration into the United States, so the number of migrants was determined more by the desire of people to come than anything else. Particularly important as a stimulus to migration, of course, was the drop in the death rate in Europe during the 19th century, which launched a long period of population growth. Free migration from Europe to the temperate zones of the world—especially North and South America and Oceania—represents one of the most significant movements of people across international boundaries in history. The social, cultural, economic, and demographic impacts of this migration were enormous (Davis 1974).

The opening up of new land in the United States coincided with a variety of political and economic problems in Europe and that helped to generate a great deal of labor migration to the United States in the 19th century, particularly after the end of the Civil War. The United States responded with its first immigration restrictions in the form of the Immigration Act of 1882. This law levied a head tax of 50 cents on each immigrant and blocked the entry of idiots, lunatics, convicts, and persons likely to become public charges (U.S. Immigration and Naturalization Service 1991).

Meanwhile, the discovery of gold in California had prompted a demand for labor—for railroad building and farming—that had been met in part by the migration of indentured Chinese laborers. However, in 1869, after the completion of the transcontinental railroad, American workers came west more readily and the Chinese showed up in the east on several occasions as strikebreakers. Resentment against the Chinese built to the point that in 1882 Congress was willing to break a recently signed treaty with China and suspend Chinese immigration for 10 years (Stephenson 1964). The law was challenged unsuccessfully in the courts and, over time, restrictions on the Chinese, even those residing in the United States, were tightened (indeed, the Chinese Exclusion Acts, as they were called, that were passed in 1882 were not repealed until 1943). The exclusion of the Chinese led to an increase in the 1880s and 1890s of Japanese immigration, but by the turn of the century hostility was building against them, too (the Japanese Exclusion Act was passed in 1924), and against several other immigrant groups.

I discussed in Chapter 7 the fact that by the late 19th century the ethnic distribution of immigrants to North America had shifted away from the predominance of northern and western Europeans toward an increase in migration from southern Europe, especially Italy. In 1890, 86 percent of all foreign-born persons in the United States were of European origin, but only 2 percent were from southern Europe—almost exclusively Italy. Only 30 years later, in 1920, it was still true that 86 percent of the foreign-born were Europeans, but 14 percent were southern Europeans—a sevenfold increase (U.S. Bureau of the Census 1975).

The immigrant processing center in New York City was moved to Ellis Island in 1892 to aid in screening people entering the United States from foreign countries (Stephenson 1964), since the changing mix of ethnicity had led to public demands for greater control over who could enter the country. In 1891 Congress legislated that aliens were not to be allowed into the country if they suffered from "a loathsome or dangerous contagious disease" (Auerbach 1961:5) or if they were criminals. Insanity was added to the unacceptable list in 1903, and tuberculosis joined it in 1907. In 1917 a highly controversial provision was passed that established a literacy requirement, thus excluding aliens over age 16 who were unable to read.

Recognition of the problems created by the still relatively free migration led to a new era of restrictions right after World War I. Europe was unsettled and in the midst of economic chaos, and there was a widespread belief that "millions of war-torn Europeans were about to descend on the United States—a veritable flood which would completely subvert the traditional American way of life" (Divine 1957:6). The United States and Canada both passed restrictive legislation (Boyd 1976), in step with the eugenics movement which had gained popularity throughout Europe and North America in the 1920s. As it was applied to migration, the ideology was ethnic purity, and the sentiment of the time about migrants is perhaps best expressed as "not too tired, not too poor, and not too many."

In 1921 Congress passed the first act in American history to put a numeric limit on immigrants. The Quota Law of 1921 "limited the number of aliens of any nationality to three percent of foreign-born persons of that nationality who lived in the U.S. in 1910" (Auerbach 1961:9). For example, in 1910 there were 11,498 people in the United States who had been born in Bulgaria (U.S. Bureau of the Census 1975), so 3 percent of that number, or 345, would be permitted to enter each year from Bulgaria. Under the law about 350,000 people could enter the United States each year as quota immigrants, although close relatives of American citizens and people in certain professions (for example, artists, nurses, professors, and domestic servants) were not affected by the quotas. The law of 1921 remained in effect only until 1924, when it was replaced by the Immigration Quota Act. The 1924 law was even more restrictive than that of 1921, because public debate over immigration had unfortunately led to a popularization of racist theories claiming that "Nordics [people from northwestern Europe] were genetically superior to others" (Divine 1957:14).

Initially, the 1924 law established quotas on the basis of 2 percent of the foreign-born in the 1890 census, when about 70 percent of the foreign-born were from northwestern Europe, as opposed to using the 1910 census (used in the 1921 law) when only about 50 percent were from northwestern Europe (Divine, 1957). To avoid the charge that the immigration law was deliberately discriminatory, though, a new quota system—the National Origins Quota—was adopted in 1929 (Auerbach 1961). This was a complex scheme in which a special Quota Board took the percentage of each nationality group in the United States in 1790 (the first U.S. Census) and then traced "the additions to that number of subsequent immigration" (Divine 1957:28). The task was not an easy one, since by and large the necessary data did not exist, so a lot of arbitrary assumptions and questionable estimates were made in the process . Once the national origins restriction had been established, the actual number of immigrants allowed from each country each year was calculated as a proportion of 150,000, which was established as the maximum number of all immigrants. Thus if 60 percent of the population was of English origin, then 60 percent of the 150,000 immigrants, or 90,000, could be from England. The number turned out to be slightly more than 150,000, because every country was allowed a minimum of 100 visas. Furthermore, close relatives of American citizens continued to be exempt from the quotas. In Canada, a similar immigration act was passed in 1927 (Boyd 1976). Congress, of course, retained the ability to override those quotas if the need arose, as it did during and after World War II when refugees from Europe were accommodated.

In 1952, in the middle of the anti-Communist McCarthy era, another attempt was made in the United States to control immigration by increasing the "compatibility" of migrants and the established North American society. The McCarran-Walter Act, or the Immigration and Naturalization Act of 1952, retained the system of national origin quotas and added to it a system of preferences based largely on occupation (Keely 1971). The McCarran-Walter Act permitted up to 50 percent of the visas from each country to be taken by highly skilled persons whose services were urgently needed. Relatives of American citizens were ranked next, followed by people with no salable skills and no relatives who were citizens of the United States. Thus even from those countries with an advantage according to the national origins quota system, the freedom of migration into the United States was severely restricted. The same was true in Canada, which passed similar legislation in the same year.

Canada, I should note, had at least two reasons for echoing the immigration policies of the United States. In the first place, it shares much of the sociocultural heritage of the United States, and second, of course, it shares a border with the United States. This cultural similarity and close proximity would have left Canada inundated with migrants excluded from the United States had Canada not passed its own restrictive laws.

Contemporary American Immigration Policy   In the 1960s the ethnically discriminatory aspects of North America's immigration policy ended, but its restrictive aspects were maintained. The Immigration Act of 1965 ended the nearly half-century of national origins as the principal determinant of who could enter this country from non-Western Hemisphere nations. Again, related changes occurred in Canada in 1962.

Although the criterion of national origins is gone, restrictions on the numbers of immigrants remain, including a limit on immigrants from Western Hemisphere as well as non-Western Hemisphere nations. The 1965 act set an annual limit of 120,000 persons from the Western Hemisphere; in addition, there is a maximum limit of 170,000 from non-Western Hemisphere nations, with no more than 20,000 allowed from any single country. Congress again retained the right to grant exemptions from those limits for any special groups, such as the Vietnamese refugees. A system of preference was retained but modified to give first crack at immigration to relatives of American citizens. Parents of U.S. citizens could migrate regardless of the quota. In addition, a certification by the Labor Department is now required for occupational preference applicants to establish that their skills are required in the United States. In 1976 the law was amended so that parents of U.S. citizens had highest priority only if their child was at least 21 years old. The intent of that change was to eliminate the fairly frequent ploy of a pregnant woman entering the country illegally, bearing her child in the United States (the child then being a U.S. citizen), and then applying for citizenship on the basis of being a parent of a U.S. citizen.

It is almost certain that the restriction of immigration in the Western Hemisphere contributed to the number of people who have become undocumented workers or illegal immigrants to the United States, especially from Mexico. The requirement of labor certification, as well as a numeric quota on Western Hemisphere nations (dominated by Mexico), came at a time when the population of Mexico was growing much more rapidly than the Mexican economy could handle, especially in terms of jobs (as I discussed in Chapter 12). Thus there are distinct push factors in Mexico. Migration of unskilled labor to the United States from Mexico had been an alternative from the 1950s through the 1960s as part of the bracero program, which California growers had pushed as a means of obtaining cheap labor. In the 1960s, however, Mexican-Americans lobbied successfully for a halt to the bracero program, and that followed on the heels of the more restrictive labor requirements. Thus it is no surprise that illegal migration from Mexico has been high ever since.

As undocumented migration from Mexico continued, the cry was often heard that the southern border of the United States was "out of control," and it was widely believed that undocumented workers were taking jobs from U.S. citizens and were draining the welfare system. Typically, reality was less dramatic. The number of undocumented workers in the United States is less than many people imagined (as I discussed in Chapter 7), yet the widespread public perception of negative consequences of undocumented migration helped push Congress to pass the 1986 Immigration Reform and Control Act (IRCA). From the standpoint of the undocumented immigrant, this was a "good news/bad news" piece of legislation. The good news was that the law offered "amnesty"—relief from the threat of deportation and the prospect of legal resident status for undocumented workers who had been living continuously in the United States since before January 1, 1982. The bad news was that in order to curtail new workers from entering the country without documentation, it is now unlawful for an employer to knowingly hire such a person. The teeth of the law include fines for employers beginning at $250, with the prospect of much larger fines and even jail sentences for employers who are repeatedly found to be hiring undocumented workers. Notice that this law was aimed entirely at curtailing illegal immigration and had no impact on the pattern of legal migration.

In the 1990s the United States also spent millions of dollars on new border fences and walls, designed to make it harder for unauthorized people to cross from Mexico into the United States, and to make it easier to apprehend people who tried to do so. The impact has been to slow down the rate of return to Mexico of people who do make it successfully to the United States, although it is difficult to tell if the actual number of migrants is much different than it would otherwise have been (Loaeza Tovar and Martin 1997).

Over time the immigration policy of the United States, and to a lesser extent Canada, has been that of gatekeeping—keeping out those people considered undesirable. Those people generally considered most desirable are professionals: physicians, scientists, engineers, teachers, lawyers, and others. Indeed, one interesting aspect of the abolition of the national origins criterion is that the proportion of all legal immigrants who are professionals has actually increased. A major reason for this seems to be a type of "brain drain" from countries whose people were previously excluded, or nearly so, from the United States.

In 1990 the United States opened the door to legal immigration a little wider, as I mentioned in Chapter 7. The 1990 amendments to the Immigration Act increased the total number of legal immigrants to be accepted each year, and although the main thrust of the changes was to enhance family reunification, the new law specifically reserved 140,000 visas per year for immigrants with special occupational skills and an additional 10,000 visas per year for individuals with at least $500,000 to invest in a new business that creates at least ten jobs.

The increase in immigration, in conjunction with the low fertility rates in the United States, has meant that immigrants have been an increasingly important and visible part of the demographic picture. Bouvier (1992) has estimated that, at current trends, about half of the population growth between 1990 and 2050 would be due to immigrants who arrive after 1990 and their descendants. These are among the reasons why a backlash against immigrants erupted in the early 1990s (Huber and Espenshade 1997), exemplified by the passage of Proposition 187 in California (restricting the access of undocumented immigrants to tax-supported health and education services). In the same general vein, the federally sponsored Commission on Immigration Reform recommended in 1995 that the United States should cut legal immigration by a third, dropping it to 550,000 per year, back to the level of 1984.

Other Countries   Labor migration has become commonplace in the world today, as I discussed in Chapter 7, but countries still accept guest workers on the naive assumption that they can be sent home when the need arises. This almost never turns out to be the case, regardless of whether the host country is in the "north" or the "south." Germany is dealing with immigrants from eastern Europe and Turkey (and other less developed nations); the United Kingdom accepts about 50,000 immigrants a year, almost exclusively from former colonies, especially India and Pakistan (Coleman 1995). Even Italy—long a source of migrants itself—was forced in 1986 to enact its first law controlling immigration into the country (Martin 1993). However, to remind you that this is not just an issue with more developed nations, Pakistan announced in 1994 that it was going to expel nearly 1 million illegal immigrants—largely from India, Afghanistan, Bangladesh, Burma, and Iran (Wall Street Journal 1994).

Out-migration is obviously not a viable solution to relieving worldwide population pressure. Although it was suggested as a possibility at the time Americans first landed on the moon (raising speculation about establishing colonies in space), the cost and problems of social organization and technological development are too immense to give it more than a passing thought.

In earlier chapters I discussed out-migration as a common response to population growth. Europeans migrated to the Western Hemisphere, Chinese have migrated globally, Indians have migrated throughout Asia and Africa, Jamaicans have headed for Great Britain, and Mexicans continue to migrate to the United States. But in none of these instances or any others has out-migration been a response to deliberate policy. Why not? Because, as you saw in Chapter 7, migration is a highly selective process—selective often of better-educated, highly motivated people, and those are generally not the people that a society wants to lose. Nonetheless, it is no longer a complete loss to a sending country when people leave. On the contrary, increasing numbers of governments are recognizing the important economic value of the remittances that migrants send home. So, they may not encourage migration in an active way, but neither will they necessarily try to stop it (Zlotnik 1994).

History is replete with examples of forced out-migration, but generally the purpose was not to control population growth but rather to control political or religious dissent. One aspect of improving the human condition as conceptualized in the Western world is, in fact, to minimize people's fear that they will be wantonly uprooted from their homes and forced elsewhere. As a direct population policy, such a possibility seems remote.

Migration need not be forced, of course; it need only be encouraged in indirect ways. This could be accomplished by raising taxes, reducing or cutting off services to residents or businesses in particular areas, and so forth. But here again such repressive measures are likely to be responded to most readily by those with the most marketable skills who can most easily move elsewhere.

Internal Migration   Despite the political importance of international migration, at least an equally important issue is the internal redistribution of populations. The question is, how can a government (at any level) influence individual choices to achieve the most desirable pattern of population distribution through migration? In the late 1960s and early 1970s the United States was being sensitized to the issue of world population growth and at the same time the Baby Boom generation was making itself felt as a factor in American demographic change. Many communities responded by passing laws and zoning restrictions designed to limit the influx of people. One of the earliest of these limits to growth policies was established in Petaluma, California in 1972. An ordinance passed that year limited construction of large housing developments and apartment houses to 500 units a year. Although challenged in court, the ordinance was ultimately upheld by the U.S. Supreme Court. By the 1980s, however, the ordinance had become largely symbolic because the population pressure was never as great as planners had feared. Of course, it may be that the publicity over the area's concern about growth diverted potential buyers toward other communities. That does not mean that limits to growth did not continue to be a concern. Throughout the United States, the increase in the number of two-earner (and thus two-car) families since the 1970s has meant that almost every American community has been experiencing an uncomfortable rise in the volume of traffic. To the average person, more traffic means more people, and that calls for limits to growth, or at least better control over growth. Enter, then, the Growth Management Act of 1985 in Florida, which was designed to guarantee that population does not grow faster than the physical means to support it. That is but one of literally hundreds of local plans that have been drawn throughout the United States to curtail or control population growth.

The state of Oregon had also tried to isolate itself from population growth in the 1970s. In 1971, when Tom McCall was governor of that state, he said, "Visit us often, but for heaven's sake, don't move here to live" (Vicker 1982). McCall later claimed that the comment was made in jest, but state policies in the 1970s did, in fact, discourage new industry and new migrants. The result was a depressed economy, and by the 1980s Oregon's gates were again beginning to open. The dean of Oregon State University's engineering school capsulized the change in attitude: "One's perception of life changes when one's pocketbook is squeezed" (quoted by Vicker 1982:50). These same questions of the tradeoff between "boosterism" (growth is good for the local economy) and "nimbyism" (not in my backyard) are being debated in Europe, not just the United States (Barlow 1995).

As population growth in developing countries has pushed people out of agricultural regions toward the cities, most countries have responded by implementing population redistribution policies aimed at diverting rural migrants from large metropolitan areas and dispersing them to other, less densely settled areas. In the mid-1980s, for example, the Mexican government announced a policy to encourage people to move almost anywhere else in Mexico besides Mexico City and started the process itself by moving some government offices out of the capital into other cities (United Nations Population Fund 1991). Thus less developed nations are attempting to promote patterns of internal migration similar to those being pushed in the United States, where a typical problem is that faced by areas wanting to slow down the influx of new people.

As early as the 1950s the government of Indonesia began experimenting with transmigration—resettlement from the main island of Java to other, less densely settled islands. The goal has been to provide jobs for landless peasants and improve the economy of those islands (van der Wijst 1985; Widjojo 1970). Over the years the program has been periodically expanded, although to be effective in reducing population density on the main island of Java, millions more would have to be relocated. Nonetheless, the program seems to be more effective than the official statistics would indicate because an analysis of census data has suggested that governmentally assisted migrants are often unofficially followed to the new location by relatives and friends (Hugo, et al. 1987), and each female migrant of reproductive age takes not only herself, but her potential children as well.

Migration is, of course, only a temporary solution to population growth, since the empty spaces people are moving to will eventually fill up. To date, the success of attempts to limit immigration has been highly variable at best, and Massey (1996) has explained this with what he has called his "perverse laws of international migration:" (1) immigration is a lot easier to start than it is to stop; (2) actions taken to restrict immigration often have the opposite effect; (3) the fundamental causes of immigration may be outside the control of policy-makers; (4) immigrants understand immigration better than politicians and academicians; and (5) because they understand immigration better than the policy-makers, immigrants are often able to circumvent policies aimed at stopping them.

In the final analysis the most effective means by which you can retard growth is to nip it in the bud (or indeed, before)—to limit fertility.

Limiting Fertility

Limiting fertility is the best way to slow down population growth in any population, but it is also the most complex. You will recall from Chapter 6 that Ansley

Coale has argued that there are three preconditions for a sustained decline in fertility: (1) the acceptance of calculated choice as a valid element in fertility; (2) the perception of advantages from reduced fertility; and (3) knowledge and mastery of effective techniques of control. Each of these components has implications for population policy, as you can see in Table 14.1. Some policy initiatives are aimed directly at influencing demographic behavior, while others are oriented toward trying to change social behavior, which will then indirectly have an impact on fertility. To ignore the first precondition is to invite failure in any proposed policy; to ignore the second precondition is to miss an opportunity to slow down or stop population growth more quickly. Yet, most national policies oriented toward limiting fertility have been aimed only at the third precondition—enhancing the ability of couples to control their fertility. Programs like that fall under the general heading of family planning.

Family Planning   One of the internationally most popular population policies to limit fertility is to provide each woman with the technological ability to have the number of children she wants. Family planning involves the provision of birth prevention information, services, and appliances. It also involves teaching women (sometimes men as well) about their bodies and teaching them how to prevent births, usually with contraceptives but sometimes also with abortion or sterilization.

An early assumption of family planning programs was that women were having large families because they were uninformed about birth prevention or lacked access to the means for preventing births. Even though that assumption has been increasingly questioned in the past few years, family planning programs have remained the most popular means of implementing a policy to slow down population growth. A major reason for the widespread prevalence of family planning is, as I mentioned before, the fact that it is usually associated with health programs, which are almost universally acceptable. This is an important point, because many issues surrounding reproduction are very sensitive politically, if not socially.

In 1965 only 21 countries in the world admitted to actively supporting family planning programs (Isaacs and Cook 1984). By 1974 the United Nations had counted 86 governments that provided direct support for family planning, and by 1989 the number was up to 123, covering 91 percent of the world's population (United Nations 1989). As you can see, the spread of family planning services and technology around the world has taken place fairly rapidly since about 1965:

> In the mid-1960s, developing countries began to adopt policies to support family planning as a means of slowing population growth. By the late 1960s, family planning had become a worldwide social movement that involved international organizations such as the United Nations Fund for Population Activities, government agencies such as the United States Agency for International Development, private American philanthropic groups such as the Ford and Rockefeller foundations, nonprofit organizations such as the affiliates of the International Planned Parenthood Federation, and a host of individuals, many with backgrounds in medicine and public health [Donaldson and Tsui 1990:4].

One example of a family planning approach to population policy is in Guatemala, which set out "to achieve a substantial decline in the birth rate by

Table 14.1 Examples of Policies to Limit Fertility

Precondition for Which Intervention Is Desired	Examples of Policies	
	Direct	Indirect
Rational choice	Provide full legal rights to women Increase legal age at marriage for women	Promote secular education Promote communication between spouses
Motivation for smaller families	*Incentives* Payments for not having children Priorities in jobs, housing, education for small families Community improvements for achievement of low birth rate *Disincentives* Higher taxes for each additional child Higher maternity and educational costs for each additional child ("user fees")	*Incentives* Economic development Increased educational opportunities for women Increased labor force opportunities for women Peer pressure campaigns Lower infant and child mortality rates *Disincentives* Child labor laws Compulsory education for children Peer pressure campaigns Community birth quotas
Availability of means for limiting family size	Legalize abortion Legalize steralization Legalize all other forms of fertility control Train family planning program workers Manufacture or buy contraceptive supplies Distribute birth control methods at all health clinics Make birth control methods available through local vendors Establish systems of community-based distribution	Public campaigns to promote knowledge and use of birth control Politicians speaking out in favor of birth control

Source: John R. Weeks, 1992, "How to Influence Fertility: The Experience So Far." In Lindsey Grant (ed.), The Elephants in the Volkswagen: How Many People Can Fit Into America? (New York: W. H. Freeman Company), Table 15.1.

expanding the government family planning programs to rural areas" (Nortman 1975:29). The Ministry of Health provides family planning services as part of maternal and child health care and services are also provided by the Family Planning Association of Guatemala (APROFAM), an affiliate of the International Planned Parenthood Federation. However, with a total fertility rate of 5.1 in the late 1990s, Guatemala still has a long way to go to achieve a "substantial decline" in the birth rate, and after three decades of government and private-sector support for family planning, the Demographic and Health Survey in Guatemala estimated that only 31 percent of ever-married women are using any form of contraception. But the fault seems to lie less with the family planning program itself than with the policy approach. Surveys indicate that the indigenous population, which accounts for two thirds of Guatemala's population, prefers larger families, is not very accepting of modern contraceptives, and is generally ambivalent even toward the concept of birthspacing (Terborgh, et al. 1995). Family planning programs are implicitly designed to close the gap of **unmet needs** in a population—the number of sexually active women who would prefer not to get pregnant but are nevertheless not using any method of contraception (Robey, Ross, and Bhushan 1996), the strongest indicator of which is the percentage of women who say they would prefer not to have *any* more children, but are not using any method of contraception (Westoff and Bankole 1996). In a population such as the Mayan in Guatemala there may be little unmet need because women still prefer to have several children.

Governmentally supported family planning programs have often evolved from the efforts of private citizens seeing a need for such services. In Egypt, for example, private, voluntary organizations opened family planning clinics in urban areas as early as the 1950s. These clinics distributed contraceptive foams and jellies to women as long as they met three preconditions: (1) they already had three children; (2) they had their husband's permission; and (3) they could show a health or economic reason that established the need for birth control (Gadalla 1978). As quaint as that sounds, it helped set the stage for the far more massive government effort currently underway in Egypt, in which government health clinics dispense pills, IUDs, and other contraceptives, and outreach workers distribute contraceptives in rural areas.

Since the mid-1980s, the Egyptian government has combined the provision of family planning with a concerted drive to increase the motivation of couples to use birth control. This has included speeches by President Hosni Mubarak, published interviews with Islamic scholars, mass media messages, and activities by local community volunteers (Gilbar 1994). To be sure, the total fertility rate dropped from a high of 7.1 in 1960 to 3.6 in the late 1990s, but it is not clear how much of that decline can be attributed to family efforts, per se. Fargues (1997) notes that changes in the birth rate have been very closely tied to other government policies that affect fertility less directly—especially economic policies and the effort to improve educational levels of women. Still, the lessons of Chapter 6 suggest that without a family planning program in place, the motivation to limit fertility will have less chance of being converted to lower birth rates.

Family planning programs represent the major type of population policy to retard growth in the world, so it is important to ask, as I just have, do they work? Or are they mere palliatives—policies adopted to smooth relations with more powerful industrialized nations that, after all, provide much of the funding and guidance for the programs?

Do Family Planning Programs Work?   In a 1975 review of family planning services, Mauldin concluded that "the performance of family planning programs to date has been mixed, ranging from poor to moderately good to nearly spectacular" (1975:35). But a few years later, after analyzing data for 81 countries, Cutright and Kelly (1981) concluded that family planning programs clearly do have an important effect on fertility. Lapham and Mauldin (1984) followed up with a more extensive analysis of nearly 100 countries and drew the same conclusion, as did yet another follow-up analysis in 1990 (Bongaarts, Mauldin, and Phillips 1990).

These studies used a measure of family planning *program effort* developed by Lapham and Mauldin, relating effort to the observed decline in the birth rate, while controlling for such things as the level of social and economic development. Mauldin and Ross (1994) later combined this index with measures of demographic and social trends in order to assess the prospects for fertility reduction between 1990 and 2015. In particular, they were interested in evaluating the likelihood that currently developing countries might reach replacement levels by the year 2015. They focused on those nations with 15 million people or more in 1990, accounting for a total of 3.9 billion people in 1990, and representing 91 percent of the population in less developed nations of the world. I show some of these results in Table 14.2.

The data in Table 14.2 suggest that several nations in Asia (China, North Korea, South Korea, Sri Lanka, Taiwan, and Thailand) will almost certainly reach replacement level by 2015. Only two other developing nations, Mexico and Colombia, fall in this category. Indeed, we know that fertility in China, Taiwan, and North and South Korea had already dropped below replacement level by the late 1990s. India falls into the group of countries that Mauldin and Ross felt could probably be able to reach replacement level by the year 2015. Also in this category are countries such as Indonesia, Brazil, and Egypt. Moving to countries in which it is possible, although not very likely, that replacement level fertility could be reached by 2015, we find Bangladesh and Kenya, among others. Most dismal are the prospects for nations such as Afghanistan, Pakistan, Iraq, Ghana, and Nigeria. In these countries, the data suggest that it is unlikely that they will be able even to come close to replacement level in the foreseeable future. Although family planning programs represent only one of the indicators used by Mauldin and Ross in their analysis, family planning effort is a key indicator (either as a cause or at least as a concomitant) of rapidly declining fertility.

In general, there are three categories of countries in which you will find family planning programs: those in which fertility was already low before the advent of organized programs (primarily in Europe and North America), those in which a family planning program promoted or enhanced a decline in fertility (mainly in East Asia and Latin America), and, finally, those in which family planning programs exist on a very small scale and/or have yet had little impact on the birth rate (largely in Africa, the Middle East, and South Asia). Let me provide you with some examples.

Table 14.2 Prospects for More Populous Developing
Countries Reaching Replacement Level Fertility by the Year 2015

Region	Certain	Probable	Possible	Unlikely
Asia	China	India	Bangladesh	Afghanistan
	North Korea	Indonesia	Myanmar	Nepal
	South Korea	Malaysia		Pakistan
	Sri Lanka	Philippines		
	Taiwan	Vietnam		
	Thailand			
Latin America	Colombia	Argentina	—	—
	Mexico	Brazil		
		Venezuela		
		Peru		
Middle East/North Africa	—	Egypt	Algeria	Iraq
		Morocco	Iran	Sudan
		Turkey		
Sub-Saharan Africa	—	South Africa	Kenya	Ethiopia
				Ghana
				Nigeria
				Tanzania
				Uganda
				Zaire

Source: Adapted from W. P. Mauldin and J. A. Ross, 1994, "Prospects and programs for fertility reduction, 1990–2015," *Studies in Family Planning* 25(2):77–95, Table 4.

Fertility Decline without Organized Family Planning—The United States Fertility declined in the United States up to the 1940s without a widespread family planning program. Nonetheless, the family planning or planned parenthood movement in the United States in fact has a fairly long history, dating back to the 19th century, when many of the early efforts were aimed at removing legal restrictions on the distribution and sale of contraceptives and on abortions. In 1873, after production of the condom had begun and in the midst of a general decline in U.S. fertility, Congress passed the Comstock Law prohibiting the distribution of contraceptives through the mails. According to Westoff and Westoff, "Anthony Comstock, for whom the law was named, formed the New York Society for the Suppression of Vice . . . (and) was responsible for seven hundred arrests, the suicide of a woman abortionist, and the seizure of thousands of books and contraceptives" (Westoff and Westoff 1971:47). The Comstock Law was not repealed until 1970, although it had been toned down by court order in 1936 (Jaffe 1971).

Jaffe has also noted that "it was not until 1958 that the ban on prescribing contraceptives in public hospitals was lifted in New York City and the way opened for publicly financed health institutions to provide family planning services. It was not until 1965, when the Supreme Court struck down the Connecticut statute barring the use of contraceptives, that a number of states repealed their restrictive laws" (1971:119).

Congress has provided federal funding for family planning programs in the United States since 1967, but the Family Planning Services and Population Research Act of 1970 specifically forbade funds to any organization that provided abortion services, even though several states already had liberalized abortion laws at that time. In 1973 the Supreme Court struck down restrictive abortion laws in the United States and one of the last obstacles to free choice in birth prevention was hurdled. Of course, there are ongoing attempts in the United States to reverse that decision and once again restrict the access of American women to legal abortion, but in the United States, as in other parts of the world, a decline in fertility did not wait for the legal restrictions to be removed from contraception and abortion; nor did it await government programs to develop new contraceptive technologies such as the IUD and the pill (both were privately financed), or for the government to provide free or subsidized birth control measures. Fertility declined dramatically from the 19th century to World War II, went up briefly, and then again declined dramatically. Only on the second decline was a wide range of birth prevention means readily available to women; in essence, the means were part of the rise in the standard of living. The same goal of family limitation can now be accomplished more easily and more confidently than ever before.

Fertility Decline Enhanced by a Family Planning Program—Taiwan   One of the countries that was an early showpiece of the success of family planning programs was Taiwan, which was heralded in the 1960s as a nation that had responded to a highly organized family planning program with a dramatic fertility decline. However, a close inspection of the data revealed that the general fertility rate was already declining before World War II, during the period of Japanese occupation. The withdrawal of the Japanese after the war removed most of the skilled personnel, but factories and a labor force accustomed to industrial work remained, and, as a result, the Taiwanese were able to take advantage of these resources fairly quickly. As the economy expanded, mortality fell and birth rates declined in an almost perfect replication of the demographic transition model.

Beginning in the early 1950s both the economy and the population of Taiwan began to grow quite rapidly. Between 1952 and 1967, for example, the income per person doubled, and although the population was growing, fertility was in the midst of a decline. There was a baby boom just after the war, but the birth rate has declined quite steadily again since about 1951, whereas the family planning program did not start until 1959 and only became fully organized in 1963 (Davis 1967). Not until 1968 did the government actually adopt an official policy condoning and encouraging family planning (Nortman 1975). Although the age at marriage went up, it appears that the fertility decline has been largely due to the effective use of contraception within marriage (Feeney 1991), and this has brought fertility down to below replacement level as of the late 1990s. Thus family planning has without doubt helped to sustain, and possibly to accelerate, the decline in fertility in Taiwan, which in turn was an accompaniment to rapid economic development.

Fertility Little Influenced by a Family Planning Program—India   In 1952 India began experimenting with family planning programs to keep the population "at a level consistent with requirements of the economy" (Samuel 1966:54). Although initial progress was pretty slow—the Indian government was moving cautiously—by 1961 there were about 1,500 clinics in operation, providing condoms, diaphragms, jellies, foam tablets, and other services, largely free (Demerath 1976).

In 1963 the birth rate still was not responding and the government, now with the advice of the Ford Foundation, reorganized its family planning effort, mainly in an outreach effort to spread the birth control message to more people. Still the birth rate did not go down, so in 1966 there was another reorganization, and in 1967 Dr. S. Sripati Chandrasekhar, a demographer, was named Minister of State for Health and Family Planning. Chandrasekhar boosted the use of the mass media, offered transistor radios to men undergoing vasectomy, pushed through a legalization of abortion, and generally brought about more family planning action than had occurred in all the previous years of family planning. Male sterilization, being both permanent as well as the cheapest method of birth control, has been especially pushed by the Indian government, and between 1967 and 1973 (when Chandrasekhar left his position) 13 million persons had been sterilized in India.

Did the birth rate respond accordingly? The birth rate in India may have declined ever so slightly between 1961 and 1971, but the death rate sank even more, so by the mid-1970s the Indian population was growing faster than ever before. Why had the family planning efforts failed? Because family planners had not taken into account the broader social context within which reproduction occurs. It was naive to think that family planners could, of their own accord, generate the kind of social and cultural revolution required to make small family norms an everyday practice throughout India (Demerath 1976).

By 1976 Indira Gandhi's government in India was beginning to show signs of desperation at being unable to slow the rate of population growth, and it became the first government in the world to lean definitely toward compulsory measures. Dr. D.N. Pai, director of family planning in Bombay, has reportedly said, "Ninety percent of the people have no stake in life. How do you motivate them? They have nothing to lose. The only way out of the situation is compulsion" (Rosenhause 1976). The government implemented a now famous, if not infamous, program of enforced sterilization. "Officially, there was no coercion, but the elaborate system of 'disincentives' amounted to the same thing. Government employees had to produce two or more candidates for sterilization. For such civil servants, or for anybody who was being pressured into submitting to sterilization himself, it was usually possible to hire a stand-in for about 200 rupees ($22). For those not in government service all sorts of privileges—such as licenses for guns, shops, ration cards—were denied unless the applicant could produce a sterilization certificate" (Guihati 1977).

As a result of the policy of enforced sterilization, the number of people (largely male) who were sterilized in India jumped from 13 million in 1976 to 22 million in 1977. By contrast, I should note, there were fewer than 4 million users of the IUD or the pill—only 3 percent of the total population of married women (U.S. Bureau of the Census 1978). As it turned out, however, the average male who was sterilized was in his thirties and already had several children (Guihati 1977), so the program was less effective in reducing fertility than those numbers might imply.

Although her campaign effectively raised the level of sterilization, Indira Gandhi's policies generated violent hostility, and in 1977 she was defeated in her bid for reelection. Her successor, 82-year-old Morarji Desai, professed support for family planning, but only half-facetiously suggested that self-control (that is, abstinence), rather than birth control, was the solution to India's high rate of population growth (Bhatia 1978). Gandhi was elected once again in 1979, but she kept a relatively low profile with respect to her support for family planning, right up to the time of her assassination in 1984. Interestingly enough, that was a period of time in which the birth rate did appear to be dropping, although survey results have shown that no change in desired family size took place between 1970 and 1980 (Khan and Prasad 1985).

In 1984 Indira Gandhi's son, Rajiv, was elected prime minister of India. He had been actively involved in the family planning movement during his mother's administration and in 1985 he declared that the nation was on "war footing" to reduce its rate of population growth to a two-child family norm by 2000 (Population Action Council 1985:1). A 5-year, $3.6 billion program was unveiled that included monetary rewards to women who limited their family size (see the discussion later in this chapter of these motivational schemes) and a broadening of the scope and quality of the nation's family planning services. He was voted out of office in 1989 and later assassinated, although neither event seemed to be related specifically to his views on population limitation.

It would no longer be fair to call the family planning program a failure in India, but it is apparent that the program has lacked consistent direction and has failed to offer couples the most flexible forms of birth control. Family planning in India still depends on sterilization, although the emphasis is now on tubal ligation for women. Other contraceptive methods represent only a small fraction of family planning effort in India (Pathak, Feeney, and Luther 1998).

Fertility is declining now in India, and the total fertility rate had declined to 3.4 by the end of the 1990s. This was about 2.0 children less than 20 years before, but it is not clear how much of the decline is due to organized family planning efforts. At least one fourth of the decline has been attributed to an increase in the average age at marriage from 17 to 19 (Retherford and Rele 1989), and a government policy to increase female literacy almost certainly has had an impact, as well (Jejeebhoy 1991). However, as I mentioned in Chapter 6, there is tremendous variability from state to state in India in the birth rate (as well as variation in several other indicators of social and economic well-being). This is due partly to the fact that implementation of national policy rests in the hands of the individual state governments. "This reality of the government structure in India remains a hindrance for the national population policy" (Panandiker and Umashankar 1994:101).

Why Don't They Always Work?   The conclusion to be drawn from the preceding three case studies is one that I have repeated earlier in the book: A decline in fertility is as dependent on motivation as it is on technology and government administrators. Consider also that in 1963, in discussing the rapid decline of fertility in Japan, the United Nations noted that "the important lesson to be learned from the Japanese experience is that the desire for fewer children spreads quickly if a strong motivation exists, but without this, family planning programmers are not likely to achieve their aims" (United Nations 1963:33).

A stronger and widely reproduced statement of the same theme is that by Davis, who noted that "by sanctifying the doctrine that each woman should have the number of children she wants, and by assuming that if she has only that number this will automatically curb population growth to the necessary degree, the leaders of current policies escape the necessity of asking why women desire so many children and how this desire can be influenced" (1967:733).

Again, the assumption that underlies family planning programs and that lends them importance as a policy designed to limit fertility is that millions of women in the world (including the United States) have more children than they want because they are ignorant of, or lack access to, effective methods of birth control—they have an unmet need for family planning. Furthermore, family planning is designed to improve the ability of women to have the number of children they want (whether that number is zero or eight), but it may do little to influence a couple's desire to have children. In places like Guatemala, Pakistan, Nepal, and Kenya—where fertility remains high—it is easy to argue that the problem is a weak family planning program. But it has become increasingly accepted throughout the world that a good family planning program must work in the context of a changing social environment that encourages couples to have fewer children and/or discourages them from having large families. A family planning program will not lower fertility in the absence of the desire for smaller families; to be effective, a population policy in some countries may need to go beyond family planning.

Beyond Family Planning—Engineering Social Change

In Chapters 6 and 10 I discussed the kinds of social factors that motivate people to want smaller families. How can the desire for children be altered? Since people have children at least partly because the perceived benefits exceed the costs, a policy must be aimed at raising costs and lowering benefits—a policy or set of policies that are tantamount to engineering social change. As you can see by looking back at Table 14.1, costs can be raised directly by government action imposing fines or taxes or eliminating deductions or allowances for children, or indirectly by restricting the availability of housing or keeping prices on child-oriented consumer goods artificially high. Benefits of children can be lowered directly by making child labor illegal and indirectly by lifting the pronatalist pressures that currently exist in virtually every society and lifting the penalties for antinatalist behavior that also currently exist.

Empowering Women

Lifting pronatalist pressures will especially involve a change in the gender roles taught to boys and girls, giving equal treatment to the sexes in the educational and occupational spheres. If a woman's adulthood and femininity are expressed in other ways besides childbearing, then the pressures lessen to bear children as a means of forcing social recognition. Likewise, if a man's role is viewed as less domineering, then the establishment of a family may be less essential to him as a means of forcing social recognition. Required, of course, are alternatives to children and families in general as bonds holding social relationships together, indeed as centers of everyday life. This does not require the abolition of the family, but it does involve playing down its importance, which may (perhaps I should say "will") require massive social change—a virtual revolution in the way social life is organized.

Any policy aimed at affecting motivation will, by definition, have to alter the way people perceive the social world and how they deal with their environment on an everyday basis. It will have to involve a restructuring of power relationships within the family, a reordering of priorities with respect to gender roles, a reorganization of the economic structure to enhance the participation of women, a concerted effort to raise the level of education for all people in the society, and a stability to economic and political life that allows people to plan for a future rather than just cope with today's problems of survival. History suggests that most of these changes have evolved somewhat naturally in the course of economic development, at least in Western nations. However, they are not inherently dependent on development and, as a result, could be part of a deliberate policy even before large-scale development occurs (see, for example, Das Gupta 1995). In fact, it is probable that those are the very kinds of social changes that would help accelerate economic development, leading, as they would, to a substantial improvement in the human condition, at least by Western standards.

Lifting the penalties for antinatalist behavior also starts with the redefinition of gender roles—a more positive evaluation of single or childless persons. But there are more specific items that could be mentioned with reference to almost any country, such as: (1) alter tax systems in which single people pay higher taxes than married people, (2) stop giving tax exemptions for children or for child care, (3) eliminate or reduce maternity benefits, and (4) stop providing larger allowances (such as welfare and health benefits) to people with children than to those without.

Economic Incentives—The Case of India

The flip side of lifting penalties is to introduce policies that specifically reward people for limiting the number of children they have. One of the first plans to use monetary incentives in a positive way to affect the desire for children was that proposed a few decades ago for the Indian government. In 1960 Stephen Enke calculated that in many countries, such as India, the resources invested in children (food, clothing, medical care, education, and so on) do not earn as high a return as resources invested in ordinary capital projects (a factory, an irrigation dam, and others). In those countries, Enke proposed that the government could afford to pay a money bonus to men and women for each birth permanently prevented (Enke 1960). India was a prime candidate for such a scheme, since at that time in India there were already state governments and private companies offering employees a free vasectomy plus a small bonus.

Enke argued for a larger bonus, perhaps as high as $100 (more than a year's income in India at the time), and also recommended sterilization rather than contraception, since it requires a one-time-only cost, and also for men rather than women, since vasectomy is the cheapest form of sterilization. His justification for the larger bonus was that "it may be several hundred times more advantageous to invest money in preventing births—in the ways suggested here—than in traditional development projects" (1960:346).

It was an innovative idea, but too expensive in the short-run and it was not implemented in the way he suggested. However, Enke also proposed an alternative plan that in his estimation would be less effective but had the advantage of not requiring immediate cash payments. This was a scheme to put money into a retirement trust fund for women for every year that passes without a child being born. This plan is currently being used by private employees on several tea plantations in India and has received the endorsement of the World Bank (World Bank 1984). Incentive programs do, in fact, play in important role in bringing women into clinics for sterilization, but we have already seen that they have not yet been effective in bringing fertility close to replacement level. This is probably because few of them have involved the kind of large-scale bonus envisioned by Enke. Two countries, in particular, have put into place systems of incentives and disincentives for childbearing that seem to have an impact on reproduction. These two countries are China and Singapore.

Incentives and Disincentives—The Case of China

Since the Communist revolution in 1949, the population of the People's Republic of China has more than doubled. The Chinese government realizes that the population problem is enormous, and they have implemented the largest, most ambitious, most significant, and certainly the most controversial policy to slow population growth ever undertaken in the world. The 1978 constitution of the People's Republic of China declared that "the state advocates and encourages birth planning," and the reasons for this were spelled out in 1979 by Vice Premier Chen Muhua (Chen 1979). She argued that there are three major explanations for the fact that in China population control was now "dictated and demanded by the socialist mode of production" (1979:94). These were: (1) too rapid an increase in population is detrimental to the acceleration of capital accumulation, (2) rapid population increase hinders the efforts to quickly raise the scientific and cultural level of the whole nation, and (3) rapid population growth is detrimental to the improvement of the standard of living. These arguments are, of course, basically those I outlined in Chapter 12 in discussing the relationship between population growth and economic development.

The goal of the Chinese government was, incredibly enough, to achieve ZPG by the year 2000, with the population stabilizing at 1.2 billion people. As you know from Chapter 8, to do so would require that the one-child family become the norm, because the youthful age structure in China in the 1970s placed a high proportion of people in the childbearing ages. How did they go about trying to achieve this goal? The first step was to convince women not to have a third child (third or higher-order births accounted for 30 percent of all births in 1979; Chen 1979). The second step was to promote the one-child family. These goals have been accomplished partly by increased social pressure (propaganda, party worker activism, and probably coercion as well) and partly by the increased manufacture and distribution of contraceptives. The heart of the policy, though, has been a carefully drawn system of economic incentives (rewards) for one-child families and disincentives (punishments) for larger families.

In the cities, couples with only one child who pledge that they will have no more (and are using some form of fertility limitation) may apply for a one-child certificate. The certificate entitles the couple to a monthly allowance to help with the cost of child rearing until the child reaches age 14. Furthermore, one-child couples receive preference over others in obtaining housing and are allotted the same amount of space as a two-child family; their child is given preference in school admission and job application; and when they retire they will receive a larger than average pension.

In the countryside, the incentives are a bit different. One-child rural families receive additional monthly work points (which determine the rural payments in cash and kind) until the child reaches age 14. These one-child families also get the same grain ration as a two-child family. In addition, all rural families receive the same size of plot for private cultivation regardless of family size, thus indirectly rewarding the small family. Each province in China has been encouraged by the central government to tailor specific policies to meet the particular needs of its residents, and some of the more widely implemented policies include an increasingly heavy tax on each child after the second, and the expectation that for each child after the second, parents will pay full maternity costs as well as full medical and education costs.

There is no question that fertility has fallen in China, as you can see in Figure 14.3. The crude birth rate dropped by 43 percent between 1972 and 1982 (Sherris 1985)—a decline exceeded in that time period only by Cuba and matched only by Singapore (as I discuss below). Furthermore, fertility has fallen not only in urban areas where the motivations for small families might be greatest, but in rural areas as well. Lavely (1984), for example, studied a rural commune in Sichuan Province, which was admittedly a model demonstration project, but which nonetheless had experienced a remarkable decline in fertility, with the total fertility rate going from nearly 7 to almost 1 in less than two decades. "Fertility before the decline was high and uncontrolled, corresponding to a pattern of natural fertility. The onset of the decline can be dated almost precisely in 1966, apparently in response to government efforts to promote birth planning" (Lavely 1984:365).

In 1971 the government instituted the *wan xi shao* (later, longer, fewer) campaign and, as I mentioned in Chapter 1, that helped to accelerate fertility decline in China. Thus fertility was already on its way down in many parts of China when the one-child policy was implemented in 1979. The birth rate in China went up slightly in 1981, perhaps in response to the relaxation of a law prohibiting early marriage, and the government responded in 1983 by trying to rebuild momentum for a fertility decline with a "New Mobilization" for comprehensive planned reproduction (Tien 1984).

Between 1985 and 1989 the total fertility rate in China again inched back up, from 2.2 to 2.6, and the Chinese government appeared to be torn between hardliners who wanted to reinvigorate the one-child policy and a more moderate element that wanted to maintain the "opening of small holes" (a relaxation of regulations) that would allow some couples to have a second child without penalty (Tien 1990). The government also very subtly, but importantly, changed the demographic target from *no more than* 1.2 billion people in the year 2000 to *about* 1.2 billion (Greenhalgh 1986). By the late 1990s, the total fertility rate was estimated to have fallen below the replacement level—to 1.8 children per woman, but enough families had exceeded the one-child ideal that China had already passed its 1.2 billion target.

Figure 14.3   The Birth Rate in China Has Reponded to Government Programs

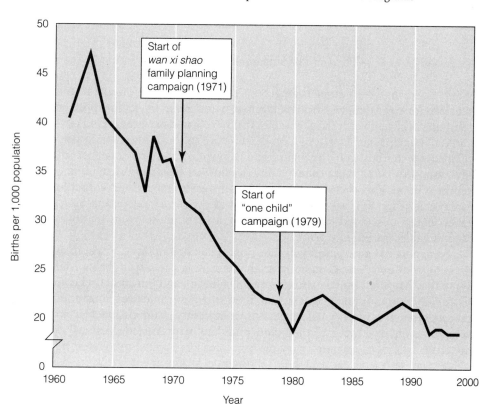

Source: J. Sherris, 1985, "The impact of family planning programs on fertility." Population Reports, J-29, Population Information Program, Johns Hopkins University, Figure 4. Used by permission; updated to 1998 by the author.

The one-child policy in China was never intended to be anything more than an interim measure that would finally put the brake on population growth in that country (Greenhalgh 1986). From the beginning, the plan had been to ease back to a two-child family after hitting the 1.2 billion level of population size. Even a quick glance at Figure 14.3 calls into question whether the one-child policy was ever as potent an instrument for fertility decline as is often thought or whether it mainly helped to publicize a fertility decline that was already well under way.

Several researchers have concluded that socioeconomic development in China has been a more important factor than government policy in motivating couples to want fewer children (see Poston and Gu 1987), although the government's attitude has certainly facilitated and enhanced those motivations (Greenhalgh, Chuzhu, and Nan 1994). It is also intriguing to note the cultural similarities between mainland China and Taiwan (which, as you know, is actually claimed by China as part of its territory) and the coincidence of rapid fertility declines in the two countries—one with a clearly defined set of incentives and disincentives, and the other with a more normal voluntary family planning program.

Disincentives Only—The Case of Singapore

While China has been emphasizing both strong incentives for small families and penalties for large families, the small (but growing) republic of Singapore began emphasizing disincentives to large families in 1969. Singapore is an island city-state on the Malaysian peninsula whose recent history was influenced by the British who developed the harbor, built the city, and controlled the area (as they did all of Malaysia) for more than a century until World War II. Japan briefly took over during the war, but after the war, as Great Britain was divesting itself of its colonies, the Malaysian Federation was formed that included both Malaysia and Singapore. In 1963 Singapore seceded from the Federation, preferring to go its own way, and in 1965 it gained its full independence.

Singapore has an ethnically diverse population of 4 million, of whom more than three fourths are ethnic Chinese, with the remaining population being Malay (who are mainly Muslim), Indian (who are mainly Hindu), and European (who are mainly Christian), and upon independence in 1965, the government of Singapore realized that its rapid rate of population growth would make it impossible to continue developing economically, since the country has few natural resources and practically no arable land of its own. Its principle resource is its population, and so population policy was seen as a form of human resource management (Drakakis-Smith and Graham 1996).

At first, the policy incorporated the family planning program approach, emphasizing the distribution of contraception through clinics and hospitals (Salaff and Wong 1978). After 4 years of family planning, however, the government grew restless at the slow pace of the decline in the birth rate. The result, in 1969, was a liberalization of the abortion law and concomitant establishment of some economic disincentives to encourage adoption of the idea that "two is enough." These measures included steeply rising maternity costs for each additional child, low school enrollment priorities for third and higher-order children, withdrawal of paid 2-month maternity leave for civil service and union women after the second child, low public housing priority for large families, and no income tax allowance for more than three children.

Have these basically coercive measures worked? Yes. Between 1966 and 1985 the total fertility rate in Singapore dropped from 4.5 children per woman to 1.4, with a bigger drop occurring among the wealthier, educated women than among poorer women, as you might expect. This classic case of differential fertility worried Singapore's prime minister Lee Kuan Yew. In 1983 he became concerned that not enough bright Singaporans were being born and he argued, "In some way or other, we must ensure that the next generation will not be too depleted of the talented. Get our educated women fully into the life cycle. Get them to replace themselves" (quoted by Leung 1983:20). There was also concern that the rapid drop in fertility would lead to an aging population in which there would be only two workers for every retired person.

Singapore flatly rejected the idea that immigration might solve that problem and instead in 1987 the government instituted the New Population Policy, in which the philosophy was "have three or more children, if you can afford them" (Yap 1995). This set of policies mainly just lifted some of the earlier disincentives, especially for university-educated and working women, although it did include innovations such as social mixers where university-educated singles could mingle (and hopefully marry). The goal was to raise the birth rate back up to replacement level by the year 1995, but that did not happen. By the late 1990s the total fertility rate had risen to 1.8, but the increase from 1.4 in the mid-1980s seemed largely to be that women who had postponed births at an earlier age had gone ahead and had them. The government has responded by lowering some of its immigration restrictions.

Indirect Policies that Influence Fertility

The centrally planned economies of China, Cuba, the former Soviet Union, and eastern Europe (prior to 1990) have shared over time the same demographic fate—the policies introduced to socialize the economy led to generally unintended declines in fertility. In China the government ultimately decided it wanted yet a lower level of fertility and enacted policies to implement that. However, Russia and the former Soviet-bloc nations of eastern Europe had a long history of dissatisfaction with low rates of population growth. There are, of course, numerous population policy lessons in the experiences of these countries, although they are by now familiar themes. The socialist centrally planned societies teach us that fertility can be indirectly influenced to decline through a combination of the following factors: (1) educating women; (2) providing women with access to the paid labor force; (3) legalizing the equality in status of males and females; (4) legalizing abortion and/or making contraceptives freely available; (5) slowly raising the standard of living, while at the same time making housing and major consumer items difficult to obtain, forcing a couple to ask themselves: "Kicsi or kosci?" (as they say in Hungarian—"A child or a car?").

Promoting or Maintaining Growth

There are two very different groups of countries that have policies to promote or maintain current rates of population growth. On the one hand are the low-fertility, slow- or nongrowing developed nations of Europe (including eastern Europe) where governments would prefer a higher birth rate, typically as a hedge against immigration and/or as a perceived stimulus to the economy. On the other hand are countries in Africa and the Middle East where population growth is viewed as too low because the population size is perceived as being too small to achieve national objectives.

Developed Countries Promoting Growth

There is not a single country in Europe that believes that its rate of growth or its level of fertility should be any lower, but in the 1990s 15 European nations believed that their birth rate was too low, as you can perhaps discern looking back at Figure 14.2. These included not only the smaller states of Estonia, Latvia, Luxembourg, and Slovenia, but the larger countries of Germany, France, Portugal, Switzerland, Russia, Ukraine, Hungary, Bulgaria, Croatia, Greece, and Romania.

Romania has the dubious distinction of having done the most in recent decades to try to raise the birth rate. In 1966 the Romanian government decided to halt the downward trend in the birth rate (which was being accomplished primarily by abortion) by establishing a policy that made abortions illegal except under extreme circumstances. The government also discouraged the use of other contraceptives by stopping their importation and by making them available only for medical reasons. The result was that the birth rate skyrocketed from 14 per 1,000 population in 1966 to 27 per 1,000 in 1967. Although it did begin to fall again after 1967 as women resorted to illegal abortions and found other means of birth control (primarily rhythm and withdrawal), by 1989 the total fertility rate of 2.3 was still higher than the 1.9 level of 1966. But 1989 brought a significant event—the change in government. In December 1989 Nicolae Ceausescu, longtime dictator of Romania, was overthrown and executed. One of the first legislative acts of the new government was to legalize abortion in order to reduce the high maternal mortality rate that had resulted from botched illegal abortions during the prior two decades (Serbanescu, et al. 1995). By the late 1990s it was estimated that the total fertility rate had dropped to 1.3, one of the world's lowest.

France, which has experienced a low birth rate for longer than almost any other existing society, has also maintained its longstanding policy of providing monthly allowances to couples who have a second or higher-order child, despite the lack of evidence that such allowances have had any measurable impact on the birth rate. Single mothers are also assured a monthly allowance, and all mothers in France have access to nursery school placement for their children by age three, so that they can return to work (Bergman 1996).

The European Parliament actually passed a resolution in 1984 calling upon member nations to consider various pronatalist policies to combat the low birth rate because, the resolution claimed, Europe's standing and influence in the world depend largely on the vitality of its population. In 1985, a government committee in Norway responded by recommending that fertility be encouraged in order to avoid a population decline. Suggestions included the usual incentives of extended maternity leave and subsidized child care (Wall Street Journal 1985).

Elsewhere in the world, including the world's largest consumer (and polluter), it is perhaps a bit harder to justify a policy that tacitly or actively encourages population growth. I discuss population policy in the United States in the essay accompanying this chapter, so let me here mention the situation in developing nations.

Developing Countries Promoting Growth

Population growth continues at high levels throughout sub-Saharan Africa because population policy has been a relatively weak agenda item in the region, although the situation is changing. In the 1980s there were still five countries—Côte d'Ivoire, Congo, Guinea, Gabon, and Equatorial Guinea—in which government policy reflected the view that fertility was too low and should be higher. By the

1990s only Gabon (which has only a bit more than a million people) remains in that category. Still, there are ten countries in sub-Saharan Africa that are satisfied with their high birth rate. These include Benin, Côte d'Ivoire, Mauritius, Togo, Djibouti, Mauritania, Somalia, Chad, Equitorial Guinea, and the Democratic Republic of Congo (Zaire), and all of these except Mauritius have rates of population growth that are well above the world average.

In the Middle Eastern nations (encompassing both northern Africa and West Asia), the countries that favor lower rates of growth are far outnumbered by those that either seek higher birth rates or are satisfied with the status quo. Algeria, Egypt, Sudan, Jordan, Morocco, Tunisia, Turkey, and Yemen and, significantly, Iran, are pursuing policies aimed at lowering fertility. Two of the Middle Eastern nations pushing policies designed to raise the birth rate are longtime adversaries—Israel and Iraq. This is, of course, easier to accomplish in Iraq, where the total fertility rate in the late 1990s was 5.7, than in Israel, where it was 2.9.

Latin America, once an area with large pockets of pronatalism, has witnessed a policy reversal since the 1970s. Brazil was once actively pronatalist, but in the early 1980s the government relaxed that stance and began to promote a program integrating health care of women and children with the provision of birth control. The government also approved legislation permitting the local production of IUDs. The birth rate has responded to these efforts; and by the late 1990s the total fertility rate had dropped to 2.5 children per woman.

In Chile during the 1970s and 80s, the government of Pinochet had been calling for a significant increase in population to protect the country from population growth in neighboring Latin American nations (Chile has had wars with both Peru and Argentina); however, events worked in the opposite direction. After overthrowing Allende's socialist government in 1973, the Pinochet regime imposed a level of austerity that cut incomes and raised the levels of unemployment. This encouraged out-migration, postponement of marriage, and the widespread use of contraception within marriage—not a great way to increase the population. Pinochet withdrew from office in 1989, and the current democratically-elected government has indicated satisfaction with the current rate of growth in Chile where the total fertility rate of 2.4 in the late 1990s was one of the lowest in South America.

World Population Conferences

Countries that want assistance with their population policy can generally find it in the form of foreign aid provided either by the United Nations Population Fund (with funding from the developed countries) and/or directly from individual countries. In the 1990s, for example, there were 17 principal international donors of population assistance. As you can see in Figure 14.4, the United States has been by far the largest donor, followed by Germany, Japan, Norway, the United Kingdom, and Sweden (note that the predominantly Catholic European countries of Italy, France, Belgium, and Austria are far less likely to contribute to population programs). Japan is the largest single donor to the UNFPA, whereas the United States prefers to spend its money directly through its Agency for International Development (USAID) and, in fact, the Reagan administration withheld all U.S. funding from the UNFPA from

Figure 14.4   Population Assistance from Major International Donors

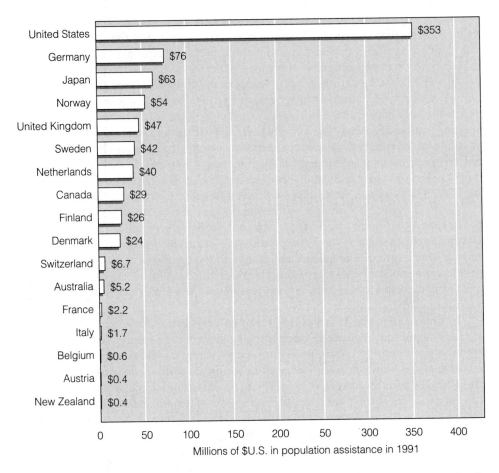

Source: Adapted from data in J. A. Ross, W. P. Mauldin, and V. C. Miller, 1993, Family Planning and Population: A Compendium of International Statistics (New York: The Population Council), Table 30.

1987–92 on the grounds that money was being used to fund abortions (a charge that was repeatedly denied by the United Nations). U.S. funding was reinstated by the Clinton administration in 1993 (Conly 1996).

In spending their money on population programs, donor countries have been guided by the list of priorities laid out at each of the last three world population conferences organized by the United Nations—1974 in Bucharest, Romania; 1984 in Mexico City; and 1994 in Cairo, Egypt. Unlike earlier world population conferences, which had been organized for professional demographers (and those kinds of meetings still take place), beginning in 1974 the United Nations-sponsored conferences brought together policy-makers, and the goal has been to develop something close to a world population policy.

The 1974 World Population Conference—Bucharest

The first of the policy-oriented World Population Conferences was held in Bucharest in 1974. After nearly 2 weeks of "often acrimonious debate" (Demeny 1994:5), the delegates to the Conference (government representatives from all over the world) agreed to a World Population Plan of Action. Table 14.3 offers you the highlights of that plan, which are important for several reasons: (1) this was the first time that an international body had established anything approximating a world population policy; (2) the 1984 and 1994 Population Conferences were designed to evaluate and alter this plan as necessary; and (3) the Plan incorporated elements from both sides of the "acrimonious debate," which, in essence, represented the U.S. perspective on the one hand (with the general support of other industrialized nations) and the perspective of the developing countries on the other (Finkle and Crane 1975).

Key government representatives of the United States came to the Bucharest meeting with a very clear perspective: developing countries should establish population growth rate targets, and those targets should be met by the provision of family planning (Donaldson 1990). Some of these ideas, in very watered-down form, can be seen in the highlights of the Conference in Table 14.3. Number 6, for example, indicates that countries are "invited to set quantitative goals," and Number 5 indicates that countries should "encourage appropriate education concerning responsible parenthood."

In contrast to the U.S. position, leaders of developing countries arrived in Bucharest with the growing conviction that they were the less advantaged participants in a world economic system organized by and for the benefit of the industrialized nations. The U.N. itself had tacitly endorsed that view just before the 1974 World Population Conference with a General Assembly resolution calling for a "New International Economic Order." Developing nations were thus interested in seeing a world plan that integrated population with economic development programs. Indeed, it was at that this Conference that a delegate from India coined the phrase "development is the best contraceptive."

The 1984 International Population Conference—Mexico City

In the 10 years between the meetings in Bucharest and Mexico City, most developing countries came to the conclusion that rapid population growth was a hindrance to their efforts to develop economically, and many of those countries implemented family planning programs to help slow down population growth. Thus the stage was set for the 1984 conference to be a better meeting of the minds. However, the United States threw cold water on that prospect by proposing, instead, that "population growth is, by itself, a neutral phenomenon." This was a "choice piece of nonsense," as Demeny (1994:10) has remarked, and it was inspired by Julian Simon's influence within the Reagan Administration (and abetted by a coalition of groups opposed to abortion). Simon's basic premise, as you recall from Chapter 12, is that population growth can help stimulate development, which, in fact, is what the developing countries have in mind as their ultimate goal. The implication is that if you are focusing your energies on population control rather than economic development

Table 14.3   Highlights of the World Population Plan of Action

In 1974 representatives from nations all over the world met at the United Nations-sponsored World Population Conference. Out of that conference came a World Population Plan of Action that became a set of benchmarks for discussion at the 1984 International Population Conference (again sponsored by the United Nations).

Some of the resolutions agreed to that are especially relevant to the issue of population growth and economic development include:

1. Governments should develop national policies and programs relating to the growth and distribution of their populations, if they have not already done so, and the rate of population change should be taken into account in development programs.

2. Countries should aim at a balance between low rather than high death rates and birth rates.

3. Highest priority should be given to the reduction of high death rates. Expectation of life should exceed 62 years in 1985 and 74 years in 2000. Where infant mortality continues high, it should be brought down to at least 120 per 1,000 live births by the year 2000.

4. Because all couples and individuals have the basic human right to decide freely and responsibly the number and spacing of their children, countries should encourage appropriate education concerning responsible parenthood and make available to persons who so desire advice and means of achieving it.

5. Family planning and related services should aim at prevention of unwanted pregnancies as well as elimination of involuntary sterility or subfecundity to enable couples to achieve their desired number of children.

6. Countries that consider their birth rates detrimental to their national purposes are invited to set quantitative goals and implement policies to achieve them by 1985.

7. Governments should ensure full participation of women in the educational, economic, social, and political life of their countries on an equal basis with men.

8. Countries that wish to increase their rate of population growth should do so through low mortality rather than high fertility, and possibly immigration.

9. To achieve the projected declines in population growth and the projected increases in life expectancy, birth rates in the developing countries should decline from the present level of 38 to 30 per 1,000 by 1985, which will require substantial national efforts and international assistance.

10. In addition to family planning, measures should be employed that affect such socioeconomic factors as reduction in infant and childhood mortality; increased education, particularly for females; improvement in the status of women; land reform; and support in old age.

11. National efforts should be intensified through expanded research programs to develop knowledge concerning the social, economic, and political interrelationships with population trends; effective means of reducing infant and childhood mortality; new and improved methods of fertility regulation to meet the varied requirements of individuals and communities, including methods requiring no medical supervision; the interrelations of health, nutrition, and reproductive biology; and methods for improving the administration, delivery, and utilization of social services, including family planning services.

12. The Plan of Action should be closely coordinated with the International Development Strategy for the Second United Nations Development Decade, reviewed in depth at 5-year intervals, and modified as appropriate.

Source: Reprinted by permission of the Population Crisis Committee.

(especially establishing free markets and democratic institutions), you may never achieve your goal of improving the well-being of your citizens.

It has been said that "in Bucharest, the Chinese had declared that overpopulation was no problem under socialism. In 1984, the Americans declared that overpopulation could be solved by capitalist development" (Donaldson 1990:130). Neither of those views is actually supportable by evidence from the real world, but the turnabout of attitudes among delegates meant that the 1984 Conference did not move the world population policy agenda along in any important way. Demeny (1994) has succinctly characterized the outcome of the Mexico City conference as follows:

> The "Recommendations for the Further Implementation of the World Population Plan of Action" adopted at Mexico City amplified the language of Bucharest but added no qualitatively novel elements. The 88 Recommendations placed added emphasis and urgency on concerns with the environment, the role and status of women, meeting "unmet needs" for family planning, and education" [Demeny 1994:12].

At the conclusion of the conference the delegates agreed to a Declaration on Population and Development, summarizing the major accomplishments of the conference, and this is shown in Table 14.4. You can see the consensus was that the World Population Plan of Action adopted at Bucharest was still valid, even if countries had not yet done very much in implementing their national population policies.

The Build-up to the Cairo Conference

In the early 1990s, as people began to anticipate the United Nation's plans for a 1994 Population Conference, the discussion splintered into the three corners of the debate about the relationship between population growth and economic development outlined in Chapter 12. Representing the view that the world has a tremendous population problem that we collectively must continue to try to resolve were members of the social and environmental sciences who study the issues, as well as people in the public and medical health service fields who deal regularly with the repercussions of rapid population growth (see, for example, Westoff 1994).

Representing the view that population growth was not the issue—that building free markets and capitalism were the important tasks—was, once again, Julian Simon and other "boomsters" (see, for example, Eberstadt 1994). From this group's perspective, there was no particular reason to have a population conference, and that view was shared by the Vatican, which has been opposed to all of the world population conferences, given their emphasis on the provision of family planning.

The newest and loudest voices added to the debate over world population policy were those of feminists, whose arguments center on the idea that abstract population targets represent additional ways in which men try to exploit women for their own purposes. In particular, the idea that "population control is a euphemism for the control of women" (Seager 1993:216) drew evidence from the

Table 14.4   Mexico City Declaration on Population and Development

The following declaration was approved by acclamation at the conclusion of the International Conference on Population in Mexico City on August 14, 1984:

1. The International Conference on Population met in Mexico City from 6 to 14 August 1984, to appraise the implementation of the World Population Plan of Action, adopted by consensus at Bucharest, ten years ago. The Conference reaffirmed the full validity of the principles and objectives of the World Population Plan of Action and adopted a set of recommendations for the further implementation of the Plan in the years ahead.

2. The world has undergone far-reaching changes in the past decade. Significant progress in many fields important for human welfare has been made through national and international efforts. However, for a large number of countries it has been a period of instability, increased unemployment, mounting external indebtedness, stagnation and even decline in economic growth. The number of people living in absolute poverty has increased.

3. Economic difficulties and problems of resource mobilization have been particularly serious in the developing countries. Growing international disparities have further exacerbated already serious problems in social and economic terms. Firm and widespread hope was expressed that increasing international co-operation will lead to a growth in welfare and wealth, their just and equitable distribution and minimal waste in use of resources, thereby promoting development and peace for the benefit of the world's population.

4. Population growth, high mortality and morbidity, and migration problems continue to be causes of great concern requiring immediate action.

5. The Conference confirms that the principal aim of social, economic and human development, of which population goals and policies are integral parts, is to improve the standards of living and quality of life of the people. This Declaration constitutes a solemn undertaking by the nations and international organizations gathered in Mexico City to respect national sovereignty to combat all forms of racial discrimination including *apartheid,* and to promote social and economic development, human rights and individual freedom.

* * * *

6. Since Bucharest the global population growth rate has declined from 2.03 to 1.67 percent per year. In the next decade the growth rate will decline more slowly. Moreover, the annual increase in numbers is expected to continue and may reach 90 million by the year 2000. Ninety percent of that increase will occur in developing countries and at that time 6.1 billion people are expected to inhabit the Earth.

7. Demographic differences between developed and developing countries remains striking. The average life expectancy at birth, which has increased almost everywhere, is 73 years in developed countries, while in developing countries it is only 57 years and families in developing countries tend to be much larger than elsewhere. This gives cause for concern since social and population pressures may contribute to the continuation of the wide disparity in welfare and the quality of life between developing and developed countries.

8. In the past decade, population issues have been increasingly recognized as a fundamental element in development planning. To be realistic, development policies, plans and programmes must reflect the inextricable links between population, resources, environment and development. Priority should be given to action programmes integrating all essential population and development factors, taking fully into account the need for rational utilization of natural resources and protection of the physical environment and preventing its further deterioration.

Table 14.4 (continued)

9. The experience with population policies in recent years is encouraging. Mortality and morbidity rates have been lowered, although not to the desired extent. Family planning programmes have been successful in reducing fertility at relatively low cost. Countries which consider that their population growth rate hinders their national development plans should adopt appropriate population policies and programmes. Timely action could avoid the accentuation of problems such as overpopulation, unemployment, food shortages, and environmental degradation.

10. Population and development policies reinforce each other when they are responsive to individual, family and community needs. Experience from the past decade demonstrates the necessity of the full participation by the entire community and grass-roots organizations in the design and implementation of policies and programmes. This will ensure that programmes are relevant to local needs and in keeping with personal and social values. It will also promote social awareness of demographic problems.

11. Improving the status of women and enhancing their role is an important goal in itself and will also influence family life and size in a positive way. Community support is essential to bring about the full integration and participation of women into all phases and functions of the development process. Institutional, economic and cultural barriers must be removed and broad and swift action taken to assist women in attaining full equality with men in the social, political and economic life of their communities. To achieve this goal, it is necessary for men and women to share jointly responsibilities in areas such as family life, child-caring and family planning. Governments should formulate and implement concrete policies which would enhance the status and role of women.

12. Unwanted high fertility adversely affects the health and welfare of individuals and families, especially among the poor, and seriously impedes social and economic progress in many countries. Women and children are the main victims of unregulated fertility. Too many, too close, too early and too late pregnancies are a major cause of maternal, infant and childhood mortality and morbidity.

13. Although considerable progress has been made since Bucharest, millions of people still lack access to safe and effective family planning methods. By the year 2000 some 1.6 billion women will be of childbearing age, 1.3 billion of them in developing countries. Major efforts must be made now to ensure that all couples and individuals can exercise their basic human rights to decide freely, responsibly and without coercion, the number and spacing of their children and to have the information, education and means to do so. In exercising this right, the best interests of their living and future children as well as the responsibility towards the community should be taken into account.

14. Although modern contraceptive technology has brought considerable progress into family planning programmes, increased funding is required in order to develop new methods and to improve the safety, efficacy and acceptability of existing methods. Expanded research should also be undertaken in human reproduction to solve problems of infertility and subfecundity.

15. As part of the overall goal to improve the health standards for all people, special attention should be given to maternal and child health services within a primary health care system. Through breast-feeding, adequate nutrition, clean water, immunization programmes, oral rehydration therapy and birth spacing, a virtual revolution in child survival could be achieved. The impact would be dramatic in humanitarian and fertility terms.

16. The coming decades will see rapid changes in population structures with marked regional variations. The absolute numbers of children and youth in developing countries will continue to rise so rapidly that special programmes will be necessary to respond to their needs and aspirations, including productive employment. Aging of populations is a phenomenon which many countries will experience. This issue requires attention

Table 14.4   (continued)

particularly in developed countries in view of its social implications and the active contribution the aged can make to the social, cultural and economic life in their countries.

17. Rapid urbanization will continue to be a salient feature. By the end of the century, 3 billion people, 48 percent of the world's population, might live in cities, frequently very large cities. Integrated urban and rural development strategies should therefore be an essential part of population policies. They should be based on a full evaluation of the costs and benefits to individuals, groups and regions involved, should respect basic human rights and use incentives rather than restrictive measures.

18. The volume and nature of international migratory movements continue to undergo rapid changes. Illegal or undocumented migration and refugee movements have gained particular importance; labour migration of considerable magnitude occurs in all regions. The outflow of skills remains a serious human resource problem in many developing countries. It is indispensable to safeguard the individual and social rights of the persons involved and to protect them from exploitation and treatment not in conformity with basic human rights; it is also necessary to guide these different migration streams. To achieve this, the co-operation of countries of origin and destination and the assistance of international organizations are required.

19. As the years since 1974 have shown, the political commitment of Heads of State and other leaders and the willingness of Governments to take the lead in formulating population programmes and allocating the necessary resources are crucial for the further implementation of the World Population Plan of Action. Governments should attach high priority to the attainment of self-reliance in the management of such programmes, strengthen their administrative and managerial capabilities, and ensure co-ordination of international assistance at the national level.

20. The years since Bucharest have also shown that international co-operation in the field of population is essential for the implementation of recommendations agreed upon by the international community and can be notably successful. The need for increased resources for population activities is emphasized. Adequate and substantial international support and assistance will greatly facilitate the efforts of Governments. It should be provided wholeheartedly and in a spirit of universal solidarity and enlightened self-interest. The United Nations family should continue to perform its vital responsibilities.

21. Non-governmental organizations have a continuing important role in the implementation of the World Population Plan of Action and deserve encouragement and support from Governments and international organizations. Members of Parliament, community leaders, scientists, the media and others in influential positions are called upon to assist in all aspects of population and development work.

* * * *

22. At Bucharest, the world was made aware of the gravity and magnitude of the population problems and their close interrelationship with economic and social development. The message of Mexico City is to forge ahead with effective implementation of the World Population Plan of Action aimed at improving standards of living and quality of life for all peoples of this planet in promotion of their common destiny in peace and security.

23. IN ISSUING THIS DECLARATION, ALL PARTICIPANTS AT THE INTERNATIONAL CONFERENCE ON POPULATION REITERATE THEIR COMMITMENT AND REDEDICATE THEMSELVES TO THE FURTHER IMPLEMENTATION OF THE PLAN.

Source: United Nations Fund for Population Activities, 1984, Document No. E/CONF.76/L.4.

coercive policies put into place in China. Although in agreement on this point, the feminist movement split into two differing views on the importance of population issues and thus on the relevance of a population conference. One view was an essentially neo-Marxist position that the problems in the world are caused not by population growth but by inequities in the distribution of power and resources (Hartmann 1995). The other feminist perspective did not deny the importance of population growth as a constraint to human well-being, but argued that the most important issue for the world is to improve the levels of women's rights and reproductive health on a global basis (see, for example, Chesler 1994; Dixon-Mueller 1993). In so doing, it was argued, the "solution" to the high birth rate will be effected.

The 1994 International Conference on Population and Development—Cairo

The International Conference on Population and Development was preceded by several preparatory meetings (PrepComs) sponsored by the United Nations Population Division and the United Nations Population Fund, as well as meetings sponsored by nongovernmental organizations (NGOs), and by a plethora of books, professional journal articles, magazine articles, newspaper columns and editorials, talk radio and TV, and any other imaginable medium of human communication. In the process, the World Population Plan of Action was essentially scrapped, and in its place a new Programme of Action was drafted and circulated in printed form to the delegates and made available to the rest of us on the internet.

Most of the content of the Programme of Action had been agreed to ahead of time at the PrepComs, but controversial material was bracketed for discussion by the delegates in Cairo. The most controversial issue was abortion, and the Vatican created worldwide public attention with its opinion of the matter. After five days of debate and special committee meetings, the Vatican did succeed in altering some of the language regarding safe abortions.

As approved at the conclusion of the Cairo conference, the Programme of Action covers an extremely wide range of topics, in recognition of the fact that the causes and consequences of population growth reach into almost every aspect of human existence. Some critics have complained that many of the action items are too vague to be effective—drafted that way to avoid controversy. Indeed, the conference report stayed away from China's controversial population policies by simply never mentioning China at all! Still, the conference regalvanized world interest in and concern about population growth. Nafis Sadik, Executive Director of the United Nations Population Fund, served as secretary-general of the conference, and she summarized the work of the conference as follows:

> The delegates have crafted a Programme of Action for the next 20 years which starts from the reality of the world we live in, and shows us the path to a better reality. The Programme contains highly specific goals and recommendations in the mutually reinforcing areas of infant and maternal mortality, education, and reproductive health and family planning, but its effect will be more wide-ranging than that. This Programme of Action has the potential to change the world.

Implementing the Programme of Action recognizes that healthy families are created by choice, not chance.

The Programme of Action recognizes that poverty is the most formidable enemy of choice. One of its most important effects will be to draw women into the mainstream of development. Better health and education, and freedom to plan their family's future, will widen women's economic choices; but it will also liberate their minds and spirits. As the leader of the Zimbabwe delegation put it, it will empower women, not with the power to fight, but with the power to decide. That power of decision alone will ensure many changes in the post-Cairo world [Sadik 1994:4–5].

In Table 14.5, I have summarized those action items that relate to the three components of the demographic transition—the transitions of fertility, mortality, and migration. Targets still exist but are reserved largely for improving health and lowering mortality. Perhaps most importantly, most of the concerns that feminists have had about the role of women in society were addressed very specifically in a variety of places within the Programme of Action and many have been specifically endorsed for funding by organizations such as the U.S. Agency for International Development (Daulaire 1995) and the World Bank (World Bank 1994). It is in this context of reproductive decision making that the language of the 1994 International Conference on Population and Development differs most from its predecessors, and that deserves additional comment.

The Three Conferences in the Context of Demographic Theory

All three World Population Conferences were built on the basic premise that population growth is a potential hindrance to a country's ability to develop economically, and that a plan for the improved well-being of a population needs to include a strategy for limiting population growth (even if not all delegates necessarily agreed with that premise). All three conferences further accepted the idea that mortality reduction should always be a high priority, despite the fact that reductions in mortality put added pressure on population growth. All three conferences accepted the idea that migration should be controlled, although none of the conferences has been very specific about how to do that.

The biggest problem that each conference dealt with was how to cope with high fertility. If we follow the U.S. perspective over time, we see that each of Coale's three preconditions for a fertility decline has been presented as the predominant U.S. policy prescription, except that they have been presented in reverse order. In 1974, the United States thought that family planning (Coale's third precondition) was the most important aspect of fertility on which to concentrate policy initiatives. In 1984, the emphasis on development implicitly recognized that the motivation to reduce fertility (Coale's second precondition) must accompany (indeed, precede) the demand for family planning services. Only by 1994 did the U.S. position get to the first precondition—that couples (especially women) must be empowered to take control of their own reproductive decisions. Only when that happens will the other preconditions become relevant for policy. To be sure, each of these elements was listed in the Action Plans of all three conferences, but with respect to fertility, in particular,

Table 14.5   Selected Items from the Programme of Action Adopted
at the 1994 International Conference on Population and Development in Cairo, Egypt

Action Items Relating Specifically to:	Chapter and Paragraph	Summary of Policy
Fertility transition	4.4	Empower women and eliminate inequalities between men and women
	4.17	Expand value of girl children
	4.26	Promote equal participation of women and men in all areas of life
	7.15	Support the principle of voluntary choice in family planning
	7.16	By the year 2015, provide universal access to a full range of safe and reliable family planning methods that are not against the law
Mortality transition	6.5	Aim to reduce high levels of infant, child, and maternal mortality
	8.5	Achieve by the year 2005 a life expectancy greater than 70 years and by the year 2015 greater than 75 years; and reduce mortality differentials
	8.16	Reduce infant and under age 5 years mortality by one third (or at least down to 50–70 per 1,000) by the year 2000
	8.21	Make significant reductions in maternal mortality by the 2015
	8.32	Control the AIDS pandemic
Migration transition	9.4	Adopt regional development strategies
	9.14	Increase management of urban development
	9.21	Address the causes of internal displacement
	10.3	Increase motivation to stay in one's own country
	10.17	Reduce the causes of undocumented migration
	10.23	Address the root causes of refugee movements

Note: The numbers refer to the chapter and paragraph within the Programme of Action.
Source: United Nations, 1994, Report of the International Conference on Population and Development
(New York: United Nations).

the emphasis is important because that influences what will likely be funded by international donors and put into place by countries facing rapid population growth.

Finally, then, by 1994 it seems as if the world community may have gotten it right in terms of a global perspective on population policy. Of course, setting a course is a lot easier than taking the trip, and policy making and policy implementation are both extremely difficult tasks. As Rumbaut (1995) reminds us, policy makers are always faced with a difficult task: "Condemned to try to control a future they cannot predict by reacting to a past that will not be repeated, policy makers are nonetheless faced with an imperative need to act that cannot be ignored as a practical or political matter" (p. 311).

Planning for a Bifurcated World—The Old and the Young

The next phase in world population policy planning is to prepare for a world in which the haves and have nots—the developed and developing countries—are divided not only by income and rates of population growth, but by distinctly different population structures. The legacy of several decades of low fertility and low mortality in the developing countries is an increasingly older population, whereas continued high fertility, combined with ever lower mortality, keeps the populations of areas such as sub-Saharan Africa, western Asia, and south Asia with very young population age structures.

In Figure 14.5 you can see that the United Nations projections suggest that by the year 2000 the more developed countries will have only 22 percent of their population under the age of 20, compared with 35 percent in the less developed nations. By contrast, 25 percent of the population in the more developed countries will be aged 60 and older, compared with only 11 percent in the less developed nations. The bottom row of data below the graph in Figure 14.5 reminds you, nonetheless, to keep in perspective that although 89 percent of the population younger than age 20 in the year 2020 will be living in the less developed nations, those countries will still be home to 70 percent of the population aged 60 and older.

Policy makers in North America, Europe, East Asia, and parts of South America are busily at work trying to figure out how to cope with an aging population. The kinds of changes in family size and household structure that produced such a high fraction of older people also means that the traditional family structure that in earlier generations might have taken care of the elderly no longer exists (Ogawa and Retherford 1997). The elderly in the more developed nations will have fewer children to call upon for support, but they are also growing older with unprecedentedly high levels of living, which they will, of course, hope to maintain. Have they saved enough money themselves to do that, or will they count on communal support—from taxes on younger people? (Wise 1997). Will North America, Europe, and Asia draw increasingly upon immigrants to support the older population? (Espenshade and Gurcak 1996).

The less developed nations will continue to struggle to meet the payroll, so to speak—needing to keep finding jobs for the numerous young people in their midst, and that will continue to put pressure on people in these countries to seek employment opportunities elsewhere. These young people also aspire to live better, eat better, and have a longer life—to emulate the living standards in the more developed world. Can it be done if they don't lower their fertility levels quickly and dramatically? Can it be done even if they do?

Summary and Conclusion

One of the major uses of demographic science is as a tool for shaping the future, for trying to improve the conditions, both social and material, of human existence.

Figure 14.5   The Young and the Old in the Year 2020

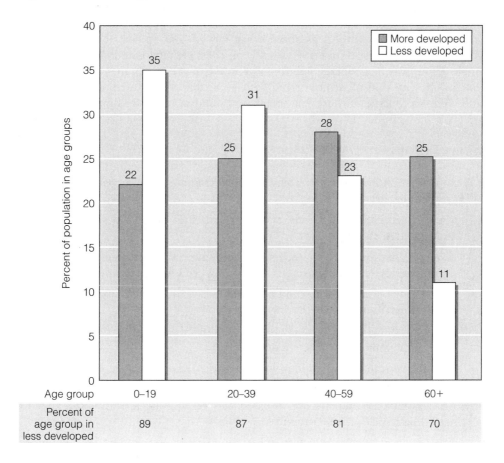

Source: Adapted from medium variant projections for the year 2020 in United Nations, 1997, The Sex and Age Distribution of the World Populations: The 1996 Revision (New York: United Nations).

To do that requires a demographic perspective—an understanding of how the causes of population change are related to the consequences. Throughout the previous 13 chapters, I have explored the ins and outs of those causes and consequences: how and why mortality, fertility, and migration change; how they affect the age and sex structure of society; how population characteristics are influenced by and in turn affect demographic changes; how women's status can influence fertility levels; how demographic changes affect the process of growing old in developed societies; and how population growth affects the urban environment, economic development, food resources, and environmental degradation. Implicit in all those discussions is the idea that an understanding of what has happened in the past and what is happening now will provide you with insights about the future, which is, of course, your first step toward advocating or promoting a population policy.

On the basis of what you now know about population change, you should be able to make some reckoning of what the future may be. If what you see upsets you and you want to advocate a policy to put the brakes on population growth, I have tried to provide you with ideas about what can be done to curtail population increase. If what you see does not worry you, then perhaps you should reread the book. Either way, I encourage you to go on to the next chapter, in which I chart for you a different kind of demographic path—one that applies demographic theory and methods to issues of business, social, and political planning strategies.

Main Points

1. A population policy is a strategy for achieving a particular pattern of demographic change.

2. Population policies are rarely ends in themselves but are designed to bring about desired social changes.

3. Policy orientations to close the gap or maintain the fit between goals and projections include retarding growth, promoting growth, and maintaining growth.

4. Growth in population size could be retarded by raising mortality—as expressed in the lifeboat and triage ethics.

5. Retarding growth is more likely to be accomplished by encouraging emigration or, at a minimum, discouraging immigration.

6. North American immigration policies are oriented toward gate-keeping—keeping out people who are not wanted.

7. Most population policies are oriented toward limiting fertility, especially by providing family planning services.

8. The evidence suggests that family planning programs are most effective when they take place in the context of economic and social change.

9. The most comprehensive way to limit fertility is by promoting the kinds of social change that alter a couple's motivation to have a large family and thereby encourage them to seek methods of family planning.

10. India has been relatively ineffective in implementing a population policy, whereas a combination of economic incentives for small families and disincentives for large families has had a significant impact in China.

11. Governments that want to promote or maintain growth can do so primarily by encouraging a further decline in the death rate and by making it difficult for women to obtain family planning services.

12. The 1974 World Population Conference produced a World Population Plan of Action that was largely reaffirmed at the 1984 International Conference on Population in Mexico City.

13. The 1994 International Conference on Population and Development produced a new Programme of Action emphasizing the importance of women's social and reproductive rights.

14. If you have been reading this book continuously at the pace of one chapter per hour, nearly 150,000 people will have been added to the world's population between the time you started reading Chapter 1 and now. That is the equivalent of nearly 500 full flights on a Boeing 777, or more than 33 flight loads of people per chapter.

Suggested Readings

1. Jason L. Finkle and C. Alison McIntosh (eds.), 1994, The New Politics of Population: Conflict and Consensus in Family Planning (Supplement to Volume 20 of Population and Development Review) (New York: Oxford University Press).

   This volume is based on a conference of experts held in Italy in 1990 and the papers contain many of the themes included in the Programme of Action from the 1994 International Conference on Population and Development.

2. John A. Ross and Elizabeth Frankenberg, 1993, Findings from Two Decades of Family Planning Research (New York: The Population Council).

   How effective are family planning programs in reducing fertility and improving the health of women, men, and their children? This volume reviews the literature to help provide answers.

3. B. Robey, P.T. Piotrow, and C. Salter, 1994, "Family planning lessons and challenges: Making programs work." Population Reports, Series J, No. 40 (Baltimore: Johns Hopkins School of Public Health, Population Information Program).

   When family planning programs do work, what is it that breeds success? This report scrupulously lays out what it takes to generate an effective program.

4. World Bank, 1994, Population and Development: Implications for the World Bank (Washington, D.C.: The World Bank).

   Because of the World Bank's important role in helping to shape economic development efforts in less developed countries, its approach to population issues is obviously critical. This volume explains a policy approach that is very much in line with the Cairo Conference.

5. A. Adlakha and J. Banister, 1995, "Demographic perspectives on China and India," Journal of Biosocial Science 27:163–178.

   No discussion of population policy can ignore India and China, and the authors, experts on the two countries, provide an excellent summary of the situation in these two demographic giants.

✸ Websites of Interest

Remember that websites are not as permanent as books and journals, so I cannot guarantee that each of the following websites still exists at the moment that you are reading this:

1. http://www.tfgi.com/software/SPEC.HTM

   Would you like to try your hand at assessing the future? The Futures Group International (tfgi), with support from the U.S. Agency for International Development, has developed a population projection program (Spectrum) that you can download at no cost from this site. The program incorporates models for assessing family planning program needs.

2. **http://www.undp.org/popin/icpd5.htm**

   The United Nations has set up a program to monitor world progress in achieving the Programme of Action from the 1994 International Conference on Population and Development (icpd) and you can check on that progress at this website, which also includes a link to the full text of the Programme of Action.

3. **http://www.populationinstitute.org**

   The Population Institute in Washington, DC is a nonprofit organization whose goal is to disseminate information about population issues and mobilize activities in support of programs that seek to lower rates of population growth and reduce the human impact on the environment.

4. **http://www.zpg.org**

   Zero Population Growth (zpg) is probably the most famous of the organizations working to build a consensus for a U.S. population policy of ZPG, and for a worldwide policy to slow down population growth and the deterioration of the environment.

5. **http://www.africa2000.com**

   Not all the voices on the internet are in favor of implementing population policies and this website offers some dissent, although in the process they also include some good material in favor of population policies, which the authors of this site then try to debunk.

CHAPTER 15
Demographics

> Our market research showed that the youthful image of the new decade had a firm basis in demographic reality. Millions of teenagers, born in the postwar baby boom, were about to surge into the national marketplace. Here was a market in search of a car. Any car that would appeal to these young customers had to have three main features: great styling, strong performance and a low price [Iacocca 1984:64].

The result was Ford's Mustang, brought out in the 1960s; and the rest, as they say, is history. Another piece of history was being written by General Motors at about the same time. The economical, mid-sized Chevrolet Nova was introduced in Mexico, with high hopes of tapping into a lucrative market. The result: appallingly low sales figures (Frost and Deakin 1983). The reason: in Spanish, No va means "it doesn't go"! The lesson: Some of the world's best business decisions have a solid demographic base; some of the worst do not.

The insights into the social world offered by the demographer are extremely valuable not only to business executives but to social planners and political strategists as well—individuals who often are less concerned with trying to change demographic trends than with using information about those trends to improve their company's profit margin, improve the provision of publicly funded services, or plot campaign and legislative strategy. It is no great exaggeration to say that every piece of information you have gleaned from the previous 14 chapters could be of great value to people who seek solutions to many practical dilemmas, and it is my purpose in this chapter to show you how to mine this treasure trove. To this end I begin with a definition of **demographics** (the popular name for applied demography) and then move to a discussion of its uses. These include planning in business (marketing, site selection, investment, and management demographics) as well as planning for social purposes (such as education, public services, health services, and criminal justice) and political planning (such as legislative analysis and campaign strategy). To help inspire you to find a useful purpose in your own life for demographic information, I conclude the chapter with a discussion of applied demography as a career choice.

Defining Demographics

Throughout this chapter I will use the terms "demographics" and "applied demography" interchangeably, even though the former term is clearly the most popular. Demographics is often used in the singular, meaning the application of demographic information and methods in business and public administration, and it is also seen in the plural, referring to the demographic information itself (Merrick and Tordella 1988). The major difference between demographics and the field of demography more generally is that the latter is concerned more with producing new knowledge and understanding of human behavior, whereas the former is concerned more with the use of existing knowledge and techniques to identify and solve problems. Thus in this chapter you will not necessarily learn anything about demographic behavior that I have not already discussed in earlier chapters, but you will learn new ways to apply that information and bring solutions to problems that have a demographic component. What might those problems be?

The Uses of Demographics

Because we live in a social world, many if not most of the decisions that have to be made about life involve people and when the issues relate to how many people there are, and where they live and work, and what are they like, then demographics becomes part of the *decision-making process* (Pol and Thomas 1997), in which you systematically lay out a strategy for achieving your goal. These goals will obviously be different depending upon whether your problems relate to business, social planning, or political planning.

There are several major uses to which demographics are regularly put by people in business, including marketing strategies, site selection, investment decisions, and management of human resources. Marketing demographics answer questions such as: Who buys my product, and where are those people located? Site selection answers the basic question of: Where should I locate my business? Investment demographics answer questions like: What are the potentially most profitable kinds of products in the future? What other countries might have a market for my products or services? Human-resource management demographics look at questions such as: Is my company hiring the proper racial and ethnic mix of employees? Will future labor force shortages prompt me to discourage older employees from retiring?

Social agencies have also recognized the value of demographics for planning. School districts ask: How many students will there be next year? Or: Do we need to redraw school attendance areas? Police departments need to ask what the changes in population growth might portend for criminal activity in their areas. Health officials plan services by identifying specific high-risk populations (people who have an unusually high need for particular services) and attempting to meet their needs: Where is the best location for a new skilled nursing facility? Or: Does the community hospital really need an additional 40 beds?

Demographics have worked their way into the political process. In the United States, of course, the constitutional basis of the Census of Population is to provide data for the apportionment of seats in the House of Representatives (as I will discuss in more detail later in the chapter). Legislators also ask questions about how population growth and distribution influence the allocation of tax dollars. Will the increase in the older population bankrupt the social security system? Would federal subsidies to inner-city areas help lower the unemployment rate? Are illegal aliens creating an undue burden on the educational and welfare systems? Politicians are not completely altruistic, of course, and have been known to use demographics for their own personal use when campaign time rolls around. What are the characteristics of a candidate's supporters and where are these voters located? Important and interesting though the uses of demographics for social and political planning may be, however, the most extensive use of demographics clearly is in business, and the literal explosion in the use of demographic data can be traced to computerized geographic information systems.

Geographic Information Systems— The Tool of Demographics

"We know who you are and we know where to find you" could be described as the motto of marketing. The combined computerization of demographic (who you are) and geographic (where to find you) data has created a revolution in the access a business has to information about potential customers, with maps, data, and computers forming the principal components of the geographic information system (GIS), as I mentioned in Chapter 2. To be useful, the data must be geographically referenced—they must have some geographic identifier (an address, ZIP code, census tract, county, state, or country, for example, going from most to least specific) that permits them to be linked to the map. Demographic data are virtually always geographically referenced (and have been in the United States since 1850), and the Geography Division of the U.S. Bureau of the Census works closely with the Population Division to make sure that data are identified for appropriate levels of "census geography," as shown in Table 15.1.

Knowledge and understanding are based on information, and our information base grows by being able to tap more deeply into rich data sources such as the census.

Table 15.1 The Census Provides Geographically Referenced Data for a Wide Range of Geographic Areas

Basic Geographic Hierarchy of Census Data:

United States

Region (9)

Division (4)

State (50, plus the *outlying areas* of American Samoa, Guam, northern Mariana Islands, Palau, Puerto Rico, and the Virgin Islands)

County (the basic administrative and legal subdivision of states)

County Subdivision (also known as minor civil divisions, including towns and townships)

Place (incorporated cities and unincorporated census-designated places)

Census Tract/Block Numbering Area (tracts are small, relatively permanent subdivisions of a county, delineated for all metropolitan areas and other densely settled areas; block numbering areas are similar to census tracts except that they are set up for nonmetropolitan areas that have not been tracted)

Block Group (a cluster of blocks within census tracts, usually containing about 400 housing units)

Block (the smallest geographic unit, usually bounded on all sides by readily identifiable features such as streets, railroad tracks, or bodies of water)

Other Commonly Used Geographic Entities that Do Not Necessarily Coincide with Census Data:

Three-Digit Zip Code (the first three digits of the Zip [Zone Improvement Program] code, representing major cities or sectional distribution centers)

Zip Code (the standard five-digit Zip Code, encompassing areas similar to, but not identical with, census tracts)

Zip Code + 4 (the Zip Code plus an additional 4-digit number unique to an address)

Areas of Dominant Influence (ADI—developed by the Arbitron Corporation)

Area Code (telephone)

Source: U.S. Bureau of the Census, 1992, Maps and More (Washington, D.C.: Government Printing Office).

GIS is an effective tool for doing this. The data in question could, of course, be from anywhere, including from a company's own database of customers. The only requirements are that your data must be geocoded (such as the addresses of all of a bank's depositors) and that you have a geographic database (such as the TIGER file, which I discussed in Chapter 2) that will find those geographic points on a map. The census, however, is the single most useful source of data for most business purposes, and a survey of readers of *Business Geographics* magazine (a publication devoted to business uses of GIS) found that "demographic analysis" was the leading area of interest of its readers, mentioned by fully 90 percent of respondents to the survey (Bryan 1994).

When geographic information systems incorporate demographic data, the resulting combination is called **geodemographics** (demographic data analyzed for specific geographic regions) that generates powerful results in a variety of applications. I have used census data for the United States (which are available to you through the internet, or you can purchase them on CD-ROM) in conjunction with one of the commercially available GIS software programs (Atlas*GIS from Environmental Systems Research Institute) to prepare the maps shown in Figure 15.1. If you were a person interested in marketing to Hispanics in San Diego, one of the first questions you would reasonably ask is, "Where in San Diego are Hispanics?" (not to be confused with "Where in the World is Carmen Sandiego?"). In the larger map in Figure 15.1, the computer has shaded those census tracts in San Diego County with a "high concentration" of Hispanics (defined for this map as census tracts with at least 1,000 Hispanics and at least 20 percent of the population in 1990 being Hispanic).

I have circled two areas on the map where there are geographic concentrations of Hispanics—one in the north of the county and one in the south, adjacent to the Mexican border. Using other features of the GIS, I was able to determine that the southern concentration has more Hispanics and that they live in greater proximity to one another than in the northern part of the county. Therefore, as a retailer it probably makes more sense to focus my interest in the southern region. Now, zooming in on that area (the smaller map), I asked the computer to shade darkly those census tracts in which 50 percent or more of the population was Hispanic, whereas the lighter shading represents tracts with 25 percent to 49 percent of the population Hispanic. At this scale the streets also show up on the map, and it seems reasonable to suggest to the retailer that a business location near either Interstate 805 or Interstate 5 would be readily accessible to a large fraction of the Hispanic population in San Diego County.

The map in Figure 15.1 illustrates only one phase in the use of GIS. Because it is a system, its value lies in the ability to work interactively with the data—obtain results such as those shown in Figure 15.1, ask additional questions, get additional results, and so on. The only shortcoming of GIS is that it does not know which questions to ask. They must come from the operator, and the rest of the chapter is devoted to helping you think about what those questions might be. What are the ways in which demographics can be most profitably exploited?

Business Planning

Making a profit in business requires, among other things, having an edge on your competitors. This may mean that your product or service is better or at least different in some meaningful way, or it may mean that you have been able to find potential customers that your competition has ignored, or it may mean that you have found a better location for your business than have your competitors. Profits are also maximized by investing in the businesses with the greatest opportunities for growth, as well by being able to employ and keep the best possible labor force. In each instance, demographic knowledge can be one of the keys to generating higher profits.

Marketing Demographics

One of the most impressive (and profitable) ways in which demographics is employed in business is for marketing. A prime example of this is the use of demographics to help a company properly segment and target the market for a product.

Segmenting and Targeting Populations   **Segmentation** refers to the manufacturing and packaging of products or the provision of services that appeal to specific so-

Figure 15.1   Combining Demographic and Geographic Information to Locate Potential Hispanic Consumers in San Diego County

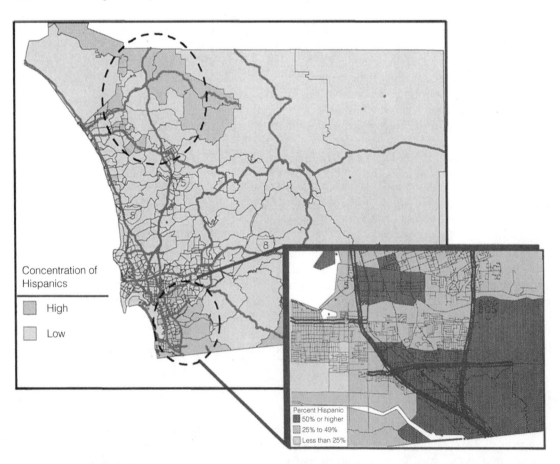

ciodemographically identifiable groups within the population. "Consumer markets are segmented on the basis of such demographic variables as geographic location, rate of product usage, income, age, sex, education, stage in the family life cycle, religion, race, and social class. Industrial markets are segmented demographically according to such variables as geographic location, kind of business, rate of product usage, and size of user" (Levy, Frerichs, and Gordon 1994:177). Automobile manufacturers are most famous for segmenting the market and producing different cars to appeal to different categories of people. Indeed, Chrysler designed its Neon car specifically to appeal to "Generation X" (Lavin 1993). Retailing chains often do the same thing by tailoring their inventory (such as clothing styles) to meet the demographic characteristics of the customers in a given area.

Closely related to segmenting (sometimes indistinguishable in the literature) is the concept of **targeting.** It involves picking out particular sociodemographic characteristics of people who might purchase what you have to offer, then appealing to the consumer tastes and behavior reflected in those particular characteristics. This is a little trickier than segmenting the market because you do not necessarily know in advance how appealing your product or service might be to specific sociodemographic groups.

A first step in targeting would be to examine data from the Consumer Expenditure Survey, conducted regularly by the U.S. Bureau of Labor Statistics. Some of these data are published in the Statistical Abstract of the United States and are available over the internet, but in order to get greater detail you have to either buy the data tapes from the government, or go to someone else who has purchased the data. Summaries of these data are also published by private companies, such as the Official Guide to Household Spending (Tran 1995). Let's suppose that you were interested in starting a business that might allow you to make money from your lifelong interest in floor coverings (why not?). Who might the potential customers be? Table 15.2 shows the age breakdown of consumer units purchasing floor coverings according to the 1994 Consumer Expenditure Survey. In 1994 the average

Table 15.2 What Is the Market for Floor Coverings in the United States?

| | Total | \multicolumn{6}{c}{Age of reference person in household:} | | | | | |
		Under 25	25–34	35–44	45–54	55–64	65–74
Number of consumer units (thousands of households)	93,263	7,093	19,540	23,440	18,633	12,624	11,933
Average annual expenditure on floor coverings ($)	123	38	85	142	165	167	86
Aggregate annual expenditures by all consumer units ($million)	11,468	270	1,661	3,328	3,074	2,108	1,026
Market share (%)	100	2	14	29	27	18	9

Source: Adapted from U.S. Bureau of Labor Statistics, Consumer Expenditure Survey, 1994–95, accessed from the website of the Consumer Expenditure Survey.

household in the United States spent $123 on floor coverings, ranging from a high of $167 for people aged 55 to 64 (people who are at their peak earnings and are perhaps upgrading their house prior to retirement) to a low of $31 for households where the reference person is under 25 (people who have a greater likelihood of being renters and are thus less likely to invest very much in carpets, linoleum, wood, and tile flooring). None of those numbers is very high on a per household basis, but you can see that the total market for floor coverings in the United States in the mid-1990s added up to $11.5 billion. Important to note, however, is the fact that people aged 35 to 44 (the baby boomers) had the largest share of the market (29 percent), despite the fact that they were not spending as much per household as were people who were older. Still, the boomers were entering the "nesting" age where they were most likely to be remodeling their homes (Blumenthal 1996), and this is why the boomers get catered to—they have disproportionate clout in the consumer market.

The consumer expenditure survey data give a sense of the general market for floor coverings and other categories of products and services, but they do not tell you specifically which consumers will be drawn to your specific product or service. The link between your product and the demographic profile of your potential or targeted consumers is called **psychographic research.** Researchers use a variety of methods, including surveys and focus groups to uncover the characteristics of people to whom particular products or advertising campaigns will be appealing (Piirto 1991; Weinstein 1994).

Armed with that information (and keeping in mind that behavior changes over time, so the research must be updated regularly), the next task is to find out where such people live and where they shop (remembering that these also change over time) so that they can be tracked as potential customers. It is rare, of course, for a single person (demographer or otherwise) to do all of this research. Much of it is handled by teams of people at one of the many demographic companies that have sprung up in the United States. A directory published annually by *American Demographics* lists more than 100 American firms offering demographic products and services. *Business Geographics* also publishes an annual list of firms offering geographic information system services to business clients, and most of those companies are also on the *American Demographics* list, underscoring the increasing inseparability of demographics and GIS.

Targeting by Demographic Characteristics   One of the most important tasks a business can undertake is to know its clientele in order to maximize its appeal. In Chapters 8 through 11, I emphasized the ways in which sociodemographic characteristics can influence behavior. It is this relationship that is of interest to business. The kinds of products you buy and how much you will spend on those products is dictated, at least in part, by factors such as age, gender, education, income, socioethnic identity, household structure and living arrangements, and whether or not you are living in a city.

Age and Gender   People at different ages have different needs and tastes for products and differing amounts of money to spend. Companies catering to the youngest age group have to keep track of the number of births (their potential market) as well as the characteristics of the parents and grandparents (who spend the money on behalf of the babies). The baby market has seen some wild fluctuations in recent decades in the United States, as you can see in Figure 15.2. The number of babies being born each year plummeted during the 1960s and 1970s, rebounded in the 1980s—peaking in 1990, slacked off in the 1990s, but is projected to rise steadily into the first part of the 21st century. Riding that wave has been tricky. Johnson & Johnson has flourished by convincing adults that baby shampoo is as good for football players as it is for babies and by diversifying its product line in a demographically relevant way, including acquiring the ownership of Ortho Pharmaceuticals (the largest U.S. manufacturer of contraceptives—helping to keep the birth rate low—and a large manufacturer of drugs to treat chronic diseases associated with aging) and the ownership of Tylenol (one of the world's most popular pain relievers).

It is dangerous business, however, to be lulled into believing that every company dealing with baby products will necessarily live or die on the peaks and troughs of

Figure 15.2   Riding the Age Wave: Births and Selected Age Groups in the United States

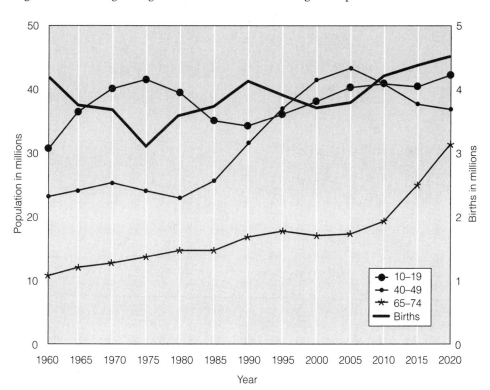

Sources: Data on births 1960–90 are from U.S. Bureau of the Census, 1996, Statistical Abstract of the United States (Washington, DC: Government Printing Office): Table 90; age data for 1960–90 are from the United Nations, 1997, The Sex and Age Distribution of the World Populations: The 1996 Revisions (New York: United Nations); birth and age data for 1995 through 2020 are from Jennifer C. Day, 1996, Population Projections of the United States by Age, Sex, Race, and Hispanic Origin: 1995–2050, U.S. Bureau of the Census Current Population Reports P20-1130: Tables 1 and 2.

birth cycles. In 1988 there were 345,000 fewer births than in 1958, but there were 455,000 more first births in 1988 than in 1958. If you have hung around new parents and new grandparents, you know that people react differently to first births than to others. In particular, they open their pocketbooks wider to pay for cribs, buggies, strollers, diapers, and every conceivable type of baby toy designed to stimulate and "improve the quality" of the babies. Businesses that cater to these needs, then, are as sensitive to birth order as they are to the absolute volume of births.

As the ups and downs in births work their way through the age structure, we can see that the age pattern of teenagers (ages 10 to 19) follows a wave that is a delayed reaction to the birth wave. Thus as the baby market dried up in the 1970s, the young adult population was expanding, only to deflate again in the 1980s, but by then the baby market was growing again. As we move into the 21st century, the crop of babies born in the 1980s will be moving through the teen and young adult ages. In the 1990s the parents of these children (sometimes called "Generation Y"—Kaufman 1997) were spending money on their education and entertainment. Later, they will be competing for space in college and hunting for a well-paying job, which will reinvigorate job-training and job-placement agencies. The number of people at each age is only part of the marketing equation, of course. The other part is how much money they have to spend, and it turns out that teenagers and young adults now have more money per person than their predecessors. This is due to a combination of factors, including a greater fraction of younger people who work part-time and an increase in the fraction of households with two earners but fewer children among whom the discretionary income must be shared. The result has been an increase in the economic clout of young people—which especially benefits the food, entertainment, sports, and apparel markets.

The middle age group (represented in Figure 15.2 by people aged 40 to 49) was relatively unchanged in size and essentially unnoticed until the Baby Boom generation began moving into it in the 1980s. Since then, it has become a new "boom" industry. Sales of lineless bifocal lenses tripled between 1985 and 1990, as did walking shoes (running shoe sales slowed to a walk). When not walking, the middle-aged baby boomers were driving their luxury or near-luxury cars and then their sport utility vehicles. It was the Baby Boom reaching middle-age that helped to fatten the non-fat market. People aged 45 to 49 are more likely than any other age to be concerned about the health effects of fat (Mogelonsky 1995) and as baby boomers began reaching that age, the market flourished for low-fat and non-fat foods. However, younger people, especially college students, have led the movement toward vegetarian meals (American Demographics 1995), and they may carry those food preferences into their middle ages at a time when the number of people 40 to 49 will dip as the baby boomers are replaced by the smaller cohort of Generation X.

The young-old population (ages 65 to 74) has been steadily increasing in numbers over time and, as you will recall from Chapter 9, has also become increasingly affluent. This segment of the population creates a market for a variety of things from leisure travel, to appliances with larger print, to door handles that are shaped to be used more easily by arthritic fingers (see Dychtwald and Flower 1989, for a long list of product adaptations for an aging population). Perhaps most importantly, the aging of the population in North America has spurred the marketing of health services and products aimed at that age group (see, for example, Phillips and Kobrinetz 1996; Pol and Thomas 1992).

The changing roles of women in society have provided opportunities for businesses to target women differently than in the past. Consider, for example, that more than half of all married women are now working, compared with one in four only a few decades ago. "Ads of the 1950s were replete with housewives whose greatest pleasure seemed to be seeing their faces reflected in well-scrubbed floors and polished furniture. In the 1970s, the housewives gave ground to images of impeccably dressed women executives dashing off to fancy French restaurants" (Bralove 1982:33). In the 1980s, the imagery changed again, to a new type of woman who is "fully dimensional . . . she cares about her home but isn't obsessive about it. If she has a career, it isn't necessarily as a TV anchor woman" (Bralove 1982:33). By the 1990s the ads were aimed at women who need a briefcase, women who need the latest technology in a cell phone, and women who need the most powerful laptop computer and presentation projector, not to mention a BMW convertible.

As the status and roles of women have changed, there have been complementary shifts in the male population requiring adjustments in marketing strategy. Increasing proportions of men are remaining single longer before marrying (and single after a divorce), with attendant problems of housekeeping and cooking, and married men are also increasingly likely to involve themselves in domestic affairs.

Sociodemographic Characteristics Education is an important sociodemographic variable, as you already know, and is a particularly potent predictor of consumer tastes. In 1990 there were 27 million households in the United States in which the head had at least some college education. That represents a market larger than the total population of most countries of the world. Magazine editorial boards watch these trends, and as the baby boomers graduated from college and starting getting good jobs, several magazines responded by upgrading their image or starting special editions "aimed at their high-demographic readers, as identified by the zip codes of prosperous communities" (Machalaba 1982:34). In 1981 *Apartment Life* magazine changed its name to *Metropolitan Home* and shifted its format to appeal to more affluent and educated consumers. Within a year its readership was dramatically different and its sales of advertising pages had risen by 8 percent.

Ethnic groups provide another identifiable target for marketers. African-Americans represent the best-educated, wealthiest, and fourth-largest (after Nigeria, Ethiopia, and South Africa) population of African-origin persons in the world—by anyone's standards a large group of consumers. Thus in 1991, Spiegel, Inc., teamed up with *Ebony* magazine to develop a fashion line and catalog aimed at African-American women (Schwadel 1991) and a national conference of African-American executives in 1990 concluded that the only problem with targeting to African-Americans was that there is not enough of it (Wynter 1990).

There is also little wonder that Hispanics have caught the attention of marketers when you consider that the number of Hispanics will have doubled between 1980 and the year 2000 and now represents more than 11 percent of the total U.S. population—virtually the same as the black population. At the same time, marketers have discovered that the Hispanic market is itself diverse—partly because of the fact that the Hispanic market spans three different major groups—Mexican-Americans (centered in Los Angeles), Puerto Ricans (primarily in New York City), and Cuban-Americans (concentrated in Miami), and partly because there exists both an English-speaking and a Spanish-speaking market within each of those groups (see, for example, Templin 1996).

Each Hispanic group has its own demographic implications for marketing. Mexican-Americans, for example, have birth rates nearly twice those of Anglos. Thus manufacturers of baby products find a denser market there than among Anglos. Puerto Ricans have high fertility, but also a high proportion of broken families and a very high incidence of poverty. Therefore, marketing is better reserved for basics such as food and clothing. Cuban-Americans, on the other hand, have low fertility and the greatest amount of discretionary income of any of the Hispanic groups, and are most likely to purchase expensive consumer durables such as large appliances and stereo equipment.

Asian-Americans represent a smaller, yet growing, ethnic market in the United States, especially along the Pacific Coast. Ethnically, it is an even more diverse group than Hispanics, and thus somewhat difficult to crack into. Asian-Americans have themselves become an important group of entrepreneurs, and these businesses often are targeted specifically to the Asian-American population (Noah 1991).

Household Structure   Some products are consumed just by individuals—things like clothes and personal care items. Other consumer goods, such as cars, furniture, appliances, and even food and beverages, are shared by members of the household. For these types of products, the characteristics of households may be as important for marketing as are the characteristics of the individual members of the household. Two Census Bureau demographers have tracked the shifts in household type over time (Green and Welniak 1991) and, as an extension of the discussion in Chapter 10, I offer you the data in Table 15.3. You can see that Green and Welniak have identified nine household types in the United States. The most common of these in 1990 was the dual-earner married couple with children. They represented 19 percent of all households in 1990, and by multiplying the number of such households by their average income, Green and Welniak were able to estimate that they accounted for 26 percent of the market share.

Greater diversity was clearly the household theme between 1970 and 1990. Especially noticeable in Table 15.3 is the drop in market share (from 25 percent to only 9 percent) of other married couples with children. This group represents the "traditional" family of one breadwinner, a spouse at home, and their children. Nearly every other type of household increased market share between 1970 and 1990 at the expense of the traditional one-earner family with children. Diversity means business opportunity for those who can spot it. People living alone, for example, are likely to have different eating patterns than a traditional family, and convenience counts for a lot. Thus the market has expanded for high-quality frozen food, and for take-out and delivery of a wide range of meals (Coleman 1995).

Household expenditures are also influenced by the life-cycle stages through which household members are passing. Wilkes (1995) has used data from the Consumer Expenditure Survey in the United States to show that household expenditures rise sharply when couples marry and form households, and stay high especially when children are present, then fall sharply as the head of household ages.

Table 15.3   The Market Share Changed for Different
Types of Households in the United States Between 1970 and 1990

Type of Household	1970	1990
Dual earner married couples with children:		
Percent of all households	20%	19%
Average household income[a]	$38,700	$49,600
Market share[b]	26%	26%
Other married couples with children:		
Percent of all households	21%	8%
Average household income[a]	$35,600	$40,000
Market share[b]	25%	9%
Childless married couples with householders under age 45 (young couples):		
Percent of all households	6%	7%
Average household income[a]	$33,100	$48,300
Market share[b]	7%	9%
Childless married couples with householders aged 45 to 64 (empty nesters):		
Percent of all households	15%	13%
Average household income[a]	$39,500	$53,800
Market share[b]	19%	19%
Childless married couples with householders aged 65 or older (mature couples):		
Percent of all households	9%	9%
Average household income[a]	$20,900	$32,300
Market share[b]	6%	8%
Single parents:		
Percent of all households	6%	9%
Average household income[a]	$17,500	$21,400
Market share[b]	4%	6%
Childless singles under age 45 (young singles):		
Percent of all households	3%	9%
Average household income[a]	$20,300	$25,200
Market share[b]	2%	6%
Childless singles aged 45 or older (older singles):		
Percent of all households	14%	6%
Average household income[a]	$11,900	$17,500
Market share[b]	5%	8%
Other multiple member (shared households):		
Percent of all households	6%	10%
Average household income[a]	$27,200	$36,900
Market share[b]	6%	10%

[a]Mean income for 1969 and 1989, in 1989 dollars.

[b]Average household income times the number of households in this category.

Source: Adapted from G. Green and E. Welniak, 1991, "The nine household markets," American Demographics (October):38.

Cluster Marketing   Knowing the sociodemographic target groups for your business does not mean that you can easily find and appeal to such people or households. However, the fact that "birds of a feather flock together" has meant that neighborhoods can be identified on the basis of a whole set of shared sociodemographic characteristics. This greatly facilitates the process of marketing to particular groups. In the 1970s

> . . . a computer scientist turned entrepreneur named Jonathan Robbin devised a wildly popular target-marketing system by matching zip codes with census data and consumer surveys. Christening his creation PRIZM (Potential Rating Index for Zip Markets), he programmed computers to sort the nation's 36,000 zips into forty "lifestyle clusters." Zip 85254 in Northeast Phoenix, Arizona, for instance, belongs to what he called the Furs and Station Wagons cluster, where surveys indicate that residents tend to buy lots of vermouth, belong to a country club, read Gourmet and vote the GOP ticket. In 02151, a Revere Beach, Massachusetts, zip designated Old Yankee Rows, tastes lean toward beer, fraternal clubs, Lakeland Boating and whoever the Democrats are supporting [Weiss 1988:xii].

The PRIZM system has made Robbin's company, Claritas Corporation, one of the more successful ones among the growing list of firms offering demographic services. It combines demographic characteristics with lifestyle variables and permits a business to home in on the specific neighborhoods where its products can be most profitably marketed (Phillips 1998). In keeping with the changing demographics of America, Claritas added two new types of neighborhood clusters to their system in the 1990s—"Latino America," concentrated in Miami, Chicago, and the southwest and representing middle-class neighborhoods with a high percentage of families that are Hispanic; and "Family Scramble" which is a set of neighborhoods concentrated along the U.S.-Mexican border (Wynter 1994).

Site Selection

One of the most common uses to which demographic data are put is in helping businesses select a site (Pol and Thomas 1997). This is an extension of the concept of targeting because businesses that rely on walk-in traffic naturally prefer a location that is closest to the maximum number of people who are likely to want to buy goods or services from them. Site selection often involves the innovative use of geographic information systems. For example, in the mid-1990s American Honda Corporation wanted to locate its motorcycle dealerships in those neighborhoods closest to the people who would likely be buying them (a thirty-something male, married, attended college, and earns about $33,200 per year; Hoerning 1996). Using geocoded census data, the likely neighborhoods were found, and then possible sites for the businesses were located and these properties were evaluated according to the number of potential customers living within a given radius of each site. Demographics do not represent the sole measure of potential success—access to the site, and visibility of the location are also important, and a firm has to be able to buy or lease the property for a price that allows it to make a profit, but the population component is the single most important factor for many retail businesses (Albert 1998; Fenker 1996).

Manufacturing and service firms also make use of demographics in site selection. Knowledge of the optimal sociodemographic characteristics for a firm's workers (for instance, males or females, certain preferred racial or ethnic groups, certain required level of educational attainment or background) can be crucial in helping a company decide where it should locate a new plant or facility. These decisions can then feed back to influence the community's long-term demographic structure if firms that need similar kinds of employees wind up clustering together in the same geographic region, such as the Silicon Valley in Northern California.

Investment Demographics

To invest is to put your money to use for the purpose of securing a gain in income. Companies invest in themselves by ploughing their profits back into the company in order to develop new products or expand or change markets; individuals (and companies also) invest by buying stocks or bonds of those companies that are viewed as having a profitable future, or by purchasing real property whose value is judged to rise in the future. Clearly, a wide range of factors goes into the decision to invest in one product line, a particular company, or a specific piece of property, but one very important consideration is future demographic change. Companies that plan for demographic shifts have a better chance at success than those that do not. In fact, it has been said that "Wall Street uses demographics to pinpoint opportunity" (Nordberg 1983:23).

Basically, making sound investment decisions (as opposed to lucky ones) involves peering into the future, forecasting likely scenarios, and then acting on the basis of what seems likely to happen. Because you are reading this book, you already have a good feel for the shape of things to come demographically. Most people do not. In 1991, a veteran retail analyst in New York City looked at demographic trends and concluded that "two-worker families move less and end up putting more money into home furnishings and home improvement. A cohort of aging empty-nesters will do the same" (quoted in Oliver 1991:57). This led to the advice to buy stock in Home Depot, Pier 1 Imports, and Williams-Sonoma (Oliver, 1991). All three have prospered since then, although we cannot be sure how much of the gain was due specifically to demographic trends.

Even the popularity of investing in North America and Europe has something to do with demographic trends. You will recall from Chapters 8 through 10 that not only is the population aging, but that the middle-aged and young-old populations have high incomes and the highest levels of wealth. These people have been looking for ways to invest that wealth. What do the demographics suggest about investment opportunities?

The fact that 90 percent of the world's population growth in the foreseeable future will occur in the less developed nations is an important reason for the globalization of business and the internationalization of investment. Developing nations have millions of babies, and each one needs some kind of diaper. Procter & Gamble, maker of Pampers disposable diapers, has found a huge market out there. At the same time, the declining birth rates in many countries have, as I have mentioned in earlier chapters, put an increasing fraction of population into the young adult ages. From Malaysia to Argentina young adults are buying cell phones, satellite dishes, color TVs, CD players, and the perennial favorites, Levis and Coca Cola (Wysocki 1997). Companies selling in these markets are thus potential investment targets.

International investors have been particularly intrigued by the world's two most populous countries. General Motors, Chrysler, and Ford all have invested in car manufacturing in China, as have Volkswagen and Peugeot Citroen from Europe. The problem, of course, is that a huge population does not necessarily mean a huge market if most people are poor. Pizza Hut and McDonald's serve up fast food in China, "but for most of its 1.2 billion people the purchase of electronic goods, stylish clothes or even ice cream remains a distant dream" (Economist 1994:75).

The same can be said for India, which is almost as populous, but less well-off than China. As of the mid-1990s, only 7 percent of India's population had a telephone, and only 12 percent owned a television (probably the same 12 percent that owned a refrigerator) (Jordan 1996). The so-called "consuming class" in India (those with at least some discretionary income) may number as many as 200 million (The Economist 1997), but the middle class (in the American or European sense) in India represents only a tiny fraction of the total population (Bussey 1994). Nonetheless it is growing and in that growth lies opportunity.

Demographics of Human-Resource Management

Whether or not a firm uses demographics in its marketing, site selection or investment decisions, nearly all firms must cope with the fact that their employees potentially are drawn from differing sociodemographic strata and geographic areas. A recognition of these differences may well enable managers to deal more humanely and successfully with their human resources.

Sensitivity to the needs of workers is viewed by most management professionals as a prerequisite for maximizing employee morale and productivity, and the underlying demographic makeup of a company will influence what sorts of needs exist. A company that employs a large number of women may find itself dealing with the potential effects of maternity leaves, with the desirability of opening a child-care center, and with protecting employees from sexual harassment and the higher crime risk that might be associated with working late and leaving for home after dark. A company that counts on immigrants or resident aliens for its labor force will find itself needing bilingual supervisors and a set of policies recognizing that different cultural backgrounds demand different kinds of inducements to productivity.

Effective managers also must be aware of the possible impact of demographic trends on their work force. As the baby bust generation moved into the labor force in the 1980s and 1990s, these smaller-sized cohorts created greater competition among companies for the best young workers, keeping unemployment low (all other things staying the same) and leading some companies to think about ways to recruit younger workers from other states or other countries. In the meantime, the baby boomers were clogging up the middle management levels, significantly slowing the rate of promotion and leading to executive frustrations, and contributing to the popularity of "downsizing" American companies.

Social Planning

Demographics are widely used in American society to chart population movements and plan for social change. When such planning breaks down, as it often does in cities of third-world nations where rapid population growth is occurring, the result can be a certain amount of chaos, social foment, and even political upheaval. However, in the United States regional planning agencies have become commonplace in an effort to coordinate the wide range of public services that maintain a high quality of life. Laws (1994) has suggested that the overriding demographic changes to which social policy planners must respond in the United States include the transition in family and household structure, immigration, increased longevity, and the changing distribution of the population within metropolitan areas.

Education

Perhaps most obvious in its need for demographic information is the educational system. Public elementary and secondary school districts cannot readily recruit students or market their services to new prospects; they rise and fall on demographic currents that determine enrollment. Of course, not every community experiences the national trend; there is variation among and between individual school districts and they have a need for precise information, because even within a district some geographic areas may be growing while others are diminishing in the number of school-aged children.

Do you close some schools and open new ones elsewhere? If so, what do you do with the old schools, and how do you pay for the new ones? Should children be bused from one area to another? If so, how is that to be organized and paid for? Can teachers be retrained to fill classes where demand is growing? These and a host of related questions push school-district demographers to pore over questionnaires about the number of younger, pre-school-aged siblings reported by currently enrolled students; to examine birth records to see where the families are that are having babies; and to talk to realtors to assess the potential for home turnover and to evaluate the likely number of children among families moving in and out. From these data they construct enrollment projections, which are a vital part of public education planning.

> The consequences of failing to project public school enrollments accurately are clear. Underestimates may result in crowded classrooms, shortages of education personnel, and outmoded facilities. The quality of education suffers as a result. On the other hand, overestimates may lead to excess capacity in capital resources and to underutilized educators. [Espenshade and Hagemann 1977:1]

Educators must also bear in mind the important influence that economic conditions have on job opportunities, and thus on the likelihood that people will be moving in or out. This means that school district demographers increasingly pay attention to regional economic models which are designed to forecast local economic trends on the theory that people follow jobs (and different kinds of people follow different kinds of jobs), so the creation of jobs within a region will influence local demographic trends (Exter 1996).

Neither are colleges and universities immune to the effects of demographics (Fishlow 1997). The number of high school graduates in the United States declined each year between 1988 and 1992, then it began increasing again in the mid-1990s as the baby boomlet children reached college age, but demand for college has increased more rapidly than population growth because an increasing fraction of high school graduates are now attending college. In California, immigration since the 1960s has brought to the state a large number of young adults who then had children, and those kids were beginning to move through high school and into colleges in the late 1990s, further contributing to what some educators have called Tidal Wave II (the first tidal wave, of course, was brought about by the Baby Boom).

Health Services

If you had been thinking of having a baby in the 1980s, you might have considered going to Las Vegas, where in 1983 the Sunrise Hospital (one of that city's three major hospitals) cut its maternity ward fees nearly in half in order to increase business (Witt 1983). That is a far cry from the 1950s and 1960s, when babies were being delivered in the halls of nearly every medical facility in the nation because hospitals simply could not keep up with the demands placed on maternity wards. Over the years, hospitals and other health care providers have learned that they have to "reposition" themselves in a classic marketing sense in order to meet the needs of a society that is changing demographically (Rives 1997).

A demographic change that most communities in North America and Europe face is the aging of the population. In the United States people aged 65 and older are four times more likely than those aged 15 to 44 to be admitted to a hospital, and they spend an average of 3 more days each year in the hospital (National Center for Health Statistics 1994). You will recall from Chapter 4 that the ends of the age continuum are where the risks of death are highest and the pattern of death rates is a good (albeit not perfect) index of demand for health services. In 1960, 59 percent of all deaths occurred to people 65 or older, while those aged 85 or older accounted for 11 percent of all deaths. By 2010 it is estimated that 74 percent of deaths will be of people aged 65 and older (and 31 percent accounted for by people aged 85 and older) (Spencer 1989), so the supply of geriatricians needs to increase to meet this demand.

At the local level, demographic detectives are usually available to spot the health care pitfalls and opportunities. As Kinstler and Pol (1990) have pointed out, local hospitals and physicians need to track the local changes in age and sex structure (because of the implied shifts in demand from the population that is aging in place), and they must also keep an eye on in- and out-migration. As you know, migration can have a dramatic short-run impact on health care demand that may seem sudden if no one is paying attention to population trends in the area.

Unfortunately, broad health issues and problems often transcend the ability of local communities to find a remedy. Infant death resulting from low birth weight (which I discussed in Chapter 4) is a national issue, even though it must be dealt with at the local level. Likewise, the preference of older persons to remain at home even if they suffer from severe chronic illness is a national issue with repercussions at the community level. In general, the provision of health services is so costly, and the training of health personnel so lengthy and expensive, that legislative action is often required to permit the system of health services to respond adequately to demographic changes.

Criminal Justice

Crime, like health, is closely tied to the age and sex structure, as I discussed in Chapter 8, and in addition everybody knows that certain kinds of crimes—especially street crimes and drug-related crimes—are more likely to occur in neighborhoods with certain particular sociodemographic characteristics. Areas where the poverty level and unemployment are high are especially prone to these visible kinds of crime (as opposed to white-collar crime, for example) (Flowers 1989; Short 1997). Police Departments and criminal justice researchers are able to use this kind of information in conjunction with a geographic information system to track reported crimes and to better manage policing resources.

Religious Planning

"Build it and they will come" may apply to ballparks and it might have once applied to churches in America, but why leave it to chance? Religious organizations and private companies employ demographic researchers to keep track of household preferences in religion so that churches can respond most effectively to the needs of their members and can, in essence, "market" themselves to the demographically appropriate community. For example, the Presbyterian Church maintains a panel survey of its membership and in the late 1990s the median age of members was 55; 61 percent were female; 98 percent described their race/ethnicity as white; and over half lived within three miles of their church (The Presbyterian Panel 1998). This is the kind of information that can then be used in a geographic information system to help locate new sites or expand local membership (Wendelken 1996).

THE DEMOGRAPHICS OF JURIES—A CASE STUDY

The Sixth Amendment to the U.S. Constitution guarantees every American accused of a crime the right to be tried by a jury of peers. A "jury of peers" means that the jury must be chosen from a panel of people that represents a demographic cross section of the community, according to a 1975 U.S. Supreme Court ruling in the case of Taylor v. Louisiana. By that broad definition, of course, you can easily imagine that the makeup of a jury could become exceedingly complex (for example, finding the proportionate numbers of "butchers, bakers, and candlestick makers"). So,

in 1979, in Duren v. Missouri, the Supreme Court established a three-pronged test to determine whether a jury panel is a fair cross section of the population. To successfully argue that it is not, a defendant must be able to show that (1) the group alleged to be excluded or underrepresented is "distinctive" or "cognizable;" (2) the underrepresentation is not reasonable given the number of such "distinctive" persons in the community; and (3) the underrepresentation is a result of some systematic bias in the jury selection process.

(Continued)

THE DEMOGRAPHICS OF JURIES—A CASE STUDY—CONT'D

Obviously, if a defendant has no quarrel with the jury panel, it is unlikely that any existing demographic unrepresentativeness will be challenged. On the other hand, if you, as a defendant, are a member of a group that you think is either excluded from or underrepresented on the jury panel, you are more likely to challenge the jury panel, based on the assumption that people who share your sociodemographic characteristics will be more apt than others to sympathize with your plight. In such a case, the first step is to make sure that these sympathetic (you hope) people are adequately represented in the pool of potential jurors.

Over time, courts have consistently agreed that African-Americans and Hispanics in the United States represent distinctive groups, where distinctive (or "cognizable," as the court rulings usually say) is defined as "a group of citizens who share a common perspective gained precisely because of membership in the group, which perspective cannot be adequately represented by other members of the community" (Rubio v. Superior Court [California, 1979]). The demographer's job, then, becomes to examine whether or not African-Americans and/or Hispanics are underrepresented on specific jury panels and, if so, to assess whether the reason for the underrepresentation is found within the jury selection system (and thus is remediable).

Although there is some evidence that African-Americans historically had been deliberately kept out of juries in a few southern states (Benokratis 1982), one of the first major legal tests of the issue occurred not in the South but in California, in the case of People v. Rhymes. Deborah Rhymes is an African-American woman who was convicted of a crime in Pomona (a community in southern California) in 1979. She immediately appealed the conviction, however, on the grounds that she had been denied a jury of her peers—only two African-Americans had been in the pool of jurors, and none had been seated on the jury hearing her case.

The court of appeals agreed to hear the case in 1982, and her lawyer called on Edgar Butler, a social demographer in the Department of Sociology at the University of California, Riverside. He testified that the 1970 census showed African-

Americans to represent 10.8 percent of the population in the county of Los Angeles (the 1980 census data had not yet been tabulated at the time of the hearing), whereas a survey of prospective jurors that he had conducted showed that only 6.2 percent were African-American. Thus there were only 57 percent as many African-Americans in the pool of jurors as one would have expected given the demographic makeup of the community.

Why the disparity? Other defense witnesses testified that the process of selecting jurors was biased because at that time in California jurors were drawn exclusively from lists of registered voters. Yet African-Americans (as well as Hispanics) are less likely than members of most other groups in the population to be registered voters. In 1982, for example, 66 percent of whites aged 18 and older interviewed nationally in the Current Population Survey reported that they were registered to vote, compared with 59 percent of African-Americans and only 35 percent of Hispanics (U.S. Bureau of the Census 1983). The logical conclusion was that minority group members were disproportionately missing from the voter registration rolls and therefore were disproportionately less likely to be called in for jury duty. The court concluded that Rhymes had, indeed, been denied a fair trial and ordered her freed on her own recognizance pending a retrial on the same charges, but with a new jury.

In the meantime, this and several similar cases throughout California prompted the state legislature to revamp the source lists from which jurors are selected. Beginning in 1984, each judicial district in California was required to merge data from lists of people obtained not only from the Registrar of Voters but also from the Department of Motor Vehicles (from which people obtain driver's licenses and identification cards). Although that change did improve the demographic cross section of jury panels in California (as has occurred in other states where similar changes have been introduced), it did not solve all of the potential problems.

In 1988, in one of many such cases in which I have been involved, I was called to provide testimony about jury demographics in the case of

People v. Ware in the Santa Monica branch court-house of the Los Angeles Superior Court. Charles Ware, the defendant, had been charged with killing two people while burglarizing a home, and he was facing the death penalty. His lawyers were concerned that Ware, who is African-American, would be tried by a predominantly white (and possibly less sympathetic) jury because they believed that juries in Santa Monica tended to be disproportionately white and were underrepresentative of African-Americans.

A survey of 2,553 people showing up for jury duty at the Santa Monica courthouse in late 1987 and early 1988 revealed that in fact, 81 percent identified themselves as white, whereas only 5.8 percent identified themselves as African-American. To begin to unravel the mystery, you need to know that in Los Angeles County, the Office of the Jury Commissioner draws jurors from an area within a 20-mile driving radius of a particular courthouse. Analysis of the 1980 census data showed that 18.7 percent of the population aged 18 and older within a 20-mile driving radius of the Santa Monica courthouse was African-American.

The difference between 18.7 percent African-American in the community and 5.8 percent in the jury survey does not automatically lead to the conclusion that a disparity exists. Not everyone who is 18 or older is necessarily eligible to serve on a jury. To qualify, you must be a citizen and you must be adequately proficient in English. In a large, cosmopolitan area such as Los Angeles, many people are noncitizens or, especially if they are recently naturalized citizens, may not be sufficiently proficient in English to be able to serve on a jury. Using information available from the 1980 Census Public Use Microdata Sample (PUMS)—a detailed 1 percent sample of individuals counted in the 1980 census—I was able to find out how many people in each racial and ethnic category were "jury eligible"—18 years of age or older, a citizen, and a speaker of English. These data revealed that in Los Angeles County, 95 percent of whites aged 18 and older are likely to be eligible for jury duty, compared with 98 percent of African-Americans but only 52 percent of Asians and 47 percent of Hispanics. After adjusting the data for these estimates, it turned out that 21.9 percent of the jury-eligible population within a 20-mile radius of the Santa Monica courthouse was African-American. This is substantially higher than the 5.8 percent African-American population actually showing up at the courthouse. So why the big disparity this time?

With further analysis I discovered that a majority of African-Americans who lived within a 20-mile radius of the Santa Monica courthouse also lived within a 20-mile radius of the downtown courthouse in Los Angeles. Additionally, the jury commissioner's office had a policy of generally allowing people to serve in the courthouse nearest their home, and, for most African-Americans in the region, downtown was closer than Santa Monica. As a result, there were actually more African-Americans showing up for jury duty downtown (where, by the way, O.J. Simpson was tried and acquitted in the criminal case against him) than would have been expected based on the percentage of jury-eligible African-Americans within the 20-mile driving radius. It was apparent, then, that there were too few African-Americans on jury duty in Santa Monica (where, you may recall, O.J. Simpson lost the civil case against him) at least partly because there were too many African-Americans on jury duty downtown.

On hearing the evidence, the court decided that African-Americans (a "distinctive" group) were unfairly underrepresented in the Santa Monica courthouse and that the cause of the disparity was the way in which jurors were assigned to different courthouses. The court therefore ordered the jury commissioner to redirect African-American jurors from downtown Los Angeles to the Santa Monica courthouse for Mr. Ware's trial. He was subsequently convicted of murder, but spared the death penalty.

For the typical defendant coming to trial in the United States, it is doubtful that the demographic composition of the jury will make a difference in the way in which evidence is weighed, arguments are perceived, and judgment is rendered. But, for a few people, demographics could be the key to a fair trial, and it is likely that a consideration of jury demographics will remain a useful tool to help ensure justice for all.

Political Planning

The original purpose of the Census of Population in the United States was, in reality, political planning—to determine how membership in the House of Representatives should be distributed. Membership just starts the process. Then comes the task of electing the representatives to the House. This often involves a dose of demographic data. Once elected, politicians regularly confront demographic reality in the course of legislating societal change. Let us examine each of these aspects of the demography of political planning.

Congressional Reapportionment and Redistricting

The U.S. Bureau of Census was required by law to deliver total population counts for all 50 states from the 1990 census to the President on or before December 31, 1990. These data were then sent to the House of Representatives where they were used to determine the number of representatives to which each state is entitled (Poston 1990; Robey 1989). Since 1913 the total number of House seats has been fixed at 435, and the Constitution requires that every state get at least one seat. The first 50 House seats are thus used up. The question remains of how to allocate the remaining 385 seats, remembering that Congressional districts cannot cross state boundaries and there can be neither partial districts nor sharing of seats. Since 1940 this number has been based on a formula called the method of equal proportions, which rank-orders a state's priority for each of those 385 seats based on the total population of a state compared with all other states. The calculations themselves are cumbersome (see Poston 1997), but they produce an allocation of House seats that is now accepted without much criticism, although the reapportionment process has historically been marked by controversy.

The reapportionment based on the 1990 census caused no stir at all for 29 of the 50 states, who neither lost nor gained seats. However, there were changes in 21 states and, in general, the Sun Belt gained at the expense of the Rust Belt. California added seven seats (the largest gain), while New York lost three seats (the largest loss).

Once the number of seats per state has been reapportioned, the real fight begins. This involves the reconfiguration of Congressional districts (geographic areas) that each seat will represent. In the 1960s a series of Supreme Court decisions extended to the state and local levels the requirement that legislative districts be drawn in such a way as to ensure demographically equal representation. Public Law 94-171 mandates that the Census Bureau provide population counts by age (under 18 and 18 and over) and by race and ethnicity down to the block level for local communities. These data were then used to redefine legislative district boundaries (see Williams 1994).

When this information was delivered in 1991, it unleashed a flurry of demographic activity. The availability of data on computer tape and CD-ROM offered advocacy groups the opportunity to use the census data to draft their own proposals for redistricting and to lobby for their plan using the same sophisticated methods available to politicians. The new openness of data permitted "special-interest groups to achieve what political cartographers since Elbridge Gerry, who in 18th century Massachusetts lent his name to gerrymandering, have sought: political districts designed to achieve a specific outcome" (Shribman 1991:A14). The most common outcome sought was to create Congressional Districts with an African-American majority, ostensibly in order to increase the number of African-Americans elected to Congress. However, in 1996 the U.S. Supreme Court struck down several such gerrymandered districts in North Carolina and Texas, arguing that race cannot be the main factor in redistricting (Barrett 1996).

As usual, a need can create the chance for profit, and many of the more than 100 demographic data companies in the United States have profited by their ability to use geographic information systems that turn geocoded data into sets of maps demarcating groups with specific sociodemographic characteristics. This mix of demography, geography, and politics does not end with redistricting. Indeed, that is only part of the interplay. The candidates running for elected office in these and other places increasingly depend upon demographic intelligence.

Campaign Strategy

Demographics are used extensively by politicians seeking office. After researching the sociodemographic characteristics of voters in their constituency, they must either match their election platform to the political attitudes of such voters or realize that they will face an uphill battle to convince people to vote for them. "The demography of the voting-age population within a political district is a crucial ingredient in winning an election because demographic characteristics such as age, sex, race, and education are strongly related to the likelihood of voting and, to some extent, to political preferences" (Hill and Kent 1988:2).

In truth, the strategies for "packaging and selling" a candidate are not unlike those used for any other product or service. Census and survey data are used to map the demographic characteristics of voters in a candidate's area and an appropriate plan of action is prepared, quite possibly using the psychographic approach that I discussed earlier with reference to marketing. Remember, though, that to be elected, a candidate must appeal to voters, not necessarily to all people who reside in an area. Since not all residents may be eligible to vote (because they are too young, are not citizens, or are convicted felons) and because not all who are eligible will actually vote, these differences may be crucial.

Geodemographics has been credited with helping Bill Clinton get elected President in 1992, as well as assisting in the Republican takeover of Congress in 1994. The Clinton-Gore campaign team used GIS not only to target potential voters in person, by mail, and by the mass media, but also to target and recruit campaign workers (Bryan 1993). It has become increasingly clear that those candidates who pay attention to demographics have a greater chance of winning the opportunity to put their demographic literacy to work in effective legislation.

Legislative Analysis

Demographics underlie many of the major social issues that confront national and state legislators, as you know from the previous chapters. As Peter Morrison has said, "Demographic change bears directly on the formulation of social policy because it determines in large part whose wealth or income is redistributed" (Morrison 1980:85). The demographics of the baby boom, for example, probably helped fuel inflation in the United States during the 1970s. Government policies in that period were oriented toward creating new jobs for the swelling numbers of labor force entrants, directly contributing to inflation through government expenditure and indirectly having an effect because Congress' attention was turned away from combating inflation (Freund 1982).

The Baby Boom was having an impact on legislative analysis in the late 1990s, as well. The question has become: How will the country finance the retirement of the baby boomers? In fact, this is a question that many nations are facing as declining fertility and increased longevity have contributed to the prospect of substantial increases in the older population. As the older cohorts begin to squeeze the social security system, legislative action will be required to make long-run changes in the financing and benefit structure of social security if the system is to survive. Changes will be made, of course, even if their exact shape is difficult to forecast. Increased self-reliance is one proposed solution: requiring people when younger to save for their own retirement, through mandatory contributions to mutual funds and other investment instruments. It may also be that, when the time comes, taxes will be raised on younger people in order to bail out those older people who, in fact, did not save enough for their retirement. Preston (1984) has pointed out that the older population in the United States has been gaining not only in numbers, but also in economic and political clout over the years, and they may continue to use that clout to their own advantage as time goes by.

As you can see, then, demography insinuates its way into an incredibly wide range of issues in our public and private worlds. This helps to underscore why it is so important for you to be aware of the underlying influences of demographics in nearly every aspect of your life—from the probability of finding a suitable mate to the probable size of your retirement check. If you take this interest in population change seriously, you may even wish to consider a career in demographics.

Should You Pursue a Career in Demographics?

A broad awareness on the part of businesses and planning organizations that demographics represents a useful (and thus profitable) resource means that an increasing number of job opportunities exist in the field of applied demography, although in truth most such jobs are not labeled "demographer." They may be labeled corporate planner, economist, information analyst, market analyst, market researcher, research analyst, survey analyst, researcher, research scientist or social scientist, but the substance of the job will be demographic analysis.

Many large companies and organizations in the United States have demographers on their payrolls. At the Chesapeake and Potomac Telephone Companies, as at Pacific Gas and Electric and Southern California Edison, demographers project the formation of new households and the size and composition of households in order to estimate future demand for utilities. General Motors has a staff demographer to do research on GM's health benefits population (Kintner 1992; Pol and Thomas 1997), whereas Dayton-Hudson Corporation in Minneapolis uses demographic insight regularly in its planning for Dayton and Hudson stores as well as for Mervyn's, B. Dalton Bookseller, and Target Stores (Finch 1986; Pol and Thomas 1997). Many of the jobs as applied demographers will be found in the myriad companies now offering geodemographic services and they can be most easily traced by referring to *American Demographics* and *Business Geographics* magazines. Demographers are also employed by planning agencies at the federal, state, and local level, including regional agencies, counties, cities, chambers of commerce, economic development organizations, health departments, family planning agencies, and school districts.

In 1982 the Population Association of America created a Committee on Business Demography, and in 1985 the Business Demography Group and the State and Local Demography Group of the Population Association of America joined forces to begin the regular publication of a newsletter/journal called *Applied Demography*. This provides insights into the people, organizations, and methods of demographics.

What background should you have, besides reading this book, if you want to pursue a career in applied demography? The Population Association of America suggests that "a strong background in demographic techniques is recommended, especially in the areas of estimates, projections, and trend analysis. In addition, several companies are concerned with international demography, so coursework that includes the analysis of international data is helpful. A thorough understanding of census geography enables private-sector demographers to use U.S. demographic data correctly and effectively. Research methods, with an emphasis on the analysis of secondary data sources, is advocated. Finally, the study of migration theory, especially as it relates to internal migration, is also recommended" (Population Association of America 1983:3). Your employability will be especially enhanced if you have a knowledge of computer-based geographic information systems, as I have discussed throughout the chapter.

Being an applied demographer is not necessarily glamorous, but that doesn't mean that you work in obscurity. The company that owns *The Wall Street Journal* also owns *American Demographics* magazine and its former editor, Cheryl Russell— a baby boomer herself and author of books on the demographics of the Baby Boom, has been routinely quoted in the national media and not just as a demographics expert, but also as a famous personality. We know, for example, that she "wouldn't be caught dead in a minivan" (Templin 1995). Now let's see, what are the demographics of people who are quoted in the national media . . . ?

Summary and Conclusion

Until the 1980s business did little business with demography, but the computerization of census and geographic data has changed all that. Demographics, particularly GIS and geodemographics, has found a comfortable and important niche in the business world, especially in marketing, where demographic information permits more accurate segmentation of markets and the targeting of specific groups of consumers. Investors also pay attention to the way demographic trends point to future market growth and thus to potential profits. Human-resource managers are learning the value of knowing the demographics of their employees and of the labor force in general.

Before business got its hands on demographics, the major application of demographic theory and methods had been in planning—social planning, especially in education, but also in public and health services. Political planners have picked up on demographics, too, using it both for analysis of potential legislation and for strategy in election campaigns.

The applied uses of demography may help keep demography in front of you for the rest of your life. You are likely to encounter demographics in magazines, newspapers, television, and in the course of your work, no matter what it might be. As that happens, I hope that you will not fail to use your demographic perspective (your demographic literacy, if you will) in its broadest scope to keep track of local, national, and international population trends, because demographic events will contribute to nearly all of the major social changes you will witness over your lifetime, just as you will continue to contribute to those population trends with your own behavior.

Main Points

1. Demographics is the application of population theory and methods to the solution of practical problems.

2. Geographic information systems (GIS) have allowed a vast expansion in the use and analysis of demographic data for applied purposes.

3. A major use of demographics is in the marketing of products and services in the private sector.

4. Marketing demographics involve segmenting markets (tailoring products and services to a specific demographic group) and targeting (aiming the advertising of a product or service to a specific demographic group).

5. Demographics are an important component of site selection for many types of businesses.

6. A wide range of sociodemographic characteristics are used in targeting, including age and gender, education, income, race/ethnicity, and household structure.

7. Investors use demographics to pinpoint areas of potential market growth, because population is a major factor behind social change (and thus opportunity).

8. Human-resource managers use demographics to help increase their awareness of the special qualities and needs that exist among their present and prospective employees.

9. Local agencies use demographics to plan for the adequate provision of services for their communities.

10. Education and criminal justice are areas of social planning in which demographics have been used extensively.

11. Health planners pay close attention to demographics because sickness and health are affected by sociodemographic characteristics and because the demand for services shifts with demographic trends.

12. Politicians find demographics helpful in the analysis of legislation and in the strategy for their own election to office.

13. You, too, could become a demographer. See your local professor for more details.

Suggested Readings

1. Hallie J. Kintner, Thomas W. Merrick, Peter A. Morrison, and Paul R. Voss, 1997, Demographics: A Casebook for Business and Government (Santa Monica: RAND).

   In a series of case studies, practitioners shows you how demographic data can be applied to government planning, long-term corporate strategy, human resource management, and marketing.

2. Louis G. Pol and Richard K. Thomas, 1997, Demography for Decision Making (Westport, CT: Quorum Books).

   This is meant to be an all-purpose introduction to demography for business students, and the authors approach that task by using examples from business to illustrate the major concepts and methods of population analysis.

3. Steve H. Murdock and David R. Ellis, 1991, Applied Demography: An Introduction to Basic Concepts, Methods, and Data (Boulder: Westview Press).

   A detailed reference manual and background resource on the types of analyses that are typically employed in applied demography.

4. Michael J. Weiss, 1988, The Clustering of America (New York: Harper and Row).

   A journalistic but very detailed description of the PRIZM system of clustering zip codes according to demographic and life-style characteristics. The author traveled to specific zip code neighborhoods to see if the numbers seemed to fit the people. (They did.)

5. Diane Crispell, 1993, Insider's Guide to Demographic Know-How : Everything You Need to Find, Analyze, and Use Information About Your Customers, Third Edition (Ithaca, NY: American Demographics Press).

   Although the title makes this sound more like a plumbing manual, it is actually a compilation of articles from American Demographics along with a lot of other useful information to provide a certain amount of "seat of the pants" demographic training.

🌐 Websites of Interest

Remember that websites are not as permanent as books and journals, so I cannot guarantee that each of the following websites still exists at the moment that you are reading this:

1. http://www.demosphere.com

   Demosphere International is a geodemographic marketing firm that specializes in data for countries other than the United States. At their website you will find descriptions of their products for countries such as MarketMexico™ and IndiaMap™.

2. http://www.agewave.com

   Age Wave is a company organized by Ken Dychtwald to tap into the opportunities arising from the demographics of aging in the United States.

3. **http://www.claritas.com**

   Claritas is the demographics firm that developed the PRIZM cluster marketing approach (they now call it "precision" marketing) and their website offers a number of on-line examples of their products and services.

4. **http://www.natdecsys.com**

   National Decision Systems is another large demographics firm which has countered the PRIZM marketing clusters with their own Micro Vision Segments. You can check them out by playing the lifestyle game at this website.

5. **http://www.popassoc.org**

   This is the website of the Population Association of America. Contact them if you have questions about the field of demography.

APPENDIX
The Life Table,
Net Reproduction Rate,
and Standardization

In the preceding 15 chapters I have referred to the expectation of life, the probabilities of survival, the level of generational replacement, and standardization of vital rates. Although you should be familiar with the concepts, the actual measurements are somewhat complex and so I have reserved them for this appendix rather than sidetracking you (and maybe derailing you) earlier on. The expectation of life and the probabilities of survival are derived from a set of statistical calculations called a life table, while the level of generational replacement is measured by the net reproduction rate—a statistic that requires data from a life table in its calculation. Standardization is a procedure that permits us to control for the effects of other variables when we compare crude birth or death rates.

The Life Table

The life table is a statistical device for summarizing the mortality experience of a population. It is a basic demographic tool and has a wide range of applications beyond the study of mortality. Life-table techniques have been used to study marriage patterns (see Schoen and Weinick 1993), migration (see Long 1988), school dropouts, labor force participation, fertility, family planning, and other problems (see Chiang 1984; Namboodiri and Suchindran 1987; Smith 1992). It is the backbone of the insurance industry, since actuaries use life tables to estimate the likelihood of occurrence of almost any event for which an insurance company might want to issue a policy. Life table analysis has even been used by General Motors to estimate the cost of maintaining warranties given the probability of a particular car part failing ("dying") from a specific model-year "population" (Merrick and Tordella 1988).

The major goal of the life table, as it is applied to the study of mortality trends and levels, is to calculate the average remaining lifetime, or expectation of life as it is usually called. It is an index of the number of additional years beyond the current age that a typical individual can expect to live if mortality levels remain unchanged. It is an average, representing the potential experience of a hypothetical group of people.

In discussing each part of the life table, reference will be made to Tables A.1 and A.2, life tables for U.S. females and males in 1995. Life-table calculations begin with a set of age/sex-specific death rates (see Chapter 4), and then the first step in calculating expectations of life is to find the probabilities of dying during any given age interval. Note that in the following examples I will be referring to an **abridged life table**, rather than a **complete life table**. The former groups ages into 5-year categories, while the latter uses single years of age. The calculations are slightly different for the two different kinds of life tables, but the interpretation of the results is identical.

Probability of Dying

The probability of dying ($_nq_x$) between ages x and $x + n$ is obtained by converting age/sex-specific death rates ($_nM_x$) to probabilities. You will recall that such death

Table A.1 Life Table for U.S. Females (1995)

(1)	(2)	(3)	(4)	(5)	(6)	(7)	(8)	(9)	(10)
	Number of Females in the Population	Number of Deaths in the Population	Age-Specific Death Rates in the Interval	Probabilities of Death (Proportion of Persons Alive at Beginning Who Die During Interval)	Of 100,000 Hypothetical People Born Alive:		Number of Years Lived		Expectation of Life
					Number Alive at Beginning of Interval	Number Dying During Age Interval	In the Age Interval	In This and All Subsequent Age Intervals	Average Number of Years of Life Remaining at Beginning of Age Interval
Age interval									
x to $x + n$	$_nP_x$	$_nD_x$	$_nM_x$	$_nq_x$	l_x	$_nd_x$	$_nL_x$	T_x	e_x
0–1	1,878,234	12,961	0.00690	0.00680	100,000	680	99,419	7,894,694	78.9
1–5	7,687,709	2,784	0.00036	0.00143	99,320	142	396,940	7,795,275	78.5
5–10	9,376,656	1,568	0.00017	0.00084	99,178	83	495,664	7,398,335	74.6
10–15	9,229,291	1,809	0.00020	0.00097	99,095	96	495,268	6,902,671	69.7
15–20	8,799,492	4,021	0.00046	0.00228	98,999	226	494,466	6,407,403	64.7
20–25	8,795,073	4,446	0.00051	0.00253	98,773	250	493,251	5,912,937	59.9
25–30	9,475,578	6,109	0.00064	0.00322	98,523	317	491,839	5,419,686	55.0
30–35	10,965,646	9,810	0.00089	0.00449	98,206	441	489,978	4,927,847	50.2
35–40	11,177,707	14,148	0.00127	0.00636	97,765	622	487,380	4,437,869	45.4
40–45	10,228,329	17,991	0.00176	0.00879	97,143	854	483,739	3,950,489	40.7
45–50	8,889,062	23,014	0.00259	0.01285	96,289	1,237	478,577	3,466,750	36.0
50–55	7,008,047	29,064	0.00415	0.02050	95,052	1,949	470,690	2,988,173	31.4
55–60	5,767,355	37,793	0.00655	0.03229	93,103	3,006	458,443	2,517,483	27.0
60–65	5,319,671	55,429	0.01042	0.05095	90,097	4,590	439,678	2,059,040	22.9
65–70	5,422,136	84,932	0.01567	0.07567	85,507	6,470	412,207	1,619,362	18.9
70–75	4,994,933	121,957	0.02442	0.11587	79,037	9,158	373,466	1,207,155	15.3
75–80	3,960,862	151,144	0.03816	0.17576	69,879	12,282	320,139	833,689	11.9
80–85	2,854,412	181,625	0.06364	0.27695	57,597	15,951	249,221	513,550	8.9
85+	2,611,279	378,436	0.11940	1.00000	41,646	41,646	264,329	264,329	6.3

Source: National Center for Health Statistics, unpublished tables, Abridged Life Tables, United States, 1995 (**http://www.cdc.gov/nchswww/datawh/statab/ unpubd/mortabs.htm;** accessed 1998).

Table A.2 Life Table for U.S. Males (1995)

(1)	(2)	(3)	(4)	(5)	(6)	(7)	(8)	(9)	(10)
					Of 100,000 Hypothetical People Born Alive:		Number of Years Lived		Expectation of Life
Age interval	Number of Males in the Population	Number of Deaths in the Population	Age-Specific Death Rates in the Interval	Probabilities of Death (Proportion of Persons Alive at Beginning Who Die During Interval)	Number Alive at Beginning of Interval	Number Dying During Age Interval	In the Age Interval	In This and All Subsequent Age Intervals	Average Number of Years of Life Remaining at Beginning of Age Interval
x to $x + n$	$_nP_x$	$_nD_x$	$_nM_x$	$_nq_x$	l_x	$_nd_x$	$_nL_x$	T_x	e_x
0–1	1,969,872	16,622	0.00844	0.00831	100,000	831	99,283	7,254,216	72.5
1–5	8,055,333	3,609	0.00045	0.00177	99,169	176	396,268	7,154,933	72.1
5–10	9,843,300	2,212	0.00022	0.00113	98,993	112	494,659	6,758,665	68.3
10–15	9,685,241	3,007	0.00031	0.00153	98,881	151	494,130	6,264,006	63.3
15–20	9,265,025	11,068	0.00119	0.00594	98,730	586	492,339	5,769,876	58.4
20–25	9,087,045	14,709	0.00162	0.00806	98,144	791	488,785	5,277,537	53.8
25–30	9,529,765	16,572	0.00174	0.00865	97,353	842	484,637	4,788,752	49.2
30–35	10,902,150	25,254	0.00232	0.01154	96,511	1,114	479,803	4,304,115	44.6
35–40	11,071,207	32,339	0.00292	0.01462	95,397	1,395	473,674	3,824,312	40.1
40–45	9,990,476	37,792	0.00378	0.01882	94,002	1,769	465,893	3,350,638	35.6
45–50	8,559,836	42,609	0.00498	0.02458	92,233	2,267	455,919	2,884,745	31.3
50–55	6,621,815	48,313	0.00730	0.03580	89,966	3,221	442,290	2,428,826	27.0
55–60	5,317,251	58,848	0.01107	0.05398	86,745	4,682	422,678	1,986,536	22.9
60–65	4,726,807	83,442	0.01766	0.08489	82,063	6,966	393,799	1,563,858	19.1
65–70	4,505,822	119,415	0.02651	0.12480	75,097	9,372	352,934	1,170,059	15.6
70–75	3,836,272	154,586	0.04031	0.18399	65,725	12,093	299,083	817,125	12.4
75–80	2,720,385	164,372	0.06044	0.26347	53,632	14,130	232,960	518,042	9.7
80–85	1,609,321	155,036	0.09636	0.38762	39,502	15,312	158,348	285,082	7.2
85+	1,016,875	182,823	0.17984	1.00000	24,190	24,190	126,734	126,734	5.2

Source: National Center for Health Statistics, unpublished tables, Abridged Life Tables, United States, 1995 (**http://www.cdc.gov/nchswww/datawh/statab/ unpubd/mortabs.htm**; accessed 1998).

rates relate deaths in a single year (shown as $_nD_x$ in column 3 of Table A.1) to the total (usually midyear) population in the given age group (shown as $_nP_x$ in column 2 of Table A.1). Thus

$$_nM_x = \frac{_nD_x}{_nP_x}.$$

However, a probability of death relates the number of deaths during any given number of years (that is, between any given exact ages) to the number of people who started out being alive and at risk of dying. For most age groups, except the very youngest and oldest, for which special adjustments are made, death rates ($_nM_x$) for a given sex for ages x to $x + n$ may be converted to probabilities of dying according to the following formula:

$$_nq_x = \frac{(n)\,(_nM_x)}{1 + (a)\,(n)\,(_nM_x)}$$

I should point out that this formula is only an estimate of the actual probability of death. Rarely do we have the data that would permit an exact calculation, but the difference between the estimation and the "true" number will seldom be significant. The principal difference between reality and estimation is the fraction that, in the above equation, is shown as a, where a is usually 0.5. This fraction implies that deaths are distributed evenly over an age interval and thus the average death occurs halfway through that interval. This is a good estimate for every age above 4, regardless of race or sex (Chiang 1984). At the younger ages, however, death tends to occur earlier in the age interval. The more appropriate fraction for ages 0 to 1 is 0.85 and for ages 1 to 4 (shown in Tables A.1 and A.2 as 1 to 5) it is 0.60. Another special case is the oldest age group (85+ in Tables A.1 and A.2) because it is open-ended, going to the highest age at which people might die. Thus it is obvious that the probability of death in this interval is 1.0000—death is certain.

In Tables A.1 and A.2 the age-specific death rates for 1995 are given in column (4). In column (5) they have been converted to probabilities of death from exact age x (for example, 10) to exact age $x + n$ (for example, $10 + 5 = 15$).

Deaths in the Life Table

Once the probabilities of death have been calculated, the number of deaths that would occur to the hypothetical life-table population is calculated. The life table assumes an initial population of 100,000 live births, which are then subjected to the specific mortality schedule. These 100,000 babies represent what is called the *radix* (l_0). During the first year, the number of babies dying is equal to the radix (100,000) times the probability of death. Subtracting the babies who died ($_1d_0$) gives the number of people still alive at the beginning of the next age interval (l_1). These calculations are shown in columns (7) and (6) of Tables A.1 and A.2, respectively. In general:

$$_nd_x = (_nq_x)\,(l_x)$$

and

$$l_{x+n} = l_x - {_n}d_x$$

Although we have not yet calculated the expectation of life, we already have some very useful information in column (6). This column indicates how many of the original 100,000 people are still alive at the beginning of any given age interval. Thus since in Table A.1 93,103 women (out of the original 100,000) are still alive at age 55, we can say that the probability of surviving to age 55 was 93 percent for U.S. females in 1995. In other words, a female baby born in that year would have a 93 percent chance of living to age 55 if her mortality risks throughout her lifetime were exactly those experienced by the female population of the United States in 1995.

Number of Years Lived

The final two columns that lead to the calculation of expectation of life are related to the concept of number of years lived. During the 5-year period, for example, between the fifth and the tenth birthdays, each person lives 5 years. If there were 98,000 people sharing their tenth birthdays, then they all would have lived a total of $5 \times 98,000 = 490,000$ years between their fifth and tenth birthdays. Of course, if a person died after the fifth but before the tenth birthday, then we add in only those years that were lived prior to dying. This concept is analogous to the frequently used measure "person-years" as an index of the cost of something. For example, we might be told that a huge dam required 1,000 person-years to build. That does not mean that it took 1,000 years to build the dam, but rather that it could have been built if 1,000 people worked on it for a year, or if 2,000 people worked on it for half a year, and so on. With respect to mortality, we are interested in the fact that the lower the death rates, the more people will survive through an entire age interval and the greater the number of years lived will be. The number of years lived ($_nL_x$) can be estimated as follows:

$$_nL_x = n \, (l_x - a \, {_n}d_x)$$

Again I must comment that the fraction a is 0.50 for all age groups except 0 to 1 (for which we should use 0.85) and 1 to 5 (for which we should use 0.60). Furthermore, this formula will not work for the oldest, open age interval (85+), since there are no survivors at the end of that age interval and we have no data as to how many years each person will live before finally dying. All is not lost, however. We can estimate the number of years lived by dividing the number of survivors to that oldest cohort (l_{85}) by the death rate at the oldest age (M_{85}):

$$L_{85} = \frac{l_{85}}{M_{85}}$$

The results of these calculations are shown in column (8). Then we add up the years lived, cumulating from the oldest to the youngest ages. These calculations are shown in column (9) and represent T_x, the total number of years lived in a given age interval and all older age intervals. At the oldest ages (85+), T_x is just equal to $_nL_x$. But at each successively younger age (e.g., 80 to 85), T_x is equal to T_x at all older ages (e.g., 85+, which is T_{85}) plus the number of person-years lived between ages x and $x + n$ (e.g., between ages 80 and 85, which is $_5L_{80}$). Thus at any given age:

$$T_x = T_{x + n} + {_nL_x} .$$

Expectation of Life

Finally we arrive at the expectation of life (e_x), or average remaining lifetime. It is the total years remaining to be lived at exact age x and is found by dividing T_x by the number of people alive at that exact age (l_x). In column (10), we find that

$$e_x = \frac{T_x}{l_x}$$

Thus for U.S. females in 1995, the expectation of life at birth (e_0) was 7,894,694/100,000 = 78.9, while at age 55 a female could expect to live an additional 27.0 years. For males the comparable numbers are 7,254,216/100,000 = 72.5 as the expectation of life at birth in 1995; and the expectation of life at age 55 in 1995 was 22.9.

Although it has required some work, we now have a sophisticated single index that summarizes the level of mortality prevailing in a given population at a particular time. I should warn you that the formulas that I have provided will not produce results that are identical to those in Tables A.1 and A.2. This is because the U.S. National Center for Health Statistics has access to single-year of age data for all people who died, and so they do not need to make some of the simplifying assumptions that we employed in the preceding formulas. However, the differences between the calculations based on the formulas given here and the numbers shown in the tables are very small.

Other Applications of the Life Table

The life table as a measure of mortality is actually a measure of duration—of the length of human life. We can apply the same techniques to the measurement of the duration of anything else—marriage, widowhood, employment, or contraceptive use, to use examples close at hand. However, by adding to the mathematical complexity of the table, we can gain additional insights into some of these phenomena by separating out specific events or by combining events. In the former instance we produce a *multiple-decrement* table and in the latter case we get a *multistate* table (Smith 1992).

A multiple-decrement table can, for example, isolate the impact of a specific cause of death on overall mortality levels. In a classic study, Preston, Keyfitz, and Schoen (1972) found that in 1964 the expectation of life for all females in the United States was 73.8 years. Deleting heart disease as a cause of death would have raised life expectancy by 17.1 years, to 90.9; whereas deleting cancer as a cause of death would have produced a gain of 2.6 years in life expectancy, to 76.3. In a different type of analysis, we might use a multiple-decrement table to ask what are the probabilities at each age of leaving the labor force before death? Of dying before your spouse? Or of discontinuing a contraceptive method and becoming pregnant?

Multistate tables move us to the level of sophistication that allows for a succession of possible contingencies—several migrations; entering and leaving the labor force, including periods of unemployment and retirement; school enrollment, with dropouts and reentries; moving through different marital statuses; and moving from one birth parity to another. Elaborate discussion of these techniques is, of course, beyond the scope of this book, but you should be aware that such methods do, in fact, exist to help untangle the mysteries of the social world (see, for example, Keyfitz 1985; Smith 1992).

Net Reproduction Rate and Mean Length of Generation

The net reproduction rate (NRR) is a measure of generational replacement, as I discussed in Chapter 5. Its lowest possible value is 0, meaning that there will be no next generation if things remain as they are. A value of 1 means that the next generation will be exactly replaced, no more and no less. Values greater than 1, of course, indicate that the next generation will be larger in size than the current one. More specifically, the NRR measures the average number of surviving female children that will be born to the female babies born in a given year, assuming no change in the age-specific fertility and mortality rates (and ignoring the effect of migration).

To calculate the NRR for the United States for 1995, for example, we need the age-specific birth rates, the proportion of births that are females, and a life table for 1995 (we will use the data from Table A.1). From the life table we need to know the probability that a female baby will survive through the reproductive ages.

We begin the calculation of the net reproduction rate by calculating the gross reproduction rate (GRR), as you can see in Table A.3. As you will recall from Chapter 5, the GRR is the number of female births that a woman would have in her lifetime if current age-specific birth rates remained unchanged. From the vital statistics data, we obtain the number of live births according to the age of the mother (column 4). What we actually need are only the female births, but typically we do not have the sex ratio at each age of the mother, so we approximate this by applying the overall proportion of female births. In the United States in 1995, for example, 48.78 percent of all births were females (a sex ratio of 105 male births per 100 female births). Since there is very little variation from age to age in the sex ratio at birth, we can apply this percentage to all ages, as shown in column 5. We then divide that number of female births at each age by the total number of women in the population in that age group (column 3—obtained from Table A.1), to produce the age-specific rates of female births (column 6).

The age-specific birth rate refers to only a single year and we want to assume that this hypothetical (synthetic) cohort of women will experience this rate of childbearing for the entire 5-year interval (for example, from ages 15 through 19), so we have to multiply the birth rates in column 6 by 5 to obtain the female birth rate

Table A.3 Calculation of Gross Reproduction Rate (GRR), Net Reproduction Rate (NRR), and Mean Length of Generation, United States, 1995

(1)	(2)	(3)	(4)	(5)	(6)	(7)	(8)	(9)	(10)
Age Group	Midpoint of Interval	Number of Women in Age Group	Number of Births to Women in Age Group	Number of Female Births to Women in Age Group	Female Births Per Woman	GRR: Female Birth Rate During 5-Year Interval	Proportion of Female Babies Surviving to Midpoint of Age Interval	NRR: Surviving Daughters per Woman during 5-Year Interval	Mean Length of Generation (T) Column (2) × Column (9)
10–14	12.5	9,229,291	12,242	5,972	0.00065	0.00324	0.9905	0.00320	0.04006
15–19	17.5	8,799,492	499,873	243,840	0.02771	0.13855	0.9889	0.13702	2.39785
20–24	22.5	8,795,073	965,547	470,999	0.05355	0.26776	0.9865	0.26415	5.94334
25–29	27.5	9,475,578	1,063,539	518,800	0.05475	0.27376	0.9837	0.26929	7.40542
30–34	32.5	10,965,646	904,666	441,300	0.04024	0.20122	0.9800	0.19719	6.40856
35–39	37.5	11,177,707	383,745	187,193	0.01675	0.08373	0.9748	0.08162	3.06080
40–44	42.5	10,228,329	67,250	32,805	0.00321	0.01604	0.9675	0.01551	0.65938
45–49	47.5	8,889,062	2,727	1,330	0.00015	0.00075	0.9572	0.00072	0.03402
						0.98505 = GRR		0.96870 = NRR	25.94942 divided by NRR = 26.79 = mean length of generation

Sources: Columns 3 and 8 are from Table A.1; column 4 data are from Stephanie J. Ventura, Joyce A. Martin, Sally C. Curtin, and T.J. Matthews, 1997, Report of Final Natality Statistics, 1995, Monthly Vital Statistics Report, 45(11): Table 2.

during the 5-year interval, and this result is shown in column 7. The sum of column 7 is the gross reproduction rate (GRR), which, as I mentioned in Chapter 5, is the female total fertility rate. In 1995, the GRR was .98505, which is slightly below replacement level.

To calculate the net reproduction rate, we must take mortality into account and we do that by using data from the life table (Table A.1) to calculate the probability of a female surviving from birth to the midpoint (column 2) of each interval. This probability is found by dividing the $_nL_x$ value in the life table for each age by 5 times the radix, or 500,000. Thus the proportion of female babies surviving to the midpoint of each age interval (column 8) is:

$$\text{Proportion of surviving females} = {}_nL_x / 5l_0$$

This number is then applied to the 5-year female birth rate to estimate the number of babies born who will survive to the midpoint of that age group. The sum of this column (column 9) is equal to the net reproduction rate. In 1995 this was 0.9687, or just slightly below replacement level.

By adding one more column to Table A.3 we are able to provide another useful index called the *mean length of generation,* or the average age at childbearing (typically labeled *T*) (Palmore and Gardner 1983). Column 10 illustrates the calculation. You multiply the midpoint of each age interval (column 2) by the surviving daughters per woman for that age interval (column 8) and then you divide the sum of those calculations by the net reproduction rate (the sum of column 8), yielding a figure of 26.79 years.

Standardization

In discussing crude death and birth rates in Chapters 4 and 5, respectively, I noted that they were "crude" because they failed to take the age and sex structure into account. I then went on to suggest other measures, such as the life table and the total fertility rate, which allow us to correct for these shortcomings, as long as the necessary data are available. The issue does not die there, however, because we can conceive of several other factors besides age and sex that could affect overall death and birth rates (and, yes, the puns were intended).

We know from Chapter 4 that death rates can be influenced not only by age and sex, but also by urban/rural residence, education, occupation, income, and marital status. Fertility rates are affected by these same variables, as you learned in Chapter 6. One way to take these or any other characteristics into account is through **standardization**—a set of calculations that statistically control for the effects of an external factor. In demography, the variable most often taken into account with standardization is, in fact, age. So I will use it in the following example, which illustrates the use of the standardization procedure as applied to the crude death rate. Remember, however, that an analogous set of procedures could be applied to any variable that the researcher thinks may be muddying up a comparison of overall crude rates.

There are two common ways of computing standardized rates. "One consists of applying different age-specific rates to a standard population; this is called (for no very good reason) in English and American usage, 'direct standardization.' The other consists of applying a standard set of rates to different populations by age; this is called 'indirect standardization.' In both types, the object is to calculate the number of deaths to be expected in one population, on the basis of some information from another population. This number of 'expected deaths' is used in calculating the standardized death rate" (Barclay 1958:161). Both the direct and indirect methods are described in Barclay (1958) and Shryock and associates (1976). In the example here, shown in Table A.4, I use the direct standardization procedure to calculate the age-standardized (also called the age-adjusted) death rate for Mexican females in 1990, compared with U.S. females in the same year, using the 1990 female population of the United States as the standard population.

In 1990 the crude death rate for females in Mexico was 4.4 deaths per 1,000 population (the sum of column 2 in Table A.4 divided by the sum of column 3). In the United States in the same year, the crude death rate for females was 8.1 per 1,000 (column 7 divided by column 5). Although Mexico has been developing economically, you would probably suspect that the overall risk of death was still greater in Mexico in 1990 than it was in the United States. Because of the higher fertility in Mexico, the age structure in that country is younger than in the United States, putting disproportionate numbers of people in the relatively low-risk ages, even though the actual risk of dying at each specific age was higher. Age-standardization permits us to see how much of the difference in the crude death rate between Mexico and the United States was due to the differences in those two nations' age structures.

In Table A.4, you can see that the process of standardization begins with the calculation of age-specific death rates for Mexican females in 1990 (column 4), by dividing column 2 by column 3 at each age. We then apply those rates to the standard population (column 5) to see how many deaths would result in the standard population if they were exposed to the age-specific risk of death among Mexican females in 1990. This produces column 6, which is the expected number of deaths at each age in the standard population, given Mexico's age-specific death rates. Our interest is in the sum of this column—the total number of expected deaths, which we divide by the total number of people in the standard population (column 5) to yield our age-adjusted or age-standardized death rate. The result is an age-adjusted death rate of 9.8 deaths per 1,000 among Mexican females when applied to the U.S. age structure. Thus we can conclude that since the Mexican crude death rate (unadjusted) was lower than that in the United States, whereas the age-adjusted rate is higher, the observed difference in the crude death rates between the two countries is attributable entirely to the younger age structure in Mexico than in the United States.

Table A.4 Age-Standardized Death Rates for Mexico (Females, 1990) Using the 1985 U.S. Population as the Standard

(1)	(2)	(3)	(4)	(5)	(6)	(7)
Age Group	Deaths to Females in Mexico in 1990	Number of Females in Mexico in 1990	Age-Specific Death Rate for Females Mexico 1990 (Per 1,000)	Standard Population (U.S. Females, 1990)	"Expected" Deaths in Standard Population Based on Death Rates in Mexico (Column 4 × Column 5)	Actual Deaths in Standard Population
0–4	38,281	5,066,071	7.56	8,962,034	67,720	19,457
5–9	2,892	5,256,003	0.55	8,836,652	4,862	1,632
10–14	2,157	5,190,086	0.42	8,347,082	3,470	1,677
15–19	2,916	4,934,605	0.59	8,651,317	5,112	4,040
20–24	3,136	4,116,137	0.76	9,344,716	7,120	4,820
25–29	3,213	3,374,496	0.95	10,617,109	10,108	6,785
30–34	3,384	2,826,118	1.20	10,985,954	13,153	9,249
35–39	4,210	2,383,084	1.77	10,060,874	17,775	11,492
40–44	4,522	1,803,757	2.51	8,923,802	22,372	14,608
45–49	5,570	1,528,609	3.64	7,061,976	25,733	18,759
50–54	6,760	1,239,475	5.45	5,835,775	31,828	25,195
55–59	8,757	981,606	8.92	5,497,386	49,040	37,338
60–64	10,633	846,563	12.56	5,669,120	71,208	60,712
65–69	12,493	619,790	20.16	5,579,428	112,465	88,977
70–74	12,759	435,653	29.29	4,585,517	134,293	112,905
75–79	15,423	314,922	48.97	3,721,601	182,263	143,241
80–84	16,257	223,374	72.78	2,567,645	186,870	162,792
85+	29,806	215,327	138.42	2,222,467	307,640	311,208
TOTALS	183,169	41,355,676		127,470,455	1,253,033	1,034,887
Crude death rates:	4.4				9.8	8.1

Source: Adapted from data in the U.S. Bureau of the Census International Data Base

GLOSSARY

This glossary contains words or terms that appeared in boldface type in the text. I have tried to include terms that are central to an understanding of the study of population. The chapter notation in parentheses refers to the chapter in which the term is first discussed in detail.

abortion the expulsion of a fetus prematurely; a miscarriage—may be either induced or spontaneous (Chapter 5).

abridged life table a life table (see definition) in which ages are grouped into categories (usually five-year age groupings) (Appendix).

accidental death loss of life unrelated to disease of any kind but attributed to the physical, social, or economic environment (Chapter 4).

acculturation a process undergone by immigrants in which they adopt the host language, bring their diet more in line with host culture, and participate in cultural activities of the host society (Chapter 7).

achieved characteristics those sociodemographic characteristics such as education, occupation, income, marital status, and labor force participation, over which we do have some degree of control (Chapter 10).

adaptation a process undergone by immigrants in which they adjust to the new physical and social environment of the host society (Chapter 7).

administrative data information derived from administrative records including tax returns, utility records, school enrollment, and participation in government programs (Chapter 2).

age/sex pyramid graph of the number of people in a population by age and sex (Chapter 8).

age/sex structure the number (or percent) of people in a population distributed by age and sex (Chapter 8).

age/sex-specific death rate the number of people of a given age and sex who died in a given year divided by the total (average midyear) number of people of that age and sex (Chapter 4).

age-specific fertility rate the number of children born to women of a given age divided by the total number of women that age (Chapter 5).

age stratification the assignment of social roles and social status on the basis of age (Chapter 8).

age structure the distribution of people in a population by age; a "young" population is one with more than 35 percent of the population under age 15; an "old" population is one with more than 10 percent aged 65 and older (Chapter 8).

Agricultural Revolution change that took place roughly 10,000 years ago when humans first began to domesticate plants and animals, thereby making it easier to settle in permanent establishments (Chapter 1).

alien a person born in, or belonging to, another country who has not acquired citizenship by naturalization—distinguished from citizen (Chapter 7).

Alzheimer's disease a disease involving a change in the brain's neurons, producing behavioral shifts; a major cause of organic brain disorder among older persons (Chapter 9).

ambivalence state of being caught between competing pressures and thus being uncertain about how to behave properly (Chapter 6).

amenorrhea   temporary absence or suppression of the menstrual discharge (Chapter 5).

anovulatory   pertaining to a menstrual cycle in which no ovum (egg) is released (Chapter 5).

antinatalist   based on an ideological position that discourages childbearing (Chapter 3).

arable   land that is suitable for agricultural purposes (Chapter 13).

ascribed characteristics   sociodemographic characteristics such as gender, race, and ethnicity, with which we are born and over which we have essentially no control (Chapter 10).

assimilation   a process undergone by immigrants in which they not only accept the outer trappings of the host culture, but also assume the behaviors and attitudes of the host culture (Chapter 7).

asylees   people who have been forced out of their country of nationality and are seeking refuge in the new country in which they are now living (Chapter 7).

atmosphere   the whole mass of air surrounding the earth (Chapter 13).

average age of a population   one measure of the age distribution of a population—may be calculated as either the mean or the median (Chapter 8).

Baby Boom   the dramatic rise in the birth rate in the United States following World War II—people born between 1946 and 1964 (Chapter 6).

base year   the beginning year of a population projection (Chapter 8).

biosphere   the zone of Earth where life is found (Chapter 13).

birth control   regulation of the number of children you have through the deliberate prevention of conception; note that birth control does not necessarily include abortion, although it is often used in the same sense as the terms contraception, family planning, and fertility control (Chapter 5).

birth rate   see crude birth rate (Chapter 5).

boomsters   a nickname given to people who believe that population growth stimulates economic development (Chapter 12).

bracero   a Mexican laborer admitted legally into the United States for a short time to perform seasonal, usually agricultural, labor (Chapter 7).

capital   a stock of goods used for the production of other goods rather than for immediate enjoyment; anything invested to yield income in the future (Chapter 12).

capitalism   an economic system in which the means of production, distribution, and exchange of wealth are maintained chiefly by private individuals or corporations, as contrasted to government ownership (Chapter 3).

cardiovascular disease   a disease of the heart or blood vessels (Chapter 4).

carrying capacity   the size of population that could theoretically be maintained indefinitely at a given level of living with a given type of economic system (Chapter 1).

celibacy   permanent nonmarriage, implying as well that a celibate is a person who never enters a sexual union (Chapter 5).

census of population   an official enumeration of an entire population, usually with details as to age, sex, occupation, and other population characteristics (Chapter 2).

chain migration   the process whereby migrants are part of an established flow from a common origin to a prepared destination where others have previously migrated (Chapter 7).

child control   the practice of controlling family size after the birth of children, through the mechanisms of infanticide, fosterage, and orphanage (Chapter 6).

child–woman ratio   a census-based measure of fertility, calculated as the ratio of children aged 0–4 to the number of women aged 15–49 (Chapter 5).

chronic disease   disease that continues for a long time or recurs frequently (as opposed to acute)—often associated with degeneration (Chapter 4).

cohabitation   living together without benefit of legal marriage (Chapter 10).

cohort   persons who share something in common; in demography this is most often the year (or grouped years) of birth (Chapter 5).

cohort component method of population projection   a population projection made by applying age-specific survival rates, age-specific-fertility rates, and age-specific measures of migration to the base year population in order to project the population to the target year (Chapter 8).

cohort flow   the movement through time of a group of people born in the same year (Chapter 8).

Columbian exchange the exchange of food, products, people, and diseases between Europe and the Americas as a result of explorations by Columbus and others (Chapter 3).

communicable disease (also called infectious disease) a disease capable of being communicated or transmitted from person to person (Chapter 4).

complete life table a life table (see definition) based on data by single years of age (Appendix).

components of change method of migration estimation a method of measuring net migration between two dates by comparing total population estimates with that which would have resulted solely from the components of birth and death, with the residual being attributed to migration (Chapter 7).

components of growth a method of estimating and/or projecting population size by adding births, subtracting deaths, and adding net migration occurring in an interval of time, then adding the result to the population at the beginning of the interval (Chapter 8).

condom a thin sheath, usually of rubber, worn over the penis during sexual intercourse to prevent conception or venereal disease (Chapter 5).

consolidated metropolitan statistical areas (CMSAs) groupings of the very largest Metropolitan Statistical Areas in the United States (Chapter 11).

content error an inaccuracy in the data obtained in a census; possibly an error in reporting, editing, or tabulating (Chapter 2).

contraception the prevention of conception or impregnation by any of various techniques or devices; birth control (Chapter 5).

contraceptive prevalence the percentage of people (usually measured in terms of couples) currently using a particular contraceptive method (Chapter 5).

coverage error the percentage of a particular group (or total population) that is inadvertently not counted in a census (Chapter 2).

crowding the gathering of a large number of people closely together; the number of people per space per unit of time (Chapter 11).

crude birth rate the number of births in a given year divided by the total midyear population in that year (Chapter 5).

crude death rate the number of deaths in a given year divided by the total midyear population in that year (Chapter 4).

crude net migration rate a measure of migration calculated as the number of in-migrants minus the number of out-migrants divided by the total midyear population (Chapter 7).

death rate see crude death rate (Chapter 4).

de facto population the people actually in a given territory on the census day (Chapter 2).

degeneration the biological deterioration of a body (Chapter 4).

de jure population the people who legally "belong" in a given area whether or not they are there on the census day (Chapter 2).

demographic analysis (DA) a method of evaluating the accuracy of a census by estimating the demographic components of change since the previous census and comparing it with the new census count (Chapter 2).

demographic balancing equation the formula that shows that the population at time 2 is equal to the population at time 1 plus the births between time 1 and 2 minus the deaths between time 1 and 2 plus the in-migrants between time 1 and 2 minus the out-migrants between time 1 and 2 (Chapter 2).

demographic change and response the theory that the response made by individuals to population pressures is determined by the means available to them (Chapter 3).

demographic characteristics indicators of a person's social and economic position or status in society, including education, occupation, income, and race and ethnicity (see population characteristics) (Chapter 10).

demographic overhead the general cost of adding people to a population caused by the necessity of providing goods and services (Chapter 12).

demographic perspective a way of relating basic information to theories about how the world operates demographically (Chapter 3).

demographics the application of demographic science to practical problems; any applied use of population statistics (Chapter 15).

demographic structure the organized pattern of demographic behavior that characterizes a population at a particular time (Chapter 1).

demographic transition the process whereby a country moves from high birth and high death rates to low birth and low death rates with an interstitial spurt in population growth (Chapter 3).

demography   the science of population (Chapter 1).

density   the ratio of people to physical space (Chapter 11).

dependency ratio   the ratio of people of dependent age (0–14 and 65+) to people of economically active ages (15–64) (Chapter 8).

diaphragm   a barrier method of fertility control—a thin, dome-shaped device, usually of rubber, inserted in the vagina and worn over the uterine cervix to prevent conception during sexual intercourse (Chapter 5).

doctrine   a principle laid down as true and beyond dispute (Chapter 3).

donor area   the area from which migrants come (Chapter 7).

doomster   a nickname given to people who believe that population growth retards economic development (Chapter 12).

douche   (as a contraceptive) washing of the vaginal area after intercourse to prevent conception (Chapter 5).

dual system estimation (DSE)   a method of evaluating a census by comparing respondents in the census with respondents in a carefully selected postcensal survey (Chapter 2).

ecological footprint   the total area of productive land and water required to produce the resources for and assimilate the waste from a given population (Chapter 13).

economic development   a rise in the average standard of living associated with economic growth; a rise in per capita income (Chapter 12).

economic growth   an increase in the total amount of income produced by a nation without regard to the total number of people (Chapter 12).

ecosphere   all of the earth's ecosystems; see also biosphere (Chapter 13).

ecosystem   a system formed by the interaction of a community of organisms with their environment (Chapter 13).

edge cities   cities that have been created in suburban, often unincorporated, areas and that replicate most of the functions of the older central city (Chapter 11).

emigrant   a person who leaves one country to settle in another (Chapter 7).

endogenous factors   those things that are within the scope of (internal to) one's own control (Chapter 6).

epidemiological transition   the pattern of long-term shifts in health and disease patterns as mortality moves from high to low levels (Chapter 4).

error theory   a theory of human aging based on the notion that random error occurs in the synthesis of protein, leading to irreversible damage (and thus aging) of a cell (Chapter 9).

ethnic group   a group of people of the same race or nationality who share a common and distinctive culture while living within a larger society (Chapter 10).

ethnicity   the ancestral origins of a particular group (Chapter 10).

ethnocentric   characterized by a belief in the inherent superiority of one's own group and culture accompanied by a feeling of contempt for other groups and cultures (Chapter 3).

exogenous factors   those things that are beyond the control of (external to) the average person (Chapter 6).

exponentially   (increasing) in a compound fashion (Chapter 1).

extensification of agriculture   increasing agricultural output by putting more land into production (Chapter 13).

extended family   family members beyond the nuclear family (Chapter 10).

exurbanization   nonrural population growth beyond the suburbs (Chapter 11).

family   a group of people who are related to each other by birth, marriage, or adoption (Chapter 10).

family household   a household in which the householder is living with one or more persons related to her or him by birth, marriage, or adoption (Chapter 10).

family planning   the conscious effort of individuals or couples to regulate the number and spacing of births (Chapter 5).

fecundity   the physical capacity to reproduce (Chapter 5).

fertility   reproductive performance rather than the mere capacity to do so; one of the three basic demographic processes (Chapter 5).

fertility awareness methods of birth control   methods of natural family planning that are combined with a barrier method to protect against conception during the fertile period (Chapter 5).

fertility differential a variable in which people show clear differences in fertility according to their categorization by that variable (Chapter 6).

fertility rate see general fertility rate (Chapter 5).

fertility transition the shift from natural fertility to fertility limitation (Chapter 6).

food security a term meaning that people have physical and economic access to the basic food they need in order to work and function normally (Chapter 13).

force of mortality the factors that prevent people from living to their biological maximum (Chapter 4).

forward survival method of migration estimation a method of estimating migration between two censuses by combining census data with life table probabilities of survival between the two censuses (Chapter 7).

fosterage the practice of placing a child in someone else's home (Chapter 6).

gender role a social role considered appropriate for a person of one sex or the other (Chapter 10).

general fertility rate the total number of births in a year divided by the total midyear number of women in the childbearing ages (Chapter 5).

gentrification restoration and habitation of older homes in central city areas by urban or suburban elites (Chapter 11).

geodemographics use of demographic data specific to a particular geographic region (Chapter 15).

geographic information system (GIS) a computer-based system of organizing and geographically analyzing data, including demographic variables (Chapter 2).

geo-referenced a piece of information that includes some form of geographic identification such as an address, ZIP code, census tract, county, state, or country (Chapter 2).

gestation the process of carrying a fetus in the uterus during the period from conception to delivery (Chapter 5).

global warming an increase in the global temperature caused by a build-up of greenhouse gases (Chapter 13).

greenhouse gases those atmospheric gases (especially carbon dioxide and water) that trap radiated heat from the sun and warm the surface of the earth (Chapter 13).

Green Revolution an improvement in agricultural production begun in the 1940s based on high-yield-variety strains of grain and increased use of fertilizers, pesticides, and irrigation (Chapter 13).

gross domestic product the total value of goods and services produced within the geographic boundaries of a nation in a given year, without reference to international trade (Chapter 12).

gross rate of in-migration the total number of immigrants divided by the total midyear population in the area of destination (Chapter 7).

gross rate of out-migration the total number of out-migrants divided by the total midyear population in the area of origin (Chapter 7).

gross national product (GNP) the total output of goods and services produced by a country, including income earned from abroad (Chapter 12).

gross reproduction rate the total fertility rate multiplied by the proportion of all births that are girls. It is generally interpreted as the number of female children that a female just born may expect to have in her lifetime, assuming that birth rates stay the same and ignoring her chances of survival through her reproductive years (Chapter 5).

high growth potential the first stage in the demographic transition in which a population has a pattern of high birth and death rates (Chapter 3).

high-yield varieties (HYV) dwarf types of grains that have shorter stems and produce more stalks than most traditional varieties (Chapter 13).

historical sources data derived from sources such as early censuses, genealogies, family reconstitution, grave sites, and archeological findings (Chapter 2).

host area the destination area of migrants; the area into which they migrate (Chapter 7).

household all of the people who occupy a housing unit (Chapter 10).

housing unit a house, apartment, mobile home or trailer, group of rooms (or even a single room if occupied as a separate living quarters) intended for occupancy as a separate living quarters (Chapter 10).

housing unit method a method of projecting a population by assuming that changes in the number of occupied housing units reflect underlying changes in population size (Chapter 15).

human capital   investments in individuals that can improve their economic productivity and thus their overall standard of living; including things such as education and job-training, and often enhanced by migration (Chapter 7).

hydrologic cycle   the water cycle by which ocean and land water evaporates into the air, is condensed, and then returns to the ground as precipitation (Chapter 13).

hydrosphere   the earth's water resources, including water in oceans, lakes, rivers, groundwater, and water in glaciers and ice caps (Chapter 13).

illegal migrants   migrants coming into a country without proper government approval (Chapter 7).

immigrant   a person who moves into a country of which he or she is not a native for the purpose of taking up permanent residence (Chapter 7).

impaired fecundity   a reduced ability to reproduce; same as subfecundity (Chapter 5).

implementing strategy   a possible means (such as migration) whereby a goal (such as an improvement in income) might be attained (Chapter 7).

incipient decline   the third (final) stage in the demographic transition when a country has moved from having a very high rate of natural increase to having a very low (possibly negative) rate of increase (Chapter 3).

Industrial Revolution   the totality of the changes in economic and social organization that began about 1750 in England, later in other countries; characterized chiefly by the replacement of hand tools with power-driven machines and by the concentration of industry in large establishments (Chapter 1).

infant mortality   death during the first year of life (Chapter 4).

infant mortality rate   the number of deaths to infants under 1 year of age divided by the number of live births in that year (and usually multiplied by 1,000) (Chapter 4).

infanticide   the deliberate killing or abandonment of an infant; a method of "family control" in many premodern and some modern societies (Chapter 1).

infecundity   the inability to produce offspring, synonymous with sterility (Chapter 5).

in-migrant   a person who moves into an area. This term usually refers to an internal migrant; an international migrant is an immigrant (Chapter 7).

intensification of agriculture   the process of increasing crop yield by any means—mechanical, chemical, or otherwise (Chapter 13).

intercensal years   the years between the taking of censuses (Chapter 2).

intermediate variables   means for regulating fertility; the variables through which any social factors influencing the level of fertility must operate (Chapter 5).

internal migration   permanent change in residence within national boundaries (Chapter 7).

international migration   permanent change of residence involving movement from one country to another (Chapter 7).

intervening obstacles   factors that may inhibit migration even if a person is motivated to migrate (Chapter 7).

intrauterine device (IUD)   any small, mechanical device for semipermanent insertion in the uterus as a contraceptive (Chapter 5).

late fetal mortality   fetal deaths that occur after at least 28 weeks of gestation (Chapter 4).

life expectancy   the average duration of life beyond any age, of persons who have attained that age, calculated from a life table (Chapter 4).

lifespan   the longest period over which an organism or species may live (Chapter 4).

life table   an actuarial table showing the number of persons who die at any given age, from which the expectation of life is calculated (Chapter 4).

linear growth   rate of growth of a population that assumes that the population is growing in a straight-line (linear) fashion (Chapter 8).

lithosphere   the outer shell of the earth's surface (Chapter 13).

logarithm   the exponent indicating the power to which a fixed number, the base, must be raised to produce a given number (Chapter 1).

logarithmic growth   the concept that human populations have the capacity to grow in a logarithmic fashion, increasing geometrically in size from one generation to the next (Chapter 1).

longevity   the ability to resist death, measured as the average age at death (Chapter 4).

long-term immigrant   a person whose stay in the place of destination is more than 1 year (Chapter 7).

Malthusian pertaining to the theories of Malthus, which state that population tends to increase at a geometric rate, while the means of subsistence increase at an arithmetic rate, resulting in an inadequate supply of the goods supporting life, unless a catastrophe occurs to reduce the population or the increase of population is checked by sexual restraint (Chapter 3).

marital status the state of being single, married, separated, divorced, widowed, or living in a consensual union (cohabiting) (Chapter 10).

marriage squeeze an imbalance between the numbers of males and females in the prime marriage ages (Chapter 10).

Marxist an adherent of Karl Marx or his theories, which assert that throughout history the state has been a device for the exploitation of the masses by a dominant class, leading to a rejection of Malthusian theory (Chapter 3).

maternal mortality the death of a woman as a result of pregnancy or childbearing (Chapter 4).

maternal mortality rate the number of deaths to women due to pregnancy and childbirth divided by the number of live births in a given year (and usually multiplied by 100,000) (Chapter 4).

mega-city a term used by the United Nations to denote any urban agglomeration with more than 8 million people (Chapter 11).

megalopolis see urban agglomeration (Chapter 11).

menarche the first menstrual period; the establishment of menstruation (Chapter 5).

menopause the time when menstruation ceases permanently, usually occurring between the ages of 45 and 50 (Chapter 5).

mercantilism the view that a nation's wealth depended upon its store of precious metals and that generating this kind of wealth was facilitated by population growth (Chapter 3).

metropolitan area see standard metropolitan statistical area.

metropolitan statistical area (MSA) a "standalone" metropolitan area composed of a county with an urbanized area of at least 50,000 people (Chapter 11).

migrant a person who makes a permanent change of residence substantial enough in distance to involve a shift in that individual's round of social activities (Chapter 7).

migration the process of changing residence from one geographic location to another; one of the three basic demographic processes (Chapter 7).

migration effectiveness the crude net migration rate divided by the total migration rate (Chapter 7).

migration ratio the ratio of the net number of migrants (in-migrants minus out-migrants) to the difference between the number of births and deaths—measuring the contribution that migration makes to overall population growth (Chapter 7).

migration transition the shift of people from rural to urban places, and the shift to higher levels of international migration (Chapter 7).

migration turnover rate the total migration rate divided by the crude net migration rate (Chapter 7).

model stable population a population whose age and sex structure is implied by a given set of mortality rates and a particular rate of population growth (Chapter 8).

modernization the process of societal development involving urbanization, industrialization, rising standards of living, better education, and improved health that is typically associated with a "Western" lifestyle and world view (Chapter 3).

momentum of population growth the potential for future increase in population size that is inherent in any present age/sex structure even if fertility levels were to drop immediately to replacement level (Chapter 8).

moral restraint according to Malthus, the avoidance of sexual intercourse prior to marriage and the delay of marriage until a man can afford all the children his wife might bear; a desirable preventive check on population growth (Chapter 3).

mortality deaths in a population; one of the three basic demographic processes (Chapter 4).

mortality transition the epidemiological transition, as it is usually called—the shift from high to low mortality (Chapter 4).

mover a person who moves within the same county and thus, according to the U.S. Bureau of the Census definitions, has not moved far enough to become a migrant (Chapter 7).

multiple crop to grow more than one crop per year on the same plot of ground (Chapter 13).

nationalist a person seeking freedom for his or her country from economic and political exploitation by more powerful nations (Chapter 12).

natural fertility   fertility levels that exist in the absence of deliberate, or at least modern, fertility control (Chapter 5).

natural increase   the excess of births over deaths; the difference between the crude birth rate and the crude death rate is the rate of natural increase (Chapter 1).

neo-Malthusian   a person who accepts the basic Malthusian premise that population growth tends to outstrip resources, but (unlike Malthus) believes that birth control measures are appropriate checks to population growth (Chapter 3).

neo-Marxist   a person who accepts the basic principle of Marx that problems in society are created by an unjust and inequitable distribution of resources of any and all kinds, without necessarily believing that communism is the answer to those problems (Chapter 12).

neonatal   pertaining to the first 28 days after birth (Chapter 4).

net migration   the difference between those who move in and those who move out of a particular region in a given period of time (Chapter 7).

net reproduction rate (NRR)   a measure of generational replacement; specifically, the average number of female children that will be born to the female babies who were themselves born in a given year, assuming no change in the age-specific fertility and mortality rates and ignoring the effect of migration (Chapter 5 and Appendix).

nonfamily household   one that includes people who live alone, or with nonfamily coresidents (friends living together, a single householder who rents out rooms, cohabiting couples, etc.) (Chapter 10).

nonsampling errors   errors that occur in an enumeration process as a result of missing people who should be counted, or inaccurate information provided by respondents, or inaccurate recording or processing of information (Chapter 2).

nuclear family   at least one parent and their/his/her children (Chapter 10).

optimum population size   the number of people that would provide the best balance of people and resources for a desired standard of living (Chapter 12).

opportunity costs   with respect to fertility, these represent the things that are foregone in order to have children (Chapter 6).

orphanage   the practice of abandoning children in such a way that they are likely to be cared for by strangers (Chapter 6).

out-migrant   a person who leaves an area with the intention of changing residence. This term usually refers to internal migration, whereas emigrant refers to an international migrant (Chapter 7).

overshoot   exceeding a region's carrying capacity (Chapter 1).

overpopulation   a situation in which the population has overshot a region's carrying capacity (Chapter 11).

ozone layer   the region in the earth's upper atmosphere that protects the earth from the ultraviolet rays of the sun (Chapter 13).

pathological   in biology, the term refers to disease; it also has been used to refer to deviant forms of behavior (Chapter 4).

per capita gross national product   a common measure of average income in a nation, calculated by dividing the total value of goods and services produced in a nation (GNP) by the total population size (Chapter 12).

perinatal   pertaining to the time from shortly before birth to 7 days after birth (Chapter 4).

period data   population data that refer to a particular year and represent a cross section of population at one specific time (Chapter 5).

period rates   rates referring to a specific, limited period of time, usually 1 year (Chapter 5).

physiocratic   the philosophy that the real wealth of a nation is in the land, not in the number of people (Chapter 3).

population characteristics   those demographic traits or qualities that differentiate one individual from another, including age, sex, race, ethnicity, marital status, occupation, education, income, wealth, and urban-rural residence (Chapters 8 through 11).

population explosion   a popular term referring to the rapid increase in the size of the world's population, especially the increase since World War II (Chapter 3).

population modeling   using the techniques of projecting populations to evaluate the possible consequences of different combinations of population processes (Chapter 8).

population momentum see momentum of population growth (Chapter 8).

population policy a formalized set of procedures designed to achieve a particular pattern of population change (Chapter 14).

population processes fertility, mortality, and migration; the dynamic elements of demographic analysis (Chapter 1).

population projection the calculation of the number of persons we can expect to be alive at a future date, given the number now alive and given reasonable assumptions about age-specific mortality, fertility rates, and migration (Chapter 8).

population pyramid see age/sex pyramid.

population register a vital registration system in which data on births, deaths, marriages, divorces, and geographic mobility are tracked for each individual in the country (Chapter 2).

positive checks a term used by Malthus to refer to factors (essentially mortality) that limit the size of human populations by "weakening" or "destroying the human frame" (Chapter 3).

postcoital contraception methods designed to avert pregnancy within a few days after intercourse, usually by taking a large dosage of the same hormones contained in the contraceptive pill (Chapter 5).

postneonatal pertaining to the period from 28 days after birth to 1 year of age (Chapter 4).

postpartum following childbirth (Chapter 5).

poverty index a measure of need that in the United States is based on the premise that one third of a family's income is spent on food; the cost of an economy food plan multiplied by 3. Since 1964, it has increased at the same rate as the consumer price index (Chapter 10).

preventive checks in Malthus's writings, any limits to birth, among which Malthus himself preferred moral restraint (Chapter 3).

primary metropolitan statistical area (PMSA) MSA within a CMSA (Chapter 11).

primate city a disproportionately large leading city (Chapter 11).

Principle of Population the Malthusian theory that human population increases geometrically whereas the available food supply increases only arithmetically, leading constantly to "misery" (Chapter 3).

programmed time clock theory a theory of aging based on the idea that each person has a built-in biological time clock; barring death from accidents or disease, cells still die because each is programmed to reproduce itself only a fixed number of times (Chapter 9).

pronatalist an attitude, doctrine, or policy that favors a high birth rate; also known as "populationist" (Chapter 3).

proximate determinants of fertility a renaming of the intermediate variables (defined previously) with an emphasis on age at entry into marriage and proportions married, use of contraception, use of abortion, and prevalence of breast-feeding (Chapter 5).

prudential restraint a Malthusian concept referring to delaying marriage without necessarily avoiding premarital intercourse (Chapter 3).

psychographic research description of a person's lifestyle, values, attitudes, and personality traits—used in conjunction with marketing demographics (Chapter 15).

purchasing power parity (PPP) a refinement of GDP and defined as the number of units of a country's currency required to buy the same amounts of goods and services in the domestic market as one dollar would buy in the United States (Chapter 12).

push–pull theory a theory of migration that says some people move because they are pushed out of their former location, whereas others move because they have been pulled, or attracted, to another location (Chapter 7).

race a group of people characterized by a more or less distinctive combination of inheritable physical traits (Chapter 10).

racial genocide the deliberate extermination of people who belong to a particular racial group (Chapter 10).

rank-size rule a hypothesis derived from studies of city-systems that says that the population size of a given city within a country will be approximately equal to the population of the largest city divided by the city's rank in the city-system (Chapter 11).

rational choice theory any theory based on the idea that human behavior is the result of individuals making calculated cost-benefit analyses about how to act (Chapter 3).

relative income hypothesis the perspective that fertility is influenced less by absolute levels of income than by relative levels of well-being (Chapter 3).

refugee a person who has been forced out of his or her country of nationality (Chapter 7)

religious pluralism the existence of two or more religious groups side by side in society, without any domination of one group over any others (Chapter 10).

replacement-level fertility the level of childbearing at which a cohort of women is having just enough children to replace themselves and their partners; best measured by a net reproduction rate of 1.0, but also estimated by a total fertility rate of 2.1 (Chapter 5).

residential mobility the process of changing residence, whether it be over a short or long distance (Chapter 7).

rural of, or pertaining to, the countryside. Rural populations are generally defined as those that are nonurban in character (Chapter 11).

sample surveys a method of collecting data by obtaining information from a sample of the total population, rather than by a complete census (Chapter 2).

sampling error error that occurs in sampling due to the fact that a sample is rarely identical in every way to the population from which it was drawn (Chapter 2).

secularization a spirit of autonomy from other worldly powers; a sense of responsibility for one's own well-being (Chapter 3).

segmentation the manufacturing and packaging of products or the provision of services that appeal to specific sociodemographically identifiable groups within the population (Chapter 15).

semen the male reproductive fluid (Chapter 5).

senescence a decline in physical viability accompanied by a rise in vulnerability to disease (Chapter 9).

senility an outdated term implying a loss of mental abilities (Chapter 9).

sex ratio the number of males per the number of females in a population (usually multiplied by 100 to get rid of the decimal point) (Chapter 8).

sex structure the distribution of people in a population according to sex (Chapter 8).

social class a group of people in a society who have more or less similar levels of income, education, and occupational prestige (Chapter 10).

social capital the network of family, friends, and acquaintances that increases a person's chances of success in life (Chapter 7).

social institutions sets of procedures (norms, laws, etc.) that organize behavior in society in fairly predictable and ongoing ways (Chapter 3).

socialism an economic system whereby the community as a whole (i.e., the government) owns the means of production and a social system that minimizes social stratification (Chapter 3).

socialization the process of learning the behavior appropriate to particular social roles (Chapter 8).

social roles the set of obligations and expectations that characterize a particular position within society (Chapter 8).

social status relative position or standing in society (Chapter 8).

social stratification a system of social inequality through which members of society are ranked in status from high to low (Chapter 8).

socioeconomic status a person's status in society as determined by that individual's combination of social and economic characteristics (Chapter 10).

sojourner an international mover seeking paid employment in another country but never really setting up a permanent residence in the new location (Chapter 7).

spermicide a chemical agent with the ability to kill sperm (Chapter 5).

sponge a disposable contraceptive containing a spermicide that acts as a chemical barrier after insertion in the vagina (Chapter 5).

stable population a population in which the percentage of people at each age and sex does not change over time because age-specific rates of fertility, mortality, and migration remain constant over a long period of time (Chapter 8).

standard metropolitan statistical area (SMSA) a term previously used by the U.S. Bureau of the Census to define a county or group of contiguous counties that contain at least one city of 50,000 inhabitants or more and any contiguous counties that are socially and economically integrated with the central city or cities (Chapter 11).

standardization   a set of procedures for controlling for the effects of external factors influencing overall crude rates such as the crude birth or death rate (Appendix).

stationary population   a type of stable population in which the birth rate equals the death rate and so the number of people remains the same as does the age/sex distribution (Chapter 8).

step migration   the process whereby a migrant moves in stages progressively farther away from the place of origin (Chapter 7).

sterile   biologically incapable of reproduction. This term is synonymous with infecund (Chapter 5).

sterilization   the process of rendering a person sterile (either voluntarily—surgically—or involuntarily) (Chapter 5).

structural economic mobility   the situation in which most, if not all, people in an entire society experience an improvement in living levels, even though some people may be improving faster than others (Chapter 10).

subfecundity   a reduced ability to reproduce; same as impaired fecundity (Chapter 5).

suburban   pertaining to populations in low-density areas close to and integrated with central cities (Chapter 11).

suburbanize   to become suburban—a city suburbanizes by growing in its outer rings (Chapter 11).

Sun Belt   the area in the southern part of the United States that is a popular destination for migrants; in particular, California, Arizona, Texas, and Florida (Chapter 7).

supply-demand framework   a version of neoclassical economics in which it is assumed that couples attempt to maintain a balance between the potential supply of and the demand for children, taking into account the costs of fertility regulation (Chapter 6).

surgical contraception   permanent methods of contraception including tubal ligation and hysterectomy for women, and vasectomy for men (Chapter 5).

symptothermal method   method of fertility control based on avoiding intercourse during the fertile period, as determined by observation of changes in a woman's cervical mucus along with the taking of basal body temperature, or the use of other indices of the menstrual cycle (Chapter 5).

synthetic cohort   a measurement obtained by treating period data as though they represented a cohort (Chapter 5).

targeting   a marketing technique of picking out particular sociodemographic characteristics and appealing to differences in consumer tastes and behavior reflected in those particular characteristics (Chapter 15).

target year   the year to which we project a population forward in time (Chapter 8).

theoretical effectiveness   with respect to birth control methods, the probability of preventing a pregnancy if a method is used exactly as it should be (Chapter 5).

theory   a system of assumptions, accepted principles, and rules of procedure devised to analyze, predict, or otherwise explain a set of phenomena (Chapter 3).

theory of relative income   see relative income hypothesis (Chapter 3).

total fertility rate   an estimate of the average number of children that would be born to each woman if the current age-specific birth rates remained constant (Chapter 5).

total migration rate   the sum of in-migrants plus out-migrants, divided by the total midyear population (Chapter 7).

transnational migrant   an international migrant who maintains close ties in both the country of origin and of destination (Chapter 7).

transitional growth   the second (middle) stage of the demographic transition when death rates have dropped but birth rates are still high. During this time, population size is increasing steadily—this is the essence of the "population explosion" (Chapter 3).

troposphere   that portion of the atmosphere which is closest to (within about 11 miles of) the earth's surface (Chapter 13).

underdeveloped   describes those nations with very low average levels of income. This is sometimes a euphemism for a poor country, but the term also implies that the country has the potential to raise its level of living (Chapter 12).

unmet need   as applied to family planning, it refers to the number of sexually active women who would prefer not to get pregnant but are nevertheless not using any method of contraception (Chapter 14).

urban   describes a spatial concentration of people whose lives are organized around nonagricultural activities (Chapter 11).

urban agglomeration   according to the United Nations, the population contained within the contours of contiguous territory inhabited at urban levels of residential density without regard to administrative boundaries (Chapter 11).

urban transition   see urbanization (Chapter 11).

urbanism   the changes that occur in life-style and social interaction as a result of living in urban places (Chapter 11).

urbanization   the process whereby the proportion of people in a population who live in urban places increases (Chapters 2 and 11).

usual residence   the concept of including people in the census on the basis of where they usually reside (Chapter 2).

use effectiveness   the actual pregnancy prevention performance associated with the use of a particular fertility control measure (Chapter 5).

vasectomy   a technique of male sterilization in which each vas deferens is cut and tied, thus preventing sperm from being ejaculated during intercourse (Chapter 5).

vital statistics   data referring to the so-called vital events of life, especially birth and death, but usually also including marriage and divorce and sometimes abortion (Chapter 2).

wealth flow   a term coined by John Caldwell to refer to the intergenerational transfer of income (Chapter 3).

wealth of a nation   the sum of known natural resources and our human capacity to transform those resources into something useful (Chapter 12).

wear and tear theory   the theory of aging that argues that humans are like machines that eventually wear out due to the stresses and strains of constant use (Chapter 9).

withdrawal   a form of fertility control that requires the male to withdraw his penis from his partner's vagina prior to ejaculation; also called coitus interruptus (Chapter 5).

world-systems theory   the theory that since the 16th century the world market has developed into a set of core nations and a set of peripheral countries that are dependent on the core (Chapter 7).

xenophobia   fear of strangers (Chapter 7).

zero population growth (ZPG)   a situation in which a population is not changing in size from year to year, as a result of the combination of births, deaths, and migration (Chapter 8).

BIBLIOGRAPHY

Abbott, C. 1993. *The Metropolitan Frontier: Cities in the Modern American West.* Tucson: The University of Arizona Press.

Abernethy, V. 1979. *Population Pressure and Cultural Adjustment.* New York: Human Sciences Press.

Abma, J.A. Chandra, W. Mosher, L. Peterson, and L. Piccinino. 1997. "Fertility, Family Planning, and Women's Health: New Data From the 1995 National Survey of Family Growth." *National Center for Health Statistics Vital Health Statistics* 23 (19).

Abramovitz, J.N. 1998. "Sustaining the World's Forests." In *State of the World 1998*, edited by L. Brown, C. Flavin, and H. French. New York: W.W. Norton & Company.

Acevedo, D. and T.J. Espenshade. 1992. "Implications of a North American Free Trade Agreement for Mexican Migration into the United States." *Population and Development Review* 18 (4):729–44.

Adamchak, D., A. Wilson, A. Nyanguru, and J. Hampson. 1991. "Elderly Support and Intergenerational Transfer in Zimbabwe: An Analysis by Gender, Marital Status, and Place of Residence." *The Gerontologist* 31:505–13.

Adelman, C. 1982. "Saving Babies With a Signature." *Wall Street Journal* 28 July.

Adlakha, A. and J. Banister. 1995. "Demographic Perspectives on China and India." *Journal of Biosocial Science* 27:163–178.

Adlakha, A. and D. Kirk. 1974. "Vital Rates in India 1961–71 Estimated From 1971 Census Data." *Population Studies* 28 (3):381–400.

Aguirre, B.E., R. Saenz, J. Edmiston, N. Yang, E. Agramonte, and D. Stuart. 1993. "The Human Ecology of Tornadoes." *Demography* 30 (4):623–33.

Ahlburg, D. and M. Schapiro. 1984. "Socioeconomic Ramifications of Changing Cohort Size: An analysis of U.S. Post-War Suicide Rates by Age and Sex." *Demography* 21 (1):97–105.

Ahmed, L. 1992. *Women and Gender in Islam: Historical Roots of a Modern Debate.* New Haven: Yale University Press.

Ahn, N. and A. Shariff. 1994. "A Comparative Study of Socioeconomic and Demographic Determinants of Fertility in Togo and Uganda." *International Family Planning Perspectives* 20:14–17.

Ahonsi, B. 1991. "Report on the Seminar on Anthropological Studies Relevant to the Sexual Transmission of HIV, Sonderborg, Denmark, 1990." *IUSSP Newsletter* 41:79–103.

Akin, J., R. Bilsbarrow, D. Guilkey, B. Popkin, D. Benoit, P. Cantrelle, M. Garenne, and P. Levi. 1981. "The Determinants of Breast-Feeding in Sri Lanka." *Demography* 18 (3):287–308.

Alba, R. and J.R. Logan. 1991. "Variations on Two Themes: Racial and Ethnic Patterns in the Attainment of Suburban Residence." *Demography* 28:431–53.

Alba, R. and V. Nee. 1997. "Rethinking Assimilation Theory for a New Era of Immigration." *International Migration Review* 31 (4):826–74.

Albert, S. 1998. "Site Selection's Essential Ingredients." *Business Geographics* (May):44.

Albert, S.M. and M.G. Cattell. 1994. *Old Age in Global Perspective.* New York: G.K. Hall & Co.

Alchon, S.A. 1997. "The Great Killers in Precolumbian America: A Hemispheric Perspective." *Latin American Population History Bulletin* (27):2–11.

Alonso, W. and P. Starr. 1982. "The Political Economy of National Statistics." *Social Science Research Council Items* 36 (3):29–35.

Amato, P.R. and A. Booth. 1995. "Changes in Gender Role Attitudes and Perceived Marital Quality." *American Sociological Review* 60 (1):58–66.

American Demographics. 1995. "Meet the New Vegetarian." *American Demographics* (January):9.

Anderson, D. 1994. *Toward a More Effective Policy Response to AIDS.* Liege: International Union for the Scientific Study of Population.

Anderson, R.N., K.D. Kochanek, and S.L. Murphy. 1997. "Report of Final Mortality Statistics 1995." *Monthly Vital Statistics Report* 45 (11, Supplement 2).

Angel, R.J. and J.L. Angel. 1993. *Painful Inheritance: Health and the New Generation of Fatherless Families*. Madison, WI: The University of Wisconsin Press.

Anker, Richard. 1998. *Gender and Jobs: Sex Segregation of Occupations in the World*. Geneva: International Labour Office.

Ankrah, E.M. 1991. "AIDS and the social side of health." *Social Science and Medicine* 32 (967–80).

Aries, P. 1962. *Centuries of Childhood*. New York: Vintage Books.

Arnold, F. 1988. "The Effect of Sex Preferences on Fertility and Family Planning: Empirical Evidence." *Population Bulletin of the United Nations* 23/24:44–56.

Arriaga, E. 1970. *Mortality Decline and Its Demographic Effects in Latin America*. Berkeley, CA: University of California Institute of International Studies.

———. 1982. "Changing Trends in Mortality in Developing Countries." Paper presented at the annual meeting of the Population Association of America. San Diego.

Associated Press. 1980. "Study on Old Age Released in China." in *San Diego Union*. San Diego, CA.

Atchley, R. 1996. *Social Forces and Aging: An Introduction to Social Gerontology, 8th Edition*. Belmont, CA: Wadsworth Publishing Co.

Auerbach, F. 1961. *Immigration Laws of the United States, 2nd Edition*. New York: Bobbs-Merrill.

Avdeev, A. and A. Monnier. 1995. "A Survey of Modern Russian Fertility." *Population: An English Selection* 7:1–38.

Axelrod, P. 1990. "Cultural and Historical Factors in the Population Decline of the Parsis of India." *Population Studies* 44:401–20.

Axinn, W.G. and A. Thornton. 1992. "The Relationship Between Cohabitation and Divorce: Selectivity or Causal Influence?" *Demography* 29 (3):357–74.

Bachrach, C. and M. Horn. 1988. "Sexual Activity Among U.S. Women of Reproductive Age." *American Journal of Public Health* 78 (3):320.

Bachu, A. 1993. "Fertility of American Women: June 1992." *Current Population Reports* Series P20 (No. 470).

Bachu, A. 1997. "Fertility of American Women, June 1995 (Update)." *Current Population Reports, Series* P-20 (499).

Baird, D.D., C.R. Weinberg, L.F. Voigt, and J.R. Daling. 1996. "Vaginal Douching and Reduced Fertility." *American Journal of Public Health* 86 (6): 844–850.

Balakrishnan, T.R., E. Lapierre-Adamczyk, and K. Krotki. 1993. *Family and Childbearing in Canada: A Demographic Analysis*. Toronto: University of Toronto Press.

Balakrishnan, T.R. and J. Chen. 1988. *Religiosity, Nuptiality and Reproduction in Canada*. London, Ontario, Canada.

Banister, J. 1987. *China's Changing Population*. Stanford, CA: Stanford University Press.

Banks, J.A. 1954. *Prosperity and Parenthood: A Study of Family Planning Among the Victorian Middle Classes*. London: Routledge & Kegan Paul.

Banks, J.A. and O. Banks. 1964. *Feminism and Family Planning in Victorian England*. Liverpool: Liverpool University Press.

Barclay, G.W. 1958. *Techniques of Population Analysis*. New York: John Wiley & Sons.

Barlow, J. 1995. "The Politics of Urban Growth: 'Boosterism' and 'Nimbyism' in European Boom Regions." *International Journal of Urban and Regional Research* 19 (1):129–44.

Barrett, P.M. 1996a. "High-Court Ruling on Census Sets Back Some Cities That Disputed '90 Numbers." *Wall Street Journal* 21 March, p. B6.

———. 1996b. "Minority Voting Districts Struck Down by High Court for Lack of 'Compactness'." *The Wall Street Journal* 14 June.

Basu, A.M. 1993. "How Pervasive Are Sex Differentials in Childhood Nutritional Levels in South Asia?" *Social Biology* 40 (1–2):25–37.

———. 1997. "The 'Politicization' of Fertility to Achieve Non-Demographic Objectives." *Population Studies* 51 (1):5–18.

Bean, F. and G. Swicegood. 1985. *Mexican American Fertility Patterns*. Austin, TX: University of Texas Press.

Bean, F. and M. Tienda. 1987. *The Hispanic Population of the United States*. New York: Russell Sage Foundation.

Bean, F., G. Vernez, and C. Keely. 1989. *Opening and Closing the Doors: Evaluating Immigration Reform and Control*. Washington, D.C.: The Urban Institute.

Becker, G. 1960. "An Economic Analysis of Fertility." in *Demographic Change and Economic Change in Developed Countries*, edited by National Bureau of Economic Research. Princeton: Princeton University Press.

Becker, J. 1997. *Hungry Ghosts: Mao's Secret Famine*. New York: Free Press.

Behrman, J. and P. Taubman. 1989. "A Test of the Easterlin Fertility Model Using Income For Two Generations and a Comparison with the Becker Model." *Demography* 26:117–23.

Belmont, L. and F. Marolla. 1973. "Birth Order, Family Size, and Intelligence." *Science* 182:1096–1101.

Benedict, B. 1972. "Social Regulation of Fertility." in *The Structure of Human Population*, edited by G. A. Harrison and A. J. Boyce. Oxford: Clarendon Press.

Benenson, A. 1990. *Control of Communicable Disease in Man*. Washington, D.C.: American Public Health Association.

Benjamin, B. 1969. *Demographic Analysis*. New York: Praeger.

Benokratis, N. 1982. "Racial Exclusion in Juries." *Journal of Applied Behavioral Science* 18 (1):29–47.

Bergman, B.R. 1996. *Saving Our Children From Poverty: What the United States Can Learn From France*. New York: Russell Sage Foundation.

Berry, B., C. Goodwin, R. Lake, and K. Smith. 1976. "Attitudes Toward Integration: The Role of Status." in *The Changing Face of the Suburbs*, edited by B. Schwartz. Chicago: University of Chicago Press.

Berry, B.J.L. 1993. "Transnational Urbanward Migration, 1830–1980." *Annals of the Association of American Geographers* 83 (3):389–405.

Bertrand, J. and D. Bogue. 1977. "A Research-Based System for Improving Family Planning Adoption: The Guatemala Study." *Intercom* 5:8–9.

Bhatia, P. 1978. "Abstinence—Desai Solution for Baby Boom." *San Francisco Chronicle* 11 June.

Bianchi, S. 1981. *Household Composition and Racial Inequality*. New Brunswick: Rutgers University Press.

———. 1995. "Changing Economic Roles of Women and Men." in *State of the Union: America in the 1990s, Volume One, Economic Trends*, edited by R. Farley. New York: Russell Sage Foundation.

Biraben, J.-N. 1979. "Essai sur l'evolution du nombre des hommes." *Population* 34.

Bishop, D.B., J.S. Roesler, B.R. Zimmerman, and D.J. Ballard. 1993. "Diabetes." in *Chronic Disease Epidemiology and Control*, edited by R. C. Brownson, R. L. Remington, and J. R. Davis. Washington, D.C.: American Public Health Association.

Blake, J. 1967. "Family Size in the 1960s—A Baffling Fad." *Eugenics Quarterly* 14:60–74.

———. 1972. "Coercive Pronatalism and American Population Policy." in *Social and Demographic Aspects of Population Growth and the American Future*, edited by C. Westoff and R. Parke. Washington, D.C.: Government Printing Office.

———. 1974. "Can We Believe Recent Data in the United States?" *Demography* 11 (1):25–44.

———. 1981. "The Only Child in America: Prejudice Versus Performance." *Population and Development Review* 7 (1):43–54.

———. 1989. *Family Size and Achievement*. Berkeley: University of California Press.

Blanchet, D. 1991. "On Interpreting Observed Relationships Between Population Growth and Economic Growth: A Graphical Exposition." *Population and Development Review* 17 (1):105–14.

Blayo, Y. 1992. "Political Events and Fertility in China Since 1950." *Population: An English Selection* 4:209–32.

Bledsoe, C., A.G. Hill, U. D'Alessandro, and P. Langerock. 1994. "Constructing Natural Fertility: The Use of Western Contraceptive Technologies in Rural Gambia." *Population and Development Review* 20 (1):81–113.

Bloom, D. and R. Freeman. 1986. "The Effects of Rapid Population Growth on Labor Supply and Employment in Developing Countries." *Population and Development Review* 12 (3):381–414.

Bloom, D. and J. Trussell. 1984. "What Are the Determinants of Delayed Childbearing and Permanent Childlessness in the United States?" *Demography* 21 (4):591–612.

Bloom, D.E. and A. Brender. 1993. "Labor and the Emerging World Economy." *Population Bulletin* 48 (2).

Blossfeld, H. and A. de Rose. 1992. "Educational Expansion and Changes in Entry Into Marriage and Motherhood, the Experience of Italian Women." *Genus* XLVIII (3–4):73–91.

Blumenthal, K. 1996. "Boomers Entering the Peak Home-Remodeling Stage." *The Wall Street Journal* (28 June).

Bollen, K.A. and S.J. Appold. 1993. "National Industrial Structure and the Global System." *American Sociological Review* 58:283–301.

Bolton, C. and J.W. Leasure. 1979. "Evolution Politique et Baisse de la Fecondite en Occident." *Population* 34:825–44.

Bongaarts, J. 1978. "A Framework for Analyzing the Proximate Determinants of Fertility." *Population and Development Review* 4 (1):105–32.

———. 1982. "The Fertility-Inhibiting Effects of the Intermediate Fertility Variables." *Studies in Family Planning* 13:179–89.

———. 1986. "Contraceptive Use and Annual Acceptors Required for Fertility Transition: Results of a Projection Model." *Studies in Family Planning* 17 (209–16).

———. 1993. "The Supply-Demand Framework for the Determinants of Fertility: An Alternative Implementation." *Population Studies* 47:437–456.

———. 1996. "Global Trends in AIDS Mortality." *Population and Development Review* 22 (1):21–46.

Bongaarts, J., W.P. Mauldin, and J. Phillips. 1990. "The Demographic Impact of Family Planning Programs." *Studies in Family Planning* 21:299–310.

Boserup, E. 1965. *The Conditions of Agricultural Growth*. Chicago: Aldine.

———. 1970. *Woman's Role in Economic Development*. New York: St. Martin's Press.

———. 1981. *Population and Technological Change: A Study of Long-Term Trends*. Chicago: University of Chicago Press.

Bourne, K. and G. Walker. 1991. "The Differential Effect of Mother's Education on Mortality of Boys and Girls in India." *Population Studies* 45:203–20.

Bouvier, L., D.L. Poston, and N.B. Zhai. 1997. "Population Growth Impacts of Zero Net International Migration." *International Migration Review* 31 (2):294–311.

Bouvier, L. and S.L.H. Rao. 1975. *Socioreligious Factors in Fertility Decline*. Cambridge: Ballinger.

Bouvier, L. and J. van der Tak. 1976. "Infant Mortality—Progress and Problems." *Population Bulletin* 31 (2).

Bouvier, L.F. 1992. *Peaceful Invasions: Immigration and Changing America*. Lanham MD: University Press of America.

Boyd, M. 1976. "Immigration Policies and Trends: A Comparison of Canada and the United States." *Demography* 13:83–104.

Bracher, M., G. Santow, S.P. Morgan, and J. Trussell. 1993. "Marriage Dissolution in Australia: Models and Explanations." *Population Studies* 47 (3):403–26.

Brackett, J. 1967. "The Evolution of Marxist Theories of Population: Marxism Recognizes the Population Problem." in *Annual Meeting of the Population Association of America*. Cincinnati, Ohio.

Bradshaw, B. and W.P. Frisbie. 1992. "Mortality of Mexican Americans and Mexican Immigrants: Comparisons With Mexico." in *Demographic Dynamics Along the U.S.-Mexico Border*, edited by J. R. Weeks and R. Ham-Chande. El Paso: Texas Western Press of the University of Texas at El Paso.

Bradshaw, Y. 1987. "Urbanization and Underdevelopment: A Global Study of Modernization, Urban Bias, and Economic Dependency." *American Sociological Review* 52 (2):224–39.

Bradshaw, Y. and E. Fraser. 1989. "City Size, Economic Development, and Quality of Life in China: New Empirical Evidence." *American Sociological Review* 54: 986–1003.

Bralove, M. 1982. "Advertising World's Portrayal of Women is Starting to Shift." *The Wall Street Journal* 28 October.

Brandes, S.H. 1975. *Migration, Kinship, and Community*. New York: Academic Press.

———. 1990. "Ritual Eating and Drinking in Tzintzuntzan: A Contribution to the Study of Mexican Foodways." *Western Folklore* 49:163–75.

Brauchli, M.W. 1993. "Migrants Fuel China's Economic Growth." *The Wall Street Journal* 17 August.

Brentano, L. 1910. "The Doctrine of Malthus and the Increase of Population During the Last Decade." *Economic Journal* September.

Breslin, J. 1982. *Forsaking All Ohters*. New York: Simon & Schuster.

Brodie, J.F. 1994. *Contraception and Abortion in Nineteenth-Century America*. Ithaca: Cornell University Press.

Brown, C. 1994. *American Standards of Living, 1918–1988*. Cambridge, MA: Blackwell.

Brown, L. 1981a. "World Food Resources and Population: The Narrowing Gap." *Population Bulletin* 36 (3).

———. 1995. "Nature's Limits." in *State of the World 1995*, edited by L. Brown and Associates. New York: W.W. Norton & Company.

———. 1998. "The Future of Growth." in *State of the World 1998*, edited by L. Brown, C. Flavin, and H. French. New York: W.W. Norton & Company.

Brown, L. and Associates. 1991. *State of the World 1991*. New York: W.W. Norton & Company.

Brown, L., C. Flavin, and H. French. 1997. *State of the World 1997*. New York: W.W. Norton & Company.

Brown, L., H. Kane, and D. Roodman. 1994. *Vital Signs 1994: The Trends That Are Shaping the Future*. New York: W.W. Norton & Company.

Brown, L. and J. Mitchell. 1998. "Building a New Economy." in *State of the World 1998*, edited by L. Borwn, C. Flavin, and H. French. New York: W.W. Norton & Company.

Brown, L.A. 1981b. *Innovation Diffusion: A New Perspective*. London and New York: Methuen.

Brownson, R.C., J.S. Reif, M.C. Alavanja, and D.G. Bal. 1993. "Cancer." in *Chronic Disease Epidemiology and Control*, edited by R. C. Brownson, P. L. Remington, and J. R. Davis. Washington, D.C.: American Public Health Association.

Bryan, N.S. 1993. "GIS Helps Produce Clinton's Victory." *GIS World* (March/April):28–32.

———. 1994. "Exposed . . . the BG Reader." *Business Geographics* (July/August):10.

Bulatao, R. and R. Lee. 1983. *Determinants of Fertility in Developing Countries, Volume 1, Supply and Demand for Children*. New York: Academic Press.

Bumpass, L.L. and J.A. Sweet. 1995. "Cohabitation, Marriage, Nonmarital Childbearing, and Union Stability: Preliminary Findings From HSFH2." Presented at the annual meeting of the Population Association of America. San Francisco.

Bumpass, L.L., J.A. Sweet, and A. Cherlin. 1991. "The Role of Cohabitation in Declining Rates of Marriage." *Journal of Marriage and the Family* 53:913–27.

Burch, T. 1975. "Theories of Fertility as Guides to Population Policy." *Social Forces* 54 (1):126–39.

Bussey, J. 1994. "India's Market Reform Requires Perspective." *The Wall Street Journal* 6 June.

Bustamante, J. 1997. "Mexico-United States Labor Migration Flows." *International Migration Review* 31 (4):1112–1121.

Butz, W. and M. Ward. 1979. "Will U.S. Fertility Remain Low? A New Economic Interpretation." *Population and Development Review* 5:663–88.

Cain, M. 1981. "Risk and Fertility in India and Bangladesh." *Population and Development Review* 7:435–74.

Caldwell, J. 1976. "Toward a Restatement of Demographic Transition Theory." *Population and Development Review* 2 (3–4):321–66.

———. 1980. "Mass Education as a Determinant of the Timing of Fertility Decline." *Population and Development Review* 6 (2):225–56.

———. 1982. *Theory of Fertility Decline*. New York: Academic Press.

———. 1986. "Routes to Low Mortality in Poor Countries." *Population and Development Review* 12 (2):171–220.

Caldwell, J., P.H. Reddy, and P. Caldwell. 1988. *The Causes of Demographic Change: Experimental Research in South Asia*. Madison, WI: The University of Wisconsin Press.

Caldwell, J.C. and P. Caldwell. 1990. "High Fertility in Sub-Saharan Africa." *Scientific American* 269 (May): 118–25.

Calhoun, J. 1962. "Population Density and Social Pathology." *Scientific American* (May):118–25.

California Department of Finance. 1994. "Recession Slows California's Growth." *News Release of the Demographic Research Unit* 7 February.

———. 1998. "New Report Shows State Grew by 574,000 Over 12 Months." *New Release of the Demographic Research Unit* 29 January.

Campani, G. 1995. "Women Migrants: From Marginal Subjects to Social Actors." in *The Cambridge Survey of World Migration*, edited by R. Cohen. Cambridge: Cambridge University Press.

Campbell, A. 1969. "Changing Patterns of Childbearing in the United States." in *The Family in Transition*, edited by National Institutes of Health. Washington D.C.: Government Printing Office.

Campbell, P. 1996. *Population Projections for States by Age, Sex, Race, and Hispanic Origin: 1995 to 2025*. Washington, D.C.: U.S. Bureau of the Census, Population Division.

Cantillon, R. 1755 (1964). *Essai sur la Nature du Commerce en General, Edited with an English Translation by Henry Higgs*. New York: A.M. Kelley.

Carlson, E. 1981. "Governments Struggle to Keep Farms Away From Developers." *The Wall Street Journal* 23 June.

Carlson, E. 1980. "Divorce Rate Fluctuations as a Cohort Phenomenon." *Population Studies* 33:523–36.

Carlson, E. and S. Tsvetarsky. 1992. "Concentration of Rising Bulgarian Mortality Among Manual Workers." *Sociology and Social Research* 76 (2):81–84.

Carlson, E. and M. Watson. 1990. "The Family and the State: Rising Hungarian Death Rates." in *Aiding and Aging: The Coming Crisis in Support for the Elderly by Kin and State*, edited by J. Mogey. Westport, CT: Greenwood Press.

Carnes, B.A. and S.J. Olshansky. 1993. "Evolutionary Perspectives on Human Senescence." *Population and Development Review* 19 (4):793–806.

Carpenter, J. and L. Lees. 1995. "Gentrification in New York, London and Paris: An International Comparison." *International Journal of Urban and Regional Research* 19 (2):286–304.

Carr-Saunders, A.M. 1936. *World Population: Past Growth and Present Trends*. Oxford: Clarendon Press.

Caselli, G. and V. Egidi. 1991. "A New Insight Into Morbidity and Mortality in Italy." *Genus* XLVII (3–4): 1–30.

Casper, L.M., S. McLanahan, and I. Garfinkel. 1994. "The Gender-Poverty Gap: What Can We Learn From Other Countries?" *American Sociological Review* 59: 594–605.

Cassedy, J. 1969. *Demography in Early America: Beginnings of the Statistical Mind, 1600–1800*. Cambridge: Harvard University Press.

Catasús Cervera, S. and J.C.A. Fraga. 1996. "The Fertility Transition in Cuba." in *The Fertility Transition in Latin America*, edited by J. M. Guzmán, S. Singh, G. Rodríguez, and E. Pantiledes. Oxford: Clarendon Press.

Chamie, J. 1981. "Religion and Fertility." Cambridge: Cambridge University Press.

Chandler, R.F. 1971. "The Scientific Basis for the Increased Yield Potential of Rice and Wheat." in *Food, Population and Employments: The Impact of the Green Revolution*, edited by T. Poleman and D. Freebain. New York: Praeger.

Chandler, T. and G. Fox. 1974. *3000 Years of Urban Growth*. New York: Academic Press.

Chandra, A. and E.H. Stephen. 1998. "Impaired Fecundity in the United States: 1982–95." *Family Planning Perspectives* 30 (1):34–42.

Chandrasekhar, S. 1979. *"A Dirty, Filthy Book"—The Writings of Charles Knowlton and Annie Besant on Reproductive Physiology and Birth Control and an Account of the Bradlaugh-Besant Trial*. Berkeley: University of California Press.

Chang, S.-d. 1996. "The Floating Population: An Informal Process of Urbanisation in China." *International Journal of Population Geography* 2:197–214.

Chant, S. and S. Radcliffe. 1992. "Migration and Development: The Importance of Gender." in *Gender and Migration in Developing Countries*, edited by S. Chant. London: Bellhaven Press.

Charles, E. 1936. *The Twilight of Parenthood*. London: Watt's and Co.

Charlton, S. 1984. *Women in Third World Development*. Boulder, CO: Westview Press.

Chase-Dunn, C.K. 1989. *Global Formation: Structure of the World-Economy*. Cambridge, MA: Blackwell.

Chase-Dunn, C.K. and T.D. Hall. 1997. *Rise and Demise: Comparing World Systems*. Boulder, CO: Westview Press.

Chelala, C. 1998. "A Critical Move Against Female Genital Mutilation." *Populi* 25 (1):13–15.

Chen, L.C., F. Wittgenstein, and E. McKeon. 1996. "The Upsurge of Mortality in Russia: Causes and Policy Implications." *Population and Development Review* 22 (3):457–482.

Chen, M. 1979. "Birth Planning in China." *International Family Planning Perspectives* 5 (3):92–101.

Cherfas, J. 1980. "The World Fertility Survey Conference: Population Bomb Revisited." *Science* 80 (1; November):11–14.

Cherlin, A. 1992. *Marriage, Divorce, Remarriage: Revised and Enlarged Edition*. Cambridge, MA: Harvard University Press.

Chesler, E. 1994. "No, the First Priority Is: Stop Coercing Women." *The New York Times Magazine* 6 February.

Chesnais, J.-C. 1996. "Fertility, Family, and Social Policy in Contemporary Western Europe." *Population and Development Review* 22 (4):729–740.

Chiang, C. 1984. *The Life Table and Its Applications*. Melbourne, FL: Krieger.

Choi, N. 1991. "Racial Differences in the Determinants of Living Arrangements of Widowed and Divorced Elderly Women." *The Gerontologist* 34:496–504.

Choinière, R. 1993. "Les Inégalités Socio-Économiques et Culturelles de la Mortalité à Montréal à la Fin des Années 1980." *Cahiers Québécois de Démographie* 22 (2):339–362.

Choldin, H. 1978. "Urban Density and Pathology." *Annual Review of Sociology* 4:91–113.

———. 1994. *Looking For the Last Percent: The Controversy Over the Census Undercounts*. New Brunswick, NJ: Rutgers University Press.

Choucri, N. 1984. "Perspectives on Population and Conflict." in *Multidisciplinary Perspectives on Population and Conflict*, edited by N. Choucri. Syracuse: Syracuse University Press.

Cicourel, A. 1974. *Theory and Method in a Study of Argentina*. New York: Wiley.

Cincotta, R.P. and R. Engelman. 1997. *Economics and Rapid Change: The Influence of Population Growth*. Washington, D.C.: Population Action International.

Cipolla, C. 1965. *The Economic History of World Population*. London: Penguin.

———. 1969. *Literacy and Development in the West*. Baltimore: Penguin Books.

———. 1981. *Fighting the Plague in Seventeenth-Century Italy*. Madison: University of Wisconsin Press.

Clark, C. 1967. *Population Growth and Land Use*. New York: St. Martin's Press.

Clark, W. 1991. "Residential Preferences and Neighborhood Racial Segregation: A Test of the Schelling Segregation Model." *Demography* 28:1–19.

Clarke, S.C. 1995. "Advance Report of Final Marriage Statistics, 1989 and 1990." *Monthly Vital Statistics Report* 43 (12, Supplement).

Cleland, J. and C. Wilson. 1987. "Demand Theories of the Fertility Transition: An Iconoclastic View." *Population Studies* 41:5–30.

Cliff, A.D. and P. Haggett. 1996. "The Impact of GIS on Epidemiological Mapping and Modelling." in *Spatial Analysis: Modelling in a GIS Environment*, edited by P. Longley and M. Batty. New York: John Wiley & Sons.

Clough, S. 1968. *European Economic History*. New York: Walker.

Coale, A. 1973. "The Demographic Transition." in *International Population Conference*, vol. I. Liege, Belgium, pp. 53–72.

———. 1974. "The History of the Human Population." *Scientific American* 231 (3):41–51.

———. 1978. "Population Growth and Economic Development: The Case of Mexico." *Foreign Affairs* 56 (2): 5415–29.

———. 1986a. "Population Trends and Economic Development." in *World Population and U.S. Population Policy: The Choice Ahead*, edited by J. Menken. New York: W.W. Norton & Co.

———. 1986b. "Preface." in *The Decline of Fertility in Europe*, edited by A. Coale and S. C. Watkins. Princeton: Princeton University Press.

Coale, A. and J. Bannister. 1994. "Five Decades of Missing Females in China." *Demography* 31 (3):459–80.

Coale, A. and P. Demeny. 1983. *Regional Model Life Tables and Stable Populations*. New York: Academic Press.

Coale, A. and E. Hoover. 1958. *Population Growth and Economic Development in Low-Income Countries*. Princeton: Princeton University Press.

Coale, A. and J. Trussell. 1974. "Model Fertility Schedules: Variations in the Age Structure of Child-Bearing in Human Populations." *Population Index* 40 (2): 185–256.

Coale, A. and M. Zelnick. 1963. *Estimates of Fertility and Population in the United States*. Princeton: Princeton University Press.

Cohen, J. 1995. *How Many People Can the Earth Support?* New York: W. W. Norton.

Cohen, L. and M. Felson. 1979. "Social Change and Crime Rate Trends." *American Sociological Review* 44 (4):588–607.

Cohen, L. and K. Land. 1987. "Age Structure and Crime." *American Sociological Review* 52 (2):170–83.

Cohen, M.N. 1977. *The Food Crisis in Prehistory: Overpopulation and the Origins of Agriculture*. New Haven: Yale University Press.

Cole, W.A. and P. Deane. 1965. "The Growth of National Incomes." in *The Cambridge Economic History of Europe, Vol 6, Part 1*, edited by H. H. Habakkuk and M. Postan. Cambridge: Cambridge University Press.

Coleman, C. 1995. "The Unseemly Secrets of Eating Alone." *The Wall Street Journal* 6 July.

Coleman, D. 1995. "International Migration: Demographic and Socioeconomic Consequences in the United Kingdom and Europe." *International Migration Review* 29 (1):155–206.

Coleman, J. and T.J. Fararo. 1992. "Rational Choice Theory: Advocacy and Critique." Newbury Park: Sage Publications.

Coleman, R. and L. Rainwater. 1978. *Social Standing in America*. New York: Basic Books.

College Board. 1998. "College Bound Senior 1997-Table 1." http://www.collegeboard.org/press/senior97/table 01.html (accessed 1998).

Colvin, T. 1982. "Students Halt 19-Year Slide in Test Scores." *San Diego Union* 22 September.

Commission on Environmental Quality, 1980. *The Global 2000 Report to the President, Volume 1.* Washington, D.C.: Government Printing Office.

Commission on Population Growth and the American Future. 1972. *Population and the American Future.* Washington, D.C.: Government Printing Office.

Commoner, B. 1972. "The Environmental Cost of Economic Growth." *Chemistry in Britain* 8 (2):52–65.

———. 1994. "Population, Development and the Environment: Trends and Key Issues in the Developed Countries." in *Population, Environment and Development: Proceedings of the United Nations Expert Group Meeting on Populations, Environment and Development,* edited by United Nations Population Division. New York: United Nations.

Condorcet, M.J.A.N.d.C. 1795 (1955). *Sketch for an Historical Picture of the Progress of the Human Mind,* translated from the French by June Baraclough. London: Weidenfield and Nicholson.

Conly, S.R. 1996. *Taking the Lead: The United Nations and Population Assistance.* Washington, D.C.: Population Action International.

Coombs, L. and T. Sun. 1981. "Familial Values in a Developing Society: A Decade of Change in Taiwan." *Social Forces* 59 (4):1229–55.

Cornelison, A. 1980. *Strangers and Pilgrims.* New York: Holt, Rinehart & Winston.

Cornelius, W. and P. Martin. 1993. "The Uncertain Connection: Free Trade and Rural Mexican Migration to the United States." *International Migration Review* 27 (3):484–512.

Corona Vásquez, R. 1991. "Confiabilidad de los Resultados Preliminares del XI Censo General de Población y Vivienda de 1990." *Estudios Demográficos y Urbanos* 6 (1):33–68.

Courbage, Y. 1992. "Demographic Transition Among Muslims in Eastern Europe." *Population: An English Selection* 4:161–86.

Cowgill, D. 1979. "Aging and Modernization: A Revision of the Theory." in *Late Life: Communities and Environmental Policy,* edited by J. Gubrium. Springfield, IL: Charles C. Thomas, Publisher.

Cramer, J. 1980. "Fertility and Female Employment: Problems of Causal Direction." *American Sociological Review* 45 (2):167–190.

———. 1987. "Social Factors and Infant Mortality: Identifying High-Risk Groups and Proximate Causes." *Demography* 24:299–322.

Crane, B. and J. Finkle. 1989. "The United States, China, and the United Nations Population Fund: Dynamics of U.S. Policymaking." *Population and Development Review* 15 (1):23–60.

Crenshaw, E.M., A.Z. Ameen, and M. Christenson. 1997. "Population Dynamics and Economic Development: Age-Specific Population Growth Rates and Economic Growth in Developing Countries, 1965 to 1990." *American Sociological Review* 62:974–984.

Critchfield, R. 1994. *The Villagers: Changed Values, Altered Lives: The Closing of the Urban-Rural Gap.* New York: Anchor Books.

Crosby, A.W. 1972. *The Columbian Exchange: Biological and Cultural Consequences of 1492.* Westport, Conn: Greenwood Press.

———. 1989. *America's Forgotten Pandemic: The Influenza of 1918.* New York: Cambridge University Press.

Cross, A., W. Obungu, and P. Kizito. 1991. "Evidence of a Transition to Lower Fertility in Kenya." *International Family Planning Perspectives* 17:4–7.

Cutright, P. and W. Kelly. 1981. "The Role of Family Planning Programs in Fertility Declines in Less Developed Countries, 1958–1977." *International Family Planning Perspectives* 7 (4):145–51.

Dalphin, J. 1981. *The Persistence of Social Inequality in America.* Cambridge, MA: Schenkman Publishers.

Daly, H. 1971. "A Marxian-Malthusian View of Poverty and Development." *Population Studies* 15 (1).

———. 1986. "Review of 'Population and Growth and Economic Development: Policy Questions'." *Population and Development Review* 12 (3):582–5.

Danziger, S. and P. Gottschalk. 1993. "Uneven Tides: Rising Inequality in America." New York: Russell Sage Foundation.

Darnell, A. and D. Sherkat. 1997. "The Impact of Protestant Fundamentalism on Educational Attainment." *American Sociological Review* 62:306–315.

Das Gupta, M. 1995. "Fertility Decline in Punjab, India: Parallels with Historical Europe." *Population Studies* 49 (3):481–500.

Daulaire, N. 1995. "The Road From Cairo: Perspectives From Participants." in *Panel Session at the Annual Meeting of the Population Association of America.* San Francisco.

DaVanzo, J. 1976. *Why Families Move: A Model of the Geographic Mobility of Married Couples.* Santa Monica, CA: RAND Corporation.

DaVanzo, J. and A. Chan. 1994. "Living Arrangements of Older Malaysians: Who Coresides With Their Adult Children?" *Demography* 31 (1):95–113.

Davis, C., C. Haub, and J. Willette. 1983. "U.S. Hispanics: Changing the Face of America." *Population Bulletin* 38 (3).

Davis, K. 1945. "The World Demographic Transition." *The Annals of the American Academy of Political and Social Science* 237 (January):1–11.

———. 1949. *Human Society.* New York: Macmillan.

———. 1951. *The Population of India and Pakistan*. Princeton: Princeton University Press.

———. 1955. "Malthus and the Theory of Population." in *The Language of Social Research*, edited by P.L.a.M. Rosenberg. New York: Free Press.

———. 1963. "The Theory of Change and Response in Modern Demographic History." *Population Index* 29 (4):345–66.

———. 1967. "Population Policy: Will Current Programs Succeed?" *Science* 158:730–39.

———. 1972a. "The American Family in Relation to Demographic Change." in *U.S. Commission on Population Growth and the American Future, Volume 1, Demographic and Social Aspects of Population Growth*, edited by C. Westoff and R. Parke. Washington, D.C.: Government Printing Office.

———. 1972b. *World Urbanization 1950–1970, Volume 2: Analysis of Trends, Relationships, and Development*. Berkeley, CA: Institute of International Studies, University of California.

———. 1973. *Cities and Mortality*. Liège: IUSSP.

———. 1974. "The Migration of Human Populations." *Scientific American* 231:92–105.

———. 1984a. "Demographic Imbalances and International Conflict." Presented at the annual meeting of the American Sociological Association, San Antonio.

———. 1984b. "Wives and Work: The Sex Role Revolution and its Consequences." *Population and Development Review* 10 (3):397–418.

———. 1986. "Low Fertility in Evolutionary Perspective." in *Below-Replacement Fertility in Industrial Societies: Causes, Consequences, Policies*, Supplement to Population and Development Review, Volume 12, edited by K. Davis, M. S. Bernstam, and R. Ricardo-Campbell.

———. 1988. "Social Science Approaches to International Migration." in *Population and Resources in Western Intellectual Tradition*, edited by M. Teitelbaum and J. Winter. New York: The Population Council.

———. 1991. "Population and Resources: Fact and Interpretation." in *Resources, Environment, and Population: Present Knowledge, Future Options*, edited by K. Davis and M. S. Bernstam. New York: Oxford University Press.

Davis, K. and J. Blake. 1956. "Social Structure and Fertility: An Analytic Framework." *Economic Development and Cultural Change* 4:211–35.

Davis, K. and P. van den Oever. 1982. "Demographic Foundations of New Sex Roles." *Population and Development Review* 8 (3):495–511.

Davis, K. 1998. "Ancient Byproduct of War Returns: Foreigners Pay to Free Sudanese Slaves." *The San Diego Union-Tribune* 8 February

Day, J.C. 1996. "Population Projections of the United States by Age, Sex, Race, and Hispanic Origin: 1995–2050." *U.S. Bureau of the Census Current Population Reports P20-1130*.

Day, J.C. and A. Curry. 1997. "Educational Attainment in the United States: March 1996 (Update)." *Current Population Reports P20-493*.

De Santis, S. 1996. "Canada Modifies 1996 Census Form After Angry Homemaker's Campaign." *Wall Street Journal*, 9 August, p. A5C.

DeAre, D. 1990. "Longitudinal Migration Data From the Survey of Income and Program Participation." *Current Population Reports, Series P20-166*.

Deevey, E. 1960. "The Human Population." *Scientific American* 203 (3):194–205.

DeJong, G. and J. Fawcett. 1981. "Motivations for Migration: An Assessment on a Value-Expectancy Model." in *Migration Decision Making*, edited by G. DeJong and R. Gardner. New York: Pergamon Press.

DellaPergola, S. 1980. "Patterns of American Jewish Fertility." *Demography* 17 (3):261–73.

Demeny, P. 1968. "Early Fertility Decline in Austria-Hungary: A Lesson in Demographic Transition." *Daedalus* 97 (2):502–22.

———. 1994. *Population and Development*. Liège: International Union for the Scientific Study of Population.

Demerath, N. 1976. *Birth Control and Foreign Policy: The Alternatives to Family Planning*. New York: Harper & Row.

Demos, J. 1965. "Notes on Life in Plymouth Colony." *William and Mary Quarterly* 22:264–86.

Denton, N. and D. Massey. 1991. "Patterns of Neighborhood Transition in a Multiethnic World: U.S. Metropolitan Areas 1970–80." *Demography* 28:41–63.

Desmond, A., F. Wellemeyer, and F. Lorimer. 1962. "How Many People Have Ever Lived on Earth?" *Population Bulletin* 18 (1).

Devaney, B. 1983. "An Analysis of Variations in U.S. Fertility and Female Labor Force Participation Trends." *Demography* 20 (2):147–62.

Dharmalingam, A. 1994. "Old Age Support: Expectations and Experiences in a South Indian Village." *Population Studies* 48:5–19.

Dharmalingam, A. and S.P. Morgan. 1996. "Women's Work, Autonomy, and Birth Control: Evidence From Two South Indian Villages." *Population Studies* 50 (2): 187–202.

Diaz-Briquets, S. 1986. "Conflict in Central America: The Demographic Dimension." *Population Trends and Public Policy Series of the Population Reference Bureau* 10 (February).

Diaz-Briquets, S. and L. Perez. 1981. "Cuba: The Demography of Revolution." *Population Bulletin* 36 (1).

Dietz, T.L. 1995. "Patterns of Intergenerational Assistance Within the Mexican American Family." *Journal of Family Issues* 16 (3):344–56.

Divine, R. 1957. *American Immigration Policy, 1924–1952*. New Haven, CT: Yale University Press.

Dixon, J.A. and K. Hamilton. 1996. *Expanding the Measure of Wealth*, edited by The World Bank.

Dixon-Mueller, R. 1976. "The Roles of Rural Women: Female Seclusion, Economic Production, and Reproductive Choice." in *Population and Development: The Search for Selective Intervention*, edited by R. Ridker. Baltimore: The Johns Hopkins University Press.

———. 1993. *Population Policy and Women's Rights: Transforming Reproductive Choice*. Westport, CT: Praeger.

Djerassi, C. 1981. *The Politics of Contraception*. San Francisco: Freeman.

Dogan, M. and J. Kasarda. 1988. "Introduction: How Giant Cities Will Multiply and Grow." in *The Metropolis Era, Volume 1, A World of Giant Cities*, edited by M. Dogan and J. Kasarda. Newbury Park, CA: Sage Publications.

Domschke, E. and D. Goyer. 1986. *The Handbook of National Population Censuses: Africa and Asia*. Westport, CT: Greenwood Press.

Donaldson, P. and A.O. Tsui. 1990. "The International Family Planning Movement." *Population Bulletin* 45 (3).

Donaldson, P.J. 1990. *Nature Against Us: The United States and the World Population Crisis, 1965–1980*. Chapel Hill, NC: The University of North Carolina Press.

Donato, K.M., J. Durand, and D.S. Massey. 1992. "Stemming the Tide: Assessing the Deterrent Effects of the Immigration Reform and Control Act." *Demography* 29 (2):139–57.

Doornbus, G. and D. Kromhout. 1991. "Educational Level and Mortality in a 32-Year Follow-Up Study of 18-Year-Old Men in the Netherlands." *International Journal of Epidemiology* 19:374–9.

Douglas, M. 1966. "Population Control in Primitive Groups." *British Journal of Sociology* 17:263–73.

Downey, D. 1995. "When Bigger is Not Better: Family Size, Parental Resources, and Children's Educational Performance." *American Sociological Review* 60:746–761.

Drakakis-Smith, D. and E. Graham. 1996. "Shaping the Nation State: Class and the New Population Policy in Singapore." *International Journal of Population Geography* 2 (1):69–90.

Dublin, L., A. Lotka, and M. Spielgeman. 1949. *Length of Life: A Study of the Life Table*. New York: Ronald Press.

Duff, C. 1997. "Plans for Census 'Sampling' Anger GOP in Congress." *Wall Street Journal*, 12 June, p. A20.

Duleep, H. 1989. "Measuring Socioeconomic Mortality Differentials Over Time." *Demography* 26:345–51.

Dumond, D. 1975. "The Limitation of Human Population: A Natural History." *Science* 232:713–20.

Dumont, A. 1890. *Depopulation et Civilisation: Etude Demographique*. Paris: Lecrosnier & Babe.

Durand, J.D. 1967. "The Modern Expansion of World Population." *Proceedings of the American Philosophical Society* 3 (June):137–140.

Durkheim, E. 1933. *The Division of Labor in Society, translated by George Simpson*. Glencoe: Free Press.

Dychtwald, K. and J. Flower. 1989. *Age Wave: The Challenges and Opportunities of an Aging America*. Los Angeles: Jeremy P. Tarcher, Inc.

Dyson, T. 1991. "On the Demography of South Asian Famines." *Population Studies* 45:5–25.

———. 1996. *Population and Food: Global Trends and Future Prospects*. London: Routledge.

Easterlin, R.A. and E. Crimmins. 1985. *The Fertility Revolution: A Supply-Demand Analysis*. Chicago: University of Chicago Press.

Easterlin, R.A., C.M. Schaeffer, and D.J. Macunovich. 1993. "Will the Baby Boomers Be Less Well Off Than Their Parents? Income, Wealth, and Family Circumstances Over the Life Cycle in the United States." *Population and Development Review* 19 (3):497–522.

Easterlin, R.A. 1968. *Population, Labor Force, and Long Swings in Economic Growth*. New York: National Bureau of Economic Research.

———. 1978. "The Economics and Sociology of Fertility: A Synthesis." in *Historical Studies of Changing Fertility*, edited by C. Tilly. Princeton: Princeton University Press.

East-West Center. 1995. "New Survey Finds Fertility Decline in India." *Asia-Pacific Population and Policy*. Number 32 (January-February).

Eaton, J. and A. Mayer. 1954. *Man's Capacity to Reproduce*. Glencoe: Free Press.

Eberstadt, N. 1994. "Demographic Shocks After Communism: Eastern Germany, 1989–93." *Population and Development Review* 20 (1):137–52.

Edmonston, B. and J.S. Passel. 1994. "The Future Immigrant Population of the United States." in *Immigration and Ethnicity: The Integration of America's Newest Arrivals*, edited by B. Edmonston and J. S. Passel. Washington, D.C.: The Urban Institute Press.

Edmonston, B. and C. Schultze. 1995. "Modernizing the U.S. Census." in *Panel on Census Requirements in the Year 2000 and Beyond, Committee on National Statistics, National Research Council*. Washington, D.C.: National Academy Press.

Edwards, J.N., T.D. Fuller, S. Vorakitphokatorn, and S. Sermsri. 1994. *Household Crowding and Its Consequences*. Boulder, CO: Westview Press.

Ehrlich, P. 1968. *The Population Bomb*. New York: Ballantine Books.

———. 1971. *The Population Bomb, Second Edition*. New York: Sierra Club/Ballantine Books.

Ehrlich, P. and A. Ehrlich. 1972. *Population, Resources, and Environment, 2nd Edition*. San Francisco: Freeman.

———. 1990. *The Population Explosion*. New York: Simon & Schuster.

Eller, T.J. 1994. "Household Wealth and Asset Ownership: 1991." *Current Population Reports P70-34*.

Eller, T.J. and W. Fraser. 1995. "Asset Ownership of Households: 1993." *Current Population Reports P70-47.*

Ellertson, C. 1996. "History and Efficacy of Emergency Contraception: Beyond Coca-Cola." *International Family Planning Perspectives* 22 (2):52–56.

Enchautegui, M.E. 1996. *Gender and the Determinants of Interstate Migration.* Washington, D.C.: The Urban Institute.

Engels, F. 1844. *Outlines of a Critique of Political Economy, Reprinted in R. L. Meek, 1953, Marx and Engels on Malthus.* London: Lawrence and Wishart.

Englehart, R. 1994. "Alcohol plays a part in many bike accidents." *Wall Street Journal* 15 April.

Enke, S. 1960. "The Economics of Government Payments to Limit Population." *Economic Development and Cultural Change* 8 (4):339–48.

Espenshade, T.J. 1994. "Does the Threat of Border Apprehension Deter Undocumented U.S. Immigration?" *Population and Development Review* 20 (4):871–92.

Espenshade, T.J. and J.C. Gurcak. 1996. "Are More Immigrants the Answer to U.S. Population Aging?" *Population Today* (December):4–5.

Espenshade, T.J. and R. Hagemann. 1977. *Economic Aspects of Enrollment Trends.* Tallahassee, FL: Center for the Study of Population, Florida State University.

Estrella, G. 1992. "The Floating Population of the Border." in *Demographic Dynamics of the U.S.-Mexico Border*, edited by J. R. Weeks and R. Ham-Chande. El Paso: Texas Western Press, University of Texas, El Paso.

Eversley, D. 1959. *Social Theories of Malthus and the Malthusian Debate.* Oxford: Clarendon Press.

Ewald, P.W. 1994. *Evolution of Infectious Disease.* New York: Oxford University Press.

Exter, T.G. 1996. "Buying Demographics in the Late 1990s." *Business Geographics* (March):18.

Ezeh, A.C. 1993. "The Influence of Spouses Over Each Other's Contraceptive Attitudes in Ghana." *Studies in Family Planning* 24 (3):163–74.

Falkenmark, M. and C. Widstrand. 1992. "Population and Water Resources: A Delicate Balance." *Population Bulletin* 47 (3).

Faour, M. 1991. "The Demography of Lebanon: A Reappraisal." *Middle Eastern Studies* 27 (4):631–41.

Fargues, P. 1994. "Demographic Explosion or Social Upheaval." in *Democracy Without Democrats? The Renewal of Politics in the Muslim World*, edited by G. Salamé. London: I.B. Taurus Publishers.

———. 1995. "Changing Hierarchies of Gender and Generation in the Arab World." in *Family, Gender, and Population in the Middle East*, edited by C. M. Obermeyer. Cairo: The American University in Cairo.

———. 1997. "State Policies and Birth Rate in Egypt: From Socialism to Liberalism." *Population and Development Review* 23 (1):115–138.

Farley, R. 1970. *Growth of the Black Population.* Chicago: Markham.

———. 1976. "Components of Suburban Population Growth." in *The Changing Face of the Suburbs*, edited by B. Schwartz. Chicago: University of Chicago Press.

———. 1984. *Blacks and Whites: Narrowing the Gap?* Cambridge, MA: Harvard University Press.

Farley, R. and W.H. Frey. 1994. "Changes in the Segregation of Whites From Blacks During the 1980s: Small Steps Toward a More Integrated Society." *American Sociological Review* 59:23–45.

Faulkingham, R. and P. Thorbahn. 1975. "Population Dynamics and Drought: A Village in Niger." *Population Studies* 29:463–71.

Feeney, G. 1991. "Fertility Decline in Taiwan: A Study Using Parity Progression Ratios." *Demography* 28 (3):467–79.

Feinleib, S., P. Cunningham, and P. Short. 1994. *Use of Nursing and Personal Care Homes by the Civilian Population, 1987.* Rockville, MD: Public Health Services.

Felmlee, D. 1984. "A Dynamic Analysis of Women's Employment Exits." *Demography* 21 (2):171–91.

Fenker, R.M. 1996. "The Art and Science of Building Site Evaluation Models." *Business Geographics* (July/August):28–31.

Fenwick, R. and C.M. Barresi. 1981. "Health Consequences of Marital-Status Change Among the Elderly: A Comparison of Cross-Sectional and Longitudinal Analyses." *Journal of Health and Social Behavior* 22: 106–16.

Feshbach, M. and A. Friendly. 1992. *Ecocide in the USSR: Health and Nature Under Siege.* New York: Basic Books.

Field, D. 1997. "Looking Back, What Period of Your Life Brought You the Most Satisfaction?" *International Journal of Aging and Human Development* 45 (3):169–94.

Filer, R. 1990. "What We Really Know About the Homeless." *The Wall Street Journal* 10 April.

Findlay, A.M. 1995. "Skilled Transients: The Invisible Phenomenon." in *The Cambridge Survey of World Migration*, edited by R. Cohen. Cambridge: Cambridge University Press.

Findley, S. and J. Ford. 1993. "Reversals in the Epidemiological Transition in Harlem." Paper presented at the annual meeting of the Population Association of America. Cincinnati, Ohio.

Finkle, J. and B. Crane. 1975. "The Politics of Bucharest: Population Development and the New International Economic Order." *Population and Development Review* 1 (1):87–114.

Finnäs, F. 1995. "Entry Into Consensual Unions and Marriages among Finnish Women Born Between 1938 and 1967." *Population Studies* 49 (1):57–70.

Firebaugh, G. 1979. "Structural Determinants of Urbanization in Asia and Latin America, 1950–1970." *American Sociological Review* 44:199–215.

———. 1982. "Population Density and Fertility in 22 Indian Villages." *Demography* 19 (4):481–94.

Fischer, C. 1976. *The Urban Experience*. New York: Harcourt Brace Jovanovich.

———. 1981. "The Public and Private Worlds of City Life." *American Sociological Review* 46:306–16.

———. 1988. *The Urban Experience, 2nd Edition*. San Diego: Harcourt Brace Jovanovich.

Fishlow, A. 1965. *American Railroads and the Transformation of the Antebellum Economy*. Cambridge, MA: Harvard University Press.

Fishlow, H. 1997. "Enrollment Projection in a Multicampus University System." in *Demographics: A Casebook for Business and Government*, edited by H. Kintner, J., T.W. Merrick, P.A. Morrison, and P.R. Voss. Santa Monica, CA: RAND.

Fleck, S. and C. Sorrentino. 1994. "Employment and Unemployment in Mexico's Labor Force." *Monthly Labor Review* 117 (11):3–31.

Flowers, R.B. 1989. *Demographics and Criminality*. Westport, CT: Greenwood Press.

Foner, A. 1975. "Age in Society: Structure and Change." *American Behavioral Scientist* 19 (2):144–65.

Food and Agricultural Organization. 1996. *The Sixth World Food Survey*. Rome: Food and Agricultural Organization.

Foot, D.K. 1982. *Canada's Population Outlook: Demographic Futures and Economic Challanges*. Toronto: Canadian Institute for Economic Policy.

———. 1996. *Boom, Bust & Echo: How to Profit From the Coming Demographic Shift*. Toronto: Macfarlane Walter & Ross.

Fortney, J., J. Harper, and M. Potts. 1986. "Oral Contraceptives and Life Expectancy." *Studies in Family Planning* 17 (3):117–25.

Foster, G. 1967. *Tzintzuntzan: Mexican Peasants in a Changing World*. Boston: Little, Brown.

Francese, P. 1979. "The 1980 Census: The Counting of America." *Population Bulletin* 34 (4).

Freedman, A. 1991. "Nestlé to Restrict Baby-Food Supply to Third World." *Wall Street Journal* 30 January.

Freeman, O.L. 1992. "Perspectives and Prospects." *Agricultural History* 66 (2):3–11.

Freeman, R. 1976. *The Overeducated American*. New York: The Free Press.

Frejka, T. and L. Atkin. 1996. "The Role of Induced Abortion in the Fertility Transition of Latin America." in *The Fertility Transition in Latin America*, edited by J. M. Guzmán, S. Singh, G. Rodríguez, and E.A. Pantelides. Oxford: Oxford University Press.

Freund, W. 1982. "The Looming Impact of Population Changes." *The Wall Street Journal* (6 April).

Frey, W. 1990. "Metropolitan America: Beyond the Transition." *Population Bulletin* 45 (2).

———. 1995. "The New Geography of Population Shifts." in *State of the Union: America in the 1990s,* *Volume Two: Social Trends*, edited by R. Farley. New York: Russell Sage Foundation.

———. 1996. "Immigration, Domestic Migration, and Demographic Balkanization in America: New Evidence for the 1990s." *Population and Development Review* 22 (4):741–63.

Frey, W.H. and R. Farley. 1996. "Latino, Asian, and Black Segregation in U.S. Metropolitan Areas: Are Multiethnic Metros Different?" *Demography* 33 (1): 35–50.

Friedlander, D. 1983. "Demographic Responses and Socioeconomic Structure: Population Processes in England and Wales in the Nineteenth Century." *Demography* 20 (3):249–72.

Frisch, R. 1978. "Population, Food Intake, and Fertility." *Science* 199:22–30.

Frost, D. and M. Deakin. 1983. *David Frost's Book of the World's Worst Decisions*. New York: Crown Publishing.

Fuguitt, G. and D. Brown. 1990. "Residential Preferences and Population Redistribution in 1972–88." *Demography* 27:289–300.

Fuguitt, G. and J. Zuiches. 1975. "Residential Preferences and Population Distribution." *Demography* 27: 589–600.

Furstenberg, F. 1998. "Review Symposium: Families According to Research: Relative Risks: What Is The Family Doing To Our Children?" *Contemporary Sociology* 27 (3):223–225.

Furstenberg, F. and A.J. Cherlin. 1991. *Divided Families: What Happnes to Children When Parents Part?* Cambridge, MA: Harvard University Press.

Gadalla, S. 1978. *Is There Hope? Fertility and Family Planning in a Rural Community in Egypt*. Chapel Hill, NC: Carolina Population Center, University of North Carolina.

Gage, T.B. 1994. "Population Variation in Cause of Death: Level, Gender, and Period Effects." *Demography* 31 (2):271–94.

Galle, O., W. Gove, and J. McPherson. 1972. "Population Density and Pathology: What Are The Relations for Man?" *Science* 176:23–30.

Galloway, P. 1984. *Long Term Fluctuations in Climate and Population in the Pre-Industrial Era*. Berkeley: University of California.

Garcia y Griego, M., J.R. Weeks, and R. Ham-Chande. 1990. "Mexico." in *Handbook on International Migration*, edited by W.J. Serow, C.B. Nam, D.F. Sly, and R.H. Weller. New York: Greenwood Press.

Gardner, G. 1996. *Shrinking Fields: Cropland Loss in a World of Eight Billion*, edited by J.A. Peterson. Washington D.C.: World Watch Institute.

Garreau, J. 1991. *Edge City: Life on the New Frontier*. New York: Doubleday.

Gartner, R. 1990. "The Victims of Homicide: A Temporal and Cross-National Comparison." *American Sociological Review* 55:92–106.

Gayle, H.D. 1998. "An Important Perspective" in *Overseas AZT Trials: Finding Solutions for Mother-to-Infant HIV Transmission in the Developing World*, edited by the National Center for HIV, STD, and TB Prevention. Atlanta, Georgia: Center for Disease Control and Prevention, **www.cdc.gov/nchstp/od/overseas_azttrials.htm** (accessed 1998).

Gelbspan, R. 1997. *The Heat is On*. Reading, MA: Addison-Wesley.

Gendell, M. and J.S. Siegel. 1996. "Trends in Retirement Age in the United States, 1955–1993, by Sex and Race." *Journal of Gerontology: Social Sciences* 51B: S132–S139.

Gerontology Center Newsletter. 1988. *Why Do Women Live Longer Than Men?* University Park: Pennsylvania State University College of Health and Human Development.

Gertler, P.J. and J.W. Molyneaux. 1994. "How Economic Development and Family Planning Programs Combined to Reduce Indonesian Fertility." *Population and Development Review* 31 (1):33–64.

Gibson, C. 1976. "The U.S. Fertility Decline, 1961–1975: The Contribution of Changes in Marital Status and Marital Fertility." *Family Planning Perspectives* 8: 249–52.

Gilbar, G. 1994. "Population Growth and Family Planning in Egypt, 1985–92." *Middle East Contemporary Survey* 16:335–48.

Gillis, J.R., L.A. Tilly, and D. Levine. 1992. "The European Experience of Declining Fertility: A Quiet Revolution 1850–1970." Oxford: Basil Blackwell.

Glass, D.V. 1953. *Introduction to Malthus*. New York: Wiley.

Global Aging Report. 1997. "Guaranteeing the Rights of Older People: China Takes a Great Leap Forward." *Global Aging Report* 2 (5):3.

Gober, P. 1993. "Americans on the Move." *Population Bulletin* 48 (3).

Godwin, W. 1793 (1946). *Enquiry Concerning Political Justice and Its Influences on Morals and Happiness*. Toronto: University of Toronto Press.

Goldman, N. 1984. "Changes in Widowhood and Divorce and Expected Durations of Marriage." *Demography* 21 (3):297–308.

Goldscheider, C. 1971. *Population, Modernization, and Social Structure*. Boston: Little, Brown.

———. 1989. "The Demographic Embeddedness of the Arab-Jewish Conflict in Israeli Society." *Middle East Review* 21 (3):15–24.

———. 1996. *Israel's Changing Society: Population, Ethnicity, and Development*. Boulder, CO: Westview Press.

Goldscheider, C. and W. Mosher. 1991. "Patterns of Contraceptive Use in the United States: The Importance of Religious Factors." *Studies in Family Planning* 22 (2):102–15.

Goldscheider, F. and C. Goldscheider. 1993. *Leaving Home Before Marriage: Ethnicity, Familism, and Generational Relationships*. Madison, WI: The University of Wisconsin Press.

———. 1994. "Leaving and Returning Home in 20th Century America." *Population Bulletin* 48 (4).

Goldstein, A. and W. Feng. 1996. "China: The Many Facets of Demographic Change." Boulder, CO: Westview Press.

Goldstein, S. 1988. "Levels of Urbanization in China." in *The Metropolis Era: A World of Giant Cities*, edited by M. Dogan and J.D. Kasarda. Newbury Park, CA: Sage Publications.

Goldstein, S. and A. Goldstein. 1981. "The Impact of Migration on Fertility: An 'Own Children' Analysis For Thailand." *Population Studies* 35 (2):265–84.

Goldstone, J.A. 1986. "State Breakdown in the English Revolution: A New Synthesis." *American Sociological Review* 92:257–322.

———. 1991. *Revolution and Rebellion in the Early Modern World*. Berkeley: University of California Press.

Goliber, T. 1985. "Sub-Saharan Africa: Population Pressures on Development." *Population Bulletin* 40 (1).

———. 1989. "Africa's Expanding Population: Old Problems, New Policies." *Population Bulletin* 44 (3).

———. 1997. "Population and Reproductive Health in Sub-Saharan Africa." *Population Bulletin* 52 (4).

Goode, W.J. 1993. *World Changes in Divorce Patterns*. New Haven: Yale University Press.

Goodkind, D. 1997. "The Vietnamese Double Marriage Squeeze." *International Migration Review* 31 (1): 108–127.

———. 1991. "Creating New Traditions in Modern Chinese Populations: Aiming For Birth in the Year of the Dragon." *Population and Development Review* 17 (4):663–86.

———. 1995a. "The Significance of Demographic Triviality: Minority Status and Zodiacal Fertility Timing Among Chinese Malaysians." *Population Studies* 49: 45–55.

———. 1995b. "Vietnam's One-or-Two Child Policy in Action." *Population and Development Review* 21 (1): 85–112.

Gordon, M. 1975. *Agriculture and Population*. Washington, D.C.: Government Printing Office.

Goudie, A. 1990. *The Human Impact on the Natural Environment, 3rd Edition*. Cambridge, MA: MIT Press.

Goudsmit, J. 1997. *Viral Sex: The Nature of AIDS*. New York: Oxford University Press.

Gould, P. 1993. *The Slow Plague: A Geography of the AIDS Pandemic*. Oxford: Blackwell.

Gould, W.T.S. and A.M. Findlay. 1994. "Population Migration and the Changing World Order." New York: Chichester.

Gove, W. 1973. "Sex, Marital Status, and Mortality." *American Journal of Sociology* 79 (1):45–67.

Grabill, W., C. Kiser, and P. Whelpton. 1958. *The Fertility of American Women.* New York: Wiley.

Grad, S. 1996. *Income of the Population 55 or older, 1994.* Washington, D.C.: Social Security Administration.

Graebner, W. 1980. *A History of Retirement.* New Haven: Yale University Press.

Graunt, J. 1662 (1939). *Natural and Political Observations Made Upon the Bills of Mortality, Edited with an Introduction by Walter F. Willcox.* Baltimore: The Johns Hopkins University Press.

Green, G. and E. Welniak. 1991. "The Nine Household Markets." *American Demographics* (October):36–40.

Greene, M.E. and V. Rao. 1995. "The Marriage Squeeze and the Rise in Informal Marriage in Brazil." *Social Biology* 42 (1–2):65–82.

Greenhalgh, S. 1986. "Shifts in China's Population Policy 1984–86: Views From the Central, Provincial, and Local Levels." *Population and Development Review* 12 (3):491–516.

Greenhalgh, S., Z. Chuzhu, and L. Nan. 1994. "Restraining Population Growth in Three Chinese Villages, 1988–93." *Population and Development Review* 20 (2):365–96.

Gregorio, D.I., S.J. Walsh, and D. Paturzo. 1997. "The Effects of Occupation-Based Social Position on Mortality in a Large American Cohort." *American Journal of Public Health* 87 (9):1472–5.

Groat, H.T., J.W. Wicks, A. Neal, and G. Hendershot. 1982. *Working Women and Childbearing.* Hyattsville, MD: National Center for Health Statistics.

Grossbard-Schechtman, S. 1985. "Marriage Squeezes and the Marriage Market." in *Contemporary Marriage: Comparative Perspectives on a Changing Institution*, edited by K. Davis and S. Grossbard-Schechtman. New York: Russell Sage Foundation.

———. 1993. *On the Economics of Marriage: A Theory of Marriage, Labor, and Divorce.* Boulder, CO: Westview Press.

Grulich, A.E., J.M. Kaldor, O. Hendry, K. Luo, N.J. Bodsworth, and D.A. Cooper. 1997. "Risk of Kaposi's Sarcoma and Oroanal Sexual Contact." *American Journal of Epidemiology* 145 (8):673–679.

Guest, A.M., G. Almgren, and J.M. Hussey. 1998. "The Ecology of Race and Socioeconomic Distress: Infant and Working-Age Mortality in Chicago." *Demography* 35 (1):23–34.

Guihati, K. 1977. "Compulsory Sterilization: The Change in India's Population Policy." *Science* 195:1300–1305.

Guilmoto, C.Z. 1998. "Institutions and Migrations: Short-Term Versus Long-Term Moves in Rural West Africa." *Population Studies* 52:85–103.

Guttentag, M. and P. Secord. 1983. *Too Many Women?* Beverly Hills, CA: Sage Publications.

Guz, D. and J. Hobcraft. 1991. "Breastfeeding and Fertility: a Comparative Analysis." *Population Studies* 45: 91–108.

Haas, W.H. and W.J. Serow. 1993. "Amenity Retirement Migration Process: A Model and Preliminary Evidence." *The Gerontologist* 33 (2):212–220.

Hahn, B.A. 1993. "Marital Status and Women's Health: The Effects of Economic Marital Adjustments." *Journal of Marriage and the Family* 55:495–504.

Halberstein, R. 1973. "Historical-Demographic Analysis of Indian Populations in Tlaxcala, Mexico." *Social Biology* 20 (1):40–50.

Hall, D.R. and J.Z. Zhao. 1995. "Cohabitation and Divorce in Canada: Testing the Selectivity Hypothesis." *Journal of Marriage and the Family* 57:421–27.

Hall, R. and P. White. 1995. "Population Change on the Eve of the Twenty-First Century." in *Europe's Population: Towards the Next Century*, edited by R. Hall and P. White. London: University College London Press.

Ham-Chande, R. 1993. "México: País en Proceso de Envejecimiento." *Comercio Exterior (Mexico)* 43 (7): 688–96.

Hampshire, S. 1955. "Introduction." in *Sketch for an Historical Picture of the Progress of the Human Mind*, edited by M.J.A.N.d.C. Condorcet. London: Weidenfield and Nicholson.

Hampson, J. 1983. *Old Age: A Study of Aging in Zimbabwe.* Gweru, Zimbabwe: Mambo Press.

———. 1989. "Social Support For the Rural Elderly in Zimbabwe: The Transition." in *An Aging World: Dilemmas and Challenges for Law and Social Policy*, edited by J. Eekelaar and D. Pearl. Oxford: Clarendon Press.

Han, S.S. and S.T. Wong. 1994. "The Influence of Chinese Reform and Pre-Reform Policies on Urban Growth in the 1980s." *Urban Geography* 15 (6):537–64.

Handwerker, P. 1986. *Culture and Reproduction: An Anthropological Critique of Demographic Transition Theory.* Boulder: Westview Press.

Hanley, S. and K. Yamamura. 1977. *Economic and Demographic Changes in Preindustrial Japan, 1600–1868.* Princeton: Princeton University Press.

Hansen, J. 1970. *The Population Explosion: How Sociologists View It.* New York: Pathfinder Press.

Hansen, K. 1994. "Geographic Mobility: March 1992 to March 1993." *Current Population Reports* P20-481.

———. 1997. "Geographical Mobility: March 1995 to March 1996." *Current Population Reports* P20-497.

Haq, M. 1984. "Age at Menarche and the Related Issue: A Pilot Study on Urban School Girls." *Family Planning Perspectives* 13 (6):559–67.

Hardin, G. 1968. "The Tragedy of the Commons." *Science* 162:1243–48.

Harevan, T.K. and P. Uhlenberg. 1995. "Transition to Widowhood and Family Support Systems in the Twentieth Century Northeastern United States." in *Aging in the Past: Demography, Society, and Old Age*, edited by D. Kertzer and P. Laslett. Berkeley: University of California Press.

Harlan, J. 1976. "The Plants and Animals That Nourish Man." *Scientific America n* 235 (3):88–97.

Harris, M. and E.B. Ross. 1987. *Death, Sex, and Fertility: Population Regulation in Preindustrial and Developing Societies.* New York: Columbia University Press.

Harrison, P. 1993. *The Third Revolution: Population, Environment and a Sustainable World.* London: Penguin Books.

Hartmann, B. 1995. *Reproductive Rights and Wrongs: The Global Politics of Population Control.* Boston: South End Press.

Harvey, W. 1986. "Homicide Among Young Black Adults: Life in the Subculture of Exasperation." in *Homicide Among Black Americans,* edited by D. Hawkins. New York: University Press of America.

Hatcher, R.A., J. Trussell, F. Stewart, G.K. Stewart, D. Kowal, F. Guest, W. Cates, Jr., and M.S. Policar. 1994. *Contraceptive Technology, 16th Edition.* New York: Irvington.

Hatton, T.J. and J.G. Williamson. 1994. "What Drove the Mass Migrations From Europe in the Late Nineteenth Century?" *Population and Development Review* 20 (3):533–59.

Haub, C. 1995. "How many people have ever lived on earth?" *Population Today* 23 (2):4–5.

Haupt, A. 1990. "S-Night: Counting the Homeless." *Population Today* 18 (5):3–4.

Hauser, R. and P. Featherman. 1974. "Socioeconomic Achievements of U.S. Men, 1962 to 1972." *Science* 185:325–31.

Havanon, N., J. Knodel, and W. Sittitrai. 1992. "The Impact of Family Size on Wealth Accumulation in Rural Thailand." *Population Studies* 46 (1):37–52.

Hawley, A. 1972. "Population Density and the City." *Demography* 91:521–30.

Hayflick, L. 1979. "Progress in Cytogerontology." *Mechanisms of Aging and Development* 9:398–408.

Heer, D. and S.A. Grossbard-Schechtman. 1981. "The Impact of the Female Marriage Squeeze and the Contraceptive Revolution on Sex Roles and the Women's Liberation Movement in the United States, 1960 to 1975." *Journal of Marriage and the Family* 43 (1):49–65.

Heilig, G., T. Büttner, and W. Lutz. 1990. "Germany's Population: Turbulent Past, Uncertain Future." *Population Bulletin* 45 (4).

Helal, A. 1982. "Kuwaiti Push Against Illegal Immigrants Creates Shortage of Construction Workers." *Wall Street Journal* 1 December.

Heligman, L., N. Chen, and O. Babakol. 1993. "Shifts in the Structure of Population and Deaths in Less Developed Regions." in *The Epidemiological Transition: Policy and Planning Implications for Developing Countries,* edited by J.N. Gribble and S.H. Preston. Washington, D.C.: National Academy Press.

Hendry, P. 1988. "Food and Population: Beyond Five Billion." *Population Bulletin* 43 (2).

Henretta, J.C. 1997. "Changing Perspectives on Retirement." *Journal of Gerontology: Social Sciences* 52B: S1–S3.

Henry, L. 1961. "Some Data on Natural Fertility." *Eugenics Quarterly* 8:81–91.

Henshaw, S. 1990. "Induced Abortion: A World Review, 1990." *Family Planning Perspectives* 22 (2):76–89.

Herbst, S.T. 1995. *The New Food Lover's Companion, 2nd Edition.* New York: Barron's Educational Services, Inc.

Herlihy, D. and C. Klapisch-Zuber. 1985. *Tuscans and Their Families: A Study of the Florentine Catasto of 1427.* New Haven: Yale University Press.

Hern, W. 1976. "Knowledge and Use of Herbal Contraception in a Peruvian Amazon Village." *Human Organization* 35:9–19.

———. 1977. "High Fertility in a Peruvian Amazon Indian Village." *Human Ecology* 5:355–67.

———. 1992. "Polygyny and Fertility Among the Shipibo of the Peruvian Amazon." *Population Studies* 46: 53–64.

Hernandez, D. 1993. *America's Children: Resources From Family, Government and the Economy.* New York: The Russell Sage Foundation.

Heuser, R. 1976. *Fertility Tables For Birth Cohorts by Color: United States, 1917–73.* Rockville, MD: National Center for Health Statistics.

Heymann, D., J. Chin, and J. Mann. 1990. "A Global Overview of AIDS." in *Heterosexual Transmission of AIDS,* edited by N. Alexander, H. Gabelnick, and J. Spieler. New York: Wiley-Liss.

Heymann, J. 1994. "Labor Policies: Its Influence on Women's Reproductive Lives." in *Power and Decision: The Social Control of Reproduction,* edited by G. Sen and R.C. Snow. Cambridge, MA: Harvard University Press.

Higgins, M. and J.G. Williamson. 1997. "Age Structure Dynamics in Asia and Dependence on Foreign Capital." *Population and Development Review* 23 (2): 261–294.

Hill, D. and M. Kent. 1988. *Election Demographics.* Washington, D.C.: Population Reference Bureau.

Himes, N.E. 1976. *Medical History of Contraception.* New York: Schocken Books.

Himmelfarb, G. 1984. *The Idea of Poverty: England in the Early Industrial Age.* New York: Alfred A. Knopf.

Hinde, T. 1995. *The Domesday Book: England's Heritage, Then & Now.* London: Tiger Books International.

Hirschman, A. 1958. *The Strategy of Economic Development.* New Haven: Yale University Press.

Hirschman, C. and M. Butler. 1981. "Trends and Differentials in Breast Feeding: An Update." *Demography* 18 (1):39–54.

Hobbs, F.B. and B.L. Damon. 1996. "65+ in the United States." *Current Populaton Reports* P23-190.

Hobcraft, J. 1996. "Fertility in England and Wales: A Fifty Year Perspective." *Population Studies* 50 (3): 485–524.

Hodell, D., J. Curtis, and M. Brenner. 1995. "Possible Role of Climate in the Collapse of Classic Mayan Civilization." *Nature* 375:391–394.

Hoerning, R. 1996. "American Honda Jump-Starts Sales Geographically." *Business Geographics* (March):24–26.

Hogan, T.D. and D.N. Steinnes. 1998. "A Logistic Model of the Seasonal Migration Decision for Elderly Households in Arizona and Minnesota." *The Gerontologist* 38 (2):152–158.

Holdren, J. 1990. "Energy in Transition." *Scientific American* 263 (3):156–63.

Hollerbach, P. 1980. "Recent Trends in Fertility, Abortion, and Contraception in Cuba." *International Family Planning Perspectives* 6 (3):97–106.

Hollingsworth, T.H. 1969. *Historical Demography.* Ithaca: Cornell University Press.

Hope, M. and J. Young. 1990. "Review of 'Without Shelter: Homeless in the 1980s' by Peter Rossi." *Contemporary Sociology* 19:84–85.

Horwitz, T. and C. Forman. 1990. "Clashing Cultures: Immigrants to Europe From the Third World Face Racial Animosity." *Wall Street Journal* 14 August:A1.

Hout, M. 1984. "Occupational Mobility of Black Men: 1962 to 1973." *American Sociological Review* 49 (3): 308–22.

Howell, N. 1979. *Demography of the Dobe Kung.* New York: Academic Press.

Hsu, M.-L. 1994. "The Expansion of the Chinese Urban System, 1953–1990." *Urban Geography* 15 (6):514–36.

Hu, Y. and N. Goldman. 1990. "Mortality Differentials by Marital Status: An International Comparison." *Demography* 27:233–50.

Huber, G.A. and T.J. Espenshade. 1997. "Neo-Isolationism, Balanced-Budget Conservatism, and the Fiscal Impacts of Immigrants." *International Migration Review* 31:1031–1054.

Hughes, H.L. 1993. "Metropolitan Structure and the Suburban Hierarchy." *American Sociological Review* 58:417–33.

Hugo, G., T. Hull, V. Hull, and G. Jones. 1987. *The Demographic Dimension in Indonesian Development.* Singapore: Oxford University Press.

Hume, D. 1752 (1963). "Of the Populousness of Ancient Nations." in *Essays: Moral, Political and Literary,* edited by D. Hume. London: Oxford.

Hummert, M.L., T.A. Garstka, J.L. Shaner, and S. Strahm. 1995. "Judgments About Stereotypes of the Elderly." *Research on Aging* 17 (2):168–89.

Huntington, S.P. 1996. *The Clash of Civilizations and the Remaking of the World Order.* New York: Simon & Schuster.

Huttman, E., W. Blauw, and J. Saltman. 1991. "Urban Housing Segregation of Minorities in Western Europe and the United States." Durham, NC: Duke University Press.

Huzel, J. 1969. "Malthus, the Poor Law, and Population in Early Nineteenth-Century England." *Economic History Review* 22:430–52.

———. 1980. "The Demographic Impact of the Old Poor Law: More Reflexions on Malthus." *Economic History Review* 33:367–81.

———. 1984. "Parson Malthus and the Pelican Inn Protocol: A Reply to Professor Levine." *Historical Method* 17:25–27.

Iacocca, L,. 1984. *Iacocca: An Autobiography.* New York: Bantam Books.

Ingersoll, B. 1991. "Irradiation Foes Plan Media Blitz to Block Plant." *The Wall Street Journal* 26 June.

Instituto Nacional de Estadística Geografía e Informatica. 1996. *Conteo de Población y Vivienda 1995: Resultados Definitivos, Tabulados Básicos.* Aguascalientes, Mexico: INEGI.

International Labour Office. 1997. "The Fight Against Poverty: Not All Is Lost." *World of Work* 21:24–25.

Ireland Central Statistics Office. 1978. *Statistical Abstract of Ireland, 1976.* Dublin: Stationary Office.

Isaac, R.J. and V. Armat. 1990. *Madness in the Streets: How Psychiatry and the Law Abandoned the Mentally Ill.* New York: The Free Press.

Isaacs, S. and R. Cook. 1984. *Laws and Policies Affecting Fertility: A Decade of Change.* Baltimore, MD: The Johns Hopkins University.

Isbister, J. 1973. "Birth Control, Income Redistribution, and the Rate of Saving: The Case of Mexico." *Demography* 10:85–94.

Israel Central Bureau of Statistics. 1998. *Statistical Yearbook.* Tel Aviv: Internet Access: **www.cbs.gov.il/shnaton/** (accessed 1998).

Issawi, C. 1987. *An Arab Philosophy of History: Selections from the Prolegomena of Ibn Khaldun of Tunis (1332–1406).* Princeton: Princeton University Press.

Jaffe, F. 1971. "Toward the Reduction of Unwanted Pregnancy." *Science* 174:119–27.

Jain, A. and A. Adlakha. 1982. "Preliminary Estimates of Fertility Decline in India During the 1970s." *Population and Development Review* 8 (3):589–606.

James, P. 1979. *Population Malthus: His Life and Times.* London: Routledge & Kegan Paul.

Janssen, S. and R. Hauser. 1981. "Religion, Socialization, and Fertility." *Demography* 18 (4):511–28.

Jasso, G. 1985. "Marital Coital Frequency and the Passage of Time: Estimating the Separate Effects of Spouses' Ages and Marital Duration, Birth and Marriage Cohorts, and Period Influences." *American Sociological Review* 50:224–241.

Jeffery, P., R. Jeffery, and A. Lyon. 1989. *Labour Pains and Labour Power.* London: Zen Books.

Jejeebhoy, S. 1991. "Women's Status and Fertility: Successive Cross-Sectional Evidence from Tamil Nadu, India, 1970–80." *Studies in Family Planning* 20:264–72.

Jejeebhoy, S. and S. Kulkarni. 1989. "Reproductive Motivation: A Comparison of Wives and Husbands in Maharashtra, India." *Studies in Family Planning* 20: 264–72.

Jencks, C. 1994. *The Homeless*. Cambridge, MA: Harvard University Press.

Johnson, A.W. and T. Earle. 1987. *The Evolution of Human Societies: From Foraging Group to Agrarian Societies*. Stanford, CA: Stanford University Press.

Johnson, I. 1997. "China's New Containment Policy: Fighting the Rise of Megacities." *The Wall Street Journal* 11 December:A20.

Johnson, K. 1996. "The Politics of the Revival of Infant Abandonment in China, with Special Reference to Hunan." *Population and Development Review* 22:77–98.

Johnson, N. 1982. "Religious Differentials in Reproduction: The Effects of Sectarian Education." *Demography* 19 (4):495–509.

Johnston, W. 1991. "Global Work Force 2000: The New World Labor Market." *Harvard Business Review* (March-April):115–29.

Jones, L. 1981. *Great Expectations: America and the Baby Boom Generation*. New York: Ballantine.

Jones, R. 1988. "A Behavioral Model for Breastfeeding Effects on Return to Menses Postpartum in Javanese Women." *Social Biology* 35:307–23.

Jordan, M. 1996a. "India's Reforms Excite Growth, Not Voters." *The Wall Street Journal* 26 April:A10.

Jordan, M. 1996b. "Marketing Gurus Say: In India, Think Cheap, Lose the Cold Cereal." *The Wall Street Journal* 11 October.

Juárez, F. and J. Quilodrán. 1996. "Mujeres pioneras de reproducción en México." in *Nuevas Pautas Reproductivas en México*, edited by F. Juárez, J. Quilodrán, and M.E. Zavala de Cosío. Mexico City: El Colegio de Mexico.

Kahn, J. and W. Mason. 1987. "Political Alienation, Cohort Size, and the Easterlin Hypothesis." *American Sociological Review* 52:155–69.

Kalbach, W.E. and W.W. McVey. 1979. *The Demographic Bases of Canadian Society, Second Edition*. Toronto: McGraw-Hill Ryerson Limited.

Kalmijn, M. 1991. "Shifting Boundaries: Trends in Religious and Educational Homogamy." *American Sociological Review* 56:786–800.

Kannisto, V., J. Lauritsen, A.R. Thatcher, and J.W. Vaupel. 1994. "Reductions in Mortality at Advanced Ages: Several Decades of Evidence From 27 Countries." *Population and Development Review* 20 (4): 793–810.

Kaplan, C. and T. Van Valey. 1980. *Census 80: Continuing the Factfinder Tradition*. Washington, D.C.: U.S. Bureau of the Census.

Kaplan, D. 1994. "Population and Politics in a Plural Society: The Changing Geography of Canada's Linguistic Groups." *Annals of the Association of American Geographers* 84 (1):46–67.

Kasarda, J. 1995. "Industrial Restructuring and the Changing Location of Jobs." in *State of the Union: America in the 1990s, Volume One: Economic Trends*, edited by R. Farley. New York: Russell Sage Foundation.

Kaufman, J. 1997. "At Age 5, Reading, Writing and Rushing." *The Wall Street Journal* 4 February.

Kazemi, F. 1980. *Poverty and Revolution in Iran*. New York: New York University Press.

Keefe, E.F. 1962. "Self-Observation of the Cervix to Distinguish Days of Possible Fertility." *Bulletin of the Sloane Hospital for Women* 8 (4):129–36.

Keely, C. 1971. "Effects of the Immigration Act of 1965 on Selected Population Characteristics of Immigrants to the United States." *Demography* 8:157–70.

Keith, V. and D. Smith. 1988. "The Current differential in Black and White Life Expectancy." *Demography* 25: 625–32.

Kelley, A.C. 1973. "Population Growth, The Dependency Rate, and the Pace of Economic Development." *Population Studies* 27:405–14.

Kelley, A.C. and R.M. Schmidt. 1994. *Population and Income Change: Recent Evidence*. Washington, D.C.: World Bank.

Kemper, R.V. 1977. *Migration and Adaptation*. Beverly Hills, CA: Sage Publications.

———. 1991. "Migracion Sin Fronteras: El Caso del Pueblo de Tzintzuntzan, Michoacan, 1945–1990." in *Migración y Fronteras*, edited by Sociedad Mexicana de Antropología. Mexico, DF: Mesa Redonda.

Kemper, R.V. and G. Foster. 1975. "Urbanization in Mexico: The View From Tzintzuntzan." *Latin American Urban Research* 5:53–75.

Kennedy, K., R. Rivera, and A.S. McNeilly. 1989. "Consensus Statement on the Use of Breastfeeding as a Family Planning Method." *Contraception* 39:477–96.

Kephart, W. 1982. *Extraordinary Groups: The Sociology of Unconventional Life-Styles, 2nd Edition*. New York: St. Martins Press.

Kertzer, D.I. 1993. *Sacrificed for Honor: Italian Infant Abandonment and the Politics of Reproductive Control*. Boston: Beacon Press.

———. 1995. "Toward a Historical Demography of Aging." in *Aging in the Past: Demography, Society, and Old Age*, edited by D.I. Kertzer and P. Laslett. Berkeley: University of California Press.

Kertzer, D.I. and D.P. Hogan. 1989. *Family, Political Economy, and Demographic Change: The Transformation of Life in Casalecchio, Italy, 1861–1921*. Madison: University of Wisconsin Press.

Keyfitz, N. 1966. "How Many People Have Ever Lived on the Earth?" *Demography* 3 (2):581–82.

———. 1968. *Introduction to the Mathematics of Population*. Reading, MA: Addison-Wesley.

———. 1971. "On the Momentum of Population Growth." *Demography* 8 (1):71–80.

———. 1972. "Population Theory and Doctrine: A Historical Survey." in *Readings in Population*, edited by W. Petersen. New York: MacMillan.

———. 1973. "Population Theory." Chapter III in *The Determinants and Consequences of Population Trends, Volume I*, edited by United Nations. New York: United Nations.

———. 1982. "Can Knowledge Improve Forecasts?" *Population and Development Review* 8 (4):729–51.

———. 1985. *Applied Mathematical Demography, 2nd Edition*. New York: Springer-Verlag.

Khan, M. and C. Prasad. 1985. "A Comparison of 1970 and 1980 Survey Findings on Family Planning in India." *Studies in Family Planning* 16 (6):312–20.

Kiernan, K.E. 1992. "The Impact of Family Disruption in Childhood on Transitions Made in Young Adult Life." *Population Studies* 46 (2):213–34.

Kim, Y.J. and R. Schoen. 1997. "Population Momentum Expresses Population Aging." *Demography* 34 (3): 421–28.

Kinsella, K. 1995. "Aging and the Family: Present and Future Demographic Issues." in *Handbook of Aging and the Family*, edited by R. Blieszner and V. Hilkevitch Bedford. Westport, CT: Greenwood Press.

Kinstler, S. and L.G. Pol. 1990. "The Marketing Implications of Migration on Hospital and Physician Supply and Demand in the United States." *Journal of Hospital Marketing* 4 (1):119–41.

Kintner, H. 1992. "Notes From the Field: Doing Demographic Research Inside General Motors." *Applied Demography* 7 (1):8–10.

Kiser, C., W. Grabill, and A. Campbell. 1968. *Trends and Variations in Fertility in the United States*. Cambridge: Harvard University Press.

Kishor, S. 1993. "'May God Give Sons to All': Gender and Child Mortality in India." *American Sociological Review* 58:247–65.

Kisker, E. and N. Goldman. 1987. "Perils of Single Life and Benefits of Marriage." *Social Biology* 34 (3–4): 135–52.

Kitagawa, E. and P. Hauser. 1973. *Differential Mortality in the United States: A Study in Socioeconomic Epidemiology*. Cambridge: Harvard University Press.

Kitano, H. 1997. *Race Relations, 5th Edition*. New York: Prentice-Hall.

Klatsky, A.L. and G.D. Friedman. 1995. "Annotation: Alcohol and Longevity." *American Journal of Public Health* 85 (1):16–18.

Kline, J., Z. Stein, and M. Susser. 1989. *Conception to Birth: Epidemiology of Prenatal Development*. New York: Oxford University Press.

Knodel, J. 1970. "Two and a Half Centuries of Demographic History in a Bavarian Village." *Population Studies* 24:353–69.

Knodel, J. and E. van de Walle. 1979. "Lessons From the Past: Policy Implications of Historical Fertility Studies." *Population and Development Review* 5:217–45.

Knox, P. 1994. *Urbanization: An Introduction to Urban Geography*. Englewood Cliffs, NJ: Prentice-Hall.

Kobayashi, K. 1969. "Traditions and Transitions in Family Structure." in *The Family in Transition*, edited by National Institutes of Health. Washington, D.C.: Government Printing Office.

Komlos, J. 1989. "The Age at Menarche in Vienna: The Relationship Between Nutrition and Fertility." *Historical Methods* 22:158–63.

Konner, M. and C. Worthman. 1980. "Nursing Frequency, Gonadal Function, and Birth Spacing Among Kung Hunter-Gatherers." *Science* 207:788–91.

Koonin, L.M., J.C. Smith, M. Ramick, and C.A. Green. 1996. "Abortion Surveillance—United States, 1992." in *Surveillance Summaries, May 3, 1996, MMWR; 45 (No. SS-3)*, edited by Centers for Disease Control, Atlanta.

Kosinski, L. and R. Prothero. 1975. *People on the Move*. London: Methuen and Co.

Kraeger, P. 1986. "Demographic Regimes as Cultural Systems." in *The State of Population Theory: Forward From Malthus*, edited by D. Coleman and R. Schofield. Oxford: Basil Blackwell.

Kraeger, P. 1993. "Histories of Demography: A Review Article." *Population Studies* 47 (3):519–39.

Kralt, J. 1990. "Ethnic Origins in the Canadian Census, 1871–1986." in *Ethnic Demography: Canadian Immigrant, Racial and Cultural Variations*, edited by S.S. Halli, F. Trovato, and L. Driedger. Ottawa: Carleton University Press.

Kraly, E.P. 1982. "Emigration From the U.S. Among the Elderly." Paper presented at Annual Meeting of the Population Association of America. San Diego.

Kraly, E.P. and R. Warren. 1992. "Estimates of Long-Term Immigration to the United States: Moving U.S. Statistics Toward United Nations Concepts." *Demography* 29 (4):613–26.

Kritz, M. and J.M. Nogle. 1994. "Nativity Concentration and Internal Migration Among the Foreign-Born." *Demography* 31 (3):509–24.

Kunitz, S.J. 1994. *Disease and Social Diversity: The European Impact on the Health of Non-Europeans*. New York: Oxford University Press.

Kunstler, J.H. 1993. *The Geography of Nowhere: The Rise and Decline of America's Man-Made Landscape*. New York: Simon & Schuster.

Kushner, H. 1989. *Self-Destruction in the Promised Land: A Psychocultural Biology of American Suicide*. New Brunswick: Rutgers University Press.

Kuznets, S. 1965. *Economic Growth of Nations*. Cambridge, MA: Harvard University Press.

Lamison-White, L. 1997. "Poverty in the United States: 1996." *Current Population Reports P60-198*.

Landers, J. 1993. *Death and the Metropolis: Studies in the Demographic History of London 1670–1830*. Cambridge: Cambridge University Press.

Langford, C. and P. Storey. 1993. "Sex Differentials in Mortality Early in the Twentieth Century: Sri Lanka and India Compared." *Population and Development Review* 19 (2):263–82.

Lapham, R. and W.P. Mauldin. 1984. "Family Planning Program Effort and Birthrate Decline in Developing Countries." *International Family Planning Perspectives* 10 (4):109–18.

Larsen, U. 1995. "Trends in Infertility in Cameroon and Nigeria." *International Family Planning Perspectives* 21 (4):138–142.

Laslett, P. 1971. *The World We Have Lost*. London: Routledge & Kegan Paul.

———. 1991. *A Fresh Map of Life: The Emergence of the Third Age*. Cambridge, MA: Harvard University Press.

———. 1995. "Necessary Knowledge: Age and Aging in the Societies of the Past." in *Aging in the Past: Demography, Society, and Old Age*, edited by D. Kertzer and P. Laslett. Berkeley: University of California Press.

Lavely, W. 1984. "The Rural Chinese Fertility Transition: A Report From Shifang Xian, Sichuan." *Population Studies* 38 (3):365–84.

Lavin, D. 1993. "Chrysler Directs Neon Campaign at Generation X." *The Wall Street Journal* 27 August.

Laws, G. 1994. "Implications of Demographic Changes for Urban Policy and Planning." *Urban Geography* 15 (1):90–100.

Laxton, P. 1987. *London Bills of Mortality*. Cambridge, UK: Chadwyk-Healey Ltd.

Leasure, J.W. 1962. "Factors Involved in the Decline of Fertility in Spain: 1900–1950." Doctoral Thesis, Economics, Princeton University, Princeton.

———. 1982. "L Baisse de la Fecondité aux États-Unis de 1800 a 1860." *Population* 3:607–22.

———. 1989. "A Hypothesis About the Decline of Fertility: Evidence From the United States." *European Journal of Population* 5:105–17.

Lee, B. 1989. "Stability and Change in an Urban Homeless Population." *Demography* 26 (2):323–34.

Lee, B.S. and L.G. Pol. 1988. "Effect of Marital Dissolution on Fertility in Cameroon." *Social Biology* 35: 293–306.

Lee, E. 1966. "A Theory of Migration." *Demography* 3:47–57.

Lee, R. 1976. "Demographic Forecasting and the Easterlin Hypothesis." *Population and Development Review* 2:459–68.

———. 1987. "Population Dynamics in Humans and Other Animals." *Demography* 24:443–65.

Lee, R.B. 1972. "Population Growth and the Beginnings of Sedentary Life Among the Kung Bushmen." in *Population Growth: Anthropological Implications*, edited by B. Spooner. Cambridge, MA: MIT Press.

Leibenstein, H. 1957. *Economic Backwardness and Economic Growth*. New York: John Wiley & Sons.

Lengyel-Cook, M. and R. Repetto. 1982. "The Relevance of the Developing Countries to Demographic Transition Theory: Further Lessons from the Hungarian Experience." *Population Studies* 36 (1):105–28.

Leridon, H. 1990. "Cohabitation, Marriage, Separation: An Analysis of Life Histories of French Cohorts From 1968 to 1985." *Population Studies* 44 (1):127–44.

Leroux, C. 1984. "Ellis Island." *San Diego Union* 26 May.

Lesthaeghe, R.J. 1977. *The Decline of Belgian Fertility, 1800–1970*. Princeton: Princeton University Press.

———. 1980. "On the Social Control of Human Reproduction." *Population and Development Review* 6 (4): 549–80.

Lesthaeghe, R. and J. Surkyn. 1988. "Cultural Dynamics and Economic Theories of Fertility Change." *Population and Development Review* 14:1–45.

Leung, J. 1983. "Singapore's Lee Tells Educated Women: Get Married and Have Families." *The Wall Street Journal* 26 August.

Levy, S., G.R. Frerichs, and H.L. Gordon. 1994. "The Dartnell Marketing Manager's Handbook." Chicago: Dartnell Corporation.

Lewis, B. 1995. *The Middle East: A Brief History of the Last 2,000 Years*. New York: Scribner.

Li, W.L. and Y. Li. 1995. "Special Characteristics of China's Interprovincial Migration." *Geographical Analysis* 27 (2):137–51.

Lichter, D.T., D.K. McLaughlin, G. Kephart, and D.J. Landry. 1992. "Race and Retreat From Marriage: A Shortage of Marriageable Men?" *American Sociological Review* 57:781–99.

Lightbourne, R., S. Singh, and C. Green. 1982. "The World Fertility Survey." *Population Bulletin* 37 (1).

Lillard, L.A. and C.W.A. Panis. 1996. "Marital Status and Mortality: The Role of Health." *Demography* 33 (3):313–327.

Linder, F. 1959. "World Demographic Data." in *The Study of Population: An Inventory and Appraisal*, edited by P.M. Hauser and O.D. Duncan. Chicago: University of Chicago Press.

Livi-Bacci, M. 1991. *Population and Nutrition: An Essay on European Demographic History*. Cambridge: Cambridge University Press.

———. 1992. *A Concise History of World Population*. Cambridge, MA: Blackwell.

Lloyd, C. 1991. "The Contribution of the World Fertility Surveys to an Understanding of the Relationship Between Women's Work and Fertility." *Studies in Family Planning* 22 (3):144–61.

Loaeza Tovar, E.M. and S. Martin. 1997. *Migration Between Mexico & the United States: A Report of the Binational Study of Migration*. Washington, D.C.: U.S. Commission on Immigration Reform.

Loftin, C. and S.K. Ward. 1983. "A Spatial Autocorrelation Model of the Effects of Population Density on Fertility." *American Sociological Review* 48:121–128.

Logan, J.R. and R. Alba. 1993. "Locational Returns to Human Capital: Minority Access to Suburban Community Resources." *Demography* 30 (2):243–68.

London, B. 1987. "Structural Determinants of Third World Urban Change: An Ecological and Political Economic Analysis." *American Sociological Review* 52: 28–43.

Long, L. 1988. *Migration and Residential Mobility in the United States*. New York: Russell Sage Foundation.

———. 1991. "Residential Mobility Differences Among Developed Countries." *International Regional Science Review* 14 (2):133–47.

Long, L. and A. Nucci. 1996. "Convergence or Divergence in Urban-Rural Fertility Differences in the United States." Paper Presented at the annual meeting of the Population Association of America. New Orleans.

Longley, P. and M. Batty. 1996. "Analysis, Modelling, Forecasting, and GIS Technology." in *Spatial Analysis: Modelling in a GIS Environment*, edited by P. Longley and M. Batty. New York: John Wiley & Sons.

Lopez, A.D. 1983. "The Sex Mortality Differential in Developed Countries." in *Sex Differentials in Mortality: Trends, Determinants and Consequences*, edited by A.D. Lopez and L.T. Ruzick. Canberra: Australian National University.

Lorimer, F. 1959. "The Development of Demography." in *The Study of Population*, edited by P.M. Hauser and O.D. Duncan. Chicago: University of Chicago Press.

Lowell, B.L. and Z. Jing. 1994. "Unauthorized Workers and Immigration Reform: What Can We Ascertain From Employers?" *International Migration Review* 28 (3):427–46.

Lublin, J. 1984. "Suburban Population Ages, Causing Conflict and Radical Changes." *The Wall Street Journal* 1 November.

Lutz, W. 1994. "The Future of World Population." *Population Bulletin* 49 (1).

Lynch, K.A. and J.B. Greenhouse. 1994. "Risk Factors for Infant Mortality in Nineteenth-Century Sweden." *Population Studies* 48:117–33.

Machalaba, D. 1982. "More Magazines Aim for Affluent Readers, But some Worry That Shakeout is Coming." *The Wall Street Journal* 4 October.

Mackey-Smith, A. 1984. "Scholastic Aptitude Test Average Scores Rose 4 Points in 1984." *Wall Street Journal* 20 September.

Mackie, G. 1996. "Ending Footbinding and Infibulation: A Convention Account." *American Sociological Review* 61:999–1017.

MacLeod, J. and R. Gold. 1953. "The Male Factor in Fertility and Infertility: VI, Semen Quality and Certain Other Factors in Relation to the Ease of Conception." *Fertility and Sterility* 4:10–33.

Macro International Inc. 1996. "Sharp Drop in Fertility Seen in Latest Ghana Survey." *Demographic and Health Surveys Newsletter* 8 (1):3.

Makinwa-Adebusoye, P. 1994. "Report of the Seminar on Women and Demographic Change in Sub-Saharan Africa, Dakar, Senegal, 3–6 March 1993." *IUSSP Newsletter* January–April:43–48.

Malthus, T.R. 1798. *An Essay on Population (reprinted in 1965)*. New York: Augustus Kelley.

———. 1872. *An Essay on the the Principle of Population, Seventh Edition (reprinted in 1971)*. London: Reeves & Turner.

Mamdami, M. 1972. *The Myths of Population Control*. New York: Monthly Review Press.

Mandelbaum, D. 1974. *Human Fertility in India*. Berkeley: University of California Press.

Manning, W.D. 1995. "Cohabitation, Marriage, and Entry Into Motherhood." *Journal of Marriage and the Family* 57:191–200.

Manton, K.G., E. Stallard, and L. Corder. 1997. "Changes in the Age Dependence of Mortality and Disability: Cohort and Other Determinants." *Demography* 34 (1):135–157.

Marcoux, A. 1998. "The Feminization of Poverty: Claims, Facts, and Data Needs." *Population and Development Review* 24 (1):131–139.

Mare, R.D. 1991. "Five Decades of Educational Assortative Mating." *American Sociological Review* 56:15–32.

Marini, M. 1984. "Women's Educational Attainment and Parenthood." *American Sociological Review* 49 (4): 491–511.

Marini, M. and P.-L. Fan. 1997. "The Gender Gap in Earnings at Career Entry." *American Sociological Review* 62:588–604.

Markham, J.P. and J.I. Gilderbloom. 1998. "Housing Quality Among the Elderly: A Decade of Changes." *International Journal of Aging and Human Development* 46 (1):71–90.

Martin, P.L. 1993. "Comparative Migration Policies." *International Migration Review* 28 (1):164–70.

Martin, T.C. and L. Bumpass. 1989. "Recent Trends in Marital Disruption." *Demography* 26 (1):37–51.

Martine, G. 1996. "Brazil's Fertility Decline, 1965–95: A Fresh Look at Key Factors." *Population and Development Review* 22 (1):47–75.

Marx, K. 1890 (1906). *Capital: A Critique of Political Economy, translated from the Third German Edition by Samuel Moore and Edward Aveling and Edited by Frederich Engels*. New York: The Modern Library.

Masnick, G. 1981. "The Continuity of Birth-Expectation Data With Historical Trends in Cohort Parity Distributions: Implications for Fertility in the 1980s." in *Predicting Fertility*, edited by G. Henderson and P. Placek. Lexington: Lexington Books.

Mason, A. 1988. "Saving, Economic Growth, and Demographic Change." *Population and Development Review* 14 (1):113–44.

———. 1997a. "Population and the Asian Economic Miracle." *Asia-Pacific Population & Policy* (43).

Mason, K.O. 1997b. "Explaining Fertility Transitions." *Demography* 34 (4):443–454.

Massey, D. 1990. "Social Structure, Household Strategies, and the Cumulative Causation of Migration." *Population Index* 56 (1):3–26.

———. 1996a. "The Age of Extremes: Concentrated Affluence and Poverty in the Twenty-First Century." *Demography* 33 (4):395–412.

———. 1996b. "The False Legacy of the 1965 Immigration Act." *World on the Move: Newsletter of the Section on International Migration of the American Sociological Association* 2 (2):2–3.

Massey, D., R. Alarcon, J. Durand, and H. Gonzalez. 1987. *Return to Atzlan: The Social Processes of International Migration from Western Mexico.* Berkeley: University of California Press.

Massey, D., J. Arango, G. Hugo, A. Kouaouci, A. Pellegrino, and J.E. Taylor. 1993. "Theories of International Migration: A Review and Appraisal." *Population and Development Review* 19 (3):431–66.

———. 1994. "An Evaluation of International Migration Theory: The North American Case." *Population and Development Review* 20 (4):699–752.

Massey, D. and N. Denton. 1993. *American Apartheid: Segregation and the Making of the Underclass.* Cambridge, MA: Harvard University Press.

Massey, D. and K. Espinosa. 1997. "What's Driving Mexico-U.S. Migration? A Theoretical, Empirical, and Policy Analysis." *American Journal of Sociology* 102 (4):939–99.

Mathews, P., M. McCarthy, M. Young, and N. McWhirter. 1995. *The Guinness Book of World Records 1995.* New York: Bantam Books.

Matthews, W. 1980. "Inheritance Tax." *Vegetable Box* 4 (1):36.

Mauldin, W.P. 1975. "Assessment of National Family Planning Programs in Developing Countries." *Studies in Family Planning* 6 (2):30–36.

Mauldin, W.P. and J.A. Ross. 1994. "Prospects and Programs for Fertility Reduction, 1990–2015." *Studies in Family Planning* 19 (6):335–53.

Mauldin, W.P. and S. Segal. 1988. "Prevalence of Contraceptive Use: Trends and Issues." *Studies in Family Planning* 19 (6):335–53.

Maxwell, N. 1977. "Medical Secrets of the Amazon." *America* 29:2–8.

Mayhew, B. and R. Levinger. 1976. "Size and the Density of Interaction in Human Aggregates." *American Journal of Sociology* 82 (1):86–109.

McCaa, R. 1994. "Child Marriage and Complex Families Among the Nahuas of Ancient Mexico." *Latin American Population History* (26):2–11.

McCarthy, J.J. 1977. *The Sympto-Thermal Method.* Washington, D.C.: Human Life Foundation of America.

McDaniel, A. 1992. "Extreme Mortality in Nineteenth Century Africa: The Case of Liberian Immigrants." *Demography* 29 (4):581–94.

McDaniel, A. 1995. *Swing Low, Sweet Chariot: The Mortality Cost of Colonizing Liberia in the Nineteenth Century.* Chicago: University of Chicago Press.

———. 1996. "Fertility and Racial Stratification." in *Fertility in the United States: New Patterns, New Theories,* edited by J.B. Casterline, R.D. Lee, and K. Foote. New York: The Population Council.

McDevitt, T. M. 1996. "World Population Profile: 1996." U.S. Bureau of the Census Report WP/96. Washington, D.C.: U.S. Government Printing Office.

McDonald, P. 1993. "Fertility Transition Hypothesis." in *The Revolution in Asian Fertility: Dimensions, Causes, and Implications,* edited by R. Leete and I. Alam. Oxford: Clarendon Press.

McDonough, P., G.J. Duncan, D. Williams, and J. House. 1997. "Income Dynamics and Adult Mortality in the United States, 1972 through 1989." *American Journal of Public Health* 87 (9):1476–83.

McFalls, J. and M. McFalls. 1984. *Disease and Fertility.* New York: Academic Press.

McFarlan, D., N. McWhirter, M. McCarthey, and M. Young. 1991. *The Guinness Book of World Records.* New York: Bantam Books.

McGinnis, J.M. and W.H. Foege. 1993. "Actual Causes of Death in the United States." *Journal of the American Medical Association* 270 (18):2207–212.

McKeown, T. 1976. *The Modern Rise of Population.* London: Edward Arnold.

———. 1988. *The Origins of Human Disease.* Oxford: Basil Blackwell.

McKeown, T. and R. Record. 1962. "Reasons for the Decline of Mortality in England and Wales during the 19th Century." *Population Studies* 16 (2):94–122.

McLanahan, S. and L. Casper. 1995. "Growing Diversity and Inequality in the American Family." in *State of the Union: America in the 1990s, Volume Two: Social Trends,* edited by R. Farley. New York: Russell Sage Foundation.

McLanahan, S. and G. Sandefur. 1994. *Growing Up With a Single Parent: What Hurts, What Helps.* Cambridge, MA: Harvard University Press.

McNeill, W.H. 1976. *Plagues and People.* New York: Doubleday.

McNicoll, G. 1995. "On Population Growth and Revisionism: Further Questions." *Population and Development Review* 21 (2):307–340.

Meadows, D.H. 1974. *Dynamics of Growth in a Finite World*. Cambridge, MA: Wright-Allen Press.

Meadows, D.H., D.L. Meadows, J. Randers, and W. Bherens III. 1972. *The Limits to Growth*. New York: The American Library.

Meek, R. 1971. *Marx and Engels on the Population Bomb*. Berkeley, CA: Ramparts Press.

Mehta, R. 1975. "White-Collar and Blue-Collar Family Responses to Population Growth in India." in *Responses to Population Growth in India*, edited by M. Franda. New York: Praeger.

Menchik, P. 1993. "Economic Status as a Determinant of Mortality Among Black and White Older Men: Does Poverty Kill?" *Population Studies* 47:427–36.

Menken, J. 1994. "Demographic-Economic Relationships and Development." in *Population—The Complex Reality: A Report of the Population Summit of the World's Scientific Academies*, edited by F. Graham-Smith. Golden, CO: North American Press.

Mermelstein, R., B. Miller, T. Prhaska, V. Benson, and J.F. Van Nostrand. 1993. "Measures of Health." in *Health Data on Older Americans: United States, 1992, Vital and Health Statistics, Series 3: Analytic and Epidemiological Studies, No. 27*, edited by J. F. Van Nostrand, S. Furner, and R. Suzman. Hyattsville, MD: Public Health Service.

Merrick, T. and S. Tordella. 1988. "Demographics: People and Markets." *Population Bulletin* 43 (1).

Merrigan, P. and Y. St.-Pierre. 1998. "An Econometric and Neoclassical Analysis of the Timing and Spacing of Births in Canada From 1950 to 1990." *Journal of Population Economics* 11 (1).

Mertens, W. 1994. *Health and Mortality Trends Among Elderly Populations: Determinants and Consequences*. Liège: International Union for the Scientific Study of Population.

Metchnikoff, E. 1908. *The Prolongation of Life*. New York: Putnam.

Meyers, R. 1978. "An Investigation of an Alleged Centenarian." *Demography* 15 (2):235–37.

Mhloyi, M. 1994. *Status of Women, Population and Development*. Liège: International Union for the Scientific Study of Population.

Mier y Terán, M. 1991. "El Gran Cambio Demográfico." *Demos* 4:4–5.

Migration News. 1998. "Canada: Immigration, Diversity Up." *Migration News* 5 (1).

Milbank, D. 1994. "The Unwanted: Refugees from Bosnia Find Scant Welcome in Western Countries." *Wall Street Journal* 7 November:A1.

Mill, J.S. 1848 (1929). *Principles of Political Economy*. London: Longmans & Green.

———. 1873 (1924). *Autobiography*. London: Oxford University Press.

Millar, W.J., S. Wadhers, and S.K. Henshaw. 1997. "Repeat Abortions in Canada, 1975–1993." *Family Planning Perspectives* 29 (1):20–24.

Miller, G.T. 1998. *Environmental Science: Working With The Earth, 7th Edition*. Belmont, CA: Wadsworth Publishing Co.

Miller, J. 1984. "Annexation: The Outer Limits of City Growth." *American Demographics* 6 (11):30–35.

Miller, M. 1979. *Suicide After Sixty: The Final Alternative*. New York: Springer.

Moch, L.P. 1992. *Moving Europeans: Migration in Western Europe Since 1650*. Bloomington: Indiana University Press.

Mogelonsky, M. 1995. "Revenge of the Fats." *American Demographics* (July):10.

Møller, V. 1997. "AIDS: The 'Grandmother's Disease' of South Africa." *Ageing and Society* 17 (4):461–63.

Momeni, J. 1989. "Preface." in *Homelessness in the United States, Volume 1: State Surveys*, edited by J. Momeni. Westport, CT: Greenwood Press.

Montgomery, M. and B. Cohen. 1998. "From Death to Birth: Mortality Decline and Reproductive Change." in *Committee on Population, National Research Council*. Washington, D.C.: National Academy Press.

Morelos, J.B. 1994. "La Mortalidad en México: Hechos y Consensos." in *La Población en el Desarrollo Contemporáneo de México*, edited by F. Alba and G. Cabrera. Mexico City: El Colegio de México.

Moreno, L. and S. Singh. 1996. "Fertility Decline and Changes in Proximate Determinants in the Latin American Region." in *The Fertility Transition in Latin America*, edited by J.M. Guzman, S. Singh, G. Rodríguez, and E.A. Pantelides. New York: Oxford University Press.

Morgan, S.P. 1996. "Characteristic Features of Modern American Fertility." in *Fertility in the United States: New Patterns, New Theories*, edited by J.B. Casterline, R.D. Lee, and K.A. Foote. New York: The Population Council.

Morris, M., A.D. Bernhardt, and M.S. Handcock. 1994. "Economic Inequality: New Methods for New Trends." *American Sociological Review* 59:205–19.

Morrison, A.R. 1993. "Violence or Economics: What Drives Internal Migration in Guatemala?" *Economic Development and Cultural Change* 41 (4):817–32.

Morrison, P. 1980. "How Demographers Can Help Legislators." *Policy Analysis* 6 (1):85–98.

Mosher, S.W. 1983. *Broken Earth: The Rural Chinese*. New York: Free Press.

Mosher, W. and G. Hendershot. 1984. "Religion and Fertility: A Replication." *Demography* 21 (2):185–92.

Mosher, W. and W. Pratt. 1990. "Fecundity and Infertility in the United States, 1965–88." *Advance Data From Vital and Health Statistics of the National Center for Health Statistics, No. 192*.

Mosher, W., L.B. Williams, and D.P. Johnson. 1992. "Religion and Fertility in the United States: New Patterns." *Demography* 29 (2):199–214.

Muhua, C. 1979. "For the Realization of the Four Modernizations, There Must be Planned Control of Population Growth." *Excerpted in Population and Development Review* 5:723–30.

Muhuri, P.K. and S.H. Preston. 1991. "Effects of Family Composition on Mortality Differentials by Sex Among Children in Matlab, Bangladesh." *Population and Development Review* 17 (3):415–34.

Muller, P.O. 1997. "The Suburban Transformation of the Globalizing American City." *Annals of the American Association of Political and Social Science* 551:44–58.

Mumford, L. 1968. "The City: Focus and Funtion." in *International Encyclopedia of the Social Sciences*, edited by D. Sills. New York: Macmillan.

Muramatsu, M. 1971. *Country Profiles: Japan*. New York: Population Council.

Murdock, S.H. and D.R. Ellis. 1991. *Applied Demography: An Introduction to Basic Concepts, Methods, and Data*. Boulder: Westview Press.

Murray, C. 1996. "Rethinking DALYs." in *The Global Burden of Disease*, edited by C.J.L. Murray and A.D. Lopez. Cambridge: Harvard School of Public Health.

Murray, C.J.L. and A.D. Lopez. 1996. *The Global Burden of Disease: A Comprehensive Assessment of Mortality and Disability From Diseases, Injuries, and Risk Factors in 1990 and Projected to 2020*. Cambridge: Harvard University Press.

Mutchler, J.E., J.A. Burr, A.M. Pienta, and M.P. Massagli. 1997. "Pathways to Labor Force Exit: Work Transitions and Work Instability." *Journal of Gerontology: Social Sciences* 52B:S4–S12.

Myrdal, G. 1957. *Economic Theory and Underdeveloped Areas*. London: G. Duckworth.

Nag, M. 1962. *Factors Affecting Human Fertility in Non-Industrial Societies: A Cross-Cultural Study*. New Haven: Yale University Press.

Namboodiri, K. and C.M. Suchindran. 1987. *Life Table Techniques and Their Applications*. San Diego: Academic Press.

National Center for Health Statistics. 1994. *Health United States 1993*. Hyattsville, MD: U.S. Public Health Service.

———. 1996. *Vital Statistics of the United States 1992, Volume II—Mortality, Part A*. Hyattsville, MD: U.S. Department of Health and Human Services.

———. 1997. *Health, United States, 1996–97*. Hyattsville, MD: Department of Health and Human Services.

National Committee on the Status of Women. 1975. *Status of Women in India*. New Delhi, India: The Indian Council of Social Science Research.

National Research Council. 1986. *Population Growth and Economic Development: Policy Questions*. Washington, D.C.: National Academy Press.

———. 1993. *A Census That Mirrors America: Interim Report*. Washington, D.C.: National Academy Press.

Newport, F. and L. Saad. 1992. "Economy Weighs Heavily on American Minds." *Gallup Poll Monthly* 319: 30–31.

Newsweek. 1974. "How to Ease the Hunger Pangs." *Newsweek* 11 November.

Nicholls, W.H. 1970. "Development in Agrarian Economies: The Role of Agricultural Surplus, Population Pressures, and Systems of Land Tenure." in *Subsistence Agriculture and Economic Development*, edited by C. Wharton. Chicago: Aldine.

Nickerson, J. 1975. *Homage to Malthus*. Port Washington, NY: National University Publications.

Nielsen, J. 1978. *Sex in Society: Perspectives on Stratification*. Belmont, CA: Wadsworth Publishing Co.

Nizard, A. and F. Muñoz-Pérez. 1994. "Alcohol, Smoking and Mortality in France since 1950: An Evaluation of the Number of Deaths in 1986 due to Alcohol and Tobacco Consumption." *Population: An English Selection* 6:159–94.

Noah, T. 1991. "Asian-Americans Take Lead in Starting U.S. Businesses." *The Wall Street Journal* 2 August.

Nordberg, O. 1983. "Counting on Wall Street." *American Demographics* 5 (8):22–25.

Nortman, D. 1975. *Population and Family Planning Programs: A Factbook*. New York: The Population Council.

Notestein, F.W. 1945. "Population—The Long View." in *Food for the World*, edited by T.W. Schultz. Chicago: University of Chicago Press.

O'Connell, M. and C. Rogers. 1983. "Assessing Cohort Birth Expectations Data From the Current Population Survey, 1971–1981." *Demography* 29 (3):369–84.

Oeppen, J. 1993. "Back Projection and Inverse Projection: Members of a Wider Class of Constrained Projection Models." *Population Studies* 47 (2):245–68.

Ogawa, N. and R.D. Retherford. 1997. "Shifting Costs of Caring for the Elderly Back to Families in Japan: Will it Work?" *Population and Development Review* 23 (1): 41–58.

O'Hare, W.P., K.M. Pollard, T.L. Mann, and M.M. Kent. 1991. "African Americans in the 1990s." *Population Bulletin* 46 (1).

Ohlin, G. 1976. "Economic Theory Confronts Population Growth." in *Economic Factors in Population Growth*, edited by A. Coale. New York: John Wiley & Sons.

Oliver, S. 1991. "Look Homeward." *Forbes* 1 April.

Olshansky, S.J. and B.A. Carnes. 1997. "Ever Since Gompertz." *Demography* 34 (1):1–15.

Omran, A.R. 1971. "The Epidemiological Transition: A Theory of the Epidemiology of Population Change." *Milbank Memorial Fund Quarterly* 49:509–38.

———. 1977. "Epidemiologic Transition in the United States." *Population Bulletin* 32 (2).

———. 1973. "The Mortality Profile." in *Egypt: Population Problems and Prospects*, edited by A.R. Omran. Chapel Hill: Carolina Population Center.

Omran, A.R. and F. Roudi. 1993. "The Middle East Population Puzzle." *Population Bulletin* 48 (1).

Oppenheimer, V.K. 1967. "The Interaction of Demand and Supply and Its Effect on the Female Labour Force in the U.S." *Population Studies* 21 (3):239–59.

———. 1994. "Women's Rising Employment and the Future of the Family in Industrializing Societies." *Population and Development Review* 20 (2):293–342.

Oriol, W. 1982. *Aging in All Nations: A Special Report on the United Nations World Assembly on Aging*. Washington, D.C.: National Council on Aging.

Orshansky, M. 1969. "How Poverty is Measured." *Monthly Labor Review* 92 (2):37–41.

Ortiz de Montellano, B. 1975. "Empirical Aztec Medicine." *Science* 188:215–20.

Orubuloye, I., J. Caldwell, and P. Caldwell. 1991. "Sexual Networking in the Ekitia District of Nigeria." *Studies in Family Planning* 22:61–73.

Osgood, D.W., J.K. Wilson, P.M. O'Malley, J.G. Bachman, and L.D. Johnston. 1996. "Routine Activities and Individual Deviant Behavior." *American Sociological Review* 61:635–55.

Östberg, V. and D. Vägero. 1991. "Socioeconomic Differences in Mortality Among Children: Do They Persist Into Adulthood?" *Social Science and Medicine* 32: 403–10.

Overbeek, J. 1980. *Population and Canadian Society*. Toronto: Butterworths.

Pagnini, D.L. and R.R. Rindfuss. 1993. "The Divorce of Marriage and Childbearing: Changing Attitudes and Behavior in the United States." *Population and Development Review* 19 (2):331–47.

Palmore, J. and R. Gardner. 1983. *Measuring Mortality, Fertility, and Natural Increase*. Honolulu: East-West Population Institute.

Pampel, F.C. 1993. "Relative Cohort Size and Fertility: The Socio-Political Context of the Easterlin Effect." *American Sociological Review* 58:496–514.

———. 1996. "Cohort Size and Age-Specific Suicide Rates: A Contingent Relationship." *Demography* 33: 341–55.

Panandiker, V.A.P. and P.K. Umashankar. 1994. "Fertility Control and Politics in India." in *The New Politics of Population: Conflict and Consensus in Family Planning*, edited by J.L. Finkle and C.A. McIntosh. New York: Oxford University Press.

Pantaleo, G., C. Grziosi, J.F. Demerest, L. Butini, M. Montroni, C.H. Fox, J.M. Orenstein, D.P. Kotler, and A.S. Fauci. 1993. "HIV Infection is Active and Progressive in Lymphoid Tissue During the Clinically Latent Stage of Disease." *Nature* 362 (25 March): 355–58.

Pappas, G., S. Queen, W. Hadden, and G. Fisher. 1993. "The Increasing Disparity in Mortality Between Socioeconomic Groups in the United States, 1960 and 1986." *The New England Journal of Medicine* 329 (2):103–09.

Park, C.B. and N. Cho. 1995. "Consequences of Son Preference in a Low-Fertility Society: Imbalance of the Sex Ratio at Birth." *Population and Development Review* 21 (1):59–84.

Passaro, J. 1996. *The Unequal Homeless: Men on the Street, Women in Their Place*. New York: Routledge.

Passel, J.S. and B. Edmonston. 1994. "Immigration and Race: Recent trends in Immigration to the United States." in *Immigration and Ethnicity: The Integration of America's Newest Arrivals*, edited by B. Edmonston and J.S. Passel. Washington, D.C.: The Urban Institute Press.

Passel, J., C. Cowan, and K. Wolter. 1983. "Coverage of the 1980 Census." Paper presented at Annual Meeting of the Population Association of America. Pittsburgh, PA.

Pathak, K.B., G. Feeney, and N.Y. Luther. 1998. *Accelerating India's Fertility Decline: The Role of Temporary Contraceptive Methods*. Honolulu: East-West Center Program on Population.

Paul, B.K. 1991. "Health Service Resources as Determinants of Infant Death in Rural Bangladesh: An Empirical Study." *Social Science and Medicine* 32:42–49.

———. 1994. "Gender Ratios in the SMAs of Bangladesh: Is the Gap Declining?" *Urban Geography* 15 (4):345–61.

Pebley, A., H. Delgado, and E. Brinemann. 1979. "Fertility Desires and Child Mortality Experiences Among Guatemalan Women." *Studies in Family Planning* 11 (4):129–36.

Peng, P. 1996. "Population and Development in China." in *The Population Situation in China: The Insiders' View*, edited by China Population Association and the State Family Planning Commission of China. Liege, Belgium: International Union for the Scientific Study of Population.

Peter, K. 1987. *The Dynamics of Hutterite Society*. Canada: University of Alberta Press.

Petersen, W. 1975. *Population*. New York: MacMillan Publishing Co.

———. 1979. *Malthus*. Cambridge, MA: Harvard University Press.

Peyser, M. and K. Alexander 1997. "Question Time: What Will the 2000 Census Ask?" *Newsweek*, 16 June, p. 14.

Phillips, D. 1974. "The Influence of Suggestion on Suicide." *American Sociological Review* 39:340–54.

———. 1977. "Motor Vehicle Fatalities Increase Just After Publicized Suicide Stories." *Science* 196:1464–65.

———. 1978. "Airplane Accident Fatalities Increase Just After Newspaper Stories About Murder and Suicide." *Science* 201:748–50.

———. 1983. "The Impact of Mass Media Violence on U.S. Homicides." *American Sociological Review* 48 (4):560–68.

———. 1993. "Psychology and Survival." *Lancet* 342: 1142–45.

Phillips, D. and D. Smith. 1990. "Postponement of Death Until Symbolically Meaningful Occasions." *Journal of the American Medical Association* 263 (14):1947–51.

Phillips, M. 1996. "Wage Gap Based on Education Levels Off." *The Wall Street Journal* 22 July.

Phillips, N. and A.D. Kobrinetz. 1996. "Targeting Home Healthcare to Senior Citizens." *Business Geographics* (June):30–33.

Phillips, T. 1998. "Mass Media Target Customers." *Business Geographics* (May):22–23.

Piccinino, L.J. and W.D. Mosher. 1998. "Trends in Contraceptive Use in the United States: 1982–1995." *Family Planning Perspectives* 30 (1):4–10.

Piirto, R. 1991. *Beyond Mind Games: The Marketing Power of Psychographics*. Ithaca, NY: American Demographics Books.

Pimentel, D., X. Huang, A. Cardova, and P.M. 1997. "Impact of Population Growth on Food Supplies and Environment." *Population and Environment* 19 (1): 9–14.

Pimentel, D., R. Terhune, R. Dyson-Hudson, S. Rochereau, R. Samis, E. Smith, D. Denman, D. Reifschneider, and M. Shepard. 1976. "Land Degradation: Effects on Food and Energy Resources." *Science* 194:149–55.

Plane, D. and P. Rogerson. 1994. *The Geographical Analysis of Population: with Applications to Planning and Business*. New York: John Wiley & Sons.

Plato. 1960. *The Laws*, vol. (originally published circa 350 BC). New York: Dutton.

Poednak, A.P. 1989. *Racial and Ethnic Differences in Disease*. New York: Oxford University Press.

Pol, L.G. and R.K. Thomas. 1992. *The Demography of Health and Health Care*. New York: Plenum Press.

———. 1997. *Demography For Business Decision Making*. Westport, CT: Quorum Books.

Poleman, T. 1975. "World Food: A Perspective." *Science* 188:510–18.

Pollack, R.A. and S.C. Watkins. 1993. "Cultural and Economic Approaches to Fertility: Proper Marriage or Mésalliance." *Population and Development Review* 19 (3):467–96.

Pollard, K.M. 1997. "Are More Frequent Data Better? Census Bureau Tests Long Form's Possible Successor." *Population Today* 25 (9):1–2.

Pong, S. 1994. "Sex Preference and Fertility in Peninsular Malaysia." *Studies in Family Planning* 25 (3): 137–48.

Pool, D.I. 1970. "Ghana: The Attitudes of Urban Males toward Family Size and Family Limitation." *Studies in Family Planning* 60:12–17.

Popkin, B.M. 1993. "Nutritional Patterns and Transitions." *Population and Development Review* 19: 138–157.

Population Action Council. 1985. "India Seeking Two-Child Family Norm by 2000." *Popline* 7 (5):1–2.

Population Action International. 1996. *Why Population Matters*. Washington, D.C.: Population Action International.

Population Association of America. 1983. *Activities of Demographers in the Private Sector*. Washington, D.C.: Population Association of America.

Population Crisis Committee. 1990. *Cities: Life in the World's 100 Largest Metropolitan Areas*. Washington, D.C.: Population Crisis Committee.

Population Information Program. 1985. "Periodic Abstinence." *Population Reports of the Johns Hopkins University* 1 (3).

Population Institute. 1997a. "Pakistan's Future Hinges on Stabilizing Population." *PopLine* 19 (July–August):3.

———. 1997b. *Toward the 21st Century: Resource Conservation, Population and Sustainable Development*. Washington, D.C.: The Population Institute.

Population Reference Bureau. 1997. *World Population Data Sheet*. Washington D.C.: Population Reference Bureau.

———. 1998. *1998 World Population Data Sheet*. Washington, D.C.: Population Reference Bureau.

Portes, A. and R.G. Rumbaut. 1990. *Immigrant America: A Portrait*. Berkeley: University of California Press.

Postel, S. 1993. "Facing Water Scarcity." in *State of the World 1993*, edited by L. Brown. New York: W.W. Norton & Company.

———. 1996. "Forging a Sustainable Water Strategy." in *State of the World 1996*, edited by Lester Brown and Associates. New York: W.W. Norton & Company.

Poston, D.L., 1986. "Patterns of Contraceptive Use in China." *Studies in Family Planning* 17:217–27.

———. 1990. "Apportioning the Congress: A Primer." *Population Today* 18 (7/8):6–8.

———. 1997. "The U.S. Census and Congressional Apportionment." *Society* 34 (March/April):36–44.

Poston, D.L. and B. Gu. 1987. "Socioeconomic Development, Family Planning, and Fertility in China." *Demography* 24 (4):531–52.

Poston, D.L., M.X. Mao, and M.-Y. Yu. 1994. "The Global Distribution of the Overseas Chinese Around 1990." *Population and Development Review* 20 (3): 631–46.

Potter, J. 1986. "Review of 'Population Growth and Economic Development: Policy Questions'." *Population and Development Review* 12 (3):578–81.

Potter, L. 1991. "Socioeconomic Determinants of White and Black Males' Life Expectancy Differentials, 1980." *Demography* 28:303–21.

Potter, R. 1992. *Urbanisation in the Third World*. Oxford: Oxford University Press.

Prema, K. and M. Ravindranath. 1982. "The Effect of Breastfeeding Supplements on the Return of Fertility." *Studies in Family Planning* 13 (10):293–96.

Preston, S.H., 1970. *Older Male Mortality and Cigarette Smoking.* Berkeley: Institute of International Studies, University of California.

———. 1978. *The Effects of Infant and Child Mortality on Fertility.* New York: Academic Press.

———. 1984. "Children and the Elderly: Divergent Paths for America's Dependents." *Demography* 24 (4): 435–57.

———. 1986. "Are the Economic Consequences of Population Growth a Sound Basis for Population Policy?" in *World Population and U.S. Policy: The Choices Ahead*, edited by J. Menken. New York: W.W. Norton & Co.

Preston, S. and M. Guillot. 1997. "Population Dynamics in an Age of Declining Fertility." *Genus* LIII (3–4): 15–31.

Preston, S. and M.R. Haines. 1991. *Fatal Years: Child Mortality in Late Nineteenth-Century America.* Princeton: Princeton University Press.

Preston, S., N. Keyfitz, and R. Schoen. 1972. *Causes of Death: Life Tables for National Populations.* New York: Seminar Press.

Preston, S.H. and J. MacDonald. 1979. "The Incidence of Divorce Within Cohorts of American Marriages Contracted Since the Civil War." *Demography* 16: 1–25.

Pumphrey, G. 1940. *The Story of Liverpool's Public Service.* London: Hodden & Stoughton.

Putka, G. 1988. "Benefit of B.A. is Greater Than Ever." *The Wall Street Journal* 17 August.

Quinn, T. 1987. "The Global Epidemiology of the Acquired Immunodeficiency Syndrome." in *Report of the Surgeon General's Workshop on Children With HIV Infections and Their Families*, edited by B. Silverman and A. Waddell. Washington, D.C.: Department of Health and Human Services.

Rajan, S.I., U.S. Mishra, and M. Ramanathan. 1993. "The Two-Child Family in India: Is It Realistic?" *International Family Planning Perspectives* 19:125–28.

Rajulton, F. 1991. "Migrability: A Diffusion Model of Migration." *Genus* XLVII (1–2):31–48.

Ramos, L.R. 1993. "Brazil." in *Developments and Research on Aging: An International Handbook*, edited by E. Palmore. Westport, CT: Greenwood Press.

Randall, S. 1996. "Whose Reality? Local Perceptions of Fertility Versus Demographic Analysis." *Population Studies* 50 (2):221–234.

Rao, V. 1993. "Dowry 'Inflation' in Rural India: A Statistical Investigation." *Population Studies* 47 (2): 283–93.

Ravenstein, E. 1889. "The Laws of Migration." *Journal of the Royal Statistical Society* 52:241–301.

Rawlings, S. 1994. "Household and Family Characteristics: March 1993." *Current Population Reports P20-477.*

Rees, W.E. 1996a. "Ecological Footprints of the Future." *People and the Planet* http://www.ecdpm.org/patp/vol16/rees.html (accessed 1998):1–3.

———. 1996b. "Revisiting Carrying Capacity: Area-Based Indicators of Sustainability." *Population and Environment* 17 (3):195–215.

Rees, W.E. and M. Wackernagel. 1994. "Ecological Footprints and Appropriated Carrying Capacity: Measuring the Natural Capital Requirements of the Human Economy." in *Investing in Natural Capacity: The Ecological Economics Approach to Sustainability*, edited by A. Jansson, M. Hammer, C. Folke, and R. Costanza. Washington, D.C.: Island Press.

Reher, D. and R. Schofield. 1993. *Old and New Methods in Historical Demography.* Oxford: Clarendon Press.

Reider, I. and S. Pick. 1992. "El Aborto: Quién Debe Tomar la Decisión." *Demos* 5:35–36.

Reining, P., F. Camara, B. Chiñas, R. Fanale, S. Gojman de Millán, B. Lenkerd, I. Shinolhara, and I. Tinker. 1977. *Village Women: Their Changing Lives and Fertility.* Washington, D.C.: American Association for the Advancement of Science.

Reissman, L. 1964. *The Urban Process: Cities in Industrial Societies.* New York: The Free Press.

Retherford, R.D. 1975. *The Changing Sex Differentials in Mortality.* Westport: Greenwood Press.

Retherford, R.D. and J. Rele. 1989. "A Decomposition of Recent Fertility Changes in South Asia." *Population and Development Review* 15 (4):739–47.

Retherford, R.D., N. Ogawa, and S. Sakamoto. 1996. "Values and Fertility Change in Japan." *Population Studies* 50:5–26.

Retherford, R.D. and W. Sewell. 1991. "Birth Order and Intelligence: Further Test of the Confluence Model." *American Sociological Review* 56:141–58.

Revelle, R. 1984. "The Effects of Population Growth on Renewable Resources." in *Population Resources, Environment, and Development*, edited by United Nations. New York: United Nations.

Richards, A. and J. Waterbury. 1990. *A Political Economy of the Middle East.* Boulder, CO: Westview Press.

Riche, M. 1997. "Should the Census Bureau Use 'Statistical Sampling' in Census 2000?" *Insight* 18 August.

Ricketts, T.C., L.A. Savitz, W.M. Gesler, and D.N. Osborne. 1997. "Using Geographic Methods to Understand Health Issues." Agency for Health Care Policy and Research, Rockville, MD.

Riddle, J.M. 1992. *Contraception and Abortion From the Ancient World to the Renaissance.* Cambridge: Harvard University Press.

Ridley, J., M. Sheps, J. Lingner, and J. Menken. 1967. "The Effects of Changing Mortality on Natality." *Milbank Memorial Fund Quarterly* 55:77–97.

Riley, M.W. 1976a. "Age Strata in Social Systems." in *Handbook of Aging and the Social Sciences*, edited by R. Binstock and E. Shanas. New York: Van Nostrand Reinhold.

———. 1976b. "Social Gerontology and the Age Stratification of Society." in *Aging in America*, edited by C.S. Kart and B. Manard. Port Washington, NY: Alfred Publishing Co.

———. 1979. "Aging, Social Change, and Social Policy." in *Aging From Birth to Death*, edited by M.W. Riley. Boulder: Westview Press.

Riley, N. and R.W. Gardner. 1996. *China's Population: A Review of the Literature*. Liege, Belgium: International Union for the Scientific Study of Population.

Rindfuss, R.R., S.P. Morgan, and K. Offutt. 1996. "Education and the Changing Age Pattern of American Fertility: 1963–1989." *Demography* 33 (3):277–290.

Ritchey, P. and C. Stokes. 1972. "Residence Background, Migration, and Fertility." *Demography* 9:217–30.

Rives, B. 1997. "Strategic Financial Planning for Hospitals: Demographic Considerations." in *Demographics: A Casebook for Business and Government*, edited by H.J. Kintner, T.W. Merrick, P.A. Morrison, and P.R. Voss. Santa Monica, CA: RAND.

Roberts, K. 1978. "The Intrauterine Device as a Health Risk." in *Fertility and Contraception in America*, edited by House Select Committee on Population. Washington, D.C.: Government Printing Office.

Robey, B. 1983. "Achtung! Here Comes the Census." *American Demographics* 5 (10):2–4.

———. 1989. "Two Hundred Years and Counting: The 1990 Census." *Population Bulletin* 44 (1).

Robey, B., J.A. Ross, and I. Bhushan. 1996. "Meeting Unmet Need: New Strategies." *Population Reports of the Population Information Program of the The Johns Hopkins University* Series J (No. 43).

Robinson, W.C. 1986. "Another Look at the Hutterites and Natural Fertility." *Social Biology* 33:65–76.

———. 1997. "The Economic Theory of Fertility Over Three Decades." *Population Studies* 51:63–74.

Rockstein, M. and M. Sussman. 1979. *Biology of Aging*. Belmont, CA: Wadsworth Publishing Co.

Rogers, A. 1992. "Introduction." in *Elderly Migration and Population Redistribution: A Comparative Study*, edited by A. Rogers. London: Bellhaven Press.

Rogers, A., R. Rogers, and A. Belanger. 1990. "Longer Life But Worse Health? Measurement and Dynamics." *The Gerontologist* 30:640–49.

Rogers, E.M. 1995. *Diffusion of Innovations, Fourth Edition*. New York: The Free Press.

Rogers, R.G. 1992. "Living and Dying in the U.S.A.: Sociodemographic Determinants of Death Among Blacks and Whites." *Demography* 29:287–303.

———. 1995. "Sociodemographic Characteristics of Long-Lived and Healthy Individuals." *Population and Development Review* 21 (1):33–58.

Rogers, R.G. and E. Powell-Griner. 1991. "Life Expectancies of Cigarette Smokers and Nonsmokers in the United States." *Social Science and Medicine* 32: 1151–59.

Rosenhause, S. 1976. "India Taking Drastic Birth Control Step." *Los Angeles Times* 25 September.

Rosenhouse-Persson, S. and G. Sabagh. 1983. "Attitudes Toward Abortion Among Catholic Mexican-American Women: The Effects of Religiosity and Education." *Demography* 20 (1):87–98.

Rosenwaike, I. 1979. "A New Evaluation of the United States Census Data on the Extreme Aged." *Demography* 16 (2):279–88.

———. 1991. "Mortality of Hispanic Populations." New York: Greenwood Press.

Ross, C.E. and C.-l. Wu. 1995. "The Links Between Education and Health." *American Sociological Review* 60:719–745.

Ross, J.A. and E. Frankenberg. 1993. *Findings From Two Decades of Family Planning Research*. New York: The Population Council.

Ross, J.A., W.P. Mauldin, and V.C. Miller. 1993. *Family Planning and Population: A Compendium of International Statistics*. New York: The Population Council.

Rossi, P. 1955. *Why Families Move*. New York: Free Press.

———. 1989. *Without Shelter: Homelessness in the 1980s*. New York: Priority Press.

Rubinfein, E. 1989. "Boat People Arouse Japan's Xenophobia." *Wall Street Journal* 11 October:A14.

Ruggles, S. 1994a. "The Origins of African-American Family Structure." *American Sociological Review* 59: 136–51.

———. 1994b. "The Transformation of American Family Structure." *American Historical Review* 99 (1): 103–28.

———. 1997. "The Rise of Divorce and Separation in the United States, 1880–1990." *Demography* 34 (4): 455–466.

Ruggles, S. and R.R. Menard. 1995. "The Minnesota Historical Census Projects." *Historical Methods* 28 (1):1–78.

Rumbaut, R.G. 1991. "Passages to America: Perspectives on the New Immigration." in *The Recentering of America: American Society in Transition*, edited by A. Wolfe. Berkeley: University of California Press.

———. 1994. "Origins and Destinies: Immigration to the United States Since World War II." *Sociological Forum* 9(4):583–621.

———. 1995a. "The New Immigration." *Contemporary Sociology* 24 (4):307–11.

———. 1995b. "Vietnamese, Laotian, and Cambodian Americans." in *Asian Americans: Contemporary Trends and Issues*, edited by P.G. Min. Beverly Hills: Sage Publications.

———. 1997. "Assimilation and Its Discontents." *International Migration Review* 31 (4):923–60.

Rumbaut, R.G. and J.R. Weeks. 1996. "Unraveling a Public Health Enigma: Why Do Immigrants Experience Superior Perinatal Health Outcomes?" *Research in the Sociology of Health Care* 13 (B):337–391.

Russell, B. 1951. *New Hopes for a Changing World.* New York: Simon & Schuster.

Russell, C. 1992. *The Master Trend: How the Baby Boom Generation is Remaking America.* New York: Plenum Press.

Ryder, N. 1960. "The Cohort as a Concept in the Study of Social Change." *American Sociological Review* 23:843–61.

———. 1964. "Notes on the Concept of a Population." *American Journal of Sociology* 69 (5):447–63.

———. 1990. "What Is Going to Happen to American Fertility?" *Population and Development Review* 16 (3):433–54.

Sadik, N. 1994. "The Cairo Consensus: Changing the World by Choice." *Populi* 21 (9):4–5.

Sagan, C. 1989. "The Secret of the Persian Chessboard." *Parade Magazine* 5 February:14.

Salaff, J. and A. Wong. 1978. "Are Disincentives Coercive? The View From Singapore." *International Family Planning Perspectives* 4 (2):50–55.

Salter, C., H.B. Johnston, and N. Hengen. 1997. "Care for Postabortion Complications: Saving Women's Lives." *Population Reports of the Population Information Program of the The Johns Hopkins University* Series L (No. 10).

Saluter, A.F. 1993. "Marital Status and Living Arrangements: March 1993." *Current Population Reports* P20-478.

Saluter, A.F. and T.A. Lugaila. 1998. "Marital Status and Living Arrangements: March 1996." *Current Population Reports* P20-496.

Sampson, R.J. and J.H. Laub. 1990. "Crime and Deviance Over the Life Course: The Salience of Adult Social Bonds." *American Sociological Review.* 55: 609–27.

Samuel, T.J. 1966. "The Development of India's Policy of Population Control." *Milbank Memorial Fund Quarterly* 44:49–67.

———. 1994. *Quebec Separatism is Dead: Demography is Destiny.* Toronto: John Samuel and Associates.

San Diego Association of Governments. 1991. *Regional Growth Forecasts: Preliminary Series 8 Regionwide Forecast (1990–2015), Agenda Report No. RB-25.* San Diego: SANDAG.

Sanchez-Albornoz, N. 1974. *The Population of Latin America: A History.* Berkeley: University of California Press.

———. 1988. *Españoles Hacia America: La Emigración en Masa, 1880–1930.* Madrid: Alianza Editorial.

Sanderson, S.K. 1995. *Social Transformations: A General Theory of Historical Development.* Cambridge, MA: Blackwell.

Sanderson, W. 1979. "Quantitative Aspects of Marriage, Fertility, and Family Limitation in Nineteenth Century America: Another Application of the Coale Specifications." *Demography* 16 (3):336–58.

Sauvy, A. 1969. *General Theory of Population.* New York: Basic Books.

Schapera, I. 1941. *Married Life in an African Tribe.* New York: Sheridan House.

Schiffren, L. 1995. "How to Stay Married, By Those Who Have." *The Wall Street Journal* 26 June.

Schiller, N.G., L. Basch, and C.S. Blanc. 1995. "From Immigrant to Transmigrant: Theorizing Transnational Migration." *Anthropological Quarterly* 68 (1):48–63.

Schinaia, N., A. Ghirardini, F. Chiarotti, P. Mannucci, and The Italian Group. 1991. "Progress to AIDS among Italian HIV-seropositive Haemophiliacs." *AIDS* 5 (4):385–91.

Schnore, L., C. Andre, and H. Sharp. 1976. "Black Suburbanization 1930–1970." in *The Changing Face of the Suburbs,* edited by B. Schwartz. Chicago: University of Chicago Press.

Schoen, R. 1983. "Measuring the Tightness of a Marriage Squeeze." *Demography* 20 (1):61–78.

Schoen, R., Y.J. Kim, C.A. Nathanson, J. Fields, and N.M. Astone. 1997. "Why Do Americans Want Children?" *Population and Development Review* 23 (2): 333–358.

Schoen, R. and J. Kluegel. 1988. "The Widening Gap in Black and White Marriage Rates: The Impact of Population Composition and Differential Marriage Propensities." *American Sociological Review* 53: 895–907.

Schoen, R. and R.M. Weinick. 1993. "The Slowing Metabolism of Marriage: Figures From 1988 U.S. Marital Status Life Tables." *Demography* 30 (4):737–746.

Schoenbaum, M. and Waidmann. 1997. "Race, Socioeconomic Status, and Health: Accounting for Race Differences in Health." *The Journal of Gerontology Series B* 52B (Special Issue):61–73.

Schofield, R. and D. Coleman. 1986. "Introduction: The State of Population Theory." in *The State of Population Theory: Forward From Malthus,* edited by D. Coleman and R. Schofield. Oxford: Basil Blackwell.

Schofield, R. and D. Reher. 1991. "The Decline of Mortality in Europe." in *The Decline of Mortality in Europe,* edited by R. Schofield, D. Reher, and A. Bideau. Oxford: Clarendon Press.

Schuman, H. and J. Scott. 1990. "Generational Memories." *ISR Newsletter* 16 (6):4–5.

Schumpeter, J. 1942. *Capitalism, Socialism, and Democracy, Second Edition.* New York: Harper & Row.

Schwadel, F. 1991. "Spiegel, Ebony Aim to Dress Black Women." *The Wall Street Journal* 18 September.

Scrimshaw, S. 1978. "Infant Mortality and Behavior in the Regulation of Family Size." *Population and Development Review* 4 (3):383–403.

Seager, J. 1990. "The State of the Earth Atlas." New York: Simon & Schuster.

———. 1993. *Earth Follies: Coming to Feminist Terms with the Global Environmental Crisis*. New York: Routledge.

Segerberg, O. 1974. *The Immortality Factor*. New York: Dutton.

Seitz, J.L. 1995. *Global Issues: An Introduction*. Cambridge, MA: Blackwell.

Sell, R. and G. DeJong. 1983. "Deciding Whether to Move: Mobility, Wishful Thinking and Adjustment." *Sociology and Social Research* 67 (2):146–65.

Serbanescu, F., L. Morris, P. Stupp, and A. Stanescu. 1995. "The Impact of Recent Policy Changes on Fertility, Abortion, and Contraceptive Use in Romania." *Studies in Family Planning* 26 (2):76–87.

Shapiro, D. 1996. "Fertility Decline in Kinshasa." *Population Studies* 50:89–103.

Sheon, A. and C. Stanton. 1989. "Use of Periodic Abstinence and Knowledge of the Fertile Period in 12 Developing Countries." *International Family Planning Perspectives* 15:29–34.

Sheppard, R. 1982. "An Underdog-Eat-Underdog World." *Time* 14 June.

Sheps, M. 1965. "An Analysis of Reproductive Patterns in an American Isolate." *Population Studies* 19:65–80.

Sherris, J. 1985. *The Impact of Family Planning Programs on Fertility*. Baltimore, MD: The Johns Hopkins University.

Sherwood, N. 1995. "Editor's Note: The Census Debate." *Business Geographics*, May, p. 8.

Shkolnikov, V., F. Meslé, and J. Vallin. 1996. "Health Crisis in Russia: Part I, Recent Trends in Life Expectancy and Causes of Death From 1970 to 1993." *Population: An English Selection* 8:123–154.

Short, J.F.J. 1997. *Poverty, Ethnicity, and Violent Crime*. Boulder, CO: Westview Press.

Shribman, D. 1991. "With Cheaper, Better Data, Advocacy Groups Emerge as Growing Force in Redistricting Fights." *The Wall Street Journal* 29 May.

Shryock, H. 1964. *Population Mobility Within the United States*. Chicago: University of Chicago Press.

Shryock, H., J. Siegel, and associates. 1973. *The Methods and Materials of Demography*. Washington, D.C.: Government Printing Office.

Shryock, H., J. Siegel, and associates; (condensed by E. Stockwell). 1976. *The Methods and Materials of Demography*. New York: Academic Press.

Shweder, R.A. 1997. "It's Called Poor Health for a Reason." *The New York Times* 9 March.

Siegel, J.S. 1993. *A Generation of Change: A Profile of America's Older Population*. New York: Russell Sage Foundation.

———. 1974. "Estimates of Coverage of the Population by Sex, Race, and Age in the 1970 Census." *Demography* 11:1–23.

Simmel, G. 1905. "The Metropolis and Mental Life." in *Classic Essays on the Culture of Cities*, edited by R. Sennet. New York: Appleton-Century-Crofts.

Simmons, G.C., C. Smucker, S. Bernstein, and E. Jensen. 1982. "Post-Neonatal Mortality in Rural India: Implications of an Economic Model." *Demography* 19 (3): 371–89.

Simmons, I.G. 1993. *Environmental History: A Concise Introduction*. Oxford: Basil Blackwell Publishers.

Simmons, L. 1960. "Aging in Preindustrial Societies." in *Handbook of Gerontology*, edited by C. Tibbetts. Chicago: University of Chicago Press.

Simon, J.L. 1981. *The Ultimate Resource*. Princeton: Princeton University Press.

———. 1989. "On Aggregate Empirical Studies Relating Population Variables to Economic Development." *Population and Development Review* 15:323–32.

———. 1992. *Population and Development in Poor Countries: Selected Essays*. Princeton: Princeton University Press.

Singh, S. and R. Samara. 1996. "Early Marriage Among Women in Developing Countries." *International Family Planning Perspectives* 22 (4):148–157.

Singh, S. and G. Sedgh. 1997. "The Relationship of Abortion to Trends in Contraception and Fertility in Brazil, Colombia and Mexico." *International Family Planning Perspectives* 23 (1):4–14.

Sivin, I. 1989. "IUDs are Contraceptives, not Abortifacients: a Comment on Research and Belief." *Studies in Family Planning* 20:355–59.

Skinner, G.W. 1997. "Family systems and demographic processes." Pp. 53–95 in *Anthropological Demography: Toward a New Synthesis*, edited by D. Kertzer and T. Fricke. Chicago: University of Chicago Press.

Smail, J.K. 1997. "Population Growth Seems to Affect Everything But Is Seldom Held Responsible for Anything." *Politics and the Life Sciences* 16 (2):231–236.

Smil, V. 1994. "How Many People Can the Earth Feed?" *Population and Development Review* 20 (2):255–92.

Smith, C.A. and M. Pratt. 1993. "Cardiovascular Disease." in *Chronic Disease Epidemiology and Control*, edited by R.C. Brownson, P.L. Remington, and J.R. Davis. Washington, D.C.: American Public Health Association.

Smith, D.P. 1992. *Formal Demography*. New York: Plenum.

Smith, D.S. 1978. "Mortality and Family in the Colonial Chesapeake." *Journal of Interdisciplinary History* 8 (3):403–27.

Smith, G., M. Shipley, and G. Rose. 1990. "Magnitude and Causes of Socioeconomic Differentials in Mortality: Further Evidence From the Whitehall Study." *British Medical Journal* 301:429–32.

Smith, J.P. 1997. "Wealth Inequality Among Americans." *The Journal of Gerontology Series B* 52B (Special Issue):74–81.

Smith, K.R. and N.J. Waitzman. 1994. "Double Jeopardy: Interaction Efects of Marital and Poverty Status on the Risk of Mortality." *Demography* 31 (3): 487–507.

Smith, M.P. and L.E. Guarnizo. 1998. *Transnationalism From Below*. New Brunswick: Transaction Publishers.

Smith, P., S. Khoo, and S. Go. 1983. "The Migration of Women to Cities: A Comparative Perspective." in *Women in the Cities of Asia: Migration and Urban Adaptation*, edited by J. Fawcett, S. Khoo, and P. Smith. Boulder, CO: Westview Press.

Snipp, C.M. 1989. *American Indian: The First of This Land*. New York: Russell Sage Foundation.

Snow, J. 1936. *Snow on Cholera*. London: Oxford University Press.

Sörenson, A. and H. Trappe. 1995. "The Persistence of Gender Equality in Earnings in the German Democratic Republic." *American Sociological Review* 60: 398–406.

South, S.J. and G. Spitze. 1994. "Housework in Marital and Nonmarital Households." *American Sociological Review* 59:327–47.

Spar, E.J. 1998. "Demographics: End of Year (D.C. Year, That Is) Wrap-Up." *Business Geographics* (January):38–39.

Spence, A. 1989. *Biology of Human Aging*. Englewood Cliffs, NJ: Prentice-Hall.

Spencer, G. 1989. "Projections of the Population of the United States, by Age, Sex, and Race: 1988 to 2080." *Current Population Reports* P25-1018.

Spengler, J.J. 1974. *Population Change, Modernization, and Welfare*. Englewood Cliffs, NJ: Prentice-Hall.

———. 1979. *France Faces Depopulation: Postlude Edition, 1936–1976*. Durham, NC: Duke University Press.

Spooner, B. 1972. "Population Growth: Anthropological Implications." Cambridge, MA: MIT Press.

Srinivasan, K., P.H. Reddy, and K.N.M. Raju. 1978. "From One Generation to the Next: Changes in Fertility, Family Size Preferences, and Family Planning in an Indian State Between 1951 and 1975." *Studies in Family Planning* 9 (10–11):258–71.

Stack, S. 1987. "Celebrities and Suicide: A Taxonomy and Analysis." *American Sociological Review* 52: 401–12.

Stangeland, C.E. 1904. *Pre-Malthusian Doctrines of Population*. New York: Columbia University Press.

Stanton, S.S. 1992. "International Migration and Political Turmoil in the Middle East." *Population and Development Review* 18 (4):719–28.

Staples, R. 1985. "Changes in Black Family Structure: The Conflict Between Family Ideology and Structural Conditions." *Journal of Marriage and Family* 47: 1005–13.

———. 1986. "The Masculine Way of Violence." in *Homicide Among Black Americans*, edited by D. Hawkins. New York: University Press of America.

Stark, O. and R. Lucas. 1988. "Migration, Remittances, and the Family." *Economic Development and Cultural Change* 36:465–81.

Starr, P. 1987. "The Sociology of Official Statistics." in *The Politics of Numbers*, edited by W. Alonso and P. Starr. New York: Russell Sage Foundation.

Statistics Bureau of Japan. 1998. *Results of the First Basic Complete Tabulation of the 1995 Population Census*.

Statistics Canada. 1995. "Census 1996." *Focus for the Future* 10 (1):1–3.

———. 1996. "Differences Between Census Counts and Statistics Canada's Population Estimates." **http://www.statcan.ca/english/census96/petrie.htm** (accessed 1998).

———. 1997. "1996 Census: Immigration and Citizenship." **http://www.statcan.ca**.

———. 1998a. *Highlights From the Report on the Demographic Situation in Canada*. Ottawa: Statistics Canada.

———. 1998b. *Population and Growth Components*.

Statistika Centralbyran (Sweden). 1983. *Pehr Wargentin: den Svenska Statistikens Fader*. Stockholm: Statistika Centralbyran.

Steelman, L.C. and B. Powell. 1989. "Acquiring Capital for College: The Constraints of Family Configuration." *American Sociological Review* 54:844–855.

Steffensmeier, D. and M. Harer. 1987. "Is the Crime Rate Really Falling? An 'Aging' U.S. Population and Its Impact on the Nation's Crime Rate." *Journal of Research in Crime and Delinquency* 24:23–48.

Stephenson, G. 1964. *A History of American Immigration*. New York: Russell and Russell.

Stewart, J., B. Popkin, D. Guilkey, J. Akin, L. Adair, and W. Flieger. 1991. "Influences on the Extent of Breast-Feeding: A Prospective Study in the Philippines." *Demography* 28 (2):181–200.

Stokes, C. and Y.-S. Hsieh. 1983. "Female Employment and Reproductive Behavior in Taiwan, 1980." *Demography* 20 (3):313–31.

Stolnitz, G.J. 1964. "The Demographic Transition: From High to Low Birth Rates and Death Rates." Chapter 2 in *Population: The Vital Revolution*, edited by R. Freedman. Garden City: Anchor Books.

Stolzenberg, R. 1990. "Ethnicity, Geography, and Occupational Achievement of Hispanic Men in the United States." *American Sociological Review* 55:143–54.

Stolzenberg, R. and L. Waite. 1984. "Local Labor Markets, Children and Labor Force Participation of Wives." *Demography* 21 (2):157–70.

Stone, L. 1975. "On the Interaction of Mobility Dimensions in Theory on Migration Decisions." *Canadian Review of Sociology and Anthropology* 12: 95–100.

Stonich, S. 1989. "Social Processes and Environmental Destruction: A Central American Case Study." *Population and Development Review* 15 (2):269–96.

Stout, H. 1992. "SAT Scores Rise But Remain Near Lows." *Wall Street Journal* 27 August.

Straus, M. 1983. "Societal Morphogenesis and Intrafamily Violence in Cross-Cultural Perspective." in *International Perspectives on Family Violence*, edited by R. Gelles and C. Cornell. Lexington, MA: D.C. Heath.

Stroup, N.E., M.C. Dufour, E.S. Hurwitz, and J.-C. Desenclos. 1993. "Cirrhosis and Other Chronic Liver Diseases." in *Chronic Disease Epidemiology and Control*, edited by R.C. Brownson, P.L. Remington, and J.R. Davis. Washington, D.C.: American Public Health Association.

Strouse, J. 1980. "Act of Desperation." *Newsweek* 14 April.

Stycos, J.M. 1971. *Ideology, Faith and Family Planning in Latin America*. New York: McGraw-Hill.

Suharto, S. and O. Volkov. 1993. "Overview of Emerging Trends and Issues in the 1990 Round of Censuses." in *Annual Meeting of the Population Association of America.*, Cincinnati, Ohio.

Sung, K. 1995. "Measures and Dimensions of Filial Piety in Korea." *The Gerontologist* 35 (2):240–47.

Sussman, G. 1977. "Parisian Infants and Norman Wet Nurses in the Early Nineteenth Century: A Statistical Study." *Journal of Interdisciplinary History* 7 (4): 637–53.

Sutherland, I. 1963. "John Graunt: A Tercentenary Tribute." *Royal Statistical Society Journal, Series A* 126: 536–37, reprinted in K. Kammeyer 1975 (ed.), Population: Selected Essays and Research (Chicago: Rand McNally).

Suzman, R.M., D.P. Willis, and K.G. Manton. 1992. *The Oldest Old*. New York: Oxford University Press.

Tabah, L. 1980. "World Population Trends, a Stocktaking." *Population and Development Review* 6 (3): 355–90.

Ta-k'un, W. 1960. "A Critique of neo-Malthusian Theory." Excerpted in *Population and Development Review* (1979) 5 (4):699–707.

Taueber, C. 1993. "Sixty-Five Plus in America." *Current Population Reports* P23-178RV.

Taylor, W., J.S. Marks, J.R. Livengood, and J.P. Koplan. 1993. "Current Issues and Challenges in Chronic Disease Control." in *Chronic Disease Epidemiology and Control*, edited by R.C. Brownson, P.L. Remington, and J.R. Davis. Washington, D.C.: American Public Health Association.

Teitelbaum, M. 1975. "Relevance of Demographic Transition for Developing Countries." *Science* 188:420–25.

Teitelbaum, M. and J. Winter. 1988. "Introduction." in *Population and Resources in the Western Intellectual Tradition*, edited by M. Teitelbaum and J. Winter. New York: Population Council.

Templin, N. 1995. "Chrysler Retools 'Symbols of Suburbs' to Broaden Appeal." *The Wall Street Journal* 5 January.

———. 1996. "Tejano Songs Attract Hispanics Who Like Both Love and Amor." *The Wall Street Journal* 2 January.

Terborgh, A., J.E. Rosen, S. Gálvez, W. Terceros, J.T. Bertrand, and S.E. Bull. 1995. "Family Planning Among Indigenous Populations in Latin America." *International Family Planning Perspectives* 21 (4): 143–149.

———. 1994b. "What You Won't Hear at Cairo." *The Wall Street Journal* 6 September.

The Economist. 1994. "Little Match Girls." *The Economist* 15 January.

———. 1994a. "India: The Poor Get Richer." *The Economist* 5 November.

———. 1994b. "How Not to Sell 1.2 Billion Tubes of Toothpaste." *The Economist* 3 December.

———. 1995a. "Eastern European Demography: of Death and Dying." *The Economist* 7 January.

———. 1995b. "The Most Expensive Slum in the World." *The Economist* 6 May.

———. 1996a. "Cities: Fiction and Fact." *The Economist* 8 June.

———. 1996b. "Men's Traditional Culture." *The Economist* 10 August.

———. 1997. "So, Does America Want Them or Not?" *The Economist* 19 July:25.

———. 1997a. "Big MacCurrencies: Can Hamburgers Provide Hot Tips About Exchange Rates?" *The Economist* 12 April.

———. 1997b. "India: The Other Side of Prosperity." *The Economist* 21 June.

———. 1997c. "Stopping the Yangzi's Flow." *The Economist* 2 August.

———. 1998a. "China: The X-Files." *The Economist* 14 February.

———. 1998b. "How AIDS Began." *The Economist* 7 February:81–2.

The Environmental Fund. 1981. "The Perils of Vanishing Farmland." *The Other Side* No. 23.

The Presbyterian Panel. 1998. "Background Survey for the 1997–1999 Presbyterian Panel." The Presbyterian Church, Louisville, KY.

Thomas, H. 1997. *The Slave Trade: The Story of the Atlantic Slave Trade: 1440–1870*. New York: Simon & Schuster.

Thompson, W. 1929. "Population." *American Journal of Sociology* 34 (6):959–75.

Thornton, A., W.G. Axinn, and J.D. Teachman. 1995. "The Influence of School Enrollment and Accumulation on Cohabitation and Marriage in Early Adulthood." *American Sociological Review* 60:762–774.

Thornton, R. 1987. *American Indian Holocaust and Survival: A Population History Since 1492*. Norman: University of Oklahoma Press.

Tien, H. 1984. "Induced Fertility Transition: Impact of Population Planning and Socio-Economic Change in the People's Republic of China." *Population Studies* 38 (3):385–400.

———. 1990. "China's Population Policy After Tiananmen." *Population Today* 18 (9):6–8.

Tien, H.Y., P. Yu, J. Li, and Z. Liagn. 1992. "China's Demographic Dilemmas." *Population Bulletin* 47 (1).

Tietze, C. and S. Lewit. 1977. "Legal Abortion." *Scientific American* 236 (1):21–27.

Tobias, A. 1979. "The Only Article on Inflation You Need to Read: We're Getting Poorer, But We Can Do Something About It." *Esquire* 92 (5):49–55.

Tobin, G. 1976. "Suburbanization and the Development of Motor Transportation: Transportation Technology and Suburbanization Process." in *The Changing Face of the Suburbs*, edited by B. Schwartz. Chicago: University of Chicago Press.

Tolnay, S. 1989. "A New Look at the Effect of Venereal Disease on Black Fertility: The Deep South in 1940." *Demography* 26:679–90.

Torrey, E.F. 1988. *Nowhere To Go: The Tragic Odyssey of the Homeless Mentally Ill*. New York: Harper & Row.

Toulemon, L. 1997. "Cohabitation is Here To Stay." *Population: An English Selection* 9:11–46.

Trafzer, C.E. 1997. *Death Stalks the Yakama*. East Lansing: Michigan State University Press.

Tran, H.H. 1995. *The Official Guide to Household Spending: The Number One Guide to Who Spends How Much on What, Third Edition*. Ithaca, NY: New Strategist.

Trussell, J. and L. Grummer-Strawn. 1990. "Contraceptive Failure of the Ovulation Method of Periodic Abstinence." *Family Planning Perspectives* 22:65–74.

Turke, P. 1989. "Evolution and the Demand for Children." *Population and Development Review* 15: 61–90.

Turner, R.K., D. Pearce, and I. Bateman. 1993. *Environmental Economics: An Elementary Introduction*. Baltimore: The John Hopkins University Press.

Tuxill, J. and C. Bright. 1998. "Losing Strands in the Web of Life." in *State of the World 1998*, edited by L. Brown, C. Flavin, and H. French. New York: W.W. Norton & Company.

U.S. Bureau of the Census. 1975. *Historical Statistics of the United States*. Washington, D.C.: Government Printing Office.

———. 1978a. *Country Demographic Profiles: India*. Washington, D.C.: Government Printing Office.

———. 1978b. "History and Organization." U.S. Bureau of the Census, Washington, D.C.

———. 1979a. "Country Demographic Profiles: Mexico."

———. 1979b. "Fertility of American Women: June 1978." *Current Population Reports* Series P-20 (No. 341).

———. 1983. "Voting and Registration in the Election of November 1982." *Current Population Reports* P20-383.

———. 1984. "Educational Attainment in the United States: March 1981 and 1980." *Current Population Reports P20-390.*

———. 1992. "Educational Attainment in the United States: March 1991 and 1990." *Current Population Reports P20-454.*

———. 1993a. "1990 Census of Population and Housing, Technical Documentation." Government Printing Office, Washington, D.C.

———. 1993b. *We, The American Foreign Born*. Washington, D.C.: Government Printing Office.

———. 1995a. "American Housing Survey for the United States in 1993." *Current Housing Reports H150-93.*

———. 1995b. "Income, Poverty, and Valuation of Non-Cash Benefits." *Current Population Reports P60-188.*

———. 1996a. *The American Almanac: 1996–1997, Statistical Abstract of the United States*. Austin, TX: Hoover's, Inc.

———. 1996b. *World Population Profile: 1996*. Washington, D.C.: U.S. Bureau of the Census.

———. 1997a. "Asset Ownership of Households: 1993." *Current Population Reports P70-47.*

———. 1997b. "How We're Changing: Demographic State of the Nation, 1997." *Current Population Reports, Special Studies, Series P23-193.*

———. 1997c. "Money Income in the United States: 1996." *Current Population Reports P60-197.*

———. 1997d. "Report to the Congress: The Plan for Census 2000, Revised and Reissued (August)." U.S. Bureau of the Census, Washington, D.C.

———. 1997e. "Voting and Registration in the Election of November 1996." *Current Population Reports P20-504.*

———. 1998. "International Data Base.": **http://www.census.gov/ipc/www** (accessed 1998).

U.S. Department of Justice. 1971. *Crime in the United States, 1970*. Washington, D.C.: Federal Bureau of Investigation.

———. 1981. *Crime in the United States, 1980*. Washington, D.C.: Federal Bureau of Investigation.

U.S. Immigration and Naturalization Service. 1991. *An Immigrant Nation: United States Regulation of Immigration, 1798–1991*. Washington, D.C.: Government Printing Office.

———. 1992. *Immigration Reform and Control Act, Report on the Legalized Alien Population*. Washington, D.C.: Government Printing Office.

———. 1997. *1995 Statistical Yearbook of the Immigration and Naturalization Service*. Washington, D.C.: Government Printing Office.

Udry, J.R. 1994. "The Nature of Gender." *Demography* 31 (4):561–74.

Udry, J.R. and R. Cliquet. 1982. "A Cross-Cultural Examination of the Relationship Between Ages at Menarche, Marriage, and First Birth." *Demography* 19 (1): 53–64.

UNICEF. 1997. *The State of the World's Children 1997.* New York: Oxford University Press.

United Nations. 1953. *Causes and Consequences of Population Trends.* New York: United Nations.

———. 1958. *Principles and Recommendations for National Population Censuses.* New York: United Nations.

———. 1963. *Proceedings of the Asian Population Conference.* New York: United Nations.

———. 1973. *The Determinants and Consequences of Population Trends: New Summary of Findings on Interaction of Demographic, Economic and Social Factors, Volume I.* New York: United Nations.

———. 1979. *Trends and Characteristics of International Migration Since 1950.* New York: United Nations.

———. 1982. *Levels and Trends of Mortality Since 1950 (ST/ESA/SER.A/74).* New York: United Nations.

———. 1986. *Determinants of Mortality Change and Differentials in Developing Countries (ST/ESA/SER. A/94).* New York: United Nations.

———. 1988a. *Case Studies in Population Policy: Kuwait. Department of International Economic and Social Affairs, Population Studies, No. 15 (ST/ESA/SER.R/82).* New York: United Nations.

———. 1988b. *Report of the Commissioner-General of the United Nations Relief and Works Agency for Palestine Refugees in the Near East, 1 July 1987–30 June 1988.* New York: United Nations.

———. 1989. *Trends in Population Policy.* New York: United Nations.

———. 1991. *Demographic Yearbook 1989.* New York: United Nations.

———. 1992. *Demographic Yearbook 1990.* New York: United Nations.

———. 1996a. *Demographic Yearbook 1994.* New York: United Nations.

———. 1996b. *Urban Agglomerations 1996.* New York: United Nations.

———. 1997a. *Demographic Yearbook 1995.* New York: United Nations.

———. 1997b. *The Sex and Age Distribution of the World Populations: The 1996 Revision.* New York: United Nations.

United Nations Development Programme. 1997. *Human Development Report 1997.* New York: Oxford University Press.

United Nations High Commissioner for Refugees. 1997. "The 1997 Year in Review." *Refugees Magazine* 109.

United Nations Population Division. 1992. *Long-range World Population Projection: Two Centuries of Population Growth, 1950–2150.* New York: United Nations.

———. 1994. *World Population Growth from Year 0 to Stabilization* (mimeograph, 7 June 1994). New York: United Nations.

———. 1998. *World Population Projections to 2150.* New York: United Nations.

United Nations Population Fund. 1987. *1986 Report by the Executive Director.* New York: United Nations.

———. 1991a. *Population and the Environment: The Challenges Ahead.* New York: United Nations.

———. 1991b. *Population Growth and Policies in Mega-Cities: Mexico City.* New York: United Nations.

———. 1992. "India's Other Problem." *Population* 18 (2):2.

Universal Press Syndicate. 1982. "Dear Abby." 3 March.

Urquidi, V. 1992. "Perspectives on Population and Employment Along the Border." in *Demographic Dynamics of the U.S. Mexico Border,* edited by J.R. Weeks and R. Ham-Chande. El Paso: Texas Western Press of the University of Texas, El Paso.

van de Walle, E. 1975. "Foundations of the Model of Doom." *Science* 189:1077–78.

———. 1992. "Fertility Transition, Conscious Choice and Numeracy." *Demography* 29:487–502.

van de Walle, E. and J. Knodel. 1967. "Demographic Transition and Fertility Decline: The European Case." in *International Population Conference.* Sydney, Australia.

———. 1980. "Europe's Fertility Transition: New Evidence and Lessons for Today's Developing World." *Population Bulletin* 34 (6).

van der Wijst, T. 1985. "Transmigration in Indonesia: An Evaluation of a Population Redistribution Policy." *Population Research and Policy Review* 4 (1):1–30.

Vanderpool, C. 1995. *Integrated Natural Resource Systems.* East Lansing: Michigan Agricultural Experiment Station, Michigan State University.

VanLandingham, M.J., S. Suprasert, W. Sittitrai, C. Vaddhanaphuti, and N. Grandjean. 1993. "Sexual Activity Among Never-Married Men in Northern Thailand." *Demography* 30:297–314.

Ventura, S.J., J.A. Martin, S.C. Curtin, and T.J. Matthews. 1997. "Report of Final Natality Statistics, 1995." *Monthly Vital Statistics Report* 45 (11, Supplement).

Veyne, P. 1987. "The Roman Empire." in *A History of Private Life, Volume 1: From Pagan Rome to Byzantium,* edited by P. Veyne. Cambridge, MA: Harvard University Press.

Vicker, R. 1982. "Ecology-Minded Oregon, in Need of Jobs, Makes Industry Feel a Bit More Welcome." *The Wall Street Journal* 15 July.

Villeneuve-Gokalp, C. 1991. "From Marriage to Informal Union: Recent Changes in the Behavior of French Couples." *Population: An English Selection* 3:81–111.

Visaria, L. and P. Visaria. 1995. "India's Population in Transition." *Population Bulletin* 50 (3).

———. 1981. "India's Population: Second and Growing." *Population Bulletin* 36 (4).

Viscusi, W.K. 1979. *Welfare of the Elderly: An Economic Analysis and Policy Prescription.* New York: John Wiley & Sons.

Vlassoff, C. 1990. "The Value of Sons in an Indian Village." *Population Studies* 44:5–20.

Wachter, K. 1991. "Elusive Cycles: Are There Dynamically Possible Lee-Easterlin Models for U.S. Births?" *Population Studies* 45:109–35.

Wackernagel, M. and W.E. Rees. 1996. *Our Ecological Footprint: Reducing Human Impact on the Earth.* Philadelphia: New Society Publishers.

Waite, L. 1995. "Does Marriage Matter?" *Demography* 32 (4):483–508.

Waite, L., F.K. Goldscheider, and C. Witsberger. 1986. "Nonfamily Living and The Erosion of Traditional Family Orientations Among Young Adults." *American Sociological Review* 41 (2):235–51.

Waldman, P. 1991. "As Acreage Erodes, Palestinians Feel Pressure to Deal." *Wall Street Journal* 10 May.

Waldron, I. 1983. "The role of Genetic and Biological Factors in Sex Differences in Mortality." in *Sex Differentials in Mortality: Trends, Determinants and Consequences*, edited by A.D. Lopez and L.T. Ruzicka. Canberra: Australian National University.

———. 1986. "What Do We Know About Causes of Sex Differences in Mortality? A Review of the Literature." *Population Bulletin of the United States* 18-1995: 59–76.

Wall Street Journal, 1985. "Norway Moves to Boost Birthrate." *The Wall Street Journal* 16 January.

———. 1994. "Pakistan to Expel Immigrants." *The Wall Street Journal* 26 August.

Wallace, R. 1761 (1969). *A Dissertation on the Numbers of Mankind, in Ancient and Modern Times, 1st and 2nd editions, revised and corrected.* New York: Kelley.

Wallerstein, E. 1974. *The Modern World System.* New York: Academic Press.

Wallis, C. 1995. "How to Live to be 120." *Time* 6 March, p. 85.

Walsh, J. 1974. "U.N. Conference: Topping Any Agenda is the Question of Development." *Science* 185:1144.

———. 1975. "U.S. Agribusiness and Agricultural Trends." *Science* 188:531–34.

Ware, H. 1975. "The Limits of Acceptable Family Size in Western Nigeria." *Journal of Biosocial Science* 7: 273–96.

Warren, C., F. Hiyari, P. Wingo, A. Abel-Aziz, and L. Morris. 1990. "Fertility and Family Planning in Jordan: Results From the 1985 Jordan Husband's Fertility Survey." *Studies in Family Planning* 21:33–39.

Warren, C.A.B. 1998. "Aging and Identity in Premodern Times." *Research on Aging* 20 (1):11–35.

Warren, R. and E.P. Kraly. 1985. *The Elusive Exodus: Emigration from the United States.* Washington, D.C.: Population Reference Bureau.

Warwick, D. 1987. "The Indonesian Family Planning Program: Government Influence and Client Choice." *Population and Development Review* 12 (3):453–490.

Wasserman, I.M. 1984. "Imitation and Suicide: A Reexamination of the Werther Effect." *American Sociological Review* 49:427–36.

Watkins, S.C. 1986. "Conclusion." in *The Decline of Fertility in Europe*, edited by A. Coale and S. C. Watkins. Princeton: Princeton University Press.

———. 1991. *From Provinces Into Nations: Demographic Integration in Western Europe, 1870–1960.* Princeton, NJ: Princeton University Press.

Way, P. and K. Stanecki. 1994. "HIV/AIDS is Pandemic." in *World Population Profile: 1994*, edited by E. Jamison and F. Hobbs. Washington, D.C.: U.S. Bureau of the Census.

Weber, A. 1899. *The Growth of Cities in the Nineteenth Century.* New York: Columbia University Press.

Weber, P. 1994. *Net Loss: Fish, Jobs and the Marine Environment.* Washington D.C.: WorldWatch Institute.

Weeks, J.R. 1978. *Population: An Introduction to Concepts and Issues, First Edition.* Belmont, CA: Wadsworth Publishing Co.

———. 1984. *Aging: Concepts and Social Issues.* Belmont: Wadsworth Publishing Co.

———. 1988. "The demography of Islamic nations." *Population Bulletin* 43 (4):1–53.

Weeks, J.R. and K. Fuller. 1996. "Population Distribution and Migration: Components of Housing Demand and Supply." in *Human Settlements Habitat: Proceedings and Recommendations of the International Symposium on Human Settlements*, edited by D. Wozniak, T. Shuman, A. Roet, and M. Garret. San Diego: International Institute for Human Resource Development, San Diego State University.

Weeks, J.R., R.G. Rumbaut, C. Brindis, C. Korenbrot, and D. Minkler. 1989. "High Fertility Among Indochinese Refugees." *Public Health Reports* 104:143–50.

Weg, R. 1975. "Changing Physiology of Aging: Normal and Pathological." in *Aging: Scientific Perspectives and Social Issues*, edited by D. Woodruff and J. Birreu. New York: Van Nostrand.

Weinstein, A. 1994. *Market Segmentation: Using Demographics, Psychographics, and Other Niche Marketing Techniques to Predict and Model Customer Behavior, Revised Edition.* Chicago: Probus Publishing Co.

Weiss, K.H. 1973. "Demographic Models for Anthropology." *American Antiquity* 38 (2: Part II).

Weiss, K.M. 1990. "The Biodemography of Variations in Human Frailty." *Demography* 27 (2):185–206.

Weiss, M. 1988. *The Clustering of America.* New York: Harper & Row.

Weller, R., J. Macisco, and G. Martine. 1971. "Relative Importance of the Components of Urban Growth in Latin America." *Demography* 8 (2):225–32.

Wellington, A. 1994. "Accounting For the Male/Female Wage Gap Among Whites: 1976 and 1985." *American Sociological Review* 59:839–48.

Wells, R. 1971. "Family Size and Fertility Control in Eighteenth-Century America: A Study of Quaker Families." *Population Studies* 25 (1):73–82.

———. 1982. *Revolutions in Americans' Lives*. Westport, CT: Greenwood Press.

———. 1985. *Uncle Sam's Family: Issues in and Perspective on American Demographic History*. Albany: State University of New York Press.

Wendelken, S. 1996. "Religiously Mapping." *Business Geographics* (September):54.

Westing, A. 1981. "A Note on How Many Humans Have Ever Lived." *BioScience* 31 (7):523–24.

Westoff, C. 1978. "Marriage and Fertility in the Developed Countries." *Scientific American* 239 (6): 51–57.

———. 1990. "Reproductive Intentions and Fertility Rates." *International Family Planning Perspectives* 16:84–89.

———. 1994. "What's the World's Priority Task? Finally, Control Population." *The New York Times Magazine* 6 February.

———. 1997. "Population Growth: Large Problem, Low Visibility." *Politics & Life Sciences* 16 (2):226–227.

Westoff, C. and A. Bankole. 1996. "The Potential Demographic Significance of Unmet Need." *International Family Planning Perspectives* 22 (1):16–20.

Westoff, C. and R. Rindfuss. 1974. "Sex Preselection in the United States: Some Implications." *Science* 184: 633–36.

Westoff, C. and N. Ryder. 1977. "The Predictive Validity of Reproductive Intentions." *Demography* 14 (4): 431–54.

Westoff, L. and C. Westoff. 1971. *From Now to Zero*. Boston: Little, Brown.

Wharton, C. and R. Blackburn. 1988. "Lower Dose Pills." *Population Reports, Population Information Program, John Hopkins University* Series A, (Number 7,).

White, A.A. and K.F. Rust. 1997. "Preparing for the 2000 Census: Interim Report II." Washington, D.C.: National Academy Press.

Whitmore, T.M. 1992. *Disease and Death in Early Colonial Mexico*. Boulder: Westview Press.

Widjojo, N. 1970. *Population Trends in Indonesia*. Ithaca, NY: Cornell University Press.

Wilkens, J. and D. Fried. 1991. "Critics Challenge Census Findings on the Homeless." *San Diego Union* 22 April.

Wilkes, R.E. 1995. "Household Life-Cycle Stages, Transitions, and Product Expenditures." *Journal of Consumer Research* 22:27–42.

Willcox, W.F. 1936. *Natural and Political Observations Made Upon the Bills of Mortality by John Graunt*. Baltimore: The Johns Hopkins Press.

Williams, J.D. 1994. "Using Demographic Analysis to Fix Political Geography." *Applied Demography* 8 (3):1–3.

Williams, L. and B. Zimmer. 1990. "The Changing Influence of Religion on U.S. Fertility." *Demography* 27 (3): 475–81.

Williams, N. and C. Galley. 1995. "Urban-Rural Differentials in Infant Mortality in Victorian England." *Population Studies* 49:401–420.

Wilson, A.C. and R.L. Cann. 1992. "The Recent African Genesis of Humans." *Scientific American* 266 (4): 68–74.

Wilson, J.Q. 1993. *The Moral Sense*. New York: The Free Press.

Wilson, T. 1991. "Urbanism, Migration, and Tolerance: A Reassessment." *American Sociological Review* 56: 117–123.

Wilson, W. 1978. *The Declining Significance of Race*. Chicago: University of Chicago Press.

Winter, J.G. 1936. "Papyri in the University of Michigan Collection: Miscellaneous Papyri." in *Michigan Papyri, Vol. III*. Ann Arbor: University of Michigan Press.

Wirth, L. 1938. "Urbanism as a Way of Life." *American Journal of Sociology* 44 (3–24).

Wise, D.A. 1997. "Retirement Against the Demographic Trends: More Older People Living Longer, Working Less, and Saving Less." *Demography* 34 (1):83–95.

Witt, B. 1983. "Having a Baby? You Might Be Able to Save Some Money in Las Vegas." *The Wall Street Journal* 13 July.

Wittwer, S.H. 1977. "Assuring Our Food Supply—Technology, Resources and Policy." *World Development* 5 (5–7):784–95.

Wolffsohn, M. 1987. *Israel: Polity, Society, Economy, 1882–1986: An Introductory Handbook*. Atlantic Highlands, NJ: Humanities Press International.

Woodrow-Lafield, K. 1996. "Emigration from the USA: Multiplicity Survey Evidence." *Population Research and Policy Review* 15 (2):171–99.

World Bank. 1984. *World Development Report 1984*. New York: Oxford University Press.

———. 1994a. *Population and Development: Implications for the World Bank*. Washington, D.C.: World Bank.

———. 1994b. *World Development Report 1994: Infrastructure for Development*. New York: Oxford University Press.

———. 1995. *Monitoring Environmental Progress: A Report on Work in Progress*. Washington, D.C.: The World Bank.

———. 1996a. "Assessing Poverty in Kenya." *Findings: Africa Region* 55 (January):1–3.

———. 1996b. *Poverty Reduction and the World Bank*. Washington, D.C.: The World Bank.

————. 1998a. "West Africa: Community-Based Natural Resource Management." *Findings From Economic Management and Social Policy* 107 (March).

————. 1998b. *World Development Indicators*.

World Health Organization. 1995. *World Health Statistics Annual 1994*. Geneva: World Health Organization.

————. 1997. *Report on the Global HIV/AIDS Epidemic, December 1997*. Geneva: United Nations.

World Resources Institute. 1996. *World Resources 1996–97*. New York: Oxford University Press.

Wright, J. 1989a. *Address Unknown: The Homeless in America*. Hawthorne, NY: Aldine de Gruyter.

Wright, R. 1989b. "The Easterlin Hypothesis and European Fertility Rates." *Population and Development Review* 15:107–22.

Wrigley, E.A. 1974. *Population and History*. New York: McGraw-Hill.

————. 1987. *People, Cities and Wealth*. Oxford: Blackwell Publishers.

————. 1988. "The Limits to Growth." in *Population and Resources in Western Intellectual Tradition*, edited by M. Teitelbaum and J. Winter. New York: Population Council.

Wrigley, E.A. and R.S. Schofield. 1981. *The Population History of England, 1541–1871: A Reconstruction*. Cambridge: Harvard University Press.

Wynter, L.E. 1990. "Key Issue for Black Target-Marketing is Expansion, Say Industry Executives." *The Wall Street Journal* 7 May.

————. 1994. "Business & Race." *The Wall Street Journal* 20 July.

Wysocki, B.J. 1997. "The Global Mall: In Developing Nations, Many Youths Splurge, Mainly on U.S. Goods." *The Wall Street Journal* 26 June.

Yap, M.T. 1995. "Singapore's 'Three or More' Policy: The First Five Years." *Asia-Pacific Population Journal* 10 (4):39–52.

Yashin, A.I. and I.A. Iachine. 1997. "How Frailty Models Can Be Used For Evaluating Longevity Limits: Taking Advantage of an Interdisciplinary Approach." *Demography* 34 (1):31–48.

Yaukey, D. 1961. *Fertility Differences in a Modernizing Country*. Princeton: Princeton University Press.

Yoon, I. 1990. "The Changing Significance of Ethnic and Class Resources in Immigrant Business: The Case of Korean Immigrant Businesses in Chicago." *International Migration Review* 25 (2):303–32.

Young, T.K. 1994. *The Health of Native Americans: Toward a Biocultural Epidemiology*. New York: Oxford University Press.

Zajonc, R. 1976. "Family Configuration and Intelligence." *Science* 192:227–36.

Zajonc, R., B.G. Marcus, M. Berbaum, J. Bargh, and R. Moreland. 1991. "One Justified Criticism Plus Three Flawed Analyses Equal Two Unwarranted Conclusions: A Reply to Retherford and Sewell." *American Sociological Review* 56:159–65.

Zarate, A. and A. de Zarate. 1975. "On the Reconciliation of Research Findings of Migrant-Nonmigrant Fertility Differentials in Urban Areas." *International Migration Review* 9 (2):115–56.

Zavala de Cosío, M.E. 1989. "Niveles y Tendencias de la Fecundidad en México, 1960–1980." in *La Fecundidad en México: Cambios y Perspectives*, edited by B. Figueroa Campos. Mexico, DF: El Colegio de Mexico.

Zelinsky, W. 1971. "The Hypothesis of the Mobility Transition." *The Geographical Review* 61 (2):219–49.

Zhao, Z. 1994. "Demographic Conditions and Multigeneration Households in Chinese History: Results From Genealogical Research and Microsimulation." *Population Studies* 48 (3):413–26.

Zhu, T., B.T. Korber, A.J. Nahmia, E. Hooper, P.M. Sharp, and D.D. Ho. 1998. "An African HIV-1 Sequence From 1959 and Implications for the Origin of the Epidemic." *Nature* 391 (February):594–97.

Zinsser, H. 1935. *Rats, Lice and History*. Boston: Little, Brown.

Zipf, G.K. 1949. *Human Behavior and the Principle of Least Effort*. Reading, MA: Addison-Wesley.

Zlaff, V. 1973. "Ethnic Segregation in Urban Israel." *Demography* 11 (2):161–84.

Zlotnik, H. 1994. "Expert Group Meeting on Population Distribution and Migration." *International Migration Review* 28 (1):171–204.

Zou'bi, A.A., S. Poedjastoeti, and M. Ayad. 1992. *Jordan Population and Family Health Survey 1990*. Columbia, MD: MD IRD/Macro International.

GEOGRAPHIC INDEX

NAME INDEX

SUBJECT INDEX